Advances in Animal Biotechnology

Birbal Singh · Gorakh Mal ·
Sanjeev K. Gautam · Manishi Mukesh

Advances in Animal Biotechnology

Springer

Birbal Singh
ICAR-Indian Veterinary Research
Institute, Regional Station
Palampur, India

Gorakh Mal
ICAR-Indian Veterinary Research
Institute, Regional Station
Palampur, India

Sanjeev K. Gautam
Department of Biotechnology
Kurukshetra University
Kurukshetra, Haryana, India

Manishi Mukesh
Department of Animal Biotechnology
ICAR-National Bureau of Animal
Genetic Resources
Karnal, Haryana, India

ISBN 978-3-030-21308-4 ISBN 978-3-030-21309-1 (eBook)
https://doi.org/10.1007/978-3-030-21309-1

This Springer imprint is published by the registered company Springer Nature Switzerland AG
The registered company address is: Gewerbestrasse 11, 6330 Cham, Switzerland

Preface

Biotechnology is a rapidly advancing area with remarkable achievements. This book entitled, *Advances in Animal Biotechnology*, is a compilation of advances in the field of Animal Biotechnology including fishery that are not sheltered in depth in earlier publications. The book is divided broadly into different sections, *viz.*, Gut Microbiome and Nutritional Biotechnology, Assisted Reproduction Biotechnology, Livestock Genomics, Health Biotechnology, and Animal Biotechnology in Global Perspective.

The book covers the syllabi of Animal Biotechnology courses in various universities and competitive examinations at various levels. Researchers, Continuing Graduates, and Academicians, Research Institutions, and Biotech Companies will be benefited.

Palampur, India — Birbal Singh
Palampur, India — Gorakh Mal
Kurukshetra, India — Sanjeev K. Gautam
Karnal, India — Manishi Mukesh

Contents

Part IV Health Biotechnology

About the Authors

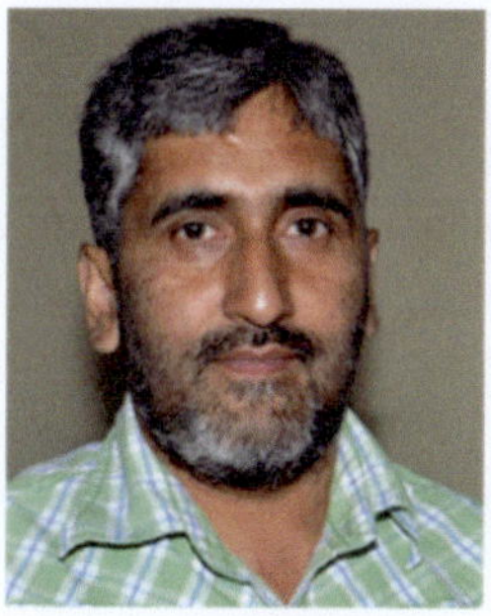

Birbal Singh, Ph.D. is Principal Scientist (Professor) at ICAR-Indian Veterinary Research Institute, Regional Station Palampur, India. He earned his Ph.D. (Animal Biotechnology) in 2007 from ICAR-National Dairy Research Institute, Karnal, India. He worked on nuclear transfer cloning in water buffaloes.

Dr. Birbal Singh is one of the leading experts in the field of animal biotechnology. He has contributed to research and teaching at post graduate levels. His expertise includes biotechnology of gut microbiome, somatic cell nuclear transfer cloning, and stem cell biology. In addition, he has worked on molecular biology, biochemical genetics, and cryopreservation of fish germplasm during his profession at ICAR-National Bureau of Fish Genetic Resources, Lucknow, India.

As a researcher, he has published more than 75 research and review papers in peer reviewed journals, more than 100 research abstracts, technical bulletins, and authored two books, namely "Textbook of Animal Biotechnology", and "Reproduction Biotechnology in Buffalo", and 20 chapters in books published by Elsevier, Springer and NOVA. He is reviewer for various international journals, and life member of assorted prestigious scientific societies.

Prof. Gorakh Mal, Ph.D. is Principal Scientist (Biochemistry), and Station Incharge at ICAR-Indian Veterinary Research Institute, Regional Station Palampur, India. He obtained his master's degree in Biochemistry from Himachal Pradesh Agriculture University Palampur, (India), and Ph.D. (Molecular Biology and Biotechnology) from University of Sheffield, South Yorkshire, UK.

Dr. Gorakh Mal has experience in molecular biology of cancer, gut microbiome and reproduction biology. He has published more than 80 scientific research papers and reviews, and more than 75 scientific abstracts in national and international journals. He has authored 2 books and contributed 15 book chapters. Dr. Mal has developed need-based technologies for value-added camel milk products during his services at ICAR-National Research Centre on Camel, Bikaner (India). He is reviewer for many prestigious international journals, and life member of various professional scientific societies.

Sanjeev K. Gautam earned his Ph.D. (Animal Biotechnology) in 2007 from National Dairy Research Institute, Karnal, India. He has worked as a research consultant at Institute for Cellular Medicine, San Jose, Costa Rica. In late 2007, he joined as Assistant Professor in Department of Biotechnology, Kurukshetra University, Kurukshetra (India). During this, he went for Postdoc in University of New Mexico, USA (2008), and Lund University, Sweden in 2010 where he worked on neuro stem cell biology. His areas of research include stem cell biology and cryopreservation.

He has published more than 60 scientific papers of national and international repute. He is co-author of "Textbook of Animal Biotechnology" (TERI, New Delhi). Dr. Gautam is working as visiting scholar in the Department of Biochemistry and Molecular Medicine in University of California, Davis, USA.

Manishi Mukesh, Ph.D. Principal Scientist (Professor) and ICAR National Fellow at ICAR-National Bureau of Animal Genetic Resources, Karnal (India). He has been actively engaged in performing research in the area of cattle and buffalo genomics.

He has made significant contribution to delineate stage-specific (heifer, lactating, and involution) transcriptomic signature of buffalo mammary gland using heterologous bovine specific microarray chip and RNASeq technologies. He also worked to establish stage specific transcriptome signature of milk derived mammary epithelial cells during lactation cycle in Sahiwal cows.

Dr. Manishi Mukesh is also engaged in thermoregulation genomics of Indian cattle and buffalo breeds, comparative physiological and transcriptional adaptation of Murrah buffaloes, Sahiwal and Karan Fries cows.

He has contributed immensely towards enriching the genomic information on indigenous cattle and buffaloes by publishing more than 80 research papers in international as well national journals, more than 100 research abstracts, several book chapters and technical bulletins. He has been involved as key resource person during several National Training Programmes.

Part I

Gut Microbiome and Nutritional Biotechnology

Metagenomics for Utilizing Herbivore Gut Potential

1

Abstract

Herbivores possess a mesmerizing gut ecosystem evolved to degrade and convert crude plant biomass and non-protein nitrogen to energy precursors and high biological value proteins. It is high time to unravel valued gut microbiota of herbivores by culture-independent molecular biological and metagenomics tools, and exploit the inferences to actually obtain microorganisms and microbial metabolites of commercial interest, and evolve strategies to improve undesirable processes associated with gut metabolism.

Highlights

- Herbivorous animals has precious gut microbiota and microbial metabolites
- Metagenomics has unravelled a lot about uncultured gut microbiota
- Many bottlenecks has to be resolved to actually utilize the potential of gut microbiota.

Keywords

Herbivores · Rumen · Metagenomics · Hydrolytic enzymes · Therapeutics · Industrial applications

1.1 Introduction

Microorganisms are the primary sources of enzymes and raw materials for industrial applications. Complex microbial ecosystems, such as extremophiles, soil (Yao et al. 2014) and compost, deep sea microbiota, and gut ecosystem of herbivores in particular, are largely underutilized genetic and biological sources of valuable microorganisms, microbial genes and biocatalysts.

The microbes inhabiting the digestive tract of herbivorous animals including vertebrates and invertebrates are of immense interest to nutrition, immunity and overall well-being of the host. As herbivores themselves lack cellulolytic enzymes to degrade plant biomass and phytometabolites therein, and synthesize proteins from non-proteins nitrogen (urea, uric acid etc.), they solely depend on metabolic activities of their gut microorganisms for energy and proteins. The system-wide investigations into microbial gut communities involved in lignocellulose degradation, detoxification of toxic phytometabolites have basic as well as applied prospects (Fig. 1.1). The gut microbial ecosystem is a potent source of valuable microbial metabolites such as enzymes, bacteriocins or antimicrobial peptides (AMPs) (Azevedo et al. 2015; Oyama et al. 2017a, b; Hatziioanou et al. 2017). Therefore, a meticulous understanding of complex gut ecosystem is of interest to microbial ecologists, livestock nutritionists, molecular biologists and industrialists.

B. Singh et al., *Advances in Animal Biotechnology*,
https://doi.org/10.1007/978-3-030-21309-1_1

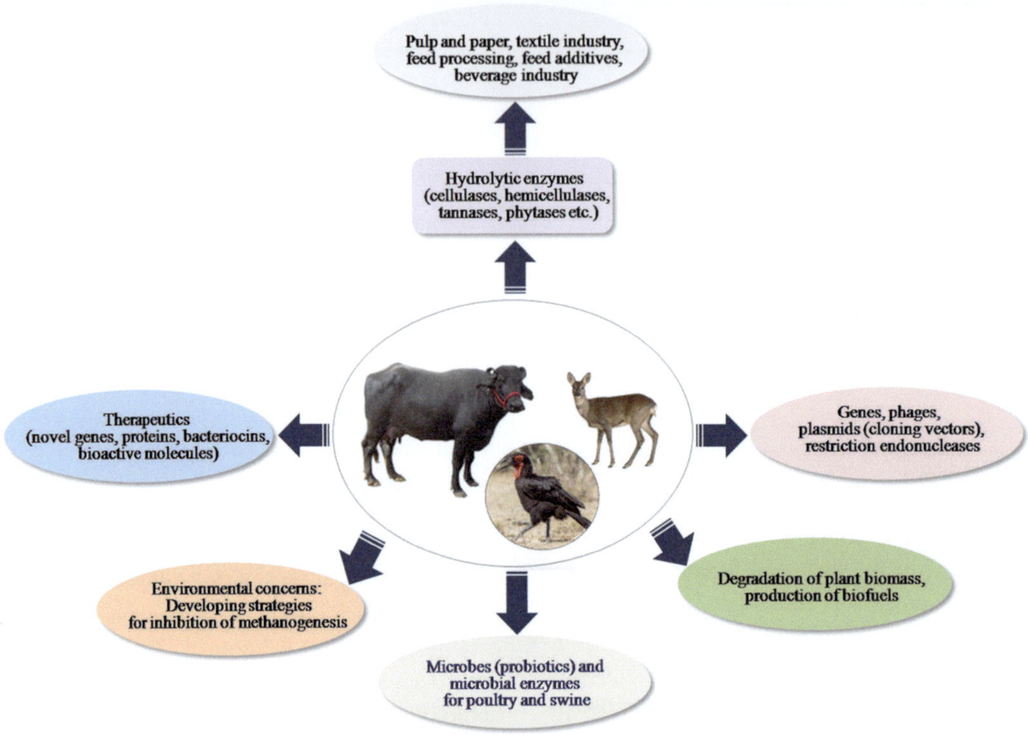

Fig. 1.1 A diagrammatic depiction of herbivore gut ecosystem as a source of well-ordered microbiome for degrading lignocelluloses-rich plant biomass and anti-nutritional plant metabolites such as tannins, oxalates, saponins, fluoroacetate, and non-protein amino acids

Handelsman et al. (1998) have suggested the term metagenomics (‘*meta*’ Greek, for transcending; more comprehensive), which constitutes a challenging domain to discover enzymes, genes and metabolic pathways from diverse microbial habitats. Soon after development of high throughput genome sequencing methodologies, and bioinformatics methods to analyze the plentiful data, the metagenomics emerged as a potential method to unravel microbial ecosystems, such as gastrointestinal (GI) tract of humans and animals (Singh et al. 2008; Hess et al. 2011; Kohl et al. 2016; Denman et al. 2018; Huws et al. 2018), skin (Chng et al. 2016), genitourinary microbes (Aagaard et al. 2012; Li et al. 2018), and epidemiological studies (Martin et al. 2018). In addition, metagenomics is used to discover extremophiles, marine microbial resources, soil, compost (Matsuzawa et al. 2015), and sewerage (Cai et al. 2018).

1.2 Why Herbivore Gut Ecosystem so Important?

The fact that food and feed processing, textile and surfactants, biofuels, and pharmaceutical industries need a sustained supply of raw materials such as enzymes, has urged the researchers to investigate microbial communities for their beneficial contributions. The herbivores by virtues of their complex and highly efficient gut microbiome very efficiently utilize the hemicelluloses-rich plant biomass, and enable them to adapt to varied agroclimatic situations. The gut microbiota of herbivores is equipped with a wide range of metabolic capabilities including synthesis of energy precursors such as short chain fatty acids (SCFAs) (White et al. 2014; Mayorga et al. 2018; Clemmons et al. 2018), increasing microbial biomass in rumen that is digested in

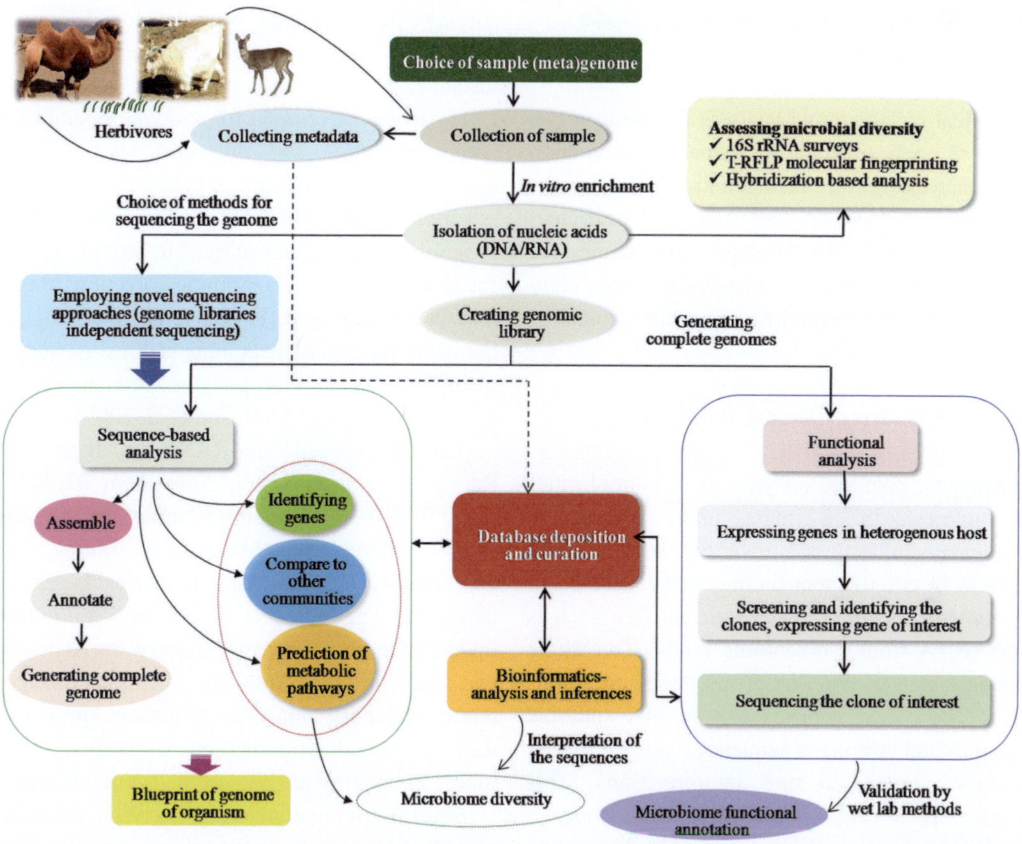

Fig. 1.2 A flowchart presentation of metagenomic profiling of herbivore GI ecosystem. Selection of the species depends on what has to be targeted. For instance wild ruminants harbour more proficient fibrolytic microbiota, while goats, rodents (e.g. Stephen's woodrats, *Neotoma stephensi*) and birds (e.g. Greater Sage-Grouse) may harbour microorganisms with ability to degrade toxic phytometabolites in plants used as forage

abomasum and lower part of GI tract, and used as a source of proteins and amino acids. In addition, some of the AMPs synthesized by bacteria confer protection against invading pathogens.

The culture-dependent microbiological techniques cannot completely describe the microbial complexity of gut ecosystem, molecular biological, and genomics methodologies are developed to expand the blueprint of microbial species inhabiting GI tract. Indeed, the very essential purpose of microbial metagenomics is to decipher non-cultivable microorganisms, their properties and eventually explore them for mercantile applications. The metagenomic analysis (Fig. 1.2), comprises of collection of samples from a niche, enrichment of microorganisms that are low in number in a population, isolation of whole genome of the microorganisms, construction, screening and analysis of metagenomic DNA/RNA libraries.

Technical advances in basic and applied sciences, technologies to construct high efficiency cloning vectors (e.g. bacterial artificial chromosomes, yeast artificial chromosomes to clone eukaryotic genes, and development of bioinformatics methods have progressed the applications of metagenomics. Figure 1.2 summarises the major technical aspects of a representative metagenomic analysis.

1.3 Metagenome Sequence Data Analysis

A typical metagenomic analysis is based on two types of analyses, namely, sequence-based analysis (screening the metagenomic libraries for nucleotide sequences), and function-based analysis (screening the metagenome clones or libraries for a particular metabolite), or a combination of both depending on requirements.

1.4 Sequence-Based Analysis

Identifying potential enzymes or a biomolecules of interest in a metagenome based on sequence-similarity is a rewarding strategy. This approach entails sequencing and analysis of genome or specific phyolegentic clusters, such as 16S rRNA from a taxonomic group. The whole genome sequencing (WGS) produces enormous data. The classic projects may involve multiple samples and billions of sequence reads for further analysis, prediction and interpretations. This requires bioinformatics and *in sillico* methods.

The metagenomic data are analysed based on sequences corresponding to a particular activity or sequences already available in database. Various software packages have been developed for analyzing the metagenome sequence data. For instance, a computer-aided program, named as DOTUR was developed for studying operational taxonomic units and microbial species in a microbial milieu (Schloss and Handelsman 2005). A series of MEGAN (MEtaGenome ANalyzers) programs, originally developed to provide as tool for investigating the taxonomic content of single dataset are available (Huson et al. 2007, 2016). Sargasso Sea data, data obtained from woolly mammoth (*Mammuthus primigennius*) bones was analyzed by MEGAN (Huson et al. 2007).

The web-based mg-RAST (metagenomic Rapid Annotation Using Subsystem Technology), facilitates processing, sharing and analysis of metagenomic data (Meyer et al. 2008).

Currently, a number of software packages are available. The readers should refer to online updates for developments in software packages and bioinformatics tools for microbial metagenomic diversity analysis, sequence similarity, functional annotation, mapping to reference genomes, and quality analysis of metagenomic data.

Several novel cellulases and glycosylases are identified from metagenomic sequences of rumen and GI microorganisms (Lopez-Cortes et al. 2007; Zhao et al. 2010) (Table 1.1).

Table 1.1 Summary of valuable microorganisms and microbial enzymes derived from metagenomics of herbivore GI tract

Microorganisms/niches/enzymes/genes	Host/origin	Inferences and recommendations (Reference)
Bacteria		
Cytophaga-Flexibacter Bacteroides phyla	Buffalo rumen	Cellulose degradation, and abundance of cellulose enzymes (Liu et al. 2009a, b)
Rumen microbes	Camel rumen	Characterization of microorganisms and microbial enzymes responsible for cellulose degradation, and generation of SCFAs. The study highlights the potential of fibrolytic microorganisms and their enzymes in food processing, and producing biofuel and fine chemicals (Gharechahi and Salekdeh 2018)
Gut microorganisms	*Rhinopithecus bieti* (primate)	Description of gut bacterial diversity, and their role in digesting plant biomass (Xu et al. 2015)
Fungi		
Fungal taxa	Murine GI tract	Identification of fungal taxa and their role in GI tract (Toyoda et al. 2009)

(continued)

Table 1.1 (continued)

Microorganisms/niches/enzymes/genes	Host/origin	Inferences and recommendations (Reference)
Methanogens		
Novel methanogens	Cattle, sheep rumen	Identification of certain groups of methanogens in rumen (Ferrer et al. 2007)
Phages		
Diverse virome	Caprids	Development of protocol to enrich and isolate DNA of rumen virome entailing viruses of various groups such as Siphoviridae, Myoviridae, Podoviridae, Mimiviridae, Microviridae, Poxviridae, Tectiviridae and Marseillevirus. The study conclude role of virome in maintaining rumen bacterial diversity (Namonyo et al. 2018)
Microbial proteins/genes		
Acetylxylan esterase family carbohydrate esterase (CE6)	Rumen	Depiction of novel rumen-origin hydrolases (Beloqui et al. 2006)
β-glucosidases	Cattle rumen	Role of enzymes in saccharification of lignocellulose (Del Pozo et al. 2012)
β-glucosidase/xylosidase (*RuBX1*) gene belonging GHF 3 β-glucosidase/ xylosidase family	Yak rumen	Demonstration of enzyme hydrolytic activity on various substrates (Zhou et al. 2012)
Cel A, *Xyl A* genes	Cow rumen	Extraction of enzymes Cel5 A and xyl A from cattle rumen (Shedova et al. 2009)
cel28a (cellulase gene)	Goat rumen	Description of gene and the enzymes, experimental validation of enzyme activity (Cheng et al. 2016)
Cellulases (Novel industrially important enzymes)	Buffalo rumen	Purification of enzyme expressed in *Escherichia coli* (Duan et al. 2009)
Genes for glycoside hydrolases (*GH*)	Yak rumen	Detection of genes *GH5, GH9 and GH10,* and their enzymes in yak hydrolases (*GH*) rumen (Dai et al. 2012). The study proposes the role of rumen enzymes in lignocellulose degradation
Glycosylases (Hybrid enzymes)	Cattle rumen	Identification of hydrolytic enzymes (Lopez-Cortes et al. 2007)
Glycosyl hydrolases (Novel enzymes)	Cattle rumen	Identification of genes producing novel glycosyl hydrolases (Zhao et al. 2010)
Glycosyl hydrolase family (GHF)	Cattle rumen	The study describes the function of GHF in digestion of ingested plant family (GHF) biomass (Ferrer et al. 2012)
Glycosyl hydrolase family 5 (GHF5)	Buffalo rumen	Revelation of GHF5 as predominant fibrolytic enzymes, possibility of family 5 (GHF5) utilizing the enzymes as feed supplements (Nguyen et al. 2012)
Glycosyl hydrolases	*Rhinopithecus bieti* fecal microbiota (Primate)	Depiction of bacterial glycoside hydrolases with ability to degrade fecal microbiota lignocelluloses (Xu et al. 2015)
Glycoside hydrolases (Protozoa origin)	Bovine rumen	Description of the enzymes (Findley et al. 2011)
KG42 xylanase (GH10 family)	Black goat rumen	Description of a novel xylanase from microbial metagenome of goat rumen. The KG42 was found to function as an endo-β-1,4-xylanase (EC 3.2.1.8), and can be used to prepare prebiotic xylooligosaccharides (Kim et al. 2018)

(continued)

Table 1.1 (continued)

Microorganisms/niches/enzymes/genes	Host/origin	Inferences and recommendations (Reference)
KG51 Bifunctional cellulase/ hemicellulase	Black goat rumen	Characterization of bifunctional enzyme cellulase (endo-β-1,4-glucanase [EC 3.2.1.4]) and hemicellulase (mannan endo-β-1,4-mannosidase [EC 3.2.1.78] and endo-β-1,4-xylanase [EC 3.2.1.8]) activities. The enzyme activity was verified through a preparation of prebiotic konjac glucomannan hydrolysates (Lee et al. 2018)
Lipolytic esterases	Sheep rumen	Identification and characterization of esterase genes, using activity-based cluster screening approach (Bayer et al. 2010)
Multiple recombinant family 26 glycohydrolase	Buffalo rumen	Characterization of multifunctional activities of the enzyme (Patel et al. 2016)
cel5A and *cel5B cellulose genes*	Cow rumen	The study involves characterization (pH optima, substrate specificity, of enzymes, and possibility of using bifunctional enzymes in cleavage of β-1,4 bonds of cellulose and hemicellulose (Rashamuse et al. 2013)
Feruloyl esterase gene	Cattle rumen	Cloning and expression of *FAE* gene encoding FAE-SH1 enzyme (*FAE*) in *Escherichia coli*, and its possible applications (Cheng et al. 2012)
Lipases-encoding genes	Cattle rumen	Biochemical and genetic characterization of lipase/esterase/and prospects (Privé et al. 2015)
RlipE1 and *RlipE2* genes their products	Cattle rumen	Purification and characterization of recombinant lipases, and their applications in rumen lipid metabolism (Liu et al. 2009a, b)
umcel3G, a gene encoding β-glucosidase	Buffalo rumen	Role of enzymes in production of biofuel (Guo et al. 2008)
Umbgl3B (β-glucosidase)	Rabbit caecum	Depiction of hydrolytic enzymes from rabbit GI tract (Feng et al. 2009)
Phytases	Buffalo rumen	Identification of phytase genes from bubaline rumen metagenome. A gene, *RPHY1* was described and expressed in *Escherichia coli* BL1. The purified enzymes was characterized for amino acid composition, molecular profile and activities (Mootapally et al. 2016)
Antimicrobial peptides (AMPs)	Rumen bacterial metagenome library	Identification and description of buwchitin, a rumen-origin AMP, and its characterization, indicating that buwchitin be effective to curtail *Enterococcus fecalis* (Oyama et al. 2017b)
Antibiotic-resistance genes	Cattle feces	Description of general microbial communities in cattle feces, identification of virulence genes, and genes associated with resistance to antibiotics and toxic compounds (Durso et al. 2011)

1.5 Function-Driven Analysis

Enzymes are the backbone of industrial development. The industrial enzyme market valued at 4.2 billion USD in 2014, is envisioned to grow at annual growth rate of 7.0% from 2015 to 2020. Food and beverage, cleaning agents, consumer goods and biofuel, and animal feed processing industry will need a sustained supply of potential biocatalysts. Bioengineering of microbial strains, and computational enzyme engineering are in increasing demand to fill the gap in demand and supply of potential biocatalysts.

Gut and faecal microorganisms serve as promising sources of useful enzymes. Identification of genes and the enzymes in a metagenome data is of paramount interest (Box 1). The functional-active screening is based on identifying the clones, characterization of positive clones by molecular or biochemical techniques. This facilitates identification of genes for producing pharmaceuticals or other commercial products.

Functional metagenomic approaches have potential for bioprospecting microbial diversity. Potentially useful enzymes are described from the metagenomic analysis of digestive tracts of herbivores. Using functional metagenomic screening, some antibiotic-resistance genes are identified (Diaz-Torres et al. 2003; Singh et al. 2012). Kim et al. (2018) have isolated a KG42 xylanase (GH10 family) from microbial metagenome of black Bengal goat. This novel xylanase was found to function as an endo-β-1,4-xylanase (EC 3.2.1.8), and was suitable for prepare prebiotic xylooligosaccharides.

Box 1. Summary of Motivating Factors that Urge Studying Gut Ecosystem

- **Sustained supply of enzymes**: Hydrolytic enzymes of rumen or herbivore gut are in demand in food and feed processing, textile and pulp industry, and production of fermentable sugars for producing biofuel.
- **Novel microbial species, genes and genetic engineering tools**: Fibrolytic microorganisms (anaerobic gut fungi, and bacteria), and fibrolytic genes are of immense importance. The rumen contains phages, and host bacteria that produce restriction enzymes (REs). Plasmids, REs, and phages can possibly serve as genetic engineering tools. The rumen bacteria and fungi have applications as microbial feed supplements.
- **Alternative therapeutics**: Bacteriocins and AMPs of rumen bacteria can be used to combat pathogens and methanogenic archaea.
- **Environmental concerns**: Preventing green house gas emissions from enteric methane emissions, using rumen phytases and cellulase enzymes to prevent or minimize release of nutrients as waste into environment from poultry, pig and fishery.

1.6 Insights from Gut Metagenomics

Certain parts of GI tract are densely populated by anaerobic microorganisms (Flint et al. 2008; Costea et al. 2017). These regions are highly active sites of the microbial metabolic activities and metabolites. The rumen is a naturally evolved ecosystem for microorganisms that are specialized in rapid hydrolysis of plant biomass for providing energy, proteins and other nutrients to the host. The rumen microbiota (bacteria, fungi, protozoa) mediate the digestion of fibrous plant cells wall, detoxification of some phytometabolites, for producing short chain fatty acids and microorganisms that are digested and assimilated in intestine (Prajapati et al. 2016). Hence, gut ecosystem of herbivores offers an inexhaustible source of microorganisms, genes and enzymes.

1.7 Deriving Commercially Important Enzymes

As stated above, the metagenomics has made remarkable advances in studying the complex microbial niches. Poly-functional glycosyl hydrolases from cow and buffalo rumen are suitable for textile and pulp industry (Ferrer et al. 2007; Duan et al. 2010). The β-glucosidase from buffalo rumen may have role in fermentative production of biofuel from lignocellose (Guo et al. 2008).

Whole genome sequence data of rumen fibrolytic bacteria, viz., *Fibrobacter succinogenes*, *Ruminococcus albus*, and *Prevotella ruminicola* could serve as source of gene sequences, and novel proteins and enzymes.

1.8 Searching Enzymes for Feed Processing

As shown above, the feral herbivores and certain domestic ruminants utilize diverse forages without apparent symptoms of adverse effects, hence can harbour genetically elite gut microbes with superior hydrolytic enzymes. Metagenomics has contributed immensely to identify microorganisms and hydrolytic microbial enzymes (Table 1.1).

KG51 bifunctional cellulase (endo-β-1,4-glucanase cellulase/hemicellulase [EC 3.2.1.4]), and hemicellulase (mannan endo-β-1,4-mannosidase [EC 3.2.1.78], and endo-β-1,4-xylanase [EC 3.2.1.8]) activities have been detected from metagenome data of Black goat rumen. The enzyme activity was verified through preparation of prebiotic konjac glucomannan hydrolysates (Lee et al. 2018). In another study, a novel KG42 xylanase (GH10 family) was also identified from metagenome of Black goat rumen. The KG42 was found to function as an endo-β-1,4-xylanase (EC 3.2.1.8), and can be used to prepare prebiotic xylooligosaccharides (Kim et al. 2018).

High milk-yielding females during early lactation due to being in negative energy balance need energy-rich diets. Rumen microorganisms, especially gut fungi and cellulolytic bacteria are the probable microbial feed supplements that can efficiently hydrolyze the plant fiber, and apply enhanced energy to these animals. Application of lactate-utilizing bacteria as microbial supplements may be useful when animals are likely to suffer from acidosis caused by high grain diets.

1.9 Microbial Detoxification of Anti-nutritional Phytometabolites

Herbivores possess a microbiome that has developed metabolic capabilities to degrade certain phytametabolies which cause toxicity when taken in excess (Singh et al. 2003). A number of bacteria and certain fungi degrading anti-nutritional dietary ingredients are isolated from the GI tract of various herbivores.

Metagenomic analysis of GI tract of various species has led to identification of microbial enzymes or consortia involved in detoxification of plant toxins. Phytase genes are noted in rumen metagenome of buffalo (Mootapally et al. 2016). Metagenomic analysis of gut contents of Stephen's woodrats (*Neotoma stephensi*) showed abundance of various microbial genes across different parts of alimentary canal. It was found that genes associated with degradation of oxalate and phenolic plant compounds were more abundant in cecum *N. stephensi* (Kohl et al. 2018).

1.10 Demand for Alternative Therapeutics

Emergence of resistance among pathogens towards routine antibiotics is a matter of concern. The situation is worsened due to slow progress in development of new antibiotics. Safer and viable options, such as genetically modified probiotics and postbiotics (Singh et al. 2017), engineered antibodies or antibody-antibiotic conjugates (Mariathasan and Tan 2017; Wagner and Maynard 2018), and AMPs (Chung and Raffatellu 2018) are proposed to prevent infectious diseases. Probiotics and postbiotics, the probiotic metabolites such as bacteriocins, bioactive AMPs are envisioned to be effective against pathogens and diseases such as cancer. The bacteria with abilities to produce antagonistic substances (organic acids, hydrogen peroxide, and bacteriocins) are reported from gut ecosystem.

Rumen microorganisms contain antibiotic resistance genes (ARGs) (resistome) that are transferable to microorganisms of human origin when rumen and human microbiota interact with each other. A deep sequence analysis of sheep rumen contents revealed abundance of ARGs therein. Abundance of resistance to antibiotic, the daptomycin, implies that ruminants might be a source of daptomycin ARGs (Hitch et al. 2018). The study emphasizes further need to discover rumen microbiome as source of clinically important ARGs.

Bacteriocinogenic microorganisms are identified from the gut of ruminants, poultry and humans. When used as alternatives to antibiotics or ionophores in the feedlot, the bacteriocins and AMPs can prevent pathogenic microorganisms. No residual bacteriocins or AMPs are expected in milk or meat as being proteins in nature, they are digested during their passage through GI tract. Azevedo et al. (2015) have shown a total of 46 bacteriocin-gene clusters in 33 rumen bacterial strains such as *Streptococcus* sp., *Ruminococcus albus*.

1.11 Environmental Concerns

Ruminants have environmental impact. Some archaea reduce H_2 and CO_2 produced during fermentation, to produce methane. According to Intergovernmental Panel on Climate Change Synthesis Report (2014), the CH_4 has a global warming potential 28-fold than that of CO_2. Hence, it is necessary to mitigate rumen methane emissions.

Understanding adaptation of methanogens to ingested dietary ingredients, and cellular and molecular level interaction between rumen archaea and protozoa, can lead to develop strategies to minimize methane emissions (Wallace et al. 2015).

1.12 Outlook and Challenges

Metagenomics provides a new way of examining microbial world. Indeed, the progress in metagenomics is attributed to ineluctable evidence that unculturable microorganisms represent majority of the environmental microbiomes, and might have important commercial applications.

Understanding metagenomicsof environmental microorganisms is an important step in discovering the role of uncultured microbiota in nutrition and well-being of humans and animals. Metatranscriptomics and metaproteomics have been developed to precisely understand the functional properties of gut microbial communities. This will help in detecting and understanding the biosynthetic pathways.

The metagenomic analysis of microbial world has certain technical limitations that restrict its wider scale use. A major problem of metagenomic analysis is that it requires sophisticated instrumentation for genome sequencing, and generates enormous sequence data. Also, technically expertise is needed to analyze and interpret the data.

Although some novel enzymes are already reported from metagenome library data, majority of them have not been authenticated by standard culture-dependent microbiological methods. Only a few gene and proteins are actually confirmed using wet lab methods.

The rumen enzymes have prospects in poultry, swine and fishery nutrition, and the environmental concerns associated with these species. Another important area for metagenomic interventions is evolving strategies to manipulate GI fermentation of fibrous residues to minimize methane emissions.

1.13 Conclusions

The microbial communities inhabiting the GI tract of mammals, particularly of herbivorous species, are the densest reservoirs. The gut microbiota of herbivores are of towering economic concern because of their metabolic merits. As culture-dependent microbiological techniques are inapt to decipher completely microbial consortia, metagenomic studies are applied to assess the microbial diversity and the metabolome of GI ecosystem. The studies should focus on investigating the potential microbiota and microbial mechanisms to maximize use of fibrous plant biomass for nutrition, and for biofuels.

References

Aagaard K, Riehle K, Ma J, Segata N, Mistretta TA, Coarfa C, Raza S, Rosenbaum S, Van den Veyver I, Milosavljevic A, Gevers D, Huttenhower C, Petrosino J, Versalovic J (2012) A metagenomic approach to characterization of the vaginal microbiome signature in pregnancy. PLoS ONE 7(6): e36466. https://doi.org/10.1371/journal.pone.0036466

Azevedo AC, Bento CB, Ruiz JC, Queiroz MV, Mantovani HC (2015) Distribution and genetic diversity of bacteriocin gene clusters in rumen microbial genomes. Appl Environ Microbiol 81(20):7290–7304. https://doi.org/10.1128/AEM.01223-15 (Epub 2015 Aug 7)

Bayer S, Kunert A, Ballschmiter M, Greiner-Stoeffele T (2010) Indication for a new lipolytic enzyme family: isolation and characterization of two esterases from a metagenomic library. J Mol Microbiol Biotechnol 18:181–187. https://doi.org/10.1159/000315459

Beloqui A, Pita M, Polaina J, Martinez-Arias A, Golyshina OV, Zumarraga M (2006) Novel polyphenol oxidase mined from a metagenome expression library of bovine rumen: biochemical properties, structural analysis, and phylogenetic relationship. J Biol Chem 281:22933–22942

Cai L, Chen TB, Zheng SW, Liu HT, Zheng GD (2018) Decomposition of lignocellulose and readily degradable carbohydrates during sewage sludge biodrying, insights of the potential role of microorganisms from a metagenomic analysis. Chemosphere 201:127–136. https://doi.org/10.1016/j.chemosphere.2018.02.177 (Epub 28 Feb 2018)

Cheng F, Sheng J, Cai T, Jin J, Liu W, Lin Y, Du Y, Zhang M, Shen L (2012) A protease-insensitive feruloyl esterase from China Holstein cow rumen metagenomic library: expression, characterization, and utilization in ferulic acid release from wheat straw. J Agric Food Chem 60:2546–2553. https://doi.org/10.1021/jf204556u

Cheng J, Huang S, Jiang H, Zhang Y, Li L, Wang J, Fan C (2016) Isolation and characterization of a non-specific endoglucanase from a metagenomic library of goat rumen. World J Microbiol Biotechnol 32(1):12. https://doi.org/10.1007/s11274-015-1957-4

Chng KR, Tay AS, Li C, Ng AH, Wang J, Suri BK, Matta SA, McGovern N, Janela B, Wong XF, Sio YY, Au BV, Wilm A, De Sessions PF, Lim TC, Tang MB, Ginhoux F, Connolly JE, Lane EB, Chew FT, Common JE, Nagarajan N (2016) Whole metagenome profiling reveals skin microbiome-dependent susceptibility to atopic dermatitis flare. Nat Microbiol 1(9): 16106. https://doi.org/10.1038/nmicrobiol.2016.106

Chung LK, Raffatellu M (2018) G.I. pros: Antimicrobial defense in the gastrointestinal tract. Semin Cell Dev Biol. 12 Feb 2018. pii: S1084-9521(17)30414-7. https://doi.org/10.1016/j.semcdb.2018.02.001 (Epub ahead of print. Review)

Clemmons BA, Voy BH, Myer PR (2018) Altering the Gut Microbiome of cattle: considerations of host-microbiome interactions for persistent microbiome manipulation. Microb Ecol 22 July 2018. https://doi.org/10.1007/s00248-018-1234-9 (Epub ahead of print)

Costea PI, Zeller G, Sunagawa S, Pelletier E, Alberti A, Levenez F, Tramontano M, Driessen M, Hercog R, Jung FE, Kultima JR, Hayward MR, Coelho LP, Allen-Vercoe E, Bertrand L, Blaut M, Brown JRM, Carton T, Cools-Portier S, Daigneault M, Derrien M, Druesne A, de Vos WM, Finlay BB, Flint HJ, Guarner F, Hattori M, Heilig H, Luna RA, van Hylckama Vlieg J, Junick J, Klymiuk I, Langella P, Le Chatelier E, Mai V, Manichanh C, Martin JC, Mery C, Morita H, O'Toole PW, Orvain C, Patil KR, Penders J, Persson S, Pons N, Popova M, Salonen A, Saulnier D, Scott KP, Singh B, Slezak K, Veiga P, Versalovic J, Zhao L, Zoetendal EG, Ehrlich SD, Dore J, Bork P (2017) Towards standards for human fecal sample processing in metagenomic studies. Nat Biotechnol 35(11):1069–1076. https://doi.org/10.1038/nbt.3960 Epub 2017 Oct 2

Climate Change Synthesis Report (2014):(https://www.ipcc.ch/site/assets/uploads/2018/05/SYR_AR5_FINAL_full_wcover.pdf)

Dai X, Zhu Y, Luo Y, Song L, Liu D, Liu L, Chen F, Wang M, Li J, Zeng X, Dong Z, Hu S, Li L, Xu J, Huang L, Dong X (2012) Metagenomic insights into the fibrolytic microbiome in yak rumen. PLoS ONE 7: e40430. https://doi.org/10.1371/journal.pone.0040430

Del Pozo MV, Fernández-Arrojo L, Gil-Martínez J, Montesinos A, Chernikova TN, Nechitaylo TY, Waliszek A, Tortajada M, Rojas A, Huws SA, Golyshina OV, Newbold CJ, Polaina J, Ferrer M, Golyshin PN (2012) Microbial beta-glucosidases from cow rumen metagenome enhance the saccharification of lignocellulose in combination with commercial cellulase cocktail. Biotechnol Biofuels 5:73. https://doi.org/10.1186/1754-6834-5-73

Diaz-Torres ML, McNab R, Spratt DA, Villedieu A, Hunt N, Wilson M, Mullany P (2003) Novel tetracycline resistance determined from the oral metagenome. Antimicrob Agents Chemother 47:1430–1432

Denman SE, Morgavi DP, McSweeney CS (Review) (2018) The application of omics to rumen microbiota function. Animal 28:1–13. https://doi.org/10.1017/s175173111800229x

Duan CJ, Liu JL, Wu X, Tang JL, Feng JX (2010) Novel carbohydrate-binding module identified in a ruminal metagenomic endoglucanase. Appl Environ Microbiol 76:4867–4870. https://doi.org/10.1128/AEM.00011-10

Duan CJ, Xian L, Zhao GC, Feng Y, Pang H, Bai XL, Tang JL, Ma QS, Feng JX (2009) Isolation and partial characterization of novel genes encoding acidic cellulases from metagenome of buffalo rumens. J Applied Microbiol 107:245–256. https://doi.org/10.1111/j.1365-2672.2009.04202.x

Durso LM, Harhay GP, Bono JL, Smith TP (2011) Virulence-associated and antibiotic resistance genes of microbial populations in cattle feces analyzed using a metagenomic approach. J Microbiol Methods 84 (2):278–282. https://doi.org/10.1016/j.mimet.2010.12.008 (Epub 2010 Dec 16)

Feng Y, Duan CJ, Liu L, Tang JL, Feng JX (2009) Properties of a metagenome-derived beta-glucosidase from the contents of rabbit cecum. Biosci Biotechnol Biochem 73:1470–1473

Ferrer M, Ghazi A, Beloqui A, Vieites JM, López-Cortés N, Marín-Navarro J, Nechitaylo TY, Guazzaroni ME, Polaina J, Waliczek A, Chernikova TN, Reva ON, Golyshina OV, Golyshin PN (2012) Functional metagenomics unveils a multifunctional glycosyl hydrolase from the family 43 catalysing the breakdown of plant polymers in the calf rumen. PLoS ONE 7: e38134. https://doi.org/10.1371/journal.pone.0038134

Ferrer M, Beloqui A, Golyshina OV, Plou FJ, Neef A, Chernikova TN (2007) Biochemical structure features of a novel cyclodextrinase from cow rumen metagenome. Biotechnol J 2:207–213

Findley SD, Mormile MR, Sommer-Hurley A, Zhang XC, Tipton P, Arnett K, Porter JH, Kerley M, Stacey G (2011) Activity-based metagenomic screening and biochemical characterization of bovine ruminal protozoan glycoside hydrolases. Appl Environ Microbiol 77(22):8106–8113. https://doi.org/10.1128/AEM.05925-11 (Epub 2011 Sep 23)

Flint HJ, Bayer EA, Rincon MT, Lamed R, White BA (2008) Polysaccharide utilization by gut bacteria: potential for new insights from genomic analysis. Nat Rev Microbiol 6(2):121–131. https://doi.org/10.1038/nrmicro1817.Review

Gharechahi J, Salekdeh GH (2018) A metagenomic analysis of the camel rumen's microbiome identifies the major microbes responsible for lignocellulose degradation and fermentation. Biotechnol Biofuels 11:216. 2 Aug 2018. https://doi.org/10.1186/s13068-018-1214-9 (eCollection 2018)

Guo H, Feng Y, Mo X, Duan C, Tang J, Feng J (2008) Cloning and expression of beta-glucosidase gene umcel3G from metagenome of buffalo rumen and characterization of the translated product. Sheng Wu Gong Cheng Xue Bao 24:232–8 (Article in Chinese, abstract in English)

Handelsman J, Rondon MR, Brady SF, Clardy J, Goodman RM (1998) Molecular biological access to the chemistry of unknown soil microbes: a new frontier for natural products. Chem Biol 5:R245–R249

Hatziioanou D, Gherghisan-Filip C, Saalbach G, Horn N, Wegmann U, Duncan SH, Flint HJ, Mayer MJ, Narbad A (2017) Discovery of a novel lantibiotic nisin O from Blautia obeum A2-162, isolated from the human gastrointestinal tract. Microbiology 163(9): 1292–1305. https://doi.org/10.1099/mic.0.000515

Hess M, Sczyrba A, Egan R, Kim TW, Chokhawala H, Schroth G, Luo S, Clark DS, Chen F, Zhang T, Mackie RI, Pennacchio LA, Tringe SG, Visel A, Woyke T, Wang Z, Rubin EM (2011) Metagenomic discovery of biomass-degrading genes and genomes from cow rumen. Science 331(6016):463–467. https://doi.org/10.1126/science.1200387

Hitch TCA, Thomas BJ, Friedersdorff JCA, Ougham H, Creevey CJ (2018) Deep sequence analysis reveals the ovine rumen as a reservoir of antibiotic resistance genes. Environ Pollut 235:571–575. https://doi.org/10.1016/j.envpol.2017.12.067 (Epub 2018 Jan 11)

Huson DH, Auch AF, Qi J, Schuster SC (2007) MEGAN analysis of metagenomic data. Genome Res 17:377–386

Huson DH, Beier S, Flade I, Górska A, El-Hadidi M, Mitra S, Ruscheweyh HJ, Tappu R (2016) MEGAN community edition—interactive exploration and analysis of large-scale microbiome sequencing data. PLoS Comput Biol 12(6):e1004957. 21 Jun 2016. https://doi.org/10.1371/journal.pcbi.1004957. eCollection 2016 Jun

Huws SA, Creevey CJ, Oyama LB, Mizrahi I, Denman SE, Popova M, Muñoz-Tamayo R, Forano E, Waters SM, Hess M, Tapio I, Smidt H, Krizsan SJ, Yáñez-Ruiz DR, Belanche A, Guan L, Gruninger RJ, McAllister TA, Newbold CJ, Roehe R, Dewhurst RJ, Snelling TJ, Watson M, Suen G, Hart EH, Kingston-Smith AH, Scollan ND, do Prado RM, Pilau EJ, Mantovani HC, Attwood GT, Edwards JE, McEwan NR, Morrisson S, Mayorga OL, Elliott C, Morgavi DP (2018) Addressing global ruminant agricultural challenges through understanding the rumen microbiome: past, present, and future. Front Microbiol 9:2161. 25 Sep 2018. https://doi.org/10.3389/fmicb.2018.02161. eCollection 2018 (Review)

Kim HB, Lee KT, Kim MJ, Lee JS, Kim KS (2018) Identification and characterization of a novel KG42 xylanase (GH10 family) isolated from the black goat rumen-derived metagenomic library. Carbohydr Res 469:1–9. https://doi.org/10.1016/j.carres.2018.08.010 (Epub 2018 Aug 22)

Kohl KD, Connelly JW, Dearing MD, Forbey JS (2016) Microbial detoxification in the gut of a specialist avian herbivore, the Greater Sage-Grouse. FEMS Microbiol Lett 363(14). pii: fnw144. https://doi.org/10.1093/femsle/fnw144 (Epub 2016 May 29)

Kohl KD, Oakeson KF, Orr TJ, Miller AW, Forbey JS, Phillips CD, Dale C, Weiss RB, Dearing MD (2018) Metagenomic sequencing provides insights into microbial detoxification in the guts of small mammalian herbivores (*Neotoma* spp.). FEMS Microbiol Ecol 94(12). 1 Nov 2018. https://doi.org/10.1093/femsec/fiy184

Lee KT, Toushik SH, Baek JY, Kim JE, Lee JS, Kim KS (2018) Metagenomic mining and functional characterization of a novel KG51 bifunctional cellulase/hemicellulase from black goat rumen. J Agric Food Chem 66(34):9034–9041. https://doi.org/10.1021/acs.jafc.8b01449 (Epub 2018 Aug 20)

Li F, Chen C, Wei W, Wang Z, Dai J, Hao L, Song L, Zhang X, Zeng L, Du H, Tang H, Liu N, Yang H, Wang J, Madsen L, Brix S, Kristiansen K, Xu X, Li J, Wu R, Jia H (2018) The metagenome of the female upper reproductive tract. Gigascience 7(10). 1 Oct 2018. https://doi.org/10.1093/gigascience/giy107

Liu K, Wang J, Bu D, Zhao S, McSweeny CS, Yu P, Li D (2009a) Isolation and biochemical characterization of two lipases from a metagenomic library of China Holstein cow rumen. Biochem Biophys Res Commun 385:605–611. https://doi.org/10.1016/j.bbrc.2009.05.110

Liu L, Tang J, Feng J (2009b) Bacterial diversity in Guangxi buffalo rumen. Wei Sheng Wu Xue Bao 49:251–56 (Article in Chinese, abstract in English)

Lopez-Cortes N, Reyes-Duarte D, Beloqui A, Polaina J, Ghazi I, Golyshina OV (2007) Catalytic role of conserved HQGE motif in the CE6 carbohydrate esterase family. FEBS Lett 581:4657–4662

Mariathasan S, Tan MW (2017) Antibody-antibiotic conjugates: a novel therapeutic platform against bacterial infections. Trends Mol Med 23(2):135–149. https://doi.org/10.1016/j.molmed.2016.12.008. Epub 2017 Jan 23 (Review)

Martin TC, Visconti A, Spector TD, Falchi M (2018) Conducting metagenomic studies in microbiology and clinical research. Appl Microbiol Biotechnol 102(20): 8629–8646. https://doi.org/10.1007/s00253-018-9209-9. Epub 2018 Aug 4 (Review)

Matsuzawa T, Kaneko S, Yaoi K (2015) Screening, identification, and characterization of a GH43 family β-xylosidase/α-arabinofuranosidase from a compost microbial metagenome. Appl Microbiol Biotechnol 99(21):8943–8954. https://doi.org/10.1007/s00253-015-6647-5 (Epub 2015 May 14)

Mayorga OL, Elliott C, Morgavi DP (2018) Addressing global ruminant agricultural challenges through understanding the rumenmicrobiome: past, present, and future. Front Microbiol 9:2161. 25 Sep 2018. https://doi.org/10.3389/fmicb.2018.02161. eCollection 2018 (Review)

Meyer F, Paarmann D, D'Souza M, Olson R, Glass EM, Kubal M, Paczian T, Rodriguez A, Stevens R, Wilke A, Wilkening J, Edwards RA (2008) The metagenomics RAST server—a public resource for the automatic phylogenetic and functional analysis of metagenomes. BMC Bioinformatics 19(9):386. https://doi.org/10.1186/1471-2105-9-386

Mootapally CS, Nathani NM, Patel AK, Jakhesara SJ, Joshi CG (2016) Mining of ruminant microbial phytase (RPHY1) from metagenomic data of mehsani buffalo breed: identification, gene cloning, and characterization. J Mol Microbiol Biotechnol 26(4):252–260. https://doi.org/10.1159/000445321 (Epub 2016 May 13)

Nguyen NH, Maruset L, Uengwetwanit T, Mhuantong W, Harnpicharnchai P, Champreda V, Tanapongpipat S, Jirajaroenrat K, Rakshit SK, Eurwilaichitr L, Pongpattanakitshote S (2012) Identification and characterization of a cellulase-encoding gene from the buffalo rumen metagenomic library. Biosci Biotechnol Biochem 76:1075–1084

Namonyo S, Wagacha M, Maina S, Wambua L, Agaba M (2018) A metagenomic study of the rumen virome in domestic caprids. Arch Virol 163(12):3415–3419. https://doi.org/10.1007/s00705-018-4022-4. Epub 2018 Sep 15. Erratum in: Arch Virol. 10 Oct 2018

Oyama LB, Crochet JA, Edwards JE, Girdwood SE, Cookson AR, Fernandez-Fuentes N, Hilpert K, Golyshin PN, Golyshina OV, Privé F, Hess M, Mantovani HC, Creevey CJ, Huws SA (2017) Buwchitin: a ruminal peptide with antimicrobial potential against *Enterococcus faecalis*. Front Chem 5:51. 12 Jul 2017b. https://doi.org/10.3389/fchem.2017.00051 (eCollection 2017)

Oyama LB, Girdwood SE, Cookson AR, Fernandez-Fuentes N, Privé F, Vallin HE, Wilkinson TJ, Golyshin PN, Golyshina OV, Mikut R, Hilpert K, Richards J, Wootton M, Edwards JE, Maresca M, Perrier J, Lundy FT, Luo Y, Zhou M, Hess M, Mantovani HC, Creevey CJ, Huws SA (2017b) The rumen microbiome: an underexplored resource for novel antimicrobial discovery. NPJ Biofilms Microbiomes. 1(3):33. https://doi.org/10.1038/s41522-017-0042-1.eCollection2017

Patel AB, Patel AK, Shah MP, Parikh IK, Joshi CG (2016) Isolation and characterization of novel multifunctional recombinant family 26 glycoside hydrolase from Mehsani buffalo rumen metagenome. Biotechnol Appl Biochem 63(2):257–265. https://doi.org/10.1002/bab.1358 (Epub 2016 Mar 14)

Prajapati VS, Purohit HJ, Raje DV, Parmar N, Patel AB, Jones OAH, Joshi CG (2016) The effect of a high-roughage diet on the metabolism of aromatic compounds by rumen microbes: a metagenomic study using Mehsani buffalo (Bubalus bubalis). Appl Microbiol Biotechnol 100(3):1319–1331. https://doi.org/10.1007/s00253-015-7239-0 (Epub 2015 Dec 28)

Privé F, Newbold CJ, Kaderbhai NN, Girdwood SG, Golyshina OV, Golyshin PN, Scollan ND, Huws SA (2015) Isolation and characterization of novel lipases/esterases from a bovine rumen metagenome. Appl Microbiol Biotechnol 99(13):5475–5485. https://doi.org/10.1007/s00253-014-6355-6 (Epub 2015 Jan 11)

Rashamuse KJ, Visser DF, Hennessy F, Kemp J, Roux-van der Merwe MP, Badenhorst J, Ronneburg T, Francis-Pope R, Brady D (2013) Characterisation of two bifunctional cellulase-xylanase enzymes isolated from a bovine rumen metagenome library. Curr Microbiol 66(2):145–151. https://doi.org/10.1007/s00284-012-0251-z (Epub 2012 Oct 20)

Schloss PD, Handelsman J (2005) Introducing DOTUR, a computer program for defining operational taxonomic units and estimating species richness. Appl Environ Microbiol 71:1501–1506

Shedova EN, Lunina NA, Berezina OV, Zverlov VV, Schwarz V, Velikodvorskaia GA (2009) Expression of the genes *celA* and *XylA* isolated from a fragment of metagenomic DNA in *Escherichia coli.* Mol Gen Mikrobiol Virusol 2:28–32 (Article in Russian, abstract in English)

Singh B, Bhat TK, Singh B (2003) Potential therapeutic applications of some antinutritional plant secondary metabolites. J Agric Food Chem 51(19):5579–5597. 10 Sep 2003 (Review)

Singh B, Gautam SK, Verma V, Kumar M, Singh B (2008) Metagenomics in animal gastrointestinal ecosystem: Potential biotechnological prospects. Anaerobe 14(3):138–44. https://doi.org/10.1016/j.anaerobe.2008.03.002. Epub 2008 Mar 26 (Review)

Singh B, Mal G, Marotta F (2017) Designer probiotics: paving the way to living therapeutics. Trends Biotechnol 35(8):679–682. https://doi.org/10.1016/j.tibtech.2017.04.001 (Epub 2017 May 5)

Singh KM, Jakhesara SJ, Koringa PG, Rank DN, Joshi CG (2012) Metagenomic analysis of virulence-associated and antibiotic resistance genes of microbes in rumen of Indian buffalo (Bubalus bubalis). Gene 507(2):146–151. https://doi.org/10.1016/j.gene.2012.07.037 (Epub 2012 Jul 28)

Toyoda A, Iio W, Mitsumori M, Minato H (2009) Isolation and identification of cellulose binding proteins from sheep rumen contents. Appl Environ Microbiol 75:1667–1673. https://doi.org/10.1128/AEM.01838-08

White BA, Lamed R, Bayer EA, Flint HJ (2014) Biomass utilization by gut microbiomes. Annu Rev Microbiol 68:279–296. https://doi.org/10.1146/annurev-micro-092412-155618. Epub 2014 Jun 16 (Review)

Xu B, Xu W, Li J, Dai L, Xiong C, Tang X, Yang Y, Mu Y, Zhou J, Ding J, Wu Q, Huang Z (2015) Metagenomic analysis of the Rhinopithecus bieti fecal microbiome reveals a broad diversity of bacterial and glycoside hydrolase profiles related to lignocellulose degradation. BMC Genom 12(16):174. https://doi.org/10.1186/s12864-015-1378-7

Wagner EK, Maynard JA (2018) Engineering therapeutic antibodies to combat infectious diseases. Curr Opin Chem Eng 19:131–141. https://doi.org/10.1016/j.coche.2018.01.007 (Epub 2018 Mar 7)

Yao J, Chen Q, Zhong G, Cao W, Yu A, Liu Y (2014) Immobilization and characterization of tannase from a metagenomic library and its use for removal of tannins from green tea infusion. J Microbiol Biotechnol 24(1):80–86

Wallace RJ, Rooke JA, McKain N, Duthie CA, Hyslop JJ, Ross DW, Waterhouse A, Watson M, Roehe R (2015) The rumen microbial metagenome associated with high methane production in cattle. BMC Genom 16:839. https://doi.org/10.1186/s12864-015-2032-0

Zhao S, Wang J, Bu D, Liu K, Zhu Y, Dong Z, Yu Z (2010) Novel glycoside hydrolases identified by screening a Chinese Holstein dairy cow rumen-derived metagenome library. Appl Environ Microbiol 76:6701–6705. https://doi.org/10.1128/AEM.00361-10

Zhou J, Bao L, Chang L, Liu Z, You C, Lu H (2012) Beta-xylosidase activity of a GH3 glucosidase/xylosidase from yak rumen metagenome promotes the enzymatic degradation of hemicellulosic xylans. Lett Appl Microbiol 54:79–87. https://doi.org/10.1111/j.1472-765X.2011.03175.x

Gut/Rumen Microbiome—A Livestock and Industrial Perspective

2

Abstract

The rumen is a complex microbial ecosystem and an active metabolic organ involved in degradation and fermentation of fibrous plant diets. The rumen microbiota is exceedingly diverse and contains representatives of all three domains—*Eucarya*, methanogenic *Archaea, and Bacteria* and phages. The digestion is carried out by microbial enzymatic, and mechanical means, i.e., mastication of ingested feed, and churning by rumen muscular movements. Short-chain fatty acids, microbial proteins, CO_2, H_2, and CH_4 are major end products of rumen digestive process.

Highlights

- Rumen hosts dense population of microorganisms: Eucarya, Archaea, and Bacteria
- Majority of the rumen microorganisms are obligatory anaerobes and unculturable
- Rumen is a potentially fertile ground for microbial ecologists, molecular microbiologists from commercial point of view.

Keywords

Rumen · Ruminants · Anaerobic fungi · Phytometabolites · Detoxification

2.1 Introduction

Herbivorous animals, especially the ruminants, contribute to global food security by producing protein-rich meat, milk, and fats. The herbivore gut, especially the rumen, is a complex ecosystem and natural habitat of a specialized microbiome that assists in degradation and fermentation of lignocellulosic plant biomass and various phytometabolites that are not degraded by humans or animals. The microbial ecosystem of rumen is exceedingly diverse possessing representatives of all three domains—*Eucarya*, *Archaea*, and *Bacteria*. The microbial population is very dense, comprising of around 10^{10} bacteria/ml, 10^6 protozoa/ml, and 10^3 fungi/ml.

The rumen is buffered over a pH range from 5.7 to 7.3, by phosphate and bicarbonate from saliva, and bicarbonate ions generated as result of rumen fermentation. Rumen temperature varies from 36 to 41 °C depending on species and physiological state of the animals. Digestion occurs primarily by microbial fermentation, and to some extent by physical breakdown during regurgitation and mastication of feed. The end products, namely SCFAs or VFAs and microbial protein, are utilized by the host. The SCFAs constitute up to 80% of the total energy requirement of the ruminants, while microbial proteins leaving the rumen account for 50–90% of the total proteins entering the duodenum or small intestine.

B. Singh et al., *Advances in Animal Biotechnology*,
https://doi.org/10.1007/978-3-030-21309-1_2

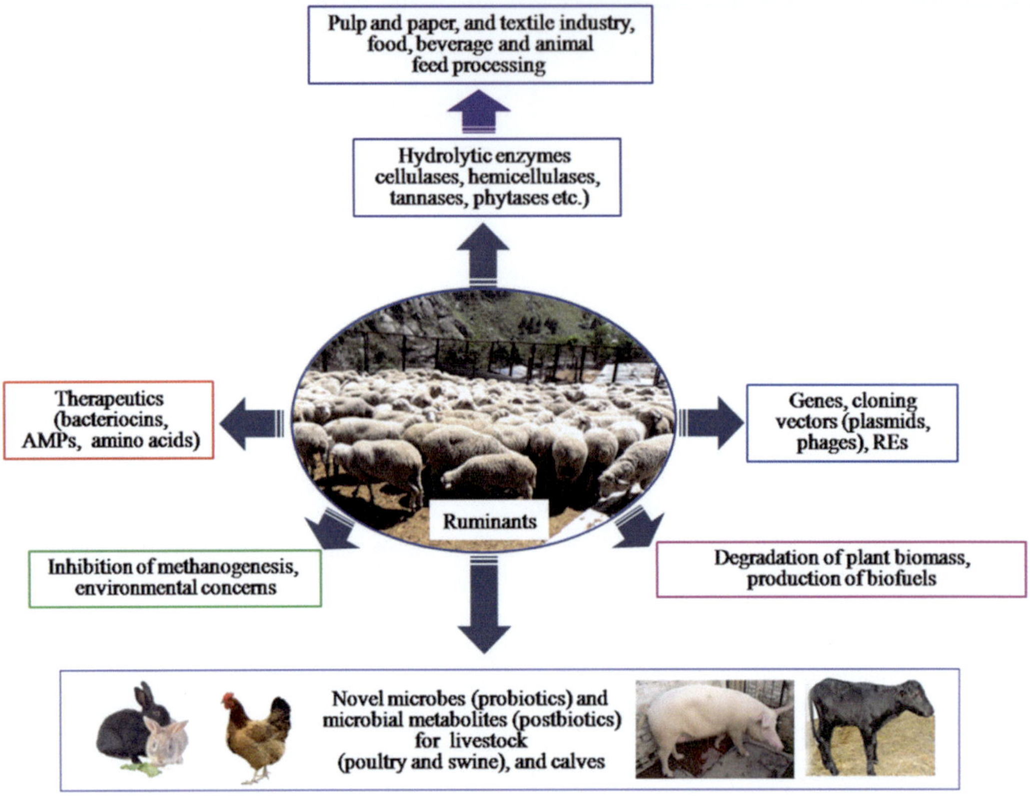

Fig. 2.1 A summary of the motive to study rumen microbiome. Hydrolytic microorganisms and enzymes from rumen microbes are the prime concern of unravelling rumen microbiome. Rumen microbes involved in digestion of forages are powerful sources of valuable enzymes that have industrial applications

In view of the importance of the rumen in the nutrition of the host, and industrial setup (Fig. 2.1), a great deal of efforts has been devoted to investigate the methods for manipulating this composite bionetwork.

2.2 Microbial Diversity in Rumen

The rumen (Fig. 2.2) is the most extensively studied digestive ecosystem of animals and is characterized by its high microbial population density and complexity of interactions among microorganisms as well as between microbes and the host. The complex microbial ecosystem includes bacteria, archaea, protozoa, yeasts and fungi. Besides, several bacteriophages are also identified form rumen. The rumen microbes are not only crucial to nutrition and livestock products quality, but also offer opportunities for industrial applications (Singh et al. 2008; Sirohi et al. 2012).

2.3 Rumen Bacteria

The bacteria are predominant being up to 10^{11} viable cells per gram of rumen contents, representing more than 200 genera. The occurrence of bacteriophages is documented to be 10^7–10^9 particles per gram of the contents. Of the rumen microbes, bacteria and fungi are primarily responsible for degradation of plant fiber, due to their fibrolytic activity and large biomass.

Much of the initial attack on ingested plant material is initiated by rumen bacteria. The majority of rumen bacteria belong to four phyla: *Bacteroidetes*, *Firmicutes*, *Proteobacteria*, and

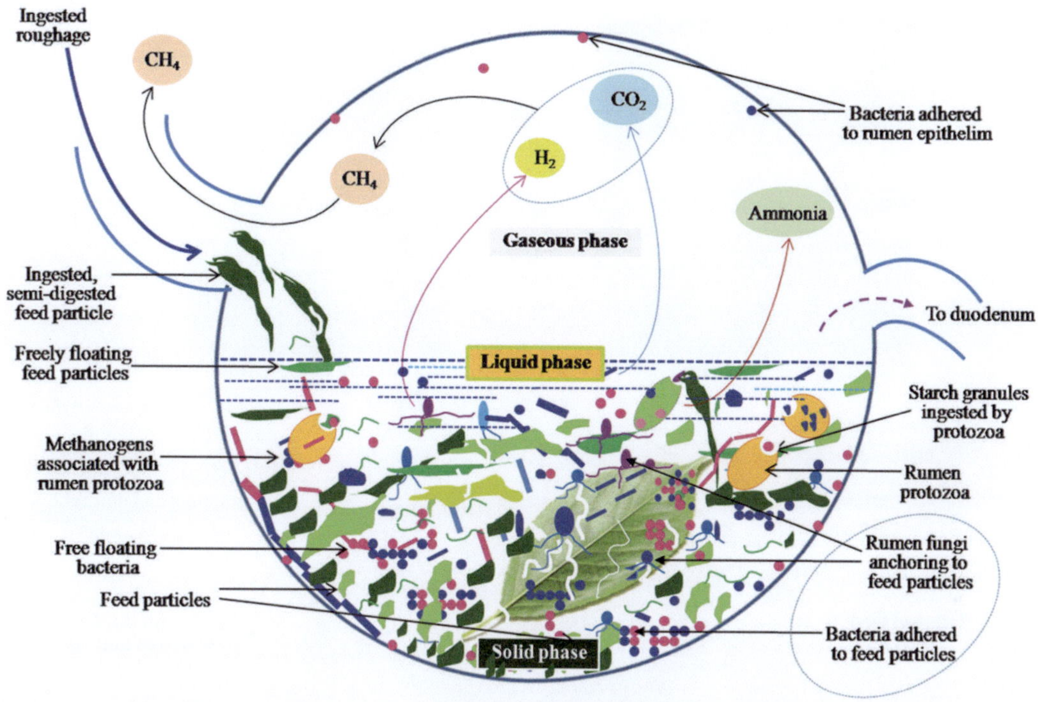

Fig. 2.2 Diagrammatic illustration of rumen, and the rumen components, viz. solid phase, liquid phase, and gaseous phase. The excessive H2 produced in rumen by rumen fungi, and protozoa is utilized by methanogens to produce methane. The methane is eructed into environment through mouth. The microbial biomass produced in rumen is digested in abomasum

Spirochaetes (Firkins and Yu 2006). Krause and Russell (1996) concluded that species diversity of rumen bacteria is much higher than expected 22 species. 16S rRNA genes of bacteria in rumen of different species have revealed a vast diversity of bacterial genera and species.

It is evident that plants synthesize various phytometabolites to ward off plant predators or deter the herbivory by insects and pests. The plant metabolites have various detrimental effects on humans as well as animals (Fig. 2.3). However, certain microbes (summarized in Table 2.1) have developed mechanisms to degrade structural polysaccharides and unavoidably present phytometabolites. Sometimes, the end products of microbial metabolism are more toxic than their precursors.

Box 1. Summary of the advancements in culturomics and genomics methods that advanced the understanding of rumen microbiota

- Development of anaerobic systems such as Hungate's roll tubes to isolate and cultivate the anaerobes (Hungate 1966)
- Developments and modification of culture media, and media supplements such as strained rumen liquor, VFAS, and reducing agents to cultivate anaerobes
- Development of anaerobic incubators or anaerobic work stations

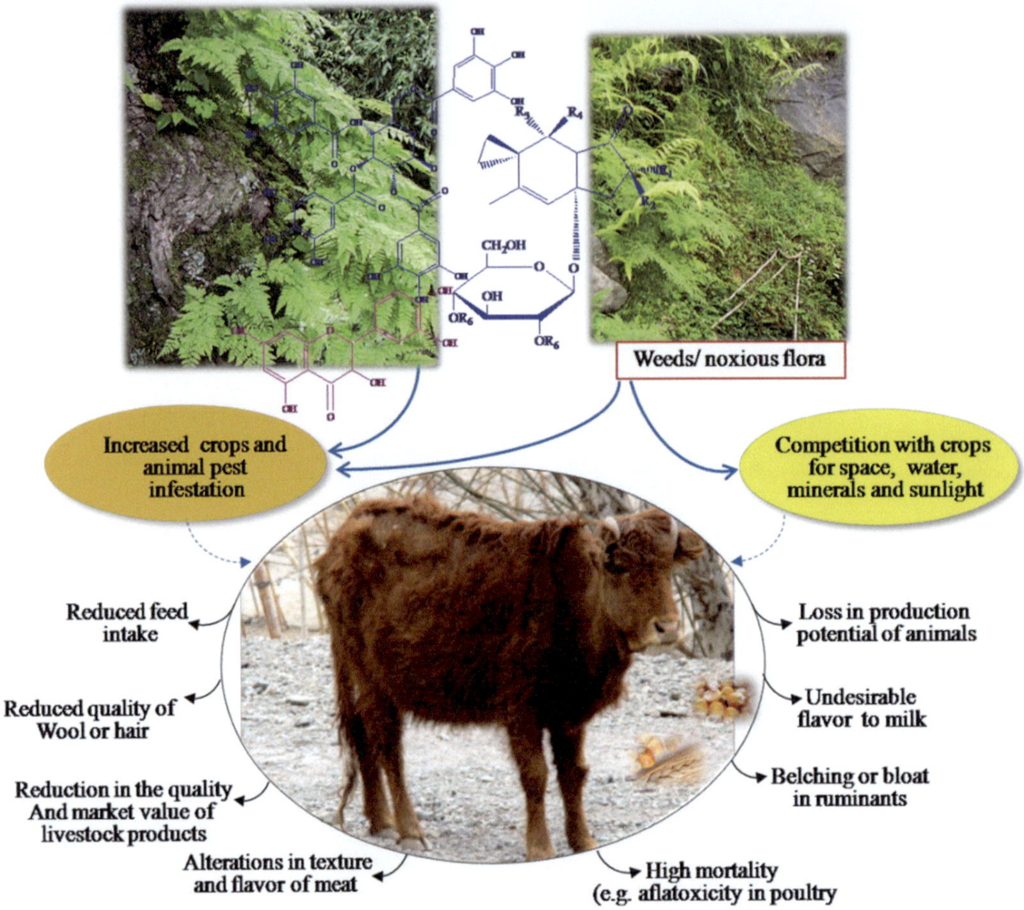

Fig. 2.3 Tree foliages, grasses, and weeds contain various metabolites that exert different effects on host, whereas phytometabolites are degraded by GI microbes; many others (e.g., mimosine) are activated and become more toxic. Plant polysaccharides are enzymatically broken down into olgo- and mono-saccharides, and ultimately produce VFAs, methane, and ethanol (Scott et al. 2011)

- Development of gel microdroplet encapsulation technique for massively parallel microbial cultivation using dilution to extinction methods (Zengler et al. 2002)
- Using improved culture medium in conjunction with a dilution method of liquid culture (Kenters et al. 2011)
- Revelations that reducing agents are not always needed in microbial culture media.

2.4 Rumen Fungi

The anaerobic fungi, which are true fungi comprising of five defined genera, are ubiquitously found in the rumen. The fungal zoospore population density ranges from 10^2 to 10^4 zoospores per gram of rumen contents. The rumen fungi occupy a unique position in the gut of large herbivores, and along with bacteria serve as primary colonizers of ingested plant material.

The gut anaerobic fungi are sensitive to oxygen; hence, they are the only identified obligatory anaerobic fungi. Their physiological,

Table 2.1 Bacterial diversity of the rumen ecosystem of domestic and wild animals, and their ability to degrade forage components

Substrate	Active microbial species
Cellulose	*Bacteroides succinogenes, Clostridium cellobioparum, Clostridium longisporum, Clostridium lochheadii, Eubacterium cellulosolvens, Cillobacterium cellulosolvens, Fibrobacter succinogenes, Ruminococcus flavefaciens*
Hemicellulose	*Bacteroides ruminicola, Eubacterium xylanophilum, E. uniformis, Butyrivibrio fibrisolvens, Prevotella bryantii, Prevotella ruminicola*
Starch	*Bacteroides amylophilus, Bacteroides ruminicola, Streptococcus bovis, Ruminobacter amylophilus, Prevetella ruminicola*
Sugars/dextrins	*Bifidobacterium globosum, B. longum, B. thermophilum, B. ruminale, B. ruminantium, Lactobacillus acidophilus, L. casei L. fermentum, L. plantarum, L. brevis, L. helveticus, Succinivibrio dextrinosolvens, Succinivibrio amylolytica, Selenomonas ruminantium, Treponema zioleckii*
Pectin	*Lachnospira multiparus, Treponema saccharophilum*
Protein degraders	*Clostridium aminophilum, Clostridium bifermentans, Clostridium sticklandii, Megasphaera elsdenii, Peptostreptococcus anaerobius, Prevotella bryantii, Prevotella ruminicola, Ruminobacter amylophilus*
Acid utilizers	*Desulphovibrio desulphuricans, Desulphatomaculum ruminis, Megasphaera elsdenii, Micrococcus lactolytica, Oxalobacter formigenes, Peptostreptococcus elsdenii, Vibrio succinogenes, Veillonella gazogenes, Veillonella alcalescens, Succiniclasticum ruminis, Wolinella succinogenes*
Bacteria producing and utilizing nitrogen	*Clostridium aminophilum, Clostridium sticklandii Peptostreptococcus anaerobius*
Fluoroacetate	*Ancylobacter dichloromethanicus* (ECPB09), *Butyrivibrio fibrisolvens* (genetically modified), *Pigmentiphaga kullae* (ECPB08), *Synergistetes* spp.
Gallic acid	*Streptococcus gallolyticus*
Mimosine degraders	*Synergistes jonesii, Streptococcus lutetiensis, Clostridium butyricum, Lactobacillus vitulinus*, and *Butyrivibrio fibrisolvens* (Derakhshani et al. 2016)
Mimosine and 3,4 DHP	*Clostridium* spp.
Nitro toxins	Clostridia, *Denitrobacterium detoxificans*
Oxalates	*Oxalobacter formigenes, Moorella thermoacetica*
Phytic acid	*Mitsuokella multiacidus* gen. nov., sp. nov., *Prevotella ruminicola*
Pyrrolizidine alkaloids	*Peptococcus heliotrinereducens* sp. nov.
Tannin-polyphenols	*Enterobacter ludwigii, Eubacterium oxidoreducens* sp. nov., *Klebsiella oxytoca*, Lactobacillus sp., *Lonepinella koalarum, Selenomonas ruminantium, Streptococcus* spp., *Streptococcus caprinus*, Fungi: *Aspergillus niger*
Methanogens	*Methanobrevibacter ruminantium, Methanobacterium formicicum, Methanosarcina barkeri, Methanomicrobium mobile*

Tables 2.1, 2.2, and 2.3 are modified from Singh et al. (2015)

biochemical, and molecular studies are hampered due to their inability to grow in vitro. Their DNA has an unusually high A + T content ranging from 72 to 87-mol%.

Described for the first time by Orpin (1976), the anaerobic fungi live in close contact with bacteria and other microorganisms. Their rhizoidal hyphae penetrate the feed particles, tear them into parts, and enhance surface area for enzymatic action. Seventeen distinct anaerobic fungi belonging to five different genera are reported so far. Interestingly, the anaerobic fungi do not possess mitochondria, but instead have hydrogenosomes, which form hydrogen and

Table 2.2 Fungal diversity of the gut ecosystem of domestic and wild animals

Species	Origin and distribution
Fibrolytic fungi	
Anaeromyces mucronatus (*Ruminomyces mucronatus*)	Cattle (Breton et al. 1990)
Caecomyces equi	Horse (Gold et al. 1988)
Neocallimastix frontalis	Cow (Orpin and Mann 1986)
N. patriciarum	Sheep (Orpin and Mann 1986)
N. hurleyensis	Cattle (Webb and Theodorou 1988)
Orpinomyces bovis	Cattle (Barr et al. 1989)
Ruminomyces elegans	Cattle (Ho and Bauchop 1990)
Piromyces communis, Piromyces mae, Piromyces dumbonica	Horse, elephant (Li et al. 1990)
Sphaeromonas communis (*Caecomyces communis*)	Cattle (Orpin 1976; Wubah and Fuller 1991)
Tannin-degrading fungi	
Aspergillus niger	Cattle fed on *Quercus leucotrichiphora* (Bhat et al. 1996)

carbon dioxide from pyruvate and malate during fermentation of carbohydrates (Ljungdahl 2008). A separate chapter has described anaerobic fungi. Some of the rumen fungi obtained by culturing in vitro are summarized in Table 2.2.

The rumen fungi are of high interest to enzymologists and microbial ecologists as they degrade lignocellulose and produce enzymes needed to hydrolyze cellulose and hemicelluloses efficiently. The *Anaeromyces mucronatus* KF8 grown in batch produces a broad range of enzymes, such as cellulase, endoglucanase, xylanase, α-xylosidase, β-xylosidase, α-glucosidase, β-glucosidase, β-galactosidase, mannosidase, cellobiohydrolase, amylase, laminarinase, pectinase, and pectate lyase.

Most of anaerobic fungi are associated with cellulosomal and polycellulosomal complexes in which the enzymes are attached through fungal dockerins to scaffolding proteins. Although cellulosomes from anaerobic fungi share many properties with cellulosomes of anaerobic cellulolytic bacteria and have comparable structures, they differ in terms of amino acid sequences.

Anaerobic fungi are generally cultured in small batch cultures. Some laboratories use milk dilution bottles for culturing rumen fungi (Fig. 2.4). The researchers have suggested different types of culture media for growing the rumen fungi.

2.5 Methanogens

Methanogens are prokaryotic microorganisms that produce methane as end product of complex biochemical pathways. They are strictly anaerobic archaea and occupy a wide variety of anoxic environments. Ruminants are among the prominent sources of methane generation. The archaeal domain in rumen includes largely methanogenic archaea belonging to phylum Euryarchaeota. The rumen archaea produce methane primarily via the hydrogenotrophic pathway (Shi et al. 2014), wherein Methanobrevibacter (Mbb.) subdivided into the SMT clade (*Mbb. smithii*, *Mbb. gottschalki*, *Mbb. millerae*, and *Mbb. thaurei*), or the RO clade (*Mbb. ruminantium* and *Mbb. olleyae*) reduces CO_2 to CH_4 (Tapio et al. 2017; Huws et al. 2018). In addition, a little bit of methane is also generated from reduction of acetate (via acetoclastic pathway), or utilization of methyl groups (methylotrophic pathway) (reviewed in Huws et al. 2018).

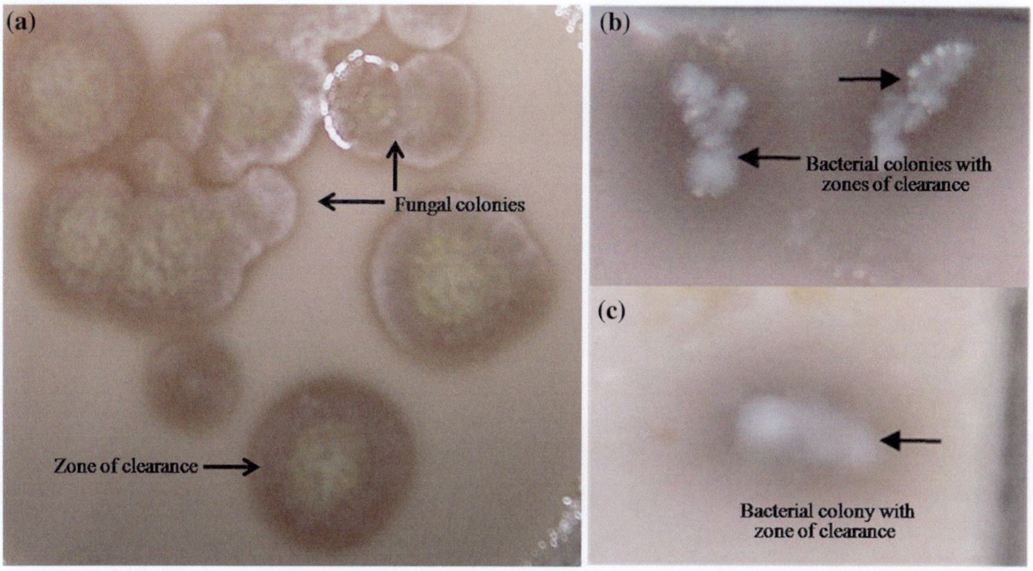

Fig. 2.4 Fungi and bacteria with ability to degrade hydrolysable tannins (HTs). **a** Gut fungal strain from of migratory *Gaddi* goats. **b** Bacterial isolates from Gaddi goats; **c** an isolate from the feces of a Indian gray langur (*Semnopithecus entellus*) that feed oak (*Quercus incana*, and *Q. leucotrichophora*) acorns. The anaerobic culture medium is supplemented with tannic acid (0.4% w/v) to make it selective for tannase activity (Singh et al. 2012)

Rumen methanogens abundance and diversity differ depending on age of ruminants, diet and geographical location, as does the methanogenesis. Interestingly, the methylotrophic methanogenesis mediated by species of the order Methanosarcinales is exclusively more active in young ruminants, while hydrogenotrophic methanogenic activity is high in adult ruminants (Friedman et al. 2017). Methane formation (Fig. 2.5), being an end product of normal process of fiber digestion, is regarded as loss of plant fiber-derived energy. However, methane formation is important as it removes excessive H_2 generated by co-existing rumen fungi and protozoa.

Lowering the methane emission is a hot topic of research. Several strategies such as dietary manipulation, enzyme inhibition of methanogens, defaunation (i.e., induced exclusion of rumen protozoa), phage therapy, and developing vaccines against rumen methanogens have been postulated. These methods target methanogens directly or indirectly. Notably, the viruses selectively active against methanogens, called as archaeophages, are detected in the rumen of some grazing animals. While formulating strategies to lower methane emissions, care must be taken into account to manage the excessive H_2 released in the rumen.

2.6 Rumen Protozoa

Rumen protozoa were first identified in 1843 (Gruby and Delafond 1843). The ciliate protozoa occur widely (10^4–10^6 per gram, distributed over 25 genera) in rumen. Among rumen protozoa (Table 2.3), the most abundant and important are the ciliates, of which the most numerous are representatives of the family *Ophryoscolecidae* (order *Entodiniomorphida*), followed by *Isotrichidae* (order *Vestibuloferida*). The former one is represented by *Entodinium caudatum*, *Eudiplodinium maggii*, *Polyplastron multivesiculatum*, *Diploplastron affine*, *Eremoplastron dilobum*, and *Epidinium caudatum*, and the latter is represented by *Dasytricha ruminantium*, *Isotricha prostoma*, and *I. intestinalis*.

The *Ophryoscolecidae* and *Isotrichidae* differ distinctly in their morphology and metabolic

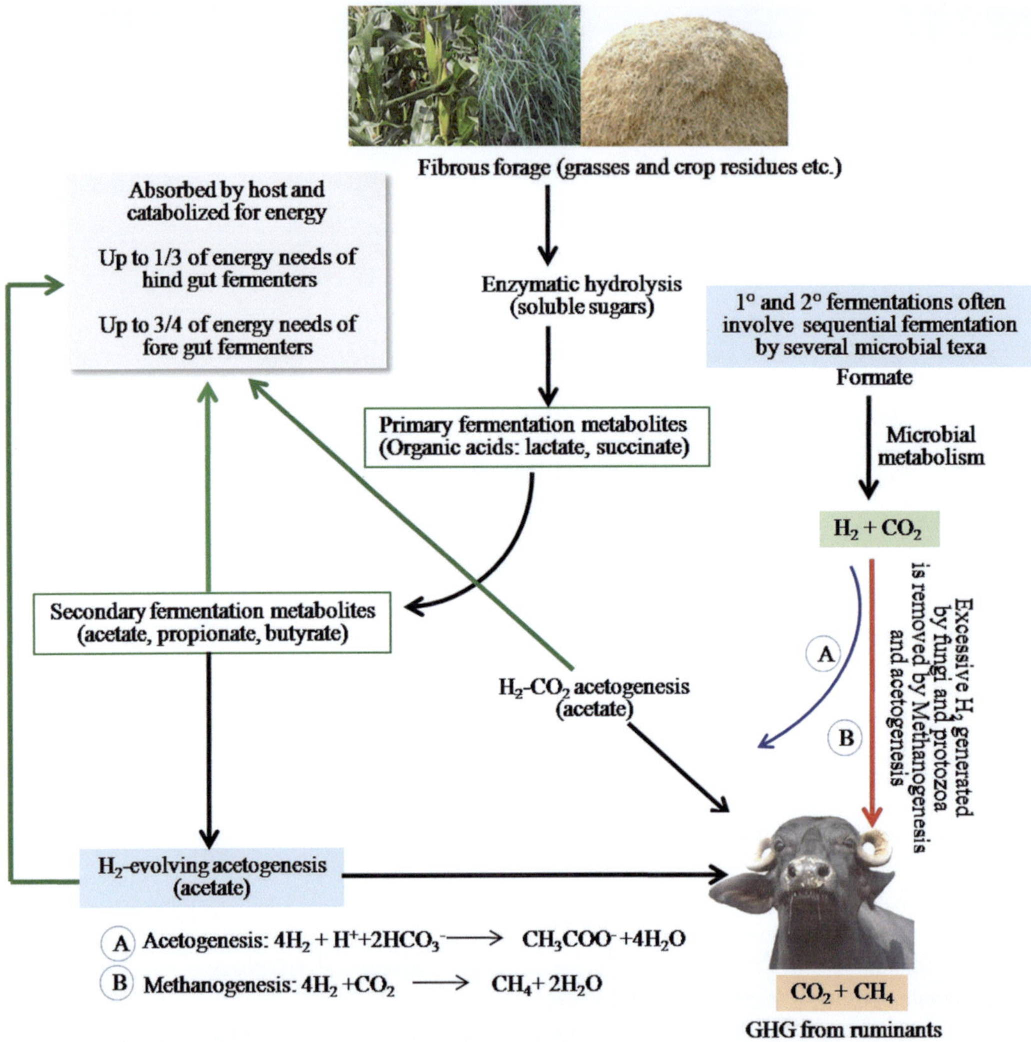

Fig. 2.5 A generalized diagrammatic presentation of different routes of methane formation in ruminants. The forages rich in fiber provide substrate for formation of CH_4 and CO_2, the major greenhouse gases (GHG)

characteristics. The *Ophryoscolecidae* are covered with a rigid pellicle and a dense ciliature at adoral ciliary zone. The caudal end of the cells ciliates bears spines and/or lobes.

Rumen has large populations ($\geq 10^6$ organisms/g of rumen contents) of ciliate protozoa (Piela et al. 2010). Rumen protozoa possess systems for hydrolyzing engulfed feed vesicles. Some bacteria engulfed along with tiny feed particles also contribute to digestive process.

The rumen protozoa contribute H_2, and therefore promote methane production. It has been shown that prokaryotic communities allied with rumen protozoa are compositionally different from their surroundings, and the relation might be as a result of explicit tropism between prokaryotes and protozoa (Levy and Jami 2018).

Despite contributing up to 50% of total microbial biomass, the role of protozoa in rumen ecosystem and function is litigious or unclear. This is because protozoa can be excluded from rumen by various methods such as plant metabolites like saponins (Patra and Saxena 2009), protozoa-specific vaccines (Williams et al.

Table 2.3 Protozoa diversity of the rumen ecosystem

Species	Origin and distribution
Holotrich protozoa	
Isotricha prostoma	Blackbuck, sheep, goats, red deer, buffalo, and zebu cattle
Isotricha intestinalis	Cattle, bison, mouse deer, and red deer
Dasytricha ruminantium	Sheep, blackbuck, buffalo
Oligoisotricha bubali	Buffalo and cattle
Entodiniomorphid protozoa	
Entodinium bovis	Yugoslavian cattle, zebu cattle, and domestic buffaloes
E. bubalum	Water buffalo
E. bursa	Cattle sheep and goat
E. caudatum	Cattle sheep and goat
E. chatterjeei	Indian goat and water buffaloes
E. longinucleatum	Cattle and water buffalo
Diplodinium dendatum	Widely distributed
D. indicum	Zebu cattle
Diploplastron affine	Domesticated ruminants
Eremoplastron asiaticus	Indian cattle
E. bubalus	Brazilian cattle and water buffalo
Eudiplodinium maggii	Widely distributed
Ostracodinium trivesiculatum	Cattle and water buffalo
Polyplastron Multivesiculatum	Cattle, sheep, goat, blackbuck
Metadinium medium	African reedbuck, cattle, water buffaloes
Epidinium caudatum	Cattle, sheep, goat
Ophryoscolex caudatus	Widely distributed
Caloscolex camelicus	Dromedary camel

2008), and antibodies (Williams et al. 2014), a process called as defaunation. The defaunated ruminants can survive for a long period (Newbold et al. 2015).

2.7 Bacteriophages in Rumen

Viruses of GI bacteria have ubiquitous distribution including rumen, GI tract of birds, and humans. Existence of bacteriophages in rumen was noted as far back as late 1960s (Adams et al. 1966; Hoogenraad et al. 1967; Paynter et al. 1969). While the research on rumen phages also continued during 1970s and 1980s, the comprehensive reports on existence of phages in rumen have come recently. The conventional methods of studying rumen phages included morphological examination by electron microscopy, genome length profiling by pulsed-field gel electrophoresis. More recent methods based on metagenomics provide insights into virome and their genetic setup.

Metagenomics has provided a comprehensive knowledge on existence of bacteriophages in the rumen of various ruminant species (Yutin et al. 2015; Anderson et al. 2017; Namonyo et al. 2018) and avian gut (Fawaz et al. 2016; Vibin et al. 2018). The rumen virophage genome encodes typical virophage major capsid proteins,

ATPase, and proteases along with a Polinton-type, protein-primed family B-DNA polymerase (Yutin et al. 2015).

Complete genome sequences of lytic phages, belonging to the viral order Caudovirales, and capable of infecting rumen *Bacteroides* sp., *Ruminococcus* sp., and *Streptococcus* sp., have been reported. In addition, co-examination of bacterial host genome revealed the existence of several genes responsible for modulating phage–host gene interaction, CRISPR/Cas elements and restriction-modifications' defense system in the phages (Gilbert et al. 2017).

Rumen viruses have both detrimental (reduction of feed utilization, transfer toxin genes) and advantageous (maintaining bacterial population balance, using them as phage therapy to control methanogens, and novel genes and enzymes) effects.

2.8 Rumen Microbial Manipulations to Enhance Animal Productivity

Rumen ecosystem can be manipulated by dietary or genetic manipulations to later the quality as well as quantity of the animal products. It is difficult to alter the genetic profile and rumen microorganisms. The rumen resists any exogenous microorganism instilled into it. That is why the genetically engineered microorganisms when introduced into rumen survive only for limited duration.

Alternatively, dietary manipulations (summarized in Box 2) are suggested to alter the microbial outputs and composition of milk.

Box 2. Summarized view of dietary manipulations to enhance rumen outputs

- **High-quality forage**: Provides essential nutrients such as fiber, carotenes, readily fermentable saccharides that promote microbial diversity and growth in the rumen that are later used as proteins for animal.
- **Saponins**: Certain phytometabolites such as saponins present in legumes inhibit rumen protozoa, minimize methane production, and promote microbial growth.
- **Tannins**: Condensed tannins at 1-3% of total DM intake are useful. They enhance the flow of rumen bypass protein by forming tannin–protein complexes, prevent oxidative stress, and suppress the growth of gastrointestinal helminths.
- **Terpineol**: Phytometabolites such as terpineol, camphor, bornyl acetate, α-pinene, thymoquinone, and thymol-induced inhibition of methane (Joch et al. 2018).
- **Antibiotics**: The antibiotics in ruminants favor the proliferation of propionate-producing bacteria. However, use of antibiotics is prohibited or restricted due to several reasons including secretion into milk, remaining in meat, or emergence of antibiotic resistance.
- **Probiotics**: Probiotics promote animal rumen efficiency. Yeasts prevent rumen acidosis and enhance feed utilization. Recombinant probiotics and postbiotics alleviate might forage toxicity, enhance nutrient utilization, and suppress pathogens (Singh et al. 2017; Huws et al. 2018).
- **Unsaturated fatty acids**: Prevent bloat, inhibit protozoa, increase general microbial population, and hamper methane emission.
- **Essential oils**: Shift in microbial population, prevent rumen acidosis, improved feed utilization, increased body weight gain in dairy calves, anti-protozoal effects and inhibition of methane (Kazemi-Bonchenari et al. 2018).
- **Vegetable oils**: Vegetable soils such as sunflower and soybean oil enhanced

trans-11 C18:1 fatty acid, *cis*-9, *trans*-11 conjugated linoleic acid in rumen (Roy et al. 2018).

2.9 Outlook and Challenges

The world population is expected to reach around 9.7 billion by 2050. The protein requirement of this gigantic population has to be satisfied primarily by animal produce (Huws et al. 2018). This implies animal production has to increase, while land availability to grow fodder crops and grains for animals will shrink. Although challenging rumen microbiome can assist meeting the goals.

The rumen microorganisms are crucial for deriving nutrients and energy precursors. Dearth of detailed understanding of rumen ecosystem on one hand, and inability to grow them in vitro, hampers utilizing their potential. Therefore, culture-independent genomics and metagenomics techniques are used to study structure complexity of rumen microorganism. Such approaches may provide detailed knowledge of ruminal enzyme systems, the absence of which has hampered rumen scientists in solving age-old challenges such as improving the digestion of low-quality forages and the efficiency of ruminal nitrogen utilization.

Possibilities to manipulate rumen ecosystem and meet the global challenges through breeding management and nutrition interventions during early life of ruminants have emerged as possible favorable interventions.

It is evident that certain specialized microorganisms establish in GI tract in response to biochemicals present in dietary components. Therefore, the ruminants feeding on fibrous roughage should be opted for isolation cellulolytic microorganisms. Likewise, browsing ruminants such as goats, and deer, and some rodents harbour microbial consortia to degrade toxic plant metabolites like tannin-polyphenols, saponins, alkaloids, nitrotoxins, and oxalates.

As the livestock sector grows in population and productivity in developing countries, it is needed that microbial–host interactions should be understood for better management of livestock as well as natural feeding resources. Besides, an important motive to study rumen microorganisms is to evolve strategies to lower methane emissions.

2.10 Conclusions

The rumen is the first and the largest chamber of the complex stomach of the ruminant ungulates. It is a highly specialized anaerobic ecosystem with very dense populations of microorganisms. The microorganisms in rumen are mostly obligate anaerobes, and contribute to digestion of plant fiber. The knowledge of rumen ecosystem is important not only to enhance feed conversion efficiency, but also to utilize the rumen microorganisms for commercial applications. It is necessary to investigate the new technologies to grow rumen and gut bacteria and utilize their beneficial traits.

References

Adams JC, Gazaway JA Jr, Brailsford MD, Hartman PA, Jacobson NL (1966) Isolation of bacteriophages from the bovine rumen. Experientia 22:717–718. https://doi.org/10.1007/BF01901335

Anderson CL, Sullivan MB, Fernando SC (2017) Dietary energy drives the dynamic response of bovine rumen viral communities. Microbiome 5(1):155. https://doi.org/10.1186/s40168-017-0374-3

Barr DJS, Kudo H, Jackober KD, Cheng KJ (1989) Morphology and development of rumen fungi: *Neocallimastix* sp., *Piromyces communis*, *Orpinomyces bovis*, gen. nov. sp. nov. Can J Bot 67:2815–2824

Bhat TK, Makkar HP, Singh B (1996) Isolation of a tannin-protein complex degrading fungus from the faeces of hill cattle. Lett Appl Mirobiol 22:257–258

Breton A, Bernalier A, Dusser M (1990) *Anaeromyces mucronatus* nov. gen, nov. sp. a new strictly anaerobic rumen fungus with polycentric thallus. FEMS Microbiol Lett 70:177–182

Derakhshani H, Corley SW, Al Jassim R (2016) Isolation and characterization of mimosine, 3,4 DHP and 2,3 DHP degrading bacteria from a commercial rumen inoculum. J Basic Microbiol 56(5):580–585. https://doi.org/10.1002/jobm.201500590

Fawaz M, Vijayakumar P, Mishra A, Gandhale PN, Dutta R, Kamble NM, Sudhakar SB, Roychoudhary P, Kumar H, Kulkarni DD, Raut AA (2016) Duck gut viral metagenome analysis captures snapshot of viral diversity. Gut Pathog 8:30. https://doi.org/10.1186/s13099-016-0113-5 (eCollection 2016)

Firkins JL, Yu Z (2006) Characterisation and quantification of the microbial populations of the rumen. In: Sejrsen K, Hvelplund T, Nielsen MO (eds) Ruminant physiology. Digestion, metabolism, and impact of nutrition on gene expression, immunology, and stress. Wageningen Academic Publishers, Wageningen, Netherlands, pp 19–54

Friedman N, Jami E, Mizrahi I (2017) Compositional and functional dynamics of the bovine rumen methanogenic community across different developmental stages. Environ Microbiol 19(8):3365–3373. https://doi.org/10.1111/1462-2920.13846 (Epub 2017 Aug 7)

Gilbert RA, Kelly WJ, Altermann E, Leahy SC, Minchin C, Ouwerkerk D, Klieve AV (2017) Toward understanding phage: host interactions in the rumen; complete genome sequences of lytic phages infecting rumen bacteria. Front Microbiol 8:2340. https://doi.org/10.3389/fmicb.2017.02340 (eCollection 2017)

Gold JJ, Heath IB, Bauchop T (1988) Ultrastructural description of a new chytrid genus of caecum anaerobe *Caecomyces equi* gen. nov., sp. nov., assigned to the Neocallimasticaceae. BioSystems 21:403–415

Gruby D, Delafond HMO (1843) Recherches ser des animalcules se de veloppant en grand nombre dans l'estomac et dans les intestins, pedant la digestion des animaux herbivores et carnivores. Compt Rend Acad Sci 17:1304–1308 (cited in)

Ho YW, Bauchop T (1990) *Ruminomyces elegans* gen. sp. nov. A polycentric anaerobic rumen fungus from cattle. Mycotaxon 38:397–405

Hoogenraad NJ, Hirk FJ, Holmes I, Millis NF (1967) Bacteriophages in rumen contents of sheep. J Gen Virol 1(4):575–576 (No abstract available)

Hungate RE (1966) The rumen and its microbes. Academic Press, New York, NY

Huws SA, Creevey CJ, Oyama LB, Mizrahi I, Denman SE, Popova M, Muñoz-Tamayo R, Forano E, Waters SM, Hess M, Tapio I, Smidt H, Krizsan SJ, Yáñez-Ruiz DR, Belanche A, Guan L, Gruninger RJ, McAllister TA, Newbold CJ, Roehe R, Dewhurst RJ, Snelling TJ, Watson M, Suen G, Hart EH, Kingston-Smith AH, Scollan ND, do Prado RM, Pilau EJ, Mantovani HC, Attwood GT, Edwards JE, McEwan NR, Morrisson S, Mayorga OL, Elliott C, Morgavi DP (2018) Addressing global ruminant agricultural challenges through understanding the rumen microbiome: past, present, and future. Front Microbiol 9:2161. doi: 10.3389/fmicb.2018.02161. eCollection 2018 (2018 Sep 25)

Kenters N, Henderson G, Jeyanathan J, Kittelmann S, Janssen PH (2011) Isolation of previously uncultured rumen bacteria by dilution to extinction using a new liquid culture medium. J Microbiol Methods 84(1): 52–60. https://doi.org/10.1016/j.mimet.2010.10.011 (Epub 2010 Oct 26)

Joch M, Mrázek J, Skřivanová E, Čermák L, Marounek M (2018) Effects of pure plant secondary metabolites on methane production, rumen fermentation and rumen bacteria populations in vitro. J Anim Physiol Anim Nutr (Berl) 102(4):869–881. https://doi.org/10.1111/jpn.12910 (Epub 2018 Apr 29)

Kazemi-Bonchenari M, Falahati R, Poorhamdollah M, Heidari SR, Pezeshki A (2018) Essential oils improved weight gain, growth and feed efficiency of young dairy calves fed 18 or 20% crude protein starter diets. J Anim Physiol Anim Nutr (Berl) 102(3): 652–661. https://doi.org/10.1111/jpn.12867 (Epub 2018 Jan 17)

Krause DO, Russell JB (1996) How many ruminal bacteria are there? J Dairy Sci 79(8):1467–1475 (Review)

Levy B, Jami E (2018) Exploring the prokaryotic community associated with the rumen ciliate protozoa population. Front Microbiol 9:2526. https://doi.org/10.3389/fmicb.2018.02526 (eCollection 2018)

Li J, Heath IB, Bauchop T (1990) *Piromyces mae* and *Piromycesdumbonica*, two new species of uniflagellate anaerobic chitridiomycete fungi from the hindgut of the horse and elephant. Can J Bot 68:1021–1033

Ljungdahl LG (2008) The cellulase/hemicellulase system of the anaerobic fungus *Orpinomyces* PC-2 and aspects of its applied use. Ann NY Acad Sci 1125:308–321

Namonyo S, Wagacha M, Maina S, Wambua L, Agaba M (2018) A metagenomic study of the rumen virome in domestic caprids. Arch Virol 163(12):3415–3419. https://doi.org/10.1007/s00705-018-4022-4 (Epub 2018 Sep 15. Erratum in: Arch Virol. 2018 Oct 10)

Newbold CJ, de la Fuente G, Belanche A, Ramos-Morales E, McEwan NR (2015) The role of ciliate protozoa in the rumen. Front Microbiol 26(6):1313. https://doi.org/10.3389/fmicb.2015.01313 . (eCollection2015. Review)

Orpin GC (1976) The characterization of the rumen bacterium Eadi's oval *Magnoovum* gen. nov. *eadii* sp. nov. Arch Microbiol 111:155–159

Orpin CG, Mann EA (1986) *Neocallimastix patriciarum*: new member of the Neocallimasticaceae inhabiting the sheep rumen. Trans Br Mycol Soc 86:178–181

Patra AK, Saxena J (2009) Dietary phytochemicals as rumen modifiers: a review of the effects on microbial populations. Antonie Van Leeuwenhoek 96(4): 363–375. https://doi.org/10.1007/s10482-009-9364-1 (Epub 2009 Jul 7. Review)

Paynter MJ, Ewert DL, Chalupa W (1969) Some morphological types of bacteriophages in bovine rumen contents. Appl Microbiol 18(5):942–943

Piela P, Michałowski T, Miltko R, Szewczyk K, Sikora R, Grzesiuk E, Sikora A (2010) Can a fermentation gas mainly produced by rumen Isotrichidae ciliates be a potential source of biohydrogen and a fuel for a chemical fuel cell? J Microbiol Biotechnol 20: 1092–1100

Roy A, Mandal GP, Patra AK (2017) Effects of different vegetable oils on rumen fermentation and conjugated linoleic acid concentration in vitro. Vet World 10(1): 11–16. https://doi.org/10.14202/vetworld.2017.11-16 (Epub 2017 Jan 8)

Scott KP, Duncan SH, Louis P, Flint HJ (2011) Nutritional influences on the gut microbiota and the consequences for gastrointestinal health. Biochem Soc Trans 39:1073–1078

Shi W, Moon CD, Leahy SC, Kang D, Froula J, Kittelmann S, Fan C, Deutsch S, Gagic D, Seedorf H, Kelly WJ, Atua R, Sang C, Soni P, Li D, Pinares-Patiño CS, McEwan JC, Janssen PH, Chen F, Visel A, Wang Z, Attwood GT, Rubin EM (2014) Methane yield phenotypes linked to differential gene expression in the sheep rumen microbiome. Genome Res 24(9):1517–1525. https://doi.org/10.1101/gr.168245.113 (Epub 2014 Jun 6)

Singh B, Bhat TK, Sharma OP, Kanwar SS, Rahi P, Gulati A (2012) Isolation of tannase-producing *Enterobacter ludwigii* GRT-1 from the rumen of migratory goats. Small Ruminant Res 102:172–176

Singh B, Gautam SK, Verma V, Kumar M, Singh B (2008) Metagenomics in animal gastrointestinal tract- potential biotechnological prospects. Anaerobe 14:138–144

Singh B, Gautam SK, Chauhan MS, Singla SK (2015) Textbook of animal biotechnology. The Energy and Resources Institute (TERI), New Delhi. ISBN-13: 978-8179933275

Singh B, Mal G, Marotta F (2017) Designer probiotics: Paving the way to living therapeutics. Trends Biotechnol 35(8):679–682. https://doi.org/10.1016/j.tibtech.2017.04.001 (Epub 2017 May 5) 2017 Aug

Sirohi SK, Singh N, Dagar SS, Puniya AK (2012) Molecular tools for deciphering the microbial community structure and diversity in rumen ecosystem. Appl Microbiol Biotechnol (in press)

Tapio I, Snelling TJ, Strozzi F, Wallace RJ (2017) The ruminal microbiome associated with methane emissions from ruminant livestock. J Anim Sci Biotechnol 19(8):7. https://doi.org/10.1186/s40104-017-0141-0. eCollection2017.Review

Webb J, Theodorou MK (1988) A rumen anaerobic fungus of the genus *Neocallimastix*: ultrastructure of the polyflagellate zoospore and young thallus. Biosystems 21:393–401

Vibin J, Chamings A, Collier F, Klaassen M, Nelson TM, Alexandersen S (2018) Metagenomics detection and characterisation of viruses in faecal samples from Australian wild birds. Sci Rep 8(1):8686. https://doi.org/10.1038/s41598-018-26851-1

Williams YJ, Rea SM, Popovski S, Skillman LC, Wright AD (2014) Technical note: protozoa-specific antibodies raised in sheep plasma bind to their target protozoa in the rumen. J Anim Sci 92(12):5757–5761. https://doi.org/10.2527/jas.2014-7873

Williams YJ, Rea SM, Popovski S, Pimm CL, Williams AJ, Toovey AF, Skillman LC, Wright AD (2008) Reponses of sheep to a vaccination of entodinial or mixed rumen protozoal antigens to reduce rumen protozoal numbers. Br J Nutr 99(1): 100–109 (Epub 2007 Aug 15)

Wubah DA, Fuller MS (1991) Studies on *Caecomyces communis*: morphology and development. Mycologia 83:303–310

Yutin N, Kapitonov VV, Koonin EV (2015) A new family of hybrid virophages from an animal gut metagenome. Biol Direct 25(10):19. https://doi.org/10.1186/s13062-015-0054-9

Zengler K, Toledo G, Rappe M, Elkins J, Mathur EJ, Short JM, Keller M (2002) Cultivating the uncultured. Proc Natl Acad Sci U S A 99(24):15681–15686 (Epub 2002 Nov 18)

3 Anaerobic Gut Fungi—A Biotechnological Perspective

Abstract

The ruminants are herbivorous animals that thrive on fibrous roughage, plant foliage, crop, and agro-industrial by-products. The microbiome present in alimentary tract of herbivores serves as an efficient system for saccharification and fermentation of ingested plant biomass and converts it into microbial proteins, ammonia, short-chain organic acids, and gases such as CO_2, H_2, and CH_4. While early studies had focused on anaerobic fungi because of their unusual metabolism and biology, in current perspective, the anaerobic fungi are viewed for potential hydrolytic genes and enzymes biotechnological applications.

Highlights

- Anaerobic gut fungi inhabit rumen and cecum of mammalian and reptile herbivores
- The gut fungi secrete enzymes that are superior to those from fibrolytic bacteria
- The enzymes of anaerobic fungi in view of their ability to utilize multiple substrates have industrial applications.

Keywords

Gut fungi · Fiber degradation · Cellulases · Tannase · Feed additives

3.1 Introduction

The way the foodstuff of herbivore differs from monogastric animals; the anatomy and mode of digestion also differ. Crop residues, fibrous agro-industrial by-products, and forages including tree leaves, grasses, plants, and bushes are fed to livestock animals. The wild herbivores also thrive primarily on various types of flora.

The digestion in herbivores includes a cascade of microbial and enzymatic processes in different parts of their gastrointestinal (GI) tract. Primary fermentation in ruminants commences in reticulo-rumen, followed by enzymatic (cellulases, hemicellulases, proteases, esterases, etc.) hydrolysis and degradation in abomasums and small intestine. Finally, fermentation completes in cecum and large intestine. When structural carbohydrates are fermented, approximately 75% of the energy in the substrate is recovered in the form of volatile fatty acids (VFAs).

3.2 Mutualism Between Fungi and Animals

The anaerobic gut fungi occupy a unique niche in the GI tract of herbivorous animals (Fig. 3.1), and act as primary colonizers of ingested plant material. Hence, rumen fungi are unique in

B. Singh et al., *Advances in Animal Biotechnology*,
https://doi.org/10.1007/978-3-030-21309-1_3

combining the resilience and invasiveness of fungi with metabolic activities of anaerobic fibrolytic bacteria. Interestingly, the anaerobic fungi do not possess mitochondria, but instead have hydrogenosomes, which form H_2 and CO_2 from pyruvate and malate during fermentation of carbohydrates (Ljungdahl 2008).

Despite the detection of zoospores of anaerobic fungi as early as in the 1910s, the proof of anaerobic fungi as an important component of herbivore digestive system was confirmed much late (Orpin 1975, 1976). The molecular biological evidences have provided valuable insights into their survival, distribution among a broad range of herbivorous animals, and adaptation to various niches (Liggenstoffer et al. 2010; Comlekcioglu et al. 2010; Kõljalg et al. 2013; Struchtemeyer et al. 2014; Makhuvele et al. 2017). Anaerobic rumen fungi first described by Orpin (1975) are evolved to co-existing bacteria and other microbes in rumen and cecum. The rumen fungi are obligatory anaerobic, but phenotypic and molecular analysis of this important group is hampered by difficulties in isolating and culturing them out of their natural habitat. The gut fungi are highly oxygen- and temperature-sensitive (Fig. 3.2). Their DNA has unusually high A + T content.

The fungi in ruminants are associated with ingested feed particles where they synthesize highly active fiber-degrading enzymes while their free-living zoospores are found in fluid phase. They possess remarkable cellulolytic and hemicellulolytic enzymatic machinery for

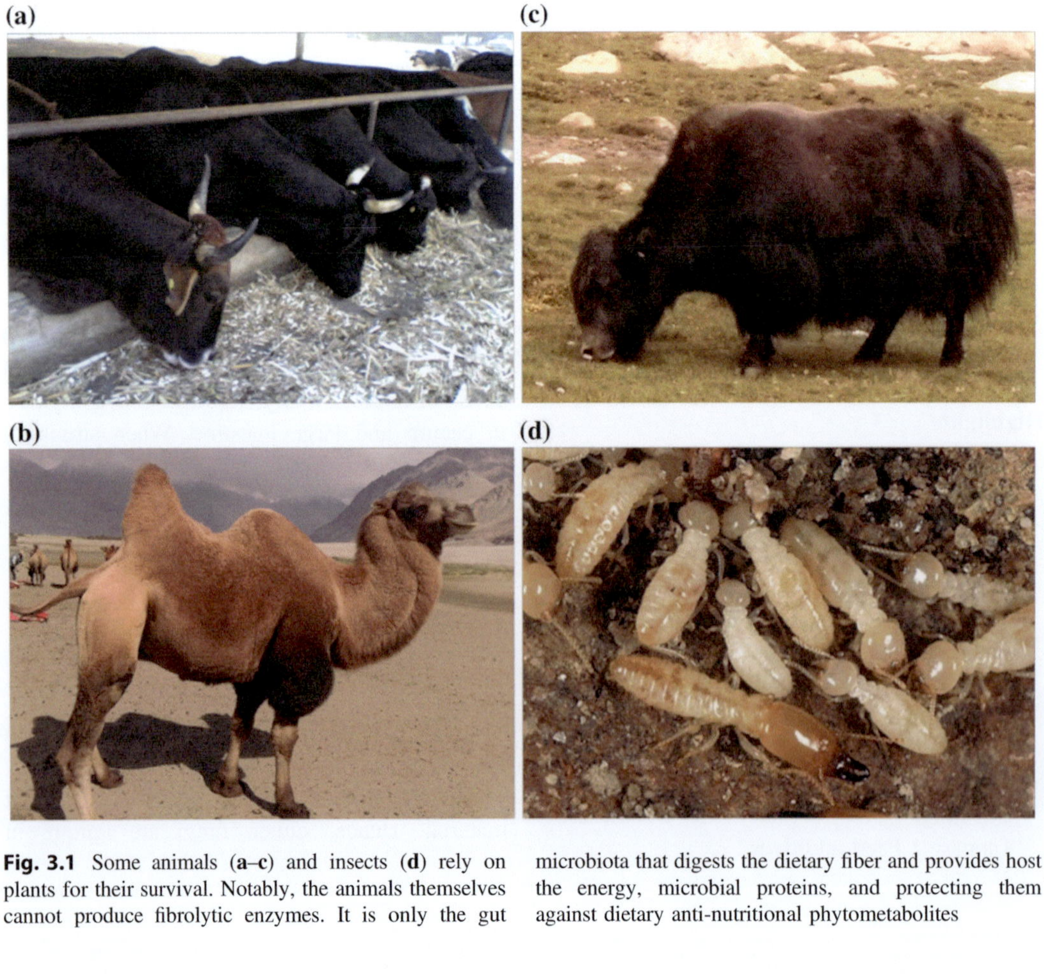

Fig. 3.1 Some animals (**a**–**c**) and insects (**d**) rely on plants for their survival. Notably, the animals themselves cannot produce fibrolytic enzymes. It is only the gut microbiota that digests the dietary fiber and provides host the energy, microbial proteins, and protecting them against dietary anti-nutritional phytometabolites

Fig. 3.2 Isolation technique for anaerobic gut fungi. **a** Enriching the fungi by providing fibrolytic substrate in Hungate's roll tubes; **b** Isolation on anaerobic agar in serum vials adapted to anaerobic environment. The enriched inocula are used to isolate fungi

sustained utilization of dietary ingredients. By colonizing to plant biomass, the fungi excrete extracellular enzymes that breakdown plant cell wall and release simpler sugar that is utilized by other gut microorganisms. They penetrate the cuticle, attack recalcitrant plant cell walls thereby weaken the forage fiber making it accessible to fibrolytic bacteria. Rumen fungi account for up to 8% of the total microbial biomass in rumen and are adapted to anaerobic environment of gut ecosystem (Youssef et al. 2013). Paul et al. (2003) have shown that compared to readily fermentable carbohydrates, the fibrous diet promotes growth and abundanceof fibrolytic fungi. The fibrolytic activity is more in fungal strains obtained from wild herbivores compared to those from domesticated ruminants (Paul et al. 2010). Seventeen distinct anaerobic fungi belonging to five different genera are reported so far from around fifty animal species (Table 3.1).

3.3 Phylogeny and Classification of Anaerobic Gut Fungi

For a long time, the rumen microbiota was considered comprising of anaerobic bacteria and protozoa. Orpin (1975) for the first time deduced that besides bacteria and protozoa, the rumen contains fungi that are obligatory anaerobic and essential to digestive system. The subsequent studies showed that gut fungi are true fungi with distribution in several species. Culture-independent molecular biological and phylogenetic studies documented global diversity of gut fungi from various herbivores including wild and undomesticated ungulates.

Based on ultrastructural properties of zoospores, the anaerobic fungi were assigned to order Spizellomycetales and the family, Neocallimasticaceae (Barr 1988). The monocentric fungus species family involves three genera, namely *Neocallimastix*, *Piromyces* (previously, Piromonas), and Caecomyces (previously, Sphaeromonas) (Gold et al. 1988). In addition, three genera comprising of polycentric fungi are Orpinomyces (Barr et al. 1989; Kudo et al. 1990), Anaeromyces (Breton et al. 1990), and Cyllamyces (Ozkose et al. 2001).

Genome-enabled mycology classifies the anaerobic GI fungi as single order (Neocallimastigales) within recently postulated phylum Neocallimastigomycota (phyla nov.) (Hibbett et al. 2007). Based on analysis of DNA sequences, six genera are found within the Neocallimastigales—Neocallimastix, Piromyces, Orpinomyces, Anaeromyces, Caecomyces, and Cyllamyces.

While studies before genomics era had focused on their atypical biology and metabolism, the majority of subsequent studies have emphasized the phylogeny and biotechnological potential of their cellulases, xylanases, and phenolic esterases. Although multiple uncharacterized isolates are reported, 8 genera and around 20 species are documented so far (Box 1) (Griffith et al. 2010).

Metagenomic studies have shown the presence of 198 fungi belonging to predominating

Table 3.1 Summary of fungal diversity of the gut ecosystem of herbivores

Fungus	Origin/host	Characteristics references
Anaeromyces mucronatus (*Ruminomyces mucronatus*)	Cattle	Fiber digestion (Breton et al. 1990)
Aspergillus niger van Tiegham	Cattle	The fungi produce tannin acyl hydrolase involved in degradation of hydrolyzable tannins (Bhat et al. 1996; Bhardwaj et al. 2003)
Neocallimastix frontalis	Cow	The study reports on existence of anaerobic fungi in rumen of cattle (in Orpin and Mann 1986)
Feramyces austinii, gen. nov., sp. nov.,	Wild Barbary sheep and fallow deer	Utilization of various sugars and plant biomass by this novel rumen fungus (Hanafy et al. 2018)
Liebetanzomyce spolymorphus, gen. et sp. nov.	Goat rumen	Description of a new anaerobic fungus resembling to other anaerobic fungi (Joshi et al. 2018)
Neocallimastigomycota phylum	Green iguana American Elk, goral, bontebok southern gerenuk, and Nile lechwe	The comprehensive study on the presence of fungi in the gut of various herbivores (Liggenstoffer et al. 2010)
Neocallimastix hurleyensis	Cattle	Fibrolytic activities (Webb and Theodorou 1988)
Neocallimastix patriciarum	Sheep	Fibrolytic activities (Orpin and Mann 1986)
N. frontalis Orpinomyces joyonii,	Yak	General description of fungi recognized yak (Wang et al. 2017)
Caecomyces sp. *Orpinomyces bovis*	Cattle	Fiber digestion (Barr et al. 1989)
Pecoramyces ruminantium, gen. nov., sp. nov	Cattle and sheep feces	Fiber degradation, i.e., cellulolytic and xylanolytic activities (Hanafy et al. 2017)
Sphaeromonas communis (*Caecomyces communis*)	Cattle	Fiber digestion (Orpin 1976; Wubah and Fuller 1991)
Piromyces communis, *P. mae*, *P. dumbonica*	Horse, elephant	Morphological description of uniflagellate fungi (Li et al. 1990)
Ruminomyces elegans	Cattle	Morphological features and procedures for zoosporogenesis for polycentric rumen fungi (Ho and Bauchop 1990)
Caecomyces equi	Horse	Fiber digestion (Gold et al. 1988)

The fungi colonize fibrous plant fragments and digest them by mechanical fissure and enzymatic hydrolysis. For simplicity of understanding, the fungi are arranged in alphabetical order
ITS-1—internal transcribed spacer region, rRNA—ribosomal RNA

phyla such as Ascomycota, Basidiomycota, and Glomeromycota, Mucoromycota, and Microsporidia (Yang et al. 2018). In addition, a total of 1739 genes including cellulase, β-glucosidase and cellulose β-1,4-cellobiosidase for cellulose degradation were also observed, indicating the existence of a complex microbial and enzyme system in GI tract of giant panda enable the animal to utilize plant-based diets (Yang et al. 2018).

Box. 1. Summarized taxonomic categorization of anaerobic got fungi

Division: Eumycota
Subdivision: Mastigomycotina
Class: Chytridiomycetes
Phylum: Neocallimastigomycota
Order: Neocallimastigales
Family: Neocallimastigaceae

Genera:

A. **Monocentric fungi** (i.e., thallus producing a single reproductive organ—sporangium or resting spore)

   (i) **Buwchfawromyces**: Zoospores with one or two flagella, thallus with a globular rhizoid
   (ii) **Neocallimastix**: Zoospores with four to twenty flagella, thallus with filamentous and branched rhizoids
   (iii) **Piromyces**: Zoospores having one to four flagella and thallus with filamentous and branched rhizoids
   (iv) **Ontomyces**: They are monocentric rhizoidal anaerobic fungi

B. **Polycentric fungi** (i.e., thallus with many centers at which reproductive organs or sporangia are developed)

   (i) **Orpinomyces:** Having multiflagellate zoospores
   (ii) **Anaeromyces**: Zoospore having one flagellum
   (iii) **Cyllamyces**: Zoospore having one to two flagella with thalloid-branched sporangiophore
   (iv) **Caecomyces**: They are bulbous monocentric fungi

3.4 Effect of Removal of Fungi on Digestion Performance of Animals

The fungi and their enzymes are more proficient to degrade structural barriers in ingested plant materials. This is evident from studies on in vitro treatment of wheat straw with aerobic fungi such as *Pleurotus* sp. (Fazaeli et al. 2004).

The gut fungi can be selectively removed from the digestive tract by antibiotics such as cycloheximide. Elimination of fungi was found to increase the number of bacteria in rumen fluid contents, thereby increasing the population of rumen protozoa, the study could not explain direct relationship between fungi and protozoa (Li and Hou 2007). When fungi were excluded from the rumen, feed intake and fiber digestibility also decreased, though total viable bacterial or ciliate protozoal populations remained unaffected (Gordon and Phillips 1993).

3.5 Anaerobic Fungi as Feed Additives in Diet of Animals

Based on review of literature on fungi, and their enzymes, it could be inferred that gut fungi are distributed in a wide variety of domesticated ruminants to wild or feral herbivores and produce hydrolytic enzymes (Kamra and Singh 2017; Solomon et al. 2016). The fibrolytic fungi in wild herbivores are more efficient degraders of lignocellulose (Paul et al. 2010). It was also inferred that anaerobic fungi could increase cellulose, hemicelluloses neutral detergent fiber digestibility of lignocellulose-enriched feed (Paul et al. 2004) and improve feed intake and body weight gain in experimental calves (Dey et al. 2004).

Induced elimination of fungi from rumen resulted in reduction in volunteer feed intake and fiber digestibility (Gordon and Phillips 1993). This highlights the role of anaerobic fungi as feed additives to improving feed and nutrient utilization. When included as feed supplements, the gut fungi might enhance quality of forages by alleviating toxic effects of hydrolyzable tannins toxicity (Singh et al. 2008; Bhat et al. 2013).

3.6 Biotechnological Potential of Anaerobic Fungi

The fungi are key sources of enzymes for industrial applications. The anaerobic fungi are of towering interest to enzymologists and microbial ecologists due to efficient fibrolytic gene pool and biocatalysts. The enzymes are promising candidates for industrial applications such as producing degradable disaccharides from cellulose and hemicelluloses-enriched municipal solid waste, animal waste, and agro-industrial wastes.

Some species such as *Anaeromyces mucronatus* KF8 is found to produce a broad range of fibrolytic enzymes, such as cellulase, endoglucanase, xylanase, α-xylosidase, β-xylosidase, α-glucosidase, β-glucosidase, β-galactosidase, mannosidase, cellobiohydrolase, amylase, laminarinase, pectinase, and pectate lyase (Kamra and Singh 2017). Notably, most of them are found as cellulosomal and polycellulosomal complexes. Although cellulosomes from anaerobic fungi share many properties with cellulosomes of aerobic cellulolytic bacteria, they differ in amino acid sequences and post-translational modification. Cheng et al. (2018) have concluded that gut-origin anaerobic fungi can decompose up to 50% of the ingested lignocellulose by means of cellulosomes and produce H_2, formate and acetate, which serve as substrates for methanogenic archaea. Hence, a co-culture of anaerobic fungi and methanogens can produce more methane in the biogas plants (Jin et al. 2011; Cheng et al. 2018).

The tannase produced from *A. niger* has prospects in beverage and feed processing industry (Singh et al. 2001, 2008; Bhardwaj et al. 2003). We foresee the use of gut fungal fibrolytic enzymes in feed processing industry and enhance nutrient utilization in poultry, fish, and swine.

3.7 Outlook and Challenges

The herbivore gut is a natural resource of microorganisms, microbial genes, and enzymes for degrading plant biomass. This is because the bacteria, protozoa and fungi together digest the plant fiber. However, degradation capacity of anaerobic fungi is higher than that confirmed in bacteria and protozoa. It is envisaged that microbial pretreatment or rumen microbial inocula may possibly increase degradation of recalcitrant substrates.

The enzyme produced by rumen fungi is unbiased in their preference because of various types of xylan-degrading enzymes produced by these fungi. Inability of gut fungi to grow in vitro impedes harvesting their actual biotechnological potential. It is necessary to develop conditions and culture media for isolating more species of gut fungi from different origins.

It is practically demanding that rumen ecosystem be manipulated to enhance utilization of fibrous feed containing some anti-nutritional plant metabolites. Therefore, studies should focus on identifying and purifying microbial enzymes that are inferred based on genome sequence data analysis, but have not been isolated in actuality.

3.8 Conclusions

Especially, the gut fungi are efficient plant biomass degraders widely distributed among reptiles and mammalian herbivores. They represent promising agents for a variety of biotechnological advances. The mechanical digestion and enzymatic digestion of ingested plants are the two features of capital interest.

References

Barr DJ (1988) How modern systematics relates to the rumen fungi. Biosystems 21(3–4):351–356

Barr DJS, Kudo H, Jackober KD, Cheng KJ (1989) Morphology and development of rumen fungi: *Neocallimastix* sp., *Piromyces communis*, *Orpinomyces bovis*, gen. nov. sp. nov. Can J Bot 67:2815–2824

Bhardwaj R, Singh B, Bhat TK (2003) Purification and characterization of tannin acyl hydrolase from *Aspergillus niger* MTCC 2425. J Basic Microbiol 43(6): 449–461

Bhat TK, Makkar HP, Singh B (1996) Isolation of a tannin-protein complex degrading fungus from the faeces of hill cattle. Lett Appl Mirobiol 22:257–258

Bhat TK, Kannan A, Singh B, Sharma OP (2013) Value addition of feed and fodder by alleviating the antinutritional effects of tannins. Agric Res 2:189–206

Breton A, Bernalier A, Dusser M, Fonty G, Gaillard-Martinie B, Guillot J (1990) *Anaeromyces mucronatus* nov. gen., nov. sp. A new strictly anaerobic rumen fungus with polycentric thallus. FEMS Microbiol Lett 58(2):177–182

Cheng Y, Shi Q, Sun R, Liang D, Li Y, Li Y, Jin W, Zhu W (2018) The biotechnological potential of anaerobic fungi on fiber degradation and methane production. World J Microbiol Biotechnol 34(10):155

Comlekcioglu U, Ozkose E, Yazdic FC, Akyol I, Ekinci MS (2010) Polysaccharidase and glycosidase production of avicel grown rumen fungus Orpinomyces sp. GMLF5. Acta Biol Hung 61(3):333–343. https://doi.org/10.1556/abiol.61.2010.3.9

Dey A, Sehgal JP, Puniya AK, Singh K (2004) Influence of an anaerobic fungal culture (*Orpinomyces* sp.) administration on growth rate, ruminal fermentation and nutrient digestion in calves. Asian-Aus. J Anim Sci 17(6):820–824

Fazaeli H, Mahmodzadeh H, Azizi A, Jelan ZA, Liang JB, Rouzbehan Y, Osman A (2004) Nutritive value of wheat straw treated with Pleurotus fungi. Asian-Aust. J Anim Sci 17:1681–1688

Gold JJ, Heath IB, Bauchop T (1988) Ultrastructural description of a new chytrid genus of caecum anaerobe, Caecomyces equi gen. nov., sp. nov., assigned to the Neocallimasticaceae. Biosystems. 1988;21(3–4):403-15

Gordon GL, Phillips MW (1993) Removal of anaerobic fungi from the rumen of sheep by chemical treatment and the effect on feed consumption and *in vivo* fibre digestion. Lett Appl Microbiol 1993(17):220–223

Griffith GW, Baker S, Fliegerova K, Liggenstoffer A, van der Giezen M, Voigt K, Beakes G (2010) Anaerobic fungi: Neocallimastigomycota. IMA Fungus 1 (2):181–185 (Epub 2010 Nov 23)

Hanafy RA, Elshahed MS, Liggenstoffer AS, Griffith GW, Youssef NH (2017) *Pecoramyces ruminantium*, gen. nov., sp. nov., an anaerobic gut fungus from the feces of cattle and sheep. Mycologia 109(2):231–243. https://doi.org/10.1080/00275514.2017.1317190 (Epub 2017 Apr 18)

Hanafy RA, Elshahed MS, Youssef NH (2018) *Feramyces austinii*, gen. nov., sp. nov., an anaerobic gut fungus from rumen and fecal samples of wild Barbary sheep and fallow deer. Mycologia 110 (3):513–525. https://doi.org/10.1080/00275514.2018.1466610 (Epub 2018 Jul 3)

Hibbett DS, Binder M, Bischoff JF, Blackwell M, Cannon PF, Eriksson OE, Huhndorf S, James T, Kirk PM, Lücking R, Thorsten Lumbsch H, Lutzoni F, Matheny PB, McLaughlin DJ, Powell MJ, Redhead S, Schoch CL, Spatafora JW, Stalpers JA, Vilgalys R, Aime MC, Aptroot A, Bauer R, Begerow D, Benny GL, Castlebury LA, Crous PW, Dai YC, Gams W, Geiser DM, Griffith GW, Gueidan C, Hawksworth DL, Hestmark G, Hosaka K, Humber RA, Hyde KD, Ironside JE, Kõljalg U, Kurtzman CP, Larsson KH, Lichtwardt R, Longcore J, Miadlikowska J, Miller A, Moncalvo JM, Mozley-Standridge S, Oberwinkler F, Parmasto E, Reeb V, Rogers JD, Roux C, Ryvarden L, Sampaio JP, Schüssler A, Sugiyama J, Thorn RG, Tibell L, Untereiner WA, Walker C, Wang Z, Weir A, Weiss M, White MM, Winka K, Yao YJ, Zhang N (2007) A higher-level phylogenetic classification of the Fungi. Mycol Res 111(Pt 5):509–547

Ho YW, Bauchop T (1990) *Ruminomyces elegans* gen. sp. nov. A polycentric anaerobic rumen fungus from cattle. Mycotaxon 38:397–405

Jin W, Cheng YF, Mao SY, Zhu WY (2011) Isolation of natural cultures of anaerobic fungi and indigenously associated methanogens from herbivores and their bioconversion of lignocellulosic materials to methane. Bioresour Technol 102(17):7925–7931. https://doi.org/10.1016/j.biortech.2011.06.026 (Epub 2011 Jun 12)

Joshi A, Lanjekar VB, Dhakephalkar PK, Callaghan TM, Griffith GW, Dagar SS (2018) *Liebetanzomycespolymorphus* gen. et sp. nov., a new anaerobic fungus (Neocallimastigomycota) isolated from the rumen of a goat. MycoKeys 40:89–110. https://doi.org/10.3897/mycokeys.40.28337 (eCollection 2018)

Kamra DN, Singh B (2017) Anaerobic gut fungi. In: Satyanarayana T, Deshmukh S, Johri BN (eds) Developments in fungal biology and applied mycology. Springer Nature, Berlin, pp 125–134. ISSBN 978-981-10-4768-8

Kõljalg U, Nilsson RH, Abarenkov K, Tedersoo L, Taylor AF, Bahram M, Bates ST, Bruns TD, Bengtsson-Palme J, Callaghan TM, Douglas B, Drenkhan T, Eberhardt U, Dueñas M, Grebenc T, Griffith GW, Hartmann M, Kirk PM, Kohout P, Larsson E, Lindahl BD, Lücking R, Martín MP, Matheny PB, Nguyen NH, Niskanen T, Oja J, Peay KG, Peintner U, Peterson M, Põldmaa K, Saag L, Saar I, Schüßler A, Scott JA, Senés C, Smith ME, Suija A, Taylor DL, Telleria MT, Weiss M, Larsson KH (2013) Towards a unified paradigm for sequence-based identification of fungi. Mol Ecol 22(21):5271–5277. https://doi.org/10.1111/mec.12481 (Epub 2013 Sep 24)

Kudo H, Jakober KD, Phillippe RC, Cheng KJ, Barr DJ, Costerton JW (1990) Isolation and characterization of cellulolytic anaerobic fungi and associated mycoplasmas from the rumen of a steer fed a roughage diet. Can J Microbiol 36(7):513–517

Li DB, Hou XZ (2007) Effect of fungal elimination on bacteria and protozoa populations and degradation of straw dry matter in the rumen of sheep and goats. Asian-Aust J Anim Sci 20(1):70–74

Li J, Heath IB, Bauchop T (1990) *Piromyces mae* and *Piromycesdumbonica*, two new species of uniflagellate anaerobic chitridiomycete fungi from the hindgut of the horse and elephant. Can J Bot 68:1021–1033

Liggenstoffer AS, Youssef NH, Couger MB, Elshahed MS (2010) Phylogenetic diversity and community structure of anaerobic gut fungi (phylum Neocallimastigomycota) in ruminant and non-ruminant herbivores. ISME J 4(10):1225–1235. https://doi.org/10.1038/ismej.2010.49 (Epub 2010 Apr 22)

Ljungdahl LG (2008) The cellulase/hemicellulase system of the anaerobic fungus *Orpinomyces* PC-2 and aspects of its applied use. Ann NY Acad Sci 1125:308–321

Makhuvele R, Ncube I, Jansen van Rensburg EL, La Grange DC (2017) Isolation of fungi from dung of wild herbivores for application in bioethanol production. Braz J Microbiol 48(4):648–655. https://doi.org/10.1016/j.bjm.2016.11.013 (Epub 2017 Jun 3)

Orpin CG (1975) J Gen Microbiol 91:249–262

Orpin GC (1976) The characterization of the rumen bacterium Eadi's oval *Magnoovum* gen. nov. *eadii* sp. nov. Arch Microbiol 111:155–159

Orpin CG, Mann EA (1986) *Neocallimastix patriciarum*: new member of the Neocallimasticaceae inhabiting the sheep rumen. Trans Br Mycol Soc 86:178–181

Ozkose E, Thomas BJ, Davies DR, Griffith GW, Theodorou MK (2001) *Cyllamyces aberensis* gen. nov., sp. nov., a new anaerobic gut fungus with branched spornagiophores isolated from cattle. Can J Bot 79:666–673

Paul SS, Kamra DN, Sastry VRB, Sahu NP, Kumar A (2003) Effect of phenolic monomers on growth and hydrolytic enzyme activities of an anaerobic fungus isolated from wild nilgai (*Baselophustrago camelus*). Letters Appl Microbiol 36:377–381

Paul SS, Kamra DN, Sastry VRB, Sahu NP, Agarwal N (2004) Effect of administration of an anaerobic gut fungus isolated from wild blue bull (*Boselaphustrago camelus*) to buffaloes (*Bubalus bubalis*) on in vivo ruminal fermentation and digestion of nutrients. Anim Feed Sci Technol 115:143–157

Paul SS, Kamra DN, Sastry VR (2010) Fermentative characteristics and fibrolytic activities of anaerobic gut fungi isolated from wild and domestic ruminants. Arch Anim Nutr. 64(4):279–292. https://doi.org/10.1080/17450391003625037

Singh B, Bhat TK, Singh B (2001) Exploiting gastrointestinal microbes for livestock and industrial development. Asian-Austr J Anim Sci 14:567–586

Singh B, Bhat TK, Kurade NP, Sharma OP (2008) Metagenomics in animal gastrointestinal ecosystem: a microbiological and biotechnological perspective. Ind J Microbiol 48:216–227

Solomon KV, Haitjema CH, Henske JK, Gilmore SP, Borges-Rivera D, Lipzen A, Brewer HM, Purvine SO, Wright AT, Theodorou MK, Grigoriev IV, Regev A, Thompson DA, O'Malley MA (2016) Early-branching gut fungi possess a large, comprehensive array of biomass-degrading enzymes. Science 351 (6278):1192–1195. https://doi.org/10.1126/science.aad1431 (Epub 2016 Feb 18)

Struchtemeyer CG, Ranganathan A, Couger MB, Liggenstoffer AS, Youssef NH, Elshahed MS (2014) Survival of the anaerobic fungus *Orpinomyces* sp. strain C1A after prolonged air exposure. Sci Rep 4:6892. https://doi.org/10.1038/srep06892

Wang X, Liu X, Groenewald JZ (2017) Phylogeny of anaerobic fungi (phylum Neocallimastigomycota), with contributions from yak in China. Antonie Van Leeuwenhoek 110(1):87–103. https://doi.org/10.1007/s10482-016-0779-1 (Epub 2016 Oct 12)

Webb J, Theodorou MK (1988) A rumen anaerobic fungus of the genus *Neocallimastix*: ultrastructure of the polyflagellate zoospore and young thallus. Biosystems. 21:393–401

Wubah DA, Fuller MS (1991) Studies on *Caecomyces communis*: morphology and development. Mycologia 83:303–310

Yang S, Gao X, Meng J, Zhang A, Zhou Y, Long M, Li B, Deng W, Jin L, Zhao S, Wu D, He Y, Li C, Liu S, Huang Y, Zhang H, Zou L (2018) Metagenomic analysis of bacteria, fungi, bacteriophages, and helminths in the gut of Giant Pandas. Front Microbiol 9:1717. https://doi.org/10.3389/fmicb.2018.01717 (eCollection 2018)

Youssef NH, Couger MB, Struchtemeyer CG, Liggenstoffer AS, Prade RA, Najar FZ, Atiyeh HK, Wilkins MR, Elshahed MS (2013) The genome of the anaerobic fungus *Orpinomyces* sp. strain C1A reveals the unique evolutionary history of a remarkable plant biomass degrader. Appl Environ Microbiol 79(15):4620–4634. https://doi.org/10.1128/aem.00821-13

Microbial Resources from Wild and Captive Animals

4

Abstract

Wild herbivores- browsers, grazers, miners and suckers depend much on their symbiotic gut microbiota for deriving nutrients and minerals from forage and tree bark. The wild animals are inhabited by microbial symbionts that impact development, physiology, ecological interaction and adaptation of host, besides their overall well being. In view of these salient merits, the microbiome of wild animals could be a promising source of valuable genes, enzymes and miscellaneous bioactive molecules. The wild and captive animals are less explored, and should therefore be explored for microbial assets in them.

Highlights

- The gut microbes play crucial role in nutrition and health of wild animals
- The gut microbes from wild animals have highly superior enzymes than those from domesticated livestock, hence may have biotechnological applications
- The gut microorganisms with therapeutic properties could be valuable microbial feed supplements.

Keywords

Ungulates · Rumen microorganisms · Fibrolytic enzymes · Herbivores · Plant metabolites

4.1 Introduction

Microorganisms are ubiquitously present in most spheres of earth. The animals including invertebrates, with a few exceptions exceedingly depend on the microbioata associated with them (Hammer et al. 2017). Indeed, the symbiotic microbial and host relationship is critical for their mutual survival. The herbivores engage gut symbionts for converting plant biomass into fermentable sugars, volatile fatty acids and microbial proteins.

Ranging from insects, including arboreal ants (*Allomerus octoarticulatus*), which depend on myrmecophytic flora for both food and shelter (Arcila Hernández et al. 2017), xylophagus beetles, and some termites (Marynowska et al. 2017; Liu et al. 2013; Scully et al. 2013), are colonized by microorganisms, variously occupying the gut and other issues. Similarly, large feral mammals depende primarily on their symbiotic microorganisms.

Evidently, the gut microbiota has increasingly large role in nutrition, health and behavioral adaptation of the wild animals. A stable, diverse and balanced gut microbial balance is indicator of healthy status of the host. The host exposed to varied diets have more diverse microbial consortia in their guts. Increased variation in diets, and dietary components increase member and varieties of gut microbial genera and species which in turn benefit the host in digesting ingested plant biomass, and protect the host

B. Singh et al., *Advances in Animal Biotechnology*,
https://doi.org/10.1007/978-3-030-21309-1_4

against pathogens and climatic stress through multiple mechanisms (Krams et al. 2017).

4.2 Diversity of Wild Herbivores

According to Hackemann and Spain (2010), there are 75.3 million wild, and 3.57 billion domesticated ruminants belonging to six families, namely *Antilocapridae*, *Bovidae*, *Cervidae*, *Giraffidae*, *Moschidae*, and *Tragulidae*. The main feral extant mammal herbivores dwelling the forests include red deer (*Cervus elaphus*), roe deer (*Careolus capreolus*), white rhinoceros (*Diceros bicornis*), giraffe (*Giraffa camelopardalis*), elephant, e.g., African bush elephant (*Loxodonta Africana*, *L. cyclotis*), Asian elephant (*Elephas maximus*), feral camels (*Camelus dromedarieus*, *C. bactrianus*), horses (*Equus ferus*), donkey (*Equus hemionus khur*), cattle (e.g. Chillingham cattle), buffalo (*Bubalus arnee*), alpaca (*Vicugna pacos*), wildebeest or gnus (*Connochaetes gnou* and *C. taurinus*), and Nilgai or blue bull (*Boselaphus tragocamelus*) (Fig. 4.1) (Table 4.1). Depending on availability of fodder and water, and threat from predators, the wild herbivores may be migratory or sedentary. There are many migratory herbivore birds such as duck, goose, blue macaw, toucan, and reptiles including iguanas and tortoises. The above animals exhibit different ranges of feeding habits.

Among small herbivores, the major genera include wild goat ibak (*Capra aegagrus*), gazelle (*Eudorcas thomsonii*), pudu (*Pudu mephistophiles*), hare (*Lepus* sp.), voles (*Microtus agretis*), rodents such as red squirrel (*Sciurus vulgaris*), wood mice (*Apodemus sylvaticus*), wood rat (*Neotoma cinerea*), and koala (*Phascolarctos cinereus*).

The domesticated livestock, particularly the *Bovidae*, is the prime focus of microbial ecologists, and wild herbivores exhibiting evolutionary adaptation to harsh climate and low-quality dietary stuff are less explored species. These animals feed on whatever plant materials available in their habitats or offered to them. Besides grasses and tree leaves and branches, animals sometimes strip the woody and fibrous bark of trees to derive more nutritious inner bark. The large ungulates are important sources of gut microorganisms with considerable biotechnological potential. Moreover, the microbes from wild

Fig. 4.1 The Nilgai or blue bull (*Boselaphus tragocamelus*), a large wild Asian antelope, grazes on roughage diets and crop residues. Being herbivorous, the *B. tragocamelus* harbours powerful fibrolytic rumen microorganisms

Table 4.1 Some prominent wild herbivores that are envisioned to be resources of microorganism with commercial potential

Species	Food habits	Likely gut microorganism	Potential applications (References)
Bovidae family			
African/Cape buffalo (*Cyncerus caffer*)	Grasses, tree foliage	Fibrolytic	Food and feed processing, textile pulp processing, biofuel production
American bison	Fibrous roughage	Fibrolytic	Food and feed processing
Antelopes	Long grasses		
Bison	Grasses and tree bark		
Blue bull or Nilcow (*Boselaphus tragocamelus*)	Grasses, herbs, woody plants, crops and crop residues	Fibrolytic	Feed processing, biofuel production (Paul et al. 2010)
Bongo (*Tragelaphus eurycerus*)	Grasses, grass and bushes, their roots, fruits, pith of rotting trees, charcoal of burnt trees	Fibrolytic, miscellaneous	Biofuel production, food processing
Bushbuck	Shrubs, herbs, Legumes, tree leaves	Fibrolytic	Biofuel production, food processing
Cattle (e.g., Gaur)	Wider variety of plants than domestic and other feral ruminants	Fibrolytic	Biofuel production, food processing
Tamaraw or Mindoro dwarf buffalo (*Bubalus mindorensis*)	European bison grasses, tree leaves bushes	Fibrolytic	Biofuel production, food processing
Saola (*Pseudoryx nghetinhensis*)	All types of plants available in forest	Fibrolytic	Biofuel production, food processing
Wisent			
Yak			
Camlidae family			
Camels	Grass, tree leaves, bushes, crop residues	Fibrolytic	Biofuel production, food processing
Cavidae family Capyra	Bushes, grass, aquatic plants, fruits, tree bark	Fibrolytic	Biofuel production
Cervidae family Deer	Weeds, grasses, herbs, tree leaves	Fibrolytic	Biofuel production, food feed processing
Reindeer or Caribou (domesticated By some tribes)	Grasses, reindeer lichen, moss, woody snow plants	Fibrolytic	Biofuel production, food feed processing
Elephantidae family			
Elephants	Tree branches and barks, fruits, grasses, shrubs	Fibrolytic	Biofuel
Equidae family			
Zebra	Grasses, tree leaves Herbs, flowers, tree Barks, twigs	Fibrolytic	Biofuel production
Girrafidae family Giraffe	Tree foliage, shrubs grasses and fruits	Fibrolytic	Biofuel production
Okapis	Tree foliage, grasses, fruits, ferns, bryophytes, and fungi		

(continued)

Table 4.1 (continued)

Species	Food habits	Likely gut microorganism	Potential applications (References)
Hippopotamidae family Hippopotamus	Aquatic plants, soft and short grasses	Fibrolytic	Biofuel production, food
Hominidae family			
Gorillas	Fruits, leaves, tender shoots, seeds	Hydrolytic and phytometabolite degraders	Misc. industrial uses
Leporidae family			
Hare	Grasses, herbs, leafy wood	Fibrolytic	Biofuel production, food
Pteropodidae family			
Megabats	Fruits, flowers, nectar	Hydrolytic	Misc. uses
Ursidae family			
Giant panda	Bamboo leaves and twigs	Fibrolytic	Misc. uses
Macropodidae family			
Kangaroos	Grasses, shrubs, tender leaves, fungi	Fibrolytic	Biofuel production, feeded processing
Phascolarctidae family			
Koalas	Eucalyptus leaves	Fibrolytic and phytometabolite degraders	Misc. industrial uses
Rhinocerotidae family			
Rhinoceros	Grasses, shoots, fruits, small tree leaves	Fibrolytic	Biofuel production, food
Trichechidae family			
Manatees (aquatic)	Mangrove foliage water seeds, algae	Hydrolytic	

animals have better cellulolytic enzymes than those reported from domestic livestock (Paul et al. 2004, 2010; Kamra and Singh 2017).

4.3 Dietary Habits of Wild Herbivores

The feral herbivores are exposed to varied dietary sources majority of which are deficient in essential nutrients such as readily fermentable energy sources, minerals, trace elements, vitamins and proteins, and have higher proportions of indigestible fiber, and various anti-nutritional or toxic phytometabolites. Herbivores have developed various mechanisms to derive nutrients from grasses and plants. The mechanisms include counteracting negative effects of tannin-polyphenols in mouth by secreting tannin-binding proline-rich salivary proteins (e.g. in monkeys, moose and mules). While volatile phytometabolites (e.g. terpenoids) are vanished during chewing, some toxins are removed as they bind eaten materials such as soil or charcoal. However, the microbial-assisted adaptation or detoxification by their gut microorganisms is important, and has commercial aspects. The toxins absorbed into blood are detoxified in liver. Prior exposure to plant toxins is correlated to increased diversity of microorganisms in the gut.

Depending on the nature of diet, the gut microorganisms establish in wild ruminant and assist the host utilize particular diet. On contrary, the captive animals undergo assorted changes such as diet, health treatment, housing, limited contact with other animals, pests, parasites and microorganisms, which restrict normal microbial diversity. In addition, the captivity leads to restriction on variety of diets which in turn plays down the diversity of their gut microorganisms (Ushida et al. 2016; Borbón-García et al. 2017; McKenzie et al. 2017).

4.4 Microbial Diversity in Wild Animals

The gut microbiota of wild animals has important contributory role in research and biomedical sciences (Viney 2018). A recent network analysis of 650 published research on gut microbiota (2009–2016) reveals that gut ecosystem of wild animals is least investigated albeit it offers under-exploited opportunities to study and utilizing them for commercial and biotechnological applications (Kohl et al. 2014; Pascoe et al. 2017). Availability of gut contents, identification of donor species, pressure from animal welfare agencies, inability of the microorganisms to grow in vitro are the major factors that impede the research on gut microorganisms in wild animals.

Using culture-independent NGS techniques, the bacteria present in fore- and distal-guts of two- and three-toed sloths were analyzed and correlated with their diets. It was noted that Proteobacteria and Fermicutes were the dominating bacterial phyla in these animals, though both types of sloths had variations in their overall gut microbial species. The differences supported hypothesis that gut microbial communities varied according to feeding habits in the species (Dill-McFarland et al. 2016).

4.5 Wild Herbivores and Their Gut Ecosystem *Vis-à-Vis* Plant Secondary Metabolites

As mammals don't synthesize enzymes for plant cell wall degradation, the symbiotic association between mammals and the microbiota in their GI tract benefit both by fulfilling mutual nutritional needs. The gut microbial composition remains almost stable despite changes in diet. As summarized in Table 4.2, it can be inferred that diversity of microorganisms with various beneficial attributes is infinite.

Ever since early reports of isolation of tannin-degrading bacteria from the gut of koalas (Osawa 1990; Osawa and Mitsuoka 1990), microorganisms with similar traits have been reported from several animals including feral and wild ruminants. The Indian northern plains gray langur (*Semnopithecus entellus*) extensively consume oak (*Quercus incana*) acorns, and harbor gut bacteria that could degrade hydrolysable tannins (Singh et al. 2008).

Some rodents thrive on vegetation, and contain gut bacteria that assist in degrading phytometabolites which when taken in excess are harmful. Desert rodents such as wood rat gut represents a useful source of novel microbial genes and enzymes involved in cellulose-degradation and detoxification of phytometabolites including phenolics, oxalates and alkaloids (Kohl et al. 2011; Miller et al. 2014).

Tannin-degrading bacteria were isolated from wood rat (*Neotoma lepidia*) and transplanted into laboratory mice (*Rattus norvegicus*). The microorganisms were found to establish stably in new host and enabled tolerance against tannin-toxicity, maintenance of body mass and reduced hepatic damage in *R. norvegicus* (Kohl et al. 2016).

Fluorocarbons are used in numerous industrial applications. Hence they are released into environment and sometimes enter food chain of humans, livestock and wild life through consumption of contaminated crops and water. The toxicity of fluorocarbons and susceptibility of animals to these compounds depend on various factors. Accumulation and behavior of some fluorocarbons, such as perfluroalkyl acids in the tissues and body fluids of ruminants (e.g. cattle) (Kowalczyk et al. 2013), is different than that reported in the tissues of non-ruminants (e.g. pig) (Numata et al. 2014). The difference might be attributed to presence of different rumen microorganisms and their action on compounds.

Table 4.2 Some valuable microorganisms from wild animals with reference to their potential metabolic attributes

Microorganisms	Host/origin	Characteristics	Inferences and recommendations (reference)
Bacteria			
Clostridium spp. *Enterobacterium agglomerans*	Koala cecum	Tannin degrader	Osawa and Sly (1991, 1992)
Clostridium sporogenes	*Neotoma albigula* feces	Oxalate-degraders	Miller et al. (2014)
Clostridium sporogenes 8-O	*Neotoma albigula* GI tract	Oxalate-degraders	Oakeson et al. (2016)
Enterococcus gallinarum	*Neotoma albigula* feces	Oxalate-degraders	Miller et al. (2014)
Enterococcus hirae	Rabbit cecum	Phytase-degrader	Marounek et al. (2009)
Eubacterium oxidoreducens	Koala cecum	Tannic acid degrader	Krumholz and Bryant (1986)
Lactobacillus gasseri, L. johnsonii, L. reuteri	*Neotoma albigula* GI tract and feces	Oxalate-degraders	Miller et al. (2014)
Lactobacillus animalis, L. murinus	*Apodemus speciiosus* (Japanese wood Mouse)	Tannin degradation	Sasaki et al. (2005), Osawa et al. (2006)
Lactobacillus apodami	Feces of *Apodemus speciiosus*	Tannase-producing	Osawa et al. (2006)
Lonepinella koalarum	Koala cecum and feces	Tannic acid	Osawa et al. (1995)
Mitsuokella multiacidus	Rumen	Phytic acid	
Selenomonas spp.	Antelope, bovine, Sheep rumen	Tannic acid	Odenyo and Osuji (1998)
Streptococcus bovis biotype I	Koala feces	Tannin–protein complex	Osawa (1990)
Streptococcus caprinus Sp. nov.,	Feral goats	Tannin resistant	Brooker et al. (1994)
Streptococcus gallolyticus	Wild Japanese rock ptarmigans (*Lagopus mutajaponica*)	Condensed tannins	Tsuchida et al. (2017)
	Apodemus specious feces	Tannins	Sasaki et al. (2005)
Fungi			
Neocallimastigo-mycota Phylum	Green iguana American Elk, goral bonte bock southern gerenuk, and Nile lechwe	Plant fiber	Liggenstoffer et al. (2010)

The fluroacetate, also called compound 1080, is a potent toxic plant metabolites found in some plants grazed by animals across the globe (Leong et al. 2017; de Oliveira Neto et al. 2017). Efforts to identify, isolate and characterize animal enzymes to detoxify fluoroacetate have not been successful. Hence, alternative methods are under consideration. Emus are resistant to fluoroacetate. Though information is scarce on gut microbiota of these birds degrading fluoroacetate, the gut microorganisms should be investigated for ability to degrade fluorocarbons.

4.6 Fibrolytic Microorganisms from the Wild Animals

Cellulose is a renewable biomaterial which can be used to produce biofuel. The resistance to microbial fermentation is a major obstacle in utilizing plant polysaccharides for energy production. The microorganisms from feral animals may serve as essential candidates to degrade plant biomass and anti-nutritional phytometabolites. A diversity of microorganisms is noted using culture-dependent as well as genomics tools from a variety of wild animals.

Especially, culture-independent molecular biological methods and metagenomics have generated plethora of information into diversity of microorganisms in gut ecosystem. A compositional and comparative analysis of rumen metetagenome of Svabard reindeer (*Rangifer terandus platyrhynchus*) have revealed deeply-branched cellulose-degrading lineages, mostly affiliated to Bacteroidetes and Fermicutes (Pope et al. 2012).

Metagenomic analyses of giant pandas was carried out wherein a total of 680 species of bacteria belonging primarily to different phyla, namely Proteobacteria, Firmicutes, Bacteroidetes, Actinobacteria, some to Cyanobacteria were detected. Among 198 fungi, the predominating phyla were Ascomycota, Basidiomycota, and Glomeromycota, Mucoromycota, and Microsporidia. In addition, 185 bacteriphages, and 45 helminths were also confirmed (Yang et al. 2018). A total of 1739 genes including cellulase, β-glucosidase and cellulose β-1,4-cellobiosodase responsible for cellulose degradation were also noted. Collectively, the data shows that a complex microbial and enzyme system is present in giant panda that enable the animal to utilize plant based diets efficiently (Yang et al. 2018).

The wild herbivores possess highly efficient cellulolytic microorganisms in their GI tract. Especially, the anaerobic rumen fungi from wild animals are more powerful degraders of fibrous diets (Paul et al. 2004; Kamra and Singh 2017). Anaerobic fungi of Neocallimastigomycota Phylum are obtained from American Elk, goral bontebock southern gerenuk, and Nile lechwe (Liggenstoffer et al. 2010). A novel rumen anaerobic fungus, named *Feramyces austinii* gen. nov., sp. nov., associated with utilization of fermentable sugars and plant biomass have been isolated from Wild Barbary sheep and fallow deer (Hanafy et al. 2018).

4.7 Birds as Resources of Microorganisms

Like mammals, the birds exposed to a broad variety of dietary resources, thereby exhibit a strong symbiosis with microorganisms colonizing their GI tract. The GI microorganisms help deriving nutrients from dietary ingredients as well as confer protection against some pathogens. Some birds depend solely on herbivory for their nutrition. These birds ingest variety of plants containing various phytometabolites.

Researchers are interested in exploring the gut microbial ecosystem of wild birds for isolating beneficial microorganisms. However, little is known about speed of digestion, plasticity of the alimentary canal that affects the microbial ecosystem in birds. Some birds, for example, Japanese rock ptarmigan (*Lagopus mutajaponica*) surviving under severe conditions of Japanese alpine region forage and wild alpine plants known to contain toxic plant metabolites such as tannins, ursolates, and saponins (Wagstaff 2008).

Ultra-deep sequencing of 16S rRNA genes of gut microbiome of the species indicated remarkable difference at phylum levels microbial species in the GI tract of wild (*Lagopus mutajaponica*) *vs.* captive Svalbard rock ptarmigans. The Firmicutes alone comprised of 80% of gut microbiota in captive ptarmigans, while Actinobacteria, Bacteroidetes, Firmicutes, and Synergistes and some unclassified genera were the major microbial phyla in *L. mutajaponica* (Ushida et al. 2016).

Carrion eaters such as vultures and alligators act as scavengers by feeding on flesh of dead animals, seem to have developed a specialized microbiome, thus exemplify a specialized host-microbe symbiosis. The functional metagenomic analysis of black vulture (*Coragyps altratus*), and turkey vulture (*Cathartes aura*) gut microbiome has revealed a remarkable conserved low diversity of gut microbiota. It has been inferred that GI tract of vultures is different from other species in terms of bacterial populations, dominated by Clostridia and Fusobacteria (Roggenbuck et al. 2014). Interestingly, a larger bacterial population is detected from the facial skin of the vultures. The more in-depth genomic studies are likely to unravel about gut microbiota of such unique birds, and their possible economic prospects.

The birds adapted to herbivory harbor fiber-degrading or similar bacteria from their GI tract. Using a species-specific primer set for each 16S ribosomal DNA i.e. ribosomal genes, a partial 16S rDNA sequence of *Fibrobacter*, and distantly related, but novel species of *Fibrobacter*, and *Ruminococcus flavifaciens* were detected in the cecum of Ostrich (*Struthio camelus*) (Matsui et al. 2010a, b). Based on these two reports, it is inferred that ostrich is a hind gut fermenter with complex microbial communities, most of which are not cultured so far. Their gut microbes might offer opportunities to have novel microbial species with biotechnological prospects.

4.8 Fish Gut and Cellulose Degradation

Fish are found in range of habitats including fresh and marine water, cold water and normal water. They are adapted to varied food habits including freshwater and marine environments. To cope with scarcity of food fish has evolved various mechanisms to fulfil their nutritional and respiration requirements.

Presumably, when many fish species stop feeding completely in scarcity of water and food, adaptation of some fish to utilize and digest offers evolutionary advantage to survive under adverse conditions (Fink and Fink 1979). Inferential evidences indicate that some loricariid catfish developed the capacity to use woody items as food source. Surgeonfish or brown tang (*Acanthurus nigrofuscus*) feed on marine plants including benthic algae. They have long intestinal tract enabling more time for microbial interaction with ingested plant materials.

The largest bacterium, named *Epulopiscium fishelsoni* have been detected from the gut of surgeonfish. Though bacterium is not cultured in vitro, it might have role in digestion and deriving nutrients from dietary ingredients (Angert et al. 1993). Cellulose-degrading bacteria and bacterial enzymes have been identified from the gut of *Panaque* sp. (Nelson et al. 1999).

4.9 Therapeutic Importance of Microorganisms from Wild Mammals

The interesting observations are established from the recent studies wherein laboratory mice were reconstituted with fecal microbiota from closely related wild mouse. The reconstituted laboratory mice exhibited low inflammation, higher resistance to influenza virus infection, inflammation and experimentally induced tumorigenesis (Rosshart et al. 2017). This study improved the concept of banking microbiota of wild animals using laboratory model animals as live reservoir.

Some primates, for example, Chimpanzee share homology with humans, and are envisaged to share similarity in their gut microbiota. The gut microorganisms, especially the lactic acid bacteria (LAB), and bifidobacteria present in the GI tract of wild or captive animals might have important role in maintaining the health of wild animals. *Bifidobacterium* species (*B. angulatum*, *B. catenulatum*, *B. dentium*, and *B. pseudocatenulatum*), and *B. pseudolongum*-like species have been have been identified from the guts of chimpanzees (reviewed in Nomoto et al. 2017). These finding have importance in future as LAB are bifidobacteria and important commensal microbes with their contributory role in gut and genitourinary health of the host.

4.10 Outlook and Challenges

Indeed majority of the Earth's terrestrial ecosystems are heavily affected and shaped by the wild herbivores. The herbivores, be terrestrial or aquatic, have developed various mechanisms to optimize nutrition from the available feed resources. In turn, the herbivores assist pollination and dispersal of plant seeds.

Thousands of herbivore species exhibit varied geographic distribution, food, living and reproduction habits. Whereas some are evolutionally adapted to extreme deserts, some are known to thrive in snow, others dwell in water. The charcoal produced due to burning of forest trees due to lightening serves as source of minerals or binds to some toxic phytomebolites ingested by the herbivores. How the speed of plant material digestion, and the plasticity of GI tract affects the gut microbial ecosystem, is an important area of scientific research. Further research will facilitate implementation of novel strategies employing superior novel microorganism from wild animals.

The field of avian gut microbiology is uneven due to variation in their dietary resources, evolutionary successful lineage, wide range of habitats across the globe, and migratory behavior. The gut microorganisms of birds should be explored for their economic traits.

Enough data is available on isolation of probiotic microorganisms from the alimentary canal or feces of the poultry birds, and promoting their health and production by means of micro-feed supplements or probiotics. The superior microorganisms their gene pool from new hosts, will be valuable assets for animal health, nutrition and industrial applications.

4.11 Conclusions

The research on host-associated microbiota and host-microbe interaction has grown swiftly. Assisted by genomics inferences, the convention microbial ecology has advanced at faster pace. However, the information is scarce on gut microbiota and their enzymes in wild animals, birds and reptiles. The gut microorganisms with detoxification properties in wild animals may serve as promising strains for agriculture, feed processing and medicine or bacteriotherapies. The studies should be carried to unravel the gut microbiota of feral animals such as wood rats and large ungulates that unlike domesticated livestock are evolved to thrive in harsh environment without proper human care and medicine.

References

Arcila Hernández LM, Sanders JG, Miller GA, Ravenscraft A, Frederickson ME (2017) Ant-plant mutualism: a dietary by-product of a tropical ant's macronutrient requirements. Ecology 98(12):3141–3151. https://doi.org/10.1002/ecy.2036

Angert ER, Clements KD, Pace NR (1993) The largest bacterium. Nature 362(6417):239–241

Borbón-García A, Reyes A, Vives-Flórez M, Caballero S (2017) Captivity shapes the gut microbiota of andean bears: insights into health surveillance. Front Microbiol 8:1316. https://doi.org/10.3389/fmicb.2017.01316 (eCollection 2017)

Brooker JD, O'Donovan LA, Skene I, Clarke K, Blackall L, Muslera P (1994) *Streptococcus caprinus* sp. nov., a tannin-resistant ruminal bacterium from feral goats. Lett Appl Microbiol 18:313–318

de Oliveira Neto TS, Riet-Correa F, Lee ST, Cook D, Sousa Barbosa FM, da Silva Neto JF, Simões SVD, Lucena RB. Poisoning in goats by the monofluoracetate-containing plant Palicourea aeneofusca (Rubiaceae). Toxicon 135:12–16. https://doi.org/10.1016/j.toxicon.2017.05.025. (2017 Sep 1)

Dill-McFarland KA, Weimer PJ, Pauli JN, Peery MZ, Suen G (2016) Diet specialization selects for an unusual and simplified gut microbiota in two- and three-toed sloths. Environ Microbiol 18(5):1391–1402. https://doi.org/10.1111/1462-2920.13022

Fink WL, Fink SV (1979) Central Amazonia and its fishes. Comp Biochem Physiol 62A:13–29

Hackmann TJ, Spain JN (2010) Invited review: ruminant ecology and evolution: perspectives useful to ruminant livestock research and production. J Dairy Sci 93(4): 1320–1334. https://doi.org/10.3168/jds.2009-2071. Review (2010 Apr)

Hammer TJ, Janzen DH, Hallwachs W, Jaffe SP, Fierer N (2017) Caterpillars lack a resident gut microbiome. Proc Natl Acad Sci USA 114(36):9641–9646. https://doi.org/10.1073/pnas.1707186114

Hanafy RA, Elshahed MS, Youssef NH (2018) *Feramyces austinii*, gen. nov., sp. nov., an anaerobic gut fungus from rumen and fecal samples of wild Barbary sheep and fallow deer. Mycologia 110 (3):513–525. https://doi.org/10.1080/00275514.2018.1466610 (Epub 2018 Jul 3)

Kamra DN, Singh B (2017) Anaerobic gut fungi. In: Satyanarayana T, Deshmukh S, Johri BN (eds) Developments in fungal biology and applied mycology. Springer Nature, Berlin, pp. 125–134. ISSBN 978-981-10-4768-8

Kohl KD, Weiss Robert B, Colin Dale M, Dearing Denise (2011) Diversity and novelty of the gut microbial community of an herbivorous rodent (*Neotoma bryanti*). Symbiosis. 54:47–54. https://doi.org/10.1007/s13199-011-0125-3

Kohl KD, Stengel A, Dearing MD (2016) Inoculation of tannin-degrading bacteria into novel hosts increases performance on tannin-rich diets. Environ Microbiol 18(6):1720–1729. https://doi.org/10.1111/1462-2920.12841

Kohl KD, Weiss RB, Cox J, Dale C, Dearing MD (2014) Gut microbes of mammalian herbivores facilitate intake of plant toxins. Ecol Lett 17(10):1238–1246. https://doi.org/10.1111/ele.12329

Kowalczyk J, Ehlers S, Oberhausen A, Tischer M, Fürst P, Schafft H, Lahrssen-Wiederholt M (2013) Absorption, distribution, and milk secretion of the perfluoroalkyl acids PFBS, PFHxS, PFOS, and PFOA by dairy cows fed naturally contaminated feed. J Agric Food Chem 61(12):2903–2912. https://doi.org/10.1021/jf304680j

Krams IA, Kecko S, Jõers P, Trakimas G, Elferts D, Krams R, Luoto S, Rantala MJ, Inashkina I, Gudrā D, Fridmanis D, Contreras-Garduño J, Grantiņa-Ieviņa L, Krama T (2017) Microbiome symbionts and diet diversity incur costs on the immune system of insect larvae. J Exp Biol pii: jeb.169227. https://doi.org/10.1242/jeb.169227

Krumholz LR, Bryant MP (1986) *Eubacterium oxidoreducens* sp. nov., requiring H2 or formate to degrade gallate, pyrogallol, phloroglucinol and quercetin. Arch Microbiol 144:8–14

Leong LEX, Khan S, Davis CK, Denman SE, McSweeney CS (2017) Fluoroacetate in plants—a review of its distribution, toxicity to livestock and microbial detoxification. J Anim Sci Biotechnol 8:55. https://doi.org/10.1186/s40104-017-0180-6 (eCollection 2017. Review)

Liggenstoffer AS, Youssef NH, Couger MB, Elshahed MS (2010) Phylogenetic diversity and community structure of anaerobic gut fungi (phylum Neocallimastigomycota) in ruminant and non-ruminant herbivores. ISME J 4(10):1225–1235. https://doi.org/10.1038/ismej.2010.49 Epub 2010 Apr 22

Liu N, Zhang L, Zhou H, Zhang M, Yan X, Wang Q, Long Y, Xie L, Wang S, Huang Y, Zhou Z (2013) Metagenomic insights into metabolic capacities of the gut microbiota in a fungus-cultivating termite (*Odontotermes yunnanensis*). PLoS One 8(7): e69184. https://doi.org/10.1371/journal.pone.0069184. Print 2013

Marounek M, Brenová N, Suchorská O, Mrázek J (2009) **Phytase** activity in rabbit cecal bacteria. Folia Microbiol (Praha) 54(2):111–114. https://doi.org/10.1007/s12223-009-0016-7 (Epub 2009 May 6)

Marynowska M, Goux X, Sillam-Dussès D, Rouland-Lefèvre C, Roisin Y, Delfosse P, Calusinska M (2017) Optimization of a metatranscriptomic approach to study the lignocellulolytic potential of the higher termite gut microbiome. BMC Genom 18 (1):681. https://doi.org/10.1186/s12864-017-4076-9

Matsui H, Ban-Tokuda T, Wakita M (2010a) Detection of fiber-digesting bacteria in the ceca of ostrich using specific primer sets. Curr Microbiol 60(2):112–116. https://doi.org/10.1007/s00284-009-9513-9 (Epub 2009 Sep 29)

Matsui H[1], Kato Y, Chikaraishi T, Moritani M, Ban-Tokuda T, Wakita M (2010b) Microbial diversity in ostrich ceca as revealed by 16S ribosomal RNA gene clone library and detection of novel Fibrobacter species. Anaerobe 16(2):83–93. https://doi.org/10.1016/j.anaerobe.2009.07.005 (Epub 2009 Jul 24)

McKenzie VJ, Song SJ, Delsuc F, Prest TL, Oliverio AM, Korpita TM, Alexiev A, Amato KR, Metcalf JL, Kowalewski M, Avenant NL, Link A, Di Fiore A, Seguin-Orlando A, Feh C, Orlando L, Mendelson JR, Sanders J, Knight R (2017) The effects of captivity on the mammalian gut microbiome. Integr Comp Biol 57 (4):690–704. https://doi.org/10.1093/icb/icx090

Miller AW, Kohl KD, Dearing MD (2014) The gastrointestinal tract of the white-throated Woodrat (Neotoma albigula) harbors distinct consortia of oxalate-degrading bacteria. Appl Environ Microbiol 80(5): 1595–1601. https://doi.org/10.1128/AEM.03742-13

Nelson JA, Wubah DA, Whitmer ME, Johnson EA, Stewart DJ (1999) Wood-eating catfishes of the genus *Panaque*: gut microflora and cellulolytic enzyme activities. J Fish Biol 54:1069–1082

Nomoto R, Takano S, Tanaka K, Tsujikawa Y, Kusunoki H, Osawa R (2017) Isolation and identification of *Bifidobacterium* species from feces of captive chimpanzees. Biosci Microbiota Food Health 36(3):91–99. https://doi.org/10.12938/bmfh.16-027

Numata J, Kowalczyk J, Adolphs J, Ehlers S, Schafft H, Fuerst P, Müller-Graf C, Lahrssen-Wiederholt M, Greiner M (2014) Toxicokinetics of seven perfluoroalkyl sulfonic and carboxylic acids in pigs fed a contaminated diet. J Agric Food Chem 62(28):6861–6870. https://doi.org/10.1021/jf405827u

Oakeson KF, Miller A, Dale C, Dearing D (2016) Draft genome sequence of an oxalate-degrading strain of *Clostridium sporogenes* from the gastrointestinal tract of the white-throated woodrat (*Neotoma albigula*). Genome Announc 4(3). pii: e00392-16. https://doi.org/10.1128/genomea.00392-16

Odenyo AA, Osuji PO (1998) Tannin-tolerant ruminal bacteria from East African ruminants. Can J Microbiol 44(9):905–909

Osawa R (1990) Formation of a clear zone on tannin-treated brain heart infusion agar by a *Streptococcus* sp. isolated from feces of koalas. Appl Environ Microbiol 56(3):829–831

Osawa R, Fujisawa T, Pukall R (2006) *Lactobacillus apodemi* sp. nov., a tannase-producing species isolated from wild mouse faeces. Int J Syst Evol Microbiol 56 (Pt 7):1693–1696

Osawa R, Mitsuoka T (1990) Selective medium for enumeration of tannin-protein complex-degrading *Streptococcus* spp. in Feces of Koalas. Appl Environ Microbiol 56(11):3609–3611

Osawa R, Rainey F, Fugisawa T, Lang E, Busse HJ, Walsh TP, Stachebrandt (1995) *Lonepinella koalarum* gen. nov., a new tannin protein complex degrading bacterium. Syst Appl Microbiol 18:56–62

Osawa R, Sly LI (1991) Phenotypic characterization of CO_2-requiring strains of *Streptococcus bovis* from koalas. Appl Environ Microbiol 57(10):3037–3039

Osawa R, Sly LI (1992) Occurrence of tannin protein complex degrading *Streptococcus* spp. in feces of various animals. Syst Appl Microbiol 15:144–147

Pascoe EL, Hauffe HC, Marchesi JR, Perkins SE (2017) Network analysis of gut microbiota literature: an overview of the research landscape in non-human animal studies. ISME J. https://doi.org/10.1038/ismej.2017.133

Paul SS, Kamra DN, Sastry VRB (2010) Fermentative characteristics and fibrolytic activities of anaerobic gut fungi isolated from wild and domestic ruminants. Arch Anim Nutr 64(4):279–292

Paul SS, Kamra DN, Sastry VRB, Sahu NP, Agarwal N (2004) Effect of administration of an anaerobic gut fungus isolated from wild blue bull (*Boselaphustrago camelus*) to buffaloes (*Bubalus bubalis*) on in vivo ruminal fermentation and digestion of nutrients. Anim Feed Sci Technol 115:143–157

Pope PB, Mackenzie AK, Gregor I, Smith W, Sundset MA, McHardy AC, Morrison M, Eijsink VG (2012) Metagenomics of the Svalbard reindeer rumen microbiome reveals abundance of polysaccharide utilization loci. PLoS One. 7(6):e38571. https://doi.org/10.1371/journal.pone.0038571 (Epub 2012 Jun 6). Erratum in: PLoS One. 2014;9(7):e104612

Roggenbuck M, Sauer C, Poulsen M, Bertelsen MF, Sørensen SJ (2014) The giraffe (Giraffa camelopardalis) rumen microbiome. FEMS Microbiol Ecol 90(1): 237–246. https://doi.org/10.1111/1574-6941.12402

Rosshart SP, Vassallo BG, Angeletti D, Hutchinson DS, Morgan AP, Takeda K, Hickman HD, McCulloch JA, Badger JH, Ajami NJ, Trinchieri G, Pardo-Manuel de Villena F, Yewdell JW, Rehermann B (2017) Wild mouse gut microbiota promotes host fitness and improves disease resistance. Cell 171(5):1015–1028. https://doi.org/10.1016/j.cell.2017.09.016

Sasaki E, Shimada T, Osawa R, Nishitani Y, Spring S, Lang E (2005) Isolation of tannin-degrading bacteria isolated from feces of the Japanese large wood mouse, *Apodemus speciosus*, feeding on tannin-rich acorns. Syst Appl Microbiol 28(4):358–365

Scully ED, Geib SM, Hoover K, Tien M, Tringe SG, Barry KW, Glavina del Rio T, Chovatia M, Herr JR, Carlson JE (2013) Metagenomic profiling reveals lignocellulose degrading system in a microbial community associated with a wood-feeding beetle. PLoS One 8(9):e73827. https://doi.org/10.1371/journal.pone.0073827 (eCollection 2013)

Singh B, Bhat TK, Sharma OP, Kurade NP (2008) Tannin-degrading bacteria from the gastrointestinal tract of Indian languor (*Semnopithecus entellus*) feeding on oak acorns. In: 49th Annual conference. International symposium on microbial biotechnology: diversiry, genomics and metagenomics. Organized by Department of Zoology, North Campus, Department of Miucrobiology, South Campus, University of Delhi, Delhi, India, p 52

Tsuchida S, Murata K, Ohkuma M, Ushida K (2017) Isolation of *Streptococcus gallolyticus* with very high degradability of condensed tannins from feces of the wild Japanese rock ptarmigans on Mt. Tateyama. J Gen Appl Microbiol 63(3):195–198. https://doi.org/10.2323/jgam.2016.09.003 (Epub 2017 Apr 7. No abstract available)

Ushida K, Segawa T, Tsuchida S, Murata K (2016) Cecal bacterial communities in wild Japanese rock ptarmigans and captive Svalbard rock ptarmigans. J Vet Med Sci 78(2):251–257. https://doi.org/10.1292/jvms.15-0313

Viney M (2018) The gut microbiota of wild rodents: challenges and opportunities. Lab Anim 23677218787538. https://doi.org/10.1177/0023677218787538 (Epub ahead of print)

Wagstaff DJ (2008) International poisonous plant checklist. CRC Press, Boca Raton

Yang S, Gao X, Meng J, Zhang A, Zhou Y, Long M, Li B, Deng W, Jin L, Zhao S, Wu D, He Y, Li C, Liu S, Huang Y, Zhang H, Zou L (2018) Metagenomic analysis of bacteria, fungi, bacteriophages, and helminths in the gut of giant pandas. Front Microbiol 9:1717. https://doi.org/10.3389/fmicb.2018.01717 (eCollection 2018)

Insect Gut—A Treasure of Microbes and Microbial Enzymes

5

Abstract

Several insects thrive on plant biomass for nutrition and habitation. Insect gut presents distinctive habitat for colonizing microbial genera and species. The insect gut microbiota contributes to digest plant materials, protecting host from parasites and pathogens, and modulation of immune system. Strategies should be evolved to unravel and exploit less known or obscure microbial diversities of these tiny creatures as sources of enzymes and raw materials having industrial applications, as well as developing insecticides by interrupting the cellulose degradation pathways in them.

Highlights

- Insect gut is distinctive and favorable niche for microbial colonization
- Some invertebrates and insects rely on gut microbiota to utilize plant biomass
- The lignocelluloytic microbial genes enzymes indicate biotechnological applications.

Keywords

Insects · Gut microbiota · Enzymes · Insect–microbiome interaction · Commercial importance

5.1 Introduction

Industrial biotechnology depends on the use of enzymes and microorganisms to produce value-added products. The microbial enzymes have fascinated the researchers since their discovery and are important in the current scenario of Industrial Revolution.

Plant materials are the prime sources for nutrition and asylum for many of terrestrial herbivorous organisms including arthropods and annelids. Various insects (Box 1) damage wood, stored forage and grasses as well as live flora. Some of them actually feed on wood or grasses, while others devastate trees and plants, and burrow into wood for shelter and making nests. The small holes are visible in wood, though in some cases the damage in not noticeable.

The diet is a major factor to regulate the variations in insect microorganisms which in turn benefit the host in digesting ingested plant biomass and protect against pests and pathogens (Krams et al. 2017). In addition, the insect microorganisms are involved in degrading certain phytometabolites that are otherwise detrimental to host. Insects, including arboreal ants, the *Allomerus octoarticulatus*, that depend on myrmecophytic flora for both food and shelter (Arcila Hernández et al. 2017), phytophagus

B. Singh et al., *Advances in Animal Biotechnology*,
https://doi.org/10.1007/978-3-030-21309-1_5

and xylophagus beetles and some termites, most are colonized by diverse genera and species of microorganisms, variously occupying their gut and some other tissues. The gut of these insects is the most efficient ecosystem to degrade lignocellulose-rich plant biomass (Pourramezan et al. 2012; Manfredi et al. 2015; Gales et al. 2018).

Box 1. Herbivorous Insects and Their Main Categories

- **Termites**
  Termites, the winged social insects, thrive on dead and decaying wood where they create tunnels by chewing the wood. The termites, especially the subterranean termites, build underground nests, while drywood termites found in structural lumber are found all over the world. Including around 2700 species across the globe, the termites are divided into seven families, namely Hodotermitidae, Kalotermitidae, Mastotermitidae, Rhinotermitidae, Serritermitidae, Termitidae, and Termopsidae (Abe et al. 2000). Although considered as agriculture and house pests, the termites and their gut microbiome play role in bioturbination and soil formation, maintaining soil fertility through breakdown of organic matter and waste vegetation.
- **Carpenter Ants**
  These are wood-eating insects and burrow deep into rotting wooden materials. Some are winged, while others are wingless.
- **Powderpost Beetles**
  Powderpost beetles are small dark or light brown insects with elongated body and strong small head. The powderpost beetles lay eggs beneath wood surface, which after hatching feed on and chew the wood and make several tiny tunnels into wooden objects. Hence, powderpost beetles are responsible for damaging furniture and molding.
- **Wood borers**
  These are the group of insects that dwell on trees, acorns, and nuts. Examples include wood wasps, wood borers, bark beetles, and longhorned beetles with various morphological, dietary, and habitat adaptations.

5.2 Insects as Multiple Beneficial Organisms

The role of insects in agriculture, animal production, industrial development, and human civilization is evident. Insect herbivory, often viewed as a negative feature of insects, has a positive facade as well. The microorganisms associated with insects enable them to utilize plant biomass and need to be viewed from a commercial biocatalysts viewpoint.

Whereas some insect larvae are sources of dietary proteins, amino acids and minerals such as calcium, in other insects and their larvae benefit humans indirectly. A number of insects play important role in carbon and nitrogen cycling in tropical ecosystems, translocating nutrients, and serve as important drivers of plant succession. Higher termites degrading lignocellulose in various stages of humification and thereby enhancing soil fertility can be used as source of microorganisms for producing bioethanol from waste plant materials such as sugarcane bagasse (Saadeddin 2014; Ben Guerrero et al. 2015).

5.3 Insect–Microbiome Interaction and Symbiosis

Symbiosis throughout plant and animal kingdom is a ubiquitous phenomenon and is vital for extending physiological and ecological capabilities of host. Insect–microbe symbiosis is quite common and permits the host to adapt to a specific environment, fitness, and co-speciation events (Peterson and Scharf 2016; Chen et al. 2018). The benefits conferred by the symbiotic microbes include converting plant polysaccharides (cellulose and hemicelluloses) into energy precursors, deriving nutrients from plants, synthesizing B vitamins (pantothenate, biotin), detoxification of ingested anti-nutritional phytometabolites, insecticides and xenobiotics, and

Table 5.1 Characteristics of microorganisms originating from various insects

Sl. No.	Name of microorganisms	Host/origin	Characteristics	Inferences and recommendations (Reference)
1	*Acinetobacter* sp.	*Microcerotermes diversus (Silvestri)*	Cellulose degradation, CMCase activities, enzymes stable at broad range of temperature and pH	Acinetobacter LB9, and Bacillus B5B had CMCase activities (1.22 U/ml, and 1.47 U/ml, respectively), enzymes envisaged to be of great commercial import (Pourramezan et al. 2012)
	Pseudomonas sp.			
	Staphylococcus sp.			
	Enterobacteriacae and Bacillaceae families			
	Actinobacteria	Reticulitermes Flavipes (Kollar)	Antagonistic activities	Actinobacteria could inhibit *Serratia marcescens*, Molds (*Trichoderma* sp. and *Metarhizium* sp.), *Candida albicans*, and basidiomycete fungi (*Gloeophyllum trabeum* (Persoon) Murrill, *Tyromyces palustris* (Berkeley & M.A. Curtis) Murrill, *Irpex lacteus* (Fries) Fries, and *Trametes versicolor* (L.) Lloyd). It is envisaged that termite microbiota could be sources of pathogen-inhibiting substances (Arango et al. 2016)
		R. tibialis Banks		
	Actinobacteria, Bacteroidetes/ Chlorobi	*Odontotermes formosanus*	Nutrition?	The study unravels co-evolutional relationship of gut microbes *vis-a-vis* host termite (Shinzato et al. 2007)
	Group, Proteobacteria			
	Domains of bacteria			
	Bacillus licheniformis	Reticulitermes chinensis Snyder	Lignin peroxidase (LiP) activities	Characterization of LiP activities in both the isolates, expression of LiP activity in *Lactococcus lactis*. The versicolor (L.) Lloyd). It is envisioned as tools to utilize lignin as energy source (Zhou et al. 2017)
	Enterobacter hormachei			
	Clostridium sp. Ne2 and Clostridia	*Nasutitermes excitiossus*		Genome sequencing revealed that isolates are closely related to *Cl. magnum* and can be used to ferment a range of sugars for producing biofuels (Wang et al. 2015a)
	Ruminoclostridium sp. Ne3, *Clostridia*	*Nasutitermes exitiosus*		The data analysis revealed that organism is hemicelluloses degrader (Wang et al. 2015b)
	Lactic acid bacteria (E. fecalis, L. lactis)	*Nasutitermes arborum, Thoracotermes macrothorax, and Anoplotermes pacificus*	Novel sugar metabolic properties	The study reveals abundance and diversity of lactic acid bacteria in the hind gut of wood and soil-feeding termites (Bauer et al. 2000)
	Protist symbionts	*Reticulitermes flavipes*	Glycosyl hydrolase family 7 (GHF7) cellulases	The GHF7 enhances lignocellulose utilization (Sethi et al. 2013)

defense against pathogens (Table 5.1). For instance, *Wolbachia* sp. in human bedbug (*Cimex lecticularis*) possesses complete biosynthetic pathway for B vitamins (B2 and B7) (Douglas 2017), and *Riesia pediculicola* in lice assists synthesis of B vitamins required essentially to insect nutrition (Boyd et al. 2017).

Notably, compared to mammals, the number and miscellany of microorganisms are low in insects, though some insects harbor large gut communities with specialized metabolic capacities. Dependable transmission routes and social behavior are important to influence gut microbial diversity. Insects such as termites, bees, and ants share microorganisms from each other, hence have more plasticity in their microbiota.

5.4 Insects as Sources of Diverse Microorganisms

The diversity and population of gut microorganisms vary depending on insect species, the morphology and part of gut, food habits, and habitat. Microbial communities reside in different body compartments, viz. bacteriocytes, haemolymph, gut, and salivary gland of the insects.

Forest-defoliating insects and caterpillars, such as pine looper moth (*Bupalus piniaria*) and pine sawfly (*Neodiprion serifer*), feed and depend on needles of native and imported *Pinus sylvestris*, *P. nigra*, *P. cembra* among others. Their gut seems to harbor bacteria specialized in degrading resin-enriched cell walls. Tree bugs, such as aphids sucking sap of plants, might have microbial consortia outside or inside their body enabling them derive nutrients from plants, boosting immunity and detoxifying plant origin toxic metabolites (Hansen and Moran 2014; Sugio et al. 2015). Aphids have attracted the attention of microbial ecologists because of their symbiotic association with broad range of endosymbionts (Arneodo and Ortego 2014). A study reports cultivable microbial diversity from various aphid species. The cultured microorganisms have been assigned 24 bacterial genera representing three phyla, namely Actinobacteria, Firmicutes, and Proteobacteria, and 3 fungal genera from two phyla, namely Ascomycota and Basidiomycota (Grigorescu et al. 2017).

Culture-independent studies and high-throughput sequencing reveal extraordinary microbial diversity residing in insect gut, and most of them are uncultured. Inability of microorganisms to grow in culture is major impediment in understanding their actual role in insects.

The termite gut comprises three major sections including foregut, midgut, and hindgut. The hindgut is the most important site for lignocellulose degradation and nutrient utilization. It is estimated that hindgut of termite may contain as much as 10^6 to 10^8 microorganisms per microliter-scale termite gut content (reviewed in Saadeddin 2014). Some termite species fix atmospheric nitrogen (N_2) with the help of diazotrophic gut bacteria (Sapountzis et al. 2016).

There are some factors that impact gut microbial diversity. Social behavior in some insects is one of the important factors. During development, many insects molt several times beginning from the larval stage of development and shedding exoskeletal lining of foregut and hindgut. This eliminates the microbial communities' associates with these niches.

Some insects such as holometabolous insects have distinctive larval, pupal, and adult stages. Remodeling of gut and other organs during metamorphosis eliminates entire larval gut contents and colonizers. Similarly, in mosquitoes such as *Anopheles punctipennis* (Say), *Culex pipiens* (L.), and *Aedes aegypti* (L.), the metamorphosis almost completely eliminates the gut bacteria, and as a result, the adult mosquitoes are almost devoid of beneficial bacteria in their gut (Moll et al. 2001).

5.5 Beneficial Attributes of Insect Gut Microorganisms

Insect gut can be regarded as small anoxic bioreactor akin to the rumen or hindgut of herbivores, wherein microorganisms and their enzymes assist degradation of dietary plant materials. The termite hindgut microorganisms enable them utilize wood. Indeed, the ability of termite gut microorganisms to convert lignocellulose to easily fermentable oligosaccharides and/or simpler sugar monomers has attracted the attention of microbial ecologists (Chaffron and Von Mering 2007; Sadeddin 2012; Gao et al. 2016).

A wood-feeding termite, *Coptomes formosanus* Shiraki, is found to possess an independent, dual cellulose-digesting systems: one comprising of endogenous celllulases in midgut, and other entailing hindgut cellulases synthesized by symbiotic flagellates (Nakashima et al. 2002).

Ex vivo lignocellulose biomass degradation potential of gut microbial consortia of various phytophagus and xylophagus insects, namely beetle (*Ergates faber*), chafer (*Potosia cuprea*), cricket (*Gryllus bimaculatus)*, cockroach (*Gromphadorrhina portentosa*), and a locust (*Locusta migratoria*), was studied by inoculating gut contents in anaerobic batch reactor, using grounded wheat straw as carbon source at neutral pH. A short duration fermentation of 8 days, drop in pH, and metabolism of substrate were observed. Reduction in pH from 7.0 to 4.5 indicates degradation of lignocelluloses biomass and production of organic acids from wheat straw. The presence of insect-dependent bacterial community structures comprising of Bacteroidia, Clostridia, and Gammaproteobacteria classes, and augmented xylanase activity was detected during fermentation. The study therefore shows that biomimetic approach can improve lignocellulose degradation in bioreactors (Gales et al. 2018).

5.6 Metagenomic Insights in Gut Microbiome

Despite several established and envisaged benefits of insect gut symbionts, the gut microbiota remains more or less unexplored. This is because culture-dependent techniques are unable to represent a blueprint of the gut microbiota. The culture-independent genomics and metagenomic approaches have provided valuable insights into structural and functional profile of insect gut microbiome (Rosenthal et al. 2013; Rossmassler et al. 2015; Peterson and Scharf 2016).

The genomics studies reveal the existence of species-specific complex microbial lineages in insects affected by microenvironmental factors (pH, available substrate, O_2, and H_2), and that most of the microbial communities cannot be grown *in vitro* (Manjula et al. 2016; Mikaelyan et al. 2017). It is emphasized that whole metagenomic sequencing could be useful in unraveling the complex microbial communities, genes, and the enzymes for industrial and biotechnological functions (Nimchua et al. 2012; Manjula et al. 2016).

So far lignin degradation is studied in white-rot basidiomycetes, and information is scarce on lignin degradation in gut microorganisms. Lignocellulose-degrading enzymes and mechanisms have been identified in some insects. It is found that hemicelluloses, lignin, and cellulose are actively degraded in the gut of Asian longhorned beetle (*Anoplophora glabripennis*), and that gut harbors microbial consortia involved in the process and enhancing nutrition capabilities of the insects. The genes encoding laccases, dye-decolorizing peroxidases, novel peroxidases and β-etherases, 36 families of glycoside hydrolases, and detoxification of some factors are identified from the gut metagenome of the *A. glabripennis* (Scully et al. 2013). Midgut microbial metagenome of *A. glabripennis* is found to be enriched in genes involved in biosynthetic pathways including synthesis of essential amino acids, sterols and vitamins, lignin degradation and ferment xylose, thereby enhancing nutritional provisioning of the host (Scully et al. 2013, 2014).

An Illumina-based study intended to infer biomass-degrading genetic makeup of free-living bacteria in the gut of a lower termite *Coptotermes gestroi* has shown that around 1460 microbial species were present in these termites. The microorganisms belonged to 12 bacterial orders entailing Actinomycetales, Bacillales, Bacteroidales, Burkholderiales, Clostridiales, Desulfovibrionales, Enterobacteriales, Lactobacillales,

Pseudomonades, Spirochaetales, Synergistales, and Xanthomonadales (Do et al. 2014). In addition, an array of ORFs or genes corresponding to plant cell wall-degrading enzymes were also recorded in the metagenome.

Large-scale metagenomic functional screening of fosmid libraries, shotgun sequencing, and biochemical assays was carried out to examine the gut microbiome of *Globitermes brachycerastes*, the wood-feeding "higher" termite. A range of cellobiose-metabolizing enzymes, namely β-glucosidases, cellobiose phosphorylases, and phopho-6-β-glucosidases was functionally verified, implying that termite gut ecosystem utilizes diverse enzyme clusters and metabolic pathways as a part of plant biomass degradation and essential metabolism (Liu et al. 2018). More information is expected in the future when more metagenome data will be available for accession and analysis.

5.7 Outlook and Challenges

Plant–insect interaction and coexistence are an intricate and diverse relationship in evolutionary history existing from past 400 million years. Insect– microbe symbiosis is an important association that enable insects to survive and proliferate in vegetation-enriched habitats, detoxification of xenobiotics and phytometabolites, and protection against pests and pathogens.

Compared to large animals such as mammals, except certain social species, majority of the insect guts has relatively few gut microbial communities. This is probably due to lack of opportunities and means of transmission of gut microbes among individuals or from parents to the next generation.

Despite ample reasons to discover insect gut microorganisms, and despite the current information on microorganisms residing inside the insect guts, a generalized perception concerning how insect gut microbial communities are organized is just in initial stage.

In addition, except some insects, for examples, termites, the insect–microbe association is less explored area primarily due to methodological and analytical limitations. Notably, the culture-independent metagenomics, metatranscriptomics, and proteomics have provided valuable insights into host–microbe associations and the role of insect gut symbionts. Though metagenomic studies have revealed abundance of diverse fibrolytic enzymes, the biotechnological potential of envisaged enzymes is not utilized. Further, majority of the gut microorganisms in insect guts are unculturable. It is necessary to develop culture media and methodologies to culture insect gut microorganism so that their enzymes can be studied, and promising ones can be selected for improvement or commercial applications.

5.8 Conclusions

The wood-feeding insects, the termites, seem to be most efficient users of lignocellulose in terms of rate and extent of cellulose degradation. The gut microbiome seems to be promising in terms of genetic resources and the enzymes for commercial applications. Ironically, the extent of insect gut microbiome association is still undefined. The studies aimed to explore the insect gut ecosystems may provide novel insights in near future.

References

Abe T, Bignell DE, Higashi M (2000) Termites: evolution, sociality, symbioses, ecology. Springer-Verlag, New York

Arango RA, Carlson CM, Currie CR, McDonald BR, Book AJ, Green F 3rd, Lebow NK, Raffa KF (2016) Antimicrobial activity of actinobacteria isolated from the guts of subterranean termites. Environ Entomol 45 (6):1415–1423. https://doi.org/10.1093/ee/nvw126

Arcila Hernández LM, Sanders JG, Miller GA, Ravenscraft A, Frederickson ME (2017) Ant-plant mutualism: a dietary by-product of a tropical ant's macronutrient requirements. Ecology 98(12):3141–3151. https://doi.org/10.1002/ecy.2036

Arneodo JD, Ortego J. Exploring the bacterial microbiota associated with native South American species of Aphis (Hemiptera: Aphididae). Environ Entomol. 2014 43(3):589–594. https://doi.org/10.1603/en13324 (Epub 2014 Apr 14)

Bauer S, Tholen A, Overmann J, Brune A (2000) Characterization of abundance and diversity of lactic acid bacteria in the hindgut of wood- and soil-feeding termites by molecular and culture-dependent techniques. Arch Microbiol 173(2): 126–137

Ben Guerrero E, Arneodo J, Bombarda Campanha R, Abrão de Oliveira P, Veneziano Labate MT, Regiani Cataldi T, Campos E, Cataldi A, Labate CA, Martins Rodrigues C, Talia P (2015) Prospection and evaluation of (Hemi) cellulolytic enzymes using untreated and pretreated biomasses in two argentinean native termites. PLoS One 10(8): e0136573. https://doi.org/10.1371/journal.pone.0136573 (eCollection 2015)

Boyd BM, Allen JM, Nguyen NP, Vachaspati P, Quicksall ZS, Warnow T, Mugisha L, Johnson KP, Reed DL (2017) Primates, lice and bacteria: speciation and genome evolution in the symbionts of hominid lice. Mol Biol Evol 34(7):1743–1757. https://doi.org/10.1093/molbev/msx117

Chaffron S, von Mering C (2007) Termites in the woodwork. Genome Biol 8:229

Chen B, Du K, Sun C, Vimalanathan A, Liang X, Li Y, Wang B, Lu X, Li L, Shao Y (2018) Gut bacterial and fungal communities of the domesticated silkworm (Bombyx mori) and wild mulberry-feeding relatives. ISME J 12(9):2252–2262. https://doi.org/10.1038/s41396-018-0174-1 (Epub 2018 Jun 12)

Do TH, Nguyen TT, Nguyen TN, Le QG, Nguyen C, Kimura K, Truong NH (2014) Mining biomass-degrading genes through Illumina-based de novo sequencing and metagenomic analysis of free-living bacteria in the gut of the lower termite Coptotermes gestroi harvested in Vietnam. J Biosci Bioeng 118(6):665–671. https://doi.org/10.1016/j.jbiosc.2014.05.010

Douglas AE (2017) The B vitamin nutrition of insects: the contributions of diet, microbiome and horizontally acquired genes. Curr Opin Insect Sci. 23:65–69. https://doi.org/10.1016/j.cois.2017.07.012 (Epub 2017 Aug 3). Review

Gales A, Chatellard L, Abadie M, Bonnafous A, Auer L, Carrère H, Godon JJ, Hernandez-Raquet G, Dumas C (2018) Screening of phytophagous and xylophagous insects guts microbiota abilities to degrade lignocellulose in bioreactor. Front Microbiol. 9:2222. https://doi.org/10.3389/fmicb.2018.02222 (eCollection 2018)

Gao G, Wang A, Gong BL, Li QQ, Liu YH, He ZM, Li G (2016) A novel metagenome-derived gene cluster from termite hindgut: Encoding phosphotransferase system components and high glucose tolerant glucosidase. Enzyme Microb Technol 84:24–31. https://doi.org/10.1016/j.enzmictec.2015.12.005

Grigorescu AS, Renoz F, Sabri A, Foray V, Hance T, Thonart P (2017) Accessing the hidden microbial diversity of aphids: an illustration of how culture-dependent methods can be used to decipher the insect microbiota. Microb Ecol. https://doi.org/10.1007/s00248-017-1092-x

Hansen AK, Moran NA (2014) The impact of microbial symbionts on host plant utilization by herbivorous insects. Mol Ecol 23(6):1473–1496. https://doi.org/10.1111/mec.12421 (Epub 2013 Aug 16)

Krams IA, Kecko S, Jõers P, Trakimas G, Elferts D, Krams R, Luoto S, Rantala MJ, Inashkina I, Gudrā D, Fridmanis D, Contreras-Garduño J, Grantiņa-Ieviņa L, Krama T (2017) Microbiome symbionts and diet diversity incur costs on the immune system of insect larvae. J Exp Biol. pii: jeb.169227. https://doi.org/10.1242/jeb.169227

Liu N, Li H, Chevrette MG, Zhang L, Cao L, Zhou H, Zhou X, Zhou Z, Pope PB, Currie CR, Huang Y, Wang Q (2018) Functional metagenomics reveals abundant polysaccharide-degrading gene clusters and cellobiose utilization pathways within gut microbiota of a wood-feeding higher termite. ISME J. https://doi.org/10.1038/s41396-018-0255-1 (Epub ahead of print)

Manfredi AP, Perotti NI, Martínez MA (2015) Cellulose degrading bacteria isolated from industrial samples and the gut of native insects from Northwest of Argentina. J Basic Microbiol 55(12):1384–1393. https://doi.org/10.1002/jobm.201500269

Manjula A, Pushpanathan M, Sathyavathi S, Gunasekaran P, Rajendhran J (2016) Comparative analysis of microbial diversity in termite gut and termite nest using ion sequencing. Curr Microbiol 72 (3):267–275. https://doi.org/10.1007/s00284-015-0947-y

Mikaelyan A, Meuser K, Brune A (2017) Microenvironmental heterogeneity of gut compartments drives bacterial community structure in wood- and humus-feeding higher termites. FEMS Microbiol Ecol. 93(1). pii: fiw210 (Epub 2016 Oct 8)

Moll RM, Romoser WS, Modrzakowski MC, Moncayo AC, Lerdthusnee K (2001) Meconial peritrophic membranes and the fate of midgut bacteria during mosquito (Diptera: Culicidae) metamorphosis. J Med Entomol 38(1):29–32

Nakashima K, Watanabe H, Saitoh H, Tokuda G, Azuma JI (2002) Dual cellulose-digesting system of the wood-feeding termite, Coptotermes formosanus Shiraki. Insect Biochem Mol Biol 32(7):777–784

Nimchua T, Thongaram T, Uengwetwanit T, Pongpattanakitshote S, Eurwilaichitr L (2012) Metagenomic analysis of novel lignocellulose-degrading enzymes from higher termite guts inhabiting microbes. J Microbiol Biotechnol 22(4):462–469

Peterson BF, Scharf ME (2016) Metatranscriptome analysis reveals bacterial symbiont contributions to lower termite physiology and potential immune functions. BMC Genom 17(1):772

Pourramezan Z, Ghezelbash GR, Romani B, Ziaei S, Hedayatkhah A (2012) Screening and identification of newly isolated cellulose-degrading bacteria from the gut of xylophagous termite *Microcerotermes diversus* (Silvestri). Mikrobiologiia 81(6):796–802

Rosenthal AZ, Zhang X, Lucey KS, Ottesen EA, Trivedi V, Choi HM, Pierce NA, Leadbetter JR

(2013) Localizing transcripts to single cells suggests an important role of uncultured deltaproteobacteria in the termite gut hydrogen economy. Proc Natl Acad Sci USA 110(40):16163–16168. https://doi.org/10.1073/pnas.1307876110

Rossmassler K, Dietrich C, Thompson C, Mikaelyan A, Nonoh JO, Scheffrahn RH, Sillam-Dussès D, Brune A (2015) Metagenomic analysis of the microbiota in the highly compartmented hindguts of six wood- or soil-feeding higher termites. Microbiome 26(3):56. https://doi.org/10.1186/s40168-015-0118-1

Saadeddin A (2014) The complexities of hydrolytic enzymes from the termite digestive system. Crit Rev Biotechnol 34(2):115–122. https://doi.org/10.3109/07388551.2012.727379 (Epub 2012 Oct 5)

Sapountzis P, de Verges J, Rousk K, Cilliers M, Vorster BJ, Poulsen M (2016) Potential for nitrogen fixation in the fungus-growing termite symbiosis. Front Microbiol 7:1993. https://doi.org/10.3389/fmicb.2016.01993 (eCollection 2016)

Scully ED, Geib SM, Hoover K, Tien M, Tringe SG, Barry KW, Glavina del Rio T, Chovatia M, Herr JR, Carlson JE (2013) Metagenomic profiling reveals lignocellulose degrading system in a microbial community associated with a wood-feeding beetle. PLoS One. 8(9):e73827. https://doi.org/10.1371/journal.pone.0073827 (eCollection 2013)

Sethi A, Kovaleva ES, Slack JM, Brown S, Buchman GW, Scharf ME (2013) A GHF7 cellulase from the protist symbiont community of Reticulitermes flavipes enables more efficient lignocellulose processing by host enzymes. Arch Insect Biochem Physiol 84(4):175–193. https://doi.org/10.1002/arch.21135 (Epub 2013 Nov 1)

Shinzato N[1], Muramatsu M, Matsui T, Watanabe Y (2007) Phylogenetic analysis of the gut bacterial microflora of the fungus-growing termite Odontotermes formosanus. Biosci Biotechnol Biochem 71(4):906–9015 (Epub 2007 Apr 7)

Sugio A, Dubreuil G, Giron D, Simon JC (2015) Plant-insect interactions under bacterial influence: ecological implications and underlying mechanisms. J Exp Bot 66(2):467–478. https://doi.org/10.1093/jxb/eru435 (Epub 2014 Nov 10)

Wang H, Lin H, Tran-Dinh N, Li D, Greenfield P, Midgley DJ (2015a) Draft genome sequence of clostridium sp. Ne2, clostridia from an enrichment culture obtained from Australian subterranean termite, Nasutitermes exitiosus. Genome Announc. 3(2). pii: e00304–15. https://doi.org/10.1128/genomea.00304-15

Wang H, Lin H, Tran-Dinh N, Li D, Greenfield P, Midgley DJ (2015b) Draft genome sequence of *Ruminoclostridium* sp. Ne3, Clostridia from an enrichment culture obtained from Australian Subterranean Termite, *Nasutitermes exitiosus*. Genome Announc.3(2). pii: e00305-15. https://doi.org/10.1128/genomea.00305-15

Zhou H, Guo W, Xu B, Teng Z, Tao D, Lou Y, Gao Y (2017) Screening and identification of lignin-degrading bacteria in termite gut and the construction of LiP-expressing recombinant Lactococcus lactis. Microb Pathog 112:63–69. https://doi.org/10.1016/j.micpath.2017.09.047

Nutraceuticals from Bioengineered Microorganisms

6

Abstract

Microbiota inhabits almost all the niches on earth and has shunned the modern world. Microbial metabolites combined with modern natural product research methodology has opened the way for a new era of nutraceuticals and therapeutic. Microbes can be engineered to produce metabolites, called as postbiotics that have health, nutrition, and industrial applications. The postbiotics confer beneficial health effects such as inhibition of gut and genitourinary microbial infections, immunomodulation, and preventing the inflammatory gut diseases and cancer.

Highlights

- Microorganisms can be engineered to do produce an array of metabolites
- The microorganism of humans or animals origin are considered to be useful
- Several recombinant microbial products are in use in humans and animals.

Keywords

Nutraceuticals · Bioengineered microbes · Recombinant enzymes · Probiotics · Postbiotics · Immunomodulation · Antimicrobial peptides

6.1 Introduction

We are in the era where each aspect of life is influenced by microorganisms and the microbial products. The market and potential for recombinant drugs are expanding as systems, and the protein production platforms are improved substantially. The bacterial strains such as strains of *Escherichia coli*, lactic acid bacteria (LAB), and yeasts are developed as protein producing units essentially because of their versatility, ease of genetic manipulation, and cultivation.

The recombinant microorganisms and microbial products have moved from plain recombinant natural products to complex protein constructs as versatile macromolecules developed from rational drug design process and computational engineering. Enormous information is available on development of engineered microorganisms for producing industrially important macromolecules. This chapter confines to the nutrition and health-related macromolecules produced from human or animals origin microorganisms. In addition, a brief preface of the microbial rich niches in the body that serve as sources of beneficial microorganism s is also cited.

B. Singh et al., *Advances in Animal Biotechnology*,
https://doi.org/10.1007/978-3-030-21309-1_6

6.2 Microbial Rich Niches in Body

Almost all the body parts are inhabited variably by microorganisms. Skin, gastrointestinal (GI) tract, genitourinary tract, and milk are the rich reservoirs of microorganisms that impact human health. Currently, the milk has emerged as one of the most dynamic microbial ecosystems comprising of LAB and bifidobacteria as predominant species in healthy lactation stages, though *Staphylococcus* and *Streptococcus* may also dominate infrequently (Hunt et al. 2011; Martín et al. 2011). Metagenomic analysis reveals that milk contains complex genomes of bacteria, archaea, viruses, fungi, and protozoa. The loss of bacterial diversity leads to infections such as mastitis (Jiménez et al. 2015).

Probiotic LAB have been reported from South African Saanen goats (Makete et al. 2016) and Brazilian goat milk (de Moraes et al. 2017). Probiotics *Lactobacillus rhamnosus* CRL 15105 isolated from goat milk could suppress *Streptococcus pneumoniae* and *Salmonella typhimurium* in mice (Salva et al. 2010). Researchers in Argentina have developed calf feed products containing a blend of freeze-dried probiotics (*L. johnsonii*, *L. murinus*, *L. salivarius*, *L. mucosae*, *L. amylovorus*, and *Enterococcus faecium*), lyoprotectors, vitamins, and minerals (Maldonado et al. 2016). The fermented milk products containing LAB (*L. murinus*, *L. mucosae*, *L. johnsonii*) were recommended as viable veterinary products for young calves owing to the beneficial effects of these products on health (low morbidity and mortality) and growth of animals (Maldonado et al. 2018).

LAB, such as *Lactococcus* sp. with probiotic properties (Fig. 6.1) isolated from genitourinary tract, have been engineered to in situ delivery of various therapeutic agents or expressing biomolecules to prevent genitourinary viral infections, such as HIV, and bacterial vaginosis (Table 6.1) (Lagenaur et al. 2011; Singh et al. 2017; Petrova et al. 2018).

6.3 The Concept and Rationale of Postbiotics

Industries rely much on natural resources for sustainable supply of raw materials and biocatalysts. It is evident that microorganisms are important sources of enzymes and other raw materials for industrial uses. However, only a few microbial niches and microorganisms are used to produce natural products of commercial importance. Hence, there is scientific and commercial interest in using probiotics and probiotic metabolites (postbiotics) to curtail pathogens, modulation of immune system, improved lactose digestion, synthesis of vitamins, lowering serum cholesterol, and increasing the availability of minerals to host. In addition, the prevention of cancer and metabolic diseases such as obesity and diabetes is among important health applications of probiotics equipped with exogenous genetic elements.

6.4 Postbiotics from Microorganisms

The synergy between postbiotics, host metabolites, and gut microbes is important to nutrition and health of the host. SCFAs produced by GI bacteria act as signaling molecule to improve regulation of lipid metabolism, glucose homeostasis, and insulin sensitivity, through activation of G-protein-coupled receptors (GPRs), therefore contributing to regulate energy balance while maintaining metabolic homoeostasis (Kimura et al. 2013; Miyamoto et al. 2016).

Postbiotics have a wide range of applications in humans and veterinary nutrition. Cell-free preparations of different probiotic bacteria are said to prevent or treat some diseases (Klein et al. 2013). For instance, Hylak@ Forte (Ratiopharm/Merckle GmbH, Germany), a bacteria-free liquid containing microbial metabolites (SCFA, lactic acid, and other unidentified metabolites) from

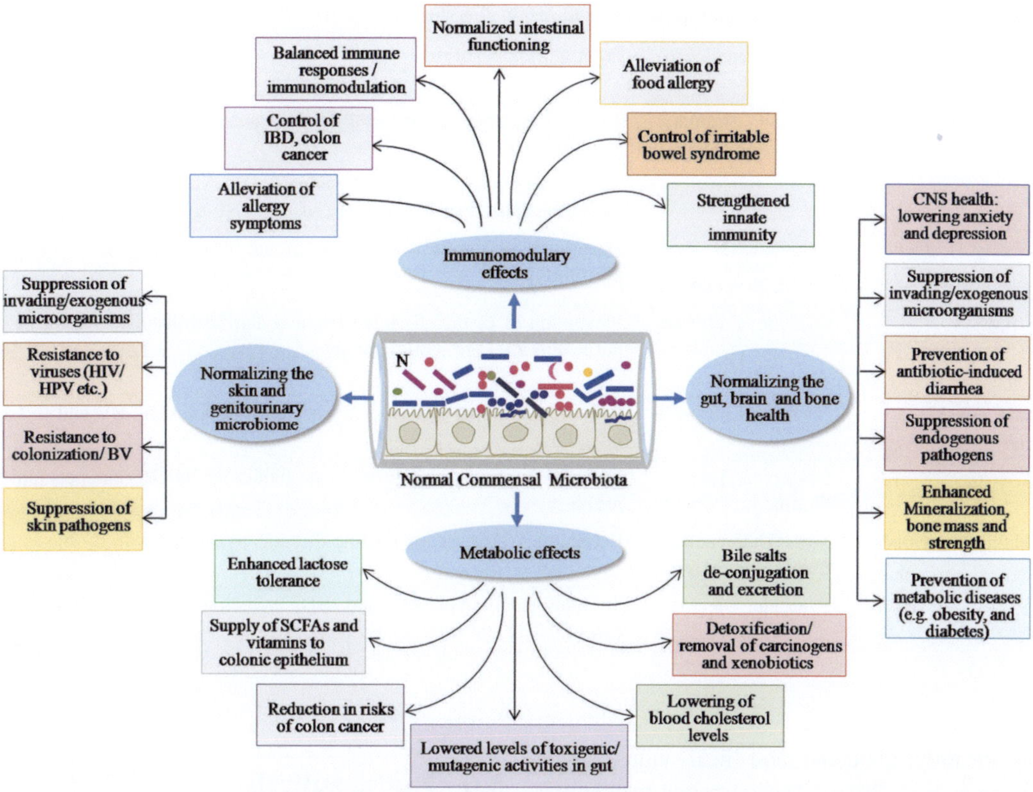

Fig. 6.1 Role of normal microbiota in homeostasis in nutrition, immunity, and health. Skin, gastrointestinal tract, and uterine mucosa are heavily colonized by diverse microbes that are further influenced by genes, ethnic origins environmental, and behavioral factors. The recombinant probiotics and their metabolites might promote normal microbiota and reduce life-threatening infections and metabolic diseases

E. coli DSM 4087, *Streptococcus faecalis* DSM 4086, *L. acidophilus* DSM 414, and *L. helveticus* DS 4183, is recommended to restore intestinal microbiota, stimulate the synthesis of epithelial cells of the intestinal wall, normalize the pH and fluid and electrolyte balance in the lumen of intestine (Shenderov 2013). CytoFlora$^{@2}$ a probiotics tincture (BioRay Inc., Laguna Hills, CA, USA), made from micronized cell wall lysates of *L. rhamnosus, B. bifidum, L. acidophilus, B. infantis, B. longum, S. thermophilus, L. plantarum, L. salivarius, L. reuteri, L. casei, L. bulgaricus, L. acidophilus DDS-1*, and *Lactobacillus sporogenes*, is used to improve intestinal dysbiosis, promote a balanced immune response, and improve symptoms in autistic kids (Ray et al. 2010). Similarly, a new commercial product Del-Immune V® (Pure Research Products, LLC, Boulder, CO, USA), a US FDA-registered formulation containing muramyl peptides, amino acids, and DNA fragments of *L. rhamnosus* V (DV strain), is suggested to support immune system (Sichel et al. 2013). This implies that postbiotics may contribute to improve health status by providing better physiological effects, albeit the exact mechanisms need scientific studies.

Improving animal health is a promising field of application, as it is evident that postbiotics influence growth of animals such as piglets and poultry (Choe et al. 2012; Kareem et al. 2016, 2017). The broilers fed with postbiotics derived from *L. plantarum* had significantly higher body weight and total weight gain compared to the broilers fed with a basal diet devoid of postbiotics. Also, postbiotics were found to increase

Table 6.1 Summary of bioengineered probiotics and metabolites for use as alternative therapeutics

Formulation (Live component)	Manufacturer	Proposed applications
ActoBiotics$^{(R)}$ (*L. lactis*)	Intrexon, Virginia USA	Targeted therapies against oral, GI tract metabolic,, allergic and autoimmune diseases
		The ActoBiotics, viz., AG013 attenuates OM; and AG014 treats IBD through localized in situ delivery of anti-TNF-α Fab
AG013 (*L. lactis* secreting hTFF1)	ActoGenix N. V. Ghent, Belgium	Developing antitumor necrosis factor, and bacterial formulation to prevent OM
CBM588 (*Clostridium butyricum*)	Osel Inc. California	Prevention of *C. difficile* infection, and diarrheal disorders, restoring gut health after antibiotic therapy
LACTIN-V (*Lactobacillus* sp.)	Osel Inc. California	Restoring feminine genitourinary health by suppressing BV and viral infections
L. lactis	Oragenics, Tampa Florida	Novel antibiotics, and proprietary probiotics for humans and pets oral health
Multiple microbes	Seres Therapeutics Cambridge, MA	Ectrobiotics$^{(R)}$ drugs for treating diseases associated with dysbiosis
SYNB1020 (*E. coli* Nissle1917)	Synlogic, UK	Treating genetic metabolic and IBDs including CD, UC, CVDs and cancer, UCD and hepatic encephalopathy (U.S. Patent #9, 487, 464).

significantly duodenal and ileal villus height (Kareem et al. 2016). The microbial metabolites (organic acids, bacteriocins, peptides) are found to interact with multiple key targets in various metabolic pathways which regulate cellular proliferation, differentiation, apoptosis, inflammation, angiogenesis, and metastasis (Kumar et al. 2013).

6.5 Bioengineering of Secondary Probiotic Metabolites

There is a genuine need to genetically modify the probiotic grade microorganisms to produce metabolites that have relevance in nutrition and health. Probiotics have been modified or engineered for treating inflammatory bowel disease (IBD). Orally administered engineered probiotics provide appreciable targeting of active biological agent at the affected site.

6.6 Nutraceuticals and Therapeutics from Bioengineered Microorganisms

Indeed, with the development of a *Lactococcus lactis* for delivery of recombinant anti-inflammatory cytokine interleukin-10 (IL-10), intended to treat inflammatory bowel disease (IBD) (Steidler et al. 2000), the field bioengineering of probiotics has advanced swiftly as a premier area of research for antibacterial and anti-viral therapies (Kumar et al. 2016; Singh et al. 2017). The *Lactococcus lactis* being non-pathogenic, non-colonizing species has been studied in a great deal for producing heterologous bioactive molecules.

The patients treated with recombinant *Lactococcus lactis* (LL-Thy 12), expressed mature human IL-10 and experienced reduced disease activity (Braat et al. 2006). Recombinant *L. lactis* producing surface piliation appendage (SpaCBA) with immunomodulating capacity, could activate Toll-like receptor 2-dependent signaling in vitro and modulate production of anti-inflammatory cytokines (TNF-α, IL-6, IL-10, and IL-12) in human dendritic cells (Von Ossowski et al. 2013). Functionality of stress-inducible controlled expression system (SICE) in *L. lactis* for delivery of proteins of health interest at mucosal surface was demonstrated in mice models (Benbouziane et al. 2013). Oral pretreatment using recombinant *L. lactis* expressing dust mite allergen Der-p2 as a mucosal vaccine, induced immune tolerance against house dust mite allergy in mice (Ai et al. Ai et al. 2014). The vaccination through oral immunization of murine models with a recombinant *Lb. gasseri* expressing conserved region of the streptococcal M6 protein (CRR6) could protect the animals against streptococcus group A infections (Mansour and Abdalaziz 2016).

6.7 Antimicrobial Peptides (AMPs)

One of the major challenges in modern medicine is a paucity of alternative therapies to combat rapidly developing multidrug resistance among microbial pathogens. In particular, the AMPs including bacteriocins, which mainly act via membrane active mechanisms, have potential to surmount the epidemic of antibiotic-resistant infections (Ong et al. 2014). This is being exerted by forming poration complexes that cross the phospholipid bilayer and causes membrane permeabilization and depletion of the proton-motive force of targeted sensitive cells. The recombinant probiotics expressing AMPs present a combined strategy that brings the benefits of AMPs and probiotics concurrently.

Recurrent urinary tract infections (UTIs) are common in women in many segments of society. Epidemiological, experimental, and clinical evidences convincingly reveal that vaginal microbiota dominated by bacteriocinogenic LAB plays a protective role against bacterial vaginosis (BV) and sexually transmitted infections including human papilloma virus (HPV) and HIV (Singh et al. 2013, 2017). A synergizing role of probiotics with anti-HIV therapy has also been envisaged (Rastmanesh et al. 2012). Recombinant microbiocides may offer a target specificity ensuring prolonged protection against sexual transmission of HIV/AIDS (Dey et al. 2013; Kumar et al. 2016).

6.8 Postbiotics for Metabolic Diseases

Some microbial metabolites are intended specifically for metabolic diseases. The microbial metabolites are either purified, or recombinant microbes are administered as such into the body to induce therapeutic effects. The recombinant microorganisms or microbial metabolites in association with resident gut microbiota confer protection against metabolic diseases such as obesity and diabetes, inflammatory responses, oxidative stress, and increasing the expansion of adhesion proteins with intestinal epithelium, for reducing intestinal permeability that increases insulin sensitivity and reduces the autoimmune responses (Gomes et al. 2014). The recombinant *Lactobacillus casei*/pSW501 inducing signal peptide SP(Usp45)-INS-specific antibodies, thereby raising the levels of IL-4 in the sera of NOD mice were found to protect them from pancreatic injury. Though further studies are warranted, this approach might be used for treating type 1 diabetes mellitus (T1DM) (Chen et al. 2007).

Recombinant probiotics expressing therapeutic factors that increase the satiety and sensitivity to adipose-derived negative feedback signals, such as leptin, could be a potential strategy (Chen et al. 2014). Oral administration of an engineered *E. coli* Nissle 1917 expressing N-acylphosphatidylethanolamines (NAPE) for eight weeks, reduced the levels of obesity in mice (Chen et al. 2014).

6.9 Recombinant Enzymes

6.9.1 Phytase

Humans and monogastric animals such as pig, poultry, rabbits, and fish are either unable to utilize certain complex nutrients or degrade them feebly. As a result, the unutilized nutrients are released into environment through feces. Among most familiar examples are phytic acid and plant cell wall polysaccharides that are not degraded in GI tract of humans and other monogastric species. The enzyme, named phytase, or recombinant probiotics producing phytase as dietary supplements enhance degradation of phytate to release phosphorus (Pakbaten et al. 2018).

Expression of native or recombinant proteins in probiotics is aimed to express and transfer these molecules to GI mucosal surface to enhance nutrient utilization. Many probiotic strains such as *Lactococcus lactis* and lactobacilli are developed to produce recombinant phytase.

6.9.2 Tannases

Tannases (EC3.1.1.20), also called as tannin acyl hydrolases, are groups of enzymes that transform hydrolysable tannins (HTs) found naturally in several crop species and forages that are fed to animals. The animals consuming plants containing an excess of HTs suffer from tannin toxicity.

A number of microorganisms including environmental fungi and bacteria (de Las Rivas et al. 2018), including those inhabiting GI tract of ruminants (Bhardwaj et al. 2003; Singh et al. 2012; Sharma et al. 2017), humans, and some monogastric animals possess tannases of varying characteristics and substrate specificities. In silico analysis of genome sequence data has revealed the presence of tannase amplicons in many bacteria and fungi of diverse origins.

Tannases has several industrial biotechnological applications such as producing gallic acid, manufacture of instant tea, processing of beverages, and food and feed processing.

Genes encoding tannase are introduced into other microorganisms either to use these microorganisms as microbial feed additives or producing high titers of tannase with finer catalytic activities. Tannase encoding genes are found in some lactic acid bacteria such as *Lactobacillus plantarum*, *L. paraplantarum*, and *L. pentosus*. Recombinant tannase produced by these strains was studied for their activities and preservation (Iwamoto et al. 2008; Matoba et al. 2013; Ueda et al. 2014).

Protocol has been described for increasing the yield of pure recombinant tannase (17 mg/ml) by affinity chromatography purification. Despite representing a novel family of tannase, the recombinant tannase showed activity against tannic acid, gallocatechin gallate, and epigallocatechin gallate, therefore making the recombinant tannase as alternative to commercially utilized fungal tannases (Curiel et al. 2009). Recombinant tannase produced from the tannase genes originally isolated from *Aspergillus niger* GH1, and expressed in engineered *Pichia pastoris* strains, comprised of 562 amino acids. The enzyme was advocated to be safer and suitable for food and beverage applications (Fuentes-Garibay et al. 2015).

6.9.3 Bile Salt Hydrolases

Bile salt hydrolases (BSH) are the enzymes produced by several lactobacilli and bifidobacteria colonizing internal body mucosal sites. Indeed, production of BSH is a desirable attribute of probiotic strains. The BSH has differential substrate spectrum, influence bile acid signaling, and sometimes are involved in bacterial pathogenesis. Deconjugated bile salts produced by BSH-like activities in a probiotic strain *Lactobacillus johnsonii* La1 had inhibitory effects against *Giardia duodenalis*, a protozoan parasite responsible for giardiasis (Travers et al. 2016; Allain et al. 2018).

The BSH in probiotics helps them reduce blood cholesterol levels of the host. The probiotics posing BSH activity has a big market potential. However, intensive research and in vivo trials are needed to investigate the negative impact of BSH activity and dynamics between probiotics, BSH activity and bile salts.

6.10 Antimicrobial Peptides (AMPs)

Emergence of resistance to broadly used antibiotics among pathogens is a major health threat among humans and animals. Antimicrobial resistance (AMR), generally regulated by the genes called, antibiotic resistance genes (ARGs), is the principle threat for effective prevention and treatment of infectious diseases caused by viruses, bacteria, yeasts, fungi and parasites. The ARGs, though widely present in diverse microorganisms, cause problem only when they are transferred to pathogenic microorganisms or are acquired by human or animal pathogens. The AMR poses a serious health threat across all sectors of societies in cancer and surgery therapies. Besides, the cost of treating the disease is increased in patients that have resistant infections compared to the patients not having drug-resistant infections. Indeed, in 2016, around 490,000 people developed multidrug resistance against tuberculosis globally. The drug resistance has caused treatment complicated against HIV and malaria (https://www.who.int/en/news-room/fact-sheets/detail/antimicrobial-resistance, accessed Jan. 22, 2019).

Environments, such as drug manufacturing industries, antibiotic-polluted places and hospitals, where frequency and evolution and transmission of antibiotic Resistance are more, have been characterized by high abundance and diversity of pathogens and horizontally transferable ARGs (Pal et al. 2016). Metagenomics offers to resolve natural or non-clinical microbial communities that serve as reservoirs of ARGs.

The antibiotic resistance poses a growing health threat and demands extensive research to develop alternative novel antimicrobial compounds that are not only effective against a broad range of pathogens, but are also safer. According to a recent review, more than 2500 AMPs have been identified and extracted from natural sources including microorganisms, plants and animals of diverse origins (Wibowo and Zhao 2018).

AMPs, as a new class of antagonistic peptides, exhibit unique and broad-spectrum antimicrobial activities against microorganisms including fungi, bacteria and viruses. The recombinant AMPs have attracted increasing attention. Unique mode of uptake of AMPs and their antimicrobial activities makes it difficult for microorganisms to develop resistance to AMPs. Hence, AMPs and AMP-carrier protein conjugates are the futuristic candidates to restrict microbial infections. AMPs and bacteriocins or antibiotic-like compounds produced by bacteria of human or animal origin are promising alternatives. Breakthroughs in computational drug designing and peptide synthesis have raised possibilities to advance the development of antimicrobial compounds and tackling the threats posed by antibiotic-resistant pathogens.

One of the major impediments in use of AMPs as antagonistic compounds is their cost of preparation.

Recent reports on the development of recombinant AMPs and their cost-effective production and purification has raised the hopes of large-scale production of AMPs for use in humans and animals. For example, a peptide ORBK (LKGCWTKSIPPKPCFK), is a cyclic cationic peptide with strong antimicrobial properties. Methods developed to express and purify ORBK sets a solid foundation for future structural and functional studies (Li et al. 2014).

A recombinant or fusion protein, named DAMP4(var)-pexiganan fusion protein, comprising of a variant of the helical biosurfactant protein DAMP4, and antimicrobial peptide pexiganan, was designed by unification of two polypeptides, at the DNA level, via an acid-sensitive cleavage site. The purified DAMP4 had antimicrobial activity comparable to the peptide synthesized by chemical peptide synthesis or biochemical engineering (Zhao et al. 2015).

Another fusion protein, named DAMP4-F-pexiganan consisting of a carrier protein DAMP4 linked to an AMP, pexiganan, through a long flexible linker is synthesized using *Escherichia coli* as host microorganism. Though further studies are needed, this novel fusion protein in terms of cost-effective production and ease of purification is expected to broaden the rational design and production of antimicrobial products based on AMPs (Sun et al. 2018).

6.11 Omega Fatty Acids

Long-chain fatty acids derived from engineered microorganism are of paramount interest.

Metabolic engineering and synthetic biology have enabled the molecular microbiologists to design and synthesize genetic circuits and introduce them into suitable vectors such as *E. coli*, and enhance quality and quantity of the long-chain fatty acids such as ω-hydroxy fatty acids (ω-HFAs). ω-3 fatty acids, such as HFAs eicosapentaenoic acid (EPA) and docosahexaenoic acid (DHA) have pro-health effects in humans (Amiri-Jami et al. 2015; Sathesh-Prabu and Lee 2015).

E. coli Nissle transformed with recombinant plasmids carrying EPA/DHA gene cluster for the synthesis of ω-3 fatty acids produced EPA. It was concluded that EPA/DHA-producing probiotic *E. coli* Nissle may be a safe, alternative and economic source for industrial and pharmaceutical production of EPA and DHA ((Amiri-Jami et al. 2015). DHA was produced by recombinant *E. coli* co-expressing *pfaABCD* from DHA-producing *Colwellia psychrerythraea* 34H, an obligatory psychrophile, with one of four *pfaE* genes from bacteria producing arachidonic acid (ARA, 20:4ω6), eicosapentaenoic acid (EPA, 20:5ω3) or DHA, respectively (Peng et al. 2016).

6.12 Outlook and Challenges

It is a time for new prescription against pathogens and metabolic diseases. A broader approach to address bacterial infection is needed as new pathogens are emerging with resistance to conventional antibiotics. A combined therapy comprising of antimicrobial property, immunomodulation, and anticancer effects of bioengineered probiotics could boost the normal GI microbiota against pathogens. Designer probiotics could radically transform the current ways of disease diagnosis and treatment of gastrointestinal disorders, cancer and genitourinary infections. Engineering the probiotics that are safer when consumed by patients, and seemingly integrate with existing clinical methods, should be a prioritized area of research.

Despite the importance of microbial natural products for human health, only a few bacterial genera are examined and utilized for novel natural products. To allow more active intervention in so-called critical care patients, i.e., patients that need acute immunosuppressive intervention, more powerful recombinant probiotics should be developed. Such new strains might utilize mechanisms similar to those used by standard non-engineered strains, but the way in which they are designed may allow for a stronger impact.

However, the recombinant microorganisms and their metabolites raise key safety concerns of human subjects per se and the biological containment of the transgenes in microorganisms. If designed with absolute attention to biological safety, the recombinant bacteria may be potential probiotics with health benefits.

Mostly, the efficiency of bioengineered microbial metabolites is from trials conducted in vitro using cell lines or model animals. The nonpathogenic bacteria within an environment such as the gut serve as reservoirs of ARGs that can transmit to other nonpathogens and pathogens. Therefore, key issues, such as strain characterization, quality control, dose optimization, lateral gene transfer from recombinant probiotics to normal microflora have to be embarked upon.

In conclusion, developing microorganisms for metabolites of economic interest is big business. The microbial metabolites from engineered microorganisms are of immense importance as therapeutics or nutraceuticals. It is desirable to discover the GI microbiota for using them as hosts for expressing recombinant proteins intended for commercial applications.

References

Ai C, Zhang Q, Ren C, Wang G, Liu X, Tian F, Zhao J, Zhang H, Chen YQ, Chen W (2014) Genetically engineered Lactococcus lactis protect against house dust mite allergy in a BALB/c mouse model. PLoS One 9(10):e109461. https://doi.org/10.1371/journal.pone.0109461 (eCollection 2014)

Allain T, Chaouch S, Thomas M, Travers MA, Valle I, Langella P, Grellier P, Polack B, Florent I,

Bermúdez-Humarán LG (2018) Bile salt hydrolase activities: a novel target to screen anti-*Giardia* lactobacilli? Front Microbiol 9:89. https://doi.org/10.3389/fmicb.2018.00089 (eCollection 2018)
Amiri-Jami M, Abdelhamid AG, Hazaa M, Kakuda Y, Griffths MW (2015) Recombinant production of omega-3 fatty acids by probiotic Escherichia coli Nissle 1917. FEMS Microbiol Lett 362(20). pii: fnv166. https://doi.org/10.1093/femsle/fnv166 (Epub 2015 Sep 13)
Benbouziane B, Ribelles P, Aubry C, Martin R, Kharrat P, Riazi A, Langella P, Bermúdez-Humarán LG (2013) Development of a Stress-Inducible Controlled Expression (SICE) system in *Lactococcus lactis* for the production and delivery of therapeutic molecules at mucosal surfaces. J Biotechnol 168(2):120–129. https://doi.org/10.1016/j.jbiotec.2013.04.019
Bhardwaj R, Singh B, Bhat TK (2003) Purification and characterization of tannin acyl hydrolase from Aspergillus niger MTCC 2425. J Basic Microbiol 43 (6):449–461
Braat H, Rottiers P, Hommes DW, Huyghebaert N, Remaut E, Remon JP, van Deventer SJ, Neirynck S, Peppelenbosch MP, Steidler L (2006) A phase I trial with transgenic bacteria expressing interleukin-10 in Crohn's disease. Clin Gastroenterol Hepatol 4(6):754–759
Chen SW, Zhong J, Huan LD (2007) Expression of human insulin in lactic acid bacteria and its oral administration in non-obese diabetic mice. Wei Sheng Wu Xue Bao. 47(6):987–991. Chinese
Chen Z, Guo L, Zhang Y, Walzem RL, Pendergast JS, Printz RL, Morris LC, Matafonova E, Stien X, Kang L, Coulon D, McGuinness OP, Niswender KD, Davies SS (2014) Incorporation of therapeutically modified bacteria into gut microbiota inhibits obesity. J Clin Invest. 124(8):3391–3406. https://doi.org/10.1172/JCI72517
Choe DW, Loh TC, Foo HL, Hair-Bejo M, Awis QS (2012) Egg production, faecal pH and microbial population, small intestine morphology, and plasma and yolk cholesterol in laying hens given liquid metabolites produced by Lactobacillus plantarum strains. Br Poult Sci 53(1):106–115. https://doi.org/10.1080/00071668.2012.659653
Curiel JA, Rodríguez H, Acebrón I, Mancheño JM, De Las Rivas B, Muñoz R (2009) Production and physicochemical properties of recombinant Lactobacillus plantarum tannase. J Agric Food Chem 57 (14):6224–6230. https://doi.org/10.1021/jf901045s
de Las Rivas B, Rodríguez H, Anguita J, Muñoz R (2018) Bacterial tannases: classification and biochemical properties. Appl Microbiol Biotechnol. https://doi.org/10.1007/s00253-018-9519-y (Epub ahead of print) Review
de Moraes GMD, de Abreu LR, do Egito AS, Salles HO, da Silva LMF, Nero LA, Todorov SD, Dos Santos KMO (2017) Functional Properties of Lactobacillus mucosae Strains Isolated from Brazilian Goat Milk Probiotics Antimicrob Proteins. 9(3):235–245. https://doi.org/10.1007/s12602-016-9244-8
Dey B, Lagenaur LA, Lusso P (2013) Protein-based HIV-1 microbicides. Curr HIV Res 11(7):576–594
Fuentes-Garibay JA, Aguilar CN, Rodríguez-Herrera R, Guerrero-Olazarán M, Viader-Salvadó JM (2015) Tannase sequence from a xerophilic Aspergillus niger Strain and production of the enzyme in Pichia pastoris. Mol Biotechnol 57(5):439–447. https://doi.org/10.1007/s12033-014-9836-z
Gomes AC, Bueno AA, de Souza RG, Mota JF (2014) Gut microbiota, probiotics and diabetes. Nutr J 17 (13):60. https://doi.org/10.1186/1475-2891-13-60
Hunt KM, Foster JA, Forney LJ, Schütte UM, Beck DL, Abdo Z, Fox LK, Williams JE, McGuire MK, McGuire MA (2011) Characterization of the diversity and temporal stability of bacterial communities in human milk. PLoS One. 6(6):e21313. https://doi.org/10.1371/journal.pone.0021313 (Epub 2011 Jun 17)
Iwamoto K, Tsuruta H, Nishitaini Y, Osawa R (2008) Identification and cloning of a gene encoding tannase (tannin acylhydrolase) from Lactobacillus plantarum ATCC 14917(T). Syst Appl Microbiol 31(4):269–277. https://doi.org/10.1016/j.syapm.2008.05.004 (Epub 2008 Jul 23)
Jiménez E, de Andrés J, Manrique M, Pareja-Tobes P, Tobes R, Martínez-Blanch JF, Codoñer FM, Ramón D, Fernández L, Rodríguez JM (2015) Metagenomic Analysis of Milk of Healthy and Mastitis-Suffering Women. J Hum Lact. 31(3):406–415. https://doi.org/10.1177/0890334415585078 (Epub 2015 May 6)
Kareem KY, Loh TC, Foo HL, Akit H, Samsudin AA (2016) Effects of dietary postbiotic and inulin on growth performance, IGF1 and GHR mRNA expression, faecal microbiota and volatile fatty acids in broilers. BMC Vet Res. 12(1):163. https://doi.org/10.1186/s12917-016-0790-9
Kareem KY, Loh TC, Foo HL, Asmara SA, Akit H (2017) Influence of postbiotic RG14 and inulin combination on cecal microbiota, organic acid concentration, and cytokine expression in broiler chickens. Poult Sci 96(4):966–975. https://doi.org/10.3382/ps/pew362
Kimura I, Ozawa K, Inoue D, Imamura T, Kimura K, Maeda T, Terasawa K, Kashihara D, Hirano K, Tani T, Takahashi T, Miyauchi S, Shioi G, Inoue H, Tsujimoto G (2013) The gut microbiota suppresses insulin-mediated fat accumulation via the short-chain fatty acid receptor GPR43. Nat Commun. 4:1829. https://doi.org/10.1038/ncomms2852
Klein G, Schanstra JP, Hoffmann J, Mischak H, Siwy J, Zimmermann K (2013) Proteomics as a Quality Control Tool of Pharmaceutical Probiotic Bacterial Lysate Products. PLoS ONE 8(6):e66682. https://doi.org/10.1371/journal.pone.0066682
Kumar M, Nagpal R, Verma V, Kumar A, Kaur N, Hemalatha R, Gautam SK, Singh B (2013) Probiotic metabolites as epigenetic targets in the prevention of colon cancer. Nutr Rev 71:23–34
Kumar M, Yadav A, Verma V, Singh B, Mal G, Nagpal R, Hemalatha R (2016) Bioengineered probiotics as a new hope for health and diseases: Potential and

prospects: An overview. Future Microbiology. 11:585–600

Lagenaur LA, Sanders-Beer BE, Brichacek B, Pal R, Liu X, Liu Y, Yu R, Venzon D, Lee PP, Hamer DH (2011) Prevention of vaginal SHIV transmission in macaques by a live recombinant Lactobacillus. Mucosal Immunol 4(6):648–657. https://doi.org/10.1038/mi.2011.30 (Epub 2011 Jul 6)

Li Y, Wang J, Yang J, Wan C, Wang X, Sun H (2014) Recombinant expression, purification and characterization of antimicrobial peptide ORBK in Escherichia coli. Protein Expr Purif 95:182–187. https://doi.org/10.1016/j.pep.2013.12.011 (Epub 2014 Jan 4)

Makete G, Aiyegoro OA, Thantsha MS (2016) Isolation, identification and screening of potential probiotic bacteria in milk from South African Saanen goats. Probiotics Antimicrob. Proteins. 9:246–254

Maldonado NC, Silva de Ruiz C, Nader-Macías ME (2016) Design of a beneficial product for newborn calves by combining Lactobacilli, minerals, and vitamins. Prep Biochem Biotechnol. 46(7):648–56. https://doi.org/10.1080/10826068.2015.1128447

Maldonado NC, Chiaraviglio J, Bru E, De Chazal L, Santos V, Nader-Macías MEF (2018) Effect of Milk Fermented with Lactic Acid Bacteria on Diarrheal Incidence, Growth Performance and Microbiological and Blood Profiles of Newborn Dairy Calves. Probiotics Antimicrob Proteins. 10(4):668–676. https://doi.org/10.1007/s12602-017-9308-4

Mansour NM, Abdelaziz SA (2016) Oral Immunization of mice with engineered *Lactobacillus gasseri* NM713 strain expressing *Streptococcus pyogenes* M6 antigen. Microbiol Immunol. https://doi.org/10.1111/1348-0421

Martín V, Mañes-Lázaro R, Rodríguez JM, Maldonado-Barragán A (2011) Streptococcus *lactarius* sp. nov., isolated from breast milk of healthy women. Int J Syst Evol Microbiol 61(Pt 5):1048–52. https://doi.org/10.1099/ijs.0.021642-0 (Epub 2010 May 28)

Matoba Y, Tanaka N, Noda M, Higashikawa F, Kumagai T, Sugiyama M (2013) Crystallographic and mutational analyses of tannase from Lactobacillus plantarum. Proteins. 81(11):2052–2058. https://doi.org/10.1002/prot.24355 (Epub 2013 Aug 23)

Miyamoto J, Kasubuchi M, Nakajima A, Irie J, Itoh H, Kimura I (2016) The role of short-chain fatty acid on blood pressure regulation. Curr Opin Nephrol Hypertens 25(5):379–383. https://doi.org/10.1097/MNH.0000000000000246

Ong ZY, Wiradharma N, Yang YY (2014) Strategies employed in the design and optimization of synthetic antimicrobial peptide amphiphiles with enhanced therapeutic potentials. Adv Drug Deliv Rev 23(78C):28–45. https://doi.org/10.1016/j.addr.2014.10.013

Pakbaten B, Majidzadeh Heravi R, Kermanshahi H, Sekhavati MH, Javadmanesh A, Mohammadi Ziarat M (2018) Production of Phytase Enzyme by a Bioengineered Probiotic for Degrading of Phytate Phosphorus in the Digestive Tract of Poultry. Probiotics Antimicrob Proteins.. https://doi.org/10.1007/s12602-018-9423-x

Pal C, Bengtsson-Palme J, Kristiansson E, Larsson DG (2016) The structure and diversity of human, animal and environmental resistomes. Microbiome. 4(1):54

Peng YF, Chen WC, Xiao K, Xu L, Wang L, Wan X (2016) DHA Production in Escherichia coli by Expressing Reconstituted Key Genes of Polyketide Synthase Pathway from Marine Bacteria. PLoS One. 11(9):e0162861. https://doi.org/10.1371/journal.pone.0162861 (eCollection 2016)

Petrova MI, van den Broek MFL, Spacova I, Verhoeven TLA, Balzarini J, Vanderleyden J, Schols D, Lebeer S (2018) Engineering Lactobacillus rhamnosus GG and GR-1 to express HIV-inhibiting griffithsin Int J Antimicrob Agents. 52(5):599–607. https://doi.org/10.1016/j.ijantimicag.2018.07.013 (Epub 2018 Jul 21)

Rastmanesh R, Catanzaro R, Bomba A, Allegri F, Marotta F (2012) Potential of prebiotics and probiotics to enhance the efficacy of HIV vaccination: a working hypothesis. Clinic Pharmacol Biopharmaceut 1:1

Ray S, Sherlock A, Wilken T, Woods T (2010) Cell wall lysed probiotic tincture decreases immune response to pathogenic enteric bacteria and improves 840 symptoms in autistic and immune compromised children. Explore 19:1–5

Salva S, Villena J, Alvarez S (2010) Immunomodulatory activity of Lactobacillus rhamnosus strains isolated from goat milk: impact on intestinal and respiratory infections Int J Food Microbiol. 141(1–2):82–89. https://doi.org/10.1016/j.ijfoodmicro.2010.03.013 (Epub 2010 Mar 18)

Sathesh-Prabu C, Lee SK (2015) Production of Long-Chain α,ω-Dicarboxylic Acids by Engineered Escherichia coli from Renewable Fatty Acids and Plant Oils J Agric Food Chem. 63(37):8199–8208. https://doi.org/10.1021/acs.jafc.5b03833 (Epub 2015 Sep 11)

Sharma D, Mal G, Kannan A, Bhar R, Sharma R, Singh B (2017) Degradation of euptox A by tannase-producing rumen bacteria from migratory goats. J Appl Microbiol 123(5):1194–1202. https://doi.org/10.1111/jam.13563

Shenderov BA (2013) Metabiotics: novel idea or natural development of probiotic conception. Microb Ecol Health Dis. 24. https://doi.org/10.3402/mehd.v24i0.20399 (eCollection 2013)

Sichel L, Timoshok NA, Pidgorskyy VS, Spivak Y (2013) Study of interferonogenous activity of the new probiotic formulation Del-Immune V®. Journal of Probiotics and Health 1:2. https://doi.org/10.4172/2329-8901.1000107

Singh B, Bhat TK, Sharma OP, Kanwar SS, Rahi P, Gulati A (2012) Isolation of tannase-producing

Enterobacter ludwigii GRT-1 from the rumen of migratory goats. Small Ruminant Research. 102: 172–176

Singh B, Mal G, Bharti D, Mohania M, Kumar M, Gautma SK, Marotta F, Yadav H, Nagpal R (2013) Probiotics in female reproductive health: paradigms, prospects and challenges. Current Women's Health Reviews. 9:236–248

Singh B, Mal G, Marotta F (2017) Designer Probiotics: Paving the Way to Living Therapeutics Trends Biotechnol. 35(8):679–682. https://doi.org/10.1016/j.tibtech.2017.04.001

Steidler L, Hans W, Schotte L, Neirynck S, Obermeier F, Falk W, Fiers W, Remaut E (2000) Treatment of murine colitis by *Lactococcus lactis* secreting interleukin-10. Science 289:1352–1355. Comment in Immunology. Therapeutic manipulation of gut flora. [Science. 2000]

Sun B, Wibowo D, Sainsbury F, Zhao CX (2018) Design and production of a novel antimicrobial fusion protein in Escherichia coli. Appl Microbiol Biotechnol 102 (20):8763–8772. https://doi.org/10.1007/s00253-018-9319-4 (Epub 2018 Aug 17)

Travers MA, Sow C, Zirah S, Deregnaucourt C, Chaouch S, Queiroz RM, Charneau S, Allain T, Florent I, Grellier P (2016) Deconjugated bile salts produced by extracellular bile-salt hydrolase-like activities from the probiotic *Lactobacillus johnsonii* La1 Inhibit *Giardia duodenalis* In vitro Growth. Front Microbiol. 7:1453 (eCollection 2016)

Ueda S, Nomoto R, Yoshida K, Osawa R (2014) Comparison of three tannases cloned from closely related lactobacillus species: *L. Plantarum*, *L. Paraplantarum*, and *L. Pentosus*. BMC Microbiol 14:87. https://doi.org/10.1186/1471-2180-14-87

von Ossowski I, Pietilä TE, Rintahaka J, Nummenmaa E, Mäkinen VM, Reunanen J, Satokari R, de Vos WM, Palva I, Palva A (2013) Using recombinant Lactococci as an approach to dissect the immunomodulating capacity of surface piliation in probiotic Lactobacillus rhamnosus GG. PLoS One 8(5):e64416. https://doi.org/10.1371/journal.pone.0064416. Print 2013 (2013 May 14)

Wibowo D, Zhao CX (2018) Recent achievements and perspectives for large-scale recombinant production of antimicrobialpeptides. Appl Microbiol Biotechnol https://doi.org/10.1007/s00253-018-9524-1

Zhao CX, Dwyer MD, Yu AL, Wu Y, Fang S, Middelberg AP (2015) A simple and low-cost platform technology for producing pexiganan antimicrobial peptide in *E. coli*. Biotechnol Bioeng 112(5):957–964. https://doi.org/10.1002/bit.25505 (Epub 2015 Jan 2)

Designer Probiotics: The Next-Gen High Efficiency Biotherapeutics

7

Abstract

The probiotic engineering is a cutting edge technology for improving disease diagnosis, treating gastrointestinal disorders and infectious diseases, and improving nutrition and ecological health. Use of bioengineered microorganisms in animals has different targets and prospects owing to differences in their anatomy, physiology, and feeding habits. In ruminants, the bioengineered microorganism is primarily aimed to enhance nutrient utilization, detoxify toxic plant metabolites, and lessen the enteric methanogenesis, while in non-ruminants, the bioengineered microorganisms are aimed to enhance nutrient utilizations, confer protection against pathogens, and inhibit infectious agents.

Highlights

- The microorganisms can be engineered to enhance their metabolic efficiency
- The bioengineered microorganisms could solve the burgeoning problem of drug-resistant pathogens
- The recombinant probiotics are promising therapeutic agents against infectious diseases.

Keywords

Designer probiotics · Live therapeutics · Recombinant AMPs · Oral vaccines · Livestock applications

7.1 Introduction

Availability of quality forage is the major problem impeding livestock production in most developing countries. As humans and animals have requirement for grains and protein-rich bioresources, the demand has increased for production of pulse crops and animals used for meat.

The long-range goal should be to develop genetically engineered microbes which are able to colonize GI tract of animals and enhance the supply of nutrients to animals. It is necessary so that nutrient requirements of high-yielding animals can be met from low-quality forages or silages, and dependence on grains or protein concentrates is minimal. In addition, genetic engineering has important applications such as detoxification of anti-nutritional phytometabolites present in native forages, crop residues, and agro-industrial byproducts used as animal feed, and minimizing methane emissions. Despite huge task to be solved, the success achieved is limited, and much remains to be done in future.

The quantity of nutrients required in high-yielding cattle or buffaloes is more than the nutrients available in conventional forages and metabolic capacity of rumen microorganisms. This is because high-yielding animals require higher amounts of readily fermentable carbohydrate and protein-rich diets. The recombinant microbes used as feed additives can fill this gap.

As the development of therapeutics to prevent bacteria, fungi, and viruses infectious diseases is

B. Singh et al., *Advances in Animal Biotechnology*,
https://doi.org/10.1007/978-3-030-21309-1_7

a routine activity of researchers, academicians, and veterinarians associated with livestock industry, this article is confined to developments in microbial therapies against infectious agents of GI, genitourinary and respiratory tract of livestock species. Importance of important body microbial niches with reference to their relevance to animal performance is highlighted.

7.2 The Era of Bioengineered Microorganisms

The bioengineering of microorganisms intended for use as feed supplements or probiotics is known to commence with the development of probiotics targeting enteric pathogens (DeGrandis et al. 1989). Applications of computational engineering, recombinant DNA technology, and synthetic biology have offered enormous possibilities to utilize microorganisms and eukaryotes for producing a variety of therapeutics and nutraceuticals targeting a range of humans and veterinary applications (Challinor and Bode 2015; Ozdemir et al. 2018; Pedrolli et al. 2018).

Studies have reported recombinant probiotics to improve nutrition (Amiri-Jami et al. 2015), and human and veterinary health (Trombert 2015; Yang et al. 2017).

Butyrivibrio fibrisolvens OB156 originating from rumen was transformed by electroporation with fluoroacetate dehalogenase gene (*H1*), from *Moraxella* sp. The genetically modified *B. fibrisolvens* OB156 could detoxify fluoroacetate, a naturally occurring plant toxin. The recombinant *B. fibrisolvens* was found to establish in the rumen of experimental sheep for longer period and protect them against fluoroacetate poisoning (Gregg et al. 1994, 1998).

With the development of a *Lactococcus lactis* for delivery of recombinant anti-inflammatory cytokine interleukin-10 (IL-10), intended to treat inflammatory bowel disease (IBD) (Steidler et al. 2000), the field bioengineering of probiotics has advanced swiftly as a premier area of research in nutrition and antibacterial and anti-viral therapies.

7.3 Recombinant Bacteria for Nutrient Utilization

Interspecies transinoculation of rumen bacteria is a well-known phenomenon to alleviate forage toxicity in animals. The recombinant probiotics are tailored to produce proteins to induce cytotoxicity or tumoricidal activity in internal tumors, or for delivering therapeutic proteins to tumors that are inaccessible for delivering therapeutics (Singh et al. 2017; Ma et al. 2018). In animals, the emphasis is on developing genetically modified probiotics to enhance nutrient utilization and confer protection against infectious agents.

Pig, poultry, and fish are the non-ruminants. They are least efficient to utilize certain feed components such as cellulose, phytic acid, saponins oxalates, etc. The unutilized components released in feces into environment act as pollutants. Therefore, efforts are made to develop recombinant microorganisms that can be used as microbial feed supplements in pig, poultry, or fish.

The plasmid carrying *Bacillus subtilis* phytase genes, i.e., (phyA) codon was introduced into *Lactobacillus acidophilus*, *L. gasseri*, and *L. gallinarum*. The phytase activity was confirmed in lactobacilli strains, which when fed to poultry birds increased body weight gain in broilers (Askelson et al. 2014). A broiler gut-origin recombinant *Lactobacillus reuteri* expressing endoglucanase gene (*celW*) of *B. subtilis* WL001, and phytase gene (phyW) of *Aspergillus fumigatus* WL002, was developed to enhance feed utilization in poultry. The recombinant L. *reuteri* improved feed conversion ratio when fed to broiler and exerted antimicrobial activities against undesirable bacteria such as *Bacteroides vulgates*, *Escherichia coli*, and *Veillonella* spp. and promoted beneficial intestinal commensals like lactobacilli and bifidobacteria (Wang et al. 2014).

Lactococcus lactis containing recombinant phytase gene (*appA2*) insert obtained from *Escherichia coli* revealed phytase activity (4U/ml) in cell extract and 19 U/ml phytase activity in culture medium supernatant. The recombinant *L. lactis* improved growth performance of broilers (Pakbaten et al. 2018). Recombinant bile salt hydrolase (BSH) from *Lactobacillus johnsonii* La1 exhibiting anti-giardial activities might have prospects to prevent widely spread, but neglected infectious giardial infections in humans and veterinary subjects (Allain et al. 2018). Some other relevant examples are summarized in Tables 7.1 and 7.2.

7.4 Beneficial Probiotic Metabolites

Bioengineered microorganisms have a pivotal role in nutrition, therapeutics, and health industry. Besides delivering drugs and gene-therapy vectors with efficiency and site-specificity, the bioengineered microorganisms have provided a lot of valuable products for diagnosis and treatment of diseases. Genome engineering and

Table 7.1 Summary of recombinant microorganisms developed for applications to livestock nutrition health augmentation

Microorganisms	Specific features/genes introduced	Possible inferences and implications
Butyrivibrio fibrisolvens	Genes for fluoroacetate dehalogenase	Developed to alleviate toxicity caused by fluoroacetate present in some forages, could establish in rumen of sheep and prevent fluoroacetate toxicity (Gregg et al. 1994, 1998)
	Neocallimastix patriciarum xylanase	Improvement in fiber digestion by recombinant bacteria (Krause et al. 2001)
	N. patriciarum xylanase	Enhanced (28.7%) degradation of neutral detergent fiber compared with normal or control strains. The study explores the feasibility of expressing rumen fungus genes in rumen bacteria to enhance fibre digestion (Gobius et al. 2002)
Escherichia coli BL-21	Phytase genes from *Selenomonas ruminantium*	The study explores commercial scale production of BL-21 recombinant phytase (Chi-Wei Lan et al. 2014)
Lactobacillus acidophilus, L. gasseri, L. gallinarum	*B. subtilis* phytase (phyA) codon	Phytase gene detected in the lactobacilli, phytase degradation activity of the recombinant strains increased due to recombinant phytase. Phytase-expressing *L. gasseri* increased body weight gain in the broilers (Askelson et al. 2014)
Lactobacillus reuteri (from broiler GI tract)	*B. subtilis* WL001 endoglucanase gene (celW), and *Aspergillus fumigatus* WL002 phytase gene (phyW) mature peptide (phyWM)	Endoglucanase and phytase activities detected in culture medium of recombinant *L. reuteri*. The recombinant *L. reuteri* improved feed conversion ratio when fed to broiler, and exerted antimicrobial activities against *Bacteroides vulgates*, *Escherichia coli*, and *Veillonella* spp. and promoted intestinal lactobacilli and bifidobacteria (Wang et al. 2014)
Lactococcus lactis	Phytase gene (*appA2*) insert of *Escherichia coli*	The transgene could express in *L. lactis* as revealed from phytase activity (4U/ml) in cell extract, and supernatant maximal phytase activity being 19 U/ml (Pakbaten et al. 2018)

Table 7.2 Summary of recombinant microorganisms developed as oral vaccines or for protection against bacterial viral infections

Microorganisms introduced	Specific features/genes	Possible inferences and implications
Escherichia coli	Expressing recombination activating gene 2 (*RAG2*) in *E. coli*	Successful expression of recombinant fusion protein, and its purification. Using purified fusion protein to produce polyclonal antibodies (Jin et al. 2017)
Lactobacillus plantatrum	Porcine epidemic diarrhea virus S gene fused to a DC-targeting peptide	The mice immunized by lavage administration of recombinant *L. plantarum* (NC8-pSIP409-pgsA'-S-DCpep) NC8 strain lead to increased secretion of cytokines, namely IFN-γ, IL-4 and IL-17, thereby indicating that recombinant strain triggered humoral immune response, and that recombinant strain could be used for producing novel oral vaccine against porcine epidemic diarrhea virus (Huang et al. 2018)
	Expression of 3M2e-HA2 influenza virus proteins in *L. plantarum*	The chicks immunized with N/pgsA'-3M2e-HA2 exhibited characteristic humoral, and T cell-mediated immune response against avian influenza virus. The study is emphasized as a novel approach and effective vaccine to promote mucosal immunity (Yang et al. 2018b)
	Expression of TGEV antigen(S) to dendritic cells (DCs) via dendritic cell-targeting peptides (DCpep)	Induction of high levels of B7 molecules on DCs, plus high levels of IgG, secretory IgA, IFN-γ, and IL-4. Expression of DC-targeted antigens could induce cellular, mucosal, and humoral immunity in murine models, indicating that recombinant *L. plantarum* could be used as oral vaccine against TGEV (Yang et al. 2018a)
	Expression of spike (S) protein originating from TGEV, fused to DC-targeting peptides (DCpep) in *L. plantarum*	The recombinant *L. plantarum* (NC8-pSIP409-pgsA-S-DCpep) expressing S fused with DCpep enhances the MHC-II$^+$CD80$^+$ B cells, and CD3$^+$, CD4$^+$ T cells of ileal lamina propria with simultaneous increase in levels of secretory IgA in feces, and IgG in serum, indicating that recombinant bacterium is suitable for use as oral vaccine against the virus (Jin et al. 2018)

synthetic biology that allow design and construction of genetic circuits, and precise fine-tuning of transgene expression, have opened up new frontiers to augment the therapeutic and nutritive potential of the probiotics.

Postbiotics, the microbial metabolites, are important factors accounting for efficacy of the probiotics. Chromatography coupled with tandem mass spectrometry and Fourier transform ion cyclotron resonance mass spectrometry with direct infusion is recommended to identify and characterize bacterial metabolites or postbiotics such as AMPs, bacteriocins, fatty acids, purines, sphingolipids, and oligosaccharides (Kok et al. 2013). Currently, fluorescent in situ hybridization (FISH), DNA pyrosequencing, microarrays (PhyloChip), and quantitative PCR assays, and analysis of conserved 16S rRNA genes for phylogenetic analysis are used to analyze complex microbial ecosystems.

7.5 Diversity of Genitourinary Microbiota

The female reproductive is specific in terms of immunological organization, epithelial barrier, sensitivity to estrogen, and microbial ecosystem (Petrova et al. 2018). The affirmation that normal genitourinary microbes protect the host against pathogens has encouraged the researchers to decipher uterine tract microbiota to manage reproductive health and infertility (Zhou et al. 2004; Martin et al. 2013).

Until recently, the microbes inhabiting urinary tract were studied by empirical culture-dependent methods that are unable to depict the complete outline of microbiota. Since the cultivable microorganisms merely represent a fraction of total microbial diversity, it is envisaged that the microorganisms that are not cultured, and therefore remain unexplored, play important role in the health of the mother and neonates (Hyman et al. 2014).

Deciphering how the microorganisms interact with each other and with uterine epithelium is fundamental for a more complete understanding of feminine reproductive health (Singh et al. 2013). These culture-independent approaches have provided information about novel, previously unidentified uterine microbiota in different endocrinological milieus, and support a shift in the microbiome during assisted reproduction and pregnancy outcomes.

In this context, it is necessary to analyze complex microbial niches, genes, and metabolic pathways therein for developing alternatives therapies that are safer and effective against antibiotic-resistant microorganisms (Thallinger et al. 2013; Cardona et al. 2015).

7.6 Recombinant Antimicrobial Peptides

One of the major challenges in modern medicine is paucity of alternative therapies against drug-resistant pathogens. Genitourinary tract infection is one of the many major causes of infertility in dairy and beef cattle. The vaginal microbiota dominated by LAB confers protection against bacterial infections. Recombinant microbiocides may offer a target specificity ensuring prolonged protection against genitourinary infections.

Infertility in livestock is a matter of serious concern. The major factors declining livestock productivity include infectious diseases, nutritional imbalance, climatic stress, pests, and parasitic infestation. Besides, the brucellosis, chlamydiosis, infectious bovine rhinotracheitis, leptospirosis, listeriosis, corynebacteriosis, campylobacteriosis, neosporosis, trichomoniasis, and viral infections are primarily responsible for causing infertility in animals.

However, little is known about microbial communities of reproductive tract of animals and aberrations or dysbiosis in cow and buffalo leading to infertility. The females unable to conceive and calving are then abandoned which pose a threat to humans by causing fatal accidents and spreading parasites and zoonotic pathogens through urine, uterine discharge, and defecation.

The information is sparse on microbiota of reproductive tract microbes and their role in reproductive health and infertility in animals (Durso et al. 2015). Culture-dependent microbial analysis and 16S rRNA sequencing of cow and ewe ectocervicovaginal lavages has revealed a large difference in their vaginal microbial composition (Swartz et al. 2014). Similar reports are available from analysis of Nellore cattle uterine metagenome (Laguardia-Nascimento et al. 2015). Antimicrobial peptides (AMPs) and bacteriocins which act via membrane active mechanisms are widely used in food industries, feed industry and also be used in personal care products, surmounting the pathogens (Ong et al. 2014; Juturu and Wu 2018). As genitourinary tract microbiota is least explored, it is important to study the reproductive ecosystem of livestock species to understand microbial causes and cure of infertility. The bioengineered probiotics producing high titers of AMPs could enhance benefits of probiotics as well as antimicrobial peptides *vis-a-vis* their role in reviving reproductive health of animals.

7.7 Health Threats to Porcine and Poultry Industry

Besides, nutritional deficiencies and toxicity, the poultry and pigs suffer from several parasitic, bacterial, and viral infections. In addition, wild and abandoned animals serve as important sources of infections to healthy animals and zoonotic diseases. Viral infections such as porcine transmissible gastroenteritis virus (TGEV), belonging to coronaviruses (CoVs) (Yang et al. 2018a; Jin et al. 2018), and porcine epidemic diarrhea virus (PEDV) (Huang et al. 2018) cause severe economic losses in pork industry worldwide. In addition to infecting poultry, the animal viruses pose a serious threat to humans (Walker et al. 2018). To prevent viral infections, therapeutics developed should be safer, and no residual remains are left in meat or milk intended for human consumption, and confer long term protection.

Viral infections such as avian influenza (AIV) are among serious threats to poultry industry (Yang et al. 2018b). Edible vaccines in the form of probiotics are recommended in view of their safety and ease of administration. Besides humans (Boesmans et al. 2018), the probiotic therapies are recommended to prevent infections in poultry (Kareem et al. 2016, 2017; Yazhini et al. 2018), calves (Maldonado et al. 2016, 2017; Plaizier et al. 2018), and porcine (Nordeste et al. 2017; Xu et al. 2018).

7.8 Recombinant Microorganisms as Oral Vaccines

Effective public health research and awareness requires a comprehensive and thorough understanding of viruses that are at risk of transmission to humans. Vaccines, antibiotics, antibodies, and antibiotics–antibody conjugates are used to prevent animals against infections.

Edible vaccines offer advantages in animals. Oral delivery of antigens by means of recombinant probiotics, such as those expressing dendritic cell (DC)-targeting peptides, fused with virus antigens is expected to confer protection in animals against viral infections (Wang et al. 2017).

Probiotic lactic acid bacteria, especially lactobacilli and lactococci, in view of their status as safe, and natural presence in humans, and animals including birds and insects are opted to express antigens of viruses of porcine and poultry species to produce cytokines and antibodies, for use as oral vaccines.

A recombinant *Lactobacillus plantarum*, named *L. plantarum* (NC8-pSIP409-pgsA-S-DCpep), expressing spike (S) protein originating from TGEV fused with DCpep enhanced MHC-II$^+$CD80$^+$ B cells, and CD3$^+$, CD4$^+$ T cells of ileal lamina propria with simultaneous increase in levels of secretory IgA in feces, and IgG in serum shows that recombinant bacterium is suitable as oral vaccine against the virus (Jin et al. 2018). In another study, 3M2e-HA2 influenza virus proteins were made to express in *L. plantarum*. The experimental chicks immunized with N/pgsA'-3M2e-HA2, exhibited characteristic humoral, and T cell-mediated immune response against avian influenza virus. The study is emphasized as a novel approach and effective vaccine to promote mucosal immunity (Yang et al. 2018a, b).

7.9 Outlook and Challenges

Development of novel antibiotics to combat drug-resistant pathogens is slow compared to rise of drug resistance. Recombinant microbes are the emerging and futuristic candidates of nutrition and health management. The bioengineered microorganisms are now making their way to improve nutrient utilization, alleviate forage toxicity, and protect animals against pathogens.

It is a time for new prescription against pathogens and metabolic diseases. A broader approach to address bacterial infection is needed as new pathogens are emerging with resistance to conventional antibiotics. A combined therapy comprising of antimicrobial property, immunomodulation, and anticancer effects of bioengineered probiotics could boost the host immune system.

Engineered probiotics that are safer upon administering or feeding should be prioritized. However, the recombinant microorganisms raise certain safety concerns. The likely probiotic-mediated induction of immune response should be carefully examined. It is likely that overproduction of AMPs by recombinant probiotics may deter the normal gut bacteria that are otherwise essential to wellbeing of the host. It is necessary to prevent proliferation of genetically modified microorganisms in environment when they come out in feces and prevent transfer of transgenes into environmental or pathogenic microorganisms. Key issues, such as strain characterization, quality control, dose optimization, lateral gene transfer from recombinant probiotics to normal microflora, should be embarked upon.

7.10 Conclusions

Exploiting microbiota is increasingly a big business. As recombinant probiotics may rebuild microbiota and restore the health with efficiency and site-specificity, the challenges and key problems associated with their use should be thoroughly investigated to enhance their applications in animals serving as part of human food chain. The research to discover normal microbiota and developing alternative microbial therapeutics should be prioritized. The therapies require a critical evaluation before being recommended for human applications.

References

Allain T, Chaouch S, Thomas M, Vallée I, Buret AG, Langella P, Grellier P, Polack B, Bermúdez-Humarán LG, Florent I (2018) Bile-salt-hydrolases from the probiotic strain *Lactobacillus johnsonii* la1 mediate anti-giardial activity. in vitro and in vivo. Front Microbiol 8:2707. https://doi.org/10.3389/fmicb.2017.02707. eCollection 2017

Amiri-Jami M, Abdelhamid AG, Hazaa M, Kakuda Y, Griffths MW (2015) Recombinant production of omega-3 fatty acids by probiotic Escherichia coli Nissle 1917. FEMS Microbiol Lett 362(20). pii: fnv166. https://doi.org/10.1093/femsle/fnv166 (Epub 2015 Sep 13)

Askelson TE, Campasino A, Lee JT, Duong T (2014) Evaluation of phytate-degrading Lactobacillus culture administration to broiler chickens. Appl Environ Microbiol 80(3):943–950. https://doi.org/10.1128/AEM.03155-13 (Epub 2013 Nov 22)

Boesmans L, Valles-Colomer M, Wang J, Eeckhaut V, Falony G, Ducatelle R, Van Immerseel F, Raes J, Verbeke K (2018) Butyrate producers as potential next-generation probiotics: safety assessment of the administration of *Butyricicoccus pullicaecorum* to healthy volunteers mSystems. 3(6). pii: e00094-18. https://doi.org/10.1128/msystems.00094-18. eCollection 2018 Nov-Dec

Cardona ST, Selin C, Gislason AS (2015) Genomic tools to profile antibiotic mode of action. Crit Rev Microbiol 41:465–472

Challinor VL, Bode HB (2015) Bioactive natural products from novel microbial sources. Ann NY Acad Sci 1354:82–97. https://doi.org/10.1111/nyas.12954. Review

Chi-Wei Lan J, Chang CK[2], Wu HS[2] (2014) Efficient production of mutant phytase (phyA-7) derived from Selenomonas ruminantium using recombinant Escherichia coli in pilot scale. J Biosci Bioeng 118(3):305–310. https://doi.org/10.1016/j.jbiosc.2014.02.024 (Epub 2014 Mar 29)

DeGrandis S, Law H, Brunton J, Gyles C, Lingwood CA (1989) Globotetraosylceramide is recognized by the pig edema disease toxin. J Biol Chem 264(21):12520–12525

Durso L, Wells JE, Kim MS (2015) Diversity of microbiomes in beef cattle. Encyclopedia of Metagenomics 129–138

Gobius KS, Xue GP, Aylward JH, Dalrymple BP, Swadling YJ, McSweeney CS, Krause DO (2002) Transformation and expression of an anaerobic fungal xylanase in several strains of the rumen bacterium Butyrivibrio fibrisolvens. J Appl Microbiol 93(1):122–133

Gregg K, Cooper CL, Schafer DJ, Sharpe H, Beard CE, Allen G, Xu J (1994) Detoxification of the plant toxin fluoroacetate by a genetically modified rumen bacterium. Biotechnology (NY). 12(13):1361–1365

Gregg K, Hamdorf B, Henderson K, Kopecny J, Wong C (1998) Genetically modified ruminal bacteria protect sheep from fluoroacetate poisoning. Appl Environ Microbiol 64(9):3496–3498

Huang KY, Yang GL, Jin YB, Liu J, Chen HL, Wang PB, Jiang YL, Shi CW, Huang HB, Wang JZ, Wang G, Kang YH, Yang WT, Wang CF (2018) Construction and immunogenicity analysis of *Lactobacillus plantarum* expressing a porcine epidemic diarrhea virus S gene fused to a DC-targeting peptide. Virus Res 2(247):84–93. https://doi.org/10.1016/j.virusres.2017.12.011 (Epub 2017 Dec 27)

Hyman RW, Fukushima M, Jiang H, Fung E, Rand L, Johnson B, Vo KC, Caughey AB, Hilton JF, Davis RW, Giudice LC (2014) Diversity of the vaginal microbiome correlates with preterm birth. Reprod. Sci. 21:32–40

Jin YB, Yang WT, Huang KY, Chen HL, Shonyela SM, Liu J, Liu Q, Feng B, Zhou Y, Zhi SL, Jiang YL, Wang JZ, Huang HB, Shi CW, Yang GL, Wang CF (2017) Expression and purification of swine RAG2 in E. coli for production of porcine RAG2 polyclonal antibodies. Biosci Biotechnol Biochem 81(8):1489–1496. https://doi.org/10.1080/09168451.2017.1340086 (Epub 2017 Jun 23)

Jin YB, Yang WT, Shi CW, Feng B, Huang KY, Zhao GX, Li QY, Xie J, Huang HB, Jiang YL, Wang JZ, Wang G, Kang YH, Yang GL, Wang CF (2018) Immune responses induced by recombinant *Lactobacillus plantarum* expressing the spike protein derived from transmissible gastroenteritis virus in piglets. Appl Microbiol Biotechnol 102(19):8403–8417. https://doi.org/10.1007/s00253-018-9205-0 (Epub 2018 Jul 18)

Juturu V, Wu JC (2018) Microbial production of bacteriocins: Latest research development and applications. Biotechnol Adv pii: S0734-9750(18)30170-8. https://doi.org/10.1016/j.biotechadv.2018.10.007

Kareem KY, Loh TC, Foo HL, Akit H, Samsudin AA (2016) Effects of dietary postbiotic and inulin on growth performance, IGF1 and GHR mRNA expression, faecal microbiota and volatile fatty acids in broilers. BMC Vet Res 12(1):163. https://doi.org/10.1186/s12917-016-0790-9

Kareem KY, Loh TC, Foo HL, Asmara SA, Akit H (2017) Influence of postbiotic RG14 and inulin combination on cecal microbiota, organic acid concentration, and cytokine expression in broiler chickens. Poult Sci 96(4):966–975. https://doi.org/10.3382/ps/pew362

Kok MG, Ruijken MM, Swann JR, Wilson ID, Somsen GW, de Jong GJ (2013) Anionic metabolic profiling of urine from antibiotic-treated rats by capillary electrophoresis-mass spectrometry. Anal Bioanal Chem. 405(8):2585–2594. https://doi.org/10.1007/s00216-012-6701-4 (Epub 2013 Jan 12)

Krause DO, Bunch RJ, Dalrymple BD, Gobius KS, Smith WJ, Xue GP, McSweeney CS (2001) Expression of a modified Neocallimastix patriciarum xylanase in *Butyrivibrio fibrisolvens* digests more fibre but cannot effectively compete with highly fibrolytic bacteria in the rumen. J Appl Microbiol 90(3):388–396

Laguardia-Nascimento M, Branco KM, Gasparini MR, Giannattasio-Ferraz S, Leite LR, Araujo FM, Salim AC, Nicoli JR, de Oliveira GC, Barbosa-Stancioli EF (2015) Vaginal microbiome characterization of nellore cattle using metagenomic analysis. PLoS One 10(11):e0143294. https://doi.org/10.1371/journal.pone.0143294. eCollection 2015

Ma S, Wang L, Huang X, Wang X, Chen S, Shi W, Qiao X, Jiang Y, Tang L, Xu Y, Li Y (2018) Oral recombinant Lactobacillus vaccine targeting the intestinal microfold cells and dendritic cells for delivering the core neutralizing epitope of porcine epidemic diarrhea virus. Microb Cell Fact 17(1):20. https://doi.org/10.1186/s12934-018-0861-7

Maldonado NC, Silva de Ruiz C, Nader-Macías ME (2016) Design of a beneficial product for newborn calves by combining Lactobacilli, minerals, and vitamins. Prep Biochem Biotechnol 46(7):648–656. https://doi.org/10.1080/10826068.2015.1128447

Maldonado NC, Chiaraviglio J, Bru E, De Chazal L, Santos V, Nader-Macías MEF (2017) Effect of milk fermented with lactic acid bacteria on diarrheal incidence, growth performance and microbiological and blood profiles of newborn dairy calves. Probiotics Antimicrob Proteins https://doi.org/10.1007/s12602-017-9308-4

Martin DH, Zozaya M, Lillis RA, Myers L, Nsuami MJ, Ferris MJ (2013) Unique vaginal microbiota that includes an unknown Mycoplasma-like organism is associated with Trichomonas vaginalis infection. J Infect Dis 207(12):1922–1931. https://doi.org/10.1093/infdis/jit100 (Epub 2013 Mar 12)

Nordeste R, Tessema A, Sharma S, Kovač Z, Wang C, Morales R, Griffiths MW (2017) Molecules produced by probiotics prevent enteric colibacillosis in pigs. BMC Vet Res 13(1):335. https://doi.org/10.1186/s12917-017-1246-6

Ong ZY, Wiradharma N, Yang YY (2014) Strategies employed in the design and optimization of synthetic antimicrobial peptide amphiphiles with enhanced therapeutic potentials. Adv Drug Deliv Rev 23 (78C):28–45. https://doi.org/10.1016/j.addr.2014.10.013

Ozdemir T, Fedorec AJH, Danino T, Barnes CP (2018) Synthetic Biology and Engineered Live Biotherapeutics: Toward Increasing System Complexity. Cell Syst 7(1):5–16. https://doi.org/10.1016/j.cels.2018.06.008. Review

Pakbaten B, Majidzadeh Heravi R, Kermanshahi H, Sekhavati MH, Javadmanesh A, Mohammadi Ziarat M (2018) Production of phytase enzyme by a bioengineered probiotic for degrading of phytate phosphorus in the digestive tract of poultry. Probiotics Antimicrob Proteins. https://doi.org/10.1007/s12602-018-9423-x

Pedrolli DB, Ribeiro NV, Squizato PN, de Jesus VN, Cozetto DA (2018) Team AQA Unesp at iGEM 2017. Engineering Microbial Living Therapeutics: The Synthetic Biology Toolbox. Trends Biotechnol. pii: S0167-7799(18)30258-0. https://doi.org/10.1016/j.tibtech.2018.09.005

Petrova MI, van den Broek MFL, Spacova I, Verhoeven TLA, Balzarini J, Vanderleyden J, Schols D, Lebeer S (2018) Engineering Lactobacillus rhamnosus GG and GR-1 to express HIV-inhibiting griffithsin. Int J Antimicrob Agents 52(5):599–607. https://doi.org/10.1016/j.ijantimicag.2018.07.013 (Epub 2018 Jul 21)

Plaizier JC, Danesh Mesgaran M, Derakhshani H, Golder H, Khafipour E, Kleen JL, Lean I, Loor J, Penner G, Zebeli Q (2018) Review: enhancing gastrointestinal health in dairy cows. Animal 24:1–20. https://doi.org/10.1017/s1751731118001921 (Epub ahead of print)

Singh B, Mal G, Bharti D, Mohania M, Kumar M, Gautma SK, Marotta F, Yadav H, Nagpal R (2013) Probiotics in female reproductive health: paradigms, prospects and challenges. Current Women's Health Reviews 9:239–248

Singh B, Mal G, Marotta F (2017) Designer probiotics: paving the way to living therapeutics. Trends Biotechnol 35(8):679–682. https://doi.org/10.1016/j.tibtech.2017.04.001 (Epub 2017 May 5)

Steidler L, Hans W, Schotte L, Neirynck S, Obermeier F, Falk W, Fiers W, Remaut E (2000) Treatment of murine colitis by *Lactococcus lactis* secreting interleukin-10. Science 289:1352–1355. Comment in Immunology. Therapeutic manipulation of gut flora. [Science. 2000]

Swartz JD, Lachman M, Westveer K, O'Neill T, Geary T, Kott RW, Berardinelli JG, Hatfield PG, Thomson JM, Roberts A, Yeoman CJ (2014) Characterization of the vaginal microbiota of ewes and cows reveals a unique microbiota with low levels of lactobacilli and near-neutral pH. Front Vet Sci 15(1):19

Thallinger B, Prasetyo EN, Nyanhongo GS, Guebitz GM (2013) Antimicrobial enzymes: an emerging strategy to fight microbes and microbial biofilms. Biotechnol J 8:97–109

Trombert A (2015) Recombinant lactic acid bacteria as delivery vectors of heterologous antigens: the future of vaccination? Benef Microbes 6(3):313–324. https://doi.org/10.3920/BM2014.0068.Review

Walker JW, Han BA, Ott IM, Drake JM (2018) Transmissibility of emerging viral zoonoses. PLoS One. 13 (11):e0206926. https://doi.org/10.1371/journal.pone.0206926. eCollection 2018

Wang L, Yang Y, Cai B, Cao P, Yang M, Chen Y (2014) Coexpression and secretion of endoglucanase and phytase genes in Lactobacillus reuteri. Int J Mol Sci 15(7):12842–12860. https://doi.org/10.3390/ijms150712842

Wang X, Wang L, Huang X, Ma S, Yu M, Shi W, Qiao X, Tang L, Xu Y, Li Y (2017) Oral delivery of probiotics expressing dendritic cell-targeting peptide fused with porcine epidemic diarrhea virus COE antigen: A Promising Vaccine Strategy Against PEDV. Viruses. 9(11). pii: E312. https://doi.org/10.3390/v9110312

Xu J, Li Y, Yang Z, Li C, Liang H, Wu Z, Pu W (2018) Yeast probiotics shape the gut microbiome and improve the health of early-weaned piglets. Front Microbiol 9:2011. https://doi.org/10.3389/fmicb.2018.02011. eCollection 2018

Yang G, Jiang Y, Tong P, Li C, Yang W, Hu J, Ye L, Gu W, Shi C, Shan B, Wang C (2017) Alleviation of enterotoxigenic Escherichia coli challenge by recombinant *Lactobacillus plantarum* expressing a FaeG- and DC-targeting peptide fusion protein. Benef Microbes. 8(3):379–391. https://doi.org/10.3920/BM2016.0116 (Epub 2017 May 15)

Yang WT, Li QY, Ata EB, Jiang YL, Huang HB, Shi CW, Wang JZ, Wang G, Kang YH, Liu J, Yang GL, Wang CF (2018a) Immune response characterization of mice immunized with *Lactobacillus plantarum* expressing spike antigen of transmissible gastroenteritis virus. Appl Microbiol Biotechnol 102(19):8307–8318. https://doi.org/10.1007/s00253-018-9238-4 (Epub 2018 Jul 28)

Yang WT, Yang GL, Zhao L, Jin YB, Jiang YL, Huang HB, Shi CW, Wang JZ, Wang G, Kang YH, Wang CF (2018b) *Lactobacillus plantarum* displaying conserved M2e and HA2 fusion antigens induces protection against influenza virus challenge. Appl Microbiol Biotechnol 102(12):5077–5088. https://doi.org/10.1007/s00253-018-8924-6 (Epub 2018 Apr 19)

Yazhini P, Visha P, Selvaraj P, Vasanthakumar P, Chandran V (2018) Dietary encapsulated probiotic effect on broiler serum biochemical parameters. Vet World 11(9):1344–1348. https://doi.org/10.14202/vetworld.2018.1344-1348 (Epub 2018 Sep 29)

Zhou X, Bent SJ, Schneider MG, Davis CC, Islam MR, Forney LJ (2004) Characterization of vaginal microbial communities in adult healthy women using cultivation-independent methods. Microbiology 150:2565–2573

Part II

Assisted Reproduction Biotechnology

8 Revolutionary Reproduction Biotechnologies in Livestock: An Overview

Abstract

Developing, conserving, and disseminating best livestock is the prime concern of reproduction biotechnology. Breakthroughs in assisted reproduction technologies (ARTs) ranging from artificial insemination to advanced transgenesis and genome editing are successfully applied to enhance production and value addition of livestock products. While the emphasis is on proliferating high-yielding breeds, these animals are susceptible to biotic and abiotic stress. Therefore, native livestock resources need due scientific attention to conserve them and utilize their genetic merit.

Highlights

- The ARTs have played a crucial role in enhancing livestock production from a domestic practice to a commercial enterprise
- High-yielding animal breeds have certain limitations that make them unfit in low-input management
- Stress-tolerance genes of native livestock are the potential sources to improve high-yielding animals to cope with imminent climatic stress.

Keywords

Reproduction biotechniques · Cryopreservation · Sex selection · Livestock · Transgenesis

8.1 Introduction

Livestock provides food (milk, meat, and eggs), fiber, manure, and power for agriculture operation, and serves as a source of income. Billions of people depend on nutrition and livelihood on livestock (Singh et al. 2009; Windig and Engelsma 2010). Domestic animals, referred to as livestock including cattle, buffaloes, pig, and poultry, occupy nearly 30% of the earth's ice-free terrestrial surface area (Steinfeld et al. 2006), and serve as global assets with a commercial value of more than $1.4 trillion (Thornton 2010).

The commercial animal production initiated with the pursuit of scientific and technical interventions in animal breeding, nutrition, and health. Revolutionary ARTs such as artificial insemination (AI), in vitro embryo production (IVEP), sex preselection, and stem cell technology have improved the production potential of domesticated animals. Animal cloning or nuclear transfer (NT) cloning is a precious technique to proliferate superior livestock, repopulate endangered mammals, and resurrect mammalian species (Hoshino et al. 2009; Selokar et al. 2014). Genomics and genetic engineering and genome editing are used to introduce new genes into animals and improve the quality of animal products. Transgenic model animals have contributed a lot to understand human and animal diseases and developmental biological processes.

B. Singh et al., *Advances in Animal Biotechnology*,
https://doi.org/10.1007/978-3-030-21309-1_8

This chapter reviews how the ARTs originally developed in a species were applied in other species and altered the practice and impact in animal farming. In addition, we have highlighted the genetic merits of some livestock species that are immensely important in view of their evolution, adaptation to prevailing extreme climates, and perform well where high-yielding animals can hardly thrive. The genes and genetic pathways of unique adaptive and production traits present in these animals are of immense biotechnological importance for incorporating into high-yielding animals.

8.2 Reproduction Biotechnologies and Their Use in Livestock

Selective breeding is the commonly used strategy in animal husbandry. The very basic purpose of manual selection is to choose the animals based on best body score or phenotypic characteristics that have economic value. Different breeds of same livestock species have different characteristics, importance, and requirements for their management. For instance, pets such as cats and dogs come in different body sizes, colors, and hairs. Cattle have been originated from a common ancestor but are present as different breeds with distinct adaptive and productive attributes. Similarly, different breeds of buffaloes, sheep, and goats have been improved for milk, meat, or both purposes. Notably, livestock breeds used as dual purpose are comparatively productive. Some countries such as India are bestowed with the immense richness of animal biodiversity that plays a predominant role in the rural and national economy.

8.3 Male-Assisted Reproduction

AI is among the initial ARTs aimed to utilize the genetic potential of superior males. In almost all the livestock species, semen is collected, diluted, and processed, and semen doses are cryopreserved and distributed for use in livestock farms or under field conditions. In pet animals also, AI is preferred to make use of the genetic merit of the males. Most of the commercial dairy farms use semen of progeny-tested bulls.

Use of AI progressed to a commercial scale with the advent of cryopreservation of semen at ultralow temperature (Polge et al. 1949), and use of antibiotics in processed semen to prevent microbial contamination (Almquist et al. 1949). Development of robust metallic cryocontainers to store liquid nitrogen proved to be extremely useful for maintaining large quantities of semen doses (reviewed in Moore and Hasler 2017).

Success in producing embryos in vitro by IVF of in vitro matured oocytes proved the suitability of cryopreserved semen in commercial livestock breeding programs (Polge 1952; Cheng 1959). Later on, several commercial companies came up with valuable accessories (disposable semen straws of assorted capacity, ampoules, catheters, automatic semen-straw marking tools, embryo cup filters, AI guns, etc.) to facilitate cryopreservation and use of the cryopreserved semen. Simultaneous advancements in electronics and instrumentation led to development of high-power microprocessor-based microscopes, high-precision automated micromanipulators, and ultrasound machines.

Cytological and molecular biological studies of oocytes in different species led to the development of novel concepts and promoting the production of embryos by innovative approaches. For instance, intracytoplasmic sperm injection (ICSI) was developed to produce embryos in particular cases from the sperms obtained from subfertile individuals or the patients suffering from testicular degeneration or impaired spermatogenesis. The technique was applied in livestock species such as equines. The zona-pellucida in equine oocyte is hard and prevents penetration of sperm into oocytes during IVF (Roels et al. 2018; Valenzuela et al. 2018). In bovines, the ICSI is found to enhance embryos production from oocytes collected from aged cows (Magata et al. 2019).

After the popularization of AI, it was realized that males are of less use in the dairy industry. The males can contribute to meat or agriculture operations. This evolved the concept of separating X- and Y-chromosome bearing sperm and only using X-chromosome sperm to produce females to launch dairy herds.

In the present scenario, the sexed semen has a great demand in dual-purpose livestock such as cattle and buffaloes. Currently, the sperm sexing by flow cytometric sorting is the only successful method to separate X- and Y-chromosome bearing sperm.

AI doses with 95% of X- or Y-chromosomes separated by Beltsville Sperm sexing technology based on flow cytometric cell sorting of DNA-stained sperm (Garner 2006) has been used to inseminate cow that has produced more than 50,000 calves of desired sex (Seidel 2009). The technology has importance in commercial dairy and beef production. However, the technology needs modification for sexing sperm in other livestock species such as pigs and horses.

So far, the sexed semen is used to produce offspring of desired sex in cattle, Mediterranean buffaloes, horses, sheep, elk, domestic cats and dogs, and dolphins (Garner et al. 2013).

Deriving germ cells and sperm from embryonic stem cells (Zhou et al. 2016; Makoolati et al. 2017), and spermatogonial stem cells (SSCs) (Deng et al. 2017), is a probable technology, and live births are already achieved using artificial gametes (Hendriks et al. 2015). However, the cost involved in producing sperm from stem cells, and susceptibility of sex-sorted sperm to cryopreservation are the major challenges that need a scientific solution.

The biomaterials during shipment or transportation are subjected to trembling and vibration that have detrimental impact on frozen sperm. Though real-time data is not available on the effect of vibrations on sperm quality, it is hypothesized that vibrations may affect semen quality. An electronic mobile sensing app (TransportLog 1.0), utilizing the built-in sensors has been developed to capture the vibrations emitted during the transportation of frozen semen. The device might measure the impact of vibrations on sperm and form the basis for developing guidelines for safe transport of cryopreserved semen (Schulze et al. 2018).

8.4 Biotechniques Utilizing Female Reproduction

8.4.1 Embryo Production and Banking

Producing embryos including through IVF, parthenogenesis or somatic cell nuclear transfer (SCNT), cryopreservation of embryos, and embryo sexing are the significant technologies in commercial livestock farms besides their use in basic and applied developmental biological studies (Fig. 8.1).

All these techniques rely on the availability of quality oocytes. Unlike sperm, the oocytes are produced in limited number during the reproductive cycle of the females. There are different ways to obtain and preserve oocytes. Abattoir-derived ovaries are convenient sources of obtaining immature cumulus–oocyte complexes (COCs) in mammals. Occasionally, the oocytes can also be obtained from live animals. Nevertheless, obtaining oocytes from live animals is expensive and needs technical expertise to administer drugs to induce multiple ovulation and collect the oocytes by laparoscopic ovum pick-up (LOPU) or ultrasound-guided aspiration of ovarian follicles.

Oocytes can also be generated from germ stem cells such as oogonia stem cells and embryonic stem cells. Germ cells have been obtained from germ stem cells in mammals (Morohaku et al. 2016; de Souza et al. 2017). The technique is successful in experimental animals such as mice but can be expected to be useful in farm animals in future. Surplus oocytes and embryos are preserved by conventional slow-freezing or vitrification.

8.5 Cryopreservation

Cryopreservation is a very important component of biomedical and veterinary sciences. Cryopreservation methods have been developed that minimize damage to cells and improve their survival. Compared to the computer-guided slow-freezing process of cryopreservation, vitrification

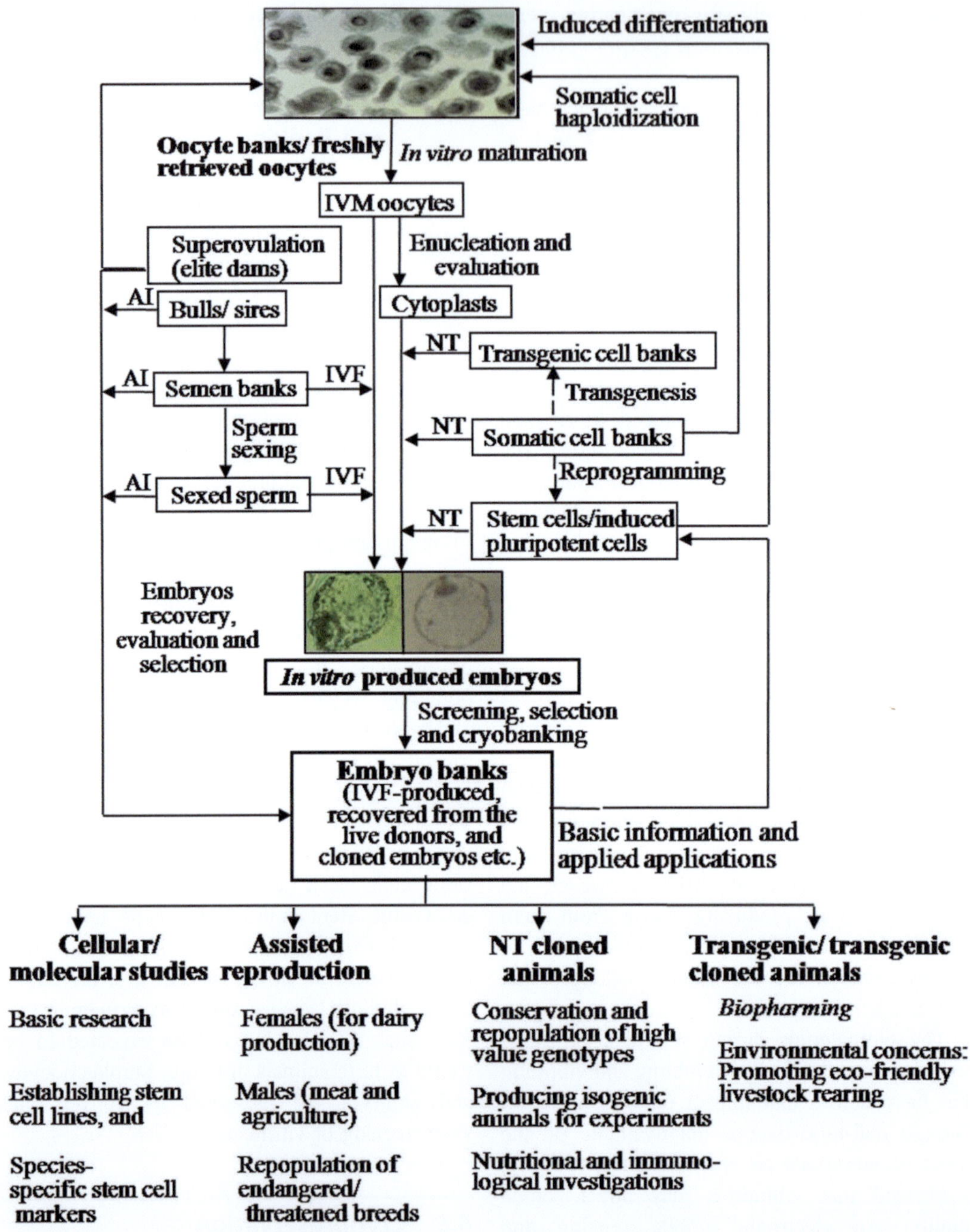

Fig. 8.1 Applications of various ARTs in buffaloes. Establishing oocytes and embryo banks should be the prioritized concern. Nuclear transfer cloning, sperm sexing, stem cells, and sperm and somatic cell banking are already established in buffaloes. The abbreviations are explained in the text

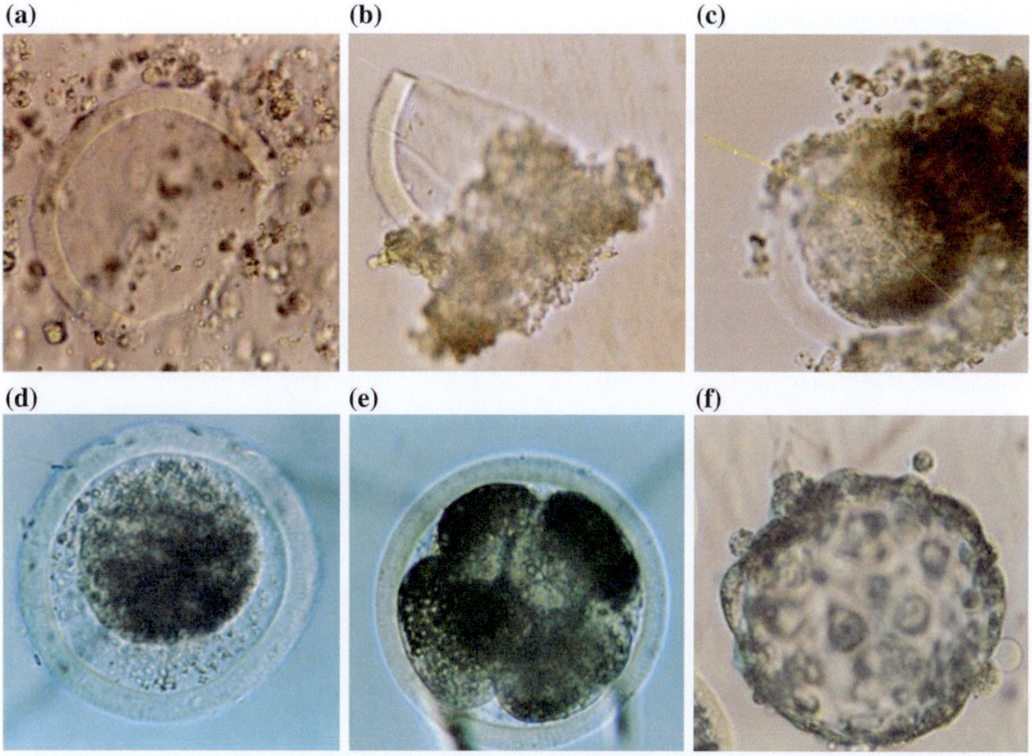

Fig. 8.2 Effect of vitrification on bubaline (*Bubalus bubalis*) oocytes and embryos. **a** and **b** Broken zona pellucida; **c** dissociation of cumulus cell mass from zona pellucida; **d** shrunken ooplasm; **e** a normal morula; **f** normal blastocyst after vitrification-warming

is simpler, and manual "flash-freezing" type of method of cryopreservation. Vitrification has been improved remarkably for storing gametes embryos and somatic cells, stem cells.

At moment, the cryopreservation is an integral component of biomedical and veterinary sciences. Cryopreservation of mammalian sperm has promoted artificial insemination and dissemination of superior males. Cryopreservation of mammalian embryos was initially shown to be feasible when Whittingham et al. (1972a, b) and Festing (1972) could successfully cryopreserve mice embryos using glycerol or dimethyl sulphoxide (DMSO). Live mouse pups were born from the frozen-thawed embryos transferred into surrogate females. Since then, the cryobiology has become an important area of research in biomedical, human and animals ARTs. As cryopreservation of oocytes, and embryos leads to various types of damages (Fig. 8.2), and genetic aberrations such as activation of apoptosis pathways in oocytes and embryos, the technique is under improvement to enhance the survival of biological materials.

Cryopreservation of immature oocytes has high demand as oocytes are needed in all the studies aimed to produce embryos and ESCs. Similarly, banking of embryos, somatic and gonadal tissues are imperative in biomedicine, reproductive biology and livestock conservation (Singh et al. 2017a, b; Do et al. 2019).

However, the major limitation of vitrification is the toxicity of high concentration of CPAs to the cells. Therefore, vitrification should be improved further to preserve cells and tissues. Success in resurrecting animals from the cells preserved without CPAs suggest that it may be possible to exclude or minimize the use of CPAs in ultralow preservation of biological material (Hoshino et al. 2009).

8.6 Sex Preselection

Gende r or sex-preselection is a big business in dairy farming. In buffaloes and cattle, the males are least preferred in view of their decreasing role in agriculture operations. The males are therefore reared for meat. Hence, it is desirable that the animals of decided sex should be produced either for meat or milk.

This is achieved by using sexed sperm in assisted reproduction program or screening the pre-implantation stage embryos based on sex. Sex preselection is viewed as an economically important technique that has more commercial relevance for dairy and beef industry. Sex preselection is possible by two approaches; either by use of sexed sperm, or sexing the in vitro-produced embryos before transferring them to estrus-synchronized females.

Molecular biological and immunological techniques are also used for sexing sperms (Alves et al. 2010; Yadav et al. 2017), and pre-implantation embryos. This allows selection of healthy embryos (Fig. 8.3), for proliferating desirable and healthy germplasm.

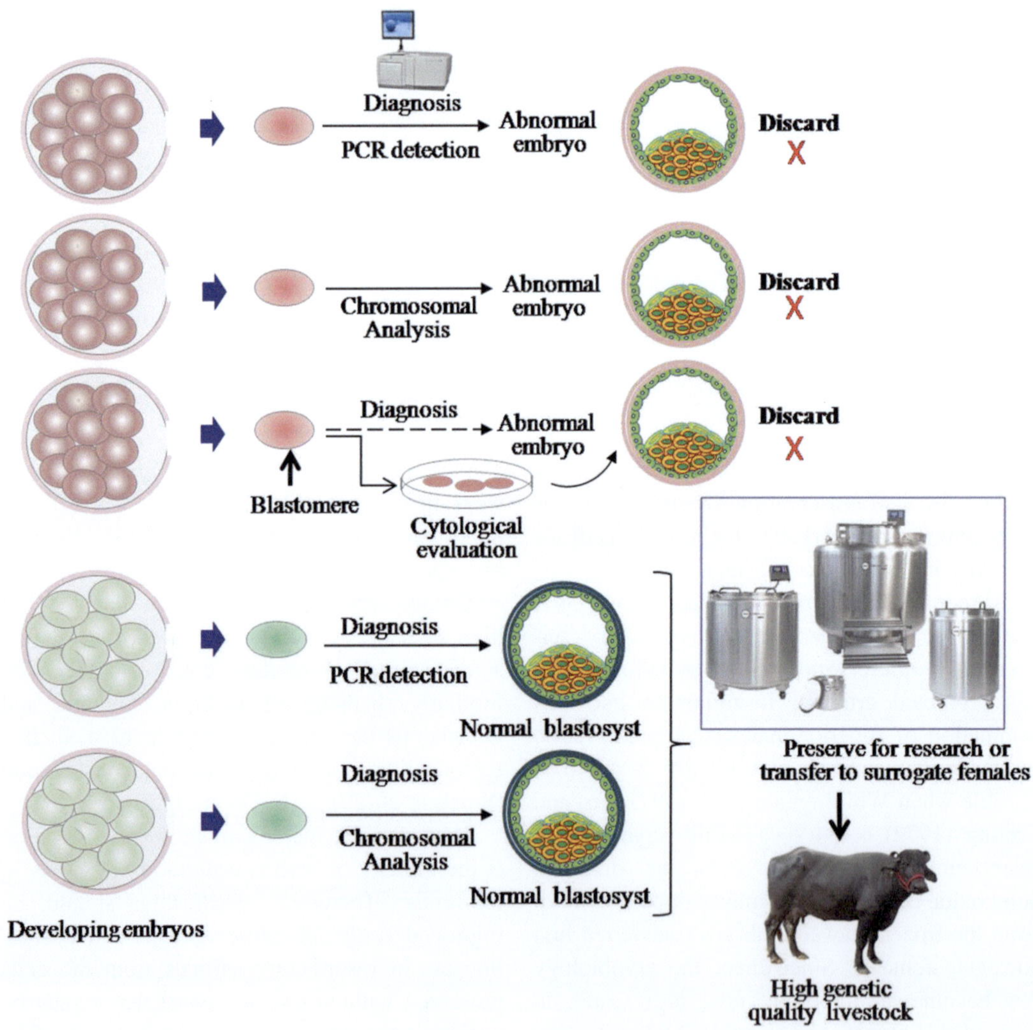

Fig. 8.3 Evaluating embryos for genetic abnormalities or sex-preselection in livestock. The abnormal embryos should be discarded. Healthy embryos, free from genetic disorders or abnormalities should be used to establish healthy animal herds

8.7 Altering the Quality of Livestock Products

Healthy livestock with superior germplasm is the backbone of agriculture. Prior to advent of genomics and development of molecular markers, the farmers are used to select livestock based on visual or phenotypic characteristics. However, the success was slow and benefits from selected livestock were low.

Embryo technologies have contributed a lot to animal reproduction, breeding and proliferating the best animals. Earlier, embryo splitting methods were the only options to produce identical twins (Willadsen 1979; Willadsen et al. 1981). Embryo splitting was used to get more embryos from individual donor animal collections (Gray et al. 1991).

Later, it was realized that nuclear transfer (NT) cloning could enhance yield of identical embryos (Prather et al. 1987). The main difference between animals produced by embryo splitting and NT was that embryo splitting involved bisecting the embryo manually and allowing them to rejuvenate into new embryos. Practically, a maximum of four embryos could be obtained from a bisected embryo.

The embryos produced by NT included the transfer of diploid cell or blastomeres into evacuated oocyte (cytoplast) followed by activation with electric stimulus, also called as electrofusion. The electrofusion and activation by certain chemicals induce reprogramming of the zygote or cell–cytoplast couplet genome to development into embryo. The technique proved to be milestone achievement in the history of embryo biotechnology, when Wilmut et al. (1997) succeeded in producing a sheep from terminally differentiated fibroblasts. This paved the way to new concepts that genome of finally differentiated cells can be reprogrammed to a state of totipotency. SCNT or mammalian cloning is the feasible way to proliferate the elite animals. Till date, a number of livestock species, and non-human primates have been cloned

Outstanding achievements of SCNT include cloning of gaur bull (*Bos gaurus*), a highly endangered large wild bovid (Lanza et al. 2000), and cloning a bull from the organs frozen without cryoprotectant in a −80 °C freezer for a decade (Hoshino et al. 2009). The live cells obtained for defrosted spermatic cord of frozen testicles were cultured and used as donor cells to produce cloned embryos, establish pregnancies and producing cloned calves. The most important conclusion of the study was that a complete genome is conserved in tissues despite long-term frozen condition (Hoshino et al. 2009). The birth of cloned buffalo calf from the epithelial cells of the cryopreserved semen of the buffalo bull that had died a decade ago paves the way to restore mammalian species that are endangered (Selokar et al. 2014).

Altering the nutritive or therapeutic value of livestock products is a decades-old concept. The recombinant proteins excreted into milk could be purified and used as therapeutics. Transgenic animals (cattle, goats, sheep, rabbits, and mice) are already in use as disease-resistant livestock (Liu et al. 2014), or as live bioreactors to produce therapeutics. Developing models of human diseases is another important motive of animal transgenesis and gene editing.

Certain native livestock species are comparatively resistant to pests and parasites, but they are less productive. Indian cattle breeds such as Tharparkar and Kankrej can perform well under ambient thermal stress. The genomics and biotechnology tools should be used to identify the genes responsible for resistance to parasites and pathogens and introduce them into more productive livestock. The genes of thermotolerance are valuable assets in indigenous tropical livestock that can be introduced into high-yielding exogenous cattle so that they can withstand thermal stress.

8.8 Adaptive Merits of Native Livestock and Role of ARTs to Conserve Them

The overall livestock sector is characterized and affected by a dichotomy involving developing and developed countries. The native farm animals, viz, cattle, buffaloes, goat, sheep, swine, equines, and poultry, camel, yak, and gayal or mithun. Livestock and poultry breeds are adapted to prevailing climatic conditions, management, and dietary resources, and maintain natural ecosystem. Less known livestock, such as mithun (*Bos frontalis*), also known as the cattle of mountains, is distributed in Northeast India, Bangladesh, Northern Myanmar and in Yunnan, China. The yak (*Bos grunniens*), a long-haired domestic bovid, churu (a cross between cattle and yak), ponies, and donkeys (Table 8.1) are important in the niche areas, though they are not high-yielding animals. The mithun is found and domesticated under free-range conditions in forests of rangelands at an altitude of 1000 to 3000 m above mean sea level. Similarly, double hump camel is found in cold arid deserts of Ladakh region of India, Tibet, China, Mongolia, Kazakhstan, and Afghanistan.

It is evident the production conditions are becoming less favorable while demand for animal products has increased worldwide. Further, the situation differs more depending on geography, economical status of the consumers, rituals, and myths.

A number of native breeds are facing genetic degradation and dilution due to the intensive production system and crossbreeding with exotic germplasm. It has resulted in the loss of precious native germplasm possessing superior germplasm for production, resistance to diseases and tolerance to ambient thermal stress.

The Indian native cattle, viz Sahiwal, Tharparkar, Haryana, Rathi, Gir, and Nagori withstand extreme climate and thermal stress. Kankrej, an important cattle breed, is known for its adaptability to high temperature, powerful draught capacity and resistance to tick-borne disease. Similarly, Chilika buffaloes are adapted to brackish water and salinity conditions (Singh et al. 2017a, b), Banni buffaloes of Kutch Gujarat (India) for drought tolerance, and Swamp buffaloes are known for draught power. Many of the indigenous animal breeds are smaller, hence they need less feed, can graze on local grasses and trees, and are comparatively more resistant to pests and parasites.

The livestock production and climate changes are affected by each other (Fig. 8.4). Climate change, especially increases in ambient temperature, will restrict the distribution and survival of high milk-yielding animal breeds (Hoffman 2010). The adaptive features of indigenous livestock enable them to survive in native forages, management practices, and agroclimatic conditions.

The animals, currently domesticated in different parts of the world, have basically originated from common ancestors but evolved differentially depending on environmental factors, genetic mutations, availability of forages, husbandry and management practices. The prioritized areas of biotechnological interventions in them include enhancing feed-utilizing efficiency by manipulating their gut ecosystem, minimizing fats and increasing proteins in meat and milk, improving reproduction efficiency, shortening gestation lengths, and age at earlier maturity, and dam and neonate health care.

8.9 Outlook and Challenges

It is clear that ARTs have noticeably changed animal rearing from a household practice to a scientific and technically managed entrepreneur. However, there are certain issues that need due attention of policy makers. It is imperative to document, conserve, and proliferate the native livestock resources that are best suited to adverse climate and low-input conditions. It should be assured that animals semen for use under field conditions must be of good quality.

Gender selection is of interest in commercial dairy and beef industry. Efficiency of sperm sorting has increased compared to that existed a decade ago. As sex-sorted sperm are susceptible to cryopreservation, their cryosurvival should be improved. It can be expected the use of sexed sperm will be adopted at field level. There is a need to invent alternative means of sorting the sperm as developing repositories of sexed embryo is expensive.

More bulls should be produced to increase the availability of quality semen to livestock owners. The focus should be on improving cryopreservation of equine and pig semen. As there is risk

Table 8.1 Milestone achievements that revolutionized the assisted reproduction, and animal production

Year	Achievements (References)
1949	Development of non-surgical methods of collecting embryos from cattle (Rowson and Dowling 1949), establishing pregnancies in cattle from transfer of embryos (Umbaugh 1949), first report on cryopreservation (−79 °C) of fowl sperm using glycerol as cryoprotectant (Polge et al. 1949), incorporating antibiotics in semen preserving media to prevent microbial contamination (Almquist et al. 1949)
1951	First birth of bovine calf from transfer of fertilized ovum (Willett et al. 1951)
1951	Stimulation of oocyte production by injecting PMSG or hCG (Rowson 1951), pioneering experiments on cryopreservation of bovine semen in solid CO_2 and alcohol (−79 °C), and survival of sperms after freeze-thawing, first births of calves from frozen-thawed semen (Polge 1952)
1954	Development of insulated liquid nitrogen containers by American Breeders service (Deforest WI), and Linde Air Products, Murray Hill, NJ (reviewed in Moore and Hasler 2017)
1958	Development of protocols for inducing superovulation by injection of FSH (Dziuk et al. 1958)
1959	Fertilization of ova in vitro, and first birth of the mammalian species (rabbit) by IVF-derived embryos (Chang 1959)
1961	First report on preserving semen in liquid nitrogen vapours (Forgason et al. 1961)
1972	Cryopreservation of mammalian (mice) embryos at −196 °C and −269 °C (Whittingham et al. 1972a, b)
1973	Cryopreservation of bovine embryos, and first report on birth of calves from transfer of frozen-thawed embryos (Wilmut and Rowson 1973a, b)
1973	Development of programmable freezers for cryopreservation of sperm (Almquist and Wiggin 1973)
1974	First attempts to produce transgenic mammalians (mouse) embryos using a virus vector (Jaenisch and Mintz 1974)
1975	First birth of a sexed-calf from transfer of embryo sexed by karyotyping (Hare et al. 1976)
1976	Development and applications of Foley catheters for non-surgical collection of oocytes from cattle (Drost et al. 1976; Elsden et al. 1976; Rowe et al. 1976)
1977	First report on IVF in cattle (Iritani and Niwa 1977)
1978	Description of microscopic parameters for determining the quality of embryos (Elsden et al. 1978)
1979	Development of embryo splitting technique (Willadsen 1979), and birth of identical twin calves from splitted embryos in cattle, sheep and horse (Willadsen et al. 1981)
1981	Birth of live calf from IVF embryos (Brackett et al. 1982), identification of ESCs from mice embryos (Evans and Kaufman 1981; Martin 1981)
1983	Flow cytometric separation of X- and Y-chromosome bearing sperm based on their DNA contents (Garner et al. 1983)
1984–85	Cutting-edge innovation of equipments for non- surgical transfer of embryos by commercial private sectors such as IMV Technologies (reviewed in Hasler 2014)
1986	First report on cloned mammalian species (sheep) (Willadson 1986), development of identical calves by nuclear transfer (NT) cloning (Willadson 1986)
1987	First report on calves produced solely by IVF and in vitro produced embryos (Lu et al. 1987), birth of NT cloned calf (Prather et al. 1987)
1988	Development of ovum-pickup (OPU) method for transrectal ultrasound-guided aspiration of oocytes from large animals (cattle) (Pieterse et al. 1988)
1989	Births of rabbit pups from sexed sperm (Johnson et al. 1989)
1998–91	Development of PCR-based assay for sexing embryos based on Y-chromosome specific probes using biopsy from embryos (Bondioli et al. 1989; Herr and Read 1991), establishing and increasing the pregnancies from the embryos multiplied by embryo splitting (Gray et al. 1991)
1993	Birth of live calves from IVF of oocytes with sexed sperm (Cran et al. 1993)

(continued)

Table 8.1 (continued)

Year	Achievements (References)
1994	Development of protocols for producing embryos as alternative to MOET, by combining OPU, IVF and ET (Looney et al. 1994)
1997	Birth of Dolly, the first mammal cloned form terminally differentiated skin fibroblasts (Wilmut et al. 1997)
1998	Birth of transgenic SCNT cloned calves (Cibelli et al. 1998)
2002	Development of parthenogenetic stem cells in non-human primates (Cibelli et al. 2002)
2006–07	Birth of transgenic cattle calves expressing human proteins (Bauman et al. 2006; Robl et al. 2007), development of pluripotency in mature somatic cells (fibroblasts) by insertional mutagenesis of defined genetic factors (Takahashi and Yamanaka 2006)
2008	Approval of United States FDA to use cloned animals (cattle, pigs and goats) for food (Tanne 2008)
2013	Announcement of CRISPR-Cas 9 System to target and cleave any DNA sequence in vitro (Cong et al. 2013)
2015	Birth of genotypes animals for studying genomic evaluation (Kasinathan et al. 2015)
2018	First cloning of non-human primate (Liu et al. 2018), birth of gene-edited monkeys (*Macaca fascicularis*) (Liu et al. 2018)
2019	Discovery of CasX, a new class of RNA-guided genome editor (Liu et al. 2019)

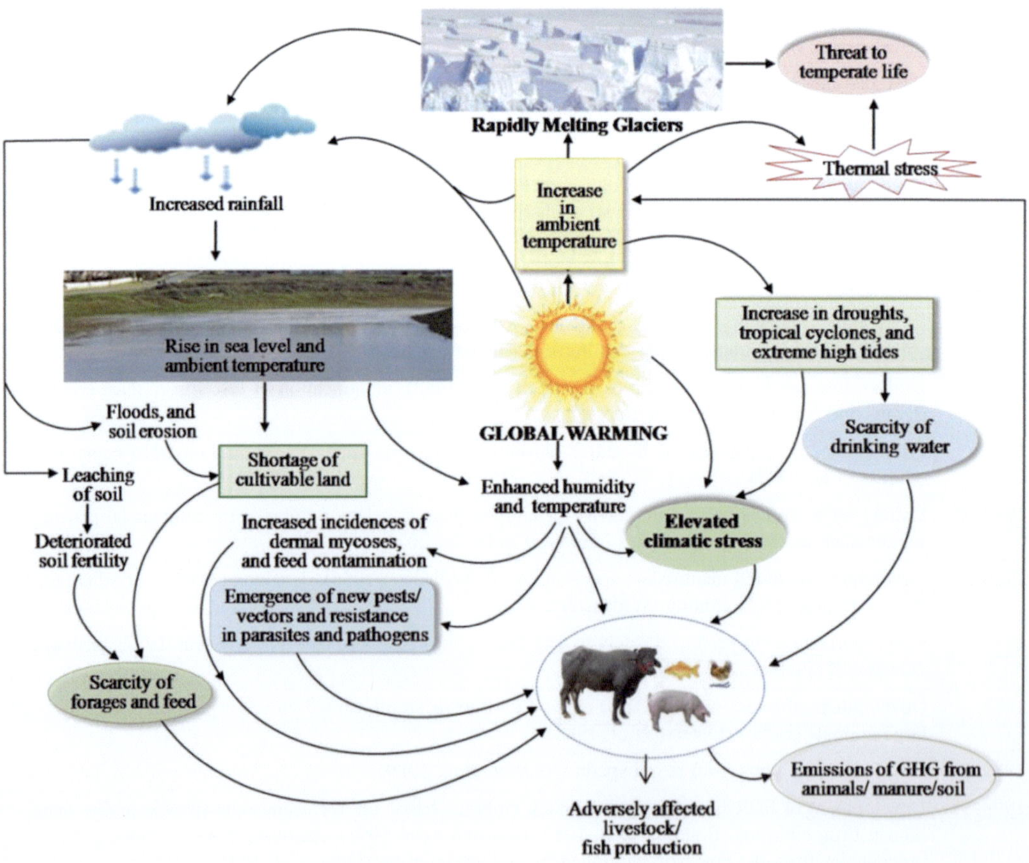

Fig. 8.4 A diagrammatic illustration of the consequences of thermal stress on livestock production. Whereas high-yielding purebred exogenous livestock is susceptible to thermal stress, some indigenous livestock can perform better. Metabolic pathways and the regulatory gene networks of evolutionarily adapted livestock may offer means to sustainable livestock production in future

of viral infections when supplements such as bovine serum albumin and fetal bovine serum are used, hence non-animal substitutes should be developed.

NT cloning has potential to multiply progeny-tested bulls as already noticed in buffaloes. The inferences obtained from a well-planned research endeavor with the goal of genetic restoration via nuclear transfer cloning are required for formulation and implementing the mammalian diversity conservation breeding programs, and proliferating the highly valuable livestock genotypes. The inferences drawn from multiple "omics" techniques should be applied to enhance the efficiency of ARTs such as IVEP.

Tendency to earn maximum benefits from selected animals has led to the depletion of native livestock resources that had genes and genetic pathways to cope with aggravating agroclimatic adversity such as resistance to thermal stress, parasites, and infectious agents. There is need to document and conserve native livestock breeds and identify the genes of economic interest or adaptation to adverse climate in them for introducing them in high-yielding breeds.

8.10 Conclusions

AI, cryopreservation of gametes, embryos and somatic cells, deriving gametes from stem cells, and proliferating animals of desired genotypes are radical-assisted reproduction techniques. Over the past few decades, cryopreservation of reproductive and somatic cells has progressed immensely. The conventional method of slow freezing requires expensive equipment, and at the same time, some cells are more sensitive to cryopreservation. Gender selection is desirable in dairy and beef livestock and likely to increase boost this sector in future. There is a need to conserve native livestock and modify the ARTs for application in different livestock species depending on their size, genetic makeup, and physiology.

References

Almquist JO, Wiggin HB (1973) Survival of bull spermatozoa frozen and thawed by different methods in plastic straws. AI Digest 21:12

Almquist JO, Glantz PJ, Shaffers HE (1949) The effect of a combination of penicillin and streptomycin upon the livability and bacterial content of bovine semen. J Dairy Sci 32:183–190

Alves BC, Hossepian de Lima VF, Moreira-Filho CA (2010) Development of Y-chromosome-specific SCAR markers conserved in taurine, zebu and bubaline cattle. Reprod Domest Anim 45(6):1047–1051. https://doi.org/10.1111/j.1439-0531.2009.01491.x

Bauman DE, Mather IH, Wall RJ, Lock AL (2006) Major advances associated with the biosynthesis of milk. J Dairy Sci 89(4):1235–1243. Review

Bondioli KR, Ellis SB, Pryor JH, Williams MW, Harpold MM (1989) The use of male-specific chromosomal DNA fragments to determine the sex of bovine preimplantation embryos. Theriogenology 31:95–104. https://doi.org/10.1016/0093-691X(89)90567-0

Brackett BG, Bousquet D, Boice ML, Donawick WJ, Evans JF, Dressel MA (1982) Normal development following in vitro fertilization in the cow. Biol Reprod 27(1):147–158

Chang MC (1959) Fertilization of rabbit ova in vitro. Nature 184(Suppl 7):466–7. No abstract available

Cibelli JB, Stice SL, Golueke PJ, Kane JJ, Jerry J, Blackwell C, Ponce de León FA, Robl JM (1998) Cloned transgenic calves produced from nonquiescent fetal fibroblasts. Science 280(5367):1256–1258

Cibelli JB, Grant KA, Chapman KB, Cunniff K, Worst T, Green HL, Walker SJ, Gutin PH, Vilner L, Tabar V, Dominko T, Kane J, Wettstein PJ, Lanza RP, Studer L, Vrana KE, West MD (2002) Parthenogenetic stem cells in nonhuman primates. Science. 295 (5556):819. No abstract available

Cong L, Ran FA, Cox D, Lin S, Barretto R, Habib N, Hsu PD, Wu X, Jiang W, Marraffini L, Zhang F (2013) Multiplex genome engineering using CRISPR/Cas systems. Science 3:1231143

Cran DG, Johnson LA, Miller NG, Cochrane D, Polge C (1993) Production of bovine calves following separation of X- and Y-chromosome bearing sperm and in vitro fertilisation. Vet Rec 132(2):40–41

de Souza GB, Costa J, da Cunha EV, Passos J, Ribeiro RP, Saraiva M, van den Hurk R, Silva J (2017) Bovine ovarian stem cells differentiate into germ cells and oocyte-like structures after culture in vitro. Reprod Domest Anim 52(2):243–250. https://doi.org/10.1111/rda.12886 (Epub 2016 Dec 7)

Deng S, Wang X, Wang Z, Chen S, Wang Y, Hao X, Sun T, Zhang Y, Lian Z, Liu Y (2017) In vitro production of functional haploid sperm cells from male germ cells of Saanen dairy goat. Theriogenology 1 (90):120–128. https://doi.org/10.1016/j.theriogenology.2016.12.002 (Epub 2016 Dec 2)

Do VH, Catt S, Kinder JE, Walton S, Taylor-Robinson AW (2019) Vitrification of invitro-derived bovine embryos: targeting enhancement of quality by refining technology and standardising procedures. Reprod Fertil Dev https://doi.org/10.1071/rd18352 (Epub ahead of print)

Drost M, Brand A, Aarts MH (1976) A device for nonsurgical recovery of bovine embryos. Theriogenology 6:503–507

Dziuk PJ, Danker JD, Nichols JR, Petersen WE (1958) Problems associated with transfer of ova between cattle. Univ Minnesota Tech Bull 222:1–75

Elsden RP, Hasler JF, Seidel GE Jr (1976) Non-surgical recovery of bovine eggs. Theriogenology 6:523–532

Elsden RP, Nelson LD, Seidel GE Jr (1978) Superovulating cows with follicle stimulating hormone and pregnant mare's serum gonadotrophin. Theriogenology 9:17–26

Evans MJ, Kaufman MH (1981) Establishment in culture of pluripotential cells from mouse embryos. Nature 292(5819):154–156

Festing M (1972) Mouse strain identifi-cation. Nature 238:351–352

Forgason JL, Berry WT Jr, Goodwin DE (1961) Freezing bull semen in liquid nitrogen vapor without instrumentation. J Anim Sci 20:970

Garner DL (2006) Flow cytometric sexing of mammalian sperm. Theriogenology. 65(5):943–57 (Epub 2005 Oct 20). Review

Garner DL, Gledhill BL, Pinkel D, Lake S, Stephenson D, Van Dilla MA, Johnson LA (1983) Quantification of the X- and Y-chromosome-bearing spermatozoa of domestic animals by flow cytometry. Biol Reprod 28 (2):312–321

Garner DL, Evans KM, Seidel GE (2013) Sex-sorting sperm using flow cytometry/cell sorting. Methods Mol Biol. 927:279–95. Review

Gray KR, Bondioli KR, Betts CL (1991) The commercial application of embryo splitting in beef cattle. Theriogenology 35:37–44

Hare WCD, Mitchell D, Betteridge KJ, Eaglesome MD, Randall GCB (1976) Sexing two-week old bovine embryos by chromosomal analysis prior to surgical transfer: preliminary methods and results. Theriogenology 5:243–253

Hasler JF (2014) Forty years of embryo transfer in cattle: a review focusing on the journal Theriogenology, the growth of the industry in North America, and personal reminisces. Theriogenology 81(1):152–169. https://doi.org/10.1016/j.theriogenology.2013.09.010

Hendriks S, Dancet EA, van Pelt AM, Hamer G, Repping S (2015) Artificial gametes: a systematic review of biological progress towards clinical application. Hum Reprod Update 21(3):285–296. https://doi.org/10.1093/humupd/dmv001

Herr CM, Reed KC (1991) Micronanipulation of bovine embryos for sex determination. Theriogenology 35:45–54

Hoffmann I (2010) Climate change and the characterization, breeding and conservation of animal genetic resources. Anim Genet 41(Suppl 1):32–46. https://doi.org/10.1111/j.1365-2052.2010.02043.x

Hoshino Y, Hayashi N, Taniguchi S, Kobayashi N, Sakai K, Otani T, Iritani A, Saeki K (2009) Resurrection of a bull by cloning from organs frozen without cryoprotectant in a −80 °C freezer for a decade. PLoS One 4(1):e4142. https://doi.org/10.1371/journal.pone.0004142 (Epub 2009 Jan 8)

Iritani A, Niwa K (1977) Capacitation of bull spermatozoa and fertilization in vitro of cattle follicular oocytes matured in culture. J Reprod Fertil 50(1): 119–121

Jaenisch R, Mintz B (1974) Simian virus 40 DNA sequences in DNA of healthy adult mice derived from preimplantation blastocysts injected with viral DNA. Proc Natl Acad Sci U S A 71(4):1250–1254

Johnson LA, Flook JP, Hawk HW (1989) Sex preselection in rabbits: live births from X and Y sperm separated by DNA and cell sorting. Biol Reprod 41 (2):199–203

Kasinathan P, Wei H, Xiang T, Molina JA, Metzger J, Broek D, Kasinathan S, Faber DC, Allan MF (2015) Acceleration of genetic gain in cattle by reduction of generation interval. Sci Rep 2(5):8674. https://doi.org/10.1038/srep08674

Lanza RP, Cibelli JB, Diaz F, Moraes CT, Farin PW, Farin CE, Hammer CJ, West MD, Damiani P (2000) Cloning of an endangered species (Bos gaurus) using interspecies nuclear transfer. Cloning 2(2): 79–90

Liu X, Wang Y, Tian Y, Yu Y, Gao M, Hu G, Su F, Pan S, Luo Y, Guo Z, Quan F, Zhang Y (2014) Generation of mastitis resistance in cows by targeting human lysozyme gene to β-casein locus using zinc-finger nucleases. Proc Biol Sci 281(1780):20133368. https://doi.org/10.1098/rspb.2013.3368. Print 2014 Apr 7

Liu Z, Cai Y, Liao Z, Xu Y, Wang Y, Wang Z, Jiang X, Li Y, Lu Y, Nie Y, Zhang X, Li C, Bian X, Poo M, Chang H, Sun Q (2018) Cloning of a gene-edited macaque monkey by somatic cell nuclear transfer. Natl Sci Rev (In oress). https://doi.org/10.1093/nsr/nwz003

Liu JJ, Orlova N, Oakes BL, Ma E, Spinner HB, Baney KLM, Chuck J, Tan D, Knott GJ, Harrington LB, Al-Shayeb B, Wagner A, Brötzmann J, Staahl BT, Taylor KL, Desmarais J, Nogales E, Doudna JA (2019) CasX enzymes comprise a distinct family of RNA-guided genome editors. Nature. https://doi.org/10.1038/s41586-019-0908-x

Looney CR, Lindsey BR, Gonseth CL, Johnson DL (1994) Commercial aspects of oocyte retrieval and in vitro fertilization (IVF) for embryo production in problem cows. Theriogenology 41:67–72

Lu KH, Gordon I, Chen HB, McGovern H (1987) In vitro culture of early bovine embryos derived from in vitro fertilization of follicular oocytes matured in vitro. In: Proceeding of Third Meeting *of the European* Embryo Transfer Association Lyon, France. Association of Embryo. Technology in Europe, Paris, France, pp 70

Magata F, Tsuchiya K, Okubo H, Ideta A (2019) Application of intracytoplasmic sperm injection to the embryo production in aged cows. J Vet Med Sci 81(1):84–90. https://doi.org/10.1292/jvms.18-0284 (Epub 2018 Nov 26)

Makoolati Z, Movahedin M, Forouzandeh-Moghadam M, Naghdi M, Koruji M (2017) Embryonic stem cell derived germ cells induce spermatogenesis after transplantation into the testes of an adult mouse azoospermia model. Clin Sci (Lond) 131(18):2381–2395. https://doi.org/10.1042/cs20171074. Print 2017 Sep 15

Martin GR (1981) Isolation of pluripotent cell line from early mouse embryos cultured in medium conditioned by teratocarcinoma stem cells. Proc Natl Acad Sci USA 78:7634–7638

Moore SG, Hasler JF (2017) A 100-year review: reproductive technologies in dairy science. J Dairy Sci 100 (12):10314–10331. https://doi.org/10.3168/jds.2017-13138.Review

Morohaku K, Tanimoto R, Sasaki K, Kawahara-Miki R, Kono T, Hayashi K, Hirao Y, Obata Y (2016) Complete in vitro generation of fertile oocytes from mouse primordial germ cells. Proc Natl Acad Sci USA 113(32):9021–9026. https://doi.org/10.1073/pnas.1603817113 (Epub 2016 Jul 25)

Pieterse MC, Kappen KA, Kruip TA, Taverne MA (1988) Aspiration of bovine oocytes during transvaginal ultrasound scanning of the ovaries. Theriogenology 30(4):751–762

Polge C (1952) Fertilizing capacity of bull spermatozoa after freezing at 79 °C. Nature 169(4302):626–627

Polge C, Smith AU, Parkes AS (1949) Revival of spermatozoa after vitrification and dehydration at low temperatures. Nature 164(4172):666

Prather RS, Barnes FL, Sims MM, Robl JM, Eyestone WH, First NL (1987) Nuclear transplantation in the bovine embryo: assessment of donor nuclei and recipient oocyte. Biol Reprod 37(4):859–866

Robl JM, Wang Z, Kasinathan P, Kuroiwa Y (2007) Transgenic animal production and animal biotechnology. Theriogenology 67(1):127–133 (Epub 2006 Oct 27)

Roels K, Smits K, Ververs C, Govaere J, D'Herde K, Van Soom A (2018) Blastocyst production after intracytoplasmic sperm injection with semen from a stallion with testicular degeneration. Reprod Domest Anim 53 (3):814–817. https://doi.org/10.1111/rda.13153 (Epub 2018 Mar 1)

Rowe RF, Del Campo MR, Eilts CL, French LR, Winch RP, Ginther OJ (1976) A single cannula technique for nonsurgical collection of ova from cattle. Theriogenology 6(5):471–483

Rowson LE (1951) Methods of inducing multiple ovulation in cattle. J Endocrinol 7(3):260–270

Rowson LE, Dowling DF (1949) An apparatus for the extraction of fertilized eggs from the living cow. Vet Rec 61:191

Schulze M, Bortfeldt R, Schäfer J, Jung M, Fuchs-Kittowski F (2018) Effect of vibration emissions during shipping of artificial insemination doses on boar semen quality. Anim Reprod Sci 192:328–334. https://doi.org/10.1016/j.anireprosci.2018.03.035

Seidel GE Jr (2009) Sperm sexing technology-the transition to commercial application. An introduction to the symposium "update on sexing mammalian sperm". Theriogenology 71(1):1–3. https://doi.org/10.1016/j.theriogenology.2008.09.015 (Epub 2008 Oct 23)

Selokar NL, Saini M, Palta P, Chauhan MS, Manik R, Singla SK (2014) Hope for restoration of dead valuable bulls through cloning using donor somatic cells isolated from cryopreserved semen. PLoS One 9 (3):e90755. https://doi.org/10.1371/journal.pone.0090755. eCollection 2014

Singh B, Chauhan MS, Singla SK, Gautam SK, Verma V, Manik RS, Singh AK, Sodhi M, Mukesh M (2009) Reproductive biotechniques in buffaloes (*Bubalus bubalis*): status, prospects and challenges. Reprod Fertil Dev 21(4):499–510. https://doi.org/10.1071/rd08172. Review

Singh B, Mal G, Singla SK (2017a) Chapter 18 vitrification: a reliable method for cryopreservation of animal embryos. Methods Mol Biol 1568:243–249. https://doi.org/10.1007/978-1-4939-6828-2_18

Singh R, Mishra SK, Rajesh C, Dash SK, Niranjan SK, Kataria RS (2017b) Chilika- a distinct registered buffalo breed of India. Int J Livest Res 7(9):259–266. https://doi.org/10.5455/ijlr.20170704044822

Steinfeld H, Gerber P, Wassenaar T, Castel V, Rosales M, de Haan C (2006) Livestock's long shadow: environmental issues and options. FAO, Rome, Italy

Takahashi K, Yamanaka S (2006) Induction of pluripotent stem cells from mouse embryonic and adult fibroblast cultures by defined factors. Cell 126(4):663–676 (Epub 2006 Aug 10)

Tanne JH (2008) FDA approves use of cloned animals for food. BMJ 336(7637):176. https://doi.org/10.1136/bmj.39468.528368.DB No abstract available

Thornton PK (2010) Livestock production: recent trends, future prospects. *Philos*. Trans R Soc Lond B Biol Sci 365:2853–2867

Umbaugh RE (1949) Superovulation and ovum transfer in cattle. Am J Vet Res 10:295–305

Valenzuela OA, Couturier-Tarrade A, Choi YH, Aubrière MC, Ritthaler J, Chavatte-Palmer P, Hinrichs K (2018) Impact of equine assisted reproductive technologies (standard embryo transfer or intracytoplasmicsperm injection (ICSI) with in vitro culture and embryo transfer) on placenta and foal morphometry and placental gene expression. Reprod Fertil Dev 30 (2):371–379. https://doi.org/10.1071/RD16536

Whittingham DG, Leibo SP, Mazur P (1972a) Survival of mouse embryos from −196 to −269 °C. Science 178:411–412

Whittingham DG, Leibo SP, Mazur P (1972b) Survival of mouse embryos frozen to −196 and −269 °C. Science 178(4059):411–414

Willadsen SM (1979) A method for culture of micromanipulated sheep embryos and its use to produce monozygotic twins. Nature 277(5694):298–300. No abstract available

Willadsen SM (1986) Nuclear transplantation in sheep embryos. Nature 320(6057):63–65

Willadsen SM, Lehn-Jensen H, Fehilly CB, Newcomb R (1981) The production of monozygotic twins of preselected parentage by micromanipulation of non-surgically collected cow embryos. Theriogenology. 15(1):23–29. No abstract available

Willett EL, Black WG, Casida LE, Stone WH, Buckner PJ (1951) Successful transplantation of a fertilized bovine ovum. Science 113(2931):247. No abstract available

Wilmut I, Rowson LE (1973a) Experiments on the low-temperature preservation of cow embryos. Vet Rec 92(26):686–90. No abstract available

Wilmut I, Rowson LE (1973b) The successful low-temperature preservation of mouse and cow embryos. J Reprod Fertil 33(2):352–353. No abstract available

Wilmut I, Schnieke AE, McWhir J, Kind AJ (1997) Campbell KH. Viable offspring derived from fetal and adult mammalian cells. Nature 385(6619):810–3. Erratum in: Nature 1997 Mar 13;386(6621):200

Windig JJ, Engelsma KA (2010) Perspectives of genomics for genetic conservation of livestock. Conserv Genet 11:635–641

Yadav SK, Gangwar DK, Singh J, Tikadar CK, Khanna VV, Saini S, Dholpuria S, Palta P, Manik RS, Singh MK, Singla SK (2017) An immunological approach of sperm sexing and different methods for identification of X- and Y-chromosome bearing sperm. Vet World 10(5):498–504. https://doi.org/10.14202/vetworld.2017.498-504

Zhou Q, Wang M, Yuan Y, Wang X, Fu R, Wan H, Xie M, Liu M, Guo X, Zheng Y, Feng G, Shi Q, Zhao XY, Sha J, Zhou Q (2016) Complete meiosis from embryonic stem cell-derived germ cells in vitro. Cell Stem Cell 18(3):330–340. https://doi.org/10.1016/j.stem.2016.01.017

9 Cryopreservation of Oocytes and Embryos

Abstract

The cryopreservation as a broad discipline entails techniques that permits freezing and freeze-thawing or vitrification-warming of biological materials with minimal loss in their biological functions. Cryopreservation is an integral component of modern biomedical sciences, assisted reproduction and conservation biology, and a fundamental subject itself. Somatic cells, stem cells, primordial germ cells, oocytes, embryos and ovarian tissue preserved aptly under ultralow temperature can endure preservation for unlimited time with almost negligible loss in metabolic and genetic eminence. Compared to conventional slow-freezing cryopreservation techniques, the emphasis is on simpler methods of cryopreservation using vitrification.

Highlights

- Cryopreservation is an integral component of biomedical sciences, agriculture, and animal-assisted reproduction
- In view of limitations of slow freezing, alternative methods are developed
- Vitrification is a cheaper and efficient method of cryopreservation.

Keywords

Cryopreservation · Vitrification · Bio-banking · Cryoprotective agents · Livestock applications

9.1 Introduction

The term "cryobiology" comes from three Greek words: "*kryos*" meaning cold, "*bios*" meaning life, and "*logos*" meaning discussion or study. Cryobiology, by definition, is the study of the effects of low temperature on biological materials from living organisms. Cryopreservation in a broader sense refers to the process of arresting biological processes and placing biological materials (cells, oocytes, sperm, embryos, tissues/organs) into a suspended state of animation at ultralow temperatures (Pegg 2015; Singh et al. 2017; Canesin et al. 2018).

The impact of cryopreservation on assisted reproduction is increasingly evident. With the invention of advanced technologies like mammalian cloning, animal transgenesis, and stem cell engineering, the conservation of biological materials through cryopreservation has become imperative (Fig. 9.1). To deal with the problem

B. Singh et al., *Advances in Animal Biotechnology*,
https://doi.org/10.1007/978-3-030-21309-1_9

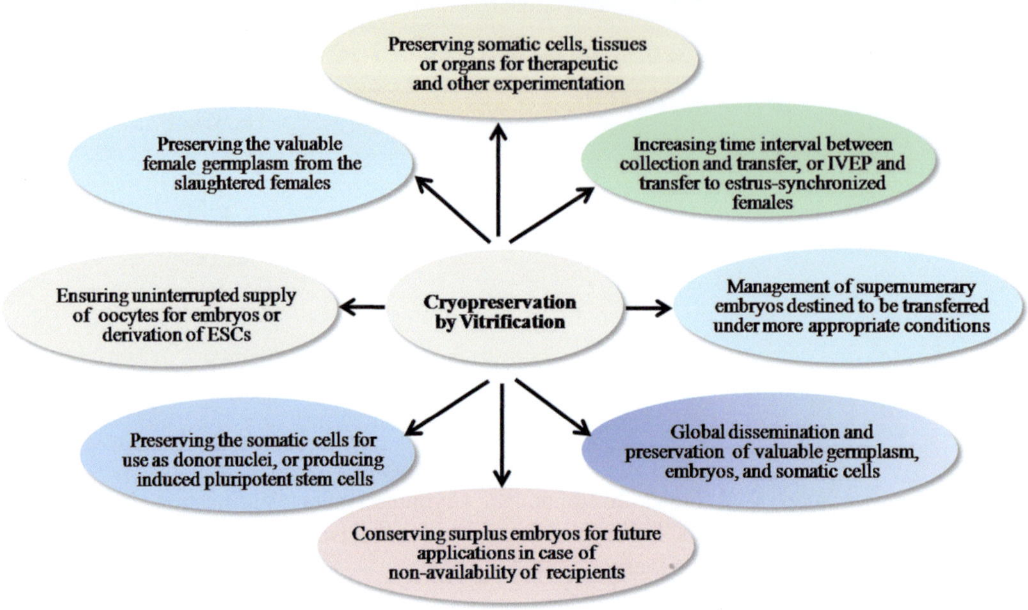

Fig. 9.1 A summarized view of the applications of cryopreservation in assisted reproduction and diversity conservation of farm animals (Singh et al. 2017). In addition, the cryopreservation aims to evolve new ways to freeze, store, and revive the cryopreserved bio-materials for clinical applications

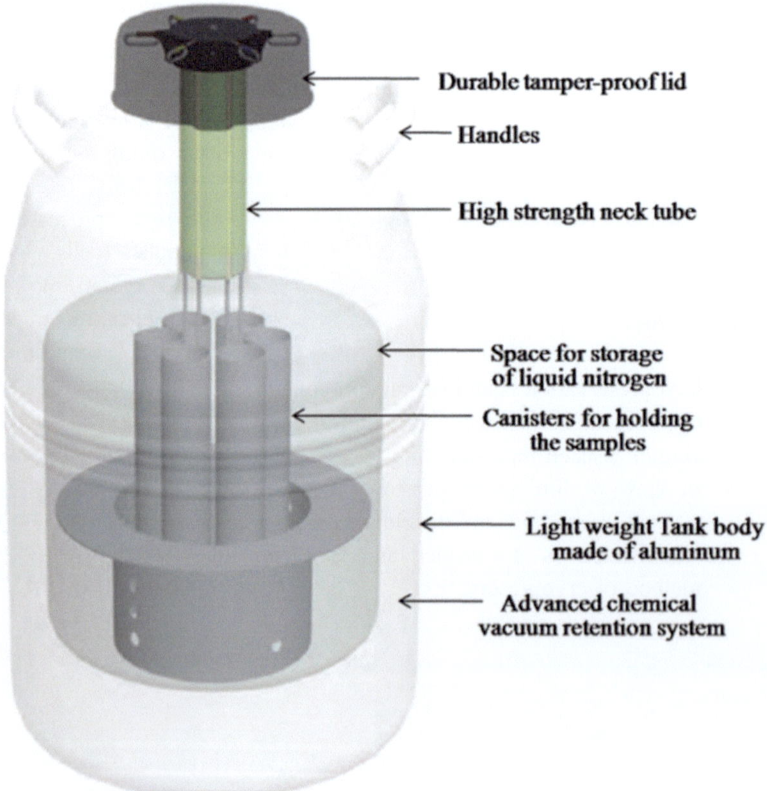

Fig. 9.2 Diagrammatic presentation of various parts of a typical commercial cryocontainer, the specialized containers used to store liquid nitrogen (LN_2). The cost, capacity, and variations of cryocontainers depend on the quantity of biological materials to be preserved

of uncertain supply and demand, the researchers have to store cells and biological materials over an extended time period. With advances in cryopreservation techniques, it is possible to preserve cells, whole organs such as blood vessels, kidney, heart, liver, and other fragile organs for transplantation with minimal damage and transport them in specialized containers called cryocontainers (Fig. 9.2) across the laboratories and humans and veterinary clinics.

9.2 General Procedural Features of Cryopreservation

Successful cryopreservation depends on several variables which include (i) the type of cell itself, (ii) the solution in which cell is suspended and whether or not the solution contains one or more specific additives called as cryoprotective agents (CPAs) (discussed below), (iii) the rate at which the cell is cooled to subzero temperatures, (iv) the minimum subzero temperature to which the cell is cooled, (v) the rate at which the cell is warmed, and (vi) the conditions under which the CPAs are removed after thawing or warming. For a given type of cell or tissue, choice of an appropriate CPAs, optimum cooling rate and warming rate play a crucial role. Depending on the suspending solution, different types of cells exhibit diverse optimum cooling rates varying from a low rate of ~0.2 °C/min to a high of 1000 °C/min. Under certain conditions, cells may even survive after being cooled at rates >100,000 °C/min.

Indeed, it is impractical to preserve various cells or tissue by using one protocol. Hence, various protocols are used for cryopreservation, and most of them have some common steps.

9.3 What Are Cryoprotectants

Whatever are the methods, certain additives known as cryoprotectants or cryoprotective agents (CPAs) are needed to avert freezing injuries due to the formation of ice crystals at ultralow temperature. Discovered accidentally by Polge et al. (1949), the CPAs are now part of most cryopreservation protocols. The CPAs in freezing solution are necessary to lower the freezing point and prevent damage to cell not only during freezing but also when cells are frozen-thawed or warmed. To be biologically acceptable, the CPAs must be non-toxic and able to permeate into cells (Pegg 2015).

In most cases, the survival of cryopreserved cells is impractical without CPAs. In addition, the CPAs make the cell membrane more elastic by affecting its lipid contents. They also ensure dehydration of the cells before cooling, thus preventing the formation of ice crystal during freezing. The CPAs are broadly classified into the following three categories:

(i) **Low molecular weight (MW) permeating cryoprotectants**

The low molecular weight permeating CPAs include ethylene glycol (EG, MW 62.07), 1, 2-propanediol (PROH) or propylene glycol (PG, MW 76.1), dimethyl sulphoxide (DMSO, MW 78.13), and glycerol (MW 92.1). The CPAs osmotically replace the intracellular water in cells before cooling and especially during slow-controlled cooling, reduce changes in the volume of cell, and prevent the formation of intracellular ice crystals. They are also believed to stabilize intracellular proteins.

(ii) **Low MW non-permeating cryoprotectants**

These include low molecular weight sugars such as galactose (MW 180.2), glucose (MW 181.1), sucrose (MW 342.3), and trehalose (MW 378.3) (Dinnyés et al. 2000; Begin et al. 2003). Trehalose, also known as mycose or tremalose, is a non-toxic saccharide occurring naturally in many organisms (mushrooms, lobsters, certain algae, and foods fermented by baker's yeast), withstanding intense low ambient temperature. Trehalose is preferred as an alternative CPA in many protocols (Stewart and He 2018).

(iii) **High MW non-permeating cryoprotectants**

Larger size molecules that cannot permeate cellular membrane represent the non-permeating category of CPAs. They increase the concentration of extracellular solutes generating an osmotic gradient across the cell membrane, which draws water out of the cell causing the cell to dehydrate before freezing. They prevent the rapid entry of water into cell subsequent to thawing during re-hydration/dilution of permeating CPAs.

Polyvinylpyrrolidone (PVP, MW 73,458), polyvinyl alcohol (PVA, MW 34,109) are large-size non-permeating CPAs. They transform the formation of ice crystal to an innocuous shape and size, and hence protect the cells from lethal mechanical injuries.

Among various classes of CPAs, presence of low MW permeating CPAs is desirable. The low MW CPAs may be combined with low MW non-permeating CPAs. The high MW CPAs are not used as such due to their failure to permeate through the cell membrane and replace cytoplasm water contents during dehydration and freezing.

9.4 Mode of Action of Cryoprotective Additives

Living cells have water and the formation of ice crystals during freezing causes damage to cell structure. The tissues are distorted, and the cells may shrink or collapse due to lethal damage on microscale. The main purpose of CPAs is to prevent damage caused by the formation of ice during freezing. Although cryoprotective actions of CPAs are not utterly understood, they have different roles and actions during cooling, thawing, or warming. When cells are first exposed to multimolar solutions of CPAs, the cells shrink because of loss of intracellular water. As the permeating CPAs enter the cells, water re-enters and the cells regain their original configuration. The CPAs enter at different diffusion rates into the cells. For example, rate of diffusion for propylene glycol into cells is relatively faster (5–7 min) compared to the diffusion of DMSO (20–30 min.) (Renard and Babinet 1984).

The cryoprotective actions of non-permeating CPAs are based on dehydration of cells prior to cooling which prevents formation of ice crystals. It is suggested that low MW non-permeating CPAs should be combined with low MW permeating CPAs to confer more protection to cells during cryopreservation. CPAs of high MW group protect the cells during freezing and warming by altering ice crystal formation to a safer form.

9.5 Mechanism of Entry of CPAs into Cells

When a cell is exposed to CPAs, it immediately contracts osmotically due to loss of water owing to differences in osmotic pressure between extracellular and intracellular milieu. At the same time, the CPA begins permeating the cell by diffusion. Finally, water begins to re-enter the cell to maintain osmotic equilibrium. At this equilibrium, the cells have an equal concentration of CPAs as that of the solution in which it is suspended. Among several variables that determine how quickly the equilibrium is established, some of the important ones include (i) permeability property of the specific solutes (e.g., ethylene glycol enters the cell faster than glycerol), (ii) concentration of CPAs (the higher the concentration, the faster the CPA will permeate into cell), (iii) temperature (the higher the temperature the faster the CPA will permeate into cell), and (iv) cellular origin, development stage, and the species.

When aqueous freezing solutions containing CPAs are frozen, the ice (water) crystals get separated, and concentration of dissolved CPAs rises. With further reduction in temperature, the concentration of CPAs and media ingredients is increased. The process continues until the temperature reaches eutectic point, and the temperature at which the entire system is solidified (Liebo 1986).

This is often, but not always, lethal to the cells (Mazur 1984). In contrast, when cells are vitrified, they remain suspended in high concentrations of CPAs, and at such high cooling rates, the intracellular ice crystals are not generated. This happens when the temperature is high enough so that molecular mobility of water is increased to a point where water molecules move and rearrange themselves from a disorderly amorphous-vitrified position to an orderly crystalline position. The phenomenon is observed well below the melting point and is therefore potentially dangerous (Rall et al. 2000).

9.6 Storage

High-quality plastic straws and ampoules are used to preserve cells and tissue. For the cells to be cryopreserved for a long time without losing viability, they must be preserved at a temperature below a glass transition temperature of cytoplasm and the suspending solution. In practice, the easy and safe way is to preserve the cells in LN_2 (−196 °C).

Storage, handling, and transportation of cryogenic materials need specialized instruments, containers, cryoware, and accessories (racks, risers, retrieval products) for high security and safety. All these are available from commercial manufacturers.

9.7 Thawing, Warming, and Post-warming

When needed, the cells cryopreserved in straws taken out of LN_2 are held in the air for around 40 s so that temperature can rise rapidly to approximately −50 °C. The straws are then transferred to water having temperature around 30 °C. Besides, it is important to prevent bursting of straws.

The optimum warming rate for a given type of cell depends on the optimum cooling rate that preceded it. Early investigators assumed that rapid warming of mammalian cells after cryopreservation was better because cells had shorter times to recrystallize and were exposed for less time to CPAs. However, freezing mouse embryos (Whittingham et al. 1972) revealed that there are exceptions to the notion. The mice embryos cryopreserved by slow cooling had greater post-thaw survival when they were warmed slowly indicating that embryo survival depends on the process of warming also. The low survival using faster warming rates might be due to osmotic effects.

9.8 Removing CPAs

Removal of CPAs is the last step of the cryopreservation. The CPAs such as DMSO could be toxic to cells, and hence should be removed before using cells for research or clinical purposes. Cells are cryopreserved in CPAs ranging in concentration from 1–8 M (1–2 M in conventional slow freezing, while 4–8 M in vitrification). In general, if cells frozen in permeating CPAs are diluted rapidly (re-hydrated), they are prone to death caused by osmotic shock. Step-wise removal of CPAs is a practical process adopted in most protocols. Although this method works well in practice, it is slow and time-consuming. A much shorter method is to use a non-permeating solute such as the sucrose as an osmotic buffer to minimize osmotic shock.

9.9 Cryopreservation by Vitrification

Cryopreservation protocols can be broadly classified as "equilibrium" (slow-freezing) or "non-equilibrium" (vitrification and ultrarapid freezing) (Rall and Fahy 1985) depending on cooling rates and the concentration of CPAs used (Fig. 9.3). Slow freezing requires a lengthy slow-controlled cooling before the sample is plunged directly into LN_2.

Vitrification, by contrast, is a rapid process of cooling of biologic samples to cryogenic temperature without detrimental phase transition of liquid to solid ice. This is achieved by directly plunging the biomaterials in the presence of highly concentrated solution of CPAs into LN_2. However, the basic concepts of these two

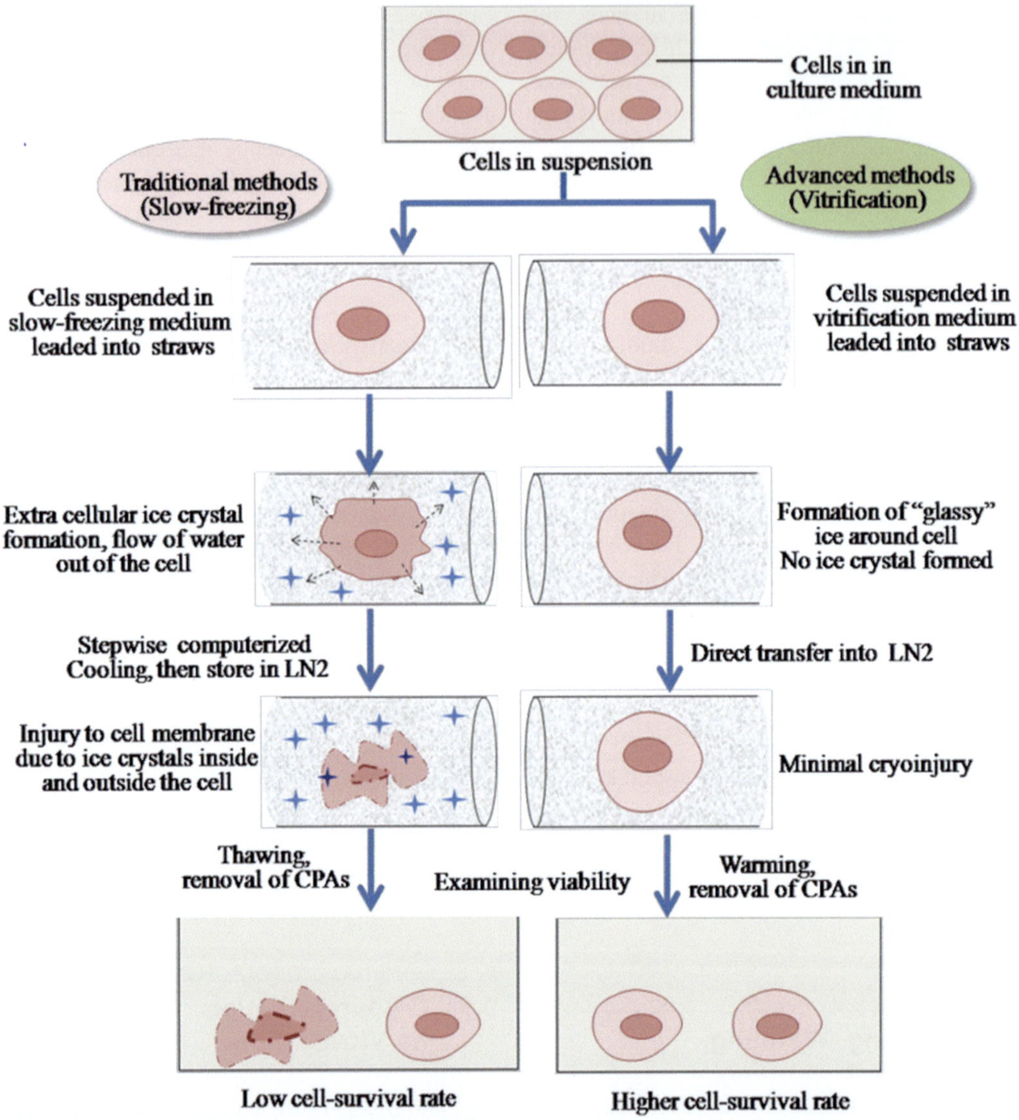

Fig. 9.3 A comparative pictorial presentation of traditional slow-freezing and vitrification cryopreservation of cells. As the cell cytoplasm is cooled to lower temperatures (below −10 °C or −15 °C), ice crystals form abruptly in the cytoplasm itself, a phenomenon referred to as intracellular nucleation. Compared to the slow-freezing method, the vitrification is simpler and leads to minimal damage to the cells cryopreserved. The straws are sealed with inert materials such as polyvinyl alcohol after loading cells into them

protocols are the same, as they both aim to protect the cells from the cryoinjuries caused by intracellular ice crystals, cellular dehydration, and drastic changes in solute concentrations at low temperature.

The radical strategy of vitrification is to completely eliminate the formation of intracellular as well as extracellular ice crystals through the imposition of ultrarapid cooling or warming rates. The protocols of vitrification are simpler,

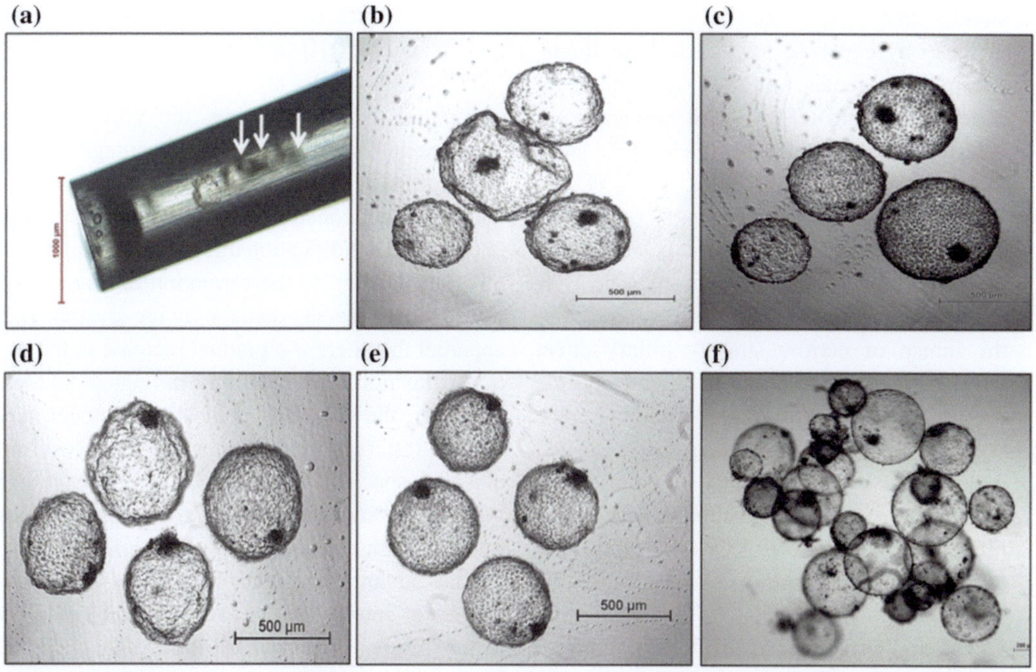

Fig. 9.4 Vitrification of animal embryos. **a** loading straw used for embryos; **b** a group of shrunken zona-free cloned buffalo blastocysts immediately after incubation with equilibration solutions; **c** re-expanded blastocysts after incubation for 5–7 min in equilibration solution; **d** a group of yet to re-expand cloned blastocysts immediately after warming; **e** a group of re-expanded cloned blastocysts after 22–24 h of culture in embryo medium; **f** re-expanded cloned blastocyst (Singh et al. 2017)

allowing the cells held directly into vitrification solutions, and then immediately transferring into LN_2. An effective way of achieving a higher cooling and warming rates would be to minimize the volume of the solution and the container used for preserving biomaterials. Vitrification protocols have emerged as promising methods of cryopreservation of animal embryos and cells (Fig. 9.4) (Singh et al. 2017).

9.10 Open and Closed Vitrification Systems

Methods of vitrification can be divided into two categories. In open system (e.g., Cryotop), the biomaterials are brought directly in contact with LN_2, whereas in the close vitrification system (e.g., CryoTip™), the cells or tissues are encapsulated in some devices and then submerged into LN_2. The open vitrification systems are the versions of OPS, Cryotop, Cryoleaf, and Cryolock systems.

The viability of vitrified warmed biomaterials by open as well as closed methods is more or less similar. The closed vitrification system provides as hygienic or aseptic alternative cryopreservation compared to open vitrification methods. The studies carried out to compare the efficacy and safety of open and closed vitrification systems have concluded that closed vitrification CryoTip system could be preferred for preserving oocytes and embryos as it eliminates possibilities of contamination without affecting survival and developmental rates in vitro and in vivo (Kuwayama et al. 2005).

1. **Open-pulled straw (OPS) vitrification**
   The 0.25 mL French mini straw (FMS) and OPS (Vajta et al. 1998), which is much

smaller than and narrower that FMS, are softened by exposing to high heat or flame, and pulled manually until the inner diameter is decreased from 1.7 to ~ 0.8 mm and from 0.15 to ~ 0.07 mm, respectively. This leads to simultaneous thinning of wall of the straws.
The heat-molten pulled straw is cut at the narrowest point with sterile blade. The cells (e.g. oocytes) along with fluid medium enter the lumen of narrow straw capillary effect when the end of the straw is dipped into the drop of vitrification solution containing cells. Alternatively, the cells can be gently aspirated 2–3 cm inside the capillary (equivalent to 1.0–2.0 μL of medium). Although it would be difficult to measure precisely rates of cooling and warming, the cooling is exceedingly fast in OPS.

2. **Cryoloop vitrification**: The cryoploop is containerless approach to increase cooling speed and cryopreservation of live cells. Cryoloop consists of a minute nylon loop (20-μm wide) mounted on a stainless steel pipe inserted into the lid of a cryovial. Oocytes or embryos are placed on loops loaded with a thin film of vitrification solution (<1 μL). The cryovial is then plunged into LN_2. The cryovial is screw-sealed. It is easy to label and store cryovial (Lane et al. 1999).
3. **Microdrops** This is another containerless method, also called microdrops or microdroplets. The technique is based on pre-equilibration in medium containing glycerol (10% v/v, before subjecting to vitrification solution (Landa and Tepla 1990). With this method, oocytes suspended in vitrification solution are aspirated into a capillary pipette and expelled directly into LN_2 as microdroplets. The 4–8 μL microdrops get vitrified without thermo-insulating device. The method is more effective than conventional straw-based vitrification for bovine embryos at early cleavage stages, which are sensitive to chilling.

9.11 Merits of Cryopreservation by Vitrification

Compared to slow-freezing, the vitrification is convenient. One of the major advantages is that vitrification avoids use of programmable freezers and minimal consumption of LN_2. The rapid cooling and solidification due to substantial rise in viscosity leading to the formation of amorphous glassy state prevents damage by ice crystals. It is apparent that there is a gradual increase in the use of vitrification in clinical and biological studies. However, the standard vitrification protocol may vary for different cells, oocytes, and embryos from assorted species. For instance, the oocytes in general, are more susceptible to cooling injuries owing to disassembling of meiotic spindle.

One advantage of vitrification is that the formation of intracellular ice can be prevented by using smaller amount of intracellular CPAs. Therefore, the low concentration of permeating CPAs seems to be desirable. Indeed, the concentration of permeating CPAs in ultrarapid vitrification is lower than that observed in conventional vitrification. In conventional vitrification, solutions with 40% (or more) permeating CPAs are used, whereas in ultrarapid vitrification, those with ~30% CPAs are more frequently used.

Box 1. Applications of Vitrification in Farm and Feral Animals

- Establishing oocyte, embryo, cell and organ banks
- Producing embryos for basic research and propagating desired animals
- Promoting conservation of endangered species through embryo biotechniques
- Extension of fertile life of oocytes
- Improving the efficiency of IVF
- Conservation of germplasm from superior animals
- Vitrification as a tool for preserving fertility.

9.12 Transplant Organ Bio-Banking

The bio-banking is an age-old phenomenon dealing with the collection and preservation of biological samples and specimens for research and academic purposes. In current clinical perspective, and in view of the accessibility of protocols and facilities for preserving biological specimens, and sustained sample viability and quality, the bio-banking or biorepository has emerged as an important area of medical and economic interest.

Whole-organ banking is not yet operational at noticeable scale despite hard efforts based on cryopreservation tools to revive the damaged tissues and organs. Vitrification is a futuristic technique for preserving organs. Innovations in cryoprotective solutions or CPAs, organ perfusion techniques, and warming strategies reveal the potential for the vitrification on a scale applicable for human and veterinary transplantation. In view of the current realism that there is a substantial scarcity of organs needed for transplantation *vis-à-vis* the number of patients on organ waiting list, the organ bio-banking has an important role. Successful approaches for organ cryopreservation (subject to the availability of organs) could enable organ banking, which would recast the entire process in which organs are recovered, allocated, preserved, and used for transplant (Finger and Bischof 2018).

In veterinary clinical perspective, the bio-banking has an important role in the collection and preservation of tissues and organs for molecular analysis of diseases and forensic cases. The organ bio-banking is important, especially when organs from humanized animal models are intended for humans or for valuable transgenic animals.

9.13 Risks and Disadvantages of Cryopreservation

Vitrification has certain limitations that should be solved to enhance the gains from this method of cryopreservation. The temperature at which biomaterials are preserved is very hurtful to cells. Reduction in temperature can result in membrane lipid phase changes and depolymerization of the cytoskelton. The increase in solute concentration may cause osmotic shrinking, and an increase in ionic concentration can have a direct effect on the membranes; this includes the solubility of the membrane proteins. Dehydration can destabilize the membrane lipid bilayers. As cryopreservation solution becomes enormously viscous, diffusion processes including osmosis are hindered. The changes in pH can denature the proteins and as cells become closely packed, and membranes may get damaged.

Effects of cryopreservation are not restricted merely to physical damage, cryopreservation also causes degeneration in cytoplasm and molecular biological process (Wang et al. 2017).. As a natural adaptive response, the cryopreservation elicits autophagic activation and increases cellular autophagy (Atg) genes (such as *Atg5*, *Atg7*, *Atg12*, *LC3a* (Map1lc3a), LC3b (Map1lc3b), and Beclin1) expression in oocytes (Bang et al. 2014). The effect is noted in immature oocytes subjected to cryopreservation. Inhibiting autophagy in immature oocytes during vitrification-warming and IVM activates apoptosis via caspase-9 and -12 pathway (Gao et al. 2017).

The studies carried out using bovine blastocysts have shown an increase in expression of apoptotic genes (*survivin*, *Fas*, *Hsp 70,* and *caspase-3*), and fragmentation of genomic DNA in frozen-thawed embryos that might impair embryo development (Park et al. 2006).

9.14 Progress in Vitrification and Its Applications in Agricultural Animals

From initial success in preservation of sperm at subzero temperature, the technique is used for preserving oocytes, embryos tissues, and organs. The interest in breeding and conserving economically important livestock species has urged the scientists to use vitrification protocols in livestock (bovine, pig, sheep, goats, and camels) (Vajta 2000; Gautam et al. 2009; Mullen and Fahy 2012; Fathi et al. 2018), as well as pet and companion animals (Abe et al. 2011; Galiguis et al. 2014; Brito et al. 2018).

The ARTs originally evolved for humans were later used for animal species. The ARTs primarily used in humans, such as PGD, sperm sexing and sex determination are increasingly used in animals such as cattle, buffalo, and pig. This will motivate the need to develop female embryos for milk production and the male embryos for meat purposes. Cryopreservation is a demanding technique in livestock and companion animals. This is because domestic pig (*Sus scrofa domesticus*) and large ruminants such as cattle (*Bos Taurus*, *B. indicus*) and buffaloes (*Bubalus bubalis*) are important meat and dairy animals.

With the increasing population, expected to reach around 9.15 billion by 2050 (Alexandratos and Bruinsma 2003), the consumption of meat is likely to rise. In addition, there is an increasing trend in producing high-yielding or transgenic animals to match the consumer demand of cells, tissues and organs, and therapeutics. As transporting live animals across large distances is expensive and tricky, hence transporting cryopreserved embryos, oocytes, and semen will increase.

It is difficult to cryopreserve embryos and oocytes in some species (such as buffaloes and pigs) due to higher intracellular lipid contents (Hazel and Williams 1990). Li et al. (2009) showed that porcine embryos treated with trypsin, or exposed to solutions of higher osmolality in vitro could partially remove lipid bilayer and facilitate their centrifugation and micromanipulation. Disrupting membrane lipid bilayer by micromanipulation, and then centrifuging embryos before vitrification had a positive outcome on their development.

Many studies are carried out to improve the efficacy of cryopreservation, development of embryos by IVF and production in livestock species as mentioned in respective chapters on various species. It was found that vitrification combined with this approach in a closed system was as successful as open-pulled straw vitrification. This could be a major step forward in cryopreservation of porcine embryos (Men et al. 2011).

9.15 Outlook and Challenges

Over the past decades, cryopreservation has emerged as an indispensable tool in biomedical and human- and animal-assisted reproduction medicine. Compared to slow freezing, the vitrification, due to being simpler and effective continues to be an area of scientific interest and improvements and developing robust repeatable methods for cryopreservation of gametes, cells, and embryos in farm animals, and genetic manipulation experiments. Improvements in vitrification methods are envisaged to have a big impact on animal husbandry.

The cryopreservation as a vital component has enabled on-demand access to supply biological materials (biomolecules, cells, tissue, and organs). The reasons for slow progress may well result from a dearth of information on how the various biophysical changes during cryopreservation regimen affect cellular functions. The research should focus to minimize adverse effects associated with cryopreservation to expedite the progression in important areas such as embryo and cell banking, cell therapy and stem cell biology.

Studies are warranted to unravel the impact of cryopreservation at molecular biological levels, viz, activation of genes implication in apoptosis, and evolving strategies to prevent activation of genes leading to apoptosis. It is crucial to improve existing protocols for their extensive use in livestock species and develop faster or effective techniques so that embryos could be made available to fulfill the field-level demands. Collaborative and multidisciplinary efforts of specialists involving biologists, physicists, and engineers may provide solutions to circumvent the damaging effects of cryopreservation.

9.16 Conclusions

The cryopreservation is an integral component of ARTs. Slow freezing needs programmable freezers that are expensive. As an alternative, the vitrification is simpler and more suitable to cells, oocyte, and embryo cryopreservation in terms of establishing pregnancies and live birth rates. The challenges such as demand for more apposite vitrification solutions that are least toxic should be resolved on priority basis so that the technique could be used at a larger scale. Research should focus on making cryopreservation economical, and more practical for bio-banking.

References

Abe Y, Suwa Y, Asano T, Ueta YY, Kobayashi N, Ohshima N, Shirasuna S, Abdel-Ghani MA, Oi M, Kobayashi Y, Miyoshi M, Miyahara K, Suzuki H (2011) Cryopreservation of canine embryos. Biol Reprod 84(2):363–368. https://doi.org/10.1095/biolreprod.110.087312 (Epub 2010 Oct 6)

Alexandratos N, Bruinsma J (2003) World agriculture: towards 2015/2030: an FAO perspective. Land Use Policy 20:375

Bang S, Shin H, Song H, Suh CS, Lim HJ (2014) Autophagic activation in vitrified-warmed mouse oocytes. Reproduction 148(1):11–19. https://doi.org/10.1530/REP-14-0036 (Epub 2014 Apr 23)

Begin I, Bhatia B, Baldassarre H, Dinnyes A, Keefer CL (2003) Cryopreservation of goat oocytes and in vivo derived 2- to 4-cell embryos using the cryoloop (CLV) and solid-surface vitrification (SSV) methods. Theriogenology 59(8):1839–1850

Brito DCC, Domingues SFS, Rodrigues APR, Figueiredo JR, Santos RR, Pieczarka JC (2018) Vitrification of domestic cat (*Felis catus*) ovarian tissue: effects of three different sugars. Cryobiology 83:97–99. https://doi.org/10.1016/j.cryobiol.2018.06.006 (Epub 2018 Jun 13)

Canesin HS, Brom-de-Luna JG, Choi YH, Pereira AM, Macedo GG, Hinrichs K (2018) Vitrification of germinal-vesicle stage equine oocytes: effect of cryoprotectant exposure time on in-vitro embryo production. Cryobiology 81:185–191. https://doi.org/10.1016/j.cryobiol.2018.01.001 (Epub 2018 Jan 3)

Dinnyés A, Dai Y, Jiang S, Yang X (2000) High developmental rates of vitrified bovine oocytes following parthenogenetic activation, in vitro fertilization, and somatic cell nuclear transfer. Biol Reprod 63 (2):513–518

Fathi M, Moawad AR, Badr MR (2018) Production of blastocysts following in vitro maturation and fertilization of dromedary camel oocytes vitrified at the germinal vesicle stage. PLoS One 13(3):e0194602. https://doi.org/10.1371/journal.pone.0194602 (eCollection 2018)

Finger EB, Bischof JC (2018) Cryopreservation by vitrification: a promising approach for transplant organ banking. Curr Opin Organ Transplant 23(3):353–360. https://doi.org/10.1097/MOT.0000000000000534

Galiguis J, Gómez MC, Leibo SP, Pope CE (2014) Birth of a domestic cat kitten produced by vitrification of lipid polarized in vitro matured oocytes. Cryobiology 68(3):459–466. https://doi.org/10.1016/j.cryobiol.2014.02.012 (Epub 2014 Mar 11)

Gao HH, Li JT, Liu JJ, Yang QA, Zhang JM (2017) Autophagy inhibition of immature oocytes during vitrification-warming and in vitro mature activates apoptosis via caspase-9 and -12 pathway. Eur J Obstet Gynecol Reprod Biol 217:89–93. https://doi.org/10.1016/j.ejogrb.2017.08.029 (Epub 2017 Aug 24)

Gautam SK, Singh B, Verma V, Palta P, Chauhan MS, Manik RS (2009) In vitro production of blastocysts from vitrified-warmed abattoir-derived buffalo (*Bubalus bubalis*) immature oocytes. Indian J Dairy Sci 62:495–502

Hazel JR, Williams EE (1990) The role of alterations in membrane lipid composition in enabling physiological adaptation of organisms to their physical environment. Prog Lipid Res 29:167–227

Kuwayama M, Vajta G, Ieda S, Kato O (2005) Comparison of open and closed methods for vitrification of human embryos and the elimination of potential contamination. Reprod Biomed Online 11(5):608–614

Landa V, Teplá O (1990) Cryopreservation of mouse 8-cell embryos in microdrops. Folia Biol (Praha) 36 (3–4):153–158

Lane M, Bavister BD, Lyons EA, Forest KT (1999) Containerless vitrification of mammalian oocytes and embryos. Nat Biotechnol 17(12):1234–1236. No abstract available

Li R, Murphy CN, Spate L, Wax D, Isom C, Rieke A et al (2009) Production of piglets after cryopreservation of embryos using a centrifugation based method for delipation without micromanipulation. Biol Reprod 80:563–571

Leibo SP (1986) Cryobiology: preservation of mammalian embryos. Basic Life Sci 37:251–272

Mazur P (1984) Freezing of living cells: mechanisms and implications. Am J Physiol 247:C-125–C142

Men H, Zhao C, Si W, Murphy CN, Spate L, Liu Y et al (2011) Birth of piglets from in vitro-produced, zona-intact porcine embryos vitrified in a closed system. Theriogenology 76:280–289

Mullen SF, Fahy GM (2012) A chronologic review of mature oocyte vitrification research in cattle, pigs, and sheep. Theriogenology 78(8):1709–1719. https://doi.org/10.1016/j.theriogenology.2012.06.008

Park SY, Kim EY, Cui XS, Tae JC, Lee WD, Kim NH, Park SP, Lim JH (2006) Increase in DNA fragmentation and apoptosis-related gene expression in frozen-thawed bovine blastocysts. Zygote 14(2):125–131

Pegg DE (2015) Principles of cryopreservation. Methods Mol Biol 1257:3–19

Polge C, Smith AU, Parkes A (1949) Revival of spermatozoa after vitrification and dehydration at low temperatures. Nature 164:166

Rall WF, Fahy GM (1985) Ice-free cryopreservation of mouse embryos at −196 degrees C by vitrification. Nature 313(6003):573–575

Rall WF, Schmidt PM, Lin X, Brown SS, Ward AC, Hansen CT (2000) Factors affecting the efficiency of embryo cryopreservation and rederivation of rat and mouse models. ILAR J 41(4):221–227 (Review)

Renard JP, Babinet C (1984) High survival of mouse embryos after rapid freezing and thawing inside plastic straws with 1-2 propanediol as cryoprotectant. J Exp Zool 230(3):443–448

Singh B, Mal G, Singla SK (2017) Chapter 18 vitrification: a reliable method for cryopreservation of animal embryos. Methods Mol Biol 1568:243–249. https://doi.org/10.1007/978-1-4939-6828-2_18

Stewart S, He X (2018) Intracellular delivery of trehalose for cell banking. Langmuir. https://doi.org/10.1021/acs.langmuir.8b02015 (Epub ahead of print)

Vajta G (2000) Vitrification of the oocytes and embryos of domestic animals. Anim Reprod Sci 60–61:357–364

Vajta G, Holm P, Kuwayama M, Booth PJ, Jacobsen H, Greve T, Callesen H (1998) Open Pulled Straw (OPS) vitrification: a new way to reduce cryoinjuries of bovine ova and embryos. Mol Reprod Dev 51:53–58

Wang N, Li CY, Zhu HB, Hao HS, Wang HY, Yan CL, Zhao SJ, Du WH, Wang D, Liu Y, Pang YW, Zhao XM (2017) Effect of vitrification on the mRNA transcriptome of bovine oocytes. Reprod Domest Anim 52(4):531–541. https://doi.org/10.1111/rda.12942 (Epub 2017 Mar 12)

Whittingham DG, Leibo SP, Mazur P (1972) Survival of mouse embryos from −196°C to −269°C. Science 178:411–412

Somatic Cell Nuclear Transfer 10

Abstract

In biological system, the term "cloning" has different meanings. Animal cloning, the somatic cell nuclear transfer (SCNT) cloning, became a buzzword when Dolly, the first mammalian species was born from the nuclear transfer of fibroblasts taken from adult sheep. The technique revolutionized the field regenerative medicine, and paved way to engineer stem cells for biomedical applications. At present, a number of animal species including cloned transgenic animals are produced. There are scopes to improve the technique for its wider use for health and welfare of humans.

Highlights

- The current nuclear transfer cloning is a more-than-two-decade-old technique that led to concept of direct reprogramming of mature cells.
- Use of technique is advocated to create animals for experiments and biomedical testing.

Keywords

SCNT · Cloned animals · Enucleation · Electrofusion · Diversity conservation · Reprogramming

10.1 Introduction

Multiplying a superior livestock, and seeing their pets as immortal, is the utmost wish of farmers and pet lovers. Cloning by SCNT (Fig. 10.1) promises to achieve the goals. Milestone breakthroughs are already achieved to disseminate superior animals, resurrecting dead animals, births from sterile animals such as mule, and cloning the pets (Table 10.1).

Nuclear transfer cloning acquired glare of publicity in 1997 with the birth of Dolly, the first mammal cloned using adult somatic cells from one animal, oocyte from other, and establishing pregnancy in third animal.

Indeed, the term "*clone*" has originated from Greek word, *klon*, meaning a twig or a small branch. The cloning as a scientific process has different connotations, protocols and outcomes in biological sciences. In plants, the cloning corresponds to a way of vegetative proliferation of flora, and asexual reproduction in some lower invertebrates such as flatworms and starfish, wherein a segment of organism after dissociation rejuvenates into sexually active adult organism. For a molecular biologist, the cloning refers to obtaining multiple copies of a DNA segment or the gene.

B. Singh et al., *Advances in Animal Biotechnology*,
https://doi.org/10.1007/978-3-030-21309-1_10

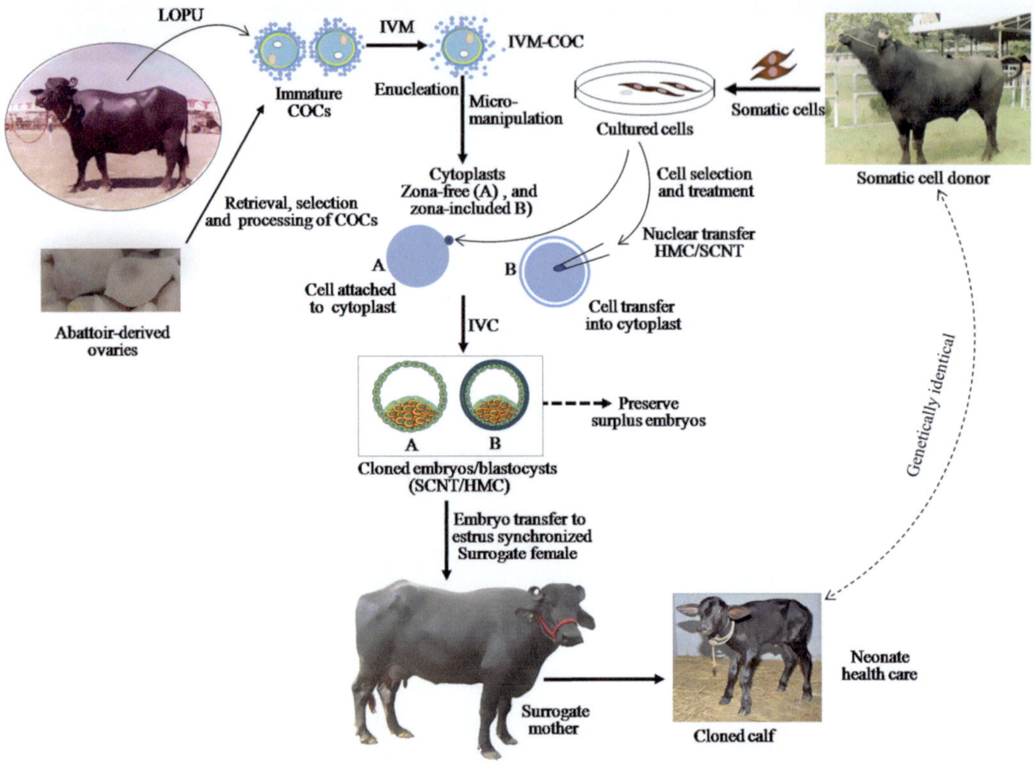

Fig. 10.1 A diagrammatic and generalized sketch of SCNT cloning. Except mitochondrial genome, the cloned offspring is identical to donor of somatic cells. Notably, the cloned calf has three mothers, but no father. *Abbreviations* LOPU laparoscopic ovum pick-up, COCs cumulus–oocyte complexes, SCNT somatic cell nuclear transfer, HMC handmade cloning; IVM in vitro maturation

Table 10.1 A comparative view of SCNT and HMC techniques in terms of cost, efficacy, and ease of use

Traditional nuclear transfer cloning (SCNT)	Handmade cloning
It is a traditional technique	It is comparatively recent
It is more invasive	Less invasive
Needs microtools, *viz*., holding capillary pipette and cutting needle	No need of microtools. It needs a specialized ultra sharp cutting blade
Requires specialized equipments and inverted microscope to prepare cytoplast and donor cell transfer	Stereo zoom microscope is needed to bisect polar body and nuclear material to prepare cytoplast
Technically complex and expensive	Simpler and cheaper
The overall operation efficiency is slow	It is comparatively faster

However, the higher vertebrates, such as mammals are unable to rejuvenate, and multiply either by parthenogenetic embryonic development or by sexual reproduction involving union of two gametes.

The principal of SCNT is that each cell in any individual possesses full genetic code for developing into an embryo, and then offspring. The SCNT helps in reprogramming the genome of finally matured cells (called donor cell), thus provides the ways of redirecting the matured cells to a totipotent embryonic stage. Hence, cloning by SCNT circumvents the processes that normally occur during gametogenesis, enabling development of embryo which develop into fetus and then live offspring. The reprogrammed cells directed to differentiate into specialized cell types. In SCNT, the recipient oocytes (cytoplasts) possess fully functional and healthier mitochondria which supports the metabolism competence, cope with the oxidative stress, and rejuvenate mitochondrial functioning of aged cells (Wu et al. 2018).

10.2 The Rationale of Nuclear Transfer Cloning

Prior to advent of current SCNT, embryo splitting was used to produce identical copies of the animals, called identical twins. The technique had some limitations. The embryos could be split into two or four parts which possess capacity to rejuvenate and develop into new embryos. The nuclear transfer cloning has paved the way to restore precious progeny-tested economically important and can be used to repopulate species that are endangered or are vulnerable to extinction.

SCNT is routinely used to produce cloned transgenic model animals for biomedical applications. The rodents (Kaminuma et al. 2017) and pigs (Lee et al. 2017) are most commonly cloned by using transgenic somatic or stem cells. A number of animal species were cloned thereafter by using SCNT or the modified forms of the original protocols. Recently, the biologists at Shanghai (China) used the SCNT to create healthy clones (Zhong Zhong and Hua Hua) of non-human primate, the macaque monkeys (*Macaca fascicularis*). More births of the cloned macaques are awaited. The scientists believe that cloning of primates combined with CRISPR-Cas9, a potent gene-editing tool, can mark the beginning of a new era of basic and biomedical sciences and will revolutionize the studies of human genetic diseases (Liu et al. 2018). It is envisaged that it might be possible to genetically engineer the primate models of human diseases by combining both the methods.

10.3 Progress in Somatic Cell Nuclear Transfer

The SCNT is an intricate technique, comprising of multiple technical steps, biological processes, and the organisms. The success depends on several factors including genomic status of diploid somatic cell, volume, and quality of cytoplasts developed from in vitro matured oocytes, cell cycle synchrony of donor nucleus, reprogramming of nuclear genome regulating pre- and post-implantation embryo development. Although many protocols are developed to improve SCNT, principles of some events such as genome epigenetic reprogramming remain unchanged. Aberrant reprogramming of genome of cloned embryos is a common problem that causes ***genetic abnormalities*** and health problems in cloned offspring. Despite this, the number and diversity of mammalian species ranging from experimental mice to primates, macaques, and livestock species have been cloned past two decades. The technique is found to be of immense importance to produce cloned gene-edited and transgenic animals.

Importantly, the SCNT is one of the most important and safer approaches to reprogram matured somatic cells into pluripotent stem cells state (Wu et al. 2018). With the efforts made during past two decades, the NT cloning is improved to some extent. However, in general, only 5–15% of transferred cloned embryos are able to develop into live offspring.

10.4 Steps Involved in the Process

Technically, the SCNT is a reproductive engineering technology used to produce cloned mammals and non-mammalian species. The nucleus is removed from in vitro matured oocyte, made as host (cytoplast) for diploid nucleus from cultured somatic cell, also called donor nucleus. The cytoplast and somatic cell upon fusion and activation in vitro, develop into a single diploid cell homologous to totipotent zygote which develop into pluripotent embryo. The technique seems to be simpler, but it comprises of a number of steps performed by using specialized equipment and cell culture techniques. Like normal embryo, the cloned mammalian embryo reached hatching blastocyst stage in 6–8 days.

10.5 Preparation of Microtools

Microtools (Fig. 10.2) are prepared from special glass capillaries. These include holding capillary pipette, cutting needle for puncturing the zona pellucida of oocyte for removing genetic materials (nucleus plus polar body), and transfer capillary pipette to transplant the specially donor nucleus cell into cytoplast. The capillary cutters or cutting needle, and capillary pipettes can be prepared in laboratories. The removal of nuclear materials is facilitated by commercially available highly efficient micromanipulators (e.g., Narishige, Japan; Eppendorf Micromanipulators, North America) attached to fluorescent inverted microscope.

A small group of alternative procedures, called handmade cloning (Vajta et al. 2001; Booth et al. 2001) or hand-guided cloning, have been developed to simplify the technique to improve the efficiency of nuclear transfer cloning.

10.6 Establishing Somatic Cell Cultures

As stated above, the somatic cells serve as donor nuclei. Researchers have emphasized the need to have authentic cell banks that should be used to produce cloned embryos (Selokar et al. 2018). The cells, be from any organ or origin, have to be cultured in vitro and prepared for subsequent use. In theory, any somatic cell or stem cells can be used as donor nuclei, but some cells such as stem cells, cumulus cells or granulosa cells, and fetal somatic cells are found to be more capable. Bone marrow stem cells are used recently to clone domestic animals. This is attributed to the ability of cell nuclear genome to be reprogrammed more accurately.

10.7 Oocytes and Their Micromanipulation

Availability of oocytes is the most important requirement. Ovaries obtained from the slaughtered females are most economical and convenient source of immature oocytes. In some cases, oocytes are aspirated from live females. Ultrasound-guided, transvaginal ovum pick-up is used to aspirate the oocytes from mature follicles on the ovaries that are overstimulated by means

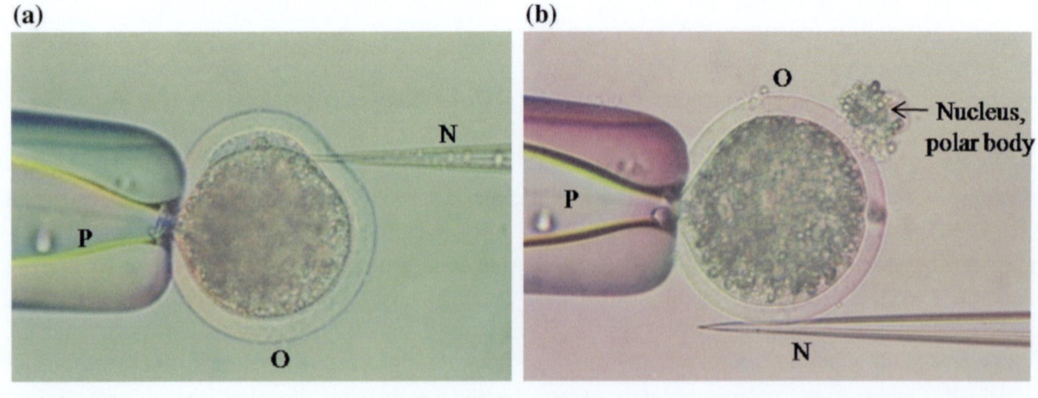

Fig. 10.2 Microsurgical removal of genetic materials (polar body and nucleus) from IVM and denuded bubaline oocyte (O). P-holding capillary pipette, N- cutting needle. The mitochondrial genome is not removed in above technique; hence it inherits mitochondrial genome to cloned offspring

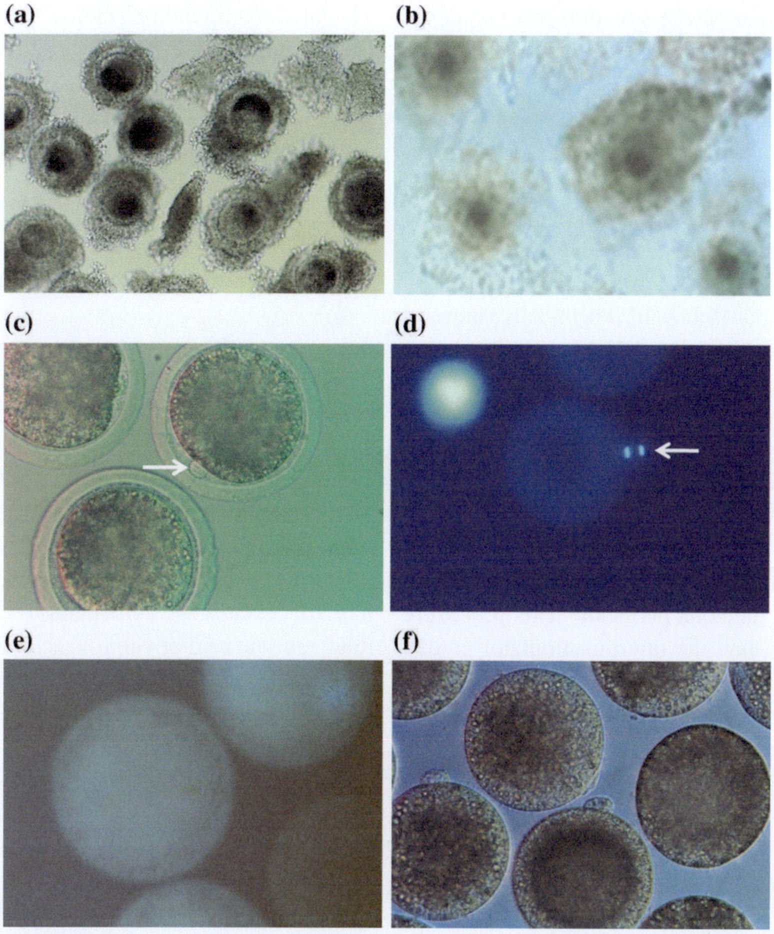

Fig. 10.3 Producing cytoplasts from the oocytes. In some species, such as pigs and buffaloes, the nuclear chromatin is embedded into the ooplasm. Demecolcine-induced extrusion of nuclear materials facilitates removal of chromatic materials. In zona-free bubaline oocytes, the extrusion cones were found to be loosely associated with the ooplasm: **a** immature buffalo oocytes; **b** IVM oocytes; **c** demecolcine-treated zona-included; and **d** zona-free IVM oocytes; **e** Zona included; and **f** zona-free cytoplasts

of hormonal injections. Often overlooked in SCNT, yet the quality of oocyte plays a critical role in the activation and subsequent development of the embryos. Variable success rates are reported using oocytes from abattoir-derived ovaries or from oocytes aspirated from ovaries of live animals. Collecting oocytes from small animals such as mice and rats is difficult due to small size of these species. Oocytes are collected after euthanizing them. In general, the oocytes or cumulus–oocyte complexes (COCs) (Fig. 10.3) with intact multilayer of cumulus cells and homogeneous ooplasm are preferred.

10.8 Enucleation

Enucleation means removal of nuclear genetic material from oocytes. The objective of enucleation is to maintain diploid status of the organisms to be cloned. Cytoplasts are prepared by removing the nuclear materials of in vitro matured oocytes. There are different modifications of the original method of enucleation of the oocytes.

Nevertheless, the oocyte lipid contents hinder removal of nuclear chromatin from IVM oocytes. Treating oocytes with certain chemicals such as demecolcine assists removal of nuclear genetic materials, thereby facilitating the preparation of cytoplasts (Vajta et al. 2005; Tani et al. 2006). The recent approaches of chemically induced enucleation are based on the phenomenon that treatment of M-II phase with some cytoskeleton-relaxing agents induces an extrusion cone over the surface of the oocyte cytoplasm, which can be removed mechanically (Hosseini et al. 2013).

In HMC, the cytoplasts are prepared by manual bisection of IVM oocytes after enzymatic hydrolysis of the zona pellucida. This procedure is convenient than the micromanipulator-guided enucleation (Table 10.1).

Enucleation is a tricky effort. Whereas one has to completely remove nuclear genetic materials including polar body, the cytoplast integrity has to be maintained. For this, sometimes oocytes are treated with chemicals such as demecolcine to facilitate removal of nuclear genetic material (Baguisi and Overstrom 2000; Gasparrini et al. 2003; Singh 2006; Tani et al. 2006; Shah et al. 2008). Care is taken that competence of oocyte is maintained to possible extent.

A calf was born from cloned embryos developed from cytoplasts developed with the help of demecolcine, implying that chemically assisted removal of genetic materials does not affect the quality of embryos (Tani et al. 2006). In addition, chemically induced enucleation is gainful as it minimizes the loss of cytoplast during enucleation and avoids the use of UV illumination which impairs competence of oocyte to develop into embryo. Cleavage rates of 63.6–70.0% and blastocyst yield of 15.5–24.2%, without significant differences between methods used to produce bovine embryos indicate that demecolcine-induced enucleation could be used to produce cloned embryos (Saraiva et al. 2015). The micromanipulation studies carried out in sika deer oocytes (Yin et al. 2014) and sheep oocytes showed that demecolcine-assisted enucleation might represent an alternative approach, especially for less skilled operators involved in nuclear transfer cloning (Iusdo et al. 2013).

10.9 Selection of Donor Nuclei

Selection of donor nucleus is an important criterion that decides the efficiency of embryo cloning. Theoretically, any somatic cell can be used as donor nucleus. The cells have to be cultured and prepared specifically before fusion with cytoplast. The treatments include serum starvation, i.e., minimizing the quantity of FBS in culture medium, treatment with certain agents like trichostatin A, cytochalasin B, or roscovitine, retinoic acid or valproic acid. However, one treatment is not applicable for all cells or species. Practically, some cells are more suitable donor nuclei. Most studies have used fibroblasts as donor nuclei.

10.10 Electrofusion of the Donor Nucleus to the Cytoplast

The electrofusion is an important technical step. Specialized equipment, the electrocell manipulator or electroporator is used to generate an AC wave for benign dielectrophoretic alignment of the cell cytoplast. The couplets are held between two electrodes of cell fusion chamber. The care has to be taken so that the cell cytoplasts are in a plane perpendicular to AC waves. With only microsecond switch-over time from AC to DC, the cell cytoplast get fused. After fusion, AC re-application keeps the cells together for the rounding-off process. The direct current (DC) square pulse may also be used as a stand-alone powerful electroporation system with a wide range of voltages and pulse lengths. The electrofusion parameters should be standardized separately different sets of experiments.

10.11 In Vitro Culture of the NT Embryos

It is necessary to exclude quiescent couplets and allow the alive embryos to grow till blastocyst stage. The hatching cloned embryos (Fig. 10.4) are either preserved by cryopreservation, or used for transfer into estrus-synchronized surrogate females. The process needs a suitable combination of culture media, supplements, and culture conditions.

10.12 Estrus Synchronization and Transfer of the Cloned Embryos

Synchronization of estrus refers to the process of scheduling the recipient females to come into estrus at approximately the same time as donor animals which have been superovulated with exogenous drugs administration. The surrogate mothers should be normal and healthy females.

Prostaglandin PGF-2α (InterVet France) and combinations of prostaglandin and gonadotropins are used to synchronize the estrus. The embryos are then transferred to surrogate mothers, but post-implantation embryo development to term, birth and survival of the cloned offspring depends on various factors.

10.13 Animals Cloned so Far

The list of animals cloned by SCNT or the modified techniques is increasing (Table 10.2). The general observation is that nuclear transfer cloning efficiency is very low due to one or other reasons. The low overall efficiency might not be an intrinsic feature of animals cloning. The selection of right procedure, viz. mechanical micromanipulation, suitable culture media, treatment of donor nuclei, post-electrofusion activation of cloned embryos, are the important factors (Singh et al. 2009; Cordova et al. 2017; Cortez et al. 2018). For instance, initially we had

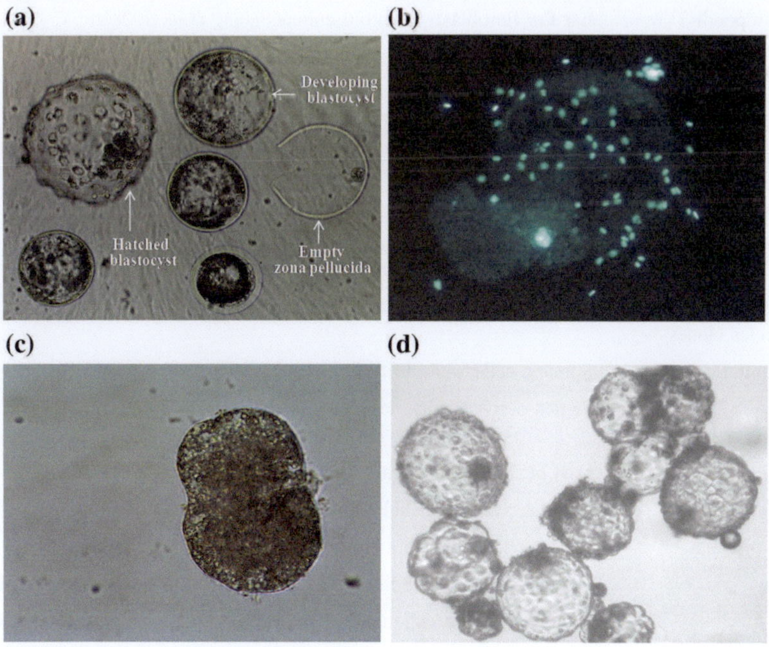

Fig. 10.4 Embryos at various stages of development in vitro. **a** morulae and hatched zona-included cloned embryos; **c** two-cell-stage zona-free HMC cloned embryo; **d** zona-free embryos; **b** the blastomeres viewed after staining with Hoeschest 32342 and viewed under UV illumination

Table 10.2 Species cloned using somatic cell nuclear transfer and modified methods of nuclear transfer cloning

Year	Species	Donor nuclei (References)
1997	Sheep (Dolly) Adult	Fibroblasts (Wilmut et al. 1997)
	Sheep (Transgenic)	Fetal fibroblasts (Schnieke et al. 1997)
1998	Cattle (transgenic)	Fetal fibroblasts (Cibelli et al. 1998)
	Cattle	Adult (Kato et al. 1998)
	Mouse	Adult (Wakayama et al. 1998)
1999	Mouse	Embryo (Wakayama et al. 1999)
	Goat	Fetal (Baguisi et al. 1999)
2000	Pig	Adult (Polejaeva et al. 2000)
	Gaur	Adult (Lanza et al. 2000)
	Mice	Adult (Wakayama et al. 2000)
2001	Mouflon	Adult (Loi et al. 2001)
2002	Cat	Adult (Shin et al. 2002)
	Rabbit	Adult (Chesné et al. 2002)
2003	Rat	Fetal (Zhou et al. 2003)
	Mule	Foetal (Woods et al. 2003)
	Horse	Adult (Galli et al. 2003)
2005	Dog	Adult (Lee et al. 2005)
2006	Ferret	Adult (Li et al. 2006a)
2007	Buffalo	Adult (Shi et al. 2007)
	Deer	Red deer from antler stem cells (Berg et al. 2007)
2008	Dog (toy poodle)	Ear Ear fibroblasts from old animals (Jang et al. 2008)
	Buffalo	Birth of water buffalo calf (Shah et al. 2009)
2009	Camel	Fetal fibroblasts (Wani et al. 2010)
	Dog	Fetal fibroblasts (Hong et al. 2009)
2011	Cattle	Somatic cells (De Bem et al. 2011)
	Anatolian grey cattle	Various somatic cells (Arat et al. 2011)
2012	Endangered grey wolf (Canis lulus)	iSCNT by using somatic cells from dead wolf, and oocytes from bitch. The bitch used as surrogate mothers (Oh et al. 2008)
	Gaur-bovine interspecies clones	Somatic cells (Srirattana et al. 2012)
	Water buffalo	Somatic cells, and vitrified-warmed HMC embryos (Saha et al. 2013)
2014	*B. bubalis, Thai swamp buffalo*	Semen-derived epithelial cells (Selokar et al. 2014)
		First birth of cloned Thai swamp buffalo (Tasripoo et al. 2014)
2015	*B. bubalis*	Urine-derived epithelial cells (Madheshiya et al. 2015)
	Korean pig	Ear fibroblasts (Hwang et al. 2015)
	Horse	Dermal fibroblasts and oocytes aspirated from live animals (Lee et al. 2015)
	Mice	Matured neurons (Mizutani et al. 2015)
2016	Pig, and sheep	Generation of cloned genome-edited transgenic pigs produced from SCNT followed by CRISPR/Cas9 microinjection (Sheets et al. 2016), cloned myostatin (MSTN) biallelic-knock out (KO), small-tailed Han sheep from transgenic fetal

(continued)

Table 10.2 (continued)

Year	Species	Donor nuclei (References)
		fibroblasts (Li et al. 2016), generation of biallelic KO sheep via gene editing and nuclear transfer cloning (Li et al. 2016)
2017	Various species	Cattle from fibroblasts (Wagner et al. 2017), camel (*C. bactreanus*) using dromedary camel as source of oocytes as well as surrogate female (Wani et al. 2017), apancreatic pigs from porcine fetal fibroblasts (Kang et al. 2017), cloned mice from proestrus, metestrus, and diestrus cells from vaginal smear (Kuwayama et al. 2017), hyper-reactive antigen-specific cloned mice (Kaminuma et al. 2017), transgenic cashmere goat from fibroblasts using piggyBac transposition system, and producing transgenic cloned goats from these cells (Bai et al. 2017)
2018	Various species	Cynomolgus monkeys (*Macaca fascicularis*), fetal fibroblasts (Liu et al. 2018), cattle using skin fibrobasts (Cortez et al. 2018), birth of viable cloned foals using bone marrow-derived MSCs as donor cells (Olivera et al. 2018), cumulus cells and fibroblasts (Azuma et al. 2018)

Table 10.3 A summary of suggestions to improve SCNT for applications in livestock

1. HMC might be efficient, and convenient rather than using traditional SCNT
2. Studies should focus on identifying molecular markers as criteria for selection of somatic cells as donor nuclei
3. Somatic cell banks of the candidate donor nuclei should be established to increase the efficiency-developing embryos
4. Emphasis should be on identifying best culture medium and supplements for improving the pre-implantation developmental competence of the cloned embryos
5. Proper healthcare of neonate cloned calves should be followed

faced difficulties in improving the SCNT efficiency in buffaloes and envisaged that it might be due to the lack of appropriate culture medium (Singh et al. 2009). Soon after shifting to another cell culture medium, the embryo development was improved, and first cloned buffalo calf was born (Shah et al. 2008, 2009). A number of cloned water buffalo calves were reported after this achievement. The suggestions to improve the efficacy of NT cloning are summarized in Tables 10.3 and 10.4.

The resurrection through cloning of proven breeding bulls that have died a long ago, by producing live offspring using somatic cells isolated from cryopreserved semen as donor cells, and birth of another cloned calf from urine-derived epithelial cells are impressive examples of HMC in buffaloes (Saini et al. 2018).

10.14 Outlook and Challenges

The Roslin Institute, at University of Edinburg, Scotland, has actually paved the way to produce cloned animals from mature somatic cells. Earlier, prior to the birth of Dolly, the scientists used to clone animals by embryo splitting or by using blastomeres to develop embryo.

Despite the fact that animal cloning is relevant to biomedical, developmental biological, and agricultural significance, the two decades of the techniques to propagate precious livestock can hardly be regarded as triumphant. In spite of massive scientific inputs and experience of cloning of around 25 mammalian species, the efficacy of SCNT is still low, whereas expertise and cost involved is high.

Table 10.4 Summary of the technical recommendations to improve efficacy of NT cloning

Activity	Explanation and inferences (reference)
Donor nuclei	Selection of the somatic cells as candidate donor nuclei; treatment of donor nuclei with pronase to achieve higher electrofusion efficiency; treating the donor nuclei with trichostatin A (Cervera et al. 2009), and roscovitine; exploring the use of pluripotent stem cells as donor nuclei Addition of chromatin remodeling factors to the oocyte to facilitate reprogramming of the cell genome (Betthauser et al. 2006) Manual bisectioning or separation of nuclear material to produce cytoplasts
Enucleation	Using handmade cloning instead of SCNT to achieve higher volume of cytoplast, minimizing the exposure to hazardous UV irradiation, increased rate of somatic cell fusion to the cytoplast within minimum time period
Electrofusion and activation	Using an appropriate combination of chemical agents for activation of the electrofused embryos; using appropriate culture medium (Singh et al. 2009); exploring alternative options for improving nuclear reprogramming (Bui et al. 2008)
Improved embryo-culture	Novel *in vitro* culture system for improving pre-implantation embryo developments, e.g., microfluidic in vitro cultivation system; Well-of-the-Well system (Vajta et al. 2005, 2008)

The most significant contribution of the NT cloning was the demonstration that adult cell genome could be reprogrammed to pluripotent embryonic state. Till date, plenty of work is carried out to reveal the plasticity of nuclear genome. On the other hand, misunderstandings and some accusations, uncorroborated fear and administrative obstacles, leading to a fiasco have discouraged the cloners to continue the tactics and overcome the associated challenges.

Widely reported health problems in cloned offspring include abnormalities of general and specific tissue growth both in utero and after birth, and impaired immunity. Abnormal expression of imprinted genes is the major cause of underlying problems. Strategies, such as selection and pre-treatment of donor nuclei, developing culture media separately for embryo development in each animal species, may hopefully improve the efficiency of animals cloning. There is need to understand the nutrition requirements of the newly constructed or cloned embryos that undergo altered reprogramming. Real-time monitoring of gene expression in pre-implantation embryos in response to nutrients supplements in cloned embryos could be helpful in developing nutritional interventions to improve the efficiency of cloning.

Concerning livestock productivity, the technology is aimed to disseminate the reproductive potential of the stud bulls, allowing breeders to keep up with the demand for semen, oocytes, and embryos. In addition, the technique heralds a new era through genomic reprogramming in the field of therapeutic cloning and regenerative medicine.

10.15 Conclusions

Nuclear transfer cloning is a valuable tool in strategic breeding programs because it can more quickly and precisely extend elite genotypes for semen production, increased breeding opportunities, and protection from untimely loss of superior animals. By cloning most elite animals with the most sought-after genetics or a higher disease-resistance, the livestock breeders can advance the quality and consistency of their livestock products. Due to complexity of the steps involved in reconstructing oocytes by nuclear transfer, novel protocols should be developed to minimize the damages inflicted during micromanipulation. Efficiency of animal cloning can possibly be improved by using optimized operational procedures, including treating the donor cells, appropriate activation of cloned couplets, use of suitable culture media for pre-implantation embryo development stage, and proper neonate health care.

References

Arat S, Caputcu AT, Akkoc T, Pabuccuoglu S, Sagirkaya H, Cirit U, Nak Y, Koban E, Bagis H, Demir K, Nak D, Senunver A, Kilicaslan R, Tuna B, Cetinkaya G, Denizci M, Aslan O (2011) Using cell banks as a tool in conservation programmes of native domestic breeds: the production of the first cloned Anatolian Grey cattle. Reprod Fertil Dev 23(8):1012–1023. https://doi.org/10.1071/RD11026

Azuma R, Miyamoto K, Oikawa M, Yamada M, Anzai M (2018) Combinational treatment of trichostatin A and vitamin C improves the efficiency of cloning mice by somatic cell nuclear transfer. J Vis Exp 26(134). https://doi.org/10.3791/57036

Baguisi A, Behboodi E, Melican DT, Pollock JS, Destrempes MM, Cammuso C, Williams JL, Nims SD, Porter CA, Midura P, Palacios MJ, Ayres SL, Denniston RS, Hayes ML, Ziomek CA, Meade HM, Godke RA, Gavin WG, Overström EW, Echelard Y (1999) Production of goats by somatic cell nuclear transfer. Nat Biotechnol 17(5):456–461

Baguisi A, Overstrom EW (2000) Induced enucleation in nuclear transfer procedures to produce cloned animals. Theriogenology 54:209 (Abstract)

Bai DP, Yang MM, Qu L, Chen YL (2017) Generation of a transgenic cashmere goat using the piggyBac transposition system. Theriogenology 15(93):1–6. https://doi.org/10.1016/j.theriogenology.2017.01.036 (Epub 2017 Jan 23)

Berg DK, Li C, Asher G, Wells DN, Oback B (2007) Red deer cloned from antler stem cells and their differentiated progeny. Biol Reprod 77(3):384–394 (Epub 2007 May 23)

Betthauser JM, Pfister-Genskow M, Xu H, Golueke PJ, Lacson JC, Koppang RW, Myers C, Liu B, Hoeschele I, Eilertsen KJ, Leno GH (2006) Nucleoplasmin facilitates reprogramming and in vivo development of bovine nuclear transfer embryos. Mol Reprod Dev 73:977–986

Booth PJ, Holm P, Callesen H (2001) Application of the zona-free manipulation technique to porcine somatic cell nuclear transfer. Cloning Stem Cells 4:191–197. https://doi.org/10.1089/15362300152725909

Bui H, Wakayama S, Kishigami S, Kim J, van Thuan N, Wakayama T (2008) The cytoplasm of mouse germinal vesicle stage oocytes can enhance somatic cell nuclear reprogramming. Development 135:2935–2945

Cervera RP, Marti-Gutierrez N, Escorihuela E, Moreno E, Stojkovic M (2009) Trichostatin A affects histone acetylation and gene expression in porcine somatic cell nuclear transfer embryos. Theriogenology 72:1097–1110

Chesné P, Adenot PG, Viglietta C, Baratte M, Boulanger L, Renard JP (2002) Cloned rabbits produced by nuclear transfer from adult somatic cells. Nat Biotechnol 20(4):366–369

Cibelli JB, Stice SL, Golueke PJ, Kane JJ, Jerry J, Blackwell C, Ponce de León FA, Robl JM (1998) Cloned transgenic calves produced from nonquiescent fetal fibroblasts. Science 280(5367):1256–1258

Cordova A, King WA, Mastromonaco GF (2017) Choosing a culture medium for SCNT and iSCNT reconstructed embryos: from domestic to wildlife species. J Anim Sci Technol 10(59):24. https://doi.org/10.1186/s40781-017-0149-1 (eCollection 2017. Review)

Cortez JV, Vajta G, Valderrama NM, Portocarrero GS, Quintana JM (2018) High pregnancy and calving rates with a limited number of transferred handmade ClonedBovine embryos. Cell Reprogram 20(1):4–8. https://doi.org/10.1089/cell.2017.0024

De Bem TH, Chiaratti MR, Rochetti R, Bressan FF, Sangalli JR, Miranda MS, Pires PR, Schwartz KR, Sampaio RV, Fantinato-Neto P, Pimentel JR, Perecin F, Smith LC, Meirelles FV, Adona PR, Leal CL (2011) Viable calves produced by somatic cell nuclear transfer using meiotic-blocked oocytes. Cell Reprogram 13(5):419–429. https://doi.org/10.1089/cell.2011.0010 (Epub 2011 Jul 8)

Galli C, Lagutina I, Crotti G, Colleoni S, Turini P, Ponderato N, Duchi R, Lazzari G (2003) Pregnancy: a cloned horse born to its dam twin. Nature 424 (6949):635 (Erratum in: Nature. 2003 Oct 16;425 (6959):680)

Gasparrini B, Gao S, Ainslie A, Fletcher J, McGarry M, Ritchie WA, Springbett AJ, Overstrom EW, Wilmut I, De Sousa PA (2003) Cloned mice derived from embryonic stem cell karyoplasts and activated cytoplasts prepared by induced enucleation. Biol Reprod 68:1259–1266

Hong SG, Jang G, Kim MK, Oh HJ, Park JE, Kang JT, Koo OJ, Kim DY, Lee BC (2009) Dogs cloned from fetal fibroblasts by nuclear transfer. Anim Reprod Sci 115(1–4):334–339. https://doi.org/10.1016/j.anireprosci.2008.12.005 (Epub 2008 Dec 6)

Hosseini SM, Hajian M, Moulavi F, Asgari V, Forouzanfar M, Nasr-Esfahani MH (2013) Cloned sheep blastocysts derived from oocytes enucleated manually using a pulled pasteur pipette. Cell Reprogram 15 (1):15–23. https://doi.org/10.1089/cell.2012.0033

Hwang IS, Kwon DJ, Oh KB, Ock SA, Chung HJ, Cho IC, Lee JW, Im GS, Hwang S (2015) Production of cloned Korean native pig by somatic cell nuclear transfer. Dev Reprod 19(2):79–84. https://doi.org/10.12717/DR.2015.19.2.079

Iuso D, Czernik M, Zacchini F, Ptak G, Loi P (2013) A simplified approach for oocyte enucleation in mammalian cloning. Cell Reprogram 15(6):490–494. https://doi.org/10.1089/cell.2013.0051 (Epub 2013 Nov 12)

Jang G, Hong SG, Oh HJ, Kim MK, Park JE, Kim HJ, Kim DY, Lee BC (2008) A cloned toy poodle produced from somatic cells derived from an aged female dog. Theriogenology 69(5):556–563. https://doi.org/10.1016/j.theriogenology.2007.11.002 (Epub 2008 Feb 19)

Kaminuma O, Katayama K, Inoue K, Saeki M, Nishimura T, Kitamura N, Shimo Y, Tofukuji S, Ishida S, Ogonuki N, Kamimura S, Oikawa M, Katoh S, Mori A, Shichijo M, Hiroi T, Ogura A (2017) Hyper-reactive cloned mice generated by direct nuclear transfer of antigen-specific $CD4_+$ T cells. EMBO Rep 18(6):885–893. https://doi.org/10.15252/embr.201643321 (Epub 2017 May 2)

Kang JD, Kim H, Jin L, Guo Q, Cui CD, Li WX, Kim S, Kim JS, Yin XJ (2017) Apancreatic pigs cloned using Pdx1-disrupted fibroblasts created via TALEN-mediated mutagenesis. Oncotarget 8(70):115480–115489. https://doi.org/10.18632/oncotarget.23301 (eCollection 2017 Dec 29)

Kato Y, Tani T, Sotomaru Y, Kurokawa K, Kato J, Doguchi H, Yasue H, Tsunoda Y (1998) Eight calves cloned from somatic cells of a single adult. Science 282(5396):2095–2098

Kuwayama H, Tanabe Y, Wakayama T, Kishigami S (2017) Birth of cloned mice from vaginal smear cells after somatic cell nuclear transfer. Theriogenology 94:79–85. https://doi.org/10.1016/j.theriogenology.2017.02.012 (Epub 2017 Feb 20)

Lanza RP, Cibelli JB, Diaz F, Moraes CT, Farin PW, Farin CE, Hammer CJ, West MD, Damiani P (2000) Cloning of an endangered species (Bos gaurus) using interspecies nuclear transfer. Cloning 2(2):79–90

Lee BC, Kim MK, Jang G, Oh HJ, Yuda F, Kim HJ, Hossein MS, Kim JJ, Kang SK, Schatten G, Hwang WS (2005) Dogs cloned from adult somatic cells. Nature 436(7051):641 (Erratum in: Nature. 2006 Oct 12;443(7112):649. Shamim, M Hossein [corrected to Hossein, M Shamim]. Nature. 2006 Mar 9;440(7081):164. Nature. 2005 Aug 25;436 (7054):1102)

Lee SE, Hyun H, Park MR, Choi Y, Son YJ, Park YG, Jeong SG, Shin MY, Ha HJ, Hong HS, Choi MK, Im GS, Park EW, Kim YH, Park C, Kim EY, Park SP (2017) Production of transgenic pig as an Alzheimer's disease model using a multi-cistronic vector system. PLoS One 12(6):e0177933. https://doi.org/10.1371/journal.pone.0177933 (eCollection 2017)

Lee W, Song K, Lee I, Shin H, Lee BC, Yeon S, Jang G (2015) Cloned foal derived from in vivo matured horse oocytes aspirated by the short disposable needle system. J Vet Sci 16(4):509–516. https://doi.org/10.4142/jvs.2015.16.4.509

Li H, Wang G, Hao Z, Zhang G, Qing Y, Liu S, Qing L, Pan W, Chen L, Liu G, Zhao R, Jia B, Zeng L, Guo J, Zhao L, Zhao H, Lv C, Xu K, Cheng W, Li H, Zhao HY, Wang W, Wei HJ (2016) Generation of biallelic knock-out sheep via gene-editing and somatic cell nuclear transfer. Sci Rep 22(6):33675. https://doi.org/10.1038/srep33675

Li Z, Sun X, Chen J, Liu X, Wisely SM, Zhou Q, Renard JP, Leno GH, Engelhardt JF. Cloned ferrets produced by somatic cell nuclear transfer. Dev Biol 293(2):439–448 (Epub 2006 Apr 3) (May 15 2006b)

Liu Z, Cai Y, Wang Y, Nie Y, Zhang C, Xu Y, Zhang X, Lu Y, Wang Z, Poo M, Sun Q (2018) Cloning of macaque monkeys by somatic cell nuclear transfer. Cell 172(4):881–887.e7. https://doi.org/10.1016/j.cell.2018.01.020 (Epub 2018 Feb 1)

Loi P, Ptak G, Barboni B, Fulka J Jr, Cappai P, Clinton M (2001) Genetic rescue of an endangered mammal by cross-species nuclear transfer using post-mortem somatic cells. Nat Biotechnol 19(10):962–964

Madheshiya PK, Sahare AA, Jyotsana B, Singh KP, Saini M, Raja AK, Kaith S, Singla SK, Chauhan MS, Manik RS, Palta P (2015) Production of a cloned buffalo (*Bubalus bubalis*) calf from somatic cells isolated from urine. Cell Reprogram 17(3):160–169. https://doi.org/10.1089/cell.2014.0097

Mizutani E, Oikawa M, Kassai H, Inoue K, Shiura H, Hirasawa R, Kamimura S, Matoba S, Ogonuki N, Nagatomo H, Abe K, Wakayama T, Aiba A, Ogura A (2015) Generation of cloned mice from adult neurons by direct nuclear transfer. Biol Reprod 92(3):81. https://doi.org/10.1095/biolreprod.114.123455 (Epub 2015 Feb 4)

Oh HJ, Kim MK, Jang G, Kim HJ, Hong SG, Park JE, Park K, Park C, Sohn SH, Kim DY, Shin NS, Lee BC (2008) Cloning endangered gray wolves (*Canis lupus*) from somatic cells collected postmortem. Theriogenology 70(4):638–647. https://doi.org/10.1016/j.theriogenology.2008.04.032 (Epub 2008 Jun 4)

Olivera R, Moro LN, Jordan R, Pallarols N, Guglielminetti A, Luzzani C, Miriuka SG, Vichera G (2018) Bone marrow mesenchymal stem cells as nuclear donors improve viability and health of cloned horses. Stem Cells Cloning 11:13–22. https://doi.org/10.2147/sccaa.s151763 (eCollection 2018)

Polejaeva IA, Chen SH, Vaught TD, Page RL, Mullins J, Ball S, Dai Y, Boone J, Walker S, Ayares DL, Colman A, Campbell KH (2000) Cloned pigs produced by nuclear transfer from adult somatic cells. Nature 407(6800):86–90

Saha A, Panda SK, Chauhan MS, Manik RS, Palta P, Singla SK (2013) Birth of cloned calves from vitrified-warmed zona-free buffalo (Bubalus bubalis) embryos produced by hand-made cloning. Reprod Fertil Dev 25(6):860–865. https://doi.org/10.1071/RD12061

Saini M, Selokar NL, Palta P, Chauhan MS, Manik RS, Singla SK (2018) An update: reproductive handmade cloning of water buffalo (Bubalus bubalis). Anim Reprod Sci pii: S0378-4320(18)30131-3. https://doi.org/10.1016/j.anireprosci.2018.08.003

Saraiva NZ, Oliveira CS, Leal CL, de Lima MR, Del Collado M, Vantini R, Monteiro FM, Niciura SC,

Garcia JM (2015) Chemically induced enucleation of activated bovine oocytes: chromatin and microtubule organization and production of viable cytoplasts. Zygote 23(6):852–862. https://doi.org/10.1017/S0967199414000537

Schnieke AE, Kind AJ, Ritchie WA, Mycock K, Scott AR, Ritchie M, Wilmut I, Colman A, Campbell KH (1997) Human factor IX transgenic sheep produced by transfer of nuclei from transfected fetal fibroblasts. Science 278(5346):2130–2133

Selokar NL, Saini M, Palta P, Chauhan MS, Manik R, Singla SK (2014) Hope for restoration of dead valuable bulls through cloning using donor somatic cells isolated from cryopreserved semen. PLoS One 9 (3):e90755. https://doi.org/10.1371/journal.pone.0090755 (eCollection 2014)

Selokar NL, Sharma P, Krishna A, Kumar D, Kumar D, Saini M, Sharma A, Vijayalakshmy K, Yadav PS (2018) Establishment of a somatic cell Bank for Indian buffalo breeds and assessing the suitability of the cryopreserved cells for somatic cell nuclear transfer. Cell Reprogram 20(3):157–163. https://doi.org/10.1089/cell.2017.0066

Shah RA, George A, Singh MK, Kumar D, Chauhan MS, Manik RS, Palta P, Singla SK (2008) Hand-made cloned buffalo (*Bubalus bubalis*) embryos: comparison of different culture media and culture systems. Cloning Stem Cells 10:435–442

Shah RA, George A, Singh MK, Kumar D, Anand T, Chauhan MS, Manik RS, Palta P, Singla SK (2009) Pregnancies established from handmade cloned blastocysts reconstructed using skin fibroblasts in buffalo (*Bubalus bubalis*). Theriogenology 71:1215–1219

Sheets TP, Park CH, Park KE, Powell A, Donovan DM, Telugu BP (2016) Somatic cell nuclear transfer followed by CRIPSR/Cas9 microinjection results in highly efficient genome editing in cloned pigs. Int J Mol Sci 17(12). pii: E2031

Shi D, Lu F, Wei Y, Cui K, Yang S, Wei J, Liu Q (2007) Buffalos (Bubalus bubalis) cloned by nuclear transfer of somatic cells. Biol Reprod 77(2):285–291 (Epub 2007 May 2)

Shin T, Kraemer D, Pryor J, Liu L, Rugila J, Howe L, Buck S, Murphy K, Lyons L, Westhusin M (2002) A cat cloned by nuclear transplantation. Nature 415 (6874):859 (Epub 2002 Feb 14)

Singh AK (2006) Studies on handmade cloning in buffaloes. M. V. Sc. Thesis submitted to Dairy Science College, National Dairy Research Institute (Deemed University), Karnal, India

Singh B, Chauhan MS, Singla SK, Gautam SK, Verma V, Manik RS, Singh AK, Sodhi M, Mukesh M (2009) Reproductive biotechniques in buffaloes (*Bubalus bubalis*): status, prospects and challenges. Reprod Fertil Dev 21(4):499–510. https://doi.org/10.1071/RD08172 (Review)

Srirattana K, Imsoonthornruksa S, Laowtammathron C, Sangmalee A, Tunwattana W, Thongprapai T, Chaimongkol C, Ketudat-Cairns M, Parnpai R (2012) Full-term development of gaur-bovine interspecies somatic cell nuclear transfer embryos: effect of trichostatin A treatment. Cell Reprogram 14(3):248–257. https://doi.org/10.1089/cell.2011.0099 (Epub 2012 May 11)

Tani T, Shimada H, Kato Y, Tsunoda Y (2006) Demecolcine-assisted enucleation for bovine cloning. Cloning Stem Cells 8:61–66

Tasripoo K, Suthikrai W, Sophon S, Jintana R, Nualchuen W, Usawang S, Bintvihok A, Techakumphu M, Srisakwattana K (2014) First cloned swamp buffalo produced from adult ear fibroblast cell. Animal 8(7):1139–1145. https://doi.org/10.1017/S1751731114001050

Vajta G, Korösi T, Du Y, Nakata K, Ieda S, Kuwayama M, Nagy ZP (2008) The Well-of-the-Well system: an efficient approach to improve embryo development. Reprod Biomed Online 17(1):73–81

Vajta G, Maddox-Hyttel P, Skou CT, Tesirlioglu RT, Peura TT, Lai L, Murphy CN, Prather RS, Kragh PM, Callesen H (2005) Highly efficient and reliable chemically assisted enucleation method for handmade cloning in cattle. Reprod Fertil Dev 17:791–797

Vajta G, Lewis IM, Hyttel P, Thouas GA, Trounson AO (2001) Somatic cell cloning without micromanipulators. Cloning 3:89–95. https://doi.org/10.1089/15204550152475590 [PubMed] [Cross Ref]

Wagner S, Cullum A, Wells DN, Laible G (2017) Taillessness in a cloned cow is not genetically transmitted. Cell Reprogram 19(6):331–336. https://doi.org/10.1089/cell.2017.0018 (Epub 2017 Oct 11)

Wakayama T, Perry AC, Zuccotti M, Johnson KR, Yanagimachi R (1998) Full-term development of mice from enucleated oocytes injected with cumulus cell nuclei. Nature 394(6691):369–374

Wakayama T, Rodriguez I, Perry AC, Yanagimachi R, Mombaerts P (1999) Mice cloned from embryonic stem cells. Proc Natl Acad Sci U S A 96(26):14984–14989

Wakayama T, Shinkai Y, Tamashiro KL, Niida H, Blanchard DC, Blanchard RJ, Ogura A, Tanemura K, Tachibana M, Perry AC, Colgan DF, Mombaerts P, Yanagimachi R (2000) Cloning of mice to six generations. Nature 407(6802):318–319 (No abstract available)

Wani NA, Wernery U, Hassan FA, Wernery R, Skidmore JA (2010) Production of the first cloned camel by somatic cell nuclear transfer. Biol Reprod 82 (2):373–379. https://doi.org/10.1095/biolreprod.109.081083

Wani NA, Vettical BS, Hong SB (2017) First cloned Bactrian camel (*Camelus bactrianus*) calf produced by interspecies somatic cell nuclear transfer: a step towards preserving the critically endangered wild Bactrian camels. PLoS One 12(5):e0177800. https://doi.org/10.1371/journal.pone.0177800 (eCollection 2017)

Wilmut I, Schnieke AE, McWhir J, Kind AJ, Campbell KH (1997) Viable offspring derived from fetal and adult mammalian cells. Nature 385(6619):810–813 (Erratum in: Nature 1997 Mar 13;386(6621):200)

Woods GL, White KL, Vanderwall DK, Li GP, Aston KI, Bunch TD, Meerdo LN, Pate BJ (2003) A mule cloned from fetal cells by nuclear transfer. Science 301(5636):1063 (Epub 2003 May 29. No abstract available)

Wu DY, Zhang X, Miao YL (2018) Reprogramming of aged cells into pluripotent stem cells by nuclear transfer. Methods Mol Biol. https://doi.org/10.1007/7651_2018_118 (Epub ahead of print)

Yin Y, Mei M, Zhang D, Zhang S, Fan A, Zhou H, Li Z (2014) The construction of cloned Sika deer embryos (Cervus nippon hortulorum) by demecolcine auxiliary enucleation. Reprod Domest Anim 49(1):164–169. https://doi.org/10.1111/rda.12246 (Epub 2013 Oct 21)

Zhou Q, Renard JP, Le Friec G, Brochard V, Beaujean N, Cherifi Y, Fraichard A, Cozzi J (2003) Generation of fertile cloned rats by regulating oocyte activation. Science 302(5648):1179 (Epub 2003 Sep 25. No abstract available)

11 Micromanipulation Technology in Health and Assisted Reproduction

Abstract

The research subject, "micromanipulation" covers the equipments, microtools and techniques for manipulation and microsurgery at cell levels. The micromanipulation is an essential process in assisted reproduction viz., treating and improving fertility, in vitro embryo production, nuclear transfer cloning, deriving embryonic cells from embryos, pre-implantation genetic diagnosis, cellular genetic engineering, and gene therapy. Besides, the technology has applications in tumor surgery and neurovascular pathology.

Highlights

- Micromanipulation is an integral component of advanced livestock assisted reproduction
- Besides, the micromanipulation is used in genome editing and genetic engineering.

Keywords

Micromanipulation · Microtools · Microsurgery · Genetic engineering · Nanotechnology-based micromanipulation · Cellular manipulation

11.1 Introduction

Micromanipulation is the process of using microtools and techniques for cell microsurgery, or manipulation to bring alterations at cell, gametes (oocytes or sperm) or embryos levels. Micromanipulation protocols are basic and essential components of research and development, clinical and assisted reproduction (Malter 2016). In addition, disease investigation and treatment, pre-implantation or prenatal genetic diagnosis (PGD) also depend on micromanipulation and microsurgery.

11.2 Equipments and Microtools for Micromanipulation

Microsurgical principles and techniques were introduced into human gynecology in the late 1960s and early 1970s. The tools used for micromanipulation of eggs, sperm, and embryos, and genetic engineering of cells and embryos, include fluorescent inverted microscopes equipped with micromanipulators and operating programs, compound microscopes accompanied essentially with image capturing and processing software, equipments for preparing microtools such as cell holding capillary glass pipettes, capillary microinjectors, capillary suction pipettes to implant cells into, or pull out the cells from developing embryo, electro cell manipulators, Laminar airflow hood, and high-quality CO_2 incubators. These instruments are often present in labs working on assisted reproduction or neurovascular pathology. Additional instruments such as Piezo-micromanipulator and microinjectors improve efficacy of micromanipulation operations.

B. Singh et al., *Advances in Animal Biotechnology*,
https://doi.org/10.1007/978-3-030-21309-1_11

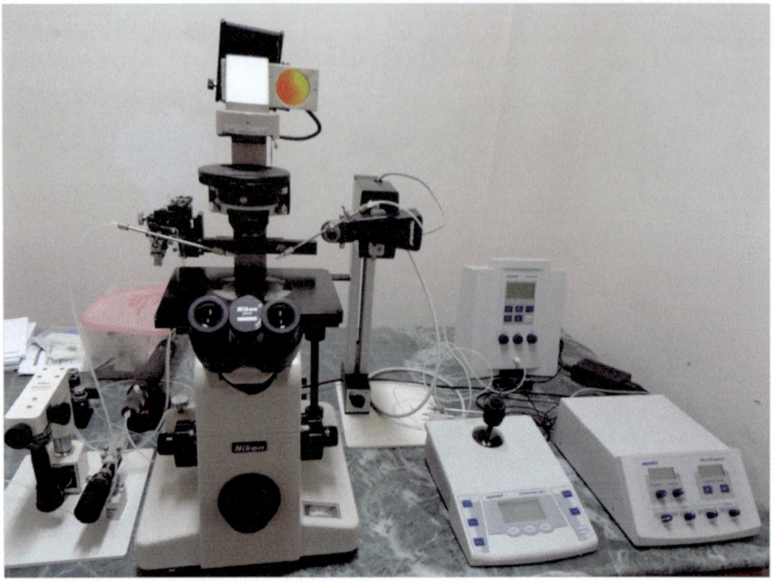

Fig. 11.1 A general purposes cell micromanipulator attached to high-resolution compound microscope

Three types of micromanipulators, based on their mode of operation, namely electronic micromanipulators (e.g., Eppendorf, Hamburg, Germany) (Fig. 11.1), hydraulic micromanipulators (Narishige, Japan), and mechanic micromanipulators (e.g., Leica, GmbH, Germany) are among the most widely used micromanipulators.

11.3 Micromanipulation in Assisted Reproduction

ARTs, especially, nuclear transfer cloning, ICSI, IVF, oocyte and embryo microsurgery and manipulation are supported mostly by micromanipulation. Injecting genes or drugs into cells, oocytes and embryo are finished by microinjection devices.

Remarkable achievements are made by using micromanipulation in the field of assisted reproduction (Fig. 11.2). Sperm fertility, IVF, and oocyte qualities are improved by using micromanipulation techniques. Micromanipulation, such as making a tiny hole in zona pellucida to facilitate adhesion of sperm to zona pellucida, followed by assisted entry into ooplasm of oocyte (Gordon and Talansky 1986), has helped in removing or alleviating the obstacles that hamper reproductive successes.

Non-ejaculated sperm obtained from the testicles or reproductive tract of men who are azoospermic can be used to fertilize normal oocyte by ICSI. Recovery of sperm from epididymis or testes and ICSI are the only hopes for fatherhood in azoospermic males. Thousands of patients have obtained offspring through this way of assisted reproduction.

In certain livestock species also, microholes are drilled into zona pellucida to facilitate entry of sperm into ooplasm to promote fertilization.

11.4 Piezo-Micromanipulator

The Piezo-micromanipulator is used to physical interact with microscopic cells and small samples under inverted microscope and is operated by means of alternating current (AC) pulse to a piezoelectric material.

Due to social reasons, many couples wish to have genetically related offspring. This poses problem when father is unable to produce enough fertile sperm. Urology, gynecology, and fertility laboratory techniques allow microsurgical retrieval of sperm in azoospermic males (Shabataev and Tal 2017).

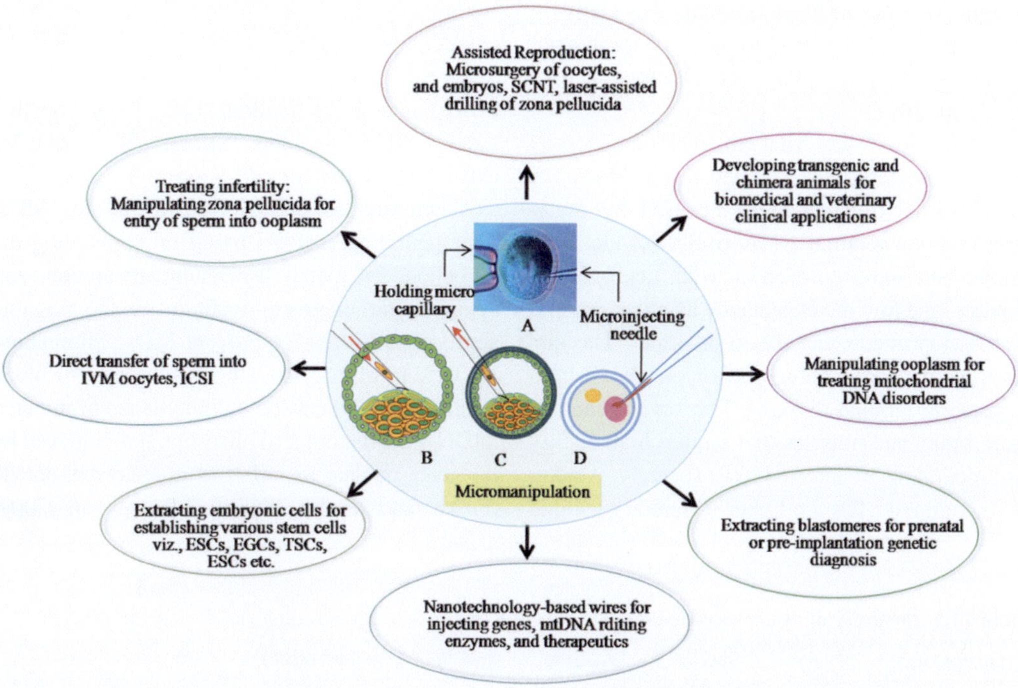

Fig. 11.2 Applications of micromanipulation in ARTs, fertility management, and common biomedical sciences. A—In vitro matured oocyte held in position with the help of microcapillary pipette, being injected genes constructs with the help of microinjection needle; B—pictorial illustration of transplanting exogenous stem cell into zona pellucida-free embryo; C—aspirating or extracting a blastomere from zona-included embryo for PGD or clinical evaluation; D—microinjection of exogenous genes into zygote pronucleus

Some leading scientific companies such as Eppendorf (Hamburg, Germany) are involved in manufacture instruments, accessories, and consumables required of microsurgical operations. The Piezo-micromanipulators, e.g., Eppendorf Transferman NK2 micromanipulator with precision menu control, provide an intuitive operation, enabling users to perform complex micromanipulation procedures with ease and precision.

The programmable FemtoJet series microinjectors (Eppendorf Cell Technology family) are perfectly suited for injecting small and medium range (femtoliters and microliters) volumes into adherent and suspension cells with high precision and repeatability (https://www.meyerinst.com/html/eppendorf/, accessed on Sept. 26, 2018).

As a therapy for treating infertility, the micromanipulation has several success stories. The Piezo-micromanipulators are used in experiments aimed to improve oocyte fertilization as well as cryopreservation. Piezo-micromanipulator assisted zona-thinning and drilling technique was found to enhance assisted hatching in human oocytes (Nakayama et al. 1998) and improved pregnancies and implantation rates in poor-prognosis infertile patients with good-quality embryo for transfer to females (Nakayama et al. 1999). Some examples of micromanipulation-assisted fertilization of oocytes are cited in other chapters in the book.

11.5 Micromanipulation for Drug Delivery into Cells

Oocyte cytoplasm has necessary components to support zygote and embryo development. However, is some rare disease conditions, the oocyte cytoplasm fails to support embryo development. With the help of micromanipulation, it is possible to improve oocytes development, especially in cases when ooplasm or cytoplasm of the oocyte is unable to support embryo development (Fulka and Fulka 2007).

Mitochondria are vital to maintain normal cellular energetic and homeostasis. Many of mitochondrial diseases (MDs) caused by mutations in mitochondrial DNA (mtDNA) are deleterious and have correlation with neurogenetic diseases, and low life expectancy as mitochondrial dysfunction affects any tissue or organ. The diseases can vary from being phenotypically silent to devastating familial diseases. The organs such as heart, brain, and muscles that require high energy for functioning are affected most severely. Among more common MDs are Leber's Hereditary Optic Neuropathy (LHON), Chronic Progressive External Ophthalmoplegia (CPEO), Kearns-Sayre Syndrome (KSS)), Mitochondrial Myopathy (MM), and Mitochondrial Encephalomyopathy, Lactic Acidosis, and Stroke like episodes (MELAS) (Radelfahr and Klopstock 2018).

There are no curative treatments for MDs. Therapeutic approaches based on stimulating the mitochondrial metabolism, supplementation, and symptomatic treatment include are the possible approaches to treat or prevent MDs. Eliminating mtDNA from oocyte is the only option to block familial diseases caused by inheritance of mutated mtDNA from mother to offspring. It is achieved by replacing nuclear materials from abnormal oocyte (Fulka and Fulka 2007; Fulka et al. 2009;

Table 11.1 Summary of examples of micromanipulation in and assisted reproduction and therapeutic sciences

Technique used	Species	Achievement and inferences (reference)
Microinjection	Sheep, pig cattle	Creating laser-assisted hole in zona pellucida, breaking plasma membrane by means of fine glass capillary inserted through hole and injecting the substance of interest back into cytoplasm. The technique found to be promising for injecting different solutions with 100% efficiency, minimal lysis and normal embryonic development (Bogliotti et al. 2016). Pronuclear microinjection procedure delivering CRISPR/Cas9 complex to zygote of hamster to develop GM hamsters (Li et al. 2018)
Gold nanowire injector (NWI)	Mice	Direct delivery of plasmids into pronucleus of mouse zygotes/embryos without using buffer, the technique is more suitable than conventional microinjection (Park et al. 2017)
Au Nanoinjector		Electric pulse-mediated gene delivery into cell, such as neuroblastoma cells, without damage to recipient cell. The technique is said to be applicable to any type of cells (Kang and Kim 2015)
Microneedle	Mouse	Around 100 times reduction in sperm number required for fertilization, the treated and IVF embryos could develop to live offspring when transferred to surrogate females, indicating that drilling is safe and effective for enhancing the efficiency of fertility upon males with low sperm counts (Gordon and Talansky 1986, 1988)
PLGA Microneedles and Fractional Ablative Laser	Pig and human skin	Provided higher transdermal delivery of methotrexate into and across the skin (Nguyen and Banga 2018)
Poly (vinyl alcohol)	Human skin	Needles made by microfolding were sharp, and strong enough to provide significantly higher transdermal delivery of doxorubicin to skin cells (Nguyen et al. 2018)
Sugar (maltose) microneedles	Human skin	The maltose microneedles fabricated by microfolding technique developed for microporation of human cadaver skin was found to be sharp, mechanically uniform, rapidly dissolvable, and promising to increase transdermal drug delivery (Nguyen and Banga 2017)

Abbreviations: PLGA—poly-(D,L-lactide-co-glycolide) acid

Patananan et al. 2016). The approach is principally based on micromanipulation and transfer of oocyte contents, intact meiotic spindle complex, nucleus, or polar body (reviewed in Patananan et al. 2016).

It has been noted that transdermal drug delivery is severely limited by skin that acts as barrier to biological compounds, drugs, and microorganisms. The micron-scale hollow needles (Table 11.1) constructed from various biological polymers and saccharides are found to be a promising technology for enhancing transdermal delivery of therapeutics such as insulin, DNA vaccines (Prausnitz 2004).

The microinjection techniques, albeit difficult and user-dependent operation, are valuable tools for assessing and studying cell gap junctions for developing therapies for multitude of genetic conditions, drug delivery, and cell channels (Fridman et al. 2016).

11.6 Micromanipulation as Tool for Genetic Engineering

The microinjection into zygote stage is most appropriate. The microinjection of genome-editing tools that create genetic modifications, siRNA, mRNAs, or blocking antibodies to investigate gene functions in pre-implantation embryos is pertinent examples. Knowles (1974) have developed a 3.0 μm internal diameter microinjection needle for transferring erythromycin-resistant mitochondria into erythromycin-sensitive *Paramecium aurelia*. Since then, the microinjection is modified for transferring genes, DNA nucleases, mtDNA-editing enzymes, and curative biomolecules into cells, zygotes, and embryos of mice and rats, and livestock species.

A modified microinjection technique was developed to transfer exogenous contents into embryos. In this technique, laser-assisted hole was created into zona pellucida of sheep and pig embryos. Plasma membrane was punctured by means of fine glass capillary inserted through hole and injecting the substances of interest back into cytoplasm. The technique was found to be promising for injecting different solutions with 100% efficiency, minimal lysis and normal embryonic development (Bogliotti et al. 2016).

The microinjection technique has certain disadvantages, such as damage to cell membrane causes mechanical stress and trauma to the cell. Due to repeated use, the microinjecting needle is prone to clogging. Besides, sophisticated instruments and technical skills are involved in it.

11.7 Nanotechnology-Based Micromanipulation and Gene-Delivery Systems

Nanotechnology-based gene-delivery system (Table 11.1) is gaining importance. Compared to animal cells which have cell membrane as barrier, transfer of exogenous materials such as genes is more difficult in zygote because of presence of zona pellucida and cytoplasmic membrane. To overcome the problem, gold (Au) nanoinjector was developed for electric pulse-mediated gene delivery into cells, such as neuroblastoma cells. The injector was safe and found to be applicable to any type of cells (Kang and Kim 2015). Another gene-delivery system, the gold nanowire injector (Au NWI) was developed for direct delivery of plasmids into pronucleus of mouse zygotes or embryos without using buffer. The delivery of plasmid was mediated by a simple electric pulse. The Au NWI offered useful alternative to conventional microinjection in terms of increased gene delivery into zygote and suppressed mosaicism (Park et al. 2017).

Micromanipulation, microinjection, optical tweezers, and confocal microscopy collectively allows quantification of membrane curvature, membrane tension, and surface density of cellular proteins. In some cases, such as traumatic brain and spinal cord injury, the mitochondria get damaged, which leads to cell demise. It is therefore necessary to transplant normal and functional mitochondria into patient cells to prevent cell death. Time-lapse confocal imaging allows examining, tracking, and cellular uptake of exogenously fluorescently tagged mitochondria injected to spinal cord cells (Gollihue et al. 2017).

11.8 Micromanipulation in Livestock Health and Infertility Treatment

Micromanipulation is used for long time in various experiments in veterinary sciences (Markert 1983; Qiu et al. 2016; Sato et al. 2018). Creating cloned transgenic animals totally depends on micromanipulation, such as opening zona pellucida to remove genetic materials to create cytoplasts, transplanting somatic cell into cytoplast to develop totipotent cloned zygote and chimera animals.

Equine blastocysts are susceptible to cryopreservation. Micromanipulation with the Piezo drill and standard sperm injection capillary pipette, and puncturing the capsule caused collapse of blastocyst that increased the vitrification cryosurvival of equine expanded blastocysts (Hinrichs and Choi 2016).

11.9 Outlook and Challenges

The micromanipulation and microsurgery are highly challenging procedures that require high degree of dexterity and manual proficiency. The embryo micromanipulation can be carried out any stage starting from pronuclear stage of zygote to hatching blastocysts.

In recent years, the genetic and molecular markers have become more important for diagnosis of several diseases. The micromanipulation techniques have wide practical applications and have helped us to answer lots of fundamental biological questions, besides their use in assisted reproduction, and generating transgenic cells and animals.

Supported by confocal microscopy and nanoengineering, the microsurgery and micromanipulation have become more competent. The questions, especially complete safety of nano-based micromanipulation must be assured when they are intended for clinical humans.

11.10 Conclusions

Micromanipulation techniques were developed for treating infertility and creating transgenic animals. Later, the techniques were modified for applications such as gene therapy, genetic engineering, and treating infertility. Many of these techniques are successfully used in laboratory and experimental animals. To overcome some limitations associated with microinjection, nanotechnology-based devices are developed, and to some extent, these devices have proved helpful.

References

Bogliotti YS, Vilarino M, Ross PJ (2016) Laser-assisted cytoplasmic microinjection in livestock zygotes. J Vis Exp (116). https://doi.org/10.3791/54465

Fridman MD, Liu J, Sun Y, Hamilton RM (2016) Microinjection technique for assessment of gap junction function. Methods Mol Biol 1437:145–154. https://doi.org/10.1007/978-1-4939-3664-9_10

Fulka H, Fulka J Jr (2007) The use of micromanipulation methods as a tool to prevention of transmission of mutated mitochondrial DNA. Curr Top Dev Biol 77:187–211

Fulka H, Langerova A, Barnetova I, Novakova Z, Mosko T, Fulka J Jr (2009) How to repair the oocyte and zygote? J Reprod Dev 55(6):583–587

Gordon JW, Talansky BE (1986) Assisted fertilization by zona drilling: a mouse model for correction of oligospermia. J Exp Zool 239(3):347–354

Gordon JW, Grunfeld L, Garrisi GJ, Talansky BE, Richards C, Laufer N (1988) Fertilization of human oocytes by sperm from infertile males after zona pellucida drilling. Fertil Steril 50(1):68–73

Gollihue JL, Patel SP, Mashburn C, Eldahan KC, Sullivan PG, Rabchevsky AG (2017) Optimization of mitochondrial isolation techniques for intraspinal transplantation procedures. J Neurosci Methods 1 (287):1–12. https://doi.org/10.1016/j.jneumeth.2017.05.023 (Epub 2017 May 26)

Hinrichs K, Choi YH (2016) Micromanipulation of equine blastocysts to allow vitrification. Reprod Fertil Dev. https://doi.org/10.1071/rd15389

Kang M, Kim B (2015) Au nanoinjectors for electrotriggered gene delivery into the cell nucleus. Methods Mol Biol 1228:55–65. https://doi.org/10.1007/978-1-4939-1680-1_6

Knowles JK (1974) An improved microinjection technique in Paramecium aurelia. Transfer of mitochondria conferring erythromycin-resistance. Exp Cell Res 88 (1):79–87

Li R, Miao J, Fan Z, Song S, Kong IK, Wang Y, Wang Z (2018) Production of genetically engineered Golden Syrian Hamsters by pronuclear injection of the CRISPR/Cas9 complex. J Vis Exp (131). https://doi.org/10.3791/56263

Markert CL (1983) Fertilization of mammalian eggs by sperm injection. J Exp Zool 228(2):195–201

Malter HE (2016) Micromanipulation in assisted reproductive technology. Reprod Biomed Online 32 (4):339–347. https://doi.org/10.1016/j.rbmo.2016.01.012 (Epub 2016 Feb 9, Review)

Nakayama T, Fujiwara H, Tastumi K, Fujita K, Higuchi T, Mori T (1998) A new assisted hatching technique using a piezo-micromanipulator. Fertil Steril 69(4):784–788

Nakayama T, Fujiwara H, Yamada S, Tastumi K, Honda T, Fujii S (1999) Clinical application of a new assisted hatching method using a piezo-micromanipulator for morphologically low-quality embryos in poor-prognosis infertile patients. Fertil Steril 71(6):1014–1018

Nguyen HX, Banga AK (2017) Fabrication, characterization and application of sugar microneedles for transdermal drug delivery. Ther Deliv 8(5):249–264. https://doi.org/10.4155/tde-2016-0096

Nguyen HX, Banga AK (2018) Delivery of methotrexate and characterization of skin treated by fabricated PLGA microneedles and fractional ablative laser. Pharm Res 35(3):68. https://doi.org/10.1007/s11095-018-2369-6

Nguyen HX, Bozorg BD, Kim Y, Wieber A, Birk G, Lubda D, Banga AK (2018) Poly (vinyl alcohol) microneedles: fabrication, characterization, and application for transdermal drug delivery of doxorubicin. Eur J Pharm Biopharm 129:88–103. https://doi.org/10.1016/j.ejpb.2018.05.017 (Epub 2018 May 22)

Park K, Kim KC, Lee H, Sung Y, Kang M, Lee YM, Ahn JY, Lim JM, Kang T, Kim B, Lee EJ (2017) Suppressing mosaicism by Au nanowire injector-driven direct delivery of plasmids into mouse embryos. Biomaterials 138:169–178. https://doi.org/10.1016/j.biomaterials.2017.05.044 (Epub 2017 May 27)

Patananan AN, Wu TH, Chiou PY, Teitell MA (2016) Modifying the mitochondrial genome. Cell Metab 23 (5):785–796. https://doi.org/10.1016/j.cmet.2016.04.004 (Review)

Prausnitz MR (2004) Microneedles for transdermal drug delivery. Adv Drug Deliv Rev 56(5):581–587

Qiu Y, Li C, Zhang S, Yang G, He M, Gao Y (2016) Systemic delivery of artemether by dissolving microneedles. Int J Pharm 508(1–2):1–9. https://doi.org/10.1016/j.ijpharm.2016.05.006 (Epub 2016 May 2)

Radelfahr F, Klopstock T (2018) Diagnostic and therapeutic approaches for mitochondrial diseases. Fortschr Neurol Psychiatr 86(9):584–591. https://doi.org/10.1055/a-0621-9255 (Epub 2018 Sep 24, German)

Sato M, Kosuke M, Koriyama M, Inada E, Saitoh I, Ohtsuka M, Nakamura S, Sakurai T, Watanabe S, Miyoshi K (2018) Timing of CRISPR/Cas9-related mRNA microinjection after activation as an important factor affecting genome editing efficiency in porcine oocytes. Theriogenology 1(108):129–138. https://doi.org/10.1016/j.theriogenology.2017.11.030 (Epub 2017 Nov 26)

Shabataev V, Tal R (2017) Artificial sperm: new horizons in procreation. Rambam Maimonides Med J 8(4). https://doi.org/10.5041/rmmj.10319

Reproduction Advances in Buffaloes 12

Abstract

The buffalo (*Bubalus bubalis*) is a prime livestock species reared for milk, meat, skin, and draught power in Asian and Mediterranean countries. The buffaloes have received the attention of reproductive molecular biologists and livestock researchers. Success has been achieved in nuclear transfer cloning, stem cell biology, genetic engineering and vitrification of gametes and embryos. The research areas which need further scientific inputs include molecular cataloguing and conservation of native breeds, developing transgenic disease-resistant herds and multiplication of the desirable genotypes.

Highlights

- Buffaloes are multipurpose milk and meat producing animals
- The buffaloes are reared in different climates and management conditions.

Keywords

Buffaloes · Milk production · Reproduction biotechniques · Nuclear transfer · Embryo biotechnology · Diversity

12.1 Introduction

12.1.1 Buffaloes as Multipurpose Livestock

Buffaloes are integral part of livestock agriculture in Asia for over 500 years. Among domestic buffaloes (*Bubalus bubalis*- Linnaeus, 1758), namely the river buffalo ($2n$ = 50) and swamp buffalo ($2n$ = 48), the river or water buffalo is an important and multipurpose livestock in Southeast Asia where nearly 97% of the global buffalo population including world's elite buffaloes (e.g., Murrah, Nili-Ravi, Surti, and Jafarabadi, etc.) are domesticated (Figs. 12.1 and 12.2). The importance of buffaloes to dairy and food industry is immense; many countries depend primarily on buffaloes for milk and meat apart from their use for agriculture operation, skin and fertilizer (manure) and employment (Singh et al. 2009). The buffalo contributes more than 50 million tons of milk and 1.43 million tons of meat (http://www-naweb.iaea.org/nafa/aph/stories/2007-india-buffaloes07…4/6/2009).

India alone has 10 recognized breeds of buffaloes including the best germplasm in the form of Murrah and Nili-Ravi buffaloes (Fig. 12.3).

B. Singh et al., *Advances in Animal Biotechnology*,
https://doi.org/10.1007/978-3-030-21309-1_12

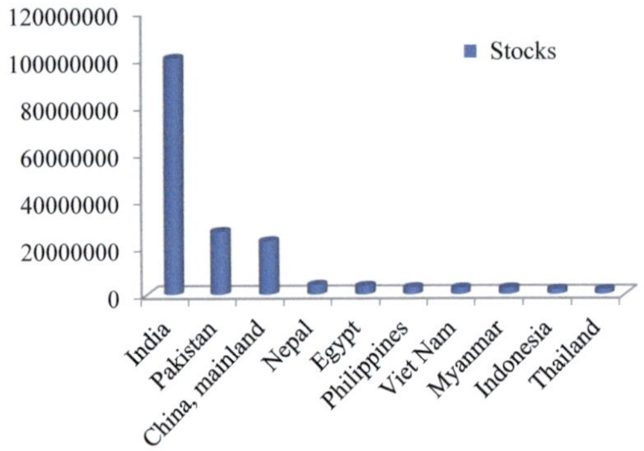

Fig. 12.1 Top ten buffaloes producing countries (Average 1994–2016). *Source* http://www.fao.org/faostat/en/#data/QA/visualize

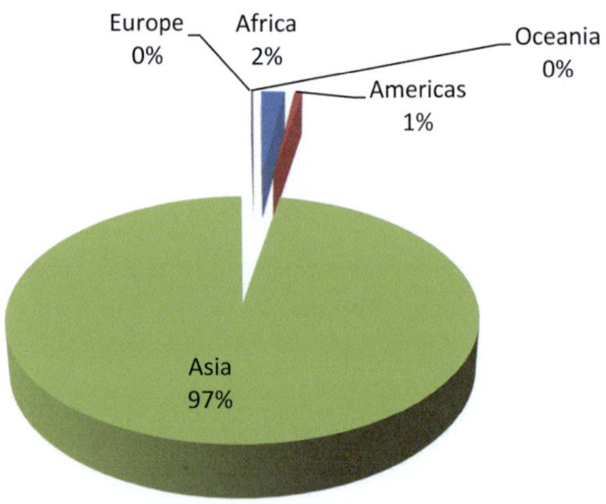

Fig. 12.2 Production share of buffaloes by regions (Average 1994–2016). *Source* http://www.fao.org/faostat/en/#data/QA/visualize

Buffaloes are found in many countries worldwide. Dissemination of buffaloes and increasing population is mainly due to peoples' increasing interest in milk and meat production. Brazil, for instance, is the largest buffalo breeding country outside Asian continent.

The buffaloes capable of consuming 6–30 kg DM roughage per day are important source of rumen microorganisms and microbial hydrolytic enzymes (cellulases, tannase, and phytase, etc.) with potential commercial applications (Singh et al. 2008; Kala et al. 2017). However, despite their evolutionary adaptation to withstand high thermal stress, considerable feed conversion efficiency, the buffalo population is declining due to poor management, emergence of new pests, parasites, and pathogens. Inherent reproductive problems (sub-estrus, seasonal estrus, long calving interval and delayed age at puberty), higher incidences of calf mortality and culling of females which fail to conceive, and lack of due attention to these issues (Nanda and Nakao 2003), are the major factor that cause decline in buffalo population and production (Fig. 12.4).

Buffalo milk is highly suitable for dairy industry for manufacturing ethnic milk products, yogurt, and cottage cheese, khoa and ghee (butter oil). Some breeds such as Bhadawari yield milk with fat contents varying from 6 to 12.5%. Compared to cattle milk, the buffalo milk is rich in calcium and has longer shelf life.

Therefore, it is necessary to evolve strategies to conserve the valuable genotypes that perform better under prevailing low-input conditions. Unique buffaloes, such as Chilika buffaloes are adapted to saline vegetation, and gregarious Toda buffalo are known to survive under hot and high humid climate.

Livestock reproduction technologies (Table 12.1) offer plentiful opportunities for repopulation and conservation of the livestock resources and producing genetically modified or transgenic buffaloes for disease-resistance and commercial-scale production of high biological value foods and therapeutic products.

12.2 Assisted Reproduction in Buffaloes

Assisted reproductive techniques (ARTs), viz., induced superovulation, recovery, and cryopreservation of oocytes, sperm and embryos, IVF, and embryo are essential components of conservation and expanding breeding stocks of livestock (Parnpai et al. 2016; Singh et al. 2016; Finger and Bischof 2018).

Artificial insemination (AI) in buffaloes was the initial innovative ART which contributed to speedy multiplication and grading up of buffaloes. Progress was then made to augmenting female reproduction efficiency through hormone-induced ovarian follicular development, collection of oocytes, evaluation and in vitro maturation, IVF

Fig. 12.3 Types of buffaloes found in North-West Himalayan Region (NWHR): **a**, **b** Gojri buffaloes; **c** non-descript buffalo; **d** Murrah and crossbred buffaloes. Native buffaloes can be maintained under low-input management and graze on poor-quality roughage. *Photo courtesy* Dr. V. Vohra, ICAR-NBAGR, Karnal (India), for providing the image of Gojri buffalo (**b**)

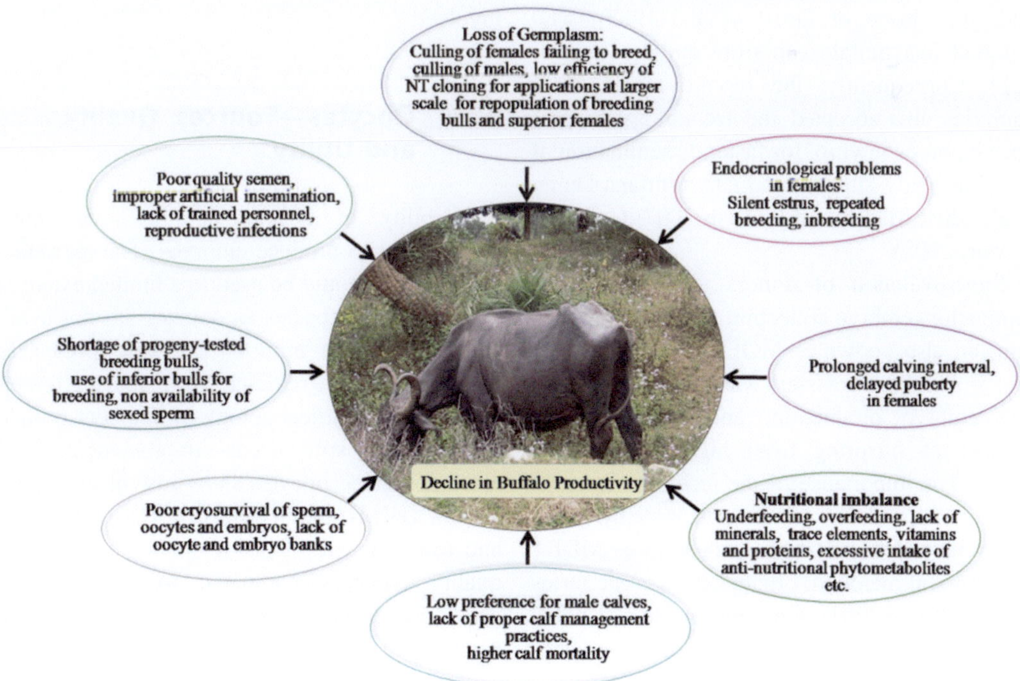

Fig. 12.4 Diagrammatic presentation of factors with adverse effects on buffaloes. Reproductive problems, lack of robust breeding and health problems due to nutritional or climatic stress are the major factors

Table 12.1 Applications of assisted reproduction and genetic engineering tools to augment buffalo production

Technologies	Prospects and applications
Oocytes and embryo banking	Production and multiplication of elite females, producing female IVF embryos using sex-sorted sperm or PGD for dairy, and the males for agricultural operations and meat production Best utilization and conservation of female germplasm (oocyte banking), repopulation of endangered breeds Increasing the genetic pool of *in situ* populations, Conserving genetic diversity and establishing founder population to manage endangered species and breeds
SCNT cloning	Repopulation of elite females, conservation and multiplication of endangered breeds, cloning stud bulls for sperm production Producing isogenic animals for nutrition and immunological and biomedical experiments Derivation of nuclear transfer stem cells for producing in vitro differentiated somatic cells as experimental models for *in vivo* studies, producing transgenic cloned animals
Stem cell technology	Using stem cells in animal cloning, generating iPSCs by cell hybridization, in vitro production of gametes (oocytes or sperms) through guided differentiation of iPSCs into gametes, clinical applications using MSCs

and embryo production, and transferring embryos to establish pregnancies.

12.3 Superovulation, Fertilization, and Embryo Banking

In vitro embryo production (IVEP) and transfer technologies in buffalo began first in the USA with the efforts of Drost et al. (1983), who reported first buffalo calf from embryo transfer (ET). Consequently, the researchers in other countries also accepted the technology to produce embryos in vitro, for faster dissemination of the genetically superior animals (Hufana-Duran et al. 2004; Madan and Prakash 2007; Galli and Lazzari 2008).

Superovulation of donors followed by insemination, embryo collection and evaluation for transfer, also known as MOET (multiple ovulation and embryo transfer) was initiated to expedite embryo production, and increasing the number of offspring from high-yielding buffaloes. The importance of the technique became evident from the work carried out by other researchers, though it was instructed that MOET is successful merely in organized scientific farms (Gandhi et al. 2007). On contrary, it was noted that overall success rate in terms of calves born per superovulation is too low for the MOET to be applied under field conditions (Perera 2008). Whereas MOET had limited success in buffalo, in vitro production of embryos in buffaloes is still low and needs optimization through improving developmental competence of the embryos (Gautam et al. 2008a, b). Further progress in the embryo production may not only lead to faster propagation of selected genotypes, but could also contribute toward establishing related technologies where embryo production is essentially involved.

12.4 Oocytes—Sources, Quality, and Utility

Availability of oocytes is the fundamental requirement to produce embryos. The reproductive potential could be used to a limited extent in most livestock species as oocytes are not available in plenty in most of livestock species. In buffaloes also, retrieval of oocytes, their applications as a technology to augment buffalo reproduction still needs improvement. The abattoir-derived oocytes show low in vitro maturation (IVM) and fertilization (IVF) capacity, and low efficiency to develop into transferable quality embryos compared to the embryos developed from oocytes that are matured and fertilized in vivo (Banwell and Thomson 2008).

The current concept of retrieving oocytes from live animals through transvaginal ovum pick-up (OPU) was based on the findings of Gasparrini (2002), who noted that OPU is a suitable method to produce transferable embryos (TE) per donor on monthly basis. While comparing in vitro developmental competence of the oocytes obtained from live animals, versus oocytes from abattoir-derived ovaries, the OPU-derived oocytes yielded higher rates of blastocysts formation (30.6 ± 4.3 *v.* 18.5 ± 1.8) (Manjunatha et al. 2008).

Collecting oocytes from live animals is technically difficult. Hence, abattoir-derived ovaries remain convenient source of oocytes (Verma et al. 2007; Gupta et al. 2017; Kumar et al. 2018).

12.5 Sex Pre-selection—Semen and Embryo Sexing

Sex selection benefits dairy producers by allowing propagation of elite dams and avoiding birth of male calves that are of less use. The demand for female calves is more, which necessitate the methods to produce female embryos. This is possible either by sexing embryos or using sexed sperm to inseminating females to produce female calves only.

Whereas sex-sorted sperm is available in cattle, efforts are underway to produce sex-sorted sperm in other species. In 2006, Murrah and Nili-Ravi buffalo bull sperm were separated by cytometric sperm-separation (Lu et al. 2006). Female twins were born from IVF-derived embryos using sexed sperm (Lu et al. 2007). Liang et al. (2008) showed that OPU-derived oocytes fertilized with sexed sperm could develop to embryo with efficiency comparable to embryos produced abattoir-derived oocytes fertilized with sexed sperm.

The major drawback of the technique is that sperm-sorting is less efficient, and sexed sperm is more susceptible to cryoinjuries (Lonergan 2007). Flow-cytometry sorting of sperm which is expensive and a proprietary method. The swim-up method was found to be effective to enrich Y-chromosome bearing sperm in Nili-Ravi bulls (Asma-Ul-Husna et al. 2017).

A study reported a 48.5% pregnancy rate in swamp or crossbred (F1 or F2) buffaloes, and 87.6% sex accuracy in the calves produced. Pregnancy rates obtained with sexed sperm were found to be higher (52.5%) from Murrah bull than that obtained with sexed sperm from Nili-Ravi bulls (being 46.1%) (Lu et al. 2015). The sexed semen technique has therefore potential for use under field conditions irrespective of breeds of animals.

However, there is need to develop alternative methods to separate sperm. Immunological inactivation of Y-chromosome bearing sperm, or phytometabolite-mediated inactivation of Y-chromosome could be a promising method, though no such evidence available. DNA amplification method, called loop-mediated isothermal amplification (LAMP) is developed for sexing bubaline embryos and sex chromosomal chimeras (Hirayama et al. 2013).

12.6 Somatic Cell Nuclear Transfer Cloning

The efforts to produce NT cloned bubaline embryos offspring initiated around two decades ago with the inferences that fibroblasts and granulosa cells could be used to produce cloned offspring (Parnpai et al. 2000). Shi et al. (2007), for the first time, reported birth of SCNT buffalo calf. The researchers at ICAR-National Dairy Research Institute, Karnal (India) and ICAR-Central Institute for Research on Buffaloes, Hisar (India) have used hand-guided cloning to clone water buffaloes using various somatic cells such as fibroblasts (Luo et al. 2013; Saini et al. 2015, 2016; Sandhu et al. 2016; Sun et al. 2016), cells from milk (Jyotsana et al. 2015), peripheral blood-derived lymphocytes (Jyotsana et al. 2016), epithelial cells from urine (Madheshiya et al. 2015), freshly collected semen (Selokar et al. 2016), and multipotent amnion mesenchymal stem cells (MSCs) (Sadeesh et al. 2016). Interspecies SCNT (iSCNT) embryos were developed using donor cells from wild buffaloes (*Bubalus arnee*), and cytoplasts from domestic water buffaloes (*B. buablis*) (Priya et al. 2014; Saini et al. 2015).

Cloned embryos were developed from epithelial cells isolated from the cryopreserved semen of four bulls, three bulls of which had already died. Pregnancies were established from cloned embryos. This paved the way to restore precious progeny-tested bulls as well as other mammalian species that are endangered or have become extinct (Selokar et al. 2014). The surviving cloned buffaloes (Fig. 12.5) are apparently healthy, and some of the cloned buffaloes have delivered healthy calf by AI. On the other hand, the cloned bulls are prepared for collecting semen. First cloned calf of Thai swamp buffalo is also reported in 2014 (Tasripoo et al. 2014).

As our understanding of epigenetics and cellular genomic reprogramming are yet partially understood, much remains to be investigated about the fundamental processes and mechanisms that can promote mammalian cloning.

12.7 Stem Cells and Transgenesis

Stem cells have multiple applications in farm animals including large ruminants, pets, and companion animals. Reports are available on a variety of stem cells in buffaloes. Ever since the first report of embryonic stem cells obtained from IVF-derived embryos (Verma et al. 2007), various reports are available on different aspects of stem cells in buffaloes (Sritnaudomchai et al. 2007; Huang et al. 2010; George et al. 2011; Kumar et al. 2014; Zandi et al. 2014; Ghosh et al. 2015) (Table 12.2).

Gene-transfected EG-like cells capable of forming chimeric embryos were reported around

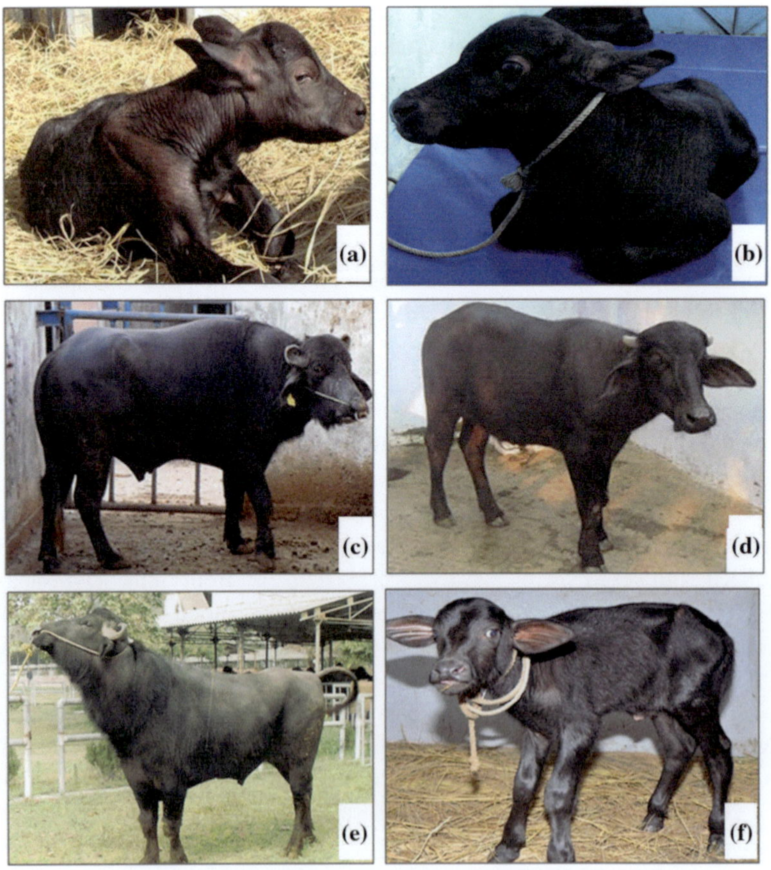

Fig. 12.5 The cloned water buffaloes at ICAR-NDRI, Karnal (India). **a** the first cloned female calf born in 2009, that survived for six days only; **b** a healthy male calf; **c** normal Murrah buffalo bull that served as source of seminal epithelial cells, and the cloned calf (**d**) developed form the bull (**c**); **e** the Murrah bull born on December 10, 1995, was used as source of cells from his cryopreserved semen; **f** the clone of bull (**e**) born on July 23. 2014. *Photo courtesy* Prof. S. K. Singla, ICAR-NDRI Karnal, India

Table 12.2 Milestone achievements in ARTs in buffaloes

Year	Achievements and remarks (reference)
1983	Birth of first buffalo calf from ET (Drost et al. 1983)
1986	Induced superovulation in buffaloes and intergenic embryostransfer involving cattle and buffalo (Drost et al. 1986)
1989	In vitro embryo production cryopreservation (Techakumphu et al. 1989)
1993	Live calf produced from transfer of frozen-thawed embryos (Kasiraj et al. 1993), sex-identification of embryos using polymerase chain reaction (Rao et al. 1993)
1997	Retrieving embryos from live animals by OPU (Galli et al. 1998), micromanipulation of oocytes and embryos (Singla et al. 1997)
1999	SCNT cloned embryos developed in buffaloes (Parnpai et al. 1999)
2001	Development of protocols for activation of SCNT embryos (Kitiyanant et al. 2001)
2003	Pregnancies from transfer of in vitro produced, and vitrified warmed embryos (Neglia et al. 2003)
2004	Vitrification of IVF-derived produced embryos, and births of calves from vitrified-warmed embryos (Hufana-Duran et al. 2004)
2005	Using sexed sperm for insemination (Presicce et al. 2005a), birth of calf from sexed semen in Mediterranean Italian buffaloes (Presicce et al. 2005b)
2006	Sexing of embryos by loop-mediated isothermal amplification (Hirayama et al. 2006)
2007	First birth of SCNT cloned buffalo calf (Shi et al. 2007), births of twin females from sexed sperm (Lu et al. 2007), first report of embryonic stem cell-like cells from IVF-produced embryos (Verma et al. 2007), and parthenogenetically produced embryos (Sritanaudomchai et al. 2007), and fetal embryogenic stem cells (Huang et al. 2007)
2008	Initiation of alternative cloning protocol i.e. HMC in water buffalo (Singh et al. 2008), production of HMC embryos (Shah et al. 2008), and transgenic IVF embryos expressing GFP (Verma et al. 2008), birth of sex-preselected calf (Liang et al. 2008)
2009	Pregnancies established from HMC embryos (Shah et al. 2009)
2010	Birth of sex-preselected buffalo calves (Lu et al. 2010)
2011	Developing cloned and transgenic embryos (George et al. 2011), production of sex-preselected bubaline calves by insemination using sexed sperm; sexing of buffalo embryos (Zoheir and Allam 2011)
2012	Establishing fetal MSCs (Dev et al. 2012)
2013	Live birth of cloned calves from vitrified-warmed HMC embryos (Saha et al. 2013)
2014	First birth of cloned Thai swamp buffalo (Tasripoo et al. 2014), isolation of epithelial cells from long term cryopreserved and frozen-thawed semen, and using the cells to produce cloned embryos, and births of calves from the embryos (Selokar et al. 2014), reprogramming fetal fibroblasts with avian egg extract (Mahapatra and Bag 2014), production of wild buffalo (B. arnee) cloned embryos by iSCNT (Priya et al. 2014), using Trichostatin A as agent to improve development of cloned embryos (Saini et al. 2014, 2016)
2015	Identification of stemness markers in amniotic-derived mesenchymal stem cells (MSCs) (Ghosh et al. 2015), establishing trophectodermal (TE) stem cells from IVF- and HMC-produced embryos (Mohapatra et al. 2015a); production of iSCNT embryos using fibroblasts from wild buffalo (*Bubalus arnee*), and cytoplasts from *B. bubalus* (Saini et al. 2015), comparative evaluation of epigeneyics status and developmental profile of cloned embryos devolved from milk epithelial cells and ear skin fibroblasts (Jyotsna et al. 2015), birth of cloned calf using urine-derived epithelial cells as donor nuclei (Madheshiya et al. 2015), validating the feasibility of use of sexed sperm for AI under field conditions (Lu et al. 2015), birth of first transgenic buffalo calves expressing enhanced green fluorescent protein (EGFP) (Meng et al. 2015), description of developmental competence, and epigenetic status, TSCs (Mohapatra et al. 2015b)
2016	Production of HMC embryos from epithelial cells isolated from a decade old frozen-thawed semen, establishing pregnancies from transfer of the embryos (Selokar et al. 2016), identification of stemness markers in amnion-derived stem cells, and using them as donor nuclei for cloned embryos (Sadeesh et al.

(continued)

Table 12.2 (continued)

Year	Achievements and remarks (reference)
	2016); production of cloned embryos from non viable somatic cells (Duah et al. 2016), and impact of Caspase-9 inhibitor on pre-implantation embryo development (Mullani et al. 2016)
2017	Oocyte-secreted factors found to promote development of buffalo embryos (Gupta et al. 2017)
2018	Establishing somatic cell banks for use in SCNT (Selokar et al. 2018), birth of cloned calves from transfer of embryos to buffaloes under field conditions, development of transgenic embryos expressing Venus fluorochrome (Kumar et al. 2018), births of six transgenic calves expressing EGFP (Lu et al. 2018)

a decade ago (Huang et al. 2008). As the buffaloes milk has high proteins, they may possibly serve as live bioreactors for producing exogenous recombinant therapeutics. While initially, the concern might be the '*biopharming*', larger-scale applications could include production of nutraceuticals or other bioactive molecules.

Innovations in intracellular delivery and integration of exogenous genes into the genome of cells or embryos through highly suitable vectors like baculoviral vectors (Chuang et al. 2007; Lufino et al. 2008; Lo et al. 2009), or multifunctional nanoparticles, liposomes, or biodegradable polymeric vectors may hopefully advance the process of genetic modifications of buffaloes.

Because the developmentally competent oocytes are restricted in number, the robust technique to derive oocytes from stem cells (Nagy et al. 2008) would facilitate investigations on the genetic, epigenetic, and environmental factors affecting embryo development. The ARTs have to be optimized to offer ample opportunities to produce fertilizable oocytes.

12.8 Outlook and Challenges

The buffaloes being efficient utilizers of fibrous roughage and adaptation to prevailing difficult tropical climate are important component of transhumant as well as commercial livestock system. The ban on slaughter of cattle in India has pushed-up domestic demand for buffaloes for buffen or carabeef. Paradoxically, the reproduction efficiency is low owing to physiological and management conditions.

Declining buffalo populations due to culling of heifers and females unable to conceive is a matter of serious concern. The general reproductive, nutritional and health management (Fig. 12.6) should be improved to increase productivity of buffaloes to ensure livelihood security of the livestock owners.

Approaches, such as establishing cell lines from diverse origins, epigenetic modifications of donor cell nuclei, post-fusion activation of cell-cytoplast couplet, and in vitro culturing of embryos at pre-implantation developmental stages have improved nuclear transfer cloning.

A limited success is achieved in producing transgenic buffalo calves. The use of sexed sperm is likely to benefit buffalo production management by producing calves of desired sex males for meat and agriculture operations, and females for milk. Delayed puberty and prolonged calving intervals should be resolved by nutritional, endocrinological, and health management interventions. Therefore, it is necessary to establish economically viable ARTs for applications under field conditions or at farmers' doorstep.

12.9 Conclusions

The buffalo farming has increasing role in national economy. The success achieved in advanced ARTs indicates that there are ample opportunities for proliferating high yielding buffaloes and progeny tested high-quality bulls. Conserving the breeds that are low-yielding, but play important role in resource-poor conditions has practical applications. Molecular biological and genomics methods should be the prioritized strategies to gain insights into biology of complex traits, and overcoming the key problems currently associated with advanced ATRs.

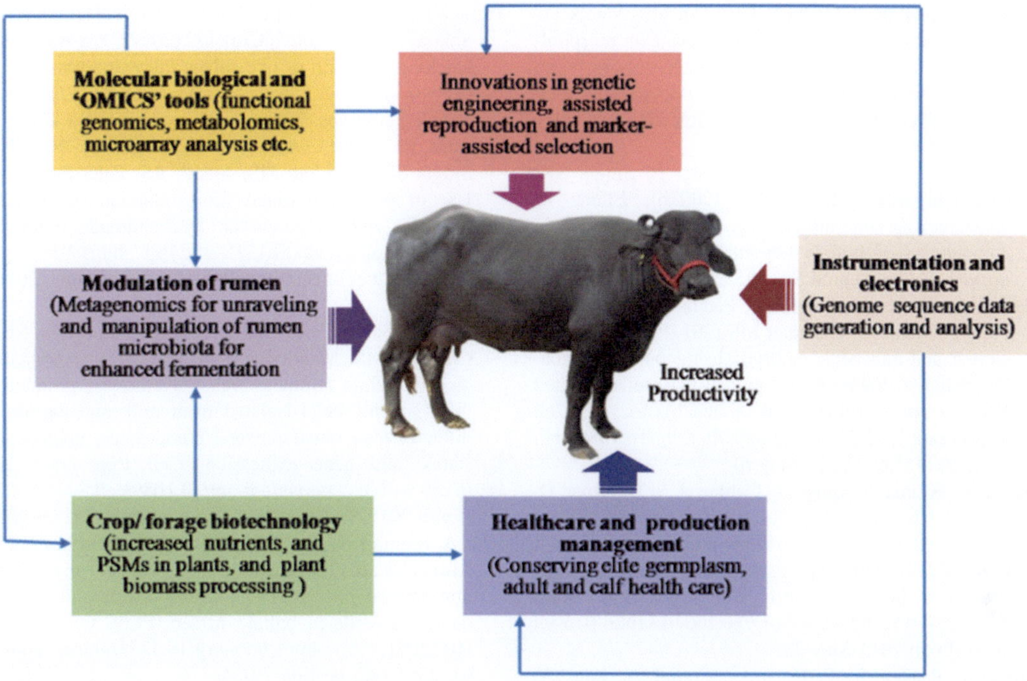

Fig. 12.6 The strategies to promote buffalo production, diversity conservations and propagation of superior bubaline germplasm. Notably, nutrition and health care of livestock are important for benefits from animal husbandry

References

Asma-Ul-Husna, Awan MA, Mehmood A, Sultana T, Shahzad Q, Ansari MS, Rakha BA, Saqlan Naqvi SM, Akhter S (2017) Sperm sexing in Nili-Ravi buffalo through modified swim up: validation using SYBR® green real-time PCR. Anim Reprod Sci 182:69–76. https://doi.org/10.1016/j.anireprosci.2017.04.011 (Epub 2017 May 6)

Banwell KM, Thomson JG (2008) In vitro maturation of mammalian oocytes: outcomes and consequences. Semin Reprod Med 26:162–174

Chuang CK, Sung LY, Hwang SM, Lo WH, Ghen HC, Hu YC (2007) Baculovirus as a new gene delivery vector for stem cell engineering and bone tissue engineering. Gene Ther 14:1417–1424

Dev K, Giri SK, Kumar A, Yadav A, Singh B, Gautam SK (2012) Expression of transcriptional factor genes (Oct-4, Nanog, and Sox-2) and embryonic stem cell-like characters in placental membrane of Buffalo (Bubalus bubalis). J Membr Biol 245(4):83–177. https://doi.org/10.1007/s00232-012-9427-5

Drost M, Wright JM Jr, Cripe WS, Richter AR (1983) Embryo transfer in water buffalo (*Bubalus bubalis*). Theriogenology 20:579–584. https://doi.org/10.1016/0093-691X(83)90082-1

Drost M, Wright JM, Elsden RP (1986) Intergenic embryo transfer between water buffalo and domestic cattle. Theriogenology 25:13–23. https://doi.org/10.1016/0093-691X(86)90180-9

Duah EK, Mohapatra SK, Sood TJ, Sandhu A, Singla SK, Chauhan MS, Manik RS, Palta P (2016) Production of hand-made cloned buffalo (Bubalus bubalis) embryos from non-viable somatic cells. Vitro Cell Dev Biol Anim 52(10):983–988 Epub 2016 Jul 15

Finger EB, Bischof JC (2018) Cryopreservation by vitrification: a promising approach for transplant organ banking. Curr Opin Organ Transplant 23(3):353–360. https://doi.org/10.1097/MOT.0000000000000534

Galli C, Duchi R, Crotti G, Lazarri G (1998) Embryo production by ovum pick up in water buffalo. Theriogenology 49:400. https://doi.org/10.1016/S0093-691X(98)90753-1

Galli C, Lazarri G (2008) The manipulation of gametes and embryos in farm animals. Reprod Domest Anim 43(Suppl 2):1–7

Gandhi RS, Singh A, Sharma AA (2007) Conceptual model for production and evaluation of Murrah bulls in multiple herds using embryo transfer. Indian J Dairy Sci 60:432–436

Gasparrini B (2002) In vitro embryo production in buffalo species: state of the art. Theriogenology 57:237–256

Gautam SK, Verma V, Palta P, Chauhan MS, Manik RS (2008a) Effect of type of cryoprotectant on morphology and developmental competence of in vitro matured buffalo (*Bubalus bubalis*) oocytes subjected to slow freezing or vitrification. Reprod Fertil Dev 20:490–496

Gautam SK, Verma V, Singh B, Palta P, Singla SK, Chauhan MS, Manik RS (2008b) Effect of slow-freezing on morphology and developmental competence of buffalo (*Bubalus bubalis*) immature oocytes. Anim Reprod Sci 105:311–318

George A, Sharma R, Singh KP, Panda SK, Singla SK, Palta P, Manik R, Chauhan MS (2011) Production of cloned and transgenic embryos using buffalo (Bubalus bubalis) embryonic stem cell-like cells isolated from in vitro fertilized and cloned blastocysts. Cell Reprogram 13(3):72–263. https://doi.org/10.1089/cell.2010.0094 (Epub 2011 May 6)

Ghosh K, Kumar R, Singh J, Gahlawat SK, Kumar D, Selokar NL, Yadav SP, Gulati BR, Yadav PS (2015) Buffalo (Bubalus bubalis) term amniotic-membrane-derived cells exhibited mesenchymal stem cells characteristics in vitro. Vitro Cell Dev Biol Anim 51(9):21–915. https://doi.org/10.1007/s11626-015-9920-0 (Epub 2015 May 28)

Gupta S, Pandey S, Parmar MS, Somal A, Paul A, Panda BSK, Bhat IA, Baiju I, Bharti MK, Saikumar G, Sarkar M, Chandra V, Sharma GT (2017) Impact of oocyte-secreted factors on its developmental competence in buffalo. Zygote 25(3):313–320. https://doi.org/10.1017/S0967199417000156

Hirayama H, Kageyama S, Moriyasu S, Sawai K, Minamihashi A (2013) Embryo sexing and sex chromosomal chimerism analysis by loop-mediated isothermal amplification in cattle and water buffaloes. J Reprod Dev 59(4):321–326 (Review)

Hirayama H, Kageyama S, Takahashi Y, Moriyasu S, Sawai K, Onoe S, Watanabe K, Kojiya S, Notomi T, Minamihashi A (2006) Rapid sexing of water buffalo (*Bubalus bubalis*) embryos using loop-mediated isothermal amplification. Theriogenology 66:1249–1256. https://doi.org/10.1016/J.THERIOGENOLOGY.2006.03.036

Huang B, Cui K, Li T, Wang XL, Lu F, Liu Q, da Silva FM, Shi D (2008) Generation of buffalo (*Bubalus bubalis*) transgenic chimeric and nuclear transfer embryos using embryonic germ-like cells expressing enhanced green fluorescent protein. Reprod Domest Anim (in press)

Huang B, Xie TS, Shi DS, Li T, Wang XL, Mo Y, Wang ZQ, Li MM (2007) Isolation and characterization of EG-like cells from Chinese swamp buffalo (*Bubalus bubalis*). Cell Biol Int 31:1079–1088

Huang B, Li T, Wang XL, Xie TS, Lu YQ, da Silva FM, Shi DS (2010) Generation and characterization of embryonic stem-like cell lines derived from in vitro fertilization Buffalo (Bubalus bubalis) embryos. Reprod Domest Anim 45(1):8–122. https://doi.org/10.1111/j.1439-0531.2008.01268.x (Epub 2008 Dec 22)

Hufana-Duran D, Pedro PB, Venturina HV, Hufana RD, Salazar AL, Duran PG, Cruz LC (2004) Post-warming hatching and birth of live calves following transfer of in vitro-derived vitrified water buffalo (*Bubalus bubalis*) embryos. Theriogenology 61:1429–1439

Jyotsana B, Sahare AA, Raja AK, Singh KP, Nala N, Singla SK, Chauhan MS, Manik RS, Palta P (2016) Use of peripheral blood for production of buffalo (*Bubalus bubalis*) embryos by handmade cloning. Theriogenology 86(5):1318–1324.e1. https://doi.org/10.1016/j.theriogenology.2016.04.073 (Epub 2016 May 5)

Jyotsana B, Sahare AA, Raja AK, Singh KP, Singla SK, Chauhan MS, Manik RS, Palta P (2015) Handmade cloned buffalo (Bubalus bubalis) embryos produced from somatic cells isolated from milk and ear skin differ in their developmental competence, epigenetic status, and gene expression. Cell Reprogram 17 (5):393–403. https://doi.org/10.1089/cell.2015.0027 (Epub 2015 Sep 2)

Kala A, Kamra DN, Kumar A, Agarwal N, Chaudhary LC, Joshi CG (2017) Impact of levels of total digestible nutrients on microbiome, enzyme profile and degradation of feeds in buffalo rumen. PLoS One 12(2): e0172051. https://doi.org/10.1371/journal.pone.0172051 (eCollection 2017)

Kasiraj R, Misra AK, Rao MM, Jaiswal RS, Rangareddi NS (1993) Successful culmination of pregnancy and live birth following transfer of frozen–thawed buffalo embryos. Theriogenology 39:1187–1192. https://doi.org/10.1016/0093-691X(93)90016-X

Kitiyanant Y, Saikhun J, Chaisalee B, White KL, Pavasuthipaisit K (2001) Somatic cell cloning in buffalo (*Bubalus bubalis*): effects of interspecies cytoplasmic recipients and activation procedures. Cloning Stem Cells 3:97–104. https://doi.org/10.1089/153623001753205052

Kumar D, Sharma P, Vijayalakshmy K, Selokar NL, Kumar P, Rajendran R, Yadav PS (2018) Generation of Venus fluorochrome expressing transgenic handmade cloned buffalo embryos using Sleeping Beauty transposon. Tissue Cell 51:49–55. https://doi.org/10.1016/j.tice.2018.02.005 (Epub 2018 Feb 24)

Kumar K, Singh R, Kumar M, Agarwal P, Mahapatra PS, Kumar A, Malakar D, Bag S (2014) Isolation and characterization of neural stem cells from buffalo. Int J Neurosci 124(6):6–450. https://doi.org/10.3109/00207454.2013.852087 (Epub 2013 Nov 7)

Liang XW, Lu YQ, Chen MT, Zhang XF, Lu SS, Zhang M, Pang CY, Huang FX, Lu KH (2008) *In vitro* embryo production in buffalo (*Bubalus bubalis*) using sexed sperm and oocytes from ovum pick up. Theriogenology 69:822–826. https://doi.org/10.1016/J.THERIOGENOLOGY.2007.11.021

Lo WH, Hwang SM, Chuang CK, Chen CY, Hu CY (2009) Development of hybrid baculoviral vector for sustained transgene expression. Mol Ther 17:658–666

Lonergan P (2007) State-of-the-art embryo technologies in cattle. Reprod Soc Fertil Suppl 64:315–325

Lu Y, Liao Y, Zhang M, Yang B, Liang X, Yang X, Lu S, Wu Z, Xu H, Liang Y, Lu K (2015) A field study on artificial insemination of swamp and crossbred buffaloes with sexed semen from river buffaloes. Theriogenology 84(6):7–862. https://doi.org/10.1016/j.theriogenology.2015.05.022

Lu F, Luo C, Li N, Liu Q, Wei Y, Deng H, Wang X, Li X, Jiang J, Deng Y, Shi D (2018) Efficient generation of transgenic buffalos (*Bubalus bubalis*) by nuclear transfer of fetal fibroblasts expressing enhanced green fluorescent protein. Sci Rep 8(1):6967. https://doi.org/10.1038/s41598-018-25120-5

Lu Y, Zhang M, Lu S, Xu D, Huang W, Meng B, Xu H, Lu K (2010) Sex-preselected buffalo (Bubalus bubalis) calves derived from artificial insemination with sexed sperm. Anim Reprod Sci 119(3–4):71–169. https://doi.org/10.1016/j.anireprosci.2010.01.001 (Epub 2010 Jan 14)

Lu YQ, Zhang M, Meng B, Lu SS, Wei YM, Lu KH (2006) Identification of X- and Y-chromosome bearing buffalo (*Bubalus bubalis*) sperm. Anim Reprod Sci 95:158–164

Lu YQ, Liang XW, Zhang M, Wang WL, Kitiyanant Y, Lu SS, Meng B, Lu KH (2007) Birth of twins after *in vitro* fertilization with flow-cytometric sorted buffalo (*Bubalus bubalis*) sperm. Anim Reprod Sci 100:192–196. https://doi.org/10.1016/J.ANIREPROSCI.2006.09.019

Lufino MM, Esder PA, Wade-Martins R (2008) Advances in high-capacity extrachromosomal vector technology: episomal maintenance, vector delivery, and transgene expression. Mol Ther 16:1525–1538

Luo C, Lu F, Wang X, Wang Z, Li X, Gong F, Jiang J, Liu Q, Shi D (2013) Treatment of donor cells with trichostatin A improves in vitro development and reprogramming of buffalo (Bubalus bubalis) nucleus transfer embryos. Theriogenology 80(8):86–878. https://doi.org/10.1016/j.theriogenology.2013.07.013 (Epub 2013 Sep 3)

Madan ML, Prakash BS (2007) Reproductive endocrinology and biotechnology applications among buffaloes. Soc Reprod Fertil Suppl 64:261–281 (Review)

Madheshiya PK, Sahare AA, Jyotsana B, Singh KP, Saini M, Raja AK, Kaith S, Singla SK, Chauhan MS, Manik RS, Palta P (2015) Production of a cloned buffalo (Bubalus bubalis) calf from somatic cells isolated from urine. Cell Reprogram 17(3):9–160. https://doi.org/10.1089/cell.2014.0097

Mahapatra PS, Bag S (2014) Reprogramming of buffalo (*Bubalus bubalis*) foetal fibroblasts with avian egg extract for generation of pluripotent stem cells. Res Vet Sci 96(2):8–292. https://doi.org/10.1016/j.rvsc.2014.02.008 (Epub 2014 Feb 18)

Manjunatha BM, Gupta PS, Ravindra JP, Devraj M, Nandi S (2008) In vitro embryo development and blastocyst hatching rates following vitrification of river buffalo embryos produced from oocytes recovered from slaughterhouse ovaries or live animals by ovum pick up. Anim Reprod Sci 104:419–426

Meng F, Li H, Wang X, Qin G, Oback B, Shi D (2015) Optimized production of transgenic buffalo embryos and offspring by cytoplasmic zygote injection. J Anim Sci Biotechnol 6:44. https://doi.org/10.1186/s40104-015-0044-x (eCollection 2015)

Mohapatra SK, Sandhu A, Neerukattu VS, Singh KP, Selokar NL, Singla SK, Chauhan MS, Manik RS, Palta P (2015a) Buffalo embryos produced by handmade cloning from oocytes selected using brilliant cresyl blue staining have better developmental competence and quality and are closer to embryos produced by in vitro fertilization in terms of their epigenetic status and gene expression pattern. Cell Reprogram 17(2):50–141. https://doi.org/10.1089/cell.2014.0077

Mohapatra SK, Sandhu A, Singh KP, Singla SK, Chauhan MS, Manik R, Palta P (2015) Establishment of trophectoderm cell lines from buffalo (Bubalus bubalis) embryos of different sources and examination of in vitro developmental competence, quality, epigenetic status and gene expression in cloned embryos derived from them. PLoS One 10(6):e0129235. https://doi.org/10.1371/journal.pone.0129235 (eCollection 2015)

Mullani N, Singh MK, Sharma A, Rameshbabu K, Manik RS, Palta P, Singla SK, Chauhan MS (2016) Caspase-9 inhibitor Z-LEHD-FMK enhances the yield of in vitro produced buffalo (Bubalus bubalis) pre-implantation embryos and alters cellular stress response. Res Vet Sci 104:4–9. https://doi.org/10.1016/j.rvsc.2015.11.008 (Epub 2015 Nov 14)

Nanda AS, Nakao T (2003) Role of buffalo in the socioeconomic development of rural Asia: current status and future prospectus. Anim Sci J 74:443–455 (Review article)

Nagy ZP, Kerkis I, Chang CC (2008) Development of artificial gametes. Reprod Biomed Online 16:539–544

Neglia G, Gasparrini B, Caracciolo di Brienza V, Di Palo R, Campanile G, Zicarelli L (2003) First pregnancies established from vitrified blastocysts entirely produced *in vitro* in Mediterranean Italian buffalo cows (*Bubalis bubalis*). Theriogenology 59:374 (Abstract). https://doi.org/10.1016/s0093-691x(02)01170-6

Parnpai R, Liang Y, Ketudat-Cairns M, Somfai T, Nagai T (2016) Vitrification of buffalo oocytes and embryos. Theriogenology 86(1):214–220. https://doi.org/10.1016/j.theriogenology.2016.04.034 (Epub 2016 Apr 21, Review)

Parnpai R, Tasrpoo K, Kamonpatana M (1999) Development of cloned swamp buffalo embryos derived from fetal fibroblast: comparison of *in vitro* cultured with or without buffalo and cattle oviductal epithelial cells. Buffalo J 15:371–384

Parnpai R, Tasripoo K, Kamonpatana M (2000) Developmental potential of cloned bovine embryos derived from quiescent and non-quiescent adult ear fibroblasts after different activation treatments. Theriogenology 53:239 (Abstract)

Perera BM (2008) Reproduction in domestic buffalo. Reprod Domest Anim 43(Suppl 2):200–206

Presicce GA, Verberckmoes S, Senatore EM, Rath D (2005a) First established pregnancies in Mediterranean Italian buffaloes (*Bubalus bubalis*) following deposition of sexed spermatozoa near the utero tubal junction. Reprod Domest Anim 40:73–75. https://doi.org/10.1111/J.1439-0531.2004.00560.X

Presicce GA, Rath D, Klinc P, Senatore EM, Pascale M (2005b) Buffalo calves born following AI with sexed semen. Reprod Domest Anim 40:349

Priya D, Selokar NL, Raja AK, Saini M, Sahare AA, Nala N, Palta P, Chauhan MS, Manik RS, Singla SK (2014) Production of wild buffalo (Bubalus arnee) embryos by interspecies somatic cell nuclear transfer using domestic buffalo (Bubalus bubalis) oocytes. Reprod Domest Anim 49(2):51–343. https://doi.org/10.1111/rda.12284 (Epub 2014 Feb 3)

Rao KB, Pawshe CH, Totey SM (1993) Sex determination of *in vitro* developed buffalo (*Bubalus bubalis*) embryos by DNA amplification. Mol Reprod Dev 36:291–296. https://doi.org/10.1002/MRD.1080360302

Sadeesh Em, Shah F, Yadav PS (2016) Differential developmental competence and gene expression patterns in buffalo (Bubalus bubalis) nuclear transfer embryos reconstructed with fetal fibroblasts and amnion mesenchymal stem cells. Cytotechnology 68 (5):1827–1848. https://doi.org/10.1007/s10616-015-9936-z (Epub 2015 Dec 14)

Saha A, Panda SK, Chauhan MS, Manik RS, Palta P, Singla SK (2013) Birth of cloned calves from vitrified-warmed zona-free buffalo (*Bubalus bubalis*) embryos produced by hand-made cloning. Reprod Fertil Dev 25(6):5–860. https://doi.org/10.1071/RD12061

Saini M, Selokar NL, Raja AK, Sahare AA, Singla SK, Chauhan MS, Manik RS, Palta P (2015) Effect of donor cell type on developmental competence, quality, gene expression, and epigenetic status of interspecies cloned embryos produced using cells from wild buffalo and oocytes from domestic buffalo. Theriogenology 84(1):101–108.e1. https://doi.org/10.1016/j.theriogenology.2015.02.018 (Epub 2015 Feb 23)

Saini M, Selokar NL, Agrawal H, Singla SK, Chauhan MS, Manik RS, Palta P (2016) Treatment of buffalo (Bubalus bubalis) donor cells with trichostatin A and 5-aza-2'-deoxycytidine alters their growth characteristics, gene expression and epigenetic status and improves the in vitro developmental competence, quality and epigenetic status of cloned embryos. Reprod Fertil Dev 28(6):37–824. https://doi.org/10.1071/RD14176

Saini M, Selokar NL, Revey T, Singla SK, Chauhan MS, Palta P, Madan P (2014) Trichostatin A alters the expression of cell cycle controlling genes and microRNAs in donor cells and subsequently improves the yield and quality of cloned bovine embryos in vitro. Theriogenology 82(7):42–1036. https://doi.org/10.1016/j.theriogenology.2014.07.027 (Epub 2014 Jul 24)

Sandhu A, Mohapatra SK, Agrawal H, Singh MK, Palta P, Singla SK, Chauhan MS, Manik RS (2016) Effect of sex of embryo on developmental competence, epigenetic status, and gene expression in buffalo (Bubalus bubalis) embryos produced by hand-made cloning. Cell Reprogram 18(5):356–365

Selokar NL, Saini M, Palta P, Chauhan MS, Manik R, Singla SK (2014) Hope for restoration of dead valuable bulls through cloning using donor somatic cells isolated from cryopreserved semen. PLoS One 9 (3):e90755. https://doi.org/10.1371/journal.pone.0090755 (eCollection 2014)

Selokar NL, Saini M, Agrawal H, Palta P, Chauhan MS, Manik R, Singla SK (2016) Buffalo (Bubalus bubalis) SCNT embryos produced from somatic cells isolated from frozen-thawed semen: effect of trichostatin A on the in vitro and in vivo developmental potential, quality and epigenetic status. Zygote 24(4):53–549. https://doi.org/10.1017/S0967199415000520 (Epub 2015 Oct 27)

Selokar NL, Sharma P, Krishna A, Kumar D, Kumar D, Saini M, Sharma A, Vijayalakshmy K, Yadav PS (2018) Establishment of a somatic cell bank for indian buffalo breeds and assessing the suitability of the cryopreserved cells for somatic cell nuclear transfer. Cell Reprogram 20(3):157–163. https://doi.org/10.1089/cell.2017.0066

Shah RA, George A, Singh MK, Kumar D, Chauhan MS, Manik RS, Palta P, Singla SK (2008) Hand-made cloned buffalo (*Bubalus bubalis*) embryos: comparison of different culture media and culture sysytems. Cloning Stem Cells 10:435–442

Shah RA, George A, Singh MK, Kumar D, Anand T, Chauhan MS, Manik RS, Palta P, Singla SK (2009) Pregnancies established from handmade cloned blastocysts reconstructed using skin fibroblasts in buffalo (*Bubalus bubalis*). Theriogenology 71:1215–1219

Shi D, Lu F, Wei Y, Cui K, Yang S, Wei J, Liu Q (2007) Buffalos (*Bubalus bubalis*) cloned by nuclear transfer of somatic cells. Biol Reprod 77:285–291. https://doi.org/10.1095/BIOLREPROD.107.060210

Singh B, Chauhan MS, Singla SK, Gautam SK, Verma V, Manik RS, Singh AK, Sodhi M, Mukesh M (2009) Reproductive biotechniques in buffaloes (*Bubalus bubalis*): status, prospects and challenges. Reprod Fertil Dev 21:499–510 (Review)

Singh B, Gautam SK, Verma V, Kumar M, Singh B (2008) Metagenomics in animal gastrointestinal ecosystem: Potential biotechnological prospects. Anaerobe 14(3):138–144. https://doi.org/10.1016/j.anaerobe.2008.03.002 (Epub 2008 Mar 26, Review)

Singh B, Mal G, Yadav PS, Singla SK (2016) Buffaloes of North-West Himalayan Region of India. In: Asian Buffalo Magazine, pp 12–16 (Jan–Dec 2016)

Singla SK, Manik RS, Madan ML (1997) Micromanipulation and cloning studies on buffalo oocytes and embryos using nucleus transfer. Indian J Exp Biol 35:1273–1283

Sritnaudomchai H, Pavasuthipaisit K, Kiyiyanant Y, Kupradinun P, Mitalipov S. Kusamran T (2007)

Characterization and multilineage differentiation of embryonic stem cells derived from a buffalo parthenogenetic embryo. Mol Reprod Dev 74:1295–1302

Sun HL, Meng LN, Zhao X, Jiang JR, Liu QY, Shi DS, Lu FH (2016) Effects of DNA methyltransferase inhibitor RG108 on methylation in buffalo adult fibroblasts and subsequent embryonic development following somatic cell nuclear transfer. Genet Mol Res 15(3). https://doi.org/10.4238/gmr.15038455

Tasripoo K, Suthikrai W, Sophon S, Jintana R, Nualchuen W, Usawang S, Bintvihok A, Techakumphu M, Srisakwattana K (2014) First cloned swamp buffalo produced from adult ear fibroblast cell. Animal 8(7):1139–1145. https://doi.org/10.1017/S1751731114001050

Techakumphu M, Lohachit C, Chantaraprateep P, Prateet P, Kobayashi G (1989) Preliminary report on cryopreservation of Thai swamp buffalo embryos: manual and automatic methods. Buffalo Bull 8:29–36

Verma V, Gautam SK, Singh B, Manik RS, Palta P, Singla SK, Gosawami SL, Chauhan MS (2007) Isolation and characterization of embryonic stem cell-like cells from in vitro-produced buffalo (*Bubalus bubalis*) embryos. Mol Reprod Dev 74:520–529

Verma V, Gautam SK, Palta P, Manik RS, Singla SK, Chauhan MS (2008) Development of a pronuclear DNA microinjection technique for production of green fluorescent protein-expressing bubaline (*Bubalus bubalis*) embryos. Theriogenology 69:655–665

Zandi M, Muzaffar M, Shah SM, Kaushik R, Singh MK, Palta P, Singla SK, Manik RS, Chauhan MS (2014) WNT3A signalling pathway in buffalo (*Bubalus bubalis*) embryonic stem cells. Reprod Fertil Dev 26 (4):551–561. https://doi.org/10.1071/RD13084

Zoheir KM, Allam AA (2011) A rapid improved method for sexing embryo of water buffalo. Theriogenology 76(1):83–87. https://doi.org/10.1016/j.theriogenology.2011.01.020 (Epub 2011 Mar 11, Erratum in: Theriogenology. 2012 May;77(8):1729)

13 Reproduction Biotechnology in Camelids

Abstract

Over the past three decades, similar to equine industry, various reproduction biotechniques, *viz.* artificial insemination, in vitro production of embryos, and nuclear transfer cloning, are developed for camelids. Inadequate understanding of reproduction physiology, incomplete genetic cataloguing of native genotypes, and lack of oocyte and proven embryo banks are the challenges in fully utilizing the reproduction potential of camels. This chapter presents an overview of advances and the bottlenecks of assisted reproduction technologies applied to camelids. Prospects of conservation of diversity of the species are discussed.

Highlights

- Camels have unique evolutionary adaptation to hard agroclimatic conditions
- Camel is turning from a beast of burden to an important farm animal with multiple benefits
- Only a few laboratories are working on reproduction expansion and biodiversity conservation of the species.

Keywords

Camels · Camelid diversity · Reproduction biotechnology · Camel semen · Embryo production

13.1 Introduction—Camel as Valued Livestock

The camelids are important livestock that play an incredible role in their cradle of domestication and original habitats through evolutionary adaptive merits and contribution to livelihood of people. The camelids contribute to the economy of various countries by thriving on native poor-quality forages, low water requirement, production of valued milk, wool/fiber and energy for transportation, and agriculture operations.

According to reports as on 2010, around 14 million camels with 90% dromedaries were existing. The dromedaries are primarily the domesticated camels mostly reared in Horn of Africa, The Sahel, Maghreb, Middle East, and South Africa (https://en.wikipedia.org/wiki/Camel, accessed on Jan. 28, 2018).

The Bactrian camel is only truly wild (as opposed to feral) species which are found in the Gobi and Taklamakan deserts of Mongolia and China. Whereas Somalia has largest populations of domestic camels, Australia is reported to have largest populations of feral camels. In addition, Syria, Egypt, Abu Dhabi, Ethiopia, Kazakhstan, and Pakistan rear camels for meat besides their use for milk and agriculture operations. Pastoralists in Kenya drink blood and milk of camels to increase stamina and immunity.

B. Singh et al., *Advances in Animal Biotechnology*,
https://doi.org/10.1007/978-3-030-21309-1_13

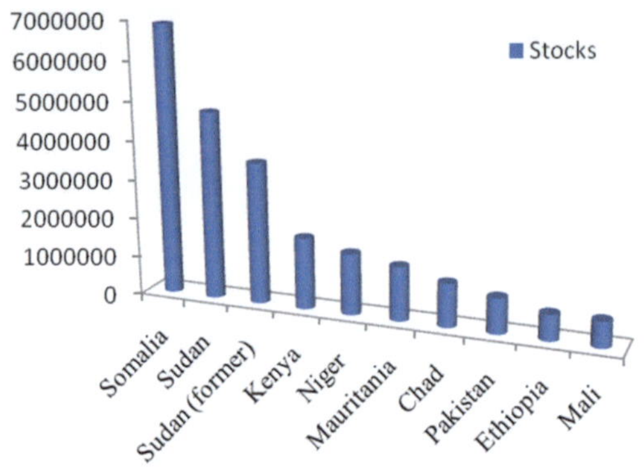

Fig. 13.1 Top 10 camel producers' country (Average 1994–2016) *Source* http://www.fao.org/faostat/en/#data/QA/visualize

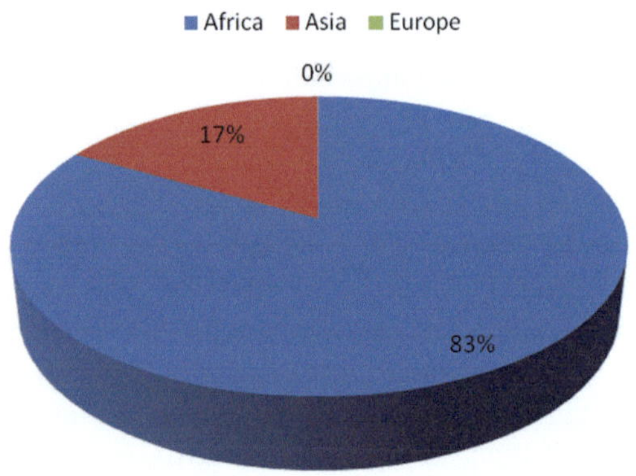

Fig. 13.2 Production share of camels by region (Average 1994–2016) *Source* http://www.fao.org/faostat/en/#data/QA/visualize

In some countries (Figs. 13.1 and 13.2), the dromedaries (Fig. 13.3) constitute the integral and major part of nomadic life serving transportation, milk, and food. Given the current scenario, and adverse conditions (Fig. 13.3) that hamper camel husbandry, some states, for instance Rajasthan in India, have declared camel (*Camelus dromedarius*) the State Animal of Rajasthan and declared the animal as critically endangered species. Whereas some countries have imposed ban on slaughter of camel, others use camel as meat animal. Some countries consider camels as competitors of forages needed for sheep. Australia, for instance, has culled hundreds of camels. In view of their peculiar evolutionary attributes (Box 1), the camel are important to several regions of the countries.

13.2 Diversity of Camelids

Camel, also known as "ship of the desert," is livestock species with unique characteristics and benefits to human health and livelihood. The convoys or caravans travelling across the deserts of the Middle East and Africa region over several years have contributed significantly to shape genetic diversity in camels (Almathen et al. 2016). Camelids appeared around 20 million years ago (Heintzman et al. 2015).

Nowadays, the camelids are divided into two categories—Old World (e.g., Bactrian and dromedary camels) and New World camelids (e.g., llama, alpaca, guanaco, and vicuna). They are thought to originate from one species in North America which around 30 million years ago segregated into two genera: Camelus and Lama (Harrison 1979). During the course of survival in varied climate and feeding resources, divergence in physiological and behavioral traits developed among New and Old World camelids.

Camels and llamas as result of their adaptation to different environments share some remarkable similarities, such as both being induced ovulators, a bicornuate uterus with left horn being larger, and an epitheliochorial placenta. Whereas New World camelid species, *viz.* llama, alpaca, guanaco, and vicuna, and Old World camelids, *viz.* Bactrian and dromedary, can hybridize or interbreed to produce viable fertile offspring, Old World camelids and New World camelids fail to interbreed owing to difference in their body sizes.

Using ARTs, hybrids of some species (e.g., Camas, the cross of camel and llama) have been produced for use as pack animals. However, interspecies mating and fertile hybrids are observed among some species. Skidmore et al. (1999) have reported hybrids of Old and New World camelids—*C. dromedarius* × *L. guanicoe* guanacos/llama and male camel.

Based on global sample set and mitochondrial genomic DNA analyses, the population structures

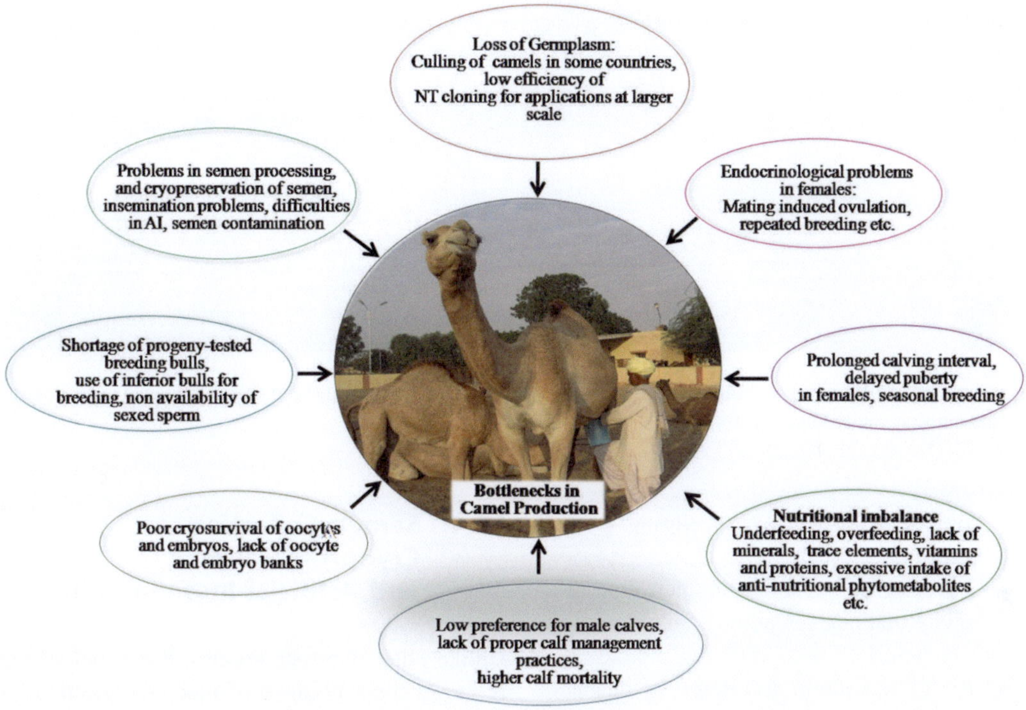

Fig. 13.3 The bottlenecks associated with low productivity of camels. Despite potential of camels, they have attracted little attention for scientific inputs from the researchers across the globe. One reason could be their confinement to particular zones of the countries

of modern dromedary camel *vis-à-vis* their extinct ancestors were studied. Phylogenetic analysis of modern and ancient dromedaries revealed restocking of modern dromedaries from ancestors inhabiting southeast coast of Arabian Peninsula (Almathen et al. 2016).

The dromedary (Fig. 13.4) is not only largest among camelids, but also largest domesticated livestock species with economic purposes. A recent phylogenetic analysis based on ancient and modern reveals a history of restocking the modern camelids from wild ancestors that have become extinct.

Box 1 Evolutionary and Adaptive Merits of Camels Enabling them as Valued Livestock

- Withstanding to water scarcity, the dromedary camel can drink once every 10 days even in extreme summer
- Unlike other ruminants, the camel red blood cells are oval which facilitates their flow even during dehydration and withstands high osmotic variation
- Ability to withstand body temperature (93 to 104 °F) fluctuations
- Minimal loss of body fluid compared to other livestock species
- Presence of *rete mirabile*, a complex network of arteries and veins that assists in cooling down the brain temperature
- Minimum sweat glands to ensure water loss through evaporation
- Ability to reabsorb the water trapped in their nostrils during exhalation
- Protection from intense solar radiations from desert because of specialized anatomical adaptations, *viz*. thick coat,

Fig. 13.4 Two types of large camels found in different habitats. **a** double-humped camel found in cold deserts; **b** single-humped camel found in tropical arid regions

long legs, presence of pedestal over sternum
- Development of thick lining in buccal cavity that enables animal chew thorny vegetation
- Ability to walk in arid desert by means of wide feet
- Efficient absorption of water in intestine and kidney. High kidney cortex to medulla ratio that reduces surface area for water/solute filtration
- Presence of additional "heavy-chain only," immunoglobulins conferring additional immunity.

13.3 Assisted Reproduction in Camel

As with other farm animals, various ARTs are developed for use in camels. Assisted reproduction is used only at scientifically organized farms. This is because the camels are seasonal breeder, and their reproduction behavior and pattern differ much from other livestock species. Both, the female and male come into heat only during breeding season. They breed by the time the females are 3–4 years old, while males are able to breed when they have become 4–5 years old. The mating occurs by natural methods.

13.4 Semen Collection, Cryopreservation, and Artificial Insemination

Unlike other livestock species, the camel semen is in gel form because of mucopolysaccharides from bulbourethral or prostate glands (Skidmore et al. 2013). AI was initiated in camelids in the early 1960s (Table 13.1). However, use of AI increased in view of improving and disseminating economically important genetic traits such as milk, meat, and wool production and racing ability as practiced in Middle East. Besides, use of germplasm of superior males for producing best offspring, reduced cost to maintain live males mating (Abraham et al. 2017), minimizing the risk and expenses involved in transportation of animals for manual breeding, and reducing risk of infections from mating, have urged the farmers to adopt AI in camel.

Semen is collected mostly by using artificial vagina (30-cm long, 5-cm internal diameter) (Bravo et al. 2000), or by electroejaculation using standard bovine ejaculator (e.g., 12 V and 180 mA, Standard Precision Electronics, Denver, CO). Electroejaculation is reported to be scratchy/uncomfortable to camel. In addition, the quality of semen collected by this method is not appreciable, hence not recommended. Artificial vagina (30-cm long, 5-cm internal diameter, 30–40 °C temperature) is also used for cattle, and

Table 13.1 Milestone achievements in reproduction biotechniques in camelids

Year	Achievements and remarks (references)
1960	First Camelid offspring from AI in Bactrian camel (Elliot 1961)
1998	Production of hybrids of Old and New World camelids (Skidmore et al. 1999)
2002	Pregnancy established from transfer of vitrified-warmed Illama (*Lama glama*) embryos (Aller et al. 2002); successful transfer of vitrified-warmed embryos in Llama (Aller et al. 2002)
2004	First offspring after transfer of frozen-thawed embryos in *C. dromedarius* (Skidmore et al. 2004)
2005	Birth of *C. dromedarius* calves from transfer of cryopreserved embryos (Nowshari et al. 2005); pregnancy/live birth from transfer of vitrified warmed *C. dromedarius* camel (Skidmore et al. 2005)
2006	First calf from IVM, IVF, and in vitro cultured abattoir-derived dromedary camel oocytes (Khatir and Anouassi 2006)
2010	Birth of "Injaz," the first SCNT cloned *C. dromedarius* camel calf, using cumulus cells as donor nuclei (Wani et al. 2010)
2016	De novo gene assembly of a domestic dromedary of North African origin (Fitak et al. 2016)
2017	Developing interspecies cloned camel embryos using oocytes from *C. bactreanus*, and fibroblasts from *C. bactreanus* and *L. glama*; first birth of SCNT cloned *C. bactreanus* using dromedary camel as source of oocytes as well as surrogate female (Wani et al. 2017)
2018	Comprehensive studies on developmental competence of SCNT cloned embryos developed from various donor cells; cumulus cells found to be more suitable donor nuclei (Wani et al. 2018), evidences that cumulus cells of antral follicles are multipotent stem cell-like cells (Saadeldin et al. 2018)
2019	First description of near-chromosome assembly of the dromedary camel (Ruvinskiy et al. 2019), diversity analysis of *C. dromedaries* females in Morocco (Boujenane et al. 2019), induction of superovulation in *C. dromedarius* (Manjinatha et al. 2019)

buffalo bulls are used for collecting semen from camel (Bravo et al. 2000). AI is reported in camelids though insemination trials are infrequent, probably due to difficulties in collecting semen as well as handling, and cryopreservation of semen.

It is difficult to cryopreserve camel semen due to gelatinous nature of seminal plasma (Mal et al. 2016; El-Bahrawy et al. 2017; Malo et al. 2017). Attempts to improve semen cryopreservation include developing capable semen extenders and their suitability in terms of maintaining sperm motility, capacitation, fertilizing ability, and improved developmental competence of embryos produced by AI or IVF (Skidmore et al. 2013).

Cryopreservation of semen is important factor determining the efficacy and utility of cryopreserved sperm. The semen is stored for short term by chilling, or long term by storing in liquid nitrogen. Variable reports exist on cryopreservation of camel semen, and embryos produced from the use of cryopreserved semen (Deen et al. 2003; Nissari-Naslajii et al. 2006, 2007). Different extenders are used to preserve semen at low temperature by chilling to 4 °C when it is intended for immediate use, or it has to be stored for indefinite time period. Using chilled semen in Green buffer fortified with 20% egg yolk, pregnancies rates of 25–30% were reported (Bravo et al. 2000), which are less than those obtained (50–55%) from inseminating the camels with freshly collected and processed semen.

The SHOTOR diluent, a Tris-based semen extender (osmolality of 330 mOsm/kg, pH 6.9), is reported to be apposite for cryopreservation of *C. bactrianus* semen (Nissari-Naslajii et al. 2007).

There are certain problems in implementing semen technologies in camel. For instance, the camel has to be trained for AV before using it as stud bulls for semen. In general, the success of assisted reproduction is low in camels (Box 2).

Box 2 Summary of the factors that slow down the success of assisted reproduction in camelids. The readers must read latest articles showing increased success rate of cryopreservation of embryos.

A. **Male factors**
   - Difficulties in collecting semen from large species
   - Possibility of contamination of semen if semen collector is not adequately trained
   - Ejaculations are obtained in fractions, each lasting for 5–20 min depending on arousal in male
   - The gelatinous nature of semen impeding the semen processing (Mal et al. 2016)
   - The sperm loss due to entrapping in viscous seminal plasma
   - Sensitivity of sperm to cryopreservation and subsequent thawing
   - Behavioral problems—aggressive nature or refusal to mate a particular female, and extended copulatory process
B. **Female factors**
   - Inability of females to ovulate without mating
   - Late onset of puberty in females, short breeding season, early embryonic mortality, long gestation period
   - Less availability of oocytes for use in IVEP, micromanipulation, and SCNT
   - Absence of zona pellucida in hatched embryos making them sensitive cryoinjuries
   - Variations in embryo size, higher lipid contents therein, and sensitivity to commonly used CPAs such as PROH and DMSO (Skidmore et al., 2004; Malo et al. 2017).

13.5 In Vitro Fertilization, Embryo Production, and Cryopreservation

Due to technical and physiological impediments, gametes and embryo cryopreservation was unable to become an integral component of embryo transfer (ET) program in camels. Over the past three decades, embryo production and transfer is practiced at commercial level in different camelids including dromedary camel and alpaca (Herrid et al. 2017). Availability of oocytes is crucial for producing embryos. As limited number of camels is slaughtered, that too in few countries, sufficient oocytes are not available so as to harness female reproduction potential.

In addition, the she-camel is induced ovulator. They ovulate only when mated, implying that ovulation should be induced before inseminating the females. Despite all constraints, some reports are available on IVF-derived embryos, cryopreservation of embryos, and establishing pregnancies and birth of camel calves following embryo transfer technologies. The efforts to preserve camel embryos under low temperature began in the mid-1990 s. The vitrification being simpler and advantageous should be used in camels (Herrid et al. 2016, 2017).

13.6 Somatic Cell Nuclear Transfer

The camels are used as dairy animal as well as source of power for transportation and agriculture in many parts of the globe. SCNT cloning is used to repopulate or multiply extinct elite livestock species (Singh et al. 2009). For the first time, fibroblasts obtained from live adult male or granulosa cells from cumulus–oocyte complex were used to produce SCNT-cloned embryos (Khatir and Anouassi 2008). In this study, 58 females were superovulated using a single dose of eCG (3500 IU), and ovaries were collected postmortem for obtaining oocytes. Both the cell types showed similar developmental competence in terms of embryos reaching blastocyst as well as hatching stages. Pregnancies were established after transferring these embryos, though none could reach term. Two years later, the SCNT-cloned embryos developed from cumulus cells (as donor nuclei) and cytoplasts from transvaginal ultrasound-guided ovum pick-up, developed to full-term fetus, and the calf named "Injaz" was born (Wani et al. 2010). More recently, a Bactrian camel (*C. bactrianus*) has been produced by interspecies SCNT wherein *C. bactrianus* was used source of donor cell as well as surrogate mother to carry on pregnancy. Though the cloned calf died shortly due to acute septicaemia, the technique was envisaged to be gifted to conserve the camels (Wani et al. 2017).

Wani et al. (2018) in studies on developmental competence of cytoplasts prepared from abattoir-derived oocytes *vis-a-vis* those derived from live animals, found that oocytes from live animals had better competence to support development of NT-cloned embryos. Also, the studies showed that all oocytes obtained from any source could support development of embryos to blastocyst stages, and development to full-term calves.

13.7 Outlook and Challenges

The camels are frugal in habits, but are ultimate pillars of national economy of many poor or developing countries, and contribute effectively to the welfare of livestock owners thriving in difficult environments. The tribe subsists on camel milk.

Modern science has started proving the worth of camelids in various aspects including food and livelihood security, sustainability, resilience, milk, and meat production under low-input management system. Despite remarkable progress observed in camel ARTs, ranging from semen cryopreservation to production of embryos in vitro and nuclear transfer cloning, the quantum of progress is low compared to other livestock species.

The countries relying on camels should therefore expedite research inputs to conserve the camel species that are evolutionarily adapted to local agroclimatic conditions and are deemed acceptable to cater the needs of local peoples. In view of some threats, such as genetic disorders, pathogenic or infectious disease, abandoning of camel rearing, there is need to establish commercially viable genes or germplasm banks of camels.

Somatic cells in the banks could be a promising source of resurrecting the threatened or extinct species through nuclear transfer cloning. Alternative strategies such as development of oocytes from stem cells should be emphasized to overcome the shortage of oocytes in camels. However, reports are scarce on this important aspect of scientific intervention.

Modifications of procedures, development, and synchrony of new procedures for cryopreservation of camelid embryos are envisaged to prove cryopreservation of camelid embryos. In addition, manipulating reproductive physiology of female camel for enhancing lifespan of corpus luteum (CL), which is generally shorter (8–10) days, appears to be advantageous to establish pregnancies.

13.8 Conclusions

The camel in view of its economic and social importance and unique milk with therapeutic attributes plays a sustainable role in many parts of the world. Without doubt, the camel might contribute more fruitfully with improved maintenance and breeding management. The ARTs in camelids need in-depth studies to improve their efficacy and suitability under field conditions in countries where camels are important to agriculture and agroclimatic conditions. The shortage of camel oocytes is the limiting factor in embryo biotechnology for commercial camel production. Also, processing and cryopreservation of camel semen is difficult. It is matter of time and development of robust technologies that may solve the problems associated with commercial camel production.

References

Abraham MC, Verdier K, Båge R, Morrell JM (2017) Semen collection methods in alpacas. Vet Rec 180 (25):613–614. https://doi.org/10.1136/vr.104074 (Review)

Aller JF, Rebuffi GE, Cancino AK, Alberio RH (2002) Successful transfer of vitrified Ilama (*Lama glama*) embryos. Anim Reprod Sci 73:121–127

Almathen F, Charruau P, Mohandesan E, Mwacharo JM, Orozco-terWengel P, Pitt D, Abdussamad AM, Uerpmann M, Uerpmann HP, De Cupere B, Magee P, Alnaqeeb MA, Salim B, Raziq A, Dessie T, Abdelhadi OM, Banabazi MH, Al-Eknah M, Walzer C, Faye B, Hofreiter M, Peters J, Hanotte O, Burger PA (2016) Ancient and modern DNA reveal dynamics of domestication and cross-continental dispersal of the dromedary. Proc Natl Acad Sci U S A 113(24):6707–6712. https://doi.org/10.1073/pnas.1519508113 (Epub 2016 May 9)

Boujenane I, El Khattaby N, Laghouaouta H, Badaoui B, Piro M (2019) Morphological diversity of female camel (*Camelus dromedarius*) populations in Morocco. Trop Anim Health Prod. https://doi.org/10.1007/s11250-019-01813-5 (Epub ahead of print)

Bravo PW, Skidmore JA, Zhao XX (2000) Reproductive aspects and storage of semen in camelidae. Anim Reprod Sci 62(1–3):173–193

Deen A, Vyas S, Sahani MS (2003) Semen collection, cryopreservation and artificial insemination in the dromedary camel. Anim Reprod Sci 77(3–4):223–233

El-Bahrawy K, Rateb S, Khalifa M, Monaco D, Lacalandra G (2017) Physical and kinematic properties of cryopreserved camel sperm after elimination of semen viscosity by different techniques. Anim Reprod Sci 187:100–108. https://doi.org/10.1016/j.anireprosci.2017.10.011 (Epub 2017 Oct 31)

Elliot FI (1961) Artificial insemination of Bactrian camel (*Camelus bactrianus*). Int Zoo Year Book 3:94

Fitak RR, Mohandesan E, Corander J, Burger PA (2016) The de novo genome assembly and annotation of a female domestic dromedary of North African origin. Mol Ecol Resour 16(1):314–324. https://doi.org/10.1111/1755-0998.12443 (Epub 2015 Jul 24)

Harrison JA (1979) Revision of the Camelidae (Artiodactyla, Tylopoda) and description of the new genus Alforjas. Paleontol Contrib 95:1–27 (University of Kansas)

Herrid M, Billah M, Malo C, Skidmore JA (2016) Optimization of a vitrification protocol for hatched blastocysts from the dromedary camel (*Camelus dromedarius*). Theriogenology 85(4):585–590. https://doi.org/10.1016/j.theriogenology.2015.09.048 (Epub 2015 Oct 24)

Heintzman PD, Zazula GD, Cahill JA, Reyes AV, MacPhee RD, Shapiro B (2015) Genomic data from extinct North American camelops revise camel evolutionary history. Mol Biol Evol 32(9):2433–2440. https://doi.org/10.1093/molbev/msv128

Herrid M, Vajta G, Skidmore JA (2017) Current status and future direction of cryopreservation of camelid embryos. Theriogenology 89:20–25. https://doi.org/10.1016/j.theriogenology.2016.10.005 (Epub 2016 Oct 13, Review)

Khatir H, Anouassi A (2006) The first dromedary (*Camelus dromedarius*) offspring obtained from in vitro matured, in vitro fertilized and in vitro cultured abattoir-derived oocytes. Theriogenology 65 (9):1727–1736 (Epub 2005 Nov 2)

Khatir H, Anouassi A (2008) Preliminary assessment of somatic cell nuclear transfer in the dromedary (Camelus dromedarius). Theriogenology 70(9):1471–7147. https://doi.org/10.1016/j.theriogenology.2008.06.094

Mal G, Vyas S, Srinivasan A, Patil NV, Pathak KM (2016) Studies on liquefaction time and proteins involved in the improvement of seminal characteristics in dromedary Camels (*Camelus dromedarius*). Scientifica (Cairo) 2016:4659358. https://doi.org/10.1155/2016/4659358 (Epub 2016 Feb 18)

Malo C, Crichton EG, Skidmore JA (2017c) Optimization of the cryopreservation of dromedary camel semen: cryoprotectants and their concentration and equilibration times. Cryobiology 74:141–147. https://doi.org/10.1016/j.cryobiol.2016.11.001 (Epub 2016 Nov 3)

Manjunatha BM, Al-Hosni A, Al-Bulushi S (2019) Simplified superovulation protocols in dromedary camels (Camelus dromedarius). Theriogenology 1(126): 214–221. https://doi.org/10.1016/j.theriogenology.2018.12.006 (Epub 2018 Dec 10)

Niasari-Naslaji A, Mosaferi S, Bahmani N, Gerami A, Gharahdaghi AA, Abarghani A, Ghanbari A (2007) Semen cryopreservation in Bactrian camel (*Camelus bactrianus*) using SHOTOR diluent: effects of cooling rates and glycerol concentrations. Theriogenology 68 (4):618–625 (Epub 2007 Jun 22)

Niasari-Naslaji A, Mosaferi S, Bahmani N, Gharahdaghi AA, Abarghani A, Ghanbari A, Gerami A (2006) Effectiveness of a tris-based extender (SHOTOR diluent) for the preservation of Bactrian camel (*Camelus bactrianus*) semen. Cryobiology 53 (1):12–21 (Epub 2006 May 2)

Nowshari MA, Ali SA, Saleem S (2005) Offspring resulting from transfer of cryopreserved embryos in camel (*Camelus dromedarius*). Theriogenology 63 (9):2513–2522 (Epub 2004 Dec 10)

Ruvinskiy D, Larkin DM, Farré M (2019) A near chromosome assembly of the dromedary camel genome. Front Genet 10:32. https://doi.org/10.3389/fgene.2019.00032 (eCollection 2019)

Saadeldin IM, Swelum AA, Elsafadi M, Mahmood A, Alfayez M, Alowaimer AN (2018) Cumulus cells of camel (Camelus dromedarius) antral follicles are multipotent stem cells. Theriogenology 15(118):233–242. https://doi.org/10.1016/j.theriogenology.2018.06.009 (Epub 2018 Jun 19)

Singh B, Chauhan MS, Singla SK, Gautam SK, Verma V, Manik RS, Singh AK, Sodhi M, Mukesh M (2009) Reproductive biotechniques in buffaloes (Bubalus bubalis): status, prospects and challenges. Reprod Fertil Dev 21(4):499–510. https://doi.org/10.1071/RD08172 (Review)

Skidmore JA, Billah M, Loskutoff NM (2005) Comparison of two different methods for the vitrification of hatched blastocysts from the dromedary camel (*Camelus dromedarius*). Reprod Fertil Dev 17 (5):523–527

Skidmore JA, Billah M, Loskutoff NM (2004) Developmental competence in vitro and in vivo of cryopreserved, hatched blastocysts from the dromedary camel (*Camelus dromedarius*). Reprod Fertil Dev 16 (6):605–609

Skidmore JA, Billah M, Binns M, Short RV, Allen WR (1999) Hybridizing Old and New World camelids: *Camelus dromedarius* × *Lama guanicoe*. Proc R Soc Lond B 266:649–656

Skidmore JA, Morton KM, Billah M (2013) Artificial insemination in dromedary camels. Anim Reprod Sci 136(3):178–186. https://doi.org/10.1016/j.anireprosci.2012.10.008

Wani NA, Wernery U, Hassan FA, Wernery R, Skidmore JA (2010) Production of the first cloned camel by somatic cell nuclear transfer. Biol Reprod 82 (2):373–379. https://doi.org/10.1095/biolreprod.109.081083

Wani NA, Hong S, Vettical BS (2018) Cytoplast source influences development of somatic cell nuclear transfer (SCNT) embryos in vitro but not their development to term after transfer to synchronized recipients in dromedary camels (*Camelus dromedarius*). Theriogenology 25(118):137–143. https://doi.org/10.1016/j.theriogenology.2018.05.021

Wani NA, Vettical BS, Hong SB (2017) First cloned Bactrian camel (*Camelus bactrianus*) calf produced by interspecies somatic cell nuclear transfer: a step towards preserving the critically endangered wild Bactrian camels. PLoS One 12(5):e0177800. https://doi.org/10.1371/journal.pone.0177800 (eCollection 2017)

Reproduction Biotechnology in Cattle 14

Abstract

Cattle are the most important domestic ungulate livestock species used for food- and agriculture-based economic gains worldwide. Several breeds of cattle are domesticated for produce milk, meat, skin, draught power, and manure. Reproduction biotechniques entailing in vitro production of embryos, cryopreservation, sperm sexing, stem cell biology, and genetic engineering to produce value-added products and nuclear transfer cloning are now used in cattle.

Highlights

- Cattle contribute to milk, meat, manure and skin, and draught power
- Genetic engineering and assisted reproduction techniques have made use of cattle to produce recombinant therapeutics.

Keywords

Cattle · Native cattle breeds · Milk and meat · Cattle reproduction · Cryopreservation · Stem cells · Transgenic cattle

14.1 Introduction

Among the livestock species, domesticated cattle are multipurpose livestock and major contributors to pastoral economies throughout the world and are recognized as important components of world biodiversity. The domesticated cattle belong to *Bos* Genus, a member of *Bovine* subfamily within the Bovidae family of order *Artiodactyla*. There are two recognized forms of domesticated cattle, the humpless *taurine* cattle of Europe, west Africa and northern Asia (*Bos taurus*) and the humped *zebu* cattle of southern Asia and Africa (*B. indicus*). The morphological and physiological differences between zebu and taurine cattle reflect adaptation to markedly different environments as *zebu* cattle generally inhabit hot arid or semi-arid regions of Asia and Africa while taurine cattle are usually confined to temperate zones with higher levels of rainfall and denser vegetation.

According to FAO, the world cattle inventory in 2016 has 1.474 billion head of cattle. Brazil has the highest population of cattle, followed by India, China, and the USA (Figs. 14.1 and 14.2).

B. Singh et al., *Advances in Animal Biotechnology*,
https://doi.org/10.1007/978-3-030-21309-1_14

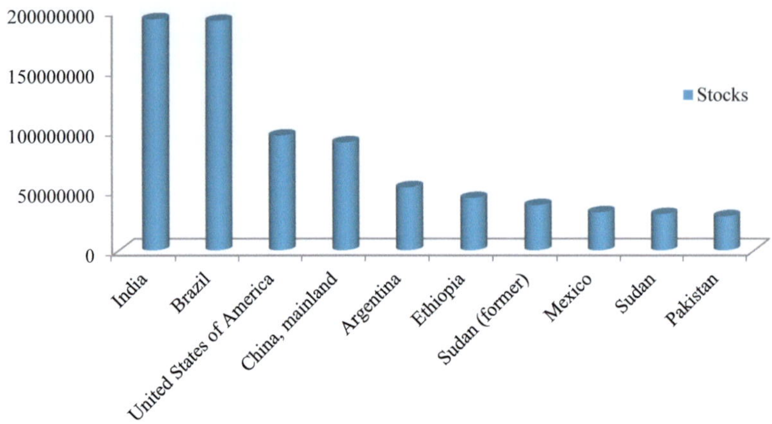

Fig. 14.1 Top ten cattle-producing countries in the world (Average 1994–2016) (Source: http://www.fao.org/faostat/en/#data/QA/visualize; accessed on Nov. 8, 2018)

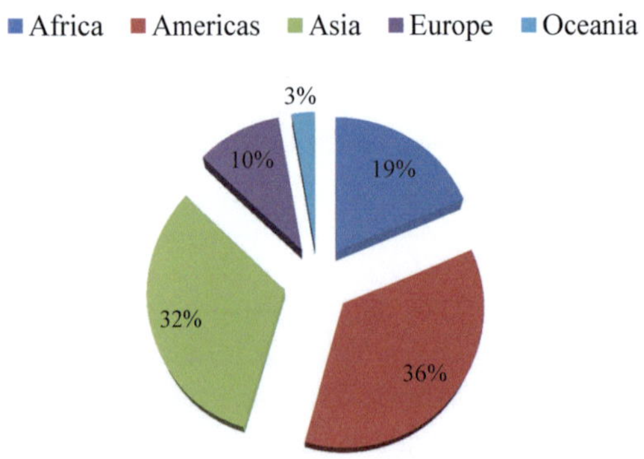

Fig. 14.2 Production share of cattle by region (Average 1994–2016); Source: http://www.fao.org/faostat/en/#data/QA/visualize; accessed on Nov. 8, 2018)

Around 104 countries have cattle inventory in excess of 1 million head (https://www.drovers.com/article/world-cattle-inventory-ranking-countries-fao, accessed on August 20, 2018).

Dairy farming, a part of agriculture for long-term production of milk, value addition of milk products, is one of the oldest practices, and a valuable tool for rural development, employment generation, and sustained income. Despite multiple factors that lower cattle production potential (Fig. 14.3), the cattle are still predominant milk species. In addition, cattle contribute to meat, hide and skin and draft power.

14.2 Biotechnology and Cattle Production

Various biotechniques are used to improve the breeding stock and productivity of cattle. These include artificial insemination (AI), oocyte collection and preservation, IVF and embryo production, stem cells techniques, somatic cell nuclear transfer cloning, and genetic engineering to produce transgenic animals (Table 14.1).

Many reproduction biotechniques were invented originally in cattle and adopted for use in other livestock species. The reproduction techniques have vividly enhanced cattle production in terms of quantity and quality.

The AI was improved with the development of methods to handle semen at low ambient temperature, semen diluents or extenders, CPAs to preserve semen at ultralow temperature and maintaining sperm viability after cryopreservation and collecting semen from disabled or high-quality bulls by means of electroejaculators.

The progress was also due to simultaneous progress in female ARTs, viz, collection of oocytes from live animals, their maturation and fertilization in vitro and developing into transferable quality embryos. This era was referred to as

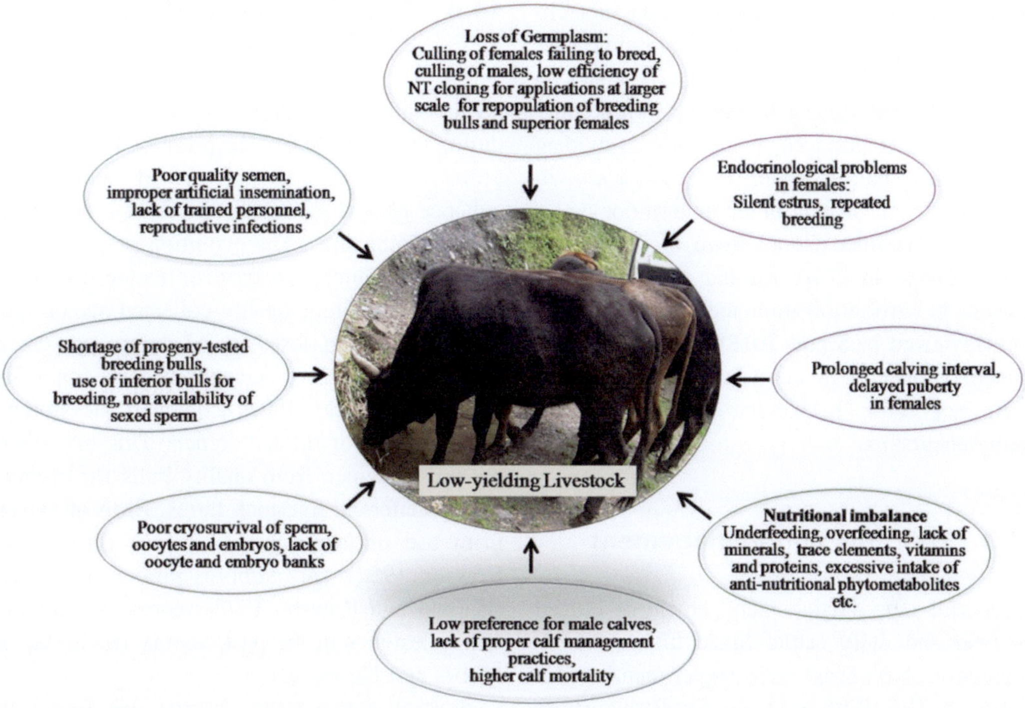

Fig. 14.3 The major factors that slow down production potential of cattle under small-scale farming system or at the levels of marginal farmers. Scarcity of robust technology for applications at farmers levels, inadequate nutrition, unhygienic livestock maintenance are predominantly responsible

Table 14.1 Applications of assisted reproduction and genetic engineering tools to augment cattle production

Technologies	Prospects and applications (references)
Male ARTs	Using AI for genetic improvement and developing cross breeds for improved milk yield, reducing management expenditure, minimizing transmission if venerable diseases through use of semen from authentic sources, propagating selectively males or females by using sexed sperm
Oocytes and embryo biotechniques	Using oocytes for producing embryos through IVF or cloning, using oocytes to produce embryos of desired sex through IVF with sexed sperm Best utilization and conservation of female germplasm (oocyte banking), repopulation of endangered cattle breeds Increasing the genetic pool of in situ populations, Conserving geneticdiversity and establishing founder population to manage endangered species and breeds
SCNT cloning	Proliferating elite females, conservation and multiplication of endangered breeds, cloning stud bulls for sperm production, producing cloned transgenic animals Producing isogenic animals for nutrition, immunological and biomedical experiments
Stem cell technology	Using stem cells in NT cloning, generating iPSCs, in vitro production of oocytes or sperms In vitro meatproduction from muscle stem cells and adipose stem cells Clinical applications- repair of mammary tissue of high yielding animals

the second generation of reproduction biotechnology in livestock (reviewed in Moore and Hasler 2017). Discovering the biology of oocytes is considered a milestone achievement in reproduction biology. Importance of the topic is evident from the fact that Sir Robert Edwards, acknowledged as pioneer worker in the field of oocyte biology was awarded Nobel Prize in Physiology and Medicine in 2010, for his contribution to advance in vitro fertilization and embryo production (reviewed in Sirard 2018). All these developments collectively contributed to transform traditional cattle husbandry to industry-level entrepreneurship.

14.3 Reproduction Management

Successful reproductive management is crucial for beef and dairy cattle. Since fifties, the reproduction biologists were experimenting for improving IVF. Birth of Louise Joy Brown (born July 25, 1978), the first human baby born by the conception of IVF embryo, was milestone breakthrough. Since then, the technology is applied to several animal species including cattle, swine, pets and small laboratory animals. The very basic objective of applying IVF in animals was to improve fertility. Later, the objectives were oriented toward improvements in genetic selection, decreasing genetic interval and improving gains from the cattle intended for milk or beef production.

Synchronization of estrus and timely AI are among most effective management tools in commercial livestock production. Many things have to go right, though many things can go wrong, in terms of a beef cow becoming pregnant and giving birth to healthy calves. For instance, AI may not be the right choice for every beef producer, but it is becoming a more attractive and cost-effective.

14.4 Artificial Insemination and Sperm Sexing

AI was initiated in cattle in Russia and Denmark during the early 1900s. The primary driving force promoting AI was the potential of technique to enhance genetic gains through semen from high genetic value sires. The evolution of using minimal sperm number to fertilize oocyte led to the concept of diluting freshly collected semen and increasing semen doses for AI. Improved semen extenders and the replacement of glass ampoules with safer plastic straws expedited the storage and handling of diluted semen. This expedited the use of semen from quality bulls for inseminating heifers in livestock farms. Birth of calves from use of frozen-thawed semen (Polge and Rowson 1952), and embryos produced *in vitro* (Wilmut and Rowson 1973) represented important milestones in the past century (reviewed in Moore and Hesler 2017).

Several native cattle breeds are found in many parts of the world that are known for their unique genetic and adaptive merits to survive and perform in difficult agroclimate. Some native (*Bos indicus*) breeds such as Tharparkar, Sahiwal (Fig. 14.4), Kankrej in tropical plane areas of India are very sturdy animals, thrive on native forages and perform well under thermal stress. They are generally low-yielding animals. In order to fulfill the milk demand and provide a livelihood to marginal farmers, crossbreeding programs were initiated to coalesce the adaptive merits of native genotypes and milk-producing traits of exogenous cattle (*Bos taurus*). The emphasis was on importing semen of superior *B. taurus* bulls. The strategy proved to be successful to increase milk production in many countries.

The semen processing and cryopreservation are now high-tech technologies. The AI has grown to the extent that more than 130 million

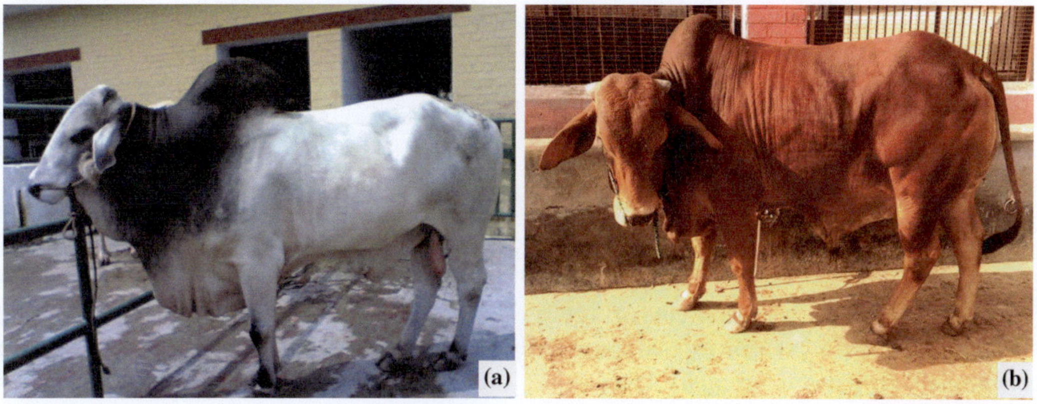

Fig. 14.4 Sturdy zebu cattle (*B. indicus*) breeds of India. **a** Tharparkar bull; **b** Sahival bull. The animals' duet to hump, and tolerance to thermal stress preferred for agriculture operations in rural areas. They perform well under thermal stress which is extremely difficult for exotic breeds. The distinct gene pool of indigenous cattle with superior adaptive traits presents an opportunity for mining specific allele(s)

cattle are inseminated annually at commercial farms and under field conditions. The bulls used as semen donors are now high-quality bulls whose merit is ascertained by genetic markers. The semen for commercial use has to pass rigorous quality tests. If they do not pass the tests, they are discarded straightforwardly. Morrel et al. (2018) have suggested that separate assays should be developed for evaluating semen in dairy and beef bulls as the semen quality of cattle used for beef and dairy differ markedly.

Literature is scarce on freeze-drying of semen and its use in cattle. However, lyophilized semen could be useful for ICSI in under certain conditions. A study has shown that freeze-dried sperm introduced into IVM oocytes by ICSI could persuade the development of blastocyst at a relatively higher proportion. Lyophilization can be used as low-cost strategy to preserve semen for biodiversity conservation of animals (Anzalone et al. 2018). One of the main advantages of preserving semen by lyophilization is that it avoids the necessity of liquid nitrogen and expensive cryocontainers.

14.5 Semen Sexing and AI

The flow cytometric separation of X- and Y-chromosome bearing sperm, IVF sex-preselected embryos, estrus synchronization and using estrus-synchronized females as surrogate mothers are among modern ARTs. Sperm sorting is an important and high-demanding technique that can enhance output from cattle and buffaloes. In addition, the emphasis is on evaluating genetic merit of the bulls based on use of high-throughput technologies to investigate into molecular markers as indicators of semen quality in terms of fertilizability (Selvaraju et al. 2018). Comparative evaluation of DNA methylation of four developmentally regulated genes (*PEG3*, *XIST*, *OCT4*, and *NANOG*) in sperms from a cloned, and a normal bull, and development of IVF blastocysts produced by semen from cloned and normal bulls, showed that sperm in both cases are equally reprogrammed during spermatogenesis and are equally good and befitting for AI programs (Yamanaka et al. 2018).

Experimental evidence reveal that imprinted and non-imprinted genes are appropriately reprogrammed during gametogenesis, hence offspring of cloned animals should be normal (Kaneda et al. 2017). The microfluidic-based separation of progressively motile cattle sperm with high genomic DNA integrity has potential for their applications in bovine industry (Nagata et al. 2018).

SexedULTRA™, a new cutting edge sex-sorting provides high-quality sorted sperm for commercial IVF or inseminating the females for producing calves of desired sex. The SexedULTRA™ semen represents a new era in semen technology that gives conception rates up to 95% of those achieved through conventional sperm. The SexedULTRA™ semen products are available from sexing technologies and the authorized distributors and are reported to be superior (in terms of fertility and conception rates) than general commercial entrepreneurs (https://mastergen.com/factsheets/sexedultra-semen/, accessed on August 19, 2018). The sexed semen is provided in 0.25 and 0.50 cc plastic straws, and aptness SexedULTRA™ sexed-sorted semen is already proven scientifically (Thomas et al. 2017).

Though the scope for errors is less while using sexed semen, there are some concerns with the semen-sexing technology. The sperm doses may have low sperm count per dose. The heifers coming into heat should be inseminated within 12 h of onset of estrus to establish pregnancy.

14.6 Embryo Biotechnologies in Cattle

Oocyte retrieval, banking, and in vitro embryo production were among the major the driving forces to harness the potential of female reproduction through ARTs. Indeed, first IVF cattle calf was produced using mid-ventral surgery for collecting immature oocytes and then transferring the in vitro produced four-cell stage morulae to oviducts (Brackett et al. 1982). Progress in sperm capacitation in vitro and use of heparin in capacitation of sperm further improved the IVF and embryo production methodologies. Drug-induced superovulation enhanced not only the quantity, but the affected quality of the competent oocytes or cumulus–oocyte complexes (COCs) retrieved from the live donor through laparoscopy. Arresting FSH for 48–72 h prior to aspiration of COCs promoted the LH-supported growth of ovarian follicles, and retrieval of developmentally more competent COCs (Blondin et al. 2002).

Cryopreservation was one of the important factors that enhanced embryo biotechnology. It was realized that embryos could be multiplied simply by bisecting them into two or more parts and allowing each one to rejuvenate to functional embryo. This way, the identical twins were produced. Over the years, the research had focussed on developing suitable culture systems for pre-implantation stage embryos and their preservation. A new era of evolving culture media and supplements initiated with the SCNT-cloned embryos. This was because cloned embryos require additional requirements to support genetic reprogramming of their genome.

14.7 Embryo Sexing

Embryo sexing is desirable to determine the sex of embryo and detecting genetic abnormalities prior to birth of animal. Sexing is done by invasive as well as non-invasive methods such as X-chromosome-linked enzymatic detection and H-Y antigen detection. Blastomeres could be used for determining the sex of embryo before transfer to recipients. The original technique for sexing the embryos involved culturing cells and examining chromosomal karyotyping, or sex chromatin. The first calf by this technique was born in 1975 (Hare et al. 1976). However, the methodology was tricky, expensive, and time-consuming based on microsurgical removal of trophoblasts from two weeks old embryos, and analyzing the karyotypes. The problem was overcome by the development of PCR assays using Y-chromosomal specific probes and a very small biopsy from blastocysts as a source of DNA (Bondioli et al. 1989; Herr and Reed 1989). Notably, in mammals, the presence of

SRY gene induces differentiation of undifferentiated gonads into testicles. In the absence of SRY genes, an alternative pathway is activated and induced the development of ovaries. Hence, SRY gene is used as a marker to determine the sex of the embryo.

14.8 Nuclear Transfer Cloning in Cattle

The use of micromanipulation of embryos has started quite early in cattle. First birth of calves from transfer of cloned embryos, developed from blastomeres taken from 2 to 32 cell stage morulae, was reported in late 1980s (Prather et al. 1987a, b). Of 19 cloned embryos transferred non-surgically, two heifers completed pregnancy gave birth to live calves (Prather et al. 1987a, b). Cibelli et al. (1998a, b), for the first time, reported SCNT transgenic cloned cattle calves from transgenic fibroblast cells. Of the 28 embryos transferred to 11 cows, 3 isogenic transgenic cloned calves were produced (Cibelli et al. 1998a).

Several milestone achievements were documented thereafter (Table 14.2) including development of SCNT-cloned transgenic bovine neurons intended for transplantation in experimental Parkinson's mice models (Zawada et al. 1998), development of transgenic ESC-like cells for transfer to pre-implantation embryos to develop chimera embryos, and birth of first bovine chimeric offspring (Cibelli et al. 1998b).

Health of cloned animals is a matter of ethical concern. The cloned animals may suffer from multiple health problems including impaired immunity, underdeveloped organs, loss of organs, and many others. Studies carried out by Lanza et al. (2001) showed that cloned cattle might be healthy and live a normal life. Subsequently, many reports were published on health status and reproductive profile of cloned animals, and the underlying causes of abnormalities. Studied were carried out on reproductive traits (viz, onset of puberty, follicular dynamics, reproductive hormone profiles, and ability to maintain pregnancy) of cloned heifers. It was shown that cloned animals produced from somatic cells irrespective of the age of donor cell have normal reproductive features (Enright et al. 2002). This was further evidenced from the study that reported birth of second-generation cloned calves from cells obtained from first-generation cloned animals. The researchers, however, failed to produce third-generation clones (Kubota et al. 2004). A recent study has reported high pregnancy rates (33% offspring per fresh HMC embryo transfer rate, compared to 25% offspring per transfer rate of IVF embryos), and live births from transfer of embryos produced by two SCNT and IVF, respectively. It was inferred that SCNT is a promising technique which needs improvements for its efficient use (Cortez et al. 2018).

Shortly after the success of SCNT in mammalian species, the commercial services were established to provide cloning of pet and livestock species. The emphasis was on producing transgenic cattle (Sano et al. 2014), animals for genomic evaluation (Kasinathan et al. 2015), gene-edited cell lines for producing cloned animals with desirable phenotypes (Carlson et al. 2016; Ishino et al. 2018).

Based on the review of published data, it is clear that up to the year 2004, about 1500 calves were obtained through cloning. The major reports were from North America, Japan, New Zealand, and Europe. Afterward, South America and Asian countries also reported births of cloned calves from different species. Moreover, compared to other ARTs, the calves produced from SCNT cloning are quite low.

The cattle are used for producing recombinant therapeutic proteins. SCNT or modified versions of the basic technique are preferred to produce transgenic animals. This is because it is comparatively easy to induce genetic changes or editing in cell genome, and use the genetically modified cells as donor nuclei to produce cloned transgenic animals.

Production of transgenic cattle expressing high levels of recombinant human RNA (rhA) expressed in their milk. The herd of transgenic cattle is reported to be an easily

Table 14.2 Milestone achievements in ARTs in cattle

Year	Achievements and remarks (reference)
1987	First birth of cloned calves from transfer of blastomeres from 2 to 32 cell bovine embryos to enucleated oocytes matured in vitro *or* in vivo, followed by electrofusion (Prather et al. 1987a, b)
1989–91	Developmentof PCR-based assays for embryo sexing by employing Y-chromosome-specific probes (Bondioli et al. 1989; Herr and Reed 1989)
1998	Birth of live calves from SCNT (Cibelli et al. 1998a, b), development of SCNT-cloned transgenic neurons intended for transplantation in Parkinson's mice models (Zawada et al. 1998), development of transgenic ESC-like cells for transfer to pre-implantation embryos, and birth of first bovine chimeric offspring, demonstrating for the first time that somatic cells can be genetically tailored and de-differentiated into ESC-like cells with possible applications in cell therapies (Cibelli et al. 1998a, b) Developmentof OPS method for vitrifying bovine embryos (Vajta et al. 1998)
2000	Development of well of the well (WOW) system for culture of zona- included or zona-free embryos (Vajta et al. 2000), birth of six clones of a17-years old Japanese Beef bull using ear skin fibroblasts, showing that fibroblasts of aged animals could serve as nucleus donor cells (Kubota et al. 2000), birth of 24 calves from 6 different cell types (cumulus cells, oviduct, uterine, female and male skin, ear and hepatic cells). The cells from young animals were found to be more suitable than those derived from aged animals
2001	Development of somatic cell cloning methods without use of micromanipulators (Vajta et al. 2001)
2002	Studied were carried out on reproductive characteristics (viz, onset of puberty, follicular dynamics, reproductive hormone profiles, follicular dynamics, and ability to maintain pregnancy) of cloned heifers. It was shown that cloned animals produced from somatic cells irrespective of the age of cells are normal (Enright et al. 2002), birth of calves from transfer of embryos produced by IVF of immature bovine oocytes preserved by open pulled vitrification (Vieira et al. 2002)
2003	Birth of cloned transgenic cattle expressing high titers of beta-casein and kappa-casein in milk (CSN2 and CSN3, respectively) (Brophy et al. 2003), birth of cloned calf from transfer of vitrified-warmed HMC embryo calf (Tecirlioglu et al. 2003)
2004	First report on serial (first and second generation) cloning of bulls (Kubota et al. 2004), birth of calf by SCNT without micromanipulators and CO_2 incubators, and development of novel submarine incubation system for in vitro development of HMC embryos (Vajta et al. 2004)
2005	Development of an efficient and reliable chemically assisted enucleation method for handmade cloning in cattle (Vajta et al. 2005), comparative evaluation of open and closed methods of vitrification, and showing evidences that CryoTip™ vitrification could eliminate the possibilities of embryo contamination during cryopreservation without compromising survival and developmental potential (Kuwayama et al. 2005)
2007	Births of prion-free transgenic cattle. The cattle envisaged as useful model for industrial bovine products (Richt et al. 2007)
2008	Microinjection of bovine and murine phospholipase C zeta1 (PLCZ1) cRNA into matured bovine oocytes to elicit sperm-like calcium oscillatory pattern and parthenogenetic activation of bovine oocytes (Ross et al. 2008)
2009	Production of transgenic cattle expressing high levels of recombinant human RNA (rhA) expressed in their milk. The herd of transgenic cattle was reported to be easily scalable and characterized source of rhA for biomedical applications (Echelard et al. 2009)
2010	Birth of cattle calves from sexed embryos (Akiyama et al. 2010)
2013	Production of antigen-specific human immunoglobulin in transchromosomic cattle (Sano et al. 2014)
2014	Studies on ICSI with flow cytometry sorted sperm revealing that ICSI can rescue damage to sperm during sorting, and can serve as an alternative approach to produce embryos with predetermined sex (Jo et al. 2014)
2015	Production of human IgG (hIgG) with hIgG1 subclass dominancy in cattle, the achievement envisaged as an important milestone for producing hpAbs for health applications (Matsushita et al. 2015)

(continued)

Table 14.2 (continued)

Year	Achievements and remarks (reference)
2016	Production of clonedembryos and calves from adipose- and amniotic fluid-derived mesenchymal stem cells (MSCs). The study opens new prospects of using new cells types as donor nuclei to produce cloned animals (da Silva et al. 2016), cloning of the African indigenous cattle breed Kenyan Boran (Yu et al. 2016)
2018	Studies were carried out to infer DNA methylation profile of some development-regulating genes in sperm of cloned *vis-a-vis* normal animals. It was found that sperm from cloned animals are adequately reprogrammed during spermatogenesis, and are comparable to sperm of normal animals, the cloned animals can therefore produce normal offspring (Yamanaka et al. 2018) Developing protocols for producing gene-edited bovine somatic cellsfor SCNT (Ishino et al. 2018), developing selection strategies for functionaland progressive sperm with high DNA integrity (Nagata et al. 2018) Studies showed that freeze-dried spermatozoaby ICSI can induce blastocyst development (Anzalone et al. 2018) Derivation of bESCs, the first stem cell line with stable genome, karyotype, expression of pluripotency-regulating genes. The cells could serve as donor nuclei for producing SCNT-clonedembryos, indicating that it will be possible to use these cells for genome engineering of livestock animals in future (Bogliotti et al. 2018) Isolation of bovine milk stem cells-like (bMSCs) from milk expressing pluripotency markers (CD90, CD73, and CD105), SOX2 and OCT4, and differentiation into various cell lineages. The bMSCs cells might have role in regenerative medicine (Pipino et al. 2018)

scalable and well-characterized source of rhA for biomedical applications (Echelard et al. 2009). This topic is discussed separately in chapter on transgenic animals.

14.9 Stem Cell Advances in Cattle

From agriculture and biomedical perspectives, the ESCs from some livestock species are important for genome engineering, genetic testing and selection and modeling human diseases (Yadav et al. 2012). The very basic purpose of studying stem cells in animals as model species is unraveling basic developmental mechanism, improving the genetics of the animal herd, enhancing reproduction efficiencies through and using stem cells for health and clinical applications. Stem cells, culture conditions, expression of stemness markers are already reported in stem cells derived from different origins, viz, embryos, fetuses, and adult cattle species (Yadav et al. 2005; Cibelli et al. 2006; Wu et al. 2016; Yuan 2018).

Despite continued efforts, the true bovine ESCs (bESCs) have not yet been successfully established so far. Expression of certain factors such as trophectoderm regulator (CDX2), expressed in ICM cells of cattle embryo, impedes the regulation of pluripotency in bESCs (Wu et al. 2016). Compared to control bESCs, the CDX2 knockdown (CDX2-KD) bESCs (bESCs-KD) were found to grow longer on feeder monolayer, express pluripotency-related genes, and down-regulation of DNA methylation on promoters of OCT4 and SOX2 (Wu et al. 2016).

The fibroblasts, generally used to develop clones of animals are short-lived cells when cultured in vitro. In contrast, the long-lasting cells such as ESCs could be more suitable for genome editing by cutting edge techniques such as CRISPR/Cas9. Induced pluripotent bovine ESCs (bESCs), the first true ESCs reported in cattle, with stable morphology, karyotype, transcriptome, expression of pluripotency marker gene expression and population doubling time are reported recently. The bESCs could serve as efficient donor cells to produce cloned embryos, thereby showing that bESCs could contribute to genomic selection, genome engineering, and production of cattle with high genetic merit (Bogliotti et al. 2018).

Mammary gland is a rich source of microbiota for neonates, and the somatic cells that are shed into milk. Bovine milk stem cells-like (bMSCs) isolated from milk showed pluripotent features viz., expression of characteristic pluripotency markers (CD90, CD73, and CD105), SOX2 and OCT4, and ability to differentiate into various cell lineages. The cells might open possibilities of their use in developmental biology and regenerative medicine (Pipino et al. 2018).

14.10 Outlook and Challenges

Cattle rearing and breeding is an age-old practice in Indian farming system and is the mainstay of agriculture economy. The commercial dairy- and meat-producing sectors have metamorphosed a lot during the past two decades. The dairy and beef industries have adopted advanced ARTs and outputs are evident (Fig. 14.5).

The present scenario calls for mining of genetic resources for identifying the genes/genomic regions associated with thermoregulation and adaptive biology in zebu cattle, and further develops functional gene resource for native cattle for better understanding of thermoregulation pathways or molecular mechanism of heat stress. Some technologies despite their envisaged huge potential are not widely used in commercial cattle production.

The SCNT technologies are improved significantly. The HMC, for example, may become more practical and useable as simpler, inexpensive, and effective alternatives to clone superior bulls or endangered animals albeit improvements are warranted.

Semen sexing has wider prospects in dairy sector where dairy heifer may give births to dairy calves, either to expand the herd for sale or as replacement of older cows. The demand for sexed semen, for instance, in largely organized dairy sector has increased rapidly, but high costs, susceptibility to cryopreservation, and low fertility rates are the major constraints. In addition, storage of semen by lyophilization has not progressed much despite its potential benefits.

The uterine metagenome should be studied for exploring its role to augment fertility and prevent uterine infections that impede fertility of the

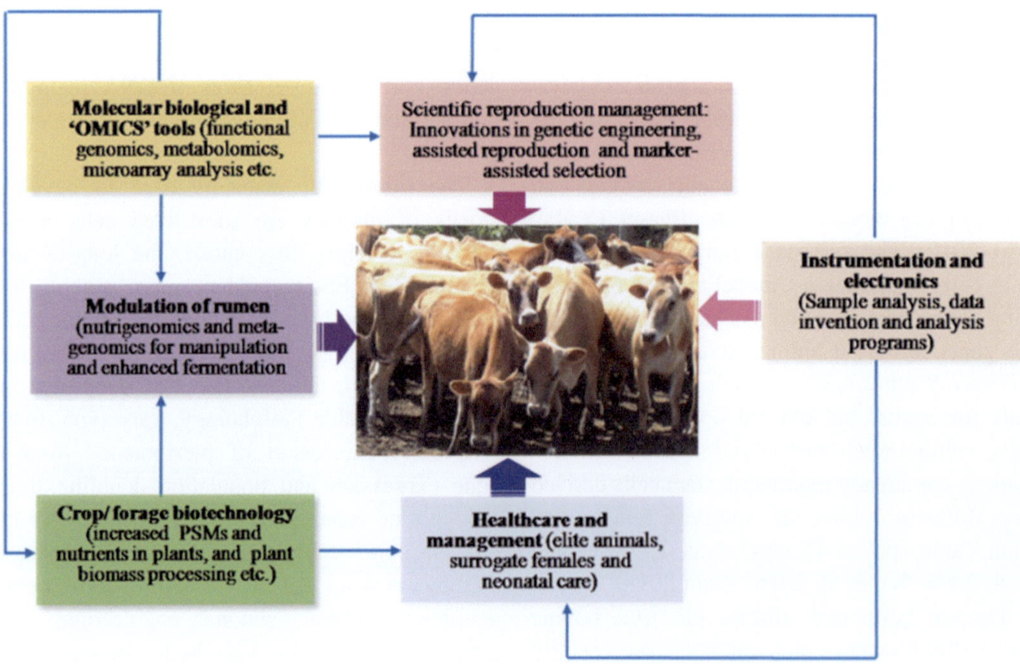

Fig. 14.5 An outline of scientific strategies for managing commercial dairy or beef cattle

animals. This is very important for countries where slaughter of cattle is banned, and animals are abandoned that cause other types of health, safety, and management problems. The sexed semen has important prospects in dairy and beef industries. The countries like India have increased demand for sexed cattle semen for producing heifers for milk. Preventing the birth of male calves is economically desirable as males have less demand.

In recent years, the use of indiscriminate crossbreeding strategies to infuse native breeds with exotic germplasm is a major cause in dilution or loss of important gene pool. Although crossbreeding may offer benefits, such as improved meat and milk production and greater draft power, it could also cause vanishing of valuable traits like superior adaptability, heat tolerance, and disease resistance developed over years of natural selection. Even though naturally evolved breeds are not as productive as their *taurine* counterparts, their loss would be a major blow to the future of cattle diversity and sustainable animal productivity. Hence, the use of locally adapted livestock and the development of appropriate intervention are effective routes to help secure livestock assets of native animal keepers.

14.11 Conclusions

Cattle are the important components of economy and agriculture. Most of the reproduction biotechniques originally developed in cattle are modified for other livestock species. While merging, cutting edge technologies such as sexed sperm, embryo sexing, and nuclear transfer cloning will show their impact later in future, flaws in more conventional approaches such as AI should be removed. It is high time to look into the prevailing problems and resolve them through scientific interventions.

References

Akiyama K, Kobayashi J, Sato Y, Sata R, Ohashi M, Sasaki E, Oda Y, Ogawa Y, Ueda S, Nabenishi H, Matoba S (2010) Calf production from vitrified bovine sexed embryos following in-straw dilution. Anim Sci J 81:461–466

Anzalone DA, Palazzese L, Iuso D, Martino G, Loi P (2018) Freeze-dried spermatozoa: an alternative biobanking option for endangered species. Anim Reprod Sci 190:85–93. https://doi.org/10.1016/j.anireprosci.2018.01.010 (Epub 2018 Jan 31)

Blondin P, Bousquet D, Twagiramungu H, Barnes F, Sirard MA (2002) Manipulation of follicular development to produce developmentally competent bovine oocytes. Biol Reprod 66(1):38–43

Bogliotti YS, Wu J, Vilarino M, Okamura D, Soto DA, Zhong C, Sakurai M, Sampaio RV, Suzuki K, Izpisua Belmonte JC, Ross PJ (2018) Efficient derivation of stable primed pluripotent embryonic stem cells from bovine blastocysts. Proc Natl Acad Sci U S A 115 (9):2090–2095. https://doi.org/10.1073/pnas.1716161115 (Epub 2018 Feb 9)

Bondioli KR, Ellis SB, Pryor JH, Williams MW, Harpold MM (1989) The use of male-specific chromosomal DNA fragments to determine the sex of bovine preimplantation embryos. Theriogenology 31:95–104

Brackett BG, Bousquet D, Boice ML, Donawick WJ, Evans JF, Dressel MA (1982) Normal development following in vitro fertilization in the cow. Biol Reprod 27(1):147–158

Brophy B, Smolenski G, Wheeler T, Wells D, L'Huillier P, Laible G (2003) Cloned transgenic cattle produce milk with higher levels of beta casein and kappa-casein. Nat Biotechnol 21(2):157–162 Epub 2003 Jan 27

Carlson DF, Lancto CA, Zang B, Kim ES, Walton M, Oldeschulte D, Seabury C, Sonstegard TS, Fahrenkrug SC (2016) Production of hornless dairy cattle from genome-edited cell lines. Nat Biotechnol 34(5):479–481. https://doi.org/10.1038/nbt.3560 (No abstract available)

Cibelli JB, Stice SL, Golueke PJ, Kane JJ, Jerry J, Blackwell C, Ponce de León FA, Robl JM (1998a) Cloned transgenic calves produced from nonquiescent fetal fibroblasts. Science 280(5367):1256–1258

Cibelli JB, Stice SL, Golueke PJ, Kane JJ, Jerry J, Blackwell C, Ponce de León FA, Robl JM (1998b) Transgenic bovine chimeric offspring produced from somatic cell-derived stem-like cells. Nat Biotechnol 16 (7):642–646

Cibelli JB, Cunniff K, Vrana KE (2006) Embryonic stem cells from parthenotes. Methods Enzymol 418:117–135

Cortez JV, Vajta G, Valderrama NM, Portocarrero GS, Quintana JM (2018) High pregnancy and calving rates with a limited number of transferred handmade cloned bovine embryos. Cell Reprogram 20(1):4–8. https://doi.org/10.1089/cell.2017.0024

da Silva CG, Martins CF, Cardoso TC, da Cunha ER, Bessler HC, Martins GH, Pivato I, Báo SN (2016) Production of bovine embryos and calves cloned by nuclear transfer using mesenchymal stem cells from amniotic fluid and adipose tissue. Cell Reprogram 18 (2):127–136. https://doi.org/10.1089/cell.2015.0064

Echelard Y, Williams JL, Destrempes MM, Koster JA, Overton SA, Pollock DP, Rapiejko KT, Behboodi E, Masiello NC, Gavin WG, Pommer J, Van Patten SM, Faber DC, Cibelli JB, Meade HM (2009) Production of recombinant albumin by a herd of cloned transgenic cattle. Transgenic Res 18(3):361–376. https://doi.org/10.1007/s11248-008-9229-9 (Epub 2008 Nov 22)

Enright BP, Taneja M, Schreiber D, Riesen J, Tian XC, Fortune JE, Yang X (2002) Reproductive characteristics of cloned heifers derived from adult somatic cells. Biol Reprod 66(2):291–296

Hare WCD, Mitchell D, Betteridge KJ, Eaglesome MD, Randall GCB (1976) Sexing two-week old bovine embryos by chromosomal analysis prior to surgical transfer: preliminary methods and results. Theriogenology 5:243–253

Herr CM, Reed KC (1989) Micronanipulation of bovine embryos for sex determination. Theriogenology 35:45–54

Ishino T, Hashimoto M, Amagasa M, Saito N, Dochi O, Kirisawa R, Kitamura H (2018) Establishment of protocol for preparation of gene-edited bovine ear-derived fibroblasts for somatic cell nuclear transplantation. Biomed Res 39(2):95–104. https://doi.org/10.2220/biomedres.39.95

Jo HT, Bang JI, Kim SS, Choi BH, Jin JI, Kim HL, Jung IS, Suh TK, Ghanem N, Wang Z, Kong IK (2014) Production of female bovine embryos with sex-sorted sperm using intracytoplasmic sperm injection: efficiency and in vitro developmental competence. Theriogenology 81(5):675–682.e1. https://doi.org/10.1016/j.theriogenology.2013.11.010 (Epub 2013 Nov 21)

Kaneda M, Watanabe S, Akagi S, Inaba Y, Geshi M, Nagai T (2017) Proper reprogramming of imprinted and non-imprinted genes in cloned cattle gametogenesis. Anim Sci J 88(11):1678–1685. https://doi.org/10.1111/asj.12846 (Epub 2017 Jun 2)

Kasinathan P, Wei H, Xiang T, Molina JA, Metzger J, Broek D, Kasinathan S, Faber DC, Allan MF (2015) Acceleration of genetic gain in cattle by reduction of generation interval. Sci Rep 2(5):8674. https://doi.org/10.1038/srep08674

Kubota C, Tian XC, Yang X (2004) Serial bull cloning by somatic cell nuclear transfer. Nat Biotechnol 22 (6):693–694 (Epub 2004 May 23)

Kubota C, Yamakuchi H, Todoroki J, Mizoshita K, Tabara N, Barber M, Yang X (2000) Six cloned calves produced from adult fibroblast cells after long-term culture. Proc Natl Acad Sci U S A. 97 (3):990–995

Lanza RP, Cibelli JB, Faber D, Sweeney RW, Henderson B, Nevala W, West MD, Wettstein PJ (2001) Cloned cattle can be healthy and normal. Science 294 (5548):1893–1894 (No abstract available)

Kuwayama M, Vajta G, Ieda S, Kato O (2005) Comparison of open and closed methods for vitrification of human embryos and the elimination of potential contamination. Reprod Biomed Online. 11(5): 608–614

Matsushita H, Sano A, Wu H, Wang Z, Jiao JA, Kasinathan P, Sullivan EJ, Kuroiwa Y (2015) Species-specific chromosome engineering greatly improves fully human polyclonal antibody production profile in cattle. PLoS One 10(6):e0130699. https://doi.org/10.1371/journal.pone.0130699 (eCollection 2015)

Moore SG, Hasler JF (2017) A 100-year review: reproductive technologies in dairy science. J Dairy Sci 100 (12):10314–10331. https://doi.org/10.3168/jds.2017-13138 (Review)

Morrell JM, Valeanu AS, Lundeheim N, Johannisson A (2018) Sperm quality in frozen beef and dairy bull semen. Acta Vet Scand 60(1):41. https://doi.org/10.1186/s13028-018-0396-2

Nagata MPB, Endo K, Ogata K, Yamanaka K, Egashira J, Katafuchi N, Yamanouchi T, Matsuda H, Goto Y, Sakatani M, Hojo T, Nishizono H, Yotsushima K, Takenouchi N, Hashiyada Y, Yamashita K (2018) Live births from artificial insemination of microfluidic-sorted bovine spermatozoa characterized by trajectories correlated with fertility. Proc Natl Acad Sci U S A 115(14):E3087–E3096. https://doi.org/10.1073/pnas.1717974115 (Epub 2018 Mar 19)

Pipino C, Mandatori D, Buccella F, Lanuti P, Preziuso A, Castellani F, Grotta L, Di Tomo P, Marchetti S, Di Pietro N, Cichelli A, Pandolfi A, Martino G (2018) Identification and characterization of a stem cell-like population in bovine milk: a potential new source for regenerative medicine in veterinary. Stem Cells Dev. https://doi.org/10.1089/scd.2018.0114

Polge C, Rowson LEA (1952) Fertilizing capacity of bull spermatozoa after freezing at −79 °C. Nature 169:626–627

Prather RS, Barnes FL, Sims MM, Robl JM, Eyestone WH, First NL (1987a) Nuclear transplantation in the bovine embryo: assessment of donor nuclei and recipient oocyte. Biol Reprod 37(4):859–866

Prather RS, Spire MF, Schalles RR (1987b) Evaluation of cryopreservation techniques for bovine embryos. Theriogenology 28(2):195–204

Richt JA, Kasinathan P, Hamir AN, Castilla J, Sathiyaseelan T, Vargas F, Sathiyaseelan J, Wu H, Matsushita H, Koster J, Kato S, Ishida I, Soto C, Robl JM, Kuroiwa Y (2007) Production of cattle lacking prion protein. Nat Biotechnol 25(1):132–138 Epub 2006 Dec 31

Ross PJ, Beyhan Z, Iager AE, Yoon SY, Malcuit C, Schellander K, Fissore RA, Cibelli JB (2008) Parthenogenetic activation of bovine oocytes using bovine and murine phospholipase C zeta. BMC Dev Biol 19(8):16. https://doi.org/10.1186/1471-213x-8-16
Sano A, Matsushita H, Wu H, Jiao JA, Kasinathan P, Sullivan EJ, Wang Z, Kuroiwa Y (2014) Physiological level production of antigen-specific human immunoglobulin in cloned transchromosomic cattle. PLoS One 8(10):e78119. https://doi.org/10.1371/journal.pone.0078119 (eCollection 2013). Erratum in: PLoS One 9(1). https://doi.org/10.1371/annotation/82d9a0e0-a898-4b29-83d7-9daf6f5c384c
Selvaraju S, Parthipan S, Somashekar L, Binsila BK, Kolte AP, Arangasamy A, Ravindra JP, Krawetz SA (2018) Current status of sperm functional genomics and its diagnostic potential of fertility in bovine(Bos taurus). Syst Biol Reprod Med 1–18. https://doi.org/10.1080/19396368.2018.1444816 (Epub ahead of print)
Sirard MA (2018) 40 years of bovine IVF in the new genomic selection context. Reproduction 156(1):R1–R7. https://doi.org/10.1530/rep-18-0008 (Epub 2018 Apr 10. Review)
Tecirlioglu RT, French AJ, Lewis IM, Vajta G, Korfiatis NA, Hall VJ, Ruddock NT, Cooney MA, Trounson AO (2003) Birth of a cloned calf derived from a vitrified hand-made cloned embryo. Reprod Fertil Dev 15(7–8):361–366
Thomas JM, Locke JWC, Vishwanath R, Hall JB, Ellersieck MR, Smith MF, Patterson DJ (2017) Effective use of SexedULTRA™ sex-sorted semen for timed artificial insemination of beef heifers. Theriogenology 98:88–93. https://doi.org/10.1016/j.theriogenology.2017.03.018 Epub 2017 Mar 23
Vajta G, Bartels P, Joubert J, de la Rey M, Treadwell R, Callesen H (2004) Production of a healthy calf by somatic cell nuclear transfer without micromanipulators and carbon dioxide incubators using the Hand-made Cloning (HMC) and the Submarine Incubation System (SIS). Theriogenology 62(8):1465–1472
Vajta G, Holm P, Kuwayama M, Booth PJ, Jacobsen H, Greve T, Callesen H (1998) Open Pulled Straw (OPS) vitrification: a new way to reduce cryoinjuries of bovine ova and embryos. Mol Reprod Dev 51 (1):53–58
Vajta G, Lewis IM, Hyttel P, Thouas GA, Trounson AO (2001) Somatic cell cloning without micromanipulators. Cloning 3(2):89–95
Vajta G, Maddox-Hyttel P, Skou CT, Tecirlioglu RT, Peura TT, Lai L, Murphy CN, Prather RS, Kragh PM, Callesen H (2005) Highly efficient and reliable chemically assisted enucleation method for handmade cloning in cattle. Reprod Fertil Dev 17(8):791–797
Vajta G, Peura TT, Holm P, Páldi A, Greve T, Trounson AO, Callesen H (2000) New method for culture of zona-included or zona-free embryos: the Well of the Well (WOW) system. Mol Reprod Dev 55(3): 256–264
Vieira AD, Mezzalira A, Barbieri DP, Lehmkuhl RC, Rubin MI, Vajta G (2002) Calves born after open pulled straw vitrification of immature bovine oocytes. Cryobiology 45(1):91–94
Wilmut I, Rowson LE (1973) Experiments on the low-temperature preservation of cow embryos. Vet. Rec. 92:686–690
Wu X, Song M, Yang X, Liu X, Liu K, Jiao C, Wang J, Bai C, Su G, Liu X, Li G (2016) Establishment of bovine embryonic stem cells after knockdown of CDX2. Sci Rep. 20(6):28343. https://doi.org/10.1038/srep28343
Yadav PS, Kues WA, Herrmann D, Carnwath JW, Niemann H (2005) Bovine ICM derived cells express the Oct4 ortholog. Mol Reprod Dev 72(2):182–190
Yadav PS, Singh RK, Singh B (2012) Animal fetal stem cells- potential health applications. Agri Res 1:67–77
Yamanaka KI, Yamashita K, Khatun H, Wada Y, Tatemoto H, Sakatani M, Takenouchi N, Takahashi M, Watanabe S (2018) Normal DNA methylation status in sperm from a somatic cell cloned bull and their fertilized embryos. Anim Sci J. https://doi.org/10.1111/asj.13086. (Epub ahead of print)
Yu M, Muteti C, Ogugo M, Ritchie WA, Raper J, Kemp S (2016) Cloning of the African indigenous cattle breed Kenyan Boran. Anim Genet 47(4):510–511. https://doi.org/10.1111/age.12441
Yuan Y (2018) Capturing bovine pluripotency. Proc Natl Acad Sci U S A 115(9):1962–1963. https://doi.org/10.1073/pnas.1800248115 (Epub 2018 Feb 14. No abstract available)
Zawada WM, Cibelli JB, Choi PK, Clarkson ED, Golueke PJ, Witta SE, Bell KP, Kane J, Ponce de Leon FA, Jerry DJ, Robl JM, Freed CR, Stice SL (1998) Somatic cell cloned transgenic bovine neurons for transplantation in parkinsonian rats. Nat Med 4(5): 569–574

Reproduction Biotechnology in Pigs 15

Abstract

Pig is an important livestock species used for pork and as valuable model animal for biomedical sciences. Adoption of reproduction biotechniques has vividly transformed the way pig has been raised for clinical research. Nuclear transfer cloning, in vitro embryo production, stem cell biology and engineering, and production of transgenic pigs for producing therapeutics, and organs for transplantation are the remarkable achievements. Pig as model animals is important for research on metabolic and lifestyle diseases and regenerative medicine.

Highlights

- Pigs are the main sources of animal origin meat and as a model animal in biomedical sciences
- Besides, genetically modified pig is used to generate cells, tissue, and organs for human xenotransplantation.

Keywords

Reproduction biotechnology · Transgenic pigs · Xenotransplantation · Biomedical applications

15.1 Introduction

Pig is an important livestock for meat (pork) and a model species for human genetic and metabolic diseases because of its prominent homology to human anatomy, physiology, and genetics. In view of low age at maturity, high fertility, short gestation period, multiple litter size per pregnancy, and tendency to rebreed shortly after parturition or farrowing make pig as animal of choice for commercial livestock rearing and marginal farmers.

Whereas pork is an essential share of diet in developed countries, many countries are global leader in pork production. People eat pork and pork-based products such as bacon, pork chops, and sausages. China is the biggest consumer of meat, and pork fulfills the demand. The statistics show that China, with the production of 54.65 million metric tons of pork per annum, is the global leader in pig production (Figs. 15.1 and 15.2).

A remarkable success is achieved in assisted reproduction in pig that has dramatically transformed the way pig is raised for pork production and biomedical sciences (Knox 2014; Zhou et al. 2015; Callesen et al. 2017). Reproductive biotechniques are utilized primarily to promote the production of pork and animals for

B. Singh et al., *Advances in Animal Biotechnology*,
https://doi.org/10.1007/978-3-030-21309-1_15

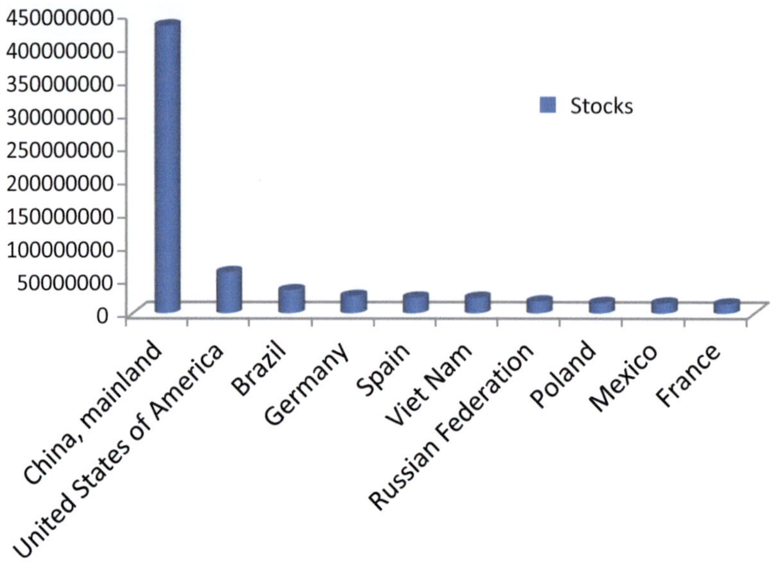

Fig. 15.1 Top ten pig producers' countries (Average 1994–2016). *Source* http://www.fao.org/faostat/en/#data/QA/visualize

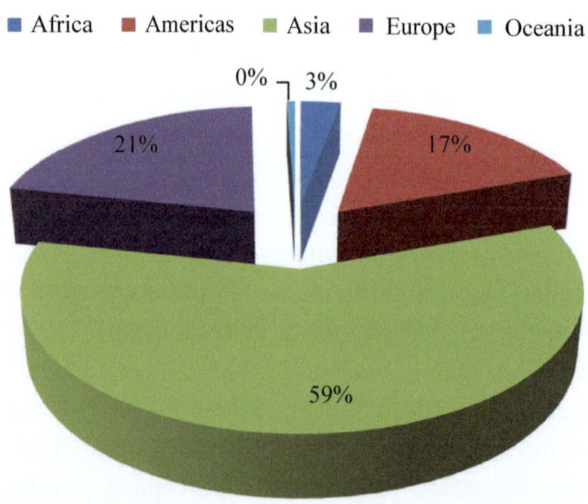

Fig. 15.2 Production share of pigs by region (Average 1994-2016). *Source* http://www.fao.org/faostat/en/#data/QA/visualize

biomedical research. However, many ARTs are restricted to particular segments of scientific organizations and commercial porcine producers. In addition, the progress in porcine genetic engineering is slow, primarily because of lack of germline-competent stem cells.

15.2 Pork as Important Food

Demand for animal food is rising in recent decades due to the increased income and fast-growing economies of the developing countries. Pork is the top-ranked food obtained from terrestrial animals. Pig is an efficient converter of normal food stuff into valuable animal proteins (Fig. 15.3). In view of the inherent adaptation to variable climates in tropical and temperate climates, ability to consume available feedstuff (maize, oats, barley, bakery waste, fish, and poultry meal, distiller's dried grain, molasses etc.), the commercial porcine breeds (e.g. Hampshire, Duroc, Chester white, Berkshire, Landrace, Poland China, etc.), and traditional or native pigs are reared under different agricultural production systems and commercial setup.

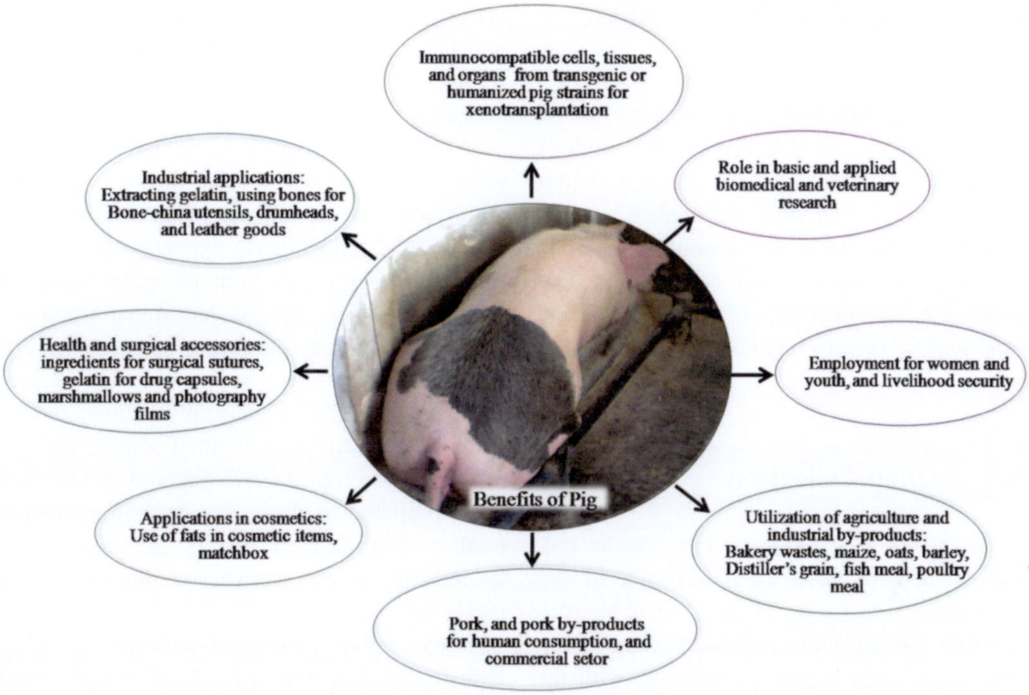

Fig. 15.3 Pig as multipurpose livestock and model animal for biomedical applications. No other animal provides benefits to human nutrition, health, and livelihood more than pig. Suede obtained from pig is used in shoes and fabric, gelatin in food, and non-food uses. Lard, the adipose tissue and fat, is used in shaving creams, soaps, makeup, and cosmetics

Pig farming is almost global, except for some regions with taboos, cultural and religious sentiments regarding consumption of pork. There are variations in pig husbandry practices in different regions of the world. Particularly, in developing countries, majorities of pigs are raised by marginally progressive farmers merely to supplement their agricultural income. On the contrary, in developed European countries and North America, the swine industry is highly advanced and scientifically managed entrepreneur integrated with crop production system.

As the majority of commercially produced pork for domestic consumption and export originates from organized farms (Knox et al. 2013), emphasis should be on safer and economical production of pork. Control on diseases, infectious agents, prohibiting antibiotics, and monensins as growth promoters should be taken care of.

15.3 Pigs as Important Model Animals in Biomedical Sciences

Pig is an indispensable model animal in diverse lines of biomedical research and fills the gap between rodent models and human patients (Flisikowska et al. 2014; Fischer et al. 2016).

Nuclear transfer cloning, transgenesis, genome editing, and stem cell technologies have enabled the researchers to create unique genotypes to develop therapeutics for curing genetic disorders, metabolic diseases, and cancer (Guitierrez et al. 2015). With advent of methods to overcome hyperacute rejection and acute vascular rejection by immune system of patients, the organs obtained from humanized pig strains could fulfill the human demand for tissues and organs (Fischer et al. 2016).

15.4 Male-Assisted Reproduction

Artificial insemination (AI) is used in commercial pig production. The benefits of AI noted in porcine-assisted reproduction include reduced cost on management of boar, better hygiene, and production of quality semen, improved fertility as each ejaculate is screened before approved for use, ease of using any boar to the sow irrespective of size, health, and temperament, thereby achieving rapid genetic improvements.

15.5 Semen Collection

It is necessary to maintain hygiene while collecting semen from boar. It is advisable to clip the long hairs on sheath to avoid their interference with the collection process. Semen is collected either using a sow in estrus or with the help of dummy sow (Fig. 15.4). The boar is allowed to mount and make a few copulation thrusts and display corkscrew activity. It is necessary to use latex-free gloves or avoid contact of any spermicidal substances while directing the penis to collect semen. After a few moments, the boar will cease thrusting and begin to ejaculate. Rhythmic pulsation should be given during the collection of semen. The gel traction should be filtered out of the ejaculate during collection using sterilized gauge putting into boucher funnel. The grip on the penis should not be loosened during ejaculation which lasts on average for 7–10 min.

AI has transformed porcine industry into an efficient production system through disseminating superior germplasm, controlling risk of disease, and reducing the expenses involved in import and export of high pedigree boars for mating the gilts. The AI has transformed the market strategy of porcine breeding from the selling of genetically superior boars to processing and selling high-quality semen. Widespread use of AI has benefitted commercial pig producers through genetic gains at economical cost compared to those expected from natural mating. European producers are the global leaders in using AI for commercial pork production.

Some threats associated with AI in swine industry include the spread of certain viral pathogens including PPR and swine viruses (Maes et al. 2016). Using disease-free semen, the countries can save resources, time, and money required to manage swine diseases such as foot and mouth diseases, swine influenza, African swine fever, classical swine fever, pseudorabies, and Japanese Encephalitis.

The progress in male-assisted reproduction was improved due to the availability of semen collecting bags, development of AI catheters,

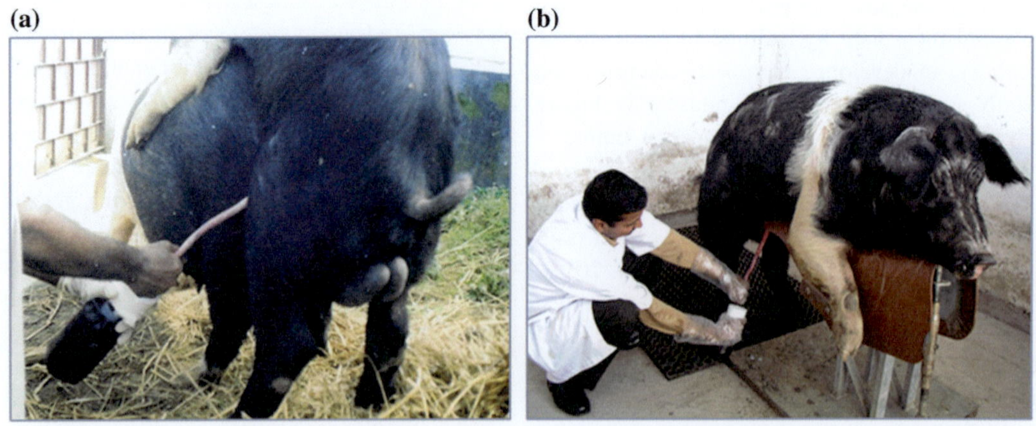

Fig. 15.4 Semen collection in boars. **a** collecting semen using a sow in estrus, **b** using dummy to collect the semen. Collecting semen from boar is tricky and requires skilled personnel during the process

extenders for long-term preservation of semen, automatic or semiautomatic semen packaging devices.

The commercial swine industry has adopted scientific methods of evaluating genetic merit of animals and production parameters. The nucleus grandparent (GGP) animals are perpetuated and maintained as purebreds and used as donors of germplasm to create F_1 animals or using semen of GGP for crossbreeding programs. The F1 animals, based on their performance, are used as future sires or dams. The problem in the use of GGP is that genetic diversity cannot be promoted, there remain chances of inbreeding depression and risk of dissemination of pathogens to new animals.

Both AI and natural service or mating are practiced in pig reproduction. The best quality boar should be selected for collecting semen. Young boars need contact with other boars to develop and main service performance. Keeping boars near sow pens is important to promote mating. Ejaculate having motility and normal morphology of sperm over 70% and total sperm count ≥ 3 billion is evaluated as a good one. The semen should be free from contamination and disease-causing agents.

Advanced methods, such as CASA, are used to evaluate and ascertain semen quality.

15.6 Sperm Sexing

Quantitative flow cytometry is used to separate X- and Y-chromosome bearing sperm. The efficiency of sperm sorting and fertilizing ability of sexed sperm is low. Applications of nanotechnology might improve sperm sorting, and certain additives such as EDTA might improve fertilizing ability of flow cytometric sex-sorted sperm (del Olmo et al. 2013). AI using sexed sperm is not a routine practice and is restricted to a few experimental uses or specific cases such as conservation of genetic resources.

Potential of sexed sperm is not fully harnessed in swine industry. There are some constraints associated with the use of liquid frozen sperm. The sperm is exposed to stress during sorting. To improve gains from AI, attention should be paid to decrease the number of sperm required per insemination dose, enhancing efficiency of sperm sexing, making available sexed sperm in commercial swine insemination programs (Roca et al. 2016).

New generation apps such as iSperm should be used for ease and accuracy of AI and production control. Nutrition management is an important aspect of pig husbandry. The pigs should be offered ample water for drinking and keeping their body cool in summer.

The biomaterials during transportation are subjected to vibrations that might affect the viability of cells. Though practice-oriented data is not available regarding general shipment procedures of frozen pig semen and the effects of vibrations on sperm quality and fertility, it is speculated that vibrations might affect semen quality. An electronic mobile sensing app (TransportLog 1.0) having built-in sensors in smartphones are developed to capture the vibrations generated during transportation of sperm containers. The device might measure the impact of vibrations on sperm and form basis for formulating recommendations or guidelines for shipping of spermatozoa (Schulze et al. 2018).

15.7 Sources of Oocytes and Embryos and Their Cryopreservation

Pigs are slaughtered routinely in most countries. Hence, ovaries can be obtained from abattoirs and used to obtain immature oocytes. In view of body size, the collection or aspirating oocytes from live animals is difficult.

Till the late 1990s, majority of the embryo preservation protocols were available for livestock species other than pig. The science of cryopreservation of pig embryos started at a faster pace since 1997. This is because pig oocytes and embryos are sensitive to hypothermia.

Delipation technology provided the proof of the concept that intracellular lipids have a correlation with hypothermic sensitivity of porcine oocytes and embryos. While the major focus of studies was on lipid content of pig embryos, little focus was on structural damage caused by cryopreservation. Dobrinsky et al. (2000) documented alterations in microfilament (MF) during and after cryopreservation of pig embryos, and utilized Cytochalasin B, a MF-inhibitor during cryopreservation by vitrification, and evaluated its effect on developmental competence of cytoskeleton-stabilized vitrified embryos. The Cytochalasin B improved survival and development of expanded and hatched blastocysts, which led to the conclusion that MF depolymerization prior to vitrification could improve embryo developmental competence after cryopreservation, and that vitrified-warmed blastocysts can be used to get healthy offspring (Dobrinsky et al. 2000).

15.8 Embryo Production, Preservation, and Transfer in Swine

Producing embryos is a common practice in clinics globally for genetic testing and stem cells for clinical applications. Many of the approaches originally invented and developed for humans are applied for animal species and vice versa.

Increasing productivity by implementing methods that are faster and lead to genetic improvement of animals is imperative. The purpose of embryo production in pig and cattle is to enhance their productivity to fulfill the milk and meat demand of humans.

Like other livestock, the pig production is also benefited from embryo biotechnology. To maximize the benefits, and reduce the generation time, the focus should be on getting embryos from females at younger stages, genetic profiling of embryos using state-of-the-art technologies such as SNP chips and sequencing (Fowler et al. 2018).

Notably, the zygotes produced in vivo are of desirable quality for genome editing and microinjection, but technical difficulties hamper in vivo production and retrieval of embryos. The embryos can be collected only after anesthetization of the animals, and the number of embryo recovered per animal is low (Gil et al. 2017a, b).

There are certain limitations that restrict the fluent use of embryo technologies in pig. The embryos and oocytes are sensitive to cryopreservation. Transfer of embryo is a surgical procedure that is difficult and expensive. A newly developed non-surgical deep intrauterine (NsDU) transfer of embryos into sow uterine horns (Martinez et al. 2004), and birth of piglets using NsDU (Cuello et al. 2005), has solved some difficulties associated with surgical transfer of embryos.

The standard methods, as developed for cryopreservation of cells and gametes in other livestock species, are applied in pig. A novel hollow fiber vitrification (HFV) (each hollow fiber devise containing 40 mice embryos or 20 porcine embryos in 2–3 mm long column in less than 0.1 μl of solution) method is developed for vitrification of cryosensitive IVM- or IVF-derived embryos (Maehara et al. 2012; Matsunari et al. 2012).

15.9 Transgenesis and Transgenic Pigs

Demand of recombinant proteins has increased in various commercial and experimental sectors. Biopharming is a technology in which transgenic organisms are generated for cheaper and bulk production of pharmaceuticals. Since its inception (Hammer et al. 1985), the animal pharming industry has promoted transgenic animals as a cost-effective method of biopharmaceutical production. Currently, a number of techniques ranging from conventional microinjection of genes into pronucleus of zygote, and virus vector-mediated gene transfer to somatic cells are available (Galli et al. 2010). The transgenic cells are used to produce cloned animals or chimera (West et al. 2015).

Pigs as model animals have applications in human biomedical testing and producing recombinant therapeutic proteins. Transgenic pigs were initially developed by microinjection of exogenous genes into zygote or early blastocyst stages of mammalian species. Then came the new era of inducing genetic modifications in somatic or stem cells, and using them to produce cloned transgenic animals.

CRISPR/Cas9 is a recent technology that facilitates the production of genetically modified or genome-edited immunocompatible animals as source of organs for xenotransplantation. The efficacy of CRISPR/Cas9 is evidenced in several cell lines and livestock species. Birth of six healthy knockout piglets by cytoplasmic microinjection of CRISPR/Cas9 vector for encoding a single-guide RNA (sgRNA) targeting exon 8 of the *GGTA1* gene into cytoplasm of pig zygotes (Petersen et al. 2016), generation of gene-modified pigs producing orthologus human mutation using CRISPR/Cas9 (Zhou et al. 2016), and birth of insulin-deficient piglets by CRISPR/Cas9-mediated genetic alterations (Cho et al. 2018), are some of the prominent achievements in biomedical sciences.

The CRISPR/Cas9-GMNN fusion gene editing was used to generate knock-in pig cell lines. In this study, humanized Cas9 was fused to N-terminal peptide of Geminin gene (GMNN), a highly conserved DNA replication-inhibiting protein found in most eukaryotes. The study paves the way to enrich specific knock-in genes in somatic cells that can be used to produce knock-in gene-edited cloned animals (Gerlach et al. 2018).

15.10 Somatic Cell Nuclear Transfer Cloning and Transgenesis

Due to the general similarities in organ sizes, physiology, and metabolism with humans, the pigs are the most preferred research model animals. Genetically modified pigs are developed for studying human diseases and producing organs for xenografting or xenotransplantation to combat the shortage of human organs.

SCNT serves as the most valuable method to resurrect the species or repopulate endangered mammals from somatic cells or viable cell nuclei. The SCNT is disgracefully intricate and inefficient in pigs. Soon after the birth of cloned sheep and goats, the SCNT procedure was modified for pigs. The year 2000 was remarkable as researchers from different laboratories announced births of healthy piglets (Betthauser et al. 2000; Onishi et al. 2000; Li et al. 2000; Polajaeva et al. 2000). The subsequent studies focused on producing somatic cell lines with altered genetic makeup or inclusion of exogenous genes in somatic cells and MSCs, and using them to produce transgenic cloned pigs.

Despite hurdles, the SCNT is used to produce cloned transgenic pigs as sources of immunocompatible tissue and organs. Majority of transgenic pigs are the cloned animals (Knox 2014; Yang et al. 2016; Zhu et al. 2018) as summarized in (Table 15.1).

15.11 Porcine Stem Cells

Almost all types of stem cells ranging from embryonic stem cells, perinatal stem cells, MSCs and reprogrammed or iPSCs are reported in pigs. ICMs cells of developing or pre-implantation embryos serve as the source of embryonic stem cells. MSCs are isolated from bone marrow, adipose tissue, perinatal or fetal and placental tissues including amniotic fluid and cord blood are the prime sources of stem cells (Bharti et al. 2016).

It is difficult to obtain MSCs from bone marrow of live animals. To overcome the hurdle, Faast et al. (2006) have suggested blood as a source of MSCs. The blood-derived MSCs were multipotent and could develop into embryo following SCNT.

The porcine uterus is an another source of MSCs. Real-time PCR showed that MSCs isolated from inner lining of porcine uterus could differentiate into adipogenic and osteogenic lineages and expressed MSC-specific markers such as CD29, CD44, CD144, CD105, and CD140b (Miernik and Karasinski 2012).

Apart from the basic description of stem cells, fundamental properties and potential, the emphasis is on utilizing the stem cells for reproduction or health applications. Some important examples are emphasized here. Sepsis, a life-threatening organ dysfunction emerging due to

Table 15.1 A chronological overview of milestone achievements in ARTs in pig

Year	Events, inferences, and recommendations
1985	Production of transgenic piglets (Hammer et al. 1985)
1986	Birth of piglets from IVF and embryo transfer
1990	Birth of piglets from transfer of frozen-thawed embryos, first report on primitive ESCs (Evans et al. 1990)
1992	Birth of transgenic pigs expressing mice Mx1 proteins that confer protection against influenza virus (Müller et al. 1992)
1997	Isolation of pluripotent stem cells from cultured porcine primordial germ cells (Shim et al. 1997)
2000	Production of transgenic pig with eco-friendly exogenous genes (phytase) addressing environmental issues (McCreath et al. 2000), Birth of four male piglets by SCNT using fetal fibroblasts, indicating that SCNT would allow production of genetically modified pigs with reduced immunogenicity for use in xenotransplantation (Betthauser et al. 2000), cloned female piglet born from microinjection of fetal fibroblast nuclei into cytoplast (Onishi et al. 2000), cloned piglet from a modified SCNT procedure (Polejaeva et al. 2000), NT cloned piglets from embryonic blastomeres from (Li et al. 2000)
2002	Birth of heterozygous knockout pigs (Lai et al. 2002)
2003	Birth of homozygous knockout pigs (Phelps et al. 2003), development of strategy for gene transfer into mammalian zygotes via lentiviruses (Hofmann et al. 2003), isolation of pluripotent cell lines from embryos of diverse origins (Li et al. 2003), Birth of cloned piglets from fetal fibroblasts transfected with enhanced GFP (Hyun et al. 2003)
2004	Development of a non-surgical deep intrauterine (NsDU) method for transfer of embryos into uterine horns (Martinez et al. 2004)
2005	Birth of live piglets from NsDU transfer (Cuello et al. 2005) Development of SCNT for producing human erythropoietin (hEPO)-transgenic cloned piglets (Lee et al. 2005)
2006	Pronucleus DNA microinjection in pigs (Kues et al. 2006), isolation of MSCs from peripheral blood, and their assessment as donor cells to produce cloned embryos (Faast et al. 2006)
2007	Development of a simplified method for cryopreservation of porcine embryos (Du et al. 2007c), birth of piglets from handmade cloning (Du et al. 2007a, b)
2008	Birth of α1,3-galactosyltransferase gene-deficient cloned piglets (Fujimura et al. 2008)
2009	Derivation and characterization of iPSCs (Wu et al. 2009)
2010	Exploring and validating therapeutic applications of MSCs in a porcine model (Dutton et al. 2010), birth of transgenic piglets by ICSI-SMGT in combination with recombinase RecA (Garcia-Vazquez et al. 2010)
2011	Birth of antibody- and B cell-deficient cloned pigs by disruption of J-region gene segment of the immunoglobulin heavy chain (Mendicino et al. 2011), birth of transgenic pigs as a model for Stargardt-like macular degeneration (Sommer et al. 2011)
2012	ICSI-mediated gene transfer for producing cloned pigs, isolation of MSCs from uterus, expressed MSC-specific markers such as CD29, CD44, CD144, CD105, and CD140b (Miernik and Karasinski 2012)
	Development of novel hollow fiber vitrification (HFV) (each hollow fiber devise containing 40 mouse embryos or 20 porcine embryos in less than 0.1 μl of solution) method for vitrification of cryosensitive IVM- or IVF-derived embryos (Maehara et al. 2012; Matsunari et al. 2012),
	Using adipose tissue-derived MSCs to regenerate damaged kidney cells in experimental renovascular disease (Eirin et al. 2012)
2013	Birth of Yorkshire pigs with phytase expressed in salivary glands (Meidinger et al. 2013), development of transgenic pig, mice, cattle embryos using SCNT and pig oocytes (Gupta et al. 2013), birth of cloned piglets from iPSCs (Fan et al. 2013), piglets expressing human α-galactosidase (Zeyland et al. 2013), birth of double knock-out piglets as model to produce organs for xenotransplantation (Lutz et al. 2013)
2014	Transgenic pig carrying a DsRed-monomer reporter gene (Chou et al. 2014), cloned transgenic pigs from recombinant cells produced by piggyBac transposase-based pmGENIE-3 plasmids (Li et al. 2014),

(continued)

Table 15.1 (continued)

Year	Events, inferences, and recommendations
	transgenic pigs produced by microinjection of sleeping beauty tranposons (Ivics et al. 2014), birth of transgenic class I MHC- null pigs (Reyes et al. 2014)
2015	Transgenic pig model expressing fungal cellulose transgenes (Lin et al. 2015), cloned transgenic miniature pigs expressing human apolipoprotein(a) in plasma (Ozawa et al. 2015), using zinc-finger nucleases-mediated knockout targeted genes in pig cells (Hisamatsu et al. 2015)
2016	Development of transgenic pigs suitable for producing organs for xenotransplantation (Fischer et al. 2016), CRISPR/Cas9-meadited generation of somatic cell reprogramming-free myostatin-mutant pigs (Tanihara et al. 2016), birth of six healthy knockout piglets by cytoplasmic microinjection of CRISPR/Cas9 into pig zygotes (Peterson et al. 2016), generation of gene-modified pigs producing orthologous human mutation using CRISPR/Cas9 (Zhou et al. 2016), generation of GGTA1-null Dinnan cloned genome-edited miniature pigs using TALENs (Cheng et al. 2016), birth of B2 m-null pigs with TALENs (Wang et al. 2016), generation of efficient method for genome editing in pig (Sheets et al. 2016)
2017	Isolation and description of stem Leydig cells reported from testicular interstitium (Yu et al. 2018), birth of piglets carrying human follistatin 344, screening the role of follistatin 344 in increasing muscle mass in pig (Chang et al. 2017), generation of Dickkopf-related protein 1 (DKK1) transgenic cloned pig (Liu et al. 2017)
2018	Isolation of tendon-derived stem cells (TDSCs) with possible applications in tendon injuries (Yang et al. 2018), development of novel strains of transgenic pigs with single-copy quad-cistronic transgene, simultaneous expression of three microbial enzymes, namely β-glucanase, xylanase, and phytase in salivary glands, thus reducing release of nutrients into environment (Zhang et al. 2018), CRISPR/Cas9-mediated generation of cloned transgenic insulin-deficient piglets (Cho et al. 2018), production of transgenic pig as central nervous system disease model (Hwang et al. 2018)
	Development of an electronic mobile sensing app (TransportLog 1.0), for measuring effects of vibrations emitted during transportation, and their effect on quality of sperm (Schulze et al. 2018)
	First-time demonstration of property of umbilical cord-derived stem cells as a promising treatment against septic shock (Laroye et al. 2018)
	Therapeutic use of MSCs to improve renal microcirculation and hemodynamics in a renovascular disease pig model. The study presents a novel therapeutic option to prevent renal metabolic syndrome (Eirin et al. 2018)

the impaired response of host to pathogenic infection, is among the leading causes of mortality in humans and animals. The treatments are scarce for preventing sepsis and septic shock. The normal MSCs (or MSCs containing exogenous genes) due to their immunomodulatory property, and conferring protection against endotoxemia, as confirmed in mice models (Mei et al. 2007), are suggested to improve organ failure. Administration of clinical grade umbilical cord-derived MSCs into a pig model of polymicrobial septic shock, significantly improved hemodynamic parameters, and alleviated organ injury, thus prevented the animal against cardiovascular failure (Laroye et al. 2018).

The adipose tissue-derived MSCs can regenerate damaged kidney cells. These cells in combination with percutaneous transluminal renal angioplasty could restore kidney function in experimental renovascular disease (Erin et al. 2012). Extracellular vesicles from autologous adipose tissue-derived MSCs, with ability to promote angiogenesis, i.e., being pro-angiogenic in nature, were found to improve renal microcirculation and hemodynamics in a renovascular disease pig model. The study presents a novel therapeutic option to prevent renal metabolic syndrome (Erin et al. 2018). Peripheral blood-derived MSCs when combined with modified demineralized bone matrix, such as granulocyte colony-stimulating factor (G-CSF), and the CXCR4 antagonist AMD3100, were found to promote cartilage repair (Zhao et al. 2018).

15.12 Outlook and Challenges

Swine industry is an important contributor to national economy and GDP. Equally important is that rearing or keeping pig in urban and peri-urban localities may also be challenging and potential health danger.

Undoubtedly, the reproduction technologies, especially AI and embryo transfer will improve pig production. There are a few problems that should be addressed. The cryopreservation of oocytes and embryos should be improved. Genetic abnormalities should be identified and the animals found to be positive for such abnormalities should be eradicated from breeding programs. Boars in AI centers or pig farms should be genetically proven and free from contagious infections. The commercial pig farming sector has already started marker-assisted selection in pig breeding programs.

Whereas in vitro embryo production has a low success rate, in vivo embryo collection is labor-intensive and requires technically skilled persons for anesthetization donor females. Also, more number of females is required to obtain enough good quality embryos.

The nuclear transfer cloning is being improved using cell lines with superior epigenetic characteristics and the MSCs or skin-derived stem cells that support embryo development. The enzymatic engineering in combination with vectors such as transposons will hopefully improve precision of genetic engineering.

Like other livestock species, cryopreservation is an important component of porcine-assisted reproduction. The pig oocytes and embryos are more sensitive to cryopreservation. There is a need to improve cryosurvival of cloned transgenic or genome-edited zygotes and embryos that are susceptible to cryopreservation and other kinds of stress compared to normal counterparts.

15.13 Conclusions

The pig reproduction biotechniques are of immense concern to make pig as efficient livestock species for food and medicine, but require considerable research inputs to resolve problems currently associated with their larger-scale applications. With the growing swine industry, there is an ever-growing demand for genetically superior females, and the males or semen from boars having high fertility and production traits. This is necessary to improve commercial swine production in developing countries.

References

Betthauser J, Forsberg E, Augenstein M, Childs L, Eilertsen K, Enos J, Forsythe T, Golueke P, Jurgella G, Koppang R, Lesmeister T, Mallon K, Mell G, Misica P, Pace M, Pfister-Genskow M, Strelchenko N, Voelker G, Watt S, Thompson S, Bishop M (2000) Production of cloned pigs from in vitro systems. Nat Biotechnol 18(10):1055–1059

Bharti D, Shivakumar SB, Subbarao RB, Rho GJ (2016) Research advancements in porcine derived mesenchymal stem cells. Curr Stem Cell Res Ther 11(1):78–93 (Review)

Callesen MM, Árnadóttir SS, Lyskjaer I, Ørntoft MW, Høyer S, Dagnaes-Hansen F, Liu Y, Li R, Callesen H, Rasmussen MH, Berthelsen MF, Thomsen MK, Schweiger PJ, Jensen KB, Laurberg S, Ørntoft TF, Elverløv-Jakobsen JE, Andersen CL (2017) A genetically inducible porcine model of intestinal cancer. Mol Oncol. 11(11):1616–1629. https://doi.org/10.1002/1878-0261.12136 (Epub 2017 Oct 10)

Chang F, Fang R, Wang M, Zhao X, Chang W, Zhang Z, Li N, Meng Q (2017) The transgenic expression of human follistatin-344 increases skeletal muscle mass in pigs. Transgenic Res 26(1):25–36. https://doi.org/10.1007/s11248-016-9985-x (Epub 2016 Oct 27)

Cheng W, Zhao H, Yu H, Xin J, Wang J, Zeng L, Yuan Z, Qing Y, Li H, Jia B, Yang C, Shen Y, Zhao L, Pan W, Zhao HY, Wang W, Wei HJ (2016) Efficient generation of GGTA1-null Diannan miniature pigs using TALENs combined with somatic cell nuclear transfer. Reprod Biol Endocrinol 14(1):77

Cho B, Kim SJ, Lee EJ, Ahn SM, Lee JS, Ji DY, Lee K, Kang JT (2018) Generation of insulin-deficient piglets by disrupting INS gene using CRISPR/Cas9 system. Transgenic Res 27(3):289–300. https://doi.org/10.1007/s11248-018-0074-1 (Epub 2018 Apr 24)

Chou CJ, Peng SY, Wu MH, Yang CC, Lin YS, Cheng WT, Wu SC, Lin YP (2014) Generation and characterization of a transgenic pig carrying a DsRed-monomer reporter gene. PLoS One 9(9): e106864. https://doi.org/10.1371/journal.pone.0106864 (eCollection 2014)

Cuello C, Berthelot F, Martinat-Botté F, Venturi E, Guillouet P, Vázquez JM, Roca J, Martínez EA (2005) Piglets born after non-surgical deep intrauterine transfer of vitrified blastocysts in gilts. Anim Reprod Sci 85(3–4):275–286

del Olmo D, Parrilla I, Gil MA, Maside C, Tarantini T, Angel MA, Roca J, Martinez EA, Vazquez JM (2013) Handling of boar spermatozoa during and after flow cytometric sex-sorting process to improve their in vitro fertilizing ability. Theriogenology 80(4):350–356. https://doi.org/10.1016/j.theriogenology.2013.04.022 (Epub 2013 Jun 7)

Dutton AQ, Choong PF, Goh JC, Lee EH, Hui JH (2010) Enhancement of meniscal repair in the avascular zone using mesenchymal stem cells in a porcine model. J Bone Joint Surg Br 92(1):169–175. https://doi.org/10.1302/0301-620X.92B1.22629

Dobrinsky JR, Pursel VG, Long CR, Johnson LA (2000) Birth of piglets after transfer of embryos cryopreserved by cytoskeletal stabilization and vitrification. Biol Reprod 62(3):564–570

Du Y, Kragh PM, Zhang Y, Li J, Schmidt M, Bøgh IB, Zhang X, Purup S, Jørgensen AL, Pedersen AM, Villemoes K, Yang H, Bolund L, Vajta G (2007a) Piglets born from handmade cloning, an innovative cloning method without micromanipulation. Theriogenology 68(8):1104–1110 (Epub 2007 Sep 21)

Du Y, Li J, Kragh PM, Zhang Y, Schmidt M, Bøgh IB, Zhang X, Purup S, Kuwayama M, Jørgensen AL, Pedersen AM, Villemoes K, Yang H, Bolund L, Vajta G (2007b) Piglets born from vitrified cloned blastocysts produced with a simplified method of delipation and nuclear transfer. Cloning Stem Cells 9(4):469–476

Du Y, Zhang Y, Li J, Kragh PM, Kuwayama M, Ieda S, Zhang X, Schmidt M, Bøgh IB, Purup S, Pedersen AM, Villemoes K, Yang H, Bolund L, Vajta G (2007c) Simplified cryopreservation of porcine cloned blastocysts. Cryobiology 54(2):181–187 (Epub 2007 Jan 27)

Eirin A, Zhu XY, Krier JD, Tang H, Jordan KL, Grande JP, Lerman A, Textor SC, Lerman LO (2012) Adipose tissue-derived mesenchymal stem cells improve revascularization outcomes to restore renal function in swine atherosclerotic renal artery stenosis. Stem Cells 30(5):1030–1041. https://doi.org/10.1002/stem.1047

Eirin A, Zhu XY, Jonnada S, Lerman A, van Wijnen AJ, Lerman LO (2018) Mesenchymal stem cell-derived extracellular vesicles improve the renal microvasculature in metabolic renovascular disease in swine. Cell Transplant 27(7):1080–1095. https://doi.org/10.1177/0963689718780942 (Epub 2018 Jun 28)

Evans MJ, Notarianni E, Laurie S, Moor RM (1990) Derivation and preliminary characterization of pluripotent cell lines from porcine and bovine blastocysts. Theriogenology. 33, 125–128

Faast R, Harrison SJ, Beebe LF, McIlfatrick SM, Ashman RJ, Nottle MB (2006) Use of adult mesenchymal stem cells isolated from bone marrow and blood for somatic cell nuclear transfer in pigs. Cloning Stem Cells 8(3):166–173

Fan N, Chen J, Shang Z, Dou H, Ji G, Zou Q, Wu L, He L, Wang F, Liu K, Liu N, Han J, Zhou Q, Pan D, Yang D, Zhao B, Ouyang Z, Liu Z, Zhao Y, Lin L, Zhong C, Wang Q, Wang S, Xu Y, Luan J, Liang Y, Yang Z, Li J, Lu C, Vajta G, Li Z, Ouyang H, Wang H, Wang Y, Yang Y, Liu Z, Wei H, Luan Z, Esteban MA, Deng H, Yang H, Pei D, Li N, Pei G, Liu L, Du Y, Xiao L, Lai L (2013) Piglets cloned from induced pluripotent stem cells. Cell Res 23(1):162–166. https://doi.org/10.1038/cr.2012.176 (Epub 2012 Dec 18. No abstract available)

Fischer K, Kraner-Scheiber S, Petersen B, Rieblinger B, Buermann A, Flisikowska T, Flisikowski K, Christan S, Edlinger M, Baars W, Kurome M, Zakhartchenko V, Kessler B, Plotzki E, Szczerbal I, Switonski M, Denner J, Wolf E, Schwinzer R, Niemann H, Kind A, Schnieke A (2016) Efficient production of multi-modified pigs for xenotransplantation by 'combineering', gene stacking and gene editing. Sci Rep. 29(6):29081. https://doi.org/10.1038/srep29081

Flisikowska T, Kind A, Schnieke A (2014) Genetically modified pigs to model human diseases. J Appl Genet 55:53–64

Fowler KE, Mandawala AA, Griffin DK, Walling GA, Harvey SC (2018) The production of pig preimplantation embryos in vitro: Current progress and future prospects. Reprod Biol 18(3):203–211. https://doi.org/10.1016/j.repbio.2018.07.001 (Epub 2018 Jul 7. Review)

Fujimura T[1], Takahagi Y, Shigehisa T, Nagashima H, Miyagawa S, Shirakura R, Murakami H (2008) Production of alpha 1,3-galactosyltransferase gene-deficient pigs by somatic cell nuclear transfer: a novel selection method for gal alpha 1,3-Gal antigen-deficient cells. Mol Reprod Dev 75(9):1372–1378. https://doi.org/10.1002/mrd.20890

Galli C, Perota A, Brunetti D, Lagutina I, Lazzari G, Lucchini F (2010) Genetic engineering including superseding microinjection: new ways to make GM pigs. Xenotransplantation 17(6):397–410. https://doi.org/10.1111/j.1399-3089.2010.00590.x (Review)

Garcia-Vazquez FA, Ruiz S, Matas C, Izquierdo-Rico MJ, Grullon LA, De Ondiz A, Vieira L, Aviles-Lopez K, Gutierrez-Adán A, Gadea J (2010) Production of transgenic piglets using ICSI-sperm-mediated gene transfer in combination with recombinase RecA. Reproduction 140:259–272

Gerlach M, Kraft T, Brenner B, Petersen B, Niemann H, Montag J (2018) Efficient knock-in of a point mutation in porcine fibroblasts using the CRISPR/Cas9-*GMNN* Fusion Gene Genes (Basel) 9(6). pii: E296. https://doi.org/10.3390/genes9060296

Gil MA, Martinez CA, Nohalez A, Parrilla I, Roca J, Wu J, Ross PJ, Cuello C, Izpisua JC, Martinez EA (2017a) Developmental competence of porcine genome-edited zygotes. Mol Reprod Dev 84(9):814–821. https://doi.org/10.1002/mrd.22829 (Epub 2017 Jun 8)

Gil MA, Martinez CA, Nohalez A, Parrilla I, Roca J, Wu J, Ross PJ, Cuello C, Izpisua JC, Martinez EA (2017) Developmental competence of porcine genome-edited zygotes. Mol Reprod Dev 84(9):814–821. https://doi.org/10.1002/mrd.22829 (Epub 2017 Jun 8. Review)

Gupta MK, Das ZC, Heo YT, Joo JY, Chung HJ, Song H, Kim JH, Kim NH, Lee HT, Ko DH, Uhm SJ (2013) Transgenic chicken, mice, cattle, and pig embryos by somatic cell nuclear transfer into pigoocytes. Cell Reprogram. 15(4):322–328. https://doi.org/10.1089/cell.2012.0074 (Epub 2013 Jun 28)

Gutierrez K, Dicks N, Glanzner WG, Agellon LB, Bordignon V (2015) Efficacy of the porcine species in biomedical research. Front Genet 6:293. https://doi.org/10.3389/fgene.2015.00293 (eCollection 2015)

Hammer RE, Pursel VG, Rexroad CE Jr, Wall RJ, Bolt DJ, Ebert KM, Palmiter RD, Brinster RL (1985) Production of transgenic rabbits, sheep and pigs by microinjection. Nature 315(6021):680–683

Hisamatsu S, Sakaue M, Takizawa A, Kato T, Kamoshita M, Ito J, Kashiwazaki N (2015) Knockout of targeted gene in porcine somatic cells using zinc-finger nuclease. Anim Sci J 86(2):132–137. https://doi.org/10.1111/asj.12259 (Epub 2014 Sep 3)

Hofmann A, Kessler B, Ewerling S, Weppert M, Vogg B, Ludwig H, Stojkovic M, Boelhauve M, Brem G, Wolf E, Pfeifer A (2003) Efficient transgenesis in farm animals by lentiviral vectors. EMBO Rep 4(11):1054–1060 (Epub 2003 Oct 17)

Hwang SU, Eun K, Yoon JD, Kim H, Hyun SH (2018) Production of transgenic pigs using a *pGFAP--CreER*$_{T2}$*/EGFP* $_{LoxP}$ inducible system for central nervous system disease models. J Vet Sci 19(3):434–445. https://doi.org/10.4142/jvs.2018.19.3.434

Hyun S, Lee G, Kim D, Kim H, Lee S, Nam D, Jeong Y, Kim S, Yeom S, Kang S, Han J, Lee B, Hwang W (2003) Production of nuclear transfer-derived piglets using porcine fetal fibroblasts transfected with the enhanced green fluorescent protein. Biol Reprod 69 (3):1060–1068 (Epub 2003 May 28)

Ivics Z, Garrels W, Mátés L, Yau TY, Bashir S, Zidek V, Landa V, Geurts A, Pravenec M, Rülicke T, Kues WA, Izsvák Z (2014) Germline transgenesis in pigs by cytoplasmic microinjection of Sleeping Beauty transposons. Nat Protoc 9(4):810–827. https://doi.org/10.1038/nprot.2014.010 (Epub 2014 Mar 13)

Knox RV (2014) Impact of swine reproductive technologies on pig and global food production. Adv Exp Med Biol 752:131–160. https://doi.org/10.1007/978-1-4614-8887-3_7 (Review)

Knox RV, Rodriguez Zas SL, Sloter NL, McNamara KA, Gall TJ, Levis DG, Safranski TJ, Singleton WL (2013) An analysis of survey data by size of the breeding herd for the reproductive management practices of North American sow farms. J Anim Sci 91(1):433–445

Kues WA, Schwinzer R, Wirth D, Verhoeyen E, Lemme E, Herrmann D, Barg-Kues B, Hauser H, Wonigeit H, Niemann H (2006) Epigenetic silencing and tissue independent expression of a novel tetracycline inducible system in double-transgenic pigs. FASEB J 20:E1–E10

Lai L, Kolber-Simonds D, Park KW, Cheong HT, Greenstein JL, Im GS, Samuel M, Bonk A, Rieke A, Day BN, Murphy CN, Carter DB, Hawley RJ, Prather RS (2002) Production of alpha-1,3-galactosyltransferase knockout pigs by nuclear transfer cloning. Science 295(5557):1089–1092 (Epub 2002 Jan 3)

Laroye C, Lemarié J, Boufenzer A, Labroca P, Cunat L, Alauzet C, Groubatch F, Cailac C, Jolly L, Bensoussan D, Reppel L, Gibot S (2018) Clinical-grade mesenchymal stem cells derived from umbilical cord improve septic shock in pigs. Intensive Care Med Exp 6(1):24. https://doi.org/10.1186/s40635-018-0194-1

Lee GS, Kim HS, Hyun SH, Lee SH, Jeon HY, Nam DH, Jeong YW, Kim S, Kim JH, Han JY, Ahn C, Kang SK, Lee BC, Hwang WS (2005) Production of transgenic cloned piglets from genetically transformed fetal fibroblasts selected by green fluorescent protein. Theriogenology 63(4):973–991

Li GP, Tan JH, Sun QY, Meng QG, Yue KZ, Sun XS, Li ZY, Wang HB, Xu LB (2000) Cloned piglets born after nuclear transplantation of embryonic blastomeres into porcine oocytes matured in vivo. Cloning 2(1): 45–52

Li M, Zhang D, Hou Y, Jiao L, Zheng X, Wang WH (2003) Isolation and culture of embryonic stem cells from porcine blastocysts. Mol Reprod Dev 65 (4):429–434

Li Z, Zeng F, Meng F, Xu Z, Zhang X, Huang X, Tang F, Gao W, Shi J, He X, Liu D, Wang C, Urschitz J, Moisyadi S, Wu Z (2014) Generation of transgenic pigs by cytoplasmic injection of piggyBac transposase-based pmGENIE-3 plasmids. Biol Reprod 90(5):93. https://doi.org/10.1095/biolreprod.113.116905 (Print 2014 May)

Lin YS, Yang CC, Hsu CC, Hsu JT, Wu SC, Lin CJ, Cheng WT (2015) Establishment of a novel, eco-friendly transgenic pig model using porcine pancreatic amylase promoter-driven fungal cellulase transgenes. Transgenic Res 24(1):61–71. https://doi.org/10.1007/s11248-014-9817-9 (Epub 2014 Jul 26)

Liu W, Wu LH, Yue M, Nashun B, Tang H, Chen Y, Chen BZ, Yuan J, Xiao D, Gu WW (2017) Generation of DKK1 transgenic Tibet minipigs by somatic cell nuclear transfer (SCNT) Oncotarget 8(43):74331–74339. https://doi.org/10.18632/oncotarget.20604 (eCollection 2017 Sep 26)

Lutz AJ, Li P, Estrada JL, Sidner RA, Chihara RK, Downey SM, Burlak C, Wang ZY, Reyes LM, Ivary B, Yin F, Blankenship RL, Paris LL, Tector AJ (2013) Double knockout pigs deficient in N-glycolylneuraminic acid and galactose α-1,3-galactose reduce the humoral barrier to xenotransplantation. Xenotransplantation 20(1):27–35. https://doi.org/10.1111/xen.12019

Maehara M, Matsunari H, Honda K, Nakano K, Takeuchi Y, Kanai T, Matsuda T, Matsumura Y, Hagiwara Y, Sasayama N, Shirasu A, Takahashi M, Watanabe M, Umeyama K, Hanazono Y, Nagashima H (2012) Hollow fiber vitrification provides a novel method for cryopreserving in vitro maturation/fertilization-derived porcine embryos. Biol Reprod 87(6):133. https://doi.org/10.1095/biolreprod.112.100339 (Print 2012 Jun)

Maes D, Van Soom A, Appeltant R, Arsenakis I, Nauwynck H (2016) Porcine semen as a vector for transmission of viral pathogens. Theriogenology 85 (1):27–38. https://doi.org/10.1016/j.theriogenology.2015.09.046 (Epub 2015 Sep 26. Review)

Martinez EA, Caamaño JN, Gil MA, Rieke A, McCauley TC, Cantley TC, Vazquez JM, Roca J, Vazquez JL, Didion BA, Murphy CN, Prather RS, Day BN (2004) Successful nonsurgical deep uterine embryo transfer in pigs. Theriogenology 61(1):137–146

Matsunari H, Maehara M, Nakano K, Ikezawa Y, Hagiwara Y, Sasayama N, Shirasu A, Ohta H, Takahashi M, Nagashima H (2012) Hollow fiber vitrification: a novel method for vitrifying multiple embryos in a single device. J Reprod Dev 58(5):599–608 (Epub 2012 Jul 4)

McCreath KJ, Howcroft J, Campbell KH, Colman A, Schnieke AE, Kind AJ (2000) Production of gene-targeted sheep by nuclear transfer from cultured somatic cells. Nature 405(6790):1066–1069 (Erratum in: Nature 2000 Nov 2;408(6808):120)

Mei SH, McCarter SD, Deng Y, Parker CH, Liles WC, Stewart DJ (2007) Prevention of LPS-induced acute lung injury in mice by mesenchymal stem cells overexpressing angiopoietin 1. PLoS Med. 4(9):e269

Meidinger RG, Ajakaiye A, Fan MZ, Zhang J, Phillips JP, Forsberg CW (2013) Digestive utilization of phosphorus from plant-based diets in the Cassie line of transgenic Yorkshire pigs that secrete phytase in the saliva. J Anim Sci 91(3):1307–1320. https://doi.org/10.2527/jas.2012-5575 (Epub 2013 Jan 7)

Mendicino M, Ramsoondar J, Phelps C, Vaught T, Ball S, LeRoith T, Monahan J, Chen S, Dandro A, Boone J, Jobst P, Vance A, Wertz N, Bergman Z, Sun XZ, Polejaeva I, Butler J, Dai Y, Ayares D, Wells K (2011) Generation of antibody- and B cell-deficient pigs by targeted disruption of the J-region gene segment of the heavy chain locus. Transgenic Res 20(3):625–641. https://doi.org/10.1007/s11248-010-9444-z (Epub 2010 Sep 26)

Miernik K, Karasinski J (2012) Porcine uterus contains a population of mesenchymal stem cells. Reproduction 143(2):203–209. https://doi.org/10.1530/REP-11-0202 (Epub 2011 Nov 7)

Müller M, Brenig B, Winnacker EL, Brem G (1992) Transgenic pigs carrying cDNA copies encoding the murine Mx1 protein which confers resistance to influenza virus infection. Gene 121(2):263–270

Onishi A, Iwamoto M, Akita T, Mikawa S, Takeda K, Awata T, Hanada H, Perry AC (2000) Pig cloning by microinjection of fetal fibroblast nuclei. Science 289 (5482):1188–1190

Ozawa M, Himaki T, Ookutsu S, Mizobe Y, Ogawa J, Miyoshi K, Yabuki A, Fan J, Yoshida M (2015) Production of cloned miniature pigs expressing high levels of human apolipoprotein(a) in plasma. PLoS One 10(7):e0132155. https://doi.org/10.1371/journal.pone.0132155 (eCollection 2015)

Phelps CJ, Koike C, Vaught TD, Boone J, Wells KD, Chen SH, Ball S, Specht SM, Polejaeva IA, Monahan JA, Jobst PM, Sharma SB, Lamborn AE, Garst AS, Moore M, Demetris AJ, Rudert WA, Bottino R, Bertera S, Trucco M, Starzl TE, Dai Y, Ayares DL (2003) Production of alpha 1,3-galactosyltransferase-deficient pigs. Science 299 (5605):411–414 (Epub 2002 Dec 19)

Petersen B, Frenzel A, Lucas-Hahn A, Herrmann D, Hassel P, Klein S, Ziegler M, Hadeler KG, Niemann H (2016) Efficient production of biallelic GGTA1 knockout pigs by cytoplasmic microinjection of CRISPR/Cas9 into zygotes. Xenotransplantation 23 (5):338–346. https://doi.org/10.1111/xen.12258 (Epub 2016 Sep 9)

Polejaeva IA, Chen SH, Vaught TD, Page RL, Mullins J, Ball S, Dai Y, Boone J, Walker S, Ayares DL, Colman A, Campbell KH (2000) Cloned pigs produced by nuclear transfer from adult somatic cells. Nature 407(6800):86–90

Reyes LM, Estrada JL, Wang ZY, Blosser RJ, Smith RF, Sidner RA, Paris LL, Blankenship RL, Ray CN, Miner AC, Tector M, Tector AJ (2014) Creating class I MHC-null pigs using guide RNA and the Cas9 endonuclease. J Immunol. 193(11):5751–5757. https://doi.org/10.4049/jimmunol.1402059 (Epub 2014 Oct 22)

Roca J, Parrilla I, Bolarin A, Martinez EA, Rodriguez-Martinez H (2016) Will AI in pigs become more efficient? Theriogenology 86(1):187–193. https://doi.org/10.1016/j.theriogenology.2015.11.026 (Epub 2015 Dec 2. Review)

Schulze M, Bortfeldt R, Schäfer J, Jung M, Fuchs-Kittowski F (2018) Effect of vibration emissions during shipping of artificial insemination doses on boar semen quality. Anim Reprod Sci. 192:328–334. https://doi.org/10.1016/j.anireprosci.2018.03.035 (Epub 2018 Mar 31)

Sheets TP, Park CH, Park KE, Powell A, Donovan DM, Telugu BP (2016) Somatic cell nuclear transfer followed by CRIPSR/Cas9 microinjection results in highly efficient genome editing in cloned pigs. Int J Mol Sci 17(12). pii: E2031

Shim H, Gutiérrez-Adán A, Chen LR, BonDurant RH, Behboodi E, Anderson GB (1997) Isolation of pluripotent stem cells from cultured porcine primordial germ cells. Biol Reprod 57(5):1089–1095

Sommer JR, Estrada JL, Collins EB, Bedell M, Alexander CA, Yang Z, Hughes G, Mir B, Gilger BC, Grob S, Wei X, Piedrahita JA, Shaw PX, Petters RM, Zhang K (2011) Production of ELOVL4 transgenic pigs: a large animal model for Stargardt-like macular degeneration. Br J Ophthalmol 95(12):1749–1754. https://doi.org/10.1136/bjophthalmol-2011-300417 (Epub 2011 Aug 26)

Tanihara F, Takemoto T, Kitagawa E, Rao S, Do LT, Onishi A, Yamashita Y, Kosugi C, Suzuki H,

Sembon S, Suzuki S, Nakai M, Hashimoto M, Yasue A, Matsuhisa M, Noji S, Fujimura T, Fuchimoto D, Otoi T (2016) Somatic cell reprogramming-free generation of genetically modified pigs. Sci Adv 2 (9):e1600803. https://doi.org/10.1126/sciadv.1600803 (eCollection 2016 Sep)

Wang Y, Du Y, Zhou X, Wang L, Li J, Wang F, Huang Z, Huang X, Wei H (2016) Efficient generation of B2 m-null pigs via injection of zygote with TALENs. Sci Rep 16(6):38854. https://doi.org/10.1038/srep38854

West FD, Terlouw SL, Dobrinsky JR, Lu Y, Jordan ET, Stice SL (2015) Generation of chimeras from porcine induced pluripotent stem cells. Meth Mol Biol 1330:153–167. https://doi.org/10.1007/978-1-4939-2848-4_14

Wu Z, Chen J, Ren J, Bao L, Liao J, Cui C, Rao L, Li H, Gu Y, Dai H, Zhu H, Teng X, Cheng L, Xiao L (2009) Generation of pig induced pluripotent stem cells with a drug-inducible system. J Mol Cell Biol. 1(1):46–54. https://doi.org/10.1093/jmcb/mjp003 (Epub 2009 Jun 3)

Yang J, Zhao Q, Wang K, Ma C, Liu H, Liu Y, Guan W (2018) Isolation, culture and biological characteristics of multipotent porcine tendon-derived stem cells. Int J Mol Med 41(6):3611–3619. https://doi.org/10.3892/ijmm.2018.3545 (Epub 2018 Mar 7)

Yang Z, Vajta G, Xu Y, Luan J, Lin M, Liu C, Tian J, Dou H, Li Y, Liu T, Zhang Y, Li L, Yang W, Bolund L, Yang H, Du Y (2016) Production of pigs by hand-made cloning using mesenchymal stem cells and fibroblasts. Cell Reprogram 18(4):256–263. https://doi.org/10.1089/cell.2015.0072

Yu H, Long W, Zhang X, Xu K, Guo J, Zhao H, Li H, Qing Y, Pan W, Jia B, Zhao HY, Huang X, Wei HJ (2018) Generation of GHR-modified pigs as Laron syndrome models via a dual-sgRNAs/Cas9 system and somatic cell nuclear transfer. J Transl Med 16(1):41. https://doi.org/10.1186/s12967-018-1409-7

Zeyland J, Gawrońska B, Juzwa W, Jura J, Nowak A, Słomski R, Smorąg Z, Szalata M, Woźniak A, Lipiński D (2013) Transgenic pigs designed to express human α-galactosidase to avoid humoral xenograft rejection. J Appl Genet. 54(3):293–303. https://doi.org/10.1007/s13353-013-0156-y (Epub 2013 Jun 19)

Zhang X, Li Z, Yang H, Liu D, Cai G, Li G, Mo J, Wang D, Zhong C, Wang H, Sun Y, Shi J, Zheng E, Meng F, Zhang M, He X, Zhou R, Zhang J, Huang M, Zhang R, Li N, Fan M, Yang J, Wu Z (2018) Novel transgenic pigs with enhanced growth and reduced environmental impact. Elife 7. pii: e34286. https://doi.org/10.7554/elife.34286

Zhao D, Li Y, Zhou X, Yang Z (2018) Peripheral blood mesenchymal stem cells combined with modified demineralized bone matrix promote pig cartilage defect repair. Cells Tissues Organs 1–9. https://doi.org/10.1159/000493210 (Epub ahead of print)

Zhou X, Wang L, Du Y, Xie F, Li L, Liu Y, Liu C, Wang S, Zhang S, Huang X, Wang Y, Wei H (2016) Efficient generation of gene-modified pigs harboring precise orthologous human mutation via CRISPR/Cas9-induced homology-directed repair in zygotes. Hum Mutat 37(1):110–118. https://doi.org/10.1002/humu.22913 (Epub 2015 Oct 23)

Zhou X, Xin J, Fan N, Zou Q, Huang J, Ouyang Z, Zhao Y, Zhao B, Liu Z, Lai S, Yi X, Guo L, Esteban MA, Zeng Y, Yang H, Lai L (2015) Generation of CRISPR/Cas9-mediated gene-targeted pigs via somatic cell nuclear transfer. Cell Mol Life Sci 72(6):1175–1184. https://doi.org/10.1007/s00018-014-1744-7 (Epub 2014 Oct 2)

Zhu XX, Zhong YZ, Ge YW, Lu KH[1], Lu SS (2018) Generation of transgenic-cloned Huanjiang Xiang pigs systemically expressing enhanced green fluorescent protein. Reprod Domest Anim. https://doi.org/10.1111/rda.13301 (Epub ahead of print)

Reproduction Biotechnology in Equines

16

Abstract

The assisted reproduction in equines has progressed to a large extent during recent two decades. Scarcity of ovaries and oocytes, lack of interest from customary equine breeders, and preference to automobiles for transportation are the main reason for tardy progress in equine ARTs. In addition, failure of mares to respond to superovulation, meager success of sperm capacitation in vitro, IVF and embryo production, and less number of laboratories working in this area are the causes of low progress. Improving oocyte maturation in vitro, IVF and enhancing developmental competence of *in vitro* produced embryos, multiplying selected progeny-tested studs and developing cell therapies for equine injuries are the prospective areas.

Highlights

- Equines are a group of different species domesticated companion animals ever since the human civilization
- Equines are diverse multipurpose livestock species.

Keywords

Equines · Companion animals · Equine reproduction · Semen cryopreservation · Embryo transfer

16.1 Introduction

Domestication of horses was one of the most important discoveries that eased the early human life. Since then, horses were used in wars, as animals of prestige and importance, in hunting and as means of transportation. Horses have benefited humans by serving as morale-booster, stress-releaser, entertainment and sources of income.

16.1.1 Equines as Companion Animals

The equines, i.e., horse, mule, donkeys, and ponies are part of human progress and civilization. The world was transformed when humans started domesticating horse for its speediness, power, and fidelity. The horse allowed men travel faster, transportation by pulling the cart, carrying load from one place to other. The horse was part of kings and emperors in games and winning the battles. The relevance of equines is evident from the fact that horse is part of human history for the past 6000 years while use of fuel-operated automobiles is merely 100 years old.

Once domesticated, the equines proved their mettle to humans by their affability and resemblance to human physiology and anatomy

B. Singh et al., *Advances in Animal Biotechnology*,
https://doi.org/10.1007/978-3-030-21309-1_16

Table 16.1 Types of equines and their features

Species	Parental lineage	Fertility status and characteristics
Donkey (*E. asinus*)	Donkey × donkey	Fertile (2n = 62), BW 80–480 kg (adult), used for pulling loads, and as pack animal for livelihood
Horse (*Equus caballus*)	Horse × mare	Fertile (2n = 64), BW 380–1000 kg (Horse); Mare, BW (200–700 kg). The horse is a famous companion animal used for riding, carrying loads, pulling carts, and as a source of livelihood
Mule (2n = 63) (*E. asinus* × *E. caballus*)	Male donkey (jack) × female horse	Hybrid and therefore sterile, mules are larger than Hinnys, BW ranges from 230 to 450 kg, They have incomplete reproductive system due to uneven number of chromosomes, breeding is rare
Hinny (*E. asinus* × *E. caballus*)	Male horse (stallion) × female donkey (jenny)	They are Hybrid (2n = 63), and sterile. Hinnys are smaller than mules, reproductive tract development is incomplete, it is difficult to obtain hinneys because of difference in the number of chromosomes of horse and the donkey

features. Among equines, the important species are—horse (*Equus caballus*), donkey (*Equus asinus*), and mule (*E. asinus* × *E. caballus*), the offspring of male donkey and female horse, ponies (Table 16.1). In addition, some native or less known equine breeds with remarkable adaptation to prevailing agroclimatic conditions are also found in various parts of countries.

Compared to ruminants, the equines require less feed and can be put to work without any impediment. Besides their use as burden carrying animals, the emphasis is on health benefits of donkey milk (Li et al. 2017; Aspri et al. 2018). Affirmation of five candidate genes, *NUMB*, *LGALS2*, *ADCY8*, *SLC25A30*, and *CA8*, associated with lactation traits in Kazakh and Yili horses, shows potential value for selecting dairy traits in horses (Liu et al. 2018).

With 50% more protein and 30% more iron contents than leanest beef, the horses offered early cave men a sustaining life (http://www.equineheritageinstitute.org/shaping-civilizations-the-role-of-the-horse-in-human-societies/, accessed on Sept. 13, 2017).

The horse production is associated with low enteric methane emissions. The horse meat, as alternative to red meat from ruminants, and its traditional products (cecina, fermented sausages, liver pates, dry-cured loin) have high nutritional value (Lorenzo et al. 2017). In addition, better uptake of polyunsaturated fatty acids by horses has positive effects on meat quality (Belaunzaran et al. 2015). Hence, in future, the horse meat could possibly be the substitute for beef.

16.1.2 Assisted Reproduction in Equines

ARTs are used expansively in modern equine reproduction and breeding management. The ARTs have helped the equine breeders to increase number of offspring produced from best quality stallions and mares by AI and IVF.

The very fundamental purpose of using ARTs is to disseminate superior genotypes and treat infertility. But, induction of drug-induced superovulation, collection of oocytes from live animals, evaluation and preservation of oocytes, producing embryos by IVF or ICSI, and nuclear transfer cloning are the prospective ARTs. The contemporary focus is to increase the success rates of ARTs for horse-rearing entrepreneurs and conserving the equine genetic diversity.

16.1.3 Male-Assisted Reproduction

Reproductive parameters viz., age at puberty, libido, testicular blood flow, serum testosterone levels, and semen quality vary for stallion and donkey jack. The parameters of puberty should be understood and defined properly. In donkeys, the sperm in semen appeared at a mean age of 18.7 months, but puberty characteristics were confirmed at 19–20 months of age (Rota et al. 2012). The stallion, in general, reaches puberty at the age of 16-18 months. Concentration of sperm in ejaculates and progressive motility of 10% are the standard parameters of confirming puberty in horse.

16.1.4 Semen Collection, Cryopreservation, and Artificial Insemination

Collecting semen and its cryopreservation is important for extending the population of animals with superior traits. Protocols have been formulated to process and preserve semen of horse and donkeys using combinations of CPAs and cryopreservation (Dorado et al. 2013; Alvarenga et al. 2016; Diaz-Jimenez et al. 2018a, b) and maintaining post-thaw quality of sperm (Al-Essawe et al. 2018a, b). Besides domesticated horse and donkeys, the efforts are made to preserve semen of some equids such as Mongolian wild Przewalski's horse (*Equus ferus przewalskii*) (Equus ferus przewaliskii) that is declared as endangered (Pukazhenthi et al. 2014).

Reproductive physiological and endocrine patterns associated with estrus cycles and pregnancy in Somali wild ass (*Equus africanus somaliensis*), an endangered equid species, have been described. The information will be useful to plan conservation and breeding strategies to conserve endangered equids (Kozlowski et al. 2018).

The semen is collected generally by standard AV, and electroejaculation methods depending on age and health of donor. As the stallions are preferred for their performance ability rather than suitability of semen to withstand cryopreservation, semen of stud stallions may or may not be of high freezing values.

Like ruminant livestock, the AI is widely used ART in horses and donkeys. The fresh or frozen, both kinds of semen are used that have increased the number of genetically elite young ones in equines and preventing proliferation of unhealthy offspring due to genetic defects or reproduction-associated infections.

Compared to other livestock species, the IVF with oocytes and sperm is poorly successful in equines. Lack of in vitro culture media to sufficiently support equine sperm capacitation is a major constraint. Under in vivo locale, the equine oviduct and associated secretions provide a suitable microenvironment to prop up sperm capacitation, oocyte-sperm interaction, fertilization and development of embryos to begin pregnancies (Leemaans et al. 2016).

Difficulties in achieving biologically effective sperm capacitation *in vitro* have hampered IVF and embryo production (reviewed in Morris 2018). The sperm capacitated *in vitro* by appropriate combination of calcium ionophore, bicarbonate, and bovine serum albumen (BSA) was used to fertilize the oocytes that resulted in live births of two foals (Palmer et al. 1991). However, despite efforts to improve capacitation and maturation of oocytes, the IVF has not improved much.

The alternative strategies used to improve or assist horse sperm penetration into oocyte, and thereby improve the IVF included partial removal of zona pellucida to facilitate entry of sperm into ooplasm (Choi et al. 1994), and drilling pores in zona pellucida, or softening the oocyte by treating with fetulin (Dell'Aquila et al. 1999). However, the attempts did not significantly improve the process of in vitro embryo production. Researchers have suggested that interaction between stallion spermatozoa and mare oolemma is complex. The interactions should be investigated to evolve strategies to improve IVF (Sessions-Bresnahan et al. 2014).

The alternative means of improving IVF by ICSI has attracted attention of horse breeders to get offspring from a subfertile male or the horse becoming aged or infertile due to testicular degeneration and decreased spermatogenesis. After development of standard practice of ICSI-derived equine embryos (Hinrichs et al. 2005) and the first report of equine ICSI using sex-sorted sperm (Clolleoni et al. 2007), the ICSI has emerged as a choice to produce embryos from sexed semen or semen of high-quality stallion.

As conventional ICSI can cause mechanical damage to oocyte, the piezo drill-assisted ICSI (PDAI) had made artificial fertilization more applicable. The technique has played important role in promoting embryo production in equines where sturdy zona pellucida prevents entry of sperm into oocytes (Smits et al. 2012). One drawback of the technique is that PDAI uses mercury which may possibly damage live cells, hence alternative and safer method, the laser-assisted ICSI (LAI) has been developed (Smits et al. 2012). Blastocyst development was noted from ICSI of intact sperm obtained from a subfertile horse, containing centriole (Roels et al. 2018).

16.1.5 Female-Assisted Reproduction

The mares, in general, reach puberty at the age of 12–15 months. Onset of puberty depends on nutrition, health status, breed, and environmental factors. Administration of hormones (e.g., hDG and GnRH) stimulates dominant follicle and improves oocyte recovery. The strategy is exclusively useful for old mares (Riera et al. 2016).

The mare is inseminated when follicle is expected to rupture and release the ovum, or in coordination with induction of ovulation by administering GnRH or LH analog (Hinrichs 2018). For determining the day of ovulation (Day 0), the preovulatory follicle dynamics are monitored after insemination.

Fertility of donkey semen in jennies is low compared to mares. A recent study shows that deep horn insemination could increase pregnancy rates in jennies, and it is advisable to go for multiple inseminations to improve fertility using frozen-thawed donkey semen (de Oliveira et al. 2016).

The embryo is flushed after day 7–8 of insemination. Various buffers and protocols are used to flush the embryos from uterine horns or uterus. The embryos are examined for quality and cryopreserved for future use or transporting to other places, labs, or equine farms.

The embryos are transferred into estrus-synchronized females by two methods—conventional method of transcervical embryo transfer and Wilsher's forceps-assisted transfer. A comparative study on efficacy of both the techniques on likelihood of establishing pregnancy indicates that Wilsher's forceps-assisted embryo transfer yielded higher pregnancy rates (Cuervo-Arango et al. 2018).

16.1.6 Oocytes—Sources, Quality, and Cryopreservation

Good quality oocytes are needed to support embryo development after IVF. As a result, the oocyte-based ARTs are gaining recognition in equine breeding business (Carnevale 2016).

There are several advantages of collecting oocytes from dominant follicle stimulated artificially by administering drugs or allowing the recruited follicles mature to a rupturing stage. The oocytes or cumulus–oocyte complexes (COCs) recovered from dominant follicles exhibit high developmental potential. Foss et al. (2013) reported as much as 70% blastocyst rate per sperm-injected oocyte from dominant follicle after ICSI vs. around 30% blastocysts from oocytes matured *in vitro*. Others have concluded the rates of 40% vs. 33% for the two types of oocytes (Jacobson et al. 2010).

The concept of equine oocyte and embryo banking by cryopreservation is desirable for many reasons. Hence, cryopreservation protocols are developed for equine oocytes and embryos. Despite fact that vitrified-warmed oocytes can mature in vitro, the developmental competence of oocytes, and success rate of vitrification of immature equine oocytes is low. Ortiz-Escrbano et al. (2017) have reported first birth of a foal from IVF of vitrified-warmed oocytes. A higher maturation rate of 37% was achieved for oocytes that were held overnight before vitrification, than the maturation rate (14%) in oocytes that were vitrified immediately (Canesin et al. 2017).

Further, the blastocysts could also be developed after ICSI from vitrified-warmed GV stage oocytes (Canesin et al. 2017, 2018).

The pregnancy rates of 50–70% are noticed from transfer of frozen-thawed embryos. The results are based on observation from a limited transfer of embryos (Squires 2016). The immature COCs surrounded by corona radiata (termed, CR oocytes) were vitrified after a short exposure to high concentration of CPAs and compared to another group of CR oocytes vitrified after long exposure to low concentration of CPAs. The warmed CR oocytes were tested for their efficacy to develop embryos after IVF. The CR oocytes vitrified by short vitrification protocol could only develop to embryos. Transfer of five of the embryos resulted in birth of one healthy foal, indicating that a short exposure time to CPAs is desirable (Ortiz-Escribano et al. 2018).

Nevertheless, the embryo production and cryopreservation have not progressed to an appreciable level in equines. This is because of several reasons including paucity of suitable protocols for inducing multiple ovulation in mare, low recovery of oocytes and embryos, technical skills, and high cost of maintaining the animals.

16.1.7 Stem Cell Technologies in Equines

Various types of stem cells are reported in equines. The finding could be noteworthy for horses as well as for other mammalian species, as well as humans. Like other livestock species, the ICM cells of developing preimplantation embryos (Paris and Stout 2010), fetal adnexa, amniotic fluid, and cord blood are the main sources of deriving MSCs (Violini et al. 2009; Iacono et al. 2017). The MSCs derived from perinatal sources such as Wharton's jelly (WJ) and amniotic membrane (AM) had rapid proliferation and were differentiated into osteogenic and chondrogenic lineages. Another study shows that MSCs from fetal sources had different characteristics such as ability to grow in culture for prolonged period, differentiation into other cell types that are necessary for therapeutic implications (Iacono et al. 2017).

Research has been extended beyond isolation and description of stem cells in equines. A study analyzed the proteome of umbilical cord-derived MSCs and observed functional relevance of activities related to primary metabolic processes, translation, production of vesicle coats, and antioxidant activities (Maia et al. 2017).

16.1.8 Stem Cells as Regenerative Medicine

The very basic purpose of regenerative medicine is to improve health conditions such as spinal cord injury and fractures cutaneous healing. The efficacy of stem cells as regenerative medicine is already proved in model animals such as mice and dog. The tendon, ligament, and bone injuries are common in horses.

Induced reprogramming of somatic cells to generate iPSCs has progressed significantly in equines. Potential of iPSCs to differentiate into various cell types, mostly toward therapeutic applications, such as skin regeneration, bone and joint healing is of paramount interest.

The iPSCs were generated from horse fibroblasts by ectopic expression of pluripotency factors (Nagy et al. 2011; Breton et al. 2013). The iPSCs have shown substantial cellular flexibility and could be differentiated into keratinocyte lineage (Aguiar et al. 2016), functional myocytes by transducing iPSC-derived mesenchymal derivatives by means of an inducible lentiviral vector coding for a human myogenic factor, MyoD (Amilon et al. 2018), and osteoblasts (Baird et al. 2018). Thus, the iPSCs may provide a source of cells for future regenerative loss or damaged skin, improving bone fractures, joint and muscular repair. Table 16.1 summarizes salient achievements of iPSCs with their proposed applications in equines.

16.1.9 Nuclear Transfer Cloning in Equine

Equines are multipurpose economically important livestock, hence nuclear transfer cloning in equine is of great interest. While equine nuclear transfer cloning is an alluring area, it has not emerged as a main field of study because of scarcity of competent oocytes. However, in view of demand to preserve high genetic value animals, the cloning is the only viable option. The success in producing cloned foals (Woods et al. 2003; Galli et al. 2003a, b) (Table 16.2) was achieved after birth of mammalian species viz., cloned sheep (Wilmut et al. 1997), cattle (Cibelli et al. 1998), and pig (Polejaeva et al. 2000).

The maturation efficiency of equine oocytes recovered from live animals by transvaginal ultrasound-guided follicle aspiration (TVFA) is higher than those obtained from abattoir-derived ovaries, but number of oocytes recovered is less.

Olivera et al. (2016) studied various parameters viz., effect of time from cytoplast-cell fusion to activation and types of donor cells to contribute to cloned equine embryos. The NT cloned embryo production efficiency using various cells was as: umbilical cord MSCs (15.6%), AF (9.3%), and FF (8.9%). Surprisingly, the microinjection of oocytes with pluripotency-inducing factors (OCT4, SOX2, MYC and KLF4, vs. AF as such as control) did not improve embryo development or pregnancy rates. It was also inferred that a

Table 16.2 A chronological overview of milestone achievements in ARTs in equines

Year	Milestone achievements and inferences (References)
1981	First report on in vitro maturation of mare oocytes (Fulka and Okolski 1981)
1987	Non-surgical recovery of follicular fluid and COCs (Palmer et al. 1987)
1989	First embryo production from IVM mare oocytes collected from abattoir (Zhang et al. 1989)
1991	First birth of foal from IVF oocytes collected from live animals on June 14, 1990 (Palmer et al. 1991)
2001	Embryo production by OPU from live donors (Galli et al. 2001)
2002	Development of ESC-like cell lines and their *in vitro* differentiation into hematopoietic and endothelial cell lineages in response to cells specific signals. The cells could have applications to treat neurological and hematopoietic disorders (Saito et al. 2002)
2003	First report on SCNT cloning of mule using fetal fibroblasts (Woods et al. 2003) and horse from adult fibroblasts (Galli et al. 2003a, b)
2005	Births of cloned foals (Lagutina et al. 2005; Hinrichs et al. 2006), demonstrating the reproducibility of the techniques in equines
2006	Horse ESC lines from proliferation of ICM cells (Li et al. 2006), studies carried out to freeze, and refreeze the stallion sperm, and using refrozen-thawed sperm to develop embryo after ICSI (Choi et al. 2006)
2007	More births of cloned horse (Hinrichs et al. 2007)
2010	Isolation of amnion-derived MSCs (Coli et al. 2010)
2011	Lyophilization of horse sperm, using lyophilized sperm to develop embryos by ICSI, resulting into birth of live foals (Choi et al. 2011), development of iPSC lines (Nagy et al. 2011) Depiction of trophoblast stem cell markers genes in ICM-derived cells from parthenogenetic embryos (Desmarais et al. 2011)
2012	Use of sex-sorted sperm in equine ET programs (Samper et al. 2012)
2013	Birth of clones fola from oocytes retrieved by transvaginal aspiration from immature follicles of live mares, supporting the possibilities of cloning horses (Choi et al. 2013), description of iPSCs from fibroblasts (Breton et al. 2013)
2014	Birth of a genomic-, and mitochondrial-DNA identical foals from oocytes from immature follicles (Choi et al. 2014)

(continued)

Table 16.2 (continued)

Year	Milestone achievements and inferences (References)
	Developing protocols for improving cryosurvival of sperm in endangered Mongolian wild Przewalski's horse (*Equus ferus przewalskii*) (Pukazhenthi et al. 2014), generation of functional neurons from keratinocyte-derived iPSCs (Sharma et al. 2014)
2015	Aspirating oocytes using a short disposable needle (14-G) from preovulatory follicles from live animals, using them to prepare SCNT embryos from skin fibroblasts, culminating in the birth of live cloned foal (Lee et al. 2015) Generation of iPSC lines using a piggyBac transposon delivery system and temporal regulation of transgene expression (Nagy and Nagy 2015), differentiating iPSCs into tendon, but compared to ESCs, the differentiation capacity of iPSCs is low (Bavin et al. 2015)
2016	Production of cloned embryos from various cells *viz*., FF, umbilical cord MSCs and iPSCs, and inferences that FF cell lines could be better nuclear donor, and a gap of 2-3 h between cytoplast-cell fusion and activation could enhance embryo development competence, and that MSCs are better donors above tested cells (Olivera et al. 2016) Development of iPSCs and their guided differentiation into keratinocyte lineage (Aguiar et al. 2016), generating MSC-like cells from iPSCs (Lepage et al. 2016), scientific assertion that placentitis causes loss in pregnancies established from SCNT embryos (Pozor et al. 2016)
2017	First birth of a foal from an improved vitrification protocol for cryopreservation of IVF-embryos oocytes (Ortiz-Escribano et al. 2017) Description of MSCs from perinatal tissues viz., Wharton's jelly, AM that might have clinical applications. The protocols are envisioned to have applications in humans and other animal species (Iacono et al. 2017) Development of embryos after ICSI of vitrified-warmed oocytes, vitrified at germinal-vesicle stage (Canesin et al. 2017) First study in global functional manner on proteomics of umbilical cord MSCs using PANTHER (Protein ANalysis THrough Evolutionary Relationships), which revealed about various metabolic activities (Maia et al. 2017)
2018	Developing protocol for improving cryopreservation and improved post-thaw quality of stallion sperm (Al-Essawe et al. 2018a, b), Improved vitrification methods for cryopreservation of donkey semen (Diaz-Jimenez et al. 2018a, b) The use of COCs surrounded by corona radiata oocytes (termed, CR oocytes), a high concentration of CPAs and shorter exposure time may enhance the developmental competence of oocytes subjected to vitrification (Ortiz-Escribano et al. 2018). After IVF, embryos could develop only from CR oocytes vitrified by short vitrification protocol. Transfer of five embryos resulted in birth of one healthy foal (Ortiz-Escribano et al. 2018) Rather than sphere method of cryopreservation, the vitrification in straws suggested to be preferable for cryopreservation and maintaining motility of donkey sperm (Diaz-Jimenez et al. 2018a, b) Production of blastocysts by ICSI of sperm from a stallion suffering from testicular degeneration with poor spermigram (Roels et al. 2018) First report on developing functional myocytes (Amilon et al. 2018) and osteocytes from equine iPSCs (Baird et al. 2018)

2–3 h interval between cytoplast-cell fusion and post-fusion activation could enhance blastocyst formation. Birth of 29 live foals from BM-derived MSCs indicate that compared to fibroblasts, the MSCs have better genome plasticity in terms of reprogramming and supporting embryo development (Olivera et al. 2018).

Hirch (2012) had concluded that the number of cloned horses was around 100–200. According to recent review (Gambini and Maserati 2017), around 370 cloned horses are produced so far. With the birth of another 29 foals (Olivera et al. 2018), we assume the number might have exceeded 400 cloned horses. As noted in other cloned foetuses, neonates and adult animals, NT cloning in equines shows epigenetic abnormalities. Ultrasonographic and histopathological examination of placenta at various stages of gestations have revealed various abnormalities and pathologies. The placentitis such as engorged allantoic vessels, enlarged umbilical vessels, and placental thickening are major problems that ultimately caused the pregnancy loss (Pozor et al. 2016).

16.2 Outlook and Challenges

Though horse is relatively immune to negative impacts of ARTs, horse owners' and breeders' trepidation regarding safety of ARTs still persists. There is need to understand the biochemical and physiological natures of oviductal secretions so that culture media could be developed to improve horse sperm capacitation, oocyte-sperm interactions, and fertilization to develop embryos.

It is desirable to document and conserve native and endangered equine breeds. The abilities to retrieve oocytes from live animals preserve and manipulate oocytes and embryos have advanced, however some aspects such as IVM of oocytes should be improved. There is need to understand the role of equine cumulus and granulosa cells, maturation events vis-à-vis developmental competence of oocytes so that artefacts that emerge after removing oocyte from ovaries could be minimized.

ESCs and iPSCs are more promising regenerative cells for deriving other cell lineages, for repairing muscle, tendon and joint, dermal healing. Another less explored area is development of germ cells, especially oogonia and oocytes from iPSCs. This will resolve the problem of scarcity of oocytes in equines. With cooperation of commercial companies, the NT cloning and stem cell therapy would progress in future.

16.3 Conclusions

The demand for cryopreservation of embryos and sperm is high in equines. The ARTs are still progressing equines. As biologically acceptable sperm capacitation, and IVF is not a satisfactorily triumphant, ICSI is used as alternative clinical procedure to produce embryos and utilize best stallion germplasm. Stem cell biology has progressed in horse in view of their importance in regenerating damaged ligaments, bones, cartilage, and skin.

References

Aguiar C, Therrien J, Lemire P, Segura M, Smith LC, Theoret CL (2016) Differentiation of equine induced pluripotent stem cells into a keratinocyte lineage. Equine Vet J 48(3):338–345. https://doi.org/10.1111/evj.12438 (Epub 2015 May 29)

Al-Essawe EM, Johannisson A, Wulf M, Aurich C, Morrell JM (2018a) Improved cryosurvival of stallion spermatozoa after colloid centrifugation is independent of the addition of seminal plasma. Cryobiology 81:145–152. https://doi.org/10.1016/j.cryobiol.2018.01.009 (Epub 2018 Feb 15)

Al-Essawe EM, Wallgren M, Wulf M, Aurich C, Macías--García B, Sjunnesson Y, Morrell JM (2018b) Seminal plasma influences the fertilizing potential of cryopreserved stallion sperm. Theriogenology 15(115): 99–107. https://doi.org/10.1016/j.theriogenology.2018.04.021

Alvarenga MA, Papa FO, Ramires Neto C (2016) Advances in stallion semen cryopreservation. Vet Clin North Am Equine Pract 32(3):521–530. https://doi.org/10.1016/j.cveq.2016.08.003 (Epub 2016 Oct 8. Review)

Amilon KR, Cortes-Araya Y, Moore B, Lee S, Lillico S, Breton A, Esteves CL, Donadeu FX (2018) Generation of functional myocytes from equine induced pluripotent stem cells. Cell Reprogram. https://doi.org/10.1089/cell.2018.0023 (Epub ahead of print)

Aspri M, Leni G, Galaverna G, Papademas P (2018) Bioactive properties of fermented donkey milk, before and after in vitro simulated gastrointestinal digestion. Food Chem 268:476–484. https://doi.org/10.1016/j.foodchem.2018.06.119 (Epub 2018 Jun 23)

Baird A, Lindsay T, Everett A, Iyemere V, Paterson YZ, McClellan A, Henson FMD, Guest DJ (2018) Osteoblast differentiation of equine induced pluripotent stem cells. Biol Open. 7(5). pii: bio033514. https://doi.org/10.1242/bio.033514

Bavin EP, Smith O, Baird AE, Smith LC, Guest DJ (2015) Equine induced pluripotent stem cells have a reduced tendon differentiation capacity compared to embryonic stem cells. Front Vet Sci 2:55. https://doi.org/10.3389/fvets.2015.00055 (eCollection 2015)

Belaunzaran X, Bessa RJ, Lavín P, Mantecón AR, Kramer JK, Aldai N (2015) Horse-meat for human consumption—current research and future opportunities. Meat Sci 108:74–81. https://doi.org/10.1016/j.meatsci.2015.05.006 (Epub 2015 May 14. Review)

Breton A, Sharma R, Diaz AC, Parham AG, Graham A, Neil C, Whitelaw CB, Milne E, Donadeu FX (2013) Derivation and characterization of induced pluripotent stem cells from equine fibroblasts. Stem Cells

Dev 22(4):611–21. https://doi.org/10.1089/scd.2012.0052 (Epub 2012 Sep 28)

Carnevale EM (2016) Advances in collection, transport and maturation of equine oocytes for assisted reproductive techniques. Vet Clin North Am Equine Pract 32(3):379–399. https://doi.org/10.1016/j.cveq.2016.07.002 (Epub 2016 Oct 8. Review)

Canesin HS, Brom-de-Luna JG, Choi YH, Ortiz I, Diaw M, Hinrichs K (2017) Blastocyst development after intracytoplasmic sperm injection of equine oocytes vitrified at the germinal-vesicle stage. Cryobiology 75:52–59. https://doi.org/10.1016/j.cryobiol.2017.02.004

Canesin HS, Brom-de-Luna JG, Choi YH, Pereira AM, Macedo GG, Hinrichs K (2018) Vitrification of germinal-vesicle stage equine oocytes: Effect of cryoprotectant exposure time on in-vitro embryo production. Cryobiology 81:185–191. https://doi.org/10.1016/j.cryobiol.2018.01.001

Choi YH, Okada Y, Hochi S, Braun J, Sato K, Oguri N (1994) In vitro fertilization rate of horse oocytes with partially removed zonae. Theriogenology 42(5):795–802

Choi YH, Ritthaler J, Hinrichs K (2014) Production of a mitochondrial-DNA identical cloned foal using oocytes recovered from immature follicles of selected mares. Theriogenology 82(3):411–417. https://doi.org/10.1016/j.theriogenology.2014.04.021 (Epub 2014 May 5)

Choi YH, Love CC, Varner DD, Hinrichs K (2006) Equine blastocyst development after **intracytoplasmic** injection of **sperm** subjected to two freeze-thaw cycles. Theriogenology 65(4):808–819 (Epub 2005 Aug 10)

Choi YH, Varner DD, Love CC, Hartman DL, Hinrichs K (2011) Production of live foals via **intracytoplasmic** injection of lyophilized **sperm** and **sperm** extract in the horse. Reproduction 142(4):529–538. https://doi.org/10.1530/REP-11-0145 (Epub 2011 Aug 16)

Choi YH, Norris JD, Velez IC, Jacobson CC, Hartman DL, Hinrichs K (2013) A viable foal obtained by equine somatic cell nuclear transfer using oocytes recovered from immature follicles of live mares. Theriogenology 79(5):791.e1–796.e1. https://doi.org/10.1016/j.theriogenology.2012.12.005 (Epub 2013 Jan 11)

Cibelli JB, Stice SL, Golueke PJ, Kane JJ, Jerry J, Blackwell C, Ponce de León FA, Robl JM (1998) Cloned transgenic calves produced from nonquiescent fetal fibroblasts. Science 280:1256–1258

Coli A, Nocchi F, Lamanna R, Iorio M, Lapi S, Urciuoli P, Scatena F, Giannessi E, Stornelli MR, Passeri S (2010) Isolation and characterization of equine amnion mesenchymal stem cells. Cell Biol Int Rep 18(1):e00011. https://doi.org/10.1042/cbr20110004

Colleoni S, Barbacini S, Necchi D, Duchi R, Lazzari G, Galli C (2007) Application of ovum pick-up, intracytoplasmic sperm injection and embryo culture in equine practice. Proc Am Assoc Equine Pract 53:554–559

Cuervo-Arango J, Claes AN, Stout TA (2018) Effect of embryo transfer technique on the likelihood of pregnancy in the mare: a comparison of conventional and Wilsher's forceps-assisted transfer. Vet Rec 183(10):323. https://doi.org/10.1136/vr.104808 (Epub 2018 May 24)

de Oliveira JV, Oliveira PV, Melo e Oña CM, Guasti PN, Monteiro GA, Sancler da Silva YF, Papa Pde M, Alvarenga MA, Dell'Aqua Junior JA, Papa FO (2016) Strategies to improve the fertility of fresh and frozen donkey semen. Theriogenology 85(7):1267–1273. https://doi.org/10.1016/j.theriogenology.2015.12.010(Epub 2015 Dec 23)

Dell'Aquila ME, De Felici M, Massari S, Maritato F, Minoia P (1999) Effects of fetuin on zona pellucida hardening and fertilizability of equine oocytes matured in vitro. Biol Reprod 61(2):533–540

Desmarais JA, Demers SP, Suzuki J Jr, Laflamme S, Vincent P, Laverty S, Smith LC (2011) Trophoblast stem cell marker gene expression in inner cell mass-derived cells from parthenogenetic equine embryos. Reproduction 141(3):321–332. https://doi.org/10.1530/REP-09-0536 (Epub 2011 Jan 5)

Diaz-Jimenez M, Dorado J, Ortiz I, Consuegra C, Pereira B, Gonzalez-De Cara CA, Aguilera R, Mari G, Mislei B, Love CC, Hidalgo M (2018a) Cryopreservation of donkey sperm using non-permeable cryoprotectants. Anim Reprod Sci 189:103–109. https://doi.org/10.1016/j.anireprosci.2017.12.013

Diaz-Jimenez M, Dorado J, Pereira B, Ortiz I, Consuegra C, Bottrel M, Ortiz E, Hidalgo M (2018b) Vitrification in straws conserves motility features better than spheres in donkey sperm. Reprod Domest Anim 53(Suppl 2):56–58. https://doi.org/10.1111/rda.13256

Dorado J, Acha D, Gálvez MJ, Ortiz I, Carrasco JJ, Díaz B, Gómez-Arrones V, Calero-Carretero R, Hidalgo M (2013) Sperm motility patterns in Andalusian donkey (Equus asinus) semen: effects of body weight, age, and semen quality. Theriogenology 79(7):1100–1109. https://doi.org/10.1016/j.theriogenology.2013.02.006 (Epub 2013 Mar 6)

Foss R, Ortis H, Hinrichs K (2013) Effect of potential oocyte transport protocols on blastocyst rates after intracytoplasmic sperm injection in the horse. Equine Vet J 45:39–43. https://doi.org/10.1111/evj.12159

Fulka J Jr, Okolski A (1981) Culture of horse oocytes in vitro. J Reprod Fertil 61:213–215

Galli C, Crotti G, Notari C, Turini P, Duchi R, Lazzari G (2001) Embryo production by ovum pick up from live donors. Theriogenology 55:1341–1357

Galli C, Lagutina I, Crotti G, Colleoni S, Turini P, Ponderato N, Duchi R, Lazzari G (2003a) Pregnancy: a cloned horse born to its dam twin. Nature 424:635

Galli C, Lagutina I, Crotti G, Colleoni S, Turini P, Ponderato N, Duchi R, Lazzari G (2003b) Pregnancy: a cloned horse born to its dam twin. Nature 424:635

Gambini A, Maserati M (2017) A journey through horse cloning. Reprod Fertil Dev 30(1):8–17. https://doi.org/10.1071/RD17374 (Review)

Hinrichs K (2018) Assisted reproductive techniques in mares. Reprod Domest Anim 53(Suppl 2):4–13. https://doi.org/10.1111/rda.13259

Hinrichs K, Choi YH, Love LB, Varner DD, Love CC, Walckenaer BE (2005) Chromatin configuration within the germinal vesicle of horse oocytes: changes post mortem and relationship to meiotic and developmental competence. Biol Reprod 72:1142–1150. https://doi.org/10.1095/biolreprod.104.036012

Hinrichs K, Choi YH, Love CC, Chung YG, Varner DD (2006) Production of horse foals via direct injection of roscovitine-treated donor cells and activation by injection of sperm extract. Reproduction 131:1063–1072

Hinrichs K, Choi YH, Varner DD, Hartman DL (2007) Production of cloned horse foals using roscovitine-treated donor cells and activation with sperm extract and/or ionomycin. Reproduction 134:319–325

Iacono E, Pascucci L, Rossi B, Bazzucchi C, Lanci A, Ceccoli M, Merlo B (2017) Ultrastructural characteristics and immune profile of equine MSCs from fetal adnexa. Reproduction 154(4):509–519. https://doi.org/10.1530/REP-17-0032

Kozlowski CP, Clawitter HL, Thier T, Fischer MT, Asa CS (2018) Characterization of estrous cycles and pregnancy in Somali wild asses (Equus africanus somaliensis) through fecal hormone analyses. Zoo Biol 37(1):35–39. https://doi.org/10.1002/zoo.21397

Jacobson CC, Choi YH, Hayden SS, Hinrichs K (2010) Recovery of mare oocytes on a fixed biweekly schedule, and resulting blastocyst formation after intracytoplasmic sperm injection. Theriogenology 73:1116–1126. https://doi.org/10.1016/j.theriogenology.2010.01.013

Lagutina I, Lazzari G, Duchi R, Colleoni S, Ponderato N, Turini P, Crotti G, Galli C (2005) Somatic cell nuclear transfer in horses: effect of oocyte morphology, embryo reconstruction method and donor cell type. Reproduction 130:559–567

Lee W, Song K, Lee I, Shin H, Lee BC, Yeon S, Jang G (2015) Cloned foal derived from in vivo matured horse oocytes aspirated by the short disposable needle system. J Vet Sci 16(4):509–516. https://doi.org/10.4142/jvs.2015.16.4.509

Leemans B, Gadella BM, Stout TA, De Schauwer C, Nelis H, Hoogewijs M, Van Soom A (2016) Why doesn't conventional IVF work in the horse? The equine oviduct as a microenvironment for capacitation/fertilization. Reproduction 152(6):R233–R245 (Epub 2016 Sep 20. Review)

Lepage SI, Nagy K, Sung HK, Kandel RA, Nagy A, Koch TG (2016) Generation, characterization, and multilineage potency of mesenchymal-like progenitors derived from equine induced pluripotent stem cells. Stem Cells Dev. 25(1):80–89. https://doi.org/10.1089/scd.2014.0409 (Epub 2015 Nov 5)

Li L, Liu X, Guo H (2017) The nutritional ingredients and antioxidant activity of donkey milk and donkey milk powder. Food Sci Biotechnol 27(2):393–400. https://doi.org/10.1007/s10068-017-0264-2 (eCollection 2018 Apr)

Li X, Zhou SG, Imreh MP, Ahrlund-Richter L, Allen WR (2006) Horse embryonic stem cell lines from the proliferation of inner cell mass cells. Stem Cells Dev. 15(4):523–531

Liu LL, Fang C, Liu WJ (2018) Identification on novel locus of dairy traits of Kazakh **horse** in Xinjiang. Gene 30(677):105–110. https://doi.org/10.1016/j.gene.2018.07.009 (Epub 2018 Jul 3)

Lorenzo JM, Munekata PES, Campagnol PCB, Zhu Z, Alpas H, Barba FJ, Tomasevic I (2017) Technological aspects of horse meat products—a review. Food Res Int 102:176–183. https://doi.org/10.1016/j.foodres.2017.09.094 (Epub 2017 Sep 30. Review)

Maia L, de Moraes CN, Dias MC, Martinez JB, Caballol AO, Testoni G, de Queiroz CM, Peña RD, Landim-Alvarenga FC, de Oliveira E (2017) A proteomic study of mesenchymal stem cells from equine umbilical cord. Theriogenology 15(100):8–15. https://doi.org/10.1016/j.theriogenology.2017.05.015

Morris LHA (2018) The development of in vitro embryo production in the horse. Equine Vet J 50(6):712–720. https://doi.org/10.1111/evj.12839 (Epub 2018 May 25. Review)

Nagy K, Nagy A (2015) Derivation of equine-induced pluripotent stem cell lines using a piggyBac transposon delivery system and temporal control of transgene expression. Methods Mol Biol 1330:79–88. https://doi.org/10.1007/978-1-4939-2848-4_8

Nagy K, Sung HK, Zhang P, Laflamme S, Vincent P, Agha-Mohammadi S, Woltjen K, Monetti C, Michael IP, Smith LC, Nagy A (2011) Induced pluripotent stem cell lines derived from equine fibroblasts. Stem Cell Rev 7 (3):693–702. https://doi.org/10.1007/s12015-011-9239-5 (Erratum in: Stem Cell Rev. 2012 Jun; 8(2):546)

Olivera R, Moro LN, Jordan R, Luzzani C, Miriuka S, Radrizzani M, Donadeu FX, Vichera G (2016) In vitro and in vivo development of horse cloned embryos generated with iPSCs, mesenchymal stromal cells and fetal or adult fibroblasts as nuclear donors. PLoS One 11(10):e0164049. https://doi.org/10.1371/journal.pone.0164049 (eCollection 2016)

Olivera R, Moro LN, Jordan R, Pallarols N, Guglielminetti A, Luzzani C, Miriuka SG, Vichera G (2018) Bone marrow mesenchymal stem cells as nuclear donors improve viability and health of cloned horses. Stem Cells Cloning 11:13–22. https://doi.org/10.2147/sccaa.s151763 (eCollection 2018)

Ortiz-Escribano N, Bogado Pascottini O, Woelders H, Vandenberghe L, De Schauwer C, Govaere J, Van den Abbeel E, Vullers T, Ververs C, Roels K, Van De Velde M, Van Soom A, Smits K (2018) An improved vitrification protocol for equine immature oocytes, resulting in a first live foal. Equine Vet J 50(3):391–397. https://doi.org/10.1111/evj.12747 (Epub 2017 Sep 21)

Ortiz-Escribano N, Bogado Pascottini O, Woelders H, Vandenberghe L, De Schauwer C, Govaere J, Van den Abbeel E, Vullers T, Ververs C, Roels K, Van De Velde M, Van Soom A, Smits K (2017) An improved vitrification protocol for equine immature oocytes, resulting in a first live foal. Equine Vet J. https://doi.org/10.1111/evj.12747

Palmer E, Duchamp G, Bezard J, Magistrini M, King WA, Bousquet D, Betteridge KJ (1987) Non-surgical recovery of follicular fluid and oocytes of mares. J Reprod Fertil Suppl 35:689–690

Palmer E, Bézard J, Magistrini M, Duchamp G (1991) In vitro fertilization in the horse. A retrospective study. J Reprod Fertil Suppl 44:375–384

Paris DB, Stout TA (2010) Equine embryos and embryonic stem cells: defining reliable markers of pluripotency. Theriogenology 74(4):516–524. https://doi.org/10.1016/j.theriogenology.2009.11.020 (Epub 2010 Jan 13. Review)

Polejaeva IA, Chen SH, Vaught TD, Page RL, Mullins J, Ball S, Dai Y, Boone J, Walker S, Ayares DL, Colman A, Campbell KHS (2000) Cloned pigs produced by nuclear transfer from adult somatic cells. Nature 407:86–90

Riera FL, Roldan JE, Gomez J, Hinrichs K (2016) Factors affecting the efficiency of foal production in a commercial oocyte transfer program. Theriogenology 85:1053–1062. https://doi.org/10.1016/j.theriogenology.2015.11.016

Pozor MA, Sheppard B, Hinrichs K, Kelleman AA, Macpherson ML, Runcan E, Choi YH, Diaw M, Mathews PM (2016) Placental abnormalities in equine pregnancies generated by SCNT from one donor horse. Theriogenology 86(6):1573–1582. https://doi.org/10.1016/j.theriogenology.2016.05.017 (Epub 2016 May 26)

Pukazhenthi BS, Johnson A, Guthrie HD, Songsasen N, Padilla LR, Wolfe BA, Coutinho da Silva M, Alvarenga MA, Wildt DE (2014) Improved sperm cryosurvival in diluents containing amides versus glycerol in the Przewalski's horse (Equus ferus przewalskii). Cryobiology 68(2):205–214. https://doi.org/10.1016/j.cryobiol.2014.01.013 (Epub 2014 Feb 6)

Roels K, Smits K, Ververs C, Govaere J, D'Herde K, Van Soom A (2018) Blastocyst production after intracytoplasmic sperm injection with semen from a stallion with testicular degeneration. Reprod Domest Anim 53 (3):814–817. https://doi.org/10.1111/rda.13153 (Epub 2018 Mar 1)

Rota A, Panzani D, Sabatini C, Camillo F (2012) Donkey jack (*Equus asinus*) semen cryopreservation: studies of seminal parameters, post breeding inflammatory response, and fertility in donkey jennies. Theriogenology 78(8):1846–1854. https://doi.org/10.1016/j.theriogenology.2012.07.015 (Epub 2012 Sep 12)

Saito S, Ugai H, Sawai K, Yamamoto Y, Minamihashi A, Kurosaka K, Kobayashi Y, Murata T, Obata Y, Yokoyama K (2002) Isolation of embryonic stem-like cells from equine blastocysts and their differentiation in vitro. FEBS Lett 531(3):389–396

Samper J, Morris L, Plough T (2012) The use of sex-sorted stallion semen in embryo transfer programs J Equine Vet Sci 32:387–389

Sessions-Bresnahan DR, Graham JK, Carnevale EM (2014) Validation of a heterologous fertilization assay and comparison of fertilization rates of equine oocytes using in vitro fertilization, perivitelline, and intracytoplasmic sperm injections. Theriogenology 82 (2):274–282. https://doi.org/10.1016/j.theriogenology.2014.04.002 (Epub 2014 Apr 16)

Sharma R, Livesey MR, Wyllie DJ, Proudfoot C, Whitelaw CB, Hay DC, Donadeu FX (2014) Generation of functional neurons from feeder-free, keratinocyte-derived equine induced pluripotent stem cells. Stem Cells Dev. 23(13):1524–1534. https://doi.org/10.1089/scd.2013.0565 (Epub 2014 Mar 25)

Smits K, Govaere J, Hoogewijs M, Piepers S, Van Soom A (2012) A pilot comparison of laser-assisted vs piezo drill ICSI for the in vitro production of horse embryos. Reprod Domest Anim 47(1):e1–e3. https://doi.org/10.1111/j.1439-0531.2011.01814.x (Epub 2011 Sep 24)

Squires EL (2016) Breakthroughs in equine embryo cryopreservation. Vet Clin North Am Equine Pract 32 (3):415–424. https://doi.org/10.1016/j.cveq.2016.07.009 (Epub 2016 Oct 8. Review)

Violini S, Ramelli P, Pisani LF, Gorni C, Mariani P (2009) Horse bone marrow mesenchymal stem cells express embryo stem cell markers and show the ability for tenogenic differentiation by in vitro exposure to BMP-12. BMC Cell Biol 22(10):29. https://doi.org/10.1186/1471-2121-10-29

Wilmut I, Schnieke AE, McWhir J, Kind AJ, Campbell KHS (1997) Viable offspring derived from fetal and adult mammalian cells. Nature 385:810–813

Woods GL, White KL, Vanderwall DK, Li GP, Aston KI, Bunch TD, Meerdo LN, Pate BJ (2003) A mule cloned from fetal cells by nuclear transfer. Science 301:1063

Zhang JJ, Boyle MS, Allen WR, Galli C (1989) Recent studies on in vivo fertilization of in vitro matured horse oocytes. Equine Vet J. 8 (Suppl.):101–104

Assisted Reproduction in Cats

17

Abstract

Among felids, the domestic cat is an imperative pet animal. The interest in assisted reproduction in cats has increased as domestic cat plays important role in reproduction of exotic and wild felids and valuable animal model for human diseases. ARTs, viz. sperm and oocyte cryopreservation, nuclear transfer cloning, stem cell technology, and transgenesis, have commercial cat rearing and human health prospects. Preserving precious germplasm or genotypes in purebred or research model catteries has raised the importance of quality cat reproduction management.

Highlights

- Petting cat is momentous for improving human mental and cognitive fitness
- The domestic cat serves as model animal for biomedical applications and assisted reproduction of wild felids.

Keywords

Domestic cat · Pets animals · Cat reproduction · Cryopreservation of oocytes · Biomedical importance

17.1 Introduction

Medical researchers and clinicians have achieved a lot to administer human health and welfare of animals. Human well-being is intertwined with the health of companion animals or pets, especially dogs and cats remaining in our vicinity (Brody 2017).

Domestic cat (*Felis domesticus*) is one of the popular companion animals. According to Dennis Turner, a leading expert on feline-human relations, "Cats are very sensitive, independent thinkers and independent actors, very elegant and beautiful animals." (http://healthypets.mercola.com/sites/healthypets/archive/2010/12/28/reasons-why-cats-make-best-pets.aspx, accessed on May 9, 2018). Besides being a pet animal, the domestic cat is an ideal model for studying long-term safety and efficacy of emergent medical modalities, such as gene and stem cell therapies for curing degenerative diseases.

Animal lovers say that cats (Fig. 17.1) are better pets due to their characteristic habits, small body size, and body language (Box 1). While kitty would not encourage to exercise, begging to go out for a hike, its purr can normalize blood

B. Singh et al., *Advances in Animal Biotechnology*,
https://doi.org/10.1007/978-3-030-21309-1_17

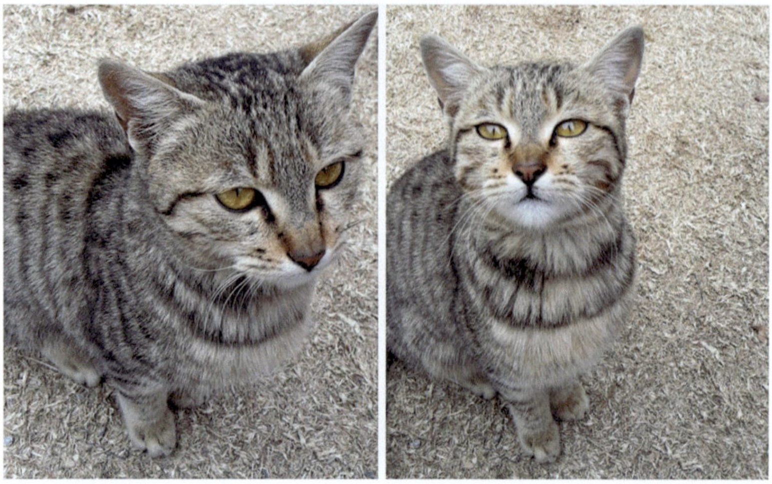

Fig. 17.1 The domestic cat is an exquisite small pet animal. In villages, the pet cats serve to control pests such as rats and mice and sometimes fight the snakes entering houses

pressure and alleviate the stress. The pedigree cat fancy has increased in many countries during past few decades.

One of the most important contribution of cat to wild animal assisted reproduction is that cat is used as donor of oocytes for preparing cytoplasts for producing cloned embryos (Van Soom et al. 2014; Gómez and Pope 2015; Moulavi et al. 2017). For instance, despite efforts, it was not possible to get viable cells from improperly frozen tissue of Asiatic cheetah (*Acinonyx jubatus venaticus*), a critically endangered wild subspecies. Then, the cell nuclei were isolated and injected directly into the cytoplasts developed form *in vitro* matured cat oocytes to produce cloned embryos.

Box 1. Reasons Why Cats are Preferred over Dogs

- Cats are better for environment. They require less food and nutrients dogs
- Cats prefer fish than corn- or beef-flavored products
- Cats control pests, rodents, and snakes that are harmful to humans
- Provide emotional support to pet lovers
- Cats are better companion to people spending solitary or lonesome life
- Cats lower mental stress and, therefore, protect against stress-induced health problems
- Cats offer better companionship, especially to forlorn women
- Cats contribute to strengthen immune system and lessen the susceptibility to allergy symptoms
- Some cats are famous as military animals and have saved lives of their owners
- Cat purring (20–140 Hz frequency) has a positive effect on bones, tendons, muscles joint mobility, and blood pressure.

17.2 Assisted Reproduction in Cats

Cattery is a business in some countries, and animal experts provide services of looking after the cats when cat owners are out of station for long time. In view of empathy of the humans toward cats, there is increasing trend in cloning their cats aging naturally.

Compared to other animals, less progress is observed in assisted reproduction technologies (ARTs) in cats because of scarcity of knowledge on their biology and behavior (Table 17.1).

Like other pet species, research on assisted reproduction is focused on ovarian biology and contraception, although progress is also noted in embryo production, interspecies nuclear transfer cloning (Gómez et al. 2006; Gómez and Pope 2015) and producing transgenic embryos for commercial or biomedical applications (Gómez et al. 2009).

Table 17.1 Chronology of some salient achievements in ARTs and their applications in cats

Year	Event and recommendations (References)
1970	Description of AI in cats (Felis catus) (Sojka et al. 1970)
1978	Semen collection by electroejaculation in domestic cats (Platz and Seager 1978), first report on birth of live kittens after AI with previously frozen semen (Platz et al. 1978)
2003	Development of IVM, and IVF, and development of IVF-derived embryos, cryopreservation, and their transfer (Gómez et al. 2003)
2002	First birth of cloned cat (Shin et al. 2002)
2005	Birth of cloned cats from adult somatic cells (Yin et al. 2005), development of cat embryos produced by ICSI (Murakami et al. 2005)
2008	Production of second generation cat clones (Yin et al. 2008), isolation of ESC-like cells from ICM cells of embryos (Yu et al. 2008)
2010	Generation of re-cloned cat (Cho et al. 2010a), confirmation of germ line transmission of transgenic genes (Cho et al. 2010b)
2014	Birth of kittens from transfer of embryos developed from vitrified oocytes (Galiguis et al. 2014)
2016	Isolation and description of cat SSCs (Powell et al. 2016), evolving strategies for improving embryo development from poor-quality cat oocytes (Ochota et al. 2016), description of SSC-markers in spermatogonia (Vansandt et al. 2016)
2017	Studies on molecular markers in spermatogonial stem cells (Bedford-Guaus et al. 2017), cryopreservation of cat sperm (Chatdarong, 2017), and oocytes by using commercially available kits and slush nitrogen technique (Fernandez-Gonzalez and Jewgenow 2017), studies on urethral catheterization and sperm cryopreservation by vitrification for developing semen banks (Swanson et al. 2017), proliferative and immunomodulatory profiling of cat MSCs (Zajic et al. 2017), studies on potential of MSCs to cure renal disorders in cats (Vidane et al. 2017)
2018	Cryopreservation of ovarian tissue (Brito et al. 2018), studies on changes in spermatozoa quality parameters subjected to cryopreservation (Cheuquemán et al. 2018)

17.3 Male-Assisted Reproduction

The male-assisted reproduction includes collection, evaluation, and cryopreservation of semen and treating infertility. The demand for semen from superior tom has increased.

Various methods are used for collecting semen in cats. Artificial vagina (AV) constructed simply by using sterile Eppendorf tube and rubber bulb, electroejaculation, urethral catheterization, and epididymis slicing are used to collect sperm from cat. Studies are carried out to evaluate semen characteristics and improve cryopreservation and dissemination. However, there are some problems in collecting semen from a tom. The small size of animal is a major problem. Compared to other pet animals, the collected semen volume is too less which makes difficult to analyze it for the quality and using the remaining aliquot for insemination or cryopreservation (Johnson 2018a).

Like other biomaterials, the cryopreservation affects the quality of semen. Cryopreservation-mediated changes include effect on total sperm motility, viability, acrosome integrity, chromatin structure, mitochondrial functions, and lipid peroxidation.

Platz et al. (1978) for the first time reported pregnancies and birth of kittens in domestic cats from intravaginal insemination of frozen-thawed semen. The success opened doors for developing semen extenders, such as Tris-egg yolk extender (TEST) and TEST-soybean lecithin extenders.

As prime objective of semen cryopreservation is to preserve sperm for future use, care should be taken to use proven quality semen, handle the semen samples carefully before and after cryopreservation, and avoid contamination at every step. Toxoplasmosis, feline leukemia virus infection, feline viral rhinotracheitis, and feline infectious peritonitis are most probable contaminants that may cause infertility in adult animals and mortality in neonates.

In cats, various factors are responsible for infertility in male as well as in female. The major ones include infections, nutrition, or management problems and age of the animal. In general, most infertility problems are complex, and their treatment depends on severity and type of the problem. The stem cells, in view of their regenerative ability, and pluripotency are viewed as promising tools to cure infertility.

The spermatogonial stem cells (SSCs) are key cells that persist throughout the life of male and mediate transmission of genetic information to offspring and therefore represent exciting new avenues for male-assisted reproduction, fertility preservation, and species conservation. Some reports on SSCs in domestic cats are worth mentioning. Expression of six SSCs markers, namely OCT4, THY-1, GPR125, GFRα1, PLFZ, and UCHL was detected in SSCs isolated from three juvenile and three adult cats (Vansandt et al. 2012). Another study conducted for expression of various SSC-specific general markers, namely KIT, CD9, ITGA6, THY-1, ADGRA3, and GFRα1, and less studied stem cell markers, namely TRA-160, TRA-1-81, SSEA-1, SSEA-4 SSCs in adult as well as prepubertal domestic cats (*Felus catus*), showed that SSCs markers identified are not expressed in domestic cat (Powell et al. 2016).

Fluorescent-activated cell sorting with antibody against epithelial cell adhesion molecules (EPCAM) was used to separate different live cells. The EPCAM (+) cells, but not the SSCs, expressed CKIT (CD117), while EPCAM(-) cells expressed high levels of OCT-4. It was concluded that SSCs might reside within the populations of EPCAM(-) cells (Bedford-Guaus et al. 2017).

We suggest that more studies are needed to define molecular markers for characterizing stemness in cat MSCs, SSCs, and germ stem cells. Recent reviews (Johnson 2018a; Quimby and Borjesson 2018) have summarized various aspects of the ARTs in cats.

17.4 Cryopreservation of Sperm

Cryopreservation is an important tool in assisted reproduction. The semen collected by urethral catheterization or by epididymal slicing is of acceptable quality for cryopreservation or inseminating the females.

It is argued that epididymal sperm cryopreserved for a prolonged periods remain motile and can undergo capacitation. It implies that technique is useful for reusing genetic materials from endangered or threatened animals or from the males even after their death (Kunkitti et al. 2016a). Epididymal sperm from corpus and cauda epididymis of seven cats were cryopreserved, frozen-thawed, and used for fertilizing IVM oocytes. The study concluded that sperm from corpus epididymis had a comparable potential to fertilize oocytes in vitro as sperm from cauda epididymis. However, it resulted in fewer embryos developing to hatching blastocysts compared to those developed form spermatozoa from the cauda (Kunkitti et al. 2016b).

Computer-assisted sperm analysis (CASA) for sperm quality parameters before and after cryopreservation showed that cryopreservation had minor effects on chromatin integrity, lipid peroxidation, and mitochondrial potential in cat sperm. However, semen collected by urethral catheterization and by epididymal slicing was of comparable excellence; hence, both methods were recommended for collecting semen for insemination or cryopreservation (Prochowska et al. 2016). Analysis of a total of 60 semen samples showed a decline in total sperm motility, mitochondrial activity, and viability from 80% in freshly obtained semen to 40% in frozen-thawed cat semen (Cheuquemán et al. 2018).

17.5 Female-Assisted Reproduction

The domestic cat is seasonal polyestrus-induced ovulator and a long-day breeder. The cats attain puberty soon; the first estrus is experienced at the age of 5–9 months on average. The queens remain in estrus unless there is interruption by pregnancy, pseudopregnancy, or undue health problems. The age at first estrus is influenced by several factors such as breed (shorter breeds reach puberty earlier than the larger breeds), season, i.e., length of daylight, and general health of the animal. Duration of light is important factor for onset of estrus; hence, breeding catteries manage to provide 12–14 h of artificial light or daylight per day to regulate or boost estrus.

As cats are induced-ovulators, for AI programs, estrus and ovulation have to be induced by using a combination of equine chorionic gonadotropin (eHG) and human chorionic gonadotropin (hCG). Care has to be taken to avoid factors that cause infertility. Insemination and ovulations should be closely synchronized. Contaminated semen should never be used for insemination.

The insemination in cats is not a routine or regular practice. This is because of practical difficulties such as small size of the animal and small volume (0.03–0.3 ml, sperm count 100–5000 × 10^6/ml) obtained from tomcat. Generally, the fresh semen is used for insemination (Tanaka et al. 2000). However, sperm can be preserved under liquid nitrogen and transported across long distance or used as and when needed (Chatdarong 2017).

Basic knowledge of anatomy of female genital tract of cat is essential for performing AI. The sperm should be deposited directly in vagina or uterus under general sedation or anesthesia. Intravaginal insemination is done by inserting a 9-cm-long 20-G needle attached to a rubber bulb on one tip (Sojka et al. 1970) or by 9-cm-long and 1.5-mm-diameter nylon probe or 9-cm-long, 2-mm-diameter stainless steel probe into anterior narrow part of non-distensible vagina (Tanaka et al. 2000). Care should be taken to keep the animals in dorsal recumbency for 20–25 min with the hind quarter elevated to ensure the entry of sufficient sperm into uterus and uterine horns. Expertise is needed to collect sperm and inseminate the females.

The ovulation commences after 20–48 h after mating (Shille et al. 1983); hence the appropriate time for insemination is second day of estrus with subsequent insemination after 48 h (Chatdarong et al. 2002).

17.6 Cryopreservation of Oocytes, Embryos, and Ovaries

Cryopreservation is important to boost female reproduction potential through oocyte and embryo banks. The preservation of cat oocytes by conventional slow-freezing and vitrification has commenced around three decades ago. Regrettably, the cryopreservation of cat oocytes is still not improved much and can be stated to be in experimental stage. Some companies have developed commercial kits for cryopreservation of feline oocytes. The kits have variable success rates in terms of cryosurvival and developmental competence of cryopreserved oocytes (Fernandez-Gonzalez and Jewgenow 2017).

Infertility in females is also an important concern. Vaginal stricture, anatomical abnormalities, renal diseases, diabetes or hyperthyroidism, and genitourinary infections could hinder conception in females. Therefore, a general and reproductive physiological examination of blood chemistry and endocrine profile of female may be followed to rule out the possibilities of infertility. Treating female infertility is a complex procedure and requires multiple tests to define the root cause of infertility.

Ever since the success achieved by using in vitro produced embryos to have live offspring, the IVF and IVEP are used to produce live cat offspring in other regions of the world (Hribal et al. 2014; Cocchia et al. 2015; Kunkitti et al. 2016a, b). However, the number of live kittens obtained is far less than those obtained in large ruminants and pigs.

The developmental competence of vitrified-warmed oocytes is low in cat. The association and communication between oocytes and cumulus cells surrounding the oocytes, mediated through connexion-based gap junctions, play important role in normal follicular development. The gap junctions remain closed generally, but may open under certain stress conditions, such as vitrification-warming. Blocking connexin 37 and connexin 43 channels with Gap26, a connexin-targeting peptide, during vitrification was found to improve maturation capacity of immature oocytes after vitrification-warming (Snoeck et al. 2018)

Cryopreservation supports preservation of genetic diversity. Unfortunately, except few reports, the cryopreservation of feline oocytes and embryos is still in experimental phases.

Introduction of new drugs marketed for cats and dogs is envisioned to impact positively the commercial cattery. Readers should refer to (Johnson 2018b; Quimby and Borjesson 2018), latest articles for updated information on various aspects of female ARTs in cats.

17.7 Nuclear Transfer Cloning

The people in view of their affection to cats want them cloned. There is trend in cloning their cats in some segments of Western society. Resurrecting pet cats by nuclear transfer cloning is a hope for pet lovers seeing their pets perpetually. The Department of Veterinary Physiology and Pharmacology, College of Veterinary Medicine, Texas A & M University, College Station, Texas, USA, reported first clone (called CC, i.e., Copy of Cat) of domestic cat, the Rainbow in 2002, inferring that methods used to clone cat might not be readily extendable to other mammalian species because of technical and molecular biological bottlenecks and interspecies barriers (Shin et al. 2002). Out of 87 implanted NT embryos, CC was one to survive. The success rate was comparable to that noted in other species at that time.

The clone and the donor of somatic cell shared same genetic code like identical twins, but were different phenotypically. This was because environmental factors including nutrition and health determine phenotype of an individual. Hence, clones differ from the mother, the donor of genetic material to cloned offspring.

Later, some studies were undertaken on various aspects of NT cloning in cats. Skrzyszowska and colleagues (2002) concluded that prolonged *in vitro* maturation of cat oocytes obstructs developmental competence of reconstructed embryos and that donor cells, whether fetal fibroblasts or oocyte-associated cumulus cells, had analogous developmental proficiency (Skrzyszowska et al. 2002).

The fetal fibroblasts and skin fibroblasts used as donor cells could be reprogrammed to develop into SCNT embryos, and when transferred into recipient queens, healthy cloned cats were produced. The study raised the hopes that adult somatic cells could be used for cloning other cats as well (Yin et al. 2005). The same group could successfully produce second generation cloned cats by SCNT using skin cells as donor nuclei from a 6-month-old G1 cloned cat. The second generation cloned kittens were born, but cloning efficiency remained low (Yin et al. 2008).

Cloning in cats offers unique commercial opportunities. Private entrepreneurs, for example, ViaGen Pets, the global leader in animals cloning and ARTs (https://viagenpets.com/cat-cloning/, accessed May 23, 2017), allow cat owners the opportunities for happiness over companionship that a beloved cat brings. The company has cloned a number of other livestock species including dogs, cows, and horses.

Moving a step ahead, SCNT in conjunction with genetic modification of donor nuclei was used to produce transgenic cat embryos. The cat fetal fibroblasts were transfected with EGFP-encoding lentiviral vectors (LV) possessing various promoters including human cytomegalovirus immediate early (hCMV-IE) promoter, the human translation elongation factor 1α (hEF-Iα) promoter, and human ubiquitin promoter. The transgenic cloned kittens produced by LV vector-mediated international integration of transgenes into the genome of donor cells expressed EGFP receptor genes in all body cells (Gómez et al. 2009). Somatic cells from a first-generation red fluorescent protein transgenic cat (first RFPTG) were used to produce re-cloned transgenic kitten (*F. catus*) (Cho et al. 2010a, b). The RFP genes were integrated into genome as revealed from the transcriptome analysis of various tissues and internal organs of the cloned animal. The study shows that RFP TG transgenic cat has normal reproductive traits such as fertility, stable germ line transmission of transgenic genes, and their expression in offspring (Cho et al. 2010a, b).These studies support evidences of generating transgenic cats as models for studying and treating human diseases.

17.8 Stem Cells in Clinical Applications

Stem cells are exciting avenues with potential in health and reproduction management and cell-based therapies in cats. MSCs, like other species, have promising health applications in cats. Bone marrow- and adipose-derived MSCs due to ease of obtaining and ability to grow in vitro are reported to be promising in cats. The cat MSCs have immunomodulatory potency comparable to that found in human MSCs and can be used for treating variety of inflammatory diseases (Quimby and Borjesson 2018).

Vidane et al. (2017) have inferred that among intrarenal (5×10^5 cells/kidney), and intravenous (2×10^6 cells $\times$ 2 treatments) infusions of allogeneic AM-derived MSCs, the intravenous injection of AMSCs were effective to confer health benefits in cats suffering from chronic renal diseases.

Studies on clinical efficacy of fresh allogeneic *vs.* fresh autologous adipose-derived MSCs for treating severe refractory feline chronic gingivostomatitis showed that fresh autologous adipose derived MSCs were more effective (Arzi et al. 2017). Allogeneic adipose-derived MSCs have been investigated for acute kidney injury (AKI) in ischemic kidney feline model. Time of treatment had a significant effect on renal function. The results, however, did not support use of allogeneic MSCs in AKI in this case. Kind of renal injury, stem cell dosage, MSC allogenicity, duration, and route or timing of administration could influence the efficacy of the cells tested (Rosselli et al. 2016). Researchers believe that stem cell therapy might be developed in future to treat cat renal diseases.

17.9 Outlook and Challenges

Human health is intertwined with the health of pets and domestic animals that share habitat and activities with humans. Like humans, the pets also suffer from common health threats and metabolic diseases. Diabetes has increased three times more in cats compared to dogs.

It is evident that humans, pets, livestock, and wildlife depend on each other and on global environment. All are impacted by biotic, abiotic, and climatic stress.

Many of the vectors-borne diseases afflict both animals and humans, so pets act as sentinels against threats to their owners (https://www.nature.com/articles/543S44a, accessed Sept. 18, 2018).

Cat assisted reproduction is a less explored area. During past three decades, appreciable progress is witnessed in understanding of behavioral, gonadal, and endocrine regime of domestic and wild feline species reproduction. The demand for high-quality semen is growing as a way to promote and preserve valuable genetic materials. AI, IVF, and embryo transfer and cryopreservation of oocytes embryos and sperm established in cats, might serve as powerful tools for increasing population of endangered felids.

There is an immediate need to develop improved regimens that would allow more predictable ovarian responses in cats. In addition, the topics that need scientific attention include a sound knowledge of female reproductive cycle, ovarian stimulation, endocrine, and behavior studies. Some inherent problems, such as low IVM which is nearly 40–80% at present, should be improved. This will take care of low IVF efficiency. As on today, stem cell therapy cannot be recommended for curing pet animals for every disease. However, if a pet has some good years of life, and budget is not the constraint, the stem cell therapy could be a viable option.

17.10 Conclusions

Cats are important pet animals and models for human disease investigations. Appropriate breeding management is a key factor to maintain good fertility in cattery. There have been greater advances in ARTs during past two decades. Cryopreservation of oocytes, sperm, and ovaries offers an important protection against loss of beloved pets and important models due to diseases or disasters. The stem cell therapy, though bit expensive at present, has potential to treat diseases in pet cats, and those used as models for investigating human diseases.

References

Arzi B, Clark KC, Sundaram A, Spriet M, Verstraete FJM, Walker NJ, Loscar MR, Fazel N, Murphy WJ, Vapniarsky N, Borjesson DL (2017) Therapeutic efficacy of fresh, allogeneic mesenchymal stem cells for severe refractory feline chronic gingivostomatitis. Stem Cells Transl Med. 6(8):1710–1722. https://doi.org/10.1002/sctm.17-0035 (Epub 2017 Jun 15)

Bedford-Guaus SJ, Kim S, Mulero L, Vaquero JM, Morera C, Adan-Milanès R, Veiga A, Raya Á (2017) Molecular markers of putative spermatogonial stem cells in the domestic cat. Reprod Domest Anim 52 (Suppl 2):177–186. https://doi.org/10.1111/rda.12819 (Epub 2016 Nov 6)

Brito DCC, Domingues SFS, Rodrigues APR, Figueiredo JR, Santos RR, Pieczarka JC (2018) Vitrification of domestic cat (Felis catus) ovarian tissue: Effects of three different sugars. Cryobiology 83:97–99. https://doi.org/10.1016/j.cryobiol.2018.06.006 (Epub 2018 Jun 13)

Brody H (2017) Animal health. Nature 543(7647):S41. https://doi.org/10.1038/543s41a (No abstract available)

Chatdarong K (2017) Retained fertilizing capability in cryopreserved feline spermatozoa. Reprod Domest Anim 52(Suppl 2):261–264. https://doi.org/10.1111/rda.12855 (Epub 2016 Nov 3)

Chatdarong K, Kampa N, Axner E, Linde-Forsberg C (2002) Investigation of cervical patency and uterine appearance in domestic cats by fluoroscopy and scintigraphy. Reprod Domest Anim 37:275–281

Cheuquemán C, Faúndez R, Sánchez R, Risopatrón J (2018) Changes in sperm function and structure after freezing in domestic cat spermatozoa. Andrologia e13080. https://doi.org/10.1111/and.13080 (Epub ahead of print)

Cho SJ, Bang JI, Yu XF, Lee YS, Kim JH, Jeon JT, Yee ST, Kong IK (2010a) Generation of a recloned transgenic cat expressing red fluorescence protein. Theriogenology 73(7):848–855. https://doi.org/10.1016/j.theriogenology.2009.09.008

Cho SJ, Lee YS, Lee JI, Bang JI, Yang J, Klassen H, Kong IK (2010b) Confirmation of germ-line transmission in the red fluorescence protein (RFP) transgenic cloned male cat. Cell Reprogram 12(6):739–747. https://doi.org/10.1089/cell.2010.0009 (Epub 2010 Sep 6)

Cocchia N, Corteggio A, Altamura G, Tafuri S, Rea S, Rosapane I, Sica A, Landolfi F, Ciani F (2015) The effects of superoxide dismutase addition to the transport medium on cumulus-oocyte complex apoptosis and IVF outcome in cats (Felis catus). Reprod Biol 15(1):56–64. https://doi.org/10.1016/j.repbio.2014.10.002 (Epub 2014 Oct 16)

Fernandez-Gonzalez L, Jewgenow K (2017) Cryopreservation of feline oocytes by vitrification using commercial kits and slush nitrogen technique. Reprod Domest Anim 52(Suppl 2):230–234. https://doi.org/10.1111/rda.12837 (Epub 2016 Nov 24)

Galiguis J, Gómez MC, Leibo SP, Pope CE (2014) Birth of a domestic cat kitten produced by vitrification of lipid polarized in vitro matured oocytes. Cryobiology 68(3):459–466. https://doi.org/10.1016/j.cryobiol.2014.02.012 (Epub 2014 Mar 11)

Gómez MC, Pope CE (2015) Cloning endangered felids by interspecies somatic cell nuclear transfer. Methods Mol Biol 1330:133–152. https://doi.org/10.1007/978-1-4939-2848-4_13

Gómez MC, Pope CE, Dresser BL (2006) Nuclear transfer in cats and its application. Theriogenology 66(1):72–81

Gómez MC, Pope CE, Kutner RH, Ricks DM, Lyons LA, Ruhe MT, Dumas C, Lyons J, Dresser BL, Reiser J (2009) Generation of domestic transgenic cloned kittens using lentivirus vectors. Cloning Stem Cells 11(1):167–176. https://doi.org/10.1089/clo.2008.0054

Gómez MC, Pope E, Harris R, Mikota S, Dresser BL (2003) Development of in vitro matured, in vitro fertilized domestic cat embryos following cryopreservation, culture and transfer. Theriogenology 60:239–251

Hribal R, Hachen A, Jewgenow K, Zahmel J, Fernandez-Gonzalez L, Braun BC (2014) The influence of recombinant feline oviductin on different aspects of domestic cat (Felis catus) IVFand embryo quality. Theriogenology 82(5):742–749. https://doi.org/10.1016/j.theriogenology.2014.06.009 (Epub 2014 Jun 16)

Johnson AK (2018) Assisted reproduction in the male cat. Vet Clin North Am Small Anim Pract 48(4):511–521. https://doi.org/10.1016/j.cvsm.2018.02.001 (Epub 2018 Apr 12. Review)

Johnson AK (2018) Assisted reproduction in the female cat. Vet Clin North Am Small Anim Pract 48(4):523–531. https://doi.org/10.1016/j.cvsm.2018.02.002 (Epub 2018 Apr 12. Review)

Kunkitti P, Bergqvist AS, Sjunnesson Y, Johannisson A, Axnér E (2016a) The tolerance of feline corpus and cauda spermatozoa to cryostress. Theriogenology 85 (3):502–508. https://doi.org/10.1016/j.theriogenology.2015.09.032

Kunkitti P, Axnér E, Bergqvist AS, Sjunnesson Y (2016b) In vitro fertilization using frozen-thawed feline epididymal spermatozoa from corpus and cauda regions. Theriogenology 86(6):1403–1408. https://doi.org/10.1016/j.theriogenology.2016.04.085 (Epub 2016 May 7)

Moulavi F, Hosseini SM, Tanhaie-Vash N, Ostadhosseini S, Hosseini SH, Hajinasrollah M, Asghari MH, Gourabi H, Shahverdi A, Vosough AD, Nasr-Esfahani MH (2017) Interspecies somatic cell nuclear transfer in Asiatic cheetah using nuclei derived from post-mortem frozen tissue in absence of cryo-protectant and in vitro matured domestic cat oocytes. Theriogenology 1 (90):197–203. https://doi.org/10.1016/j.theriogenology.2016.11.023 (Epub 2016 Nov 27)

Murakami M, Karja NW, Wongsrikeao P, Agung B, Taniguchi M, Naoi H, Otoi T (2005) Development of cat embryos produced by intracytoplasmic injection of spermatozoa stored in alcohol. Reprod Domest Anim 40(6):511–515

Ochota M, Pasieka A, Niżański W (2016) Superoxide dismutase and taurine supplementation improves in vitro blastocyst yield from poor-quality feline oocytes. Theriogenology 85(5):922–927. https://doi.org/10.1016/j.theriogenology.2015.10.042 (Epub 2015 Nov 10)

Powell RH, Galiguis J, Biancardi MN, Pope CE, Leibo SP, Wang G, Gómez MC (2016) Phenotypic and molecular characterization of domestic cat (Felis catus) spermatogonial stem cells. Biol Reprod 95 (1):20. https://doi.org/10.1095/biolreprod.115.134635

Platz CC, Seager SW (1978) Semen collection by electro—ejaculation in the domestic cat. J Am Vet Med Assoc 173:1353–1355

Platz CC, Wildt DE, Seager SW (1978) Pregnancy in the domestic cat after artificial insemination with previously frozen spermatozoa. J Reprod Fertil 52(2):279–282

Prochowska S, Niżański W, Partyka A (2016) Comparative analysis of in vitro characteristics of fresh and frozen-thawed urethral and epididymal spermatozoa from cats (Felis domesticus). Theriogenology 86(8): 2063–2072. https://doi.org/10.1016/j.theriogenology.2016.07.002 (Epub 2016 Jul 13)

Quimby JM, Borjesson DL (2018) Mesenchymal stem cell therapy in cats: Current knowledge and future potential. J Feline Med Surg. 20(3):208–216. https://doi.org/10.1177/1098612x18758590

Rosselli DD, Mumaw JL, Dickerson V, Brown CA, Brown SA, Schmiedt CW (2016) Efficacy of allogeneic mesenchymal stem cell administration in a model of acute ischemic kidney injury in cats. Res Vet Sci 108:18–24. https://doi.org/10.1016/j.rvsc.2016.07.003 (Epub 2016 Jul 5)

Shille VM, Munro C, Farmer SW, Papkoff H, Stabenfeldt GH (1983) Ovarian and endocrine responses in the cat after coitus. J Reprod Fertil 69:29–39

Shin T, Kraemer D, Pryor J, Liu L, Rugila J, Howe L, Buck S, Murphy K, Lyons L, Westhusin M (2002) A cat cloned by nuclear transplantation. Nature 415 (6874):859

Skrzyszowska M, Katska L, Ryńska B, Kania G, Smorag Z, Pieńkowski M (2002) In vitro developmental competence of domestic cat embryos after somatic cloning: a preliminary report. Theriogenology 58(8):1615–1621

Snoeck F, Szymanska KJ, Sarrazin S, Ortiz-Escribano N, Leybaert L, Van Soom A (2018) Blocking connexin channels during vitrification of immature cat oocytes improves maturation capacity after warming. Theriogenology 122:144–149. https://doi.org/10.1016/j.theriogenology.2018.09.011 (Epub ahead of print)

Sojka NJ, Jemings LL, Hamner CE (1970) Artificial insemination in the cat (Felis catus). Lab Anim Care 20:198–204

Swanson WF, Bateman HL, Vansandt LM (2017) Urethral catheterization and sperm vitrification for simplified semen banking in felids. Reprod Domest Anim 52(Suppl 2):255–260. https://doi.org/10.1111/rda.12863 (Epub 2016 Nov 3)

Tanaka A, Takagi Y, Nakagawa K, Fujimoto Y, Hori T, Tsutsui T (2000) Artificial intravaginal insemination using fresh semen in cats. J Vet Med Sci 62:1163–1167

Vansandt LM, Livesay JL, Dickson MJ, Li L, Pukazhenthi BS, Keefer CL (2016) Conservation of spermatogonial stem cell marker expression in undifferentiated felid spermatogonia. Theriogenology 86(4):1022–1035.e3. https://doi.org/10.1016/j.theriogenology.2016.03.031

Van Soom A, Rijsselaere T, Filliers M (2014) Cats and dogs: two neglected species in this era of embryo production in vitro? Reprod Domest Anim 49(Suppl 2):87–91. https://doi.org/10.1111/rda.12303 (Review)

Vansandt LM, Pukazhenthi BS, Keefer CL (2012) Molecular markers of spermatogonial stem cells in the domestic cat. Reprod Domest Anim 47(Suppl 6):256–260. https://doi.org/10.1111/rda.12079

Vidane AS, Pinheiro AO, Casals JB, Passarelli D, Hage M, Bueno RS, Martins DS, Ambrósio CE (2017) Transplantation of amniotic membrane-derived multipotent cells ameliorates and delays the progression of chronic kidney disease in cats. Reprod Domest Anim 52(Suppl 2):316–326. https://doi.org/10.1111/rda.12846 (Epub 2016 Oct 23)

Yin XJ, Lee HS, Yu XF, Kim LH, Shin HD, Cho SJ, Choi EG, Kong IK (2008) Production of second-generation cloned cats by somatic cell nuclear transfer. Theriogenology 69(8):1001–1006. https://doi.org/10.1016/j.theriogenology.2008.01.017

Yin XJ, Lee HS, Lee YH, Seo YI, Jeon SJ, Choi EG, Cho SJ, Cho SG, Min W, Kang SK, Hwang WS, Kong IK (2005) Cats cloned from fetal and adult somatic cells by nuclear transfer. Reproduction 129 (2):245–249

Yu X, Jin G, Yin X, Cho S, Jeon J, Lee S, Kong I (2008) Isolation and characterization of embryonic stem-like cells derived from in vivo-produced cat blastocysts. Mol Reprod Dev 75(9):1426–1432. https://doi.org/10.1002/mrd.20867

Zajic LB, Webb TL, Webb P, Coy JW, Dow SW, Quimby JM (2017) Comparison of proliferative and immunomodulatory potential of adipose-derived mesenchymal stem cells from young and geriatric cats. J Feline Med Surg. 19(10):1096–1102. https://doi.org/10.1177/1098612x16680703 (Epub 2016 Dec 5)

Assisted Reproduction in Dogs 18

Abstract

Dog is the first pet animal since foundation of human civilization. In modern era of human empire, the spending on pets and their care has increased. Assisted reproduction, originally confined to accepted breeds, has now extended to non-domestic canids. Cryopreservation of semen, oocytes and embryos, in vitro fertilization and somatic cell nuclear transfer cloning are used to conserve and propagate valuable genotypes. Stem cells are used as regenerative medicine or clinical health applications.

Highlights

- Human-canine association represents one of the oldest human camaraderies with animals
- With the advent of scientific animal rearing, and use of dog as biomedical model, the viewpoint to rear dog has also changed.

Keywords

Dogs · Canids · Assisted reproduction · Stem cells · Biomedical applications · Regenerative medicine

18.1 Introduction

Man and dog relation is an age-old practice and dates back around 15,000 years back since humans as hunt gatherers began to live in groups. Dog rearing assisted humans in hunting, and protecting domesticated animals from predators.

Indeed, the dogs were the first animals to be domesticated. Social adaptations of humans and dogs are quite similar, and both human and dogs can live perfectly happily with each other. It is discovered that a dog found in a grave in prehistoric era was sick for a long time and was cared for. This is said to be the oldest known grave where dead humans and dogs were buried together, and represents the earliest evidence of domestication of dogs as pet animal (Janssens et al. 2018).

Having acquired an unparalleled diversity in phenotype and behavior, the dogs are believed to descend from large gray wolf-like progenitor which is now extinct. The dogs therefore resemble wolves in some features and social behavior viz., being territorial and hunting in group. Genomics studies on origin of modern dogs have inferred that dogs were domesticated around 20,000–40,000 years ago (Ostrander et al. 2017; Parker et al. 2017).

At present, there are around 400 breeds of dogs worldwide with unique phenotypic diversity, behavior, and resistance to diseases. Genomic DNA analysis reveals that all dogs are the members of similar species, the *Canis lupus familiaris*, also known as *Canis familiaris* (Ostrander et al. 2017).

Dogs are extraordinarily attentive animals, attuned to emotional state of their owner and express contrition, if their master is annoyed. In addition, some dogs are used in biomedical

B. Singh et al., *Advances in Animal Biotechnology*,
https://doi.org/10.1007/978-3-030-21309-1_18

research and to investigate and cure human diseases, neuropathological and cognitive dysfunction, and genetic disorders (Schütt et al. 2016; Gabr et al. 2018).

The global spending on pets has raised by 14% over the past few years, and worldwide spending on animals care is estimated to be about USD109 billion. There is interest in using modern biotechniques to reproduce the dogs and maintain their health (Fig. 18.1).

18.2 Health Benefits of Dog as Pet Animal

The pets are humans' best friends. As pets, the dogs are exposed to a common environment shared by humans. Compared to rodents, body dimensions, life span, genetics, and physiology of dogs match more to humans. Dog helps release mental stress and diverts attention from stress, anxiety, and depression. People who have used to live with companion and pets have stronger immune system. According to a citation, animal rearing is beneficial to humans. Once regarded as anathema, the environmental microorganisms enrich human normal microbial diversity and activities and boost immunity and resistance against infections and allergy (Gupta 2017).

Humans develop a strong emotional affection with their pets. A lot has been discussed in literature on psychological aspects of human–pet relations. A rich microbial diversity is important for nutrition, behavior, and overall health of the host. Human infants living with pets have richer gut microbial diversity than the infants who have never come in contact with pets (Gupta 2017).

18.3 Assisted Reproduction in Dogs

History of ARTs in dogs dates back to early eighteenth century. First scientifically recorded artificial insemination (AI) was performed in Italy by Spallanzani which led to pregnancies in females and later on three pups were borne. Progress in this area could not improve, and remained slow till world's first puppy was

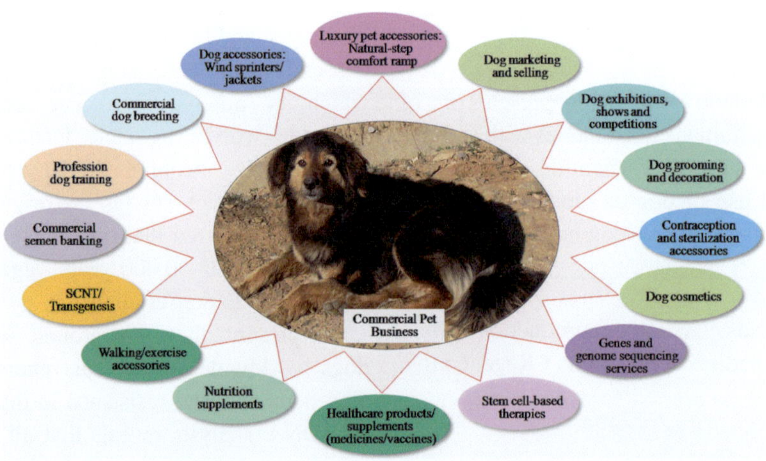

Fig. 18.1 A diagrammatic presentation of dog rearing as commercial entrepreneur. Biotechnological methods are used in nutrition (probiotics, prebiotics, and synbiotics), therapeutics (vaccines, antibodies, immunodiagnostics, and stem cells), developing biologicals for estrus synchronization, preservation of semen and embryos, and contraceptives

produced in 1969 from AI of frozen-thawed semen (cited in Farstad 2000). Lack of adequate information about reproductive physiology and sensitivity of sperms to cryopreservation are still the major constraints restrict widespread use of male ARTs in dogs.

18.4 Collection of Semen

Assisted reproduction is an important component of commercial dog breeding. Care should be taken that only healthy and high pedigree animals are used as semen donors. Once a dog is deemed fit and satisfactory, he is prepared to collect the semen. He is aroused aptly only in presence of a bitch in estrus. This is because the bitch in estrus releases pheromones that arouse the male and persuade him for mating. In absence of a bitch in estrus, the vaginal swabs should be obtained from bitch in estrus and preserved by freezing. Whenever needed, the frozen-thawed swabs should be applied onto the tail of the mail to be used as dummy. Moreover, trained donor dogs mount dummy and do not need a bitch. When mounted, the penis is directed into lumen of AV, where ejaculate is collected.

Semen is collected by different methods viz., by AV, electroejaculation, and percutaneous epididymal aspiration from cauda of epididymus. Epididymal spermatozoa are more suitable in canines. In case of sudden death, or orchiectomy, retrieval of epididymal sperms offers opportunities to collect and use the sperms to produce embryos or perform insemination (Luvoni and Morselli 2017). However, epididymal sperm are more susceptible to cryopreservation. Puppies have been born from AI with frozen-thawed epididymal spermatozoa, indicating the potential of epididymal sperm in canine-assisted reproduction.

After collection, the semen is evaluated for quality parameters and purity.

Currently, advanced methods such as auto-analyzers or CASA are available to evaluate semen and detect the contamination. Immunological and PCR methods are also used to detect contamination in semen. Semen with assured quality is chilled and should be used within 24 h, otherwise it should be cryopreserved under liquid nitrogen.

18.5 Semen Cryopreservation

Semen is cryopreserved either by slow-freezing or by vitrification. Cryopreservation by vitrification is cost-effective method that does not require specialized equipments. Intravaginal insemination with fresh and chilled epididymal semen resulted in the birth of eight puppies and one puppy, respectively (Klinc et al. 2005; Wydooghe et al. 2016). Pregnancies and birth of puppies are reported from surgical intrauterine or intratubal insemination of frozen-thawed epididymal semen.

Studies are taking place to develop expedient methods to preserve dog semen. For this, different CPAs and warming temperatures are evaluated. Among various semen extenders, based on Tris, citric acid and glucose, 20% egg yolk (TCG-20% EY), and mixtures of sucrose and/or trehalose: 250 mM sucrose; 250 mM trehalose; 125 mM sucrose + 125 mM trehalose; 250 mM sucrose + 250 mM trehalose, the extender containing CPAs either 250 mM sucrose or 250 mM trehalose, and warming at 37 °C was found to be more apposite to preserve beagle dog semen (Caturla-Sánchez et al. 2018).

On contrary, Brito et al. (2017) have concluded that slow-thawing protocols could result into sperm with better motility and binding to perivitelline membrane of chicken egg yolk, compared to fast-thawing protocols. Hence, slow-thawing protocols appear to be more suitable to cryopreserve canine sperm. Cryopreservation at ultralow temperature i.e. −196 °C, as well as semen frozen at −80 °C give comparable results, hence dog semen can also be preserved at −80 °C (Pezo et al. 2017).

18.6 Artificial Insemination

AI has come a long way ever since the first successful vaginal insemination of bitches and live birth of three puppies (cited in Mason 2018). Due to several reasons (Box 1), the AI is of interest in dogs. Major motive of AI is to proliferate the superior germplasm. In some cases, semen has to be collected before a dog dies. Sometimes semen is collected from valuable dogs that are unable to mount due to injuries or other problems.

Box 1. Summary of Advantages of Artificial Insemination and IVF in Dogs

- Avoiding transmission of genetic disorders by using semen from healthy dogs
- Increased services, i.e., more bitches can be inseminated per ejaculate from a stud dog
- Making superior semen available across distances
- Early castration of dogs that are not suitable or meant for breeding
- Using healthy, disease- and pathogen-free semen to avoid infections among females
- Overcoming the behavioral problems of females such as refusal, aggression to male
- Ease of inducing genetic alterations in spermatogonial stem cells (SSCs) and grafting the SSCs into males to produce sperms from SSCs
- Overcoming male dysfunction such as failure to mate.

Insemination should be done by a trained personnel. Semen should be deposited in front of cervix or very near to the cervix. Size of the syringe and length of pipette or tube used to transfer semen depend on body size of bitch. Bitch should not be allowed to sit or urinate for 30–45 min after insemination. Semen collected by any methods can be used for insemination depending on requirements and facilities available.

18.7 Female Reproduction Biotechnology

18.7.1 In Vitro Production of Embryos

IVF has lagged behind in dogs. Only a few reports are available on IVF-derived embryos, their cryopreservation, and live births from transfer of embryos. Nagashima et al. (2015) have reported live births from transfer of embryos in domestic dog (*C. familiaris*). Immature oocytes were collected after spaying the females. Oocytes were matured in vitro and fertilized by semen collected manually from proven-breeder (two Labrador retrievers, one beagle, and two cocker spaniels), aged 2–10 years. Embryos were cryopreserved and frozen-thawed and transferred into recipients (Nagashima et al. 2015).

Bitch oocytes show poor developmental competence which impedes in vitro embryo production. Compared to cats, the embryo development efficiency is low in dogs. Cleavage rates varying from 8 to 37% have been noted (Rodrigues et al. 2004; Otoi et al. 2006).

18.7.2 Somatic Cell Nuclear Transfer

Nuclear transfer cloning is used to clone desirable phenotypes. Domestic dogs play important role in nuclear transfer cloning of canines. They serve as donors of oocytes to prepare cypoplasts to produce cloned embryos of domestic dog as well as wild canines (Kim et al. 2007; Oh et al. 2008).

Widespread use of SCNT to clone dogs is impeded by some technical bottlenecks. Oocyte maturation efficiency is low, and fusion of cytoplast and somatic cell is problematical. Lee et al. (2005) had for the first time produced two cloned Afghan hounds. Thereafter, SCNT was improved to produce cloned transgenic dogs for biomedical applications. A short-tailed Gyeongju Donggyeong breed dog was cloned by using adult fibroblasts as donor cells (Choi et al. 2016).

Kim et al. (2017) reported birth of clones of the world's first cloned dog, the Snuppy. Adipose stem cells were used as donor cells. In another report, in vivo matured oocytes obtained using electric chemiluminescence immunoassay-predicted ovulation method were used to produce cloned embryos, which developed to term and gave birth to neonates (Lee et al. 2017). Cloned puppies were obtained from in vivo matured oocytes retrieved by using serum estradiol levels as method of choice to predict the time of ovulation (Zhao et al. 2018).

In addition to conserve the endangered dogs (Jang et al. 2009; Choi et al. 2016; Oh et al. 2016), and working dogs (Oh et al. 2009), SCNT cloning has been used to generate transgenic dogs as model animals (Hong et al. 2009b; Kim et al. 2011a, b; Oh et al. 2011).

Elite dogs are of immense importance due to their extraordinary olfactory sense. Due to their high sensing power to detect drugs, they are used in airport, quarantine areas, or rescuing humans in disasters. Elite dogs have been propagated by nuclear transfer cloning and used to perform diverse tasks (Lee et al. 2018). SCNT-cloned dogs are normal in physiology and behavior. South Korea had in 2007 introduced cloned dogs for police and military use. China has also introduced cloned dogs in Police Dog Base for training in police and forensics. Indeed, the cloning programs to establish a functional pool of outstanding military and police dogs (https://in.yahoo.com/style/china-first-cloned-police-dog-091002563.html, accessed on March 21, 2019). Table 18.1 summarizes salient achievements in ARTs in dogs.

18.7.3 Stem Cell Research

Embryonic stem cells, perinatal stem cells, adult MSCs, and iPSCs have been generated in dogs. Whereas prime motive of ESCs is to promote reproduction and unravel basic developmental mechanisms, the MSCs are used as regenerative medicine to cure injuries. Notably, the true ESCs are still not established in dogs, and only ESC-like cells are reported.

When injected around the wound, the BMSCs were found to migrate and concentrate predominantly toward the center of the wound. The study reveals that BMSCs could serve as valuable therapeutic candidates for cell-based therapies in spinal injuries in pets (Bhat et al. 2018).

The SSCs serve to treat infertility in males. It is demonstrated that SSCs grafted in testes contribute to produce sperms, thus offers avenues to generate transgenic animals (Harkey et al. 2013). Canine iPSCs (ciPSCs) have been derived from mature somatic cells, and might have use in regenerative medicines or clinical applications (Nishimura et al. 2017a, b). ciPSCs have been generated from embryonic fibroblasts and other adult stem cells. The cells were pluripotent and could be differentiated into other cell lineages such as mature megakaryocytes (MKs) and functional platelets (Nishimura et al. 2013). Another ciPSC line was developed from canine embryonic fibroblasts by means of auto-erasable replication-defective and persistent Sendai virus vector, SeVdp(KOSM)302L. The resulting ciPSCs exhibited expression of key pluripotency markers (Oct3/4, Sox2, and Nanog) and differentiated into multiple cell lineages (Tsukamoto et al. 2018). Table 18.2 summarizes the salient applications of MSCs in canines.

18.8 Outlook and Challenges

Male infertility is a problem of significant concern in pet dogs. Progress in canine-assisted reproduction has lagged behind compared to other species. This is because modern ARTs started in dogs much later than those in other species.

Table 18.1 Summary of milestone achievements in dog-assisted reproduction

Year	Achievements
1979	Surgical collection and transfer of embryos (Kinney et al. 1979)
1992	Comprehensive studies on IVM, IVF, and development of embryos (Yamada et al. 1992)
2005	First clone of dogs (Lee et al. 2005), first report on normal pregnancy and birth of puppies from intravaginal insemination with epididymal semen obtained from a dog castrated due to benign prostate hyperplasia (Klinc et al. 2005)
2006	Isolation of ESC-like cells from ICM cells of canine embryos (Hatoya et al. 2006)
2007	Dog cloned from skin fibroblasts obtained from a 2-month-old female Afghan hound (Jang et al. 2007), production of sperm from transplanted SSCs (Kim et al. 2008)
2008	Birth of cloned toy poodle produced from somatic cells obtained from an aged bitch (Jang et al. 2008)
2009	Cloning endangered Sapsaree (*Canis familiaris*), a native Korean Natural Monument canine breed (Jang et al. 2009), birth of cloned Beagle dogs (Hossein et al. 2009; Hong et al. 2009a, b), birth of offspring from cloned female dog with cloned male dog (Park et al. 2009), transgenic cloned dogs expressing red fluorescent proteins (Hong et al. 2009b), evaluating fertilizing capacity of frozen epididymal sperm (Martins et al. 2009)
2011	Studies on reproduction potential of cloned transgenic bitch. It was found that transgenic RFP bitch had normal reproduction ability, and transgenes were heritable (Hong et al. 2011), birth of re-cloned dogs from canine adipose tissue-derived MSCs (cAd-MSCs) from transgenic cloned beagle dog (Oh et al. 2011), live birth of transgenic dogs expressing GFP (Kim et al. 2011a, b)
2012	Isolation and description of MSCs from Wharton's jelly (Seo et al. 2012)
2013	Isolation, description, and transplantation of SSCs (Harkey et al. 2013), formation of functional platelets from ciPSCs (Nishimura et al. 2013).
2015	First report on live birth of seven puppies from transfer of IVF-derived, and frozen-thawed embryos (Nagashima et al. 2015)
2016	Conserving short-tailed Gyeongju Donggyeong by SCNT (Choi et al. 2016), multiplication of selected dogs by cloning (Oh et al. 2016), studies on use of acteoside to induce cell cycle synchronization of donor cells to enhance cell reprogramming to improve efficiency of SCNT in dog (Lee et al. 2016)
	Live neonates born after intravaginal insemination of bitch with chilled epididymal semen collected from dog after postmortem (Wydooghe et al. 2016)
2017	Dogs cloned by using *in vivo* matured oocytes obtained using electric chemiluminescence immunoassay-predicted ovulation technique (Lee et al. 2017), birth of clone of world's first SCNT-cloned dog (Kim et al. 2017), studies confirmed differentiation potential of MSCs from canine ovaries (Trindade et al. 2017), generation of canine-induced extraembryonic endoderm-like cell lines (Nishimura et al. 2017a), generating ciPSCs expressing doxycycline-inducible murine Oct3/4 (Pou5f1), Sox2, Klf4, and c-Myc, which were introduced using lentiviral vectors (Nishimura et al. 2017b), birth of puppies from AI using sexed sperm (Wei et al. 2017)
2018	Studies on enhancing canine oocyte maturation and embryo development in vitro by using a co-culture system comprising of oviduct epithelial cell co- culture (No et al. 2018), cloning dog using in vivo matured oocytes (Zhao et al. 2018), development of iSCNT, i.e., dog-pig interspecies cloned embryos and their developmental properties (Kim et al. 2018), studies confirmed metabolic plasticity of canine embryo-derived pluripotent cells (Tobias et al. 2018), identification of stem cells from hair follicle (de Castro et al. 2018), first report on derivation of iPSCs using auto-erasable Sendai virus vector (Tsukamoto et al. 2018)

Fortunately, SCNT has progressed significantly in canines. A number of dogs are cloned so far, besides endangered canines such as wolves. SCNT has facilitated birth of transgenic dogs as models for humans diseases.

In vitro production of embryos has lagged behind in dogs. Due to small body size and anatomy of reproductive tract of the animal, and inaccessibility of uterine tract from vagina, the embryo collection and transfer in dogs is

Table 18.2 Therapeutic application of stem cells in canines

Stem cells (origin)	Applications and inferences (references)
MSCs	Osteotomy repair after tibial tuberosity advancement with cranial cruciate ligament injury (Rocha Dos Santos et al. 2018)
MSCs (Perinatal)	Curing knee arthritis using imaging technology (Zhang et al. 2018)
MSC (human bone marrow)	Insulin-producing bone marrow-derived MSCs cured experimentally induced diabetes in dog (Gabr et al. 2018)
MSCs (Adipose tissue)	Ameliorating acute hepatitis injury (Teshima et al. 2017)
Epidermal neural crest stem cell	Feasibility of stem cell transplantation (McMahill et al. 2015)
MSCs (Placenta)	Neuroprotective properties against ischemic stroke (Yu et al. 2013)
MSCs (Bone marrow)	Applications in spinal cord injury (Bhat et al. 2018)

achieved by surgical methods. In vitro maturation efficiency is low, hence it should be improved. Only a few reports exist on births of live offspring from IVF-derived embryos. Low efficiency of canine embryo development justifies the need for further research on this topic.

Genetic testing of dogs is expanding. Dog lovers are ready to spend required amount for diagnosis of infectious diseases and genetic disorders in their pets. Some firms have come up to provide semen of proven dog pedigrees. Hence, scenario has changed a lot compared to traditional rearing of dogs. At the moment, pet genomics, genetic testing of pets governs the world top level pet canine business.

18.9 Conclusions

ARTs have gained public attention in pet animals. IVF has low success rates in dogs and needs attention to improve the same. Cloning is successful as is evident from a number of animals cloned, but it is applicable for selected genotypes or transgenic animals that are intended for human biomedical applications. SCNT might be seen as a suitable tool to conserve elite dogs and endangered canines. The ARTs should be promoted keeping in view the associated ethical and animal welfare issues.

References

Bhat IA, T B S, Somal A, Pandey S, Bharti MK, Panda BSK, B I, Verma M, J A, Sonwane A, Kumar GS, Amarpal, Chandra V, Sharma GT (2018) An allogenic therapeutic strategy for canine spinal cord injury using mesenchymal stem cells. J Cell Physiol. https://doi.org/10.1002/jcp.27086

Brito MM, Lúcio CF, Angrimani DS, Losano JD, Dalmazzo A, Nichi M, Vannucchi CI (2017) Comparison of cryopreservation protocols (single and two-steps) and thawing (fast and slow) for canine sperm. Anim Biotechnol 28(1):67–73. https://doi.org/10.1080/10495398.2016.1203797

Caturla-Sánchez E, Sánchez-Calabuig MJ, Pérez-Gutiérrez JF, Cerdeira J, Castano C, Santiago-Moreno J (2018) Vitrification of dog spermatozoa: effects of two cryoprotectants (sucrose or trehalose) and two warming procedures. Cryobiology 80:126–129. https://doi.org/10.1016/j.cryobiol.2017.11.001

Choi YB, Kim GA, Oh HJ, Kim MJ, Jo YK, Setyawan EM, Lee SH, Lee BC (2016) Cloning of the short-tailed Gyeongju Donggyeong dog via SCNT: conserving phenotypic inheritance. J Vet Med Sci 78(2):329–331. https://doi.org/10.1292/jvms.15-0107

de Castro RVG, Tavares MR, Bressan FF, Pieri NCG, Baracho Trindade Hill A, Souza AF, da R N Cruz N, Martins DS, Ambrósio CE, Meirelles FV, Garcia JM (2018) In vitro identification of a stem cell population from canine hair follicle bulge region. Tissue Cell 50:43–50. https://doi.org/10.1016/j.tice.2017.12.003

Farstad W (2000) Assisted reproductive technology in canid species. Theriogenology 53(1):175–186

Gabr MM, Zakaria MM, Refaie AF, Ismail AM, Khater SM, Ashamallah SA, Azzam MM,

Ghoneim MA (2018) Insulin-producing cells from adult human bone marrow mesenchymal stromal cells could control chemically induced diabetes in dogs: a preliminary study. Cell Transplant 27(6): 937–947. https://doi.org/10.1177/0963689718759913

Gupta S (2017) Puppy power. Nature 543:S48–S49

Harkey MA, Asano A, Zoulas ME, Torok-Storb B, Nagashima J, Travis A (2013) Isolation, genetic manipulation, and transplantation of canine spermatogonial stem cells: progress toward transgenesis through the male germ-line. Reproduction 146(1): 75–90. https://doi.org/10.1530/rep-13-0086 (Print 2013 Jul)

Hatoya S, Torii R, Kondo Y, Okuno T, Kobayashi K, Wijewardana V, Kawate N, Tamada H, Sawada T, Kumagai D, Sugiura K, Inaba T (2006) Isolation and characterization of embryonic stem-like cells from canine blastocysts. Mol Reprod Dev 73(3):298–305

Hong SG, Jang G, Kim MK, Oh HJ, Park JE, Kang JT, Koo OJ, Kim DY, Lee BC (2009a) Dogs cloned from fetal fibroblasts by nuclear transfer. Anim Reprod Sci 115(1–4):334–339. https://doi.org/10.1016/j.anireprosci.2008.12.005 (Epub 2008 Dec 6)

Hong SG, Kim MK, Jang G, Oh HJ, Park JE, Kang JT, Koo OJ, Kim T, Kwon MS, Koo BC, Ra JC, Kim DY, Ko C, Lee BC (2009b) Generation of red fluorescent protein transgenic dogs. Genesis 47(5):314–322. https://doi.org/10.1002/dvg.20504

Hong SG, Oh HJ, Park JE, Kim MJ, Kim GA, Park EJ, Koo OJ, Kang SK, Jang G, Lee BC (2011) Production of offspring from cloned transgenic RFP female dogs and stable generational transmission of the RFP gene. Genesis 49(11):835–840. https://doi.org/10.1002/dvg.20772 (Epub 2011 Aug 18)

Hossein MS, Jeong YW, Park SW, Kim JJ, Lee E, Ko KH, Hyuk P, Hoon SS, Kim YW, Hyun SH, Shin T, Hwang WS (2009) Birth of Beagle dogs by somatic cell nuclear transfer. Anim Reprod Sci 114(4): 404–414. https://doi.org/10.1016/j.anireprosci.2008.10.010

Jang G, Kim MK, Oh HJ, Hossein MS, Fibrianto YH, Hong SG, Park JE, Kim JJ, Kim HJ, Kang SK, Kim DY, Lee BC (2007) Birth of viable female dogs produced by somatic cell nuclear transfer. Theriogenology 67(5):941–947

Jang G, Hong SG, Oh HJ, Kim MK, Park JE, Kim HJ, Kim DY, Lee BC (2008) A cloned toy poodle produced from somatic cells derived from an aged female dog. Theriogenology 69(5):556–563. https://doi.org/10.1016/j.theriogenology.2007.11.002

Jang G, Hong S, Kang J, Park J, Oh H, Park C, Ha J, Kim D, Kim M, Lee B (2009) Conservation of the, using somatic cell nuclear transfer. J Vet Med Sci 71 (9):1217–1220

Janssens L, Giemsch L, Schmitz R, Street M, Dongen SV, Crombé P (2018) A new look at an old dog: Bonn-Oberkassel reconsidered. J Archaeol Sci 92:126–138

Kim MK, Jang G, Oh HJ, Yuda F, Kim HJ, Hwang WS, Hossein MS, Kim JJ, Shin NS, Kang SK, Lee BC (2007) Endangered wolves cloned from adult somatic cells. Cloning Stem Cells 9(1):130–137. Erratum in: Cloning Stem Cells. 2007 Autumn;9(3):450

Kim Y, Turner D, Nelson J, Dobrinski I, McEntee M, Travis AJ (2008) Production of donor-derived sperm after spermatogonial stem cell transplantation in the dog. Reproduction 136(6):823–831. https://doi.org/10.1530/REP-08-0226

Kim MJ, Oh HJ, Park JE, Kim GA, Hong SG, Jang G, Kwon MS, Koo BC, Kim T, Kang SK, Ra JC, Ko C, Lee BC (2011a) Generation of transgenic dogs that conditionally express green fluorescent protein. Genesis 49(6):472–478. https://doi.org/10.1002/dvg.20737

Kim MJ, Oh HJ, Park JE, Kim GA, Hong SG, Jang G, Kwon MS, Koo BC, Kim T, Kang SK, Ra JC, Ko C, Lee BC (2011) Generation of transgenic dogs that conditionally express green fluorescent protein. Genesis 49(6):472–478. https://doi.org/10.1002/dvg.20737 (Epub 2011 May 31)

Kim MJ, Oh HJ, Kim GA, Setyawan EMN, Choi YB, Lee SH, Petersen-Jones SM, Ko CJ, Lee BC (2017) Birth of clones of the world's first cloned dog. Sci Rep. 7(1):15235. https://doi.org/10.1038/s41598-017-15328-2

Kim MJ, Oh HJ, Choi YB, Lee S, Setyawan EMN, Lee SH, Lee SH, Hur TY, Lee BC (2018) Suberoylanilide hydroxamic acid during in vitro culture improves development of dog-pig interspecies cloned embryos but not dog cloned embryos. J Reprod Dev 64(3):277–282. https://doi.org/10.1262/jrd.2017-112

Kinney GM, Pennycook JW, Schriver MD, Templeton JW, Kraemer DC (1979) Surgical collection and transfer of canine embryos. Biol Reprod 20 (Suppl. 2), 170 (abstract)

Klinc P, Majdic G, Sterbenc N, Cebulj-Kadunc N, Butinar J, Kosec M (2005) Establishment of a pregnancy following intravaginal insemination with epididymal semen from a dog castrated due to benign prostatic hyperplasia. Reprod Domest Anim 40 (6):559–561

Lee BC, Kim MK, Jang G, Oh HJ, Yuda F, Kim HJ, Hossein MS, Kim JJ, Kang SK, Schatten G, Hwang WS (2005) Dogs cloned from adult somatic cells. Nature 436(7051):641. Erratum in: Nature 2006 Oct 12; 443(7112):649. Shamim, M Hossein [corrected to Hossein, M Shamim]. Nature. 2006 Mar 9;440 (7081):164. Nature. 2005 Aug 25;436(7054):1102

Lee JH, Chun JL, Kim KJ, Kim EY, Kim DH, Lee BM, Han KW, Park KS, Lee KB, Kim MK (2016) Effect of acteoside as a cell protector to produce a cloned dog. PLoS One 11(7):e0159330. https://doi.org/10.1371/journal.pone.0159330 (eCollection 2016)

Lee S, Zhao M, No J, Nam Y, Im GS, Hur TY (2017) Dog cloning with *in vivo* matured oocytes obtained using electric chemiluminescence immunoassay-predicted ovulation method. PLoS One 12(3):e0173735. https://doi.org/10.1371/journal.pone.0173735 (eCollection 2017)

Lee SH, Oh HJ, Kim MJ, Kim GA, Setyawan EMN, Ra K, Abdillah DA, Lee BC (2018) Dog cloning-no longer science fiction. Reprod Domest Anim 53(Suppl 3):133–138. https://doi.org/10.1111/rda.13358

Luvoni GC, Morselli MG (2017) Canine epididymal spermatozoa: a hidden treasure with great potential. Reprod Domest Anim 52(Suppl 2):197–201. https://doi.org/10.1111/rda.12820 (Epub 2016 Oct 18)

Martins MI, Padilha LC, Souza FF, Lopes MD (2009) Fertilizing capacity of frozen epididymal sperm collected from dogs. Reprod Domest Anim 44(Suppl 2):342–344. https://doi.org/10.1111/j.1439-0531.2009.01431.x

Mason SJ (2018) Current review of artificial insemination in Dogs. Vet Clin North Am Small Anim Pract 48 (4):567–580. https://doi.org/10.1016/j.cvsm.2018.02.005 (Epub 2018 Apr 19. Review)

McMahill BG, Spriet M, Sisó S, Manzer MD, Mitchell G, McGee J, Garcia TC, Borjesson DL, Sieber-Blum M, Nolta JA, Sturges BK (2015) Feasibility study of canine epidermal neural crest stem cell transplantation in the spinal cords of dogs. Stem Cells Transl Med. 4 (10):1173–1186. https://doi.org/10.5966/sctm.2015-0018 (Epub 2015 Aug 13)

Nagashima JB, Sylvester SR, Nelson JL, Cheong SH, Mukai C, Lambo C, Flanders JA, Meyers-Wallen VN, Songsasen N, Travis AJ (2015) Live births from domestic dog (*Canis familiaris*) embryos produced by in vitro fertilization. PLoS One 10(12):e0143930. https://doi.org/10.1371/journal.pone.0143930 (eCollection 2015)

Nishimura T, Hatoya S, Kanegi R, Sugiura K, Wijewardana V, Kuwamura M, Tanaka M, Yamate J, Izawa T, Takahashi M, Kawate N, Tamada H, Imai H, Inaba T (2013) Generation of functional platelets from canine induced pluripotent stem cells. Stem Cells Dev 22 (14):2026–2035. https://doi.org/10.1089/scd.2012.0701 (Epub 2013 Mar 15)

Nishimura T, Unezaki N, Kanegi R, Wijesekera DPH, Hatoya S, Sugiura K, Kawate N, Tamada H, Imai H, Inaba T (2017a) Generation of canine induced extraembryonic endoderm-like cell line that forms both extraembryonic and embryonic endoderm derivatives. Stem Cells Dev 26(15):1111–1120. https://doi.org/10.1089/scd.2017.0026 (Epub 2017 Apr 25)

Nishimura T, Hatoya S, Kanegi R, Wijesekera DPH, Sanno K, Tanaka E, Sugiura K, Hiromitsu Tamada NK, Imai H, Inaba T (2017b) Feeder-independent canine induced pluripotent stem cells maintained under serum-free conditions. Mol Reprod Dev 84(4):329–339. https://doi.org/10.1002/mrd.22789 (Epub 2017 Mar 13)

No J, Zhao M, Lee S, Ock SA, Nam Y, Hur TY (2018) Enhanced in vitro maturation of canine oocytes by oviduct epithelial cell co-culture. Theriogenology 1 (105):66–74. https://doi.org/10.1016/j.theriogenology.2017.09.002 (Epub 2017 Sep 6)

Oh HJ, Kim MK, Jang G, Kim HJ, Hong SG, Park JE, Park K, Park C, Sohn SH, Kim DY, Shin NS, Lee BC (2008) Cloning endangered gray wolves (*Canis lupus*) from somatic cells collected postmortem. Theriogenology 70(4):638–647. https://doi.org/10.1016/j.theriogenology.2008.04.032 (Epub 2008 Jun 4)

Oh HJ, Hong SG, Park JE, Kang JT, Kim MJ, Kim MK, Kang SK, Kim DY, Jang G, Lee BC (2009) Improved efficiency of canine nucleus transfer using roscovitine-treated canine fibroblasts. Theriogenology 2009 (72):461–470. https://doi.org/10.1016/j.theriogenology.2009.03.016

Oh HJ, Park JE, Kim MJ, Hong SG, Ra JC, Jo JY, Kang SK, Jang G, Lee BC (2011) Recloned dogs derived from adipose stem cells of a transgenic cloned beagle. Theriogenology 75(7):1221–1231. https://doi.org/10.1016/j.theriogenology.2010.11.035 (Epub 2011 Jan 8)

Oh HJ, Choi J, Kim MJ, Kim GA, Jo YK, Choi YB, Lee BC (2016) Propagation of elite rescue dogs by somatic cell nuclear transfer. Anim Sci J 87(1):21–26. https://doi.org/10.1111/asj.12402 (Epub 2015 Aug 7)

Ostrander EA, Wayne RK, Freedman AH, Davis BW. Demographic history, selection and functional diversity of the canine genome. Nat Rev Genet. 2017 Dec;18(12):705–720. https://doi.org/10.1038/nrg.2017.67 (Epub 2017 Sep 25. Review)

Otoi T, Shimizu R, Naoi H, Wongsrikeao P, Agung B, Taniguchi M (2006) Meiotic competence of canine oocytes embedded in collagen gel. Reprod Domest Anim 41(1):17–21

Park JE, Hong SG, Kang JT, Oh HJ, Kim MK, Kim MJ, Kim HJ, Kim DY, Jang G, Lee BC (2009) Birth of viable puppies derived from breeding cloned female dogs with a cloned male. Theriogenology 72(5):721–730. https://doi.org/10.1016/j.theriogenology.2009.05.007 (Epub 2009 Jul 5)

Parker HG, Dreger DL, Rimbault M, Davis BW, Mullen AB, Carpintero-Ramirez G, Ostrander EA (2017) Genomic analyses reveal the influence of geographic origin, migration, and hybridization on modern dog breed development. Cell Rep. 19(4):697–708. https://doi.org/10.1016/j.celrep.2017.03.079

Pezo F, Cheuquemán C, Salinas P, Risopatrón J (2017) Freezing dog semen using -80 °C ultra-freezer: Sperm function and in vivo fertility. Theriogenology 1 (99):36–40. https://doi.org/10.1016/j.theriogenology.2017.05.007 (Epub 2017 May 7)

Rocha Dos Santos C, da Rocha Filgueiras R, Furtado Malard P, Rodrigues da Cunha Barreto-Vianna A, Nogueira K, da Silva Leite C, Maurício Mendes de Lima E (2018) Mesenchymal stem cells in osteotomy repair after tibial tuberosity advancement in dogs with cranial cruciate ligament injury. J Exp Orthop 5(1):17. https://doi.org/10.1186/s40634-018-0130-z

Rodrigues BA, dos Santos LC, Rodrigues JL (2004) Embryonic development of in vitro matured and in vitro fertilized dog oocytes. Mol Reprod Dev 67:215–223

Schütt T, Helboe L, Pedersen LØ, Waldemar G, Berendt M, Pedersen JT (2016) Dogs with cognitive dysfunction as a spontaneous model for early Alzheimer's Disease: a translational study of neuropathological and inflammatory markers. J Alzheimers Dis 52(2):433–449. https://doi.org/10.3233/JAD-151085

Seo MS, Park SB, Kang KS (2012) Isolation and characterization of canine Wharton's jelly-derived mesenchymal stem cells. Cell Transplant 21(7):1493–1502

Teshima T, Matsumoto H, Michishita M, Matsuoka A, Shiba M, Nagashima T, Koyama H (2017) Allogenic Adipose Tissue-Derived Mesenchymal Stem Cells Ameliorate Acute Hepatic Injury in Dogs. Stem Cells Int 2017:3892514. https://doi.org/10.1155/2017/3892514 (Epub 2017 Dec 28)

Tobias IC, Isaac RR, Dierolf JG, Khazaee R, Cumming RC, Betts DH (2018) Metabolic plasticity during transition to naïve-like pluripotency in canine embryo-derived stem cells. Stem Cell Res 30:22–33. https://doi.org/10.1016/j.scr.2018.05.005 (Epub 2018 May 17)

Trindade AB, Therrien J, Garcia JM, Smith LC (2017) Mesenchymal-like stem cells in canine ovary show high differentiation potential. Cell Prolif 50(6). https://doi.org/10.1111/cpr.12391 (Epub 2017 Oct 8)

Tsukamoto M, Nishimura T, Yodoe K, Kanegi R, Tsujimoto Y, Alam ME, Kuramochi M, Kuwamura M, Ohtaka M, Nishimura K, Nakanishi M, Inaba T, Sugiura K, Hatoya S (2018) Generation of footprint-free canine induced pluripotent stem cells using auto-erasable sendai virus vector. Stem Cells Dev. https://doi.org/10.1089/scd.2018.0084

Wei YF, Chen FL, Tang SS, Mao AG, Li LG, Cheng LG, Chen C, Li FX, Wang B, Xu T, Zhang YJ, Li J, Wan JS (2017) Birth of puppies of predetermined sex after artificial insemination with a low number of sex-sorted, frozen-thawed spermatozoa in field conditions. Anim Sci J 88(8):1232–1238. https://doi.org/10.1111/asj.12763 (Epub 2017 Jan 11)

Wydooghe E, Snoeck F,Van Soom A (2016) Establishment of live birth following intravaginal artificial insemination with chilled epididymaldog semen collected post mortem: case report. In: Proceedings of 8th ISCFR with 19th EVSSAR Congress, Paris (France), p. 202, 22–25 June 2016

Yamada S, Shimazu Y, Kawaji H, Nakazawa M, Naito K, Toyoda Y (1992) Maturation, fertilization, and development of dog oocytes in vitro. Biol Reprod 46:853–858

Yu S, Tajiri N, Franzese N, Franzblau M, Bae E, Platt S, Kaneko Y, Borlongan CV (2013) Stem cell-like dog placenta cells afford neuroprotection against ischemic stroke model via heat shock protein upregulation. PLoS One 8(9):e76329. https://doi.org/10.1371/journal.pone.0076329 (eCollection 2013)

Zhang BY, Wang BY, Li SC, Luo DZ, Zhan X, Chen SF, Chen ZS, Liu CY, Ji HQ, Bai YS, Li DS, He Y (2018) Evaluation of the curative effect of umbilical cord mesenchymal stem cell therapy for knee arthritis in dogs using imaging technology. Stem Cells Int 1983025. https://doi.org/10.1155/2018/1983025 (eCollection 2018)

Zhao M, Lee S, Kim DH, No J, Nam Y, Ock SA, Ko YG, Hur TY (2018) Dog cloning with in vivo matured oocytes obtaining using serum estradiol levels for predicting time of ovulation. Theriogenology 107:109–114. https://doi.org/10.1016/j.theriogenology.2017.10.030 (Epub 2017 Oct 27)

Stem Cells and Cellular Reprogramming to Advance Livestock Industry

19

Abstract

Cellular or genomic reprogramming is an interesting phenomenon in which specialized somatic cells are made to acquire pluripotency through global resetting of the genome by epigenetic modifications. The technology has provided a promising platform to develop cells for regenerative medicine, deriving oocytes or sperm from somatic cells, and genetic engineering of multicellular organisms, especially livestock species used for human health and welfare.

Highlights

- The finally differentiated cells can be reprogrammed through epigenetic modifications of cellular genome
- The technology has revolutionized the science of regenerative medicine or cell-based therapies
- Livestock can be treated or engineered using stem cells.

Keywords

Genomic reprogramming · Induced pluripotency · Stem cells · Regenerative medicine · Livestock applications

19.1 Introduction

Pluripotency is an important characteristic of stem cells. This is the popret based on which the stem cells are able to produce other cell types. As the cells differentiate into terminal cell types, they lose the pluripotency; hence, they are unable to differentiate into other cell types.

The molecular cell biologists were interested in maintaining pluripotency, but task was not easy until, Yamanaka and colleagues in 2006, announced that they were able to amend the genome of terminally differentiated somatic cells into a pluripotent stem cell state by ectopic expression of certain genetic factors, called as Yamanaka factors (Takahashi and Yamanaka 2006).

The advent of induced pluripotency paved the way to generate stem cells from somatic cells. This created a rush among researchers to increase the efficiency of cellular genomic reprogramming, untie underlying molecular biological mechanisms of pluripotency, and develop patient-specific immunocompatible cells for cell therapies.

The induced reprogramming of the genome to establish pluripotency was alternative to establish stem cells by SCNT or by other means such as stem cell and somatic cell fusion, treating somatic cells with stem cell extracts.

B. Singh et al., *Advances in Animal Biotechnology*,
https://doi.org/10.1007/978-3-030-21309-1_19

19.2 What Is Genomic Reprogramming?

Zygote of animals is a totipotent living state which gives rise to other types of cells through development into embryo and fetus. The embryo gives rise to various cells, each with a specific function and phenotype. However, an adult somatic cell can be reprogrammed by using different approaches (Fig. 19.1), and resulting stem cells can be transformed into other cell types by guided differentiation. The art of reverting somatic cells to an embryonic pluripotent stem cells state has emerged as a unique tool to dissect the molecular events that permit the transformation of one cell type to another and provides a powerful tool for studying cell-fate transitions (Maherali and Hochedlinger 2008). Transforming somatic cells to an embryonic state-like status is a unique tool to dissect the molecular events that permit the conversion of one cell type into another.

A complex system of gene and genetic network is involved in maintaining cellular genomic reprogramming, and maintaining a delicate equilibrium between self-renewal and multilineage differentiation. There exists an extensive crosstalk among epigenetic pathways at the level

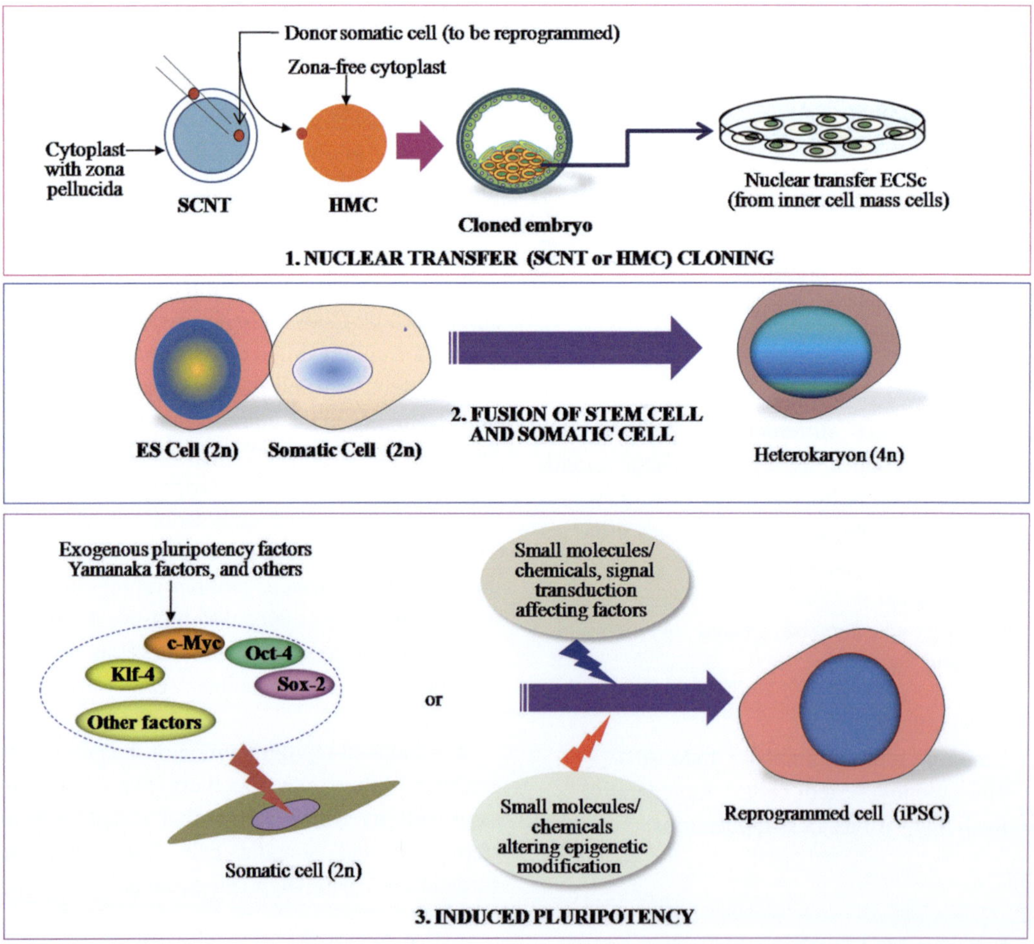

Fig. 19.1 Diagrammatic illustration of strategies to induce reprogramming of genome of somatic cells to induced pluripotent stem cell state. Nuclear transfer cloning or SCNT, cell fusion using somatic cell and ESC, and insertional mutagenesis of reprogramming genetic factors are the basic methods to induce pluripotency

of DNA, histone, and nucleosome (Li and Reinberg 2011).

The cell reprogramming is studied from two important perspectives, i.e., mechanisms of gene expression, and nuclear chromatin analysis. Somatic cell nuclei do not express Oct-4, a key transcription factor involved in cellular pluripotency (Nichols et al. 1998). The Oct-4 acts on a multitude of downstream target genes such as *Dppal-5*, *Pramel 4-7*, and *Ndp52l1*) that are expressed in pluripotent cells. Reduction in expression of Oct-4 is correlated to reduce developmental ability of embryos (Bortvin et al. 2003).

Notably, the DNA methylation is a hallmark of long-term gene silencing. The methyl groups create target sites for methyl-binding proteins, which induce transcriptional repression by recruiting co-repressors, such as histone deacetylases. Hence, DNA methylation largely contributes to gene silencing, and as such, it is essential for normal development (Morgan et al. 2005), X chromosome inactivation (Hellman and Chess 2007) and genomic imprinting (Jones and Takai 2001).

19.3 Rationale for Producing Induced Stem Cells

Deriving stem cells is a tricky practice. The embryonic stem cells (ESCs) are derived from inner cell mass cells of the pre-implantation stage embryos. For this ample quantity of embryos should be on hand. This further depends on the availability of oocytes and sperm of a particular animal species. In some cases, such as wild or endangered animals, the oocytes are not available in plenty. In addition, certain countries have imposed a ban on the slaughter of animal species. This makes it difficult to get the oocytes for producing embryos.

As an alternative strategy, it is possible to convert cultured somatic cells into pluripotent cells' stage that resembles pluripotent stem cells. Besides, the development of stem cell lines, such as pig pluripotent stem cell lines holds opportunities to provide cells for human biomedical research (Roberts et al. 2009).

19.4 Strategies for Inducing Genomic Reprogramming

Birth of Dolly, the first cloned sheep was the first evidence of reprogramming of the genome. Thereafter, strategies were developed to reprogram the cellular genome by alternative methods. It is necessary to erase the epigenetic marks' DNA methylation and histone modifications that are imposed on nuclear chromatin during development. The genomic reprogramming is based on these fundamental principles.

19.5 Reprogramming by Cell Extracts

Cell extracts initiate reprogramming of cells to multi- or pluripotent cell stage, albeit the success rate varies depending on the recipient and target cell types. The basic principle is that cell extract has regulatory components necessary to illicit genomic reprogramming upon entering the target cell. The initial barrier in the success of the approach is the entry of large cell–cytoplast molecules into target cell. This is achieved by reversible permeabilization of recipient cells by pore-forming toxin streptolysin O (SLO), exposure of the permeabilized cells to cell extract, $CaCl_2$-mediated resealing of the cells and then allowing the cells to heal themselves in cell culture system.

Cell extracts can also elicit partial differentiation in somatic or ESCs. Lysates of fetal rat cardiomyocytes promote expression of cardiomyocyte proteins and functions in human adipose MSCs (Gaustad et al. 2004). Cell-extract-based systems for reprogramming cell fate are developed to eliciting somatic cell transdifferentiation, dedifferentiation (Hansis et al. 2004;), or stem cell differentiation (Qin et al. 2005).

19.6 Reprogramming by Somatic Cell Fusion

Cell fusion combines genetic and epigenetic factors present in two different cell types, resulting in a new cell having altered genome and metabolome (Tada and Tada 2006). The resulting cell grows as binuclear cell called heterokaryon. The reprogramming factors present in stem cell modify the epigenetic state of the target somatic nucleus to a pluripotent state, While SCNT was being developed, cell fusion approaches showed that a differentiated somatic cell nucleus could be epigenetically reprogrammed to activate the expression of some genes noted in embryo (Tada et al. 2001).

Studies show that nuclei of ESCs can induce expression of Oct4 in the genome of somatic cells that can be observed as early as 24 h after fusion (Do and Schöler 2004; Pereira et al. 2008). Cell fusion-mediated genomic reprogramming shares at least some of the molecular mechanisms with in vivo reprogramming process by erasing epigenetic status.

19.7 Reprogramming by Somatic Cell Nuclear Transfer

The ability of mammalian oocyte or zygote to erase epigenetic modification is a natural process that comes into play during the period from fertilization to development of neonates. Transfer of somatic nucleus to oocyte leads to decondensation of chromatin, swelling of nucleus, and replacing methylated histones (Kikyo and Wolffe 2000). This is evidenced from the facts that development of nuclear transfer cloned embryos is influenced by methylation status of the donor cells, age, and origin.

The treatments of donor cells to minimize methylation have correlation with enhanced developmental property of cloned embryos. The process has been confirmed in a number of livestock species and cloned laboratory animals (Cibelli et al. 1998; Wakayama et al. 2001).

Reprogramming of nuclei by nuclear transplantation and cell hybridization are considered advantageous as the process does not involve exogenous genetic factors or viruses that pose threat of mutations in genome of resulting cells or embryos.

19.8 Reprogramming by Transcription Factors

Reprogramming by induced expression of mutagenic transcription factors was established when fibroblasts cultured from mice tail were transduced with stem cell-specific transcription factors (Oct3/4, Sox2, c-Myc, and Klf4), and the resulting cells reverted to an ESC-like phenotype (Takahashi and Yamanaka 2006). Interestingly, a combination of four factors (Oct4, Sox2, Klf4, and c-Myc) was able to facilitate the desired transformation of a cell to pluripotency albeit with low efficiency (Fig. 19.2).

This study heralded a new era restoring cellular pluripotency, thereby revolutionized the science of stem cells, regenerative medicine, and stem cell engineering. The discovery led to the award of Nobel Prize of 2012 in Medicine to the investigators.

The disadvantage associated with this method is the use of genetic factors raises the risk of unexpected genetic modifications or mutations. Hence, such cells cannot be safe for human applications. Hence, alternative strategies have been developed so that resulting cells are safer for humans.

19.9 Reprogramming by Using Non-genetic Factors

Possible threat of mutations in the host genome, and possibilities of tumor formation associated with stem cells carrying transgenes discourage their applications in humans. Conversely, alternative means such as mRNA or the proteins are used to develop stem cells from normal cells.

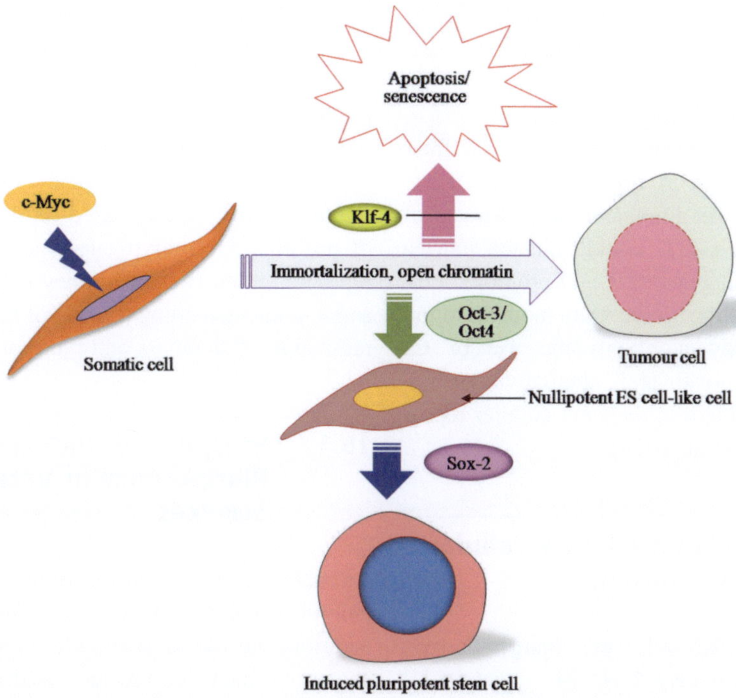

Fig. 19.2 Using transcription factors to induce pluripotency in a somatic cell. c-Myc plays a dual role; it induces apoptosis and senescence, which is possibly suppressed by Klf4. Oct3/4 and may change the cell fate from tumor cells to ESC-like pluripotent cells. Sox2 is required to sustain cellular pluripotency

The mRNA of Oct4, Sox2, Klf4, and c-Myc could produce stem cells, termed as RNA-mediated iPSCs.

It is possible to synthesize large quantities of recombinant proteins or purify the naturally existing metabolites, and translocate them into the nucleus of target cell and modifying cellular gene expression profile. The proteins fused in frame to protein transduction domains also called as cell-permeable proteins or membrane translocating sequences, and cell-penetrating peptides can be transported into the cells either for cell therapies or altered gene expression (Beerens et al. 2003).

Use of chemical approaches to produce stem cells is an emerging area. The purpose is to replace transcription factors and/or enhance the efficiency of somatic cell reprogramming (Li and Ding 2010). Development of murine iPSCs by the combined use of recombinant reprogramming proteins and the small molecules is advocated. Stable iPSCs were derived by directly delivering four reprogramming proteins (Oct4, Sox2, Klf4, and c-Myc) into somatic cell. The resulting cells called protein-induced human iPSCs (p-hiPSCs) shared homology with human ESCs (Kim et al. 2009). The finding was considered as an important step as it could develop stem cells without using mutation-inducing factors, viruses or potentially harmful chemicals. In another report, recombinant cell-penetrating reprogramming proteins were used to generate stem cells called protein-induced pluripotent stem cells (piPSCs) (Zhou et al. 2009).

19.10 Alternative Means of Inducing Reprogramming

Small molecules from various origins are found to improve reprogramming, stem cell properties, and cell regeneration. These molecules have importance as drug biomolecules. Non-steroidal anti-inflammatory drug, the Nabumetone, and an anti-cancer drug 4-hydroxytamoxifen could generate iPSCs without using Sox2, an essential inducer of reprogramming. In addition, nabumetone was able to generate iPSCs in the absence of c-Myc or Sox2, without compromising self-renewal and pluripotency properties of the resulting iPSCs (Yang et al. 2011).

Certain natural plant metabolites or their derivatives can support reprogramming of the animal cell genome. Fruit and stem extracts of *Camellia japonica* (CJ) could enhance murine and humans iPSC generation and promoted efficient wound healing in in vivo mouse wound model. The study highlights the use of CJ metabolites as novel and alternative natural metabolites capable of enhancing cellular reprogramming (Jeon et al. 2018).

19.11 MicroRNA-Mediated Cellular Reprogramming

MicroRNAs (miRNAs) are hairpin-derived RNAs, approximately 20–24 nucleotides (nt) long, which post-transcriptionally repress the expression of target genes by binding to the 30 UTR of messenger RNA (mRNA). The miRNA plays a vital role in cellular reprogramming. Ectopic introduction of ESC-specific miRNAs reverts a somatic cell back to pluripotent cell stage (Kuo and Ying 2012).

The miRNA has emerged as key regulator of gene silencing that acts by targeting specific mRNAs for degradation or by suppressing their translation. Remarkably, miRNAs are shown to directly influence virtually all biological processes, including stem cell self-renewal and lineage specification during development, as well as diseases, such as cancer. Specific miRNAs, stem cell-specific and tissue-specific regulate the expression of key transcription factors, cell cycle proteins, epigenetic modifiers, and other regulatory proteins.

Identification of miRNAs and their varied effects on stem cells have provided a far better understanding of the molecular mechanisms that fine-tune the complex gene regulatory networks controlling the proliferation and the differentiation of stem cells. Expression of ESC-specific miRNAs promotes cellular reprogramming. The murine embryonic fibroblasts were infected with retroviral vectors carrying Sox2, Oct4, and klf4, and ESC-specific cell-cycle-regulating (ESCC) miRNAs. The ESCC miRNAs increased the production of iPSCs. Among different ESCC miRNAs, miR-294 exhibited has high efficiency to induce cellular pluripotency (Mallanna and Rizzino 2010). Hence, based on the pivotal role of miRNAs in embryogenesis and somatic cell reprogramming, further studies in this area will assist in understanding the role of miRNAs in the regulation of stemness and pluripotency.

19.12 Prospects of Induced Pluripotency in Veterinary Sciences

Animal stem cells are powerful tools for studying embryo development, gene targeting, cloning. In addition, the use of stem cells in veterinary regenerative medicine has increased during recent years as highlighted in chapter on equine reproduction.

In livestock, iPSC technology has promoted stem cell science remarkably (Tables 19.1, 19.2 and 19.3). The success in establishing iPSCs in various species has raised the hopes of using iPSCs to produce transgenic animals or chimera. Despite decades of efforts, except few reports such as cattle (Bogliotti et al. 2018) and true ESCs have not been established in ruminant livestock species. Hence, the efforts should be expedited to derive true stem cell lines in livestock animals.

19.13 Derivation of Cells and Tissues for Veterinary Applications

Much of the interest in the veterinary field is centered on the use of stem cells for orthopedic injury in species like equines and camels (Richardson et al. 2007; Yadav et al. 2012; Brehm et al. 2014). The stem cells can have prospects in surgical treatment of subchondral-bone cysts, bone-fracture repair (Kraus and Kirker-Head 2006), and cartilage repair (Brehm et al. 2006;

Table 19.1 Applications of assisted reproduction and genetic engineering tools to augment cattle production

Technologies	Prospects and applications (references)
Male ARTs	
Artificial insemination	Rapid genetic improvement through AI, proliferation of superior germplasm, developing crossbreed animals for enhancing milk or meat production, minimizing management cost by using semen from superior males, minimizing transmission of infectious diseases spreading from infected males to females
Sperm sexing	Producing calves or offspring of desired sex. Females should be opted for milk, while males should be preferred for meat or draught power
ICSI	Best utilization of sperm from superior males to produce embryo or offspring
Female ARTs	
Oocytes and embryo biotechniques	Preserving oocytes from slaughtered animals for producing embryos through IVF, ICSI or nuclear transfer cloning
	Better utilization and conservation of female germplasm by oocyte banking
	Using oocytes of common species to produce interspecies cloned offspring of rare or endangered species from their somatic cells
	Conserving genetic diversity and establishing founder population of threatened or endangered animals
SCNT cloning	Conserving best livestock germplasm
	Multiplying high yielding females, conservation of endangered mammalian species, cloning progeny tested bulls for semen
	Producing cloned transgenic or genome edited animals
	Establishing groups of isogenic animals for nutrition, and immunological and studies
Stem cell technology	Generating iPSCs, *in vitro* for obtaining sperm or oocytes
	In vitro meat from muscle stem cells
	Using stem cells as regenerative medicine to cure injuries in pets and companion animals, such as equines and camels
	Repair of mammary tissue in transgenic animals

Wilke et al. 2007). Stem cell lines or iPSCs may lessen the number of live animals needed in testing of druds or toxicological studies.

19.14 Cellular Reprogramming in Assisted Reproduction

Genomic reprogramming is important in veterinary-assisted reproduction. Much emphasis is currently on developing genetically modified animals for basic research or producing therapeutic metabolites. If used as donor nuclei, the stem cells may improve the efficiency of animals cloning. The concept is supported by the birth of animals when donor cells possessing stemness features were used as donor nuclei (Zhao et al. 2010).

In vitro production of gametes is an important area where stem cells can contribute immensely (Hayashi et al. 2017). Some species of animals are endangered. Deriving oocytes from cultured stem cells can help in increasing their population. In vitro production of sperm from stem cells established from stud bulls can maximize the genetic gains from superior livestock.

19.15 Outlook and Challenges

Development of iPSCs through reprogramming of somatic cells is a breakthrough discovery in the era of stem cell biotechnology. The genome reprogramming offers ample opportunities in regenerative medicine and genetic engineering of livestock. The traditional methods for generating

Table 19.2 Milestone achievements in ARTs leading to advances in ustilizing stem cells in livestock species

Year	Achievements and remarks (reference)
1998	Birth of live calves from SCNT (Cibelli et al. 1998)
	Development of transgenic ESC-like cells for transfer to pre-implantation embryos, and birth of first chimeric offspring, demonstrating that genetically modified cells somatic cells can be genetically tailored, and de-differentiated into ESC-like cells for cell therapies (Cibelli et al. 1998)
	Confirmation of POU factor Oct4 in formation of pluripotent stem cells in mammalian embryo (Nichols et al. 1998)
2001	Development of nuclear transfer cloned embryos without using micromanipulators (Vajta et al. 2001). This paved the way to derive stem cells from cloned embryos
2006	Reprogramming of somatic cells by ectopic expression of selected transcription factors (Takahashi and Yamanaka 2006)
2012	Induction of reprogramming by ESC-specific miRNAs into somatic cells (Kuo and Ying 2013)
2018	Developing protocols for producing gene-edited bovine somatic cells for
	SCNT (Ishino et al. 2018), developing selection strategies for functional and progressive sperm with high DNA integrity (Nagata et al. 2018)
	Derivation of bESCs with stable genome, karyotype, expression of pluripotency-regulating genes. The cells could serve as donor nuclei for producing SCNT cloned embryos, indicating that it will be possible to use these cells for engineering of livestock (Bogliotti et al. 2018)
	Generating chimeric cynomolgus monkeys by means of injecting iPSCs into embryos (Kang et al. 2018)

Table 19.3 Summary of induced reprogramming to establish reprogrammed stem cells in livestock species

Species	Salient achievements and inferences
Pig	Development of iPSCs by lentiviral transduction of human (Oct4, SOX2, Klf4, and c-Myc) genes into pig fetal fibroblasts (Ezashi et al. 2009), use of drug-inducible system to generate iPSCs (Wu et al. 2009), reprogramming embryonic fibroblasts by seven reprogramming factors assembled on episomal vectors (Telugu et al. 2010), development of naïve iPSCs from embryonic fibroblasts (Fujishiro et al. 2013)
	Generating neural rosettes from iPSCs (Webb et al. 2017)
Equine	Development of iPSCs from fibroblasts using a piggyBac (PB) transposon-based method (Nagy et al. 2011); development of iPSCs from fibroblasts (Breton et al. 2013), generation of functional myocytes from equine iPSCs (Amilon et al. 2018)
	Generation and description of skeletal muscle cells (myotubes) from porcine iPSCs (Genovese et al. 2017)
Buffalo	Reprogramming of fetal fibroblasts with buffalo-specific factors (Oct4, Sox2, Klf4 and c-Myc) using retroviral vectors (Deng et al. 2012)
Bovines	Generating megakaryocytes bovine iPSCs with possible applications in regenerative therapies (Baigger et al. 2018), generation and description of single cell-derived iPSCs by piggyBac transposons with CAG promoting transcription factors (Oct3/4, Sox2, Klf4 and cMyc) (Zhao et al. 2017)
Sheep	Reprogramming of fibroblasts by mRNA (Li et al. 2011), and drug-inducible reprogramming (Bao et al. 2011; Liu et al. 2012)
Goat	iPSCs from domestic goats (Song et al. 2013; Sandmaier et al. 2015), and cashmere goats (Tai et al. 2015), generating transgenic cloned dairy goat iPSCs (tgiPSCs), and using tgiPSCs transgenic cloned dairy goat fibroblasts to produce cloned embryos (Song et al. 2016).
Rabbit	Lentiviral-mediated transformation of c-Myc, Klf4, Sox2 and Oct3/4) into rabbit hepatic and gastric cells (Honda et al. 2010; Tancos et al. 2012)
Poultry	Induction of pluripotency in chicken feather follicle cells (Kim et al. 2017)

ESCs rely on IVF embryos; the recent methods of genomic reprogramming could bypass the need for nuclear transfer, which relies on abundant availability of oocytes.

The reprogrammed cells share morphological and proliferation features with ESCs derived from naturally produced embryos. Though the technology offers unique opportunities in veterinary regenerative medicine, virus-mediated transfer of pluripotency factors to generate iPSCs hinders their use in vivo. Another disadvantage is that incomplete reprogramming of somatic cells could result in the impaired differentiation of iPSCs into the required cell types.

Regrettably, iPSCs produced by the Yamanaka factors tend to be tumorigenic, making them unsafe for clinical application. Applications of non-genetic factors to induce pluripotency are advantageous. This will provide an invaluable resource of cell types for modeling diseases, drug or toxicology screening, and patient-specific cell therapy. Some strategies seem to be more appropriate than genetic reprogramming. The chemically induced pluripotent cells may represent a simpler and valid approach.

Studying the mechanisms that prevent successful reprogramming should allow for improvements in reprogramming methods, which could have significant implications for regenerative medicine and the study of human disease. Genomics and computational chemical biology can be combined to develop new biomolecules that can induce pluripotency in somatic cells.

19.16 Conclusions

iPSCs are the multipurpose versatile tools for investigating early developmental processes and cell-based therapies in humans and veterinary setup. The progress in genomic reprogramming has offered an infinite capacity for self-renewal and unique potential of stem cells to differentiate into other cell types. With the advent of techniques to reprogram somatic cells into pluripotent stem cells, there is an increased interest in understanding the mechanisms underlying maintenance and acquisition of pluripotency. Finding alternative s to using viral vectors carrying transcription factors for reprogramming cells is now available and has solved to some extent the trepidations associated with the use of reprogrammed stem cells in medical practices.

References

Amilon KR, Cortes-Araya Y, Moore B, Lee S, Lillico S, Breton A, Esteves CL, Donadeu FX (2018) Generation of functional myocytes from equine induced pluripotent stem cells. Cell Reprogram 20 (5):275–281. https://doi.org/10.1089/cell.2018.0023 (Epub 2018 Sep 12)

Baigger A, Eicke D, Yuzefovych Y, Pogozhykh D, Blasczyk R, Figueiredo C (2018) Characterization of induced pluripotent stem cell-derived megakaryocyte lysates for potential regenerative applications. J Cell Mol Med 22(9):4545–4549. https://doi.org/10.1111/jcmm.13698 (Epub 2018 Jun 12)

Bao L, He L, Chen J, Wu Z, Liao J, Rao L, Ren J, Li H, Zhu H, Qian L, Gu Y, Dai H, Xu X, Zhou J, Wang W, Cui C, Xiao L (2011) Reprogramming of ovine adult fibroblasts to pluripotency via drug-inducible expression of defined factors. Cell Res 21(4):600–608. https://doi.org/10.1038/cr.2011.6 (Epub 2011 Jan 11)

Beerens AM, Al Hadithy AF, Rots MG, Haisma HJ (2003) Protein transduction domains and their utility in gene therapy. Curr Gene Ther 3(5):486–494

Bogliotti YS, Wu J, Vilarino M, Okamura D, Soto DA, Zhong C, Sakurai M, Sampaio RV, Suzuki K, Izpisua Belmonte JC, Ross PJ (2018) Efficient derivation of stable primed pluripotent embryonic stem cells from bovine blastocysts. Proc Natl Acad Sci USA. 115(9): 2090–2095. https://doi.org/10.1073/pnas.1716161115 (Epub 2018 Feb 9)

Bortvin A, Eggan K, Skaletsky H, Akutsu H, Berry DL, Yanagimachi R, Page DC, Jaenisch R (2003) Incomplete reactivation of Oct4-related genes in mouse embryos cloned from somatic nuclei. Development 130(8):1673–1680

Brehm W, Aklin B, Yamashita T, Rieser F, Trüb T, Jakob RP, Mainil-Varlet P (2006) Repair of superficial osteochondral defects with an autologous scaffold-free cartilage construct in a caprine model: implantation method and short-term results. Osteoarthritis Cartilage 14(12):1214–1226 (Epub 2006 Jul 3)

Brehm W, Burk J, Delling U (2014) Application of stem cells for the treatment of joint disease in horses. Methods Mol Biol 1213:215–228. https://doi.org/10.1007/978-1-4939-1453-1_18

Breton A, Sharma R, Diaz AC, Parham AG, Graham A, Neil C, Whitelaw CB, Milne E, Donadeu FX (2013) Derivation and characterization of induced pluripotent stem cells from equine fibroblasts. Stem Cells Dev.

22(4):611–621. https://doi.org/10.1089/scd.2012.0052 (Epub 2012 Sep 28)

Cibelli JB, Stice SL, Golueke PJ, Kane JJ, Jerry J, Blackwell C, Ponce de León FA, Robl JM (1998) Cloned transgenic calves produced from nonquiescent fetal fibroblasts. Science 280(5367):1256–1258

Deng Y, Liu Q, Luo C, Chen S, Li X, Wang C, Liu Z, Lei X, Zhang H, Sun H, Lu F, Jiang J, Shi D (2012) Generation of induced pluripotent stem cells from buffalo (*Bubalus bubalis*) fetal fibroblasts with buffalo defined factors. Stem Cells Dev 21(13):2485–2494. https://doi.org/10.1089/scd.2012.0018 (Epub 2012 May 14)

Do JT, Schöler HR (2004) Nuclei of embryonic stem cells reprogram somatic cells. Stem Cells 22(6):941–949

Ezashi T, Telugu BP, Alexenko AP, Sachdev S, Sinha S, Roberts RM (2009) Derivation of induced pluripotent stem cells from pig somatic cells. Proc Natl Acad Sci USA 106(27):10993–10998. https://doi.org/10.1073/pnas.0905284106 (Epub 2009 Jun 18)

Fujishiro SH, Nakano K, Mizukami Y, Azami T, Arai Y, Matsunari H, Ishino R, Nishimura T, Watanabe M, Abe T, Furukawa Y, Umeyama K, Yamanaka S, Ema M, Nagashima H, Hanazono Y (2013) Generation of naive-like porcine-induced pluripotent stem cells capable of contributing to embryonic and fetal development. Stem Cells Dev 22(3):473–482. https://doi.org/10.1089/scd.2012.0173 (Epub 2012 Oct 9)

Gaustad KG, Boquest AC, Anderson BE, Gerdes AM, Collas P (2004) Differentiation of human adipose tissue stem cells using extracts of rat cardiomyocytes. Biochem Biophys Res Commun 314(2):420–427

Genovese NJ, Domeier TL, Telugu BP, Roberts RM (2017) Enhanced development of skeletal myotubes from porcine induced pluripotent stem cells. Sci Rep 6 (7):41833. https://doi.org/10.1038/srep41833

Hansis C, Barreto G, Maltry N, Niehrs C (2004) Nuclear reprogramming of human somatic cells by xenopus egg extract requires BRG1. Curr Biol 14(16):1475–1480

Hayashi M, Kawaguchi T, Durcova-Hills G, Imai H (2017) Generation of germ cells from pluripotent stem cells in mammals. Reprod Med Biol. 17(2): 107–114. https://doi.org/10.1002/rmb2.12077 (eCollection 2018 Apr. Review)

Hellman A, Chess A (2007) Gene body-specific methylation on the active X chromosome. Science 315 (5815):1141–1143

Honda A, Hirose M, Hatori M, Matoba S, Miyoshi H, Inoue K, Ogura A (2010) Generation of induced pluripotent stem cells in rabbits: potential experimental models for human regenerative medicine. J Biol Chem 285(41):31362–31369. https://doi.org/10.1074/jbc.M110.150540 (Epub 2010 Jul 29)

Ishino T, Hashimoto M, Amagasa M, Saito N, Dochi O, Kirisawa R, Kitamura H (2018) Establishment of protocol for preparation of gene-edited bovine ear-derived fibroblasts for somatic cell nuclear transplantation. Biomed Res 39(2):95–104. https://doi.org/10.2220/biomedres.39.95

Jeon H, Kim JY, Choi JK, Han E, Song CL, Lee J, Cho YS (2018) Effects of the extracts from fruit and stem of *Camellia japonica* on induced pluripotency and wound healing. J Clin Med 7(11). pii: E449. https://doi.org/10.3390/jcm7110449

Jones PA, Takai D (2001) The role of DNA methylation in mammalian epigenetics. Science 293(5532):1068–10670

Kang Y, Ai Z, Duan K, Si C, Wang Y, Zheng Y, He J, Yin Y, Zhao S, Niu B, Zhu X, Liu L, Xiang L, Zhang L, Niu Y, Ji W, Li T (2018) Improving cell survival in injected embryos allows primed pluripotent stem cells to generate Chimeric Cynomolgus Monkeys. Cell Rep. 25(9):2563–2576. https://doi.org/10.1016/j.celrep.2018.11.001

Kikyo N, Wolffe AP (2000) Reprogramming nuclei: insights from cloning, nuclear transfer and heterokaryons. J Cell Sci 113(Pt 1):11–20

Kim D, Kim CH, Moon JI, Chung YG, Chang MY, Han BS, Ko S, Yang E, Cha KY, Lanza R, Kim KS (2009) Generation of human induced pluripotent stem cells by direct delivery of reprogramming proteins. Cell Stem Cell 4(6):472–476. https://doi.org/10.1016/j.stem.2009.05.005 (Epub 2009 May 28. No abstract available)

Kim YM, Park YH, Lim JM, Jung H, Han JY (2017) Technical note: induction of pluripotent stem cell-like cells from chicken feather follicle cells. J Anim Sci 95 (8):3479–3486. https://doi.org/10.2527/jas.2017.1418

Kraus KH, Kirker-Head C (2006) Mesenchymal stem cells and bone regeneration. Vet Surg 35(3):232–242

Kuo CH, Ying SY (2012) Advances in microRNA-mediated reprogramming technology. Stem Cells Int 2012:823709. https://doi.org/10.1155/2012/823709 (Epub 2012 Mar 28)

Kuo CH, Ying SY (2013) MicroRNA-mediated somatic cell reprogramming. J Cell Biochem 114(2):275–281. https://doi.org/10.1002/jcb.24385

Li W, Ding S (2010) Small molecules that modulate embryonic stem cell fate and somatic cell reprogramming. Trends Pharmacol Sci 31(1):36–45. https://doi.org/10.1016/j.tips.2009.10.002 (Epub 2009 Nov 4. Review)

Li G, Reinberg D (2011) Chromatin higher-order structures and gene regulation. Curr Opin Genet Dev 21 (2):175–186. https://doi.org/10.1016/j.gde.2011.01.022 (Epub 2011 Feb 20. Review)

Li M, Sancho-Martinez I, Izpisua Belmonte JC (2011) Cell fate conversion by mRNA. Stem Cell Res Ther. 2(1):5. https://doi.org/10.1186/scrt46

Liu J, Balehosur D, Murray B, Kelly JM, Sumer H, Verma PJ (2012) Generation and characterization of reprogrammed sheep induced pluripotent stem cells. Theriogenology 77(2):338–346. https://doi.org/10.1016/j.theriogenology.2011.08.006 (Epub 2011 Sep 29)

Maherali N, Hochedlinger K (2008) Guidelines and techniques for the generation of induced pluripotent stem cells. Cell Stem Cell 3(6):595–605. https://doi.org/10.1016/j.stem.2008.11.008

Mallanna SK, Rizzino A (2010) Emerging roles of microRNAs in the control of embryonic stem cells and the generation of induced pluripotent stem cells. Dev Biol 344(1):16–25. https://doi.org/10.1016/j.ydbio.2010.05.014 (Epub 2010 May 15. Review)

Morgan HD, Santos F, Green K, Dean W, Reik W (2005) Epigenetic reprogramming in mammals. Hum Mol Genet 14(Spec No 1):R47–R58

Nagata MPB, Endo K, Ogata K, Yamanaka K, Egashira J, Katafuchi N, Yamanouchi T, Matsuda H, Goto Y, Sakatani M, Hojo T, Nishizono H, Yotsushima K, Takenouchi N, Hashiyada Y, Yamashita K (2018) Live births from artificial insemination of microfluidic-sorted bovine spermatozoa characterized by trajectories correlated with fertility. Proc Natl Acad Sci USA 115(14):E3087–E3096. https://doi.org/10.1073/pnas.1717974115. Epub 2018 Mar 19

Nagy K, Sung HK, Zhang P, Laflamme S, Vincent P, Agha-Mohammadi S, Woltjen K, Monetti C, Michael IP, Smith LC, Nagy A (2011) Induced pluripotent stem cell lines derived from equine fibroblasts. Stem Cell Rev. 7(3):693–702. https://doi.org/10.1007/s12015-011-9239-5 (Erratum. In: Stem Cell Rev. 2012 Jun; 8(2):546)

Nichols J, Zevnik B, Anastassiadis K, Niwa H, Klewe-Nebenius D, Chambers I, Schöler H, Smith A (1998) Formation of pluripotent stem cells in the mammalian embryo depends on the POU transcription factor Oct4. Cell 95(3):379–391

Pereira CF, Terranova R, Ryan NK, Santos J, Morris KJ, Cui W, Merkenschlager M, Fisher AG (2008) Heterokaryon-based reprogramming of human B lymphocytes for pluripotency requires Oct4 but not Sox2. PLoS Genet 4(9):e1000170. https://doi.org/10.1371/journal.pgen.1000170

Qin M, Tai G, Collas P, Polak JM, Bishop AE (2005) Cell extract-derived differentiation of embryonic stem cells. Stem Cells 23(6):712–718

Richardson LE, Dudhia J, Clegg PD, Smith R (2007) Stem cells in veterinary medicine–attempts at regenerating equine tendon after injury. Trends Biotechnol 25(9):409–416 (Epub 2007 Aug 9. Review)

Roberts RM, Telugu BP, Ezashi T (2009) Induced pluripotent stem cells from swine (Sus scrofa): why they may prove to be important. Cell Cycle 8(19): 3078–3081 (Epub 2009 Oct 21. Review)

Sandmaier SE, Nandal A, Powell A, Garrett W, Blomberg L, Donovan DM, Talbot N, Telugu BP (2015) Generation of induced pluripotent stem cells from domestic goats. Mol Reprod Dev 82(9):709–721. https://doi.org/10.1002/mrd.22512 (Epub 2015 Jun 26)

Song H, Li H, Huang M, Xu D, Gu C, Wang Z, Dong F, Wang F (2013) Induced pluripotent stem cells from goat fibroblasts. Mol Reprod Dev 80(12):1009–1017. https://doi.org/10.1002/mrd.22266

Song H, Li H, Huang M, Xu D, Wang Z, Wang F (2016) Big animal cloning using transgenic induced pluripotent stem cells: a case study of goat transgenic induced pluripotent stem cells. Cell Reprogram 18(1):37–47. https://doi.org/10.1089/cell.2015.0035

Tada M, Tada T (2006) Epigenetic reprogramming of somatic genomes by electrofusion with embryonic stem cells. Methods Mol Biol 325:67–79

Tada M, Takahama Y, Abe K, Nakatsuji N, Tada T (2001) Nuclear reprogramming of somatic cells by in vitro hybridization with ES cells. Curr Biol 11(19): 1553–1558

Tai D, Liu P, Gao J, Jin M, Xu T, Zuo Y, Liang H, Liu D (2015) Generation of Arbas Cashmere Goat induced pluripotent stem cells through fibroblast reprogramming. Cell Reprogram. 17(4):297–305. https://doi.org/10.1089/cell.2014.0107

Takahashi K, Yamanaka S (2006) Induction of pluripotent stem cells from mouse embryonic and adult fibroblast cultures by defined factors. Cell 126(4):663–676 (Epub 2006 Aug 10)

Tancos Z, Nemes C, Polgar Z, Gocza E, Daniel N, Stout TA, Maraghechi P, Pirity MK, Osteil P, Tapponnier Y, Markossian S, Godet M, Afanassieff M, Bosze Z, Duranthon V, Savatier P, Dinnyes A (2012) Generation of rabbit pluripotent stem cell lines. Theriogenology 78(8):1774–1786. https://doi.org/10.1016/j.theriogenology.2012.06.017 (Epub 2012 Aug 24. Review)

Telugu BP, Ezashi T, Roberts RM (2010) Porcine induced pluripotent stem cells analogous to naïve and primed embryonic stem cells of the mouse. Int J Dev Biol 54(11–12):1703–1711. https://doi.org/10.1387/ijdb.103200bt

Vajta G, Lewis IM, Hyttel P, Thouas GA, Trounson AO (2001) Somatic cell cloning without micromanipulators. Cloning. 3(2):89–95

Wakayama T, Tabar V, Rodriguez I, Perry AC, Studer L, Mombaerts P (2001) Differentiation of embryonic stem cell lines generated from adult somatic cells by nuclear transfer. Science 292(5517):740–743

Webb RL, Gallegos-Cárdenas A, Miller CN, Solomotis NJ, Liu HX, West FD, Stice SL (2017) Pig induced pluripotent stem cell-derived neural rosettes parallel human differentiation into sensory neural subtypes. Cell Reprogram 19(2):88–94. https://doi.org/10.1089/cell.2016.0057

Wilke MM, Nydam DV, Nixon AJ (2007) Enhanced early chondrogenesis in articular defects following arthroscopic mesenchymal stem cell implantation in an equine model. J Orthop Res 25(7):913–925

Wu Z, Chen J, Ren J, Bao L, Liao J, Cui C, Rao L, Li H, Gu Y, Dai H, Zhu H, Teng X, Cheng L, Xiao L (2009) Generation of pig induced pluripotent stem cells with a drug-inducible system. J Mol Cell Biol 1(1):46–54. https://doi.org/10.1093/jmcb/mjp003 (Epub 2009 Jun 3)

Yadav PS, Singh RK, Singh B (2012) Fetal stem cells in farm animals: applications in health and production. Agric Res 1:67–77

Yang CS, Lopez CG, Rana TM (2011) Discovery of nonsteroidal anti-inflammatory drug and anticancer drug enhancing reprogramming and induced pluripotent stem cell generation. Stem Cells 29(10):1528–1536. https://doi.org/10.1002/stem.717

Zhao XY, Li W, Lv Z, Liu L, Tong M, Hai T, Hao J, Wang X, Wang L, Zeng F, Zhou Q (2010)

Viable fertile mice generated from fully pluripotent iPS cells derived from adult somatic cells. Stem Cell Rev 6(3):390–397. https://doi.org/10.1007/s12015-010-9160-3

Zhao L, Wang Z, Zhang J, Yang J, Gao X, Wu B, Zhao G, Bao S, Hu S, Liu P, Li X (2017) Characterization of the single-cell derived bovine induced pluripotent stem cells. Tissue Cell 49(5):521–527. https://doi.org/10.1016/j.tice.2017.05.005 (Epub 2017 May 22)

Zhou H, Wu S, Joo JY, Zhu S, Han DW, Lin T, Trauger S, Bien G, Yao S, Zhu Y, Siuzdak G, Schöler HR, Duan L, Ding S (2009) Generation of induced pluripotent stem cells using recombinant proteins. Cell Stem Cell 4(5):381–384. https://doi.org/10.1016/j.stem.2009.04.005 (Epub 2009 Apr 23. No abstract available. Erratum in: Cell Stem Cell. 2009 Jun 5;4(6):581)

Oogonia Stem Cells in Farm Animals 20

Abstract

The oogonia stem cells (OSCs) are the founding female germ stem cells that differentiate into oogonia and oocytes. The OSCs serve as the carrier of genetic information from mother to offspring. The insights from intraovarian transplantation in experimental animals reveal the capacity of OSCs to generate functional oocytes that can be fertilized and support fetal development and birth of offspring.

Highlights

- The OSCs are the founding unipotent female germ stem cells
- They generate oogonia and oocytes that can produce viable offspring.

Keywords

Oogonia stem cells · Germ cells · Oocytes · Livestock oogonia

20.1 Introduction

The oogonial stem cells (OSCs) or female germ stem cells (FGSCs) represent a subpopulation of unspecialized or undifferentiated germ stem cells in postnatal ovaries. The FGSCs can be guided to produce oocytes (Zou et al. 2019; Martin et al. 2019). The functional properties of female germ cells, also called as OSCs, are of immense interest in humans and are under investigation in animals of economic interest or endangered species.

20.2 Properties and Culturing of OSCs

The mammalian ovary is an active organ that undergoes many structural and functional alterations and contributes to produce female gametes and steroids. Folliculogenesis and the number of oocytes formed vary depending on mammalian species and health status of females. In human ovary, the oocytes development initiates during fetal life, and oocytes are enclosed within granulosa cells that develop to primordial follicles. In rodents, the follicle development commences soon after the birth (Telfer and Anderson 2019).

Some pioneering studies contributed immensely to basic research into oogenesis and germ stem cell biology. This paved the way to advance the science of female germ stem cells and their applications in reproductive cellular medicine. Johnson et al. (2004) showed that proliferative germ cells sustain oocytes and follicle development in postnatal mammalian ovaries. Later on, it was proved that FGSCs remain functional, and when transplanted into live mice, undergo oogenesis and the recipient mice could produce live offspring (Zou et al. 2009; Zhang and Wu 2016). The principles of stem cell-based

B. Singh et al., *Advances in Animal Biotechnology*,
https://doi.org/10.1007/978-3-030-21309-1_20

regenerative medicine were verified in other species such as mice and humans (Ding et al. 2016; Silvestris et al. 2018).

ESCs differentiate into ovarian follicle-like structures. The granulosa cells derived from ESCs exhibit functional characteristics such as steroidogenesis, response to FSH, and participation in development of ovarian follicles (Woods et al. 2013a).

20.3 Characteristics of OSCs

The OSCs are mitotically active germ cells, can be cultured in vitro, induced to differentiate into fertilization-competent oocytes (White et al. 2012). The FGSCs and SSCs share some common features including morphology, growth parameters, and co-highly expressed genes (Xie et al. 2014). In addition, the putative OSCs are found to express distinctive primitive germ cell-specific markers, such as DDX4, KIT, DPPA3, IFITM3, and PRDM1, and pluripotency markers (POU5F1, LIN28, and NANOG). However, oocyte or follicle-specific markers (CYP19A1, HDAC6, or ZP3) are not detected (Woods et al. 2013b; Clarkson et al. 2018).

20.4 Applications of OSCs in Livestock

The concept of neo-oogenesis has received increasing attention. Heterogeneous putative stem cells obtained from the ovary of pigs were found to express stem and germ cell-specific markers (Bui et al. 2014).

Much of the work done on OSCs is in humans and mice. Primordial germ cell-like cells (PGCLCs) (Hayashi et al. 2012) and eggs (Hayashi and Saito 2013) were obtained from murine ESCs and iPSCs. The PGCLCs differentiated into mature germinal vesicle stage oocytes that were used to produce IVF embryos and births of healthy fertile offspring (Hayashi et al. 2012).

Germ cells with stem cell-like features such as large cell nucleus, intracytoplasmic vesicle, and blebbing were obtained from 50 to 70 days of gestation stage fetal bovine ovaries (Lavoir et al. 1994). Subsequent studies showed feasibility of cryopreservation of bovine fetal ovarian germ cells (Lavoir et al. 1996). The PGCs of pig could withstand cryopreservation and supported the development of nuclear transfer embryos by fusion of karyoplasts (Liu et al. 1995).

20.5 Outlook and Challenges

Stem cells are among most widely studied niche areas in regenerative medicine, preserving fertility, and conserve the endangered species. The attempts to establish OSC lines of domestic and feral endangered animals have been less successful in part due to lack of suitable in vitro culture conditions. Most of the culture systems for animal germ stem cells are based on the inferences from murine or human germ stem cells.

Lack of stemness markers for validating pluripotency and species-specificity is one of the other causes. The field of deriving gametes from somatic cells is progressed in model animals such as mice. However, the findings provide important clues for further research on livestock oogonia and OSCs.

20.6 Conclusions

Female germ stem cells or OSCs are responsible for preserving female fertility. Currently, the knowledge about mechanisms of development and function of OSCs is in its infancy. Germ stem cells, especially the OSCs, are of immense interest in humans. The area needs attention in livestock and endangered species.

References

Bui HT, Van Thuan N, Kwon DN, Choi YJ, Kang MH, Han JW, Kim T, Kim JH (2014) Identification and characterization of putative stem cells in the adult pig ovary. Development 141(11):2235–2244. https://doi.org/10.1242/dev.104554

Clarkson YL, McLaughlin M, Waterfall M, Dunlop CE, Skehel PA, Anderson RA, Telfer EE (2018) Initial characterisation of adult human ovarian cell populations isolated by DDX4 expression and aldehyde dehydrogenase activity. Sci Rep 8(1):6953. https://doi.org/10.1038/s41598-018-25116-1

Ding X, Liu G, Xu B, Wu C, Hui N, Ni X, Wang J, Du M, Teng X, Wu J (2016) Human GV oocytes generated by mitotically active germ cells obtained from follicular aspirates. Sci Rep 30(6):28218. https://doi.org/10.1038/srep28218

Hayashi K, Ogushi S, Kurimoto K, Shimamoto S, Ohta H, Saitou M (2012) Offspring from oocytes derived from in vitro primordial germ cell-like cells in mice. Science 338(6109):971–975. https://doi.org/10.1126/science.1226889 (Epub 2012 Oct 4)

Hayashi K, Saitou M (2013) Generation of eggs from mouse embryonic stem cells and induced pluripotent stem cells. Nat Protoc 8(8):1513–1524. https://doi.org/10.1038/nprot.2013.090 (Epub 2013 Jul 11)

Johnson J, Canning J, Kaneko T, Pru JK, Tilly JL (2004) Germline stem cells and follicular renewal in the postnatal mammalian ovary. Nature 428(6979):145–150. Erratum in: Nature (2004) 430(7003):1062

Lavoir MC, Basrur PK, Betteridge KJ (1994) Isolation and identification of germ cells from fetal bovine ovaries. Mol Reprod Dev 37(4):413–424

Lavoir MC, Rumph N, Betteridge KJ, Leibo SP (1996) Cryopreservation of germ cells from bovine fetal ovaries. Reprod Fertil Dev 8(2):267–271

Liu L, Moor RM, Laurie S, Notarianni E (1995) Nuclear remodelling and early development in cryopreserved, porcine primordial germ cells following nuclear transfer into in vitro-matured oocytes. Int J Dev Biol 39(4):639–644

Martin JJ, Woods DC, Tilly JL (2019) Implications and current limitations of oogenesis from female germline or oogonial stem cells in adult mammalian ovaries. Cells 8(2):pii: E93. https://doi.org/10.3390/cells8020093

Silvestris E, Cafforio P, D'Oronzo S, Felici C, Silvestris F, Loverro G (2018) In vitro differentiation of human oocyte-like cells from oogonial stem cells: single-cell isolation and molecular characterization. Hum Reprod 33(3):464–473. https://doi.org/10.1093/humrep/dex377

Telfer EE, Anderson RA (2019) The existence and potential of germline stem cells in the adult mammalian ovary. Climacteric 1–5. https://doi.org/10.1080/13697137.2018.1543264 (Epub ahead of print)

White YA, Woods DC, Takai Y, Ishihara O, Seki H, Tilly JL (2012) Oocyte formation by mitotically active germ cells purified from ovaries of reproductive-age women. Nat Med 18(3):413–421. https://doi.org/10.1038/nm.2669

Woods DC, White YA, Tilly JL (2013a) Purification of oogonial stem cells from adult mouse and human ovaries: an assessment of the literature and a view toward the future. Reprod Sci 20(1):7–15. https://doi.org/10.1177/1933719112462632 (Epub 2012 Sep 27. Review)

Woods DC, White YA, Niikura Y, Kiatpongsan S, Lee HJ, Tilly JL (2013b) Embryonic stem cell-derived granulosa cells participate in ovarian follicle formation in vitro and in vivo. Reprod Sci 20(5):524–535. https://doi.org/10.1177/1933719113483017 (Epub 2013 Mar 27)

Xie W, Wang H, Wu J (2014) Similar morphological and molecular signatures shared by female and male germline stem cells. Sci Rep 4(4):5580. https://doi.org/10.1038/srep05580

Zhang C, Wu J (2016) Production of offspring from a germline stem cell line derived from prepubertal ovaries of germline reporter mice. Mol Hum Reprod 22(7):457–464. https://doi.org/10.1093/molehr/gaw030

Zou K, Yuan Z, Yang Z, Luo H, Sun K, Zhou L, Xiang J, Shi L, Yu Q, Zhang Y, Hou R, Wu J (2009) Production of offspring from a germline stem cell line derived from neonatal ovaries. Nat Cell Biol 11(5):631–636. https://doi.org/10.1038/ncb1869 (Epub 2009 Apr 12)

Zou K, Wang J, Bi H, Zhang Y, Tian X, Tian N, Ma W, Wu J (2019) Comparison of different in vitro differentiation conditions for murine female germline stem cells. Cell Prolif 52(1):e12530. https://doi.org/10.1111/cpr.12530 (Epub 2018 Oct 17)

Spermatogonial Stem Cells in Farm Animals

21

Abstract

Spermatogonial stem cells (SSCs) are the unipotent founding germ stem cells that originate from seminiferous tubules in testes and serve as the carrier of genetic information from father to offspring. The SSCs restore fertility and can be exploited for gene targeting and regenerative medicine. Bioengineered or gene-edited SSCs transplanted into testes generate transgenic sperm.

Highlights

- The SSCs are the adult germ stem cells that serve as basis of male fertility
- SSCs can be reprogrammed to pluripotency and differentiated into other types of cells.

Keywords

Spermatogonial stem cells · Spermatogonia · Genetic manipulation · Infertility treatment · Surrogate males

21.1 Introduction

The spermatogonial stem cells (SSCs) represent a subpopulation of unspecialized or undifferentiated germ stem cells in testes, and when removed from the testes, they display a broader developmental plasticity than they generally exhibit in vivo (Simon et al. 2010).

The SSCs originate from the primordial germ cells (PGCs), the progenitor cells for male and female germ line, which in turn are derived from epiblast cells of pre-implantation stage embryo. PGCs are the single cells that under certain culture conditions form cell colonies and resemble undifferentiated ESCs.

Indeed, the SSCs are the only adult stem cells that serve to contribute genetic material from father to the progeny. SSCs are studied in detail in model laboratory animals and humans, though they are reported in farm animals as well.

21.2 Salient Characteristics of SSCs

Like other stem cells, ability to grow in vitro, self-renewal, and differentiation into other types of cells are the notable characteristics of the SSCs (Fig. 21.1). As SSCs differentiate into spermatids, they are unipotent in nature. SSCs represent a small fraction of germ stem cells as they are overpoweringly outnumbered by other cell types such as spermatogonia, spermatocytes, spermatids, and sperm (Phillips et al. 2010).

The information on SSCs was originally based on experiments carried out in model animals such as mice, though it was not adequate for commercial and clinical applications in domestic animals. The use of SSCs into animals to cure

B. Singh et al., *Advances in Animal Biotechnology*,
https://doi.org/10.1007/978-3-030-21309-1_21

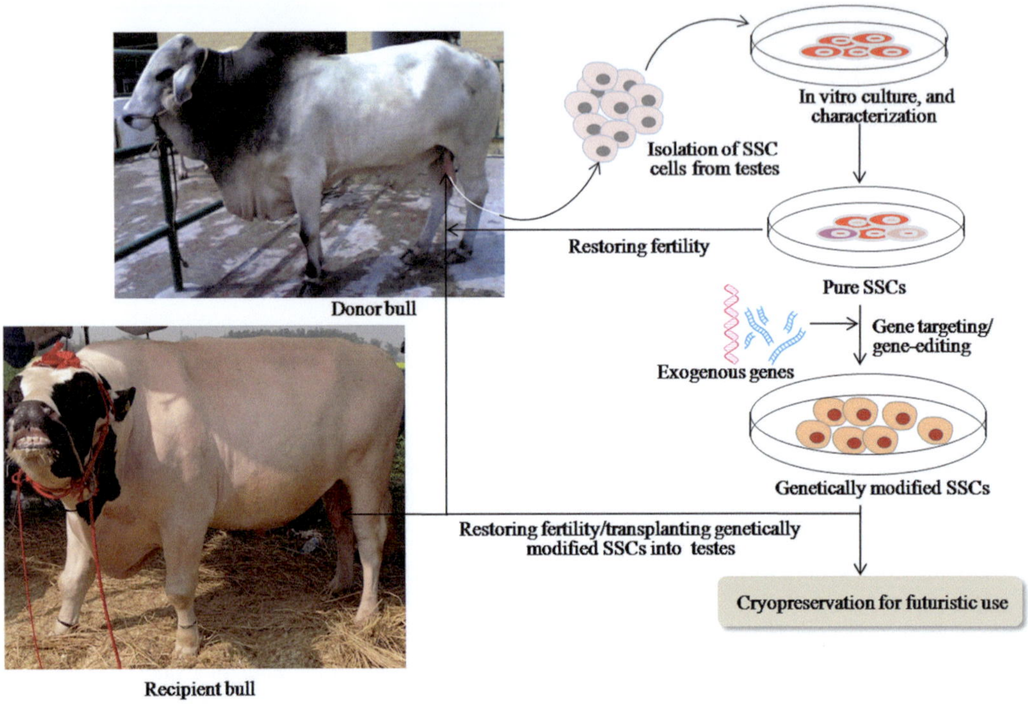

Fig. 21.1 A simplified presentation of use of SSCs to treat fertility and producing transgenic animals. Success in culturing SSCs and spermatogonia in higher-order ruminants and non-ruminants such as pig has made the SSC transplantation as a potent production tool in commercial livestock sector (Oatley 2017)

infertility initiated after pioneering work of transplanting SSCs into mice that was followed by spermatogenesis and production of viable sperms (Brinster and Avarbock 1994; Brinster and Zimmermann 1994).

As noted in rodents, the SSCs are less in number and represent only 0.03% of all germ cells in testes (Tegelenbosch and de Rooij 1993). Indeed, the majority of the information about SSCs and spermatogenic lineage is based on the inferences derived from studies in rodents as model animals. The SSCs, like other stem cells, are defined based on certain basic stemness characteristics such as ability to balance self-renewal and differentiation into other types of cells. This balance is necessary for maintaining stem pool and demand for the cells to proliferate to produce sperms.

Studies on SSCs are difficult due to their low population under natural conditions and dearth of unique species-specific characteristics to discover them. Specialized culture media, growth factors and conditions have been developed to isolate SSCs in various species such as cattle (Kim et al. 2015; Goissis et al. 2018), buffaloes (Yu et al. 2014; Sharma et al. 2019), horse (Jung et al. 2014; Costa et al. 2017), pig (Lee et al. 2014a, b; Zhao et al. 2016; Park et al. 2017a, b), goats (Heidari et al. 2012; Ma et al. 2016; Wei et al. 2018), and sheep (Qasemi-Panahi et al. 2018).

The SSCs are unipotent as they differentiate only into sperms. However, the SSCs can be reprogrammed to a state of pluripotency, so-called germline-derived pluripotent stem cells (gPSCs). Feeder-based reprogramming of spermatogonial cells is not desirable as feeder cells present an uncontrollable microenvironment and interface with growing cells. Modified feeder-free Matrigel-based in vitro cell growth system supports long-term cell proliferation and pluripotency (Lee et al. 2018).

21.3 In Vitro Maintenance of SSCs

The SSCs exist in a three dimensional (3D) microenvironment of the basement membrane of seminiferous tubule. The SSCs need similar environment for optimal growth in vitro. The SSCs can be isolated, cultured in vitro, and the sperm obtained from them can be used for IVF or ICSI of oocytes.

A proper milieu is necessary to stimulate the growth of germ cells in culture. To achieve this, various culture media and growth factors such as glial cell line-derived neurotrophic factor (GDNF), leukemia inhibitory factor (LIF), epidermal growth factor (EGF), or fibroblast growth factor 2 (FGF2) are suggested to sustain the growth and proliferation of SSCs. Bovine SSCs showed in vitro survival and growth in a specialized culture media containing growth factors such as GDNF, LIF, EGF, and FGF2 (Aponte et al. 2008). Porcine SSCs were cultured in agarose-based 3D hydrogel and in 2D culture plates either with or without feeder cells. Subsequently, the effects of 3D culture on maintenance of undifferentiated SSCs were identified by analyzing the development of cell colonies and their morphological properties, expression of stem cell-specific markers such as high activity of alkaline phosphatase (AP) and ability to self-renew. It was concluded that 3D microenvironment could effectively support the growth of porcine SSCs (Park et al. 2017a, b).

21.4 In Vitro Proliferation of SSCs

SSCs continuously undergo self-renewal and differentiation to sustain spermatogenesis throughout sexually active life of males. The SSCs are able to grow in vitro and, like other stem cells, can be manipulated genetically. Culturing of SSCs provides a basis for studying spermatogenesis and development of methods for animal genetic modification and conservation of genetic diversity.

The general concept of SSCs as cells that support formation of sperm has rather radically metamorphosed. They can also give rise to cells representing three embryonic layers in vitro, which opens opportunities for utilizing SSCs in regenerative medicine (Aponte 2015). As a result of continuous research on SSC biology, identification of SSC-specific molecular markers, the manipulation of this special class of stem cells offers novel biotechnological opportunities to manipulate and augment animal productivity.

SSCs are obtained from diverse mammalian species including humans, experimental mice, and domestic animals. However, cell lines obtained from mice SSCs have only been reprogrammed to generate pluripotent stem cells resembling ESCs. Further studies are needed to reprogram SSCs obtained from livestock species (Sahare et al. 2018).

21.5 Cryopreservation of SSCs

It is possible to cryopreserve human (Moraveji et al. 2019) and animals testicular tissues and SSCs. Cryopreservation of testicular tissue is of special interest in case of diseases such as cancer and re-establishing spermatogenesis after chemotherapy (Onofre et al. 2016). SSCs in livestock species such as cattle, pig, and horse have been cryopreserved by conventional slow-freezing and vitrification (Lee et al. 2014a, b; Kim et al. 2015; Ha et al. 2016). In animals, the cryopreservation is a valuable biotechnological approach to preserve the genetic stock of stud bulls and stallions (Costa et al. 2017).

Based on slow-freezing and vitrification using different combinations of CPAs, it was concluded that horse SSCs could be preserved without any loss in structural and functional attributes of SSCs (Costa et al. 2017).

Like other cultured cells, the cryopreserved SSCs also suffer from oxidative stress and are susceptible to apoptosis. Studies are carried out to improve cryopreservation of SSCs by incorporating catalase and α-tocopherol (α-TCP) to the basic freezing medium to prevent oxidative stress (Aliakbari et al. 2016) and including apoptosis inhibitors to protect SSCs from damage (Ha et al. 2016).

21.6 Applications of SSCs in Livestock

Stem cell transplantation is an important tool for restoring fertility in farm animals (Ezashi et al. 2009; Singh et al. 2009; Kubota and Brinster 2018). Another broader area of SSC intervention is their use in developing the gene-edited or transgenic animals (Fig. 21.1). The success of transplantation of SSCs in model animals, and ability of transplanted cells to produce sperm (Brinster and Avarbock 1994; Brinster and Zimmermann 1994), made it feasible to detect SSCs based on their functional signatures. This encouraged the researchers to pursue SSC transplanting to cure infertility in animals.

Depending on the animal species, the SSCs have different applications. In milch species, such as cattle, buffaloes, and goats, the objective is production of recombinant pharmaceutical proteins, while pigs are used to produce cells for regenerative medicine. In equines, SSCs may be used for producing progeny from geldings and therapy of infertile and subfertile stallions (Jung et al. 2014).

21.7 Genome Editing via SSCs

One of the striking differences among germ stem cells and ESCs is that germ stem cells are more stable in terms of their genetic and epigenetic status. ESCs have low genomic stability (Kanatsu-Shinohara et al. 2005). For this reason, the germ cells are more suitable for genetic modification or genome editing (Wu et al. 2015; Takashima 2018; Takashima and Shinohara 2018). The germ line cells have been modified genetically by different methods such as lipofection, electroporation, and vector-mediated gene transfer. Among main viral vectors are adenovirus- and lentivirus-mediated gene transfer (Takashima 2018; Kim et al. 2018).

SSCs modified by TALEN or CRISPR/Cas9, and transplanted into seminiferous tubules of mice established there, and formed spermatogenetic cell colonies to produce genetically modified sperm. Therefore, the gene editing can be utilized to produce genetically modified model animals (Sato et al. 2015; Sato and Ogawa 2019).

Among livestock species, the domestic pig is most widely studied animal with reference to genetic modification for producing recombinant therapeutic or cells and organs for xenotransplantation. CRISPR/Cas9 was used to edit NANOS2 genes in pig embryos to generate offspring carrying mono-allelic and bi-allelic mutations (Park et al. 2017a, b).

21.8 Generating Transgenic Animals

Germline stem cells provide opportunities to produce genetically modified animals without using embryos. SSCs can be collected from postnatal animals, cultured in vitro, and subjected to gene targeting. The genetically modified SSCs can be differentiated in vitro or in vivo into sperm that can be used to produce transgenic embryos, ESCs, and offspring. Though efficiency is low, in vitro production of sperm avoids the need to transplant SSCs into animals to produce transgenic sperm.

Bi et al. (2016) have reported gene-edited isozygous, functional myostatin (MSTN) knockout cloned pigs free of selectable marker. CRISPR/Cas9-mediated homologous recombination was used to knock out one allele of MSTN in primary cells (Bi et al. 2016).

Improving transfection efficiency of SSCs is a challenge in many species. A new method using 1-dimensional nanomaterial, called gold nanowire injector (Au nanowire injector (E-R Au NWI)), was used to transfer pEGFP-N1 plasmid carrying an exogenous gene directly into the nucleus of porcine SSCs under the influence of electric stimulus. The E-R Au NWI might be useful for efficient genetic modification of SSCs and can be used to produce transgenic pigs (Park et al. 2019).

21.9 Treating Male Infertility

There are several reasons behind infertility in males. The loss of fertility or inception of infertility is a matter of high concern. The human patients undergoing radiation and/or chemotherapy may lose fertility due to gonadal damage or damage of sperm-producing cells.

Treating infertility via SSC autotransplantation is a viable option to restore fertility (Mulder et al. 2016). However, as noted in mice, the success of transplantation of SSCs is around 12% that necessitates further improvement in this method. Co-transplantation of MSCs may enhance the homing efficiency of endogenous SSCs after transplantation (Kadam et al. 2018).

SSCs exhibited therapeutic effects in lead acetate-induced testicular damage and spermatogenesis loss. The SSCs treatment reduced expression of caspace-3 and apoptosis (Pasha et al. 2016). SSCs transplantation using germline stem cells in mature mice could generate viable sperm. When allowed to mate with female mice, the sperm produced from transplanted SSCs were able to fertilize the oocytes and produce live offspring. It was found that around 40–80 SSCs could restore fertility and enabled the male mice to mate with females (Kanatsu-Shinohara et al. 2016). It is likely that technique can be applied to males of other valued livestock species.

21.10 Opportunities and Challenges

SSCs have essential role in maintaining fertility by highly active process of spermatogenesis to produce spermatozoa and transmit genetic information to next generation.

Stem cell biology and engineering have paved the way to enhance animal health by regenerative medicine and production in terms of quality and quantity. Breakthroughs in genome-editing techniques have been applied to modify germ cells, especially the male germ cells.

Availability of methods for functional screening of SSCs has made it possible to utilize them in biomedical and livestock species. As a lot has been achieved in germ stem cell biology in rodents, the technology will have a huge impact on regenerative medicine, animal reproduction, and overall livestock productivity.

Genetic transformation of SSCs followed by differentiation into sperm enables production of transgenic animals. Minimizing damage to cells during their transfection or transformation is a major challenge. Nanotechnology-based gene transfer might be a suitable alternative to overcome the problems and enhancing gene transfer into SSCs. This is an emerging area of animal cell engineering and should be a leading focus of studies.

The imminent future advances in understanding of cellular and molecular biology of stem cells will advance our acquaintance with SSCs, regulation and genetic modifications including editing of adverse genetic mutations, enhancing male fertility and deriving gametes from somatic cells (Oatley 2017). It is envisaged that genome editing is likely to enhance genomic improvement of SSCs and their use to cure infertility or producing genetically modified animals in future.

The research areas that need a due consideration include improving feeder-free and serum-free culturing of SSCs, identifying species-specific molecular markers in economically important farm animals, and endangered animals.

21.11 Conclusions

SSCs are of immense importance in biomedical and veterinary sciences including reproductive medicine and genetic modifications of livestock species. A thorough understanding and regulation of interspecies barrier are essentially required to harness the potential of SSCs in livestock species. Though stem cell technology and spermatogenesis have progressed significantly over the last century, more work is needed to establish in vitro spermatogenesis to develop competent male germ cells from somatic cells or stem cells of various, if not all the livestock species of biomedical or agriculture importance.

References

Aliakbari F, Gilani MA, Amidi F, Baazm M, Korouji M, Izadyar F, Yazdekhasti H, Abbasi M (2016) Improving the efficacy of cryopreservation of spermatogonia stem cells by antioxidant supplements. Cell Reprogram 18(2):87–95. https://doi.org/10.1089/cell.2015.0067

Aponte PM (2015) Spermatogonial stem cells: current biotechnological advances in reproduction and regenerative medicine. World J Stem Cells 7(4):669–680. https://doi.org/10.4252/wjsc.v7.i4.669

Aponte PM, Soda T, Teerds KJ, Mizrak SC, van de Kant HJ, de Rooij DG (2008) Propagation of bovine spermatogonial stem cells in vitro. Reproduction 136(5):543–557. https://doi.org/10.1530/REP-07-0419 (Epub 2008 Jul 28)

Bi Y, Hua Z, Liu X, Hua W, Ren H, Xiao H, Zhang L, Li L, Wang Z, Laible G, Wang Y, Dong F, Zheng X (2016) Isozygous and selectable marker-free MSTN knockout cloned pigs generated by the combined use of CRISPR/Cas9 and Cre/LoxP. Sci Rep 6:31729. https://doi.org/10.1038/srep31729

Brinster RL, Avarbock MR (1994) Germline transmission of donor haplotype following spermatogonial transplantation. Proc Natl Acad Sci U S A 91(24):11303–11307

Brinster RL, Zimmermann JW (1994) Spermatogenesis following male germ-cell transplantation. Proc Natl Acad Sci U S A 91(24):11298–11302

Costa GMJ, Avelar GF, Lacerda SMSN, Figueiredo AFA, Tavares AO, Rezende-Neto JV, Martins FGP, França LR (2017) Horse spermatogonial stem cell cryopreservation: feasible protocols and potential biotechnological applications. Cell Tissue Res 370(3):489–500. https://doi.org/10.1007/s00441-017-2673-1 (Epub 2017 Aug 22)

Ezashi T, Telugu BP, Alexenko AP, Sachdev S, Sinha S, Roberts RM (2009) Derivation of induced pluripotent stem cells from the pig somatic cells. Proc Natl Acad Sci U S A 106:10993–10998

Goissis MD, Giassetti MI, Worst RA, Mendes CM, Moreira PV, Assumpção MEOA, Visintin JA (2018) Spermatogonial stem cell potential of CXCR10-positive cells from prepubertal bull testes. Anim Reprod Sci 196:219–229. https://doi.org/10.1016/j.anireprosci.2018.08.014 (Epub 2018 Aug 11)

Ha SJ, Kim BG, Lee YA, Kim YH, Kim BJ, Jung SE, Pang MG, Ryu BY (2016) Effect of antioxidants and apoptosis inhibitors on cryopreservation of murine germ cells enriched for spermatogonial stem cells. PLoS ONE 11(8):e0161372. https://doi.org/10.1371/journal.pone.0161372 (eCollection)

Heidari B, Rahmati-Ahmadabadi M, Akhondi MM, Zarnani AH, Jeddi-Tehrani M, Shirazi A, Naderi MM, Behzadi B (2012) Isolation, identification, and culture of goat spermatogonial stem cells using c-kit and PGP9.5 markers. J Assist Reprod Genet 29(10):1029–1038. https://doi.org/10.1007/s10815-012-9828-5 (Epub 2012 Jul 11)

Jung H, Roser JF, Yoon M (2014) UTF1, a putative marker for spermatogonial stem cells in stallions. PLoS ONE 9(10):e108825. https://doi.org/10.1371/journal.pone.0108825 (eCollection)

Kadam P, Ntemou E, Baert Y, Van Laere S, Van Saen D, Goossens E (2018) Co-transplantation of mesenchymal stem cells improves spermatogonial stem cell transplantation efficiency in mice. Stem Cell Res Ther 9(1):317. https://doi.org/10.1186/s13287-018-1065-0

Kanatsu-Shinohara M, Ogonuki N, Iwano T, Lee J, Kazuki Y, Inoue K, Miki H, Takehashi M, Toyokuni S, Shinkai Y, Oshimura M, Ishino F, Ogura A, Shinohara T (2005) Genetic and epigenetic properties of mouse male germline stem cells during long-term culture. Development 132(18):4155–4163 (Epub 2005 Aug 17)

Kanatsu-Shinohara M, Morimoto H, Shinohara T (2016) Fertility of male germline stem cells following spermatogonial transplantation in infertile mouse models. Biol Reprod 94(5):112. https://doi.org/10.1095/biolreprod.115.137869 (Epub 2016 Apr 6)

Kim KJ, Lee YA, Kim BJ, Kim YH, Kim BG, Kang HG, Jung SE, Choi SH, Schmidt JA, Ryu BY (2015) Cryopreservation of putative pre-pubertal bovine spermatogonial stem cells by slow. Cryobiology 70(2):175–183. https://doi.org/10.1016/j.cryobiol.2015.02.007 (Epub 2015 Feb 28)

Kim BJ, Kim YH, Oh MG, Kim KJ, Jung SE, Jin JH, Kim SU, Min KS, Ryu BY (2018) Direct modification of spermatogonial stem cells using lentivirus vectors in vivo leads to efficient generation of transgenic rats. Asian J Androl. https://doi.org/10.4103/aja.aja_80_18 (Epub ahead of print)

Kubota H, Brinster RL (2018) Spermatogonial stem cells. Biol Reprod 99(1):52–74. https://doi.org/10.1093/biolre/ioy077

Lee YA, Kim YH, Ha SJ, Kim KJ, Kim BJ, Kim BG, Choi SH, Kim IC, Schmidt JA, Ryu BY (2014a) Cryopreservation of porcine spermatogonial stem cells by slow-freezing testis tissue in trehalose. J Anim Sci 92(3):984–995. https://doi.org/10.2527/jas.2013-6843 (Epub 2014 Feb 6)

Lee WY, Lee KH, Heo YT, Kim NH, Kim JH, Kim JH, Moon SH, Chung HJ, Yoon MJ, Song H (2014b) Transcriptional coactivator undifferentiated embryonic cell transcription factor 1 expressed inspermatogonial stem cells: a putative marker of boar spermatogonia. Anim Reprod Sci 150(3–4):115–124. https://doi.org/10.1016/j.anireprosci.2014.09.010 (Epub 2014 Sep 28)

Lee SW, Wu G, Choi NY, Lee HJ, Bang JS, Lee Y, Lee M, Ko K, Schöler HR, Ko K (2018) Self-reprogramming of spermatogonial stem cells into pluripotent stem cells without microenvironment of feeder cells. Mol Cells 41(7):631–638. https://doi.org/10.14348/molcells.2018.2294 (Epub 2018 Jul 10)

Ma F, Zhou Z, Li N, Zheng L, Wu C, Niu B, Tang F, He X, Li G, Hua J (2016) Lin28a promotes self-renewal and proliferation of dairy goat spermatogonial stem cells (SSCs) through regulation of mTOR and PI3K/AKT. Sci Rep 6:38805. https://doi.org/10.1038/srep38805

Moraveji SF, Esfandiari F, Sharbatoghli M, Taleahmad S, Nikeghbalian S, Shahverdi A, Baharvand H (2019) Optimizing methods for human testicular tissue cryopreservation and spermatogonial stem cell isolation. J Cell Biochem 120(1):613–621. https://doi.org/10.1002/jcb.27419 (Epub 2018 Sep 22)

Mulder CL, Zheng Y, Jan SZ, Struijk RB, Repping S, Hamer G, van Pelt AM (2016) Spermatogonial stem cell autotransplantation and germline genomic editing: a future cure for spermatogenic failure and prevention of transmission of genomic diseases. Hum Reprod Update 22(5):561–573. https://doi.org/10.1093/humupd/dmw017 (Epub 2016 May 30. Review)

Oatley JM (2017) Recent advances for spermatogonial stem cell transplantation in livestock. Reprod Fertil Dev 30(1):44–49. https://doi.org/10.1071/RD17418

Onofre J, Baert Y, Faes K, Goossens E (2016) Cryopreservation of testicular tissue or testicular cell suspensions: a pivotal step in fertility preservation. Hum Reprod Update 22(6):744–761

Park JE, Park MH, Kim MS, Park YR, Yun JI, Cheong HT, Kim M, Choi JH, Lee E, Lee ST (2017a) Porcine spermatogonial stem cells self-renew effectively in a three dimensional culture microenvironment. Cell Biol Int 41(12):1316–1324. https://doi.org/10.1002/cbin.10844

Park KE, Kaucher AV, Powell A, Waqas MS, Sandmaier SE, Oatley MJ, Park CH, Tibary A, Donovan DM, Blomberg LA, Lillico SG, Whitelaw CB, Mileham A, Telugu BP, Oatley JM (2017b) Generation of germline ablated male pigs by CRISPR/Cas9 editing of the NANOS2 gene. Sci Rep 7:40176. https://doi.org/10.1038/srep40176

Park K, Kim MS, Kang M, Kang T, Kim B, Lee ST (2019) Successful genetic modification of porcine spermatogonial stem cells via an electrically responsive Au nanowire injector. Biomaterials 193:22–29. https://doi.org/10.1016/j.biomaterials.2018.12.005

Pasha HF, Rezk NA, Selim SA, Abd El Motteleb DM (2016) Therapeutic effect of spermatogonial stem cell on testicular damage caused by lead in rats. Gene 592(1):148–153. https://doi.org/10.1016/j.gene.2016.07.065

Phillips BT, Gassei K, Orwig KE (2010) Spermatogonial stem cell regulation and spermatogenesis. Philos Trans R Soc Lond B Biol Sci 365(1546):1663–1678. https://doi.org/10.1098/rstb.2010.0026

Qasemi-Panahi B, Movahedin M, Moghaddam G, Tajik P, Koruji M, Ashrafi-Helan J, Rafat SA (2018) Isolation and proliferation of spermatogonial cells from ghezel sheep. Avicenna J Med Biotechnol 10(2):93–97

Sahare MG, Suyatno Imai H (2018) Recent advances of in vitro culture systems for spermatogonial stem cells in mammals. Reprod Med Biol 17(2):134–142. https://doi.org/10.1002/rmb2.12087 (eCollection 2018 Apr. Review)

Sato T, Ogawa T (2019) Generating genetically engineered mice using a spermatogonial stem cell-mediated method. Methods Mol Biol 1874:87–98. https://doi.org/10.1007/978-1-4939-8831-0_5

Sato T, Sakuma T, Yokonishi T, Katagiri K, Kamimura S, Ogonuki N, Ogura A, Yamamoto T, Ogawa T (2015) Genome editing in mouse spermatogonial stem cell lines using TALEN and double-nicking CRISPR/Cas9. Stem Cell Rep 5(1):75–82. https://doi.org/10.1016/j.stemcr.2015.05.011

Sharma A, Shah SM, Saini N, Mehta P, Kumar BSB, Dua D, Singh MK, Singla SK, Palta P, Manik RS, Chauhan MS (2019) Optimization of serum-free culture conditions for propagation of putative buffalo (*Bubalus bubalis*) spermatogonial stem cells. Cell Reprogram. https://doi.org/10.1089/cell.2018.0018 (Epub ahead of print)

Simon L, Hess RA, Cooke PS (2010) Spermatogonial stem cells, in vivo transdifferentiation and human regenerative medicine. Expert Opin Biol Ther 10(4):519–530. https://doi.org/10.1517/14712591003614731

Singh B, Chauhan MS, Singla SK, Gautam SK, Verma V, Manik RS, Singh AK, Sodhi M, Mukesh M (2009) Reproductive biotechniques in buffaloes (*Bubalus bubalis*): status, prospects, and challenges. Review. Reprod. Fertil. Dev 21:499–510

Takashima S (2018) Biology and manipulation technologies of male germline stem cells in mammals. Reprod Med Biol 17(4):398–406. https://doi.org/10.1002/rmb2.12220 (eCollection 2018 Oct. Review)

Takashima S, Shinohara T (2018) Culture and transplantation of spermatogonial stem cells. Stem Cell Res 29:46–55. https://doi.org/10.1016/j.scr.2018.03.006 (Epub 2018 Mar 15)

Tegelenbosch RA, de Rooij DG (1993) A quantitative study of spermatogonial multiplication and stem cell renewal in the C3H/101 F1 hybrid mouse. Mutat Res 290(2):193–200

Wei Y, Cai S, Ma F, Zhang Y, Zhou Z, Xu S, Zhang M, Peng S, Hua J (2018) Double sex and mab-3 related transcription factor 1 regulates differentiation and proliferation in dairy goat male germline stem cells. J Cell Physiol 233(3):2537–2548. https://doi.org/10.1002/jcp.26129

Wu Y, Zhou H, Fan X, Zhang Y, Zhang M, Wang Y, Xie Z, Bai M, Yin Q, Liang D, Tang W, Liao J, Zhou C, Liu W, Zhu P, Guo H, Pan H, Wu C, Shi H, Wu L, Tang F, Li J (2015) Correction of a genetic disease by CRISPR-Cas9-mediated gene editing in mouse spermatogonialstem cells. Cell Res 25(1):67–79. https://doi.org/10.1038/cr.2014.160

Yu X, Riaz H, Dong P, Chong Z, Luo X, Liang A, Yang L (2014) Identification and IVC of spermatogonial stem cells in prepubertal buffaloes. Theriogenology 81(9):1312–1322. https://doi.org/10.1016/j.theriogenology.2014.03.002 (Epub 2014 Mar 12)

Zhao HM, Yang H, Luo FH, Li MX, Zhang S, Yang XG, Lu YQ, Lu SS, Wu YJ, Lu KH (2016) Isolation, proliferation, and induction of Bama mini-pig spermatogonial stem cells in vitro. Genet Mol Res 15(3). https://doi.org/10.4238/gmr.15038602

22 Parthenogenesis—A Potential Tool to Reproductive Biotechnology

Abstract

Parthenogenesis is a form of asexual reproduction in some organisms in which oocyte develops into an embryo without fertilization by sperm. Parthenogenesis is a natural process of reproduction in a few lower invertebrates while higher vertebrates such as humans, mice, and farm animal species have relinquished reproduction by parthenogenesis. However, oocytes can be activated or induced to develop into blastocysts by means of parthenogenetic activation albeit with remarkably low success rates. Parthenogenetic embryos serve as sources of haploid embryonic stem cells may occasionally produce live offspring.

Highlights

- Parthenogenesis is an asexual mode of reproduction in some lower animals
- Parthenogenetic embryos have applications in stem cell biology, animal biotechnology, and biomedical sciences.

Keywords

Oocyte activation · Parthenogenetic applications · Embryo development · Stem cells

22.1 Introduction

The term "parthenogenesis" refers to procreation without the direct involvement of a male. Several researchers have redefined the phenomenon. Originally derived from the Greek words (Gr. parthenos = virgin; genesis = origin) meaning, "virgin birth," the term "parthenogenesis," is a phenomenon of acknowledged biological interest which leads to the birth of young ones in a few animals such as insects and lower vertebrates. It was estimated that a proportion of animal species that reproduce only by parthenogenesis would produce only females, though these animals represent merely a fraction of the total animal population (White et al. 1977).

According to Suomalainen (1950), parthenogenesis refers to "the development of egg cell or oocyte into a new individual without fertilization," and this simple definition seems to be adequate for lower genera. Beatty (1957), modified the definition of parthenogenesis as 'the production of an embryo from a female gamete (oocyte) without involvement of a male gamete (sperm), and eventual development into an adult being. This is because embryos produced without fertilization exhibit low survival rates and undergo senescence.

B. Singh et al., *Advances in Animal Biotechnology*,
https://doi.org/10.1007/978-3-030-21309-1_22

An individual resulting from parthenogenetic activation of unfertilized oocytes is referred to as "parthenote" or "parthenogenone," "parthenogen." Of note, the parthenogenesis may initiate during early embryonic development in some mammals. However, high rates of failure in parthenogenesis in higher animals raise fundamental and as yet unanswered questions how fertilization by sperm activates a cascade of genes and genetic pathways involved in embryonic as well as post-implantation development.

Anyway, the parthenogenesis is distinct from asexual reproduction as it involves egg cells only, whereas in asexual reproduction new individuals are developed from somatic cells or tissues. In terms of physiology also, both the processes are different as the egg is a specialized reproductive cell produced by meiosis in ovaries through the process called oogenesis, while somatic cells and tissues are produced by mitosis. The parthenogenesis differs from hermaphroditism, i.e., the production of sperm and oocytes in the same organism. As the egg nucleus can be fused with the polar body, the nuclear fusion does not seem to be a bar to parthenogenesis. This process is known as "automixis," while "amphimixis" refers to the union of male and female gametes in sexual reproduction.

22.2 Parthenogenesis as Natural Process

Haploidy exists only in gametes in mammalian species. Despite not widely spread in animal kingdom, the parthenogenesis is an alternative mode of reproduction in animals such as nematodes, arthropods, insects (some bees, parasitic wasps, aphids, water fleas), cockroaches (Tanaka and Daimon 2018), and vertebrates such as sharks, crayfish (*Procambarus virginalis*) (Vogt 2018), reptiles like snakes and lizards, and occasionally in birds such as white turkeys (Savage and Harper 1986), and Chinese painted quail (Ramachandran et al. 2018). Hence, there are both disadvantages, i.e., minimizing genetic diversity and enhanced vulnerability to unfavorable mutations, as well as advantages such as the capacity to produce offspring in the absence of male in parthenogenesis.

As summarized in Fig. 22.1 and Box 1, the parthenogenesis activation may be of different types and exists in different species.

Box 1. Types of parthenogenesis

1. Natural parthenogenesis: This process completes regularly under natural conditions in certain species. The natural parthenogenesis may be of two types:

1a Complete or obligatory natural parthenogenesis: This is found in certain insects that have no sexual phase. They depend on parthenogenesis for self-reproduction.

1b Incomplete or partial or cyclic natural parthenogenesis: This is observed in insects and comprises of two generations, i.e., sexual and parthenogenetic generations that are alternate to each other. The diploid eggs produce females, while unfertilized or haploid eggs develop into males. Two types of thelytoky (diploid) are recognized. One is ameiotic parthenogenesis where diploid eggs develop into new individuals, and other is meiotic parthenogenesis where diploid eggs resulting from doubling of chromosome develop into a new individual.

2. Artificial parthenogenesis: Here, the eggs are developed into embryos and young individuals under the influence of non-natural stimuli. Various conditions such as exposing oocytes to a sudden temperature such as from 30 °C to 0–10 °C, mild electric wave shocks, exposure to UV irradiation, and pricking the oocytes by fine glass capillaries induce parthenogenesis. In addition, certain chemical treatments may also induce parthenogenesis in eggs.

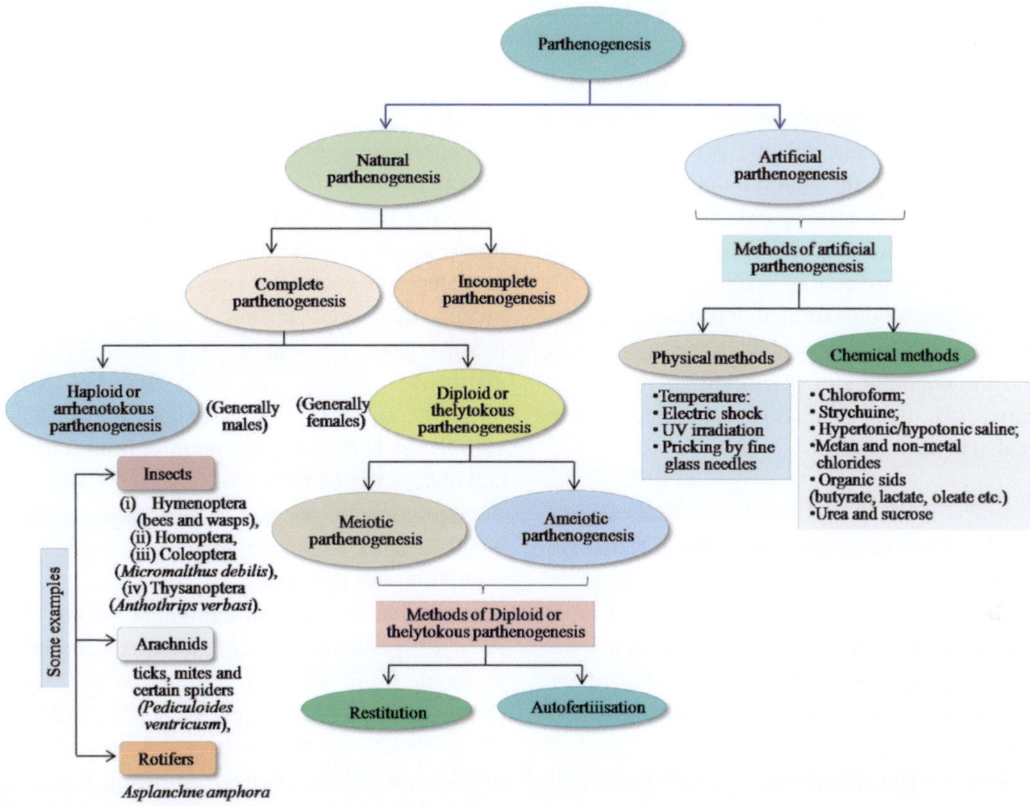

Fig. 22.1 Generalized classification of types of parthenogenesis, and the methods used to induce parthenogenesis

22.3 Induced Parthenogenesis in Mammals

The process of induced parthenogenesis began with the pioneering experiments of Pincus and Enzmann (1936) who showed that the extrusion of the polar body could be induced by contact with sperm suspension, or by heat treatment or exposure to butyric acid and hypertonic solutions. Parthenogenetic offspring are already reported in birds and mammals such as mice. Mammals have imprinted genetic regions wherein either maternal or paternal chromatin is inactivated during the normal course of development of fetus. Mammals developed by parthenogenesis may have double doses of maternally imprinted genes but lack imprinted genes from father hence may have developmental abnormalities such as defects in placental folding.

Diploid and polyploid parthenogenesis proceed via various routes. Diploid eggs may be formed by suppressing either first or second polar bodies. Alternatively, polar bodies may be extruded and subsequently fused with the egg cell. The second meiotic division may be accomplished without extrusion of the second polar body, and the resultant diploid egg cell undergoes cleavage. It is also possible, however, that the two products of the second meiotic division are separated by immediate cleavage, thus initiating haploid parthenogenesis (Witkowska 1973). If neither polar body is extruded, the resulting egg would be tetraploid, as the primary oocyte contains duplicated chromatids while preparing to undergo two meiotic divisions.

In a parthenogenetically activated oocyte, the second polar body can be removed, and a haploid parthenote develops. If the second meiotic division does not commence, a diploid parthenote is formed (Hagemann et al. 1995). To obtain diploid embryos, the matured oocytes are treated with drugs such as cytochalasin, colcimid, and demecolcine incorporated into activation medium to avoid extrusion of second polar body. The types of parthenotes formed depend on activating stimuli, intensity, postovulatory age of an oocyte, and the conditions of activation such as osmolarity, pH, and incubation temperature (King et al. 1988).

22.4 Parthenogenetic Development

Parthenogenesis is not a regular form of reproduction in mammals though mammalian oocytes, under appropriate stimuli, can be parthenogenetically activated and developed into embryos.

The experimental protocols may vary slightly for different species. In general, most protocols encompass recovery of oocytes from live animals after drug-induced superovulation or from the ovaries obtained from lately dead or slaughtered animals.

Laparoscopic ovum pick-up (LOPU) is used to obtain oocytes from live animals. The cost of retrieving oocytes from live animals is high and involves ethical concerns. In small animals, such as sheep and goats, the oocyte is collected surgically that need technical expertise. Hence, ovaries from the abattoir (Fig. 22.2) are good alternatives (Verma et al. 2007; Dev et al. 2012; Singh et al. 2012). Oocytes are collected by various methods such as aspiration from follicles and slicing the ovaries.

Aspiration of oocytes is advantageous compared to other techniques due to the rapidity of aspiration and the quality of oocytes recovered. The average recovery of culturable quality, i.e., A- and B-grade oocytes in *B. bubalis* were found to range from 0.4 to 1.76 oocytes per ovary (Totey et al. 1992; Madan et al. 1994; Nandi et al. 2000). On the contrary, sheep and goat ovaries are smaller; hence, aspirating follicular oocytes from them is tricky. Yadav et al. (2007)

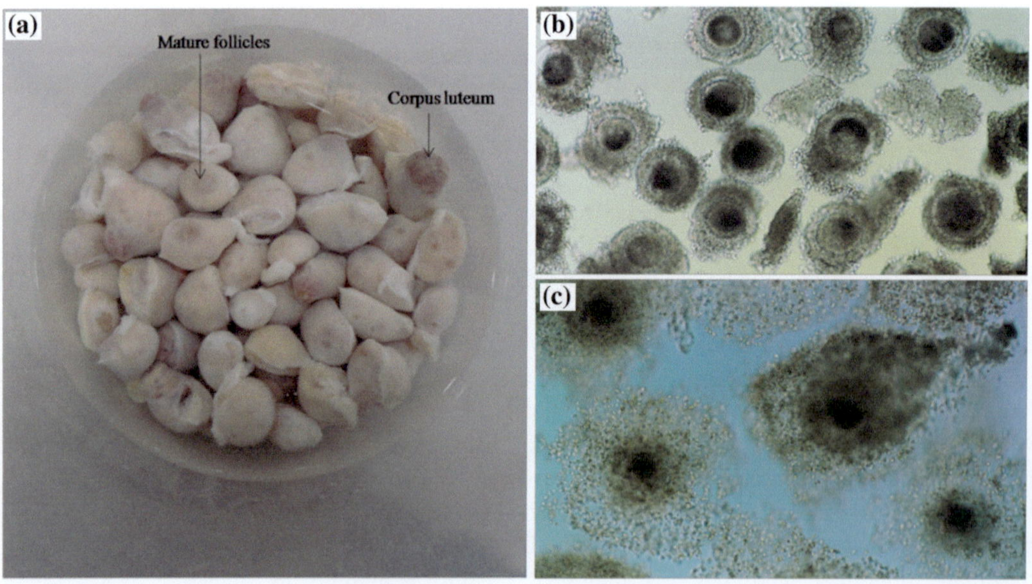

Fig. 22.2 Abattoir-derived water buffalo (*Bubalus bubalis*) ovaries (**a**); immature oocytes (**b**); and IVM oocytes (**c**). Aspiring oocytes by the 18-gauge hypodermic needle is recommended for good recovery of oocytes with compact cumulus cell mass in animals such as cattle, buffaloes, sheep, and goats

compared different techniques of collecting oocytes from sheep and goat and inferred that follicle puncture enhances recovery of oocytes. However, clogging of aspirated oocytes or cumulus–oocyte complexes (COCs) is a problem in goats and sheep that causes dissociation of cumulus cells from immature oocytes, thereby reduces the quality of COCs.

22.5 In Vitro Maturation of Oocytes

Both nuclear and ooplasmic maturation is required to insure parthenogenetic activation and embryo development in vitro. Nuclear maturation refers to the progression to the metaphase II and expansion of cumulus cells as morphological indicators for the maturation of oocytes (Fig. 22.2). Different culture media, supplements, and incubation time are needed for different species. A- and B-grade oocytes should be selected for experiments and studies.

22.6 Activation of Oocytes

The immature COCs with multiple compact layers of cumulus cells have fine in vitro maturation rate, and development capacity after parthenogenetic activation, while denuded oocytes fail to mature and develop further (Fukui and Sakuma 1980).

There is no broad range method that can support the development of oocytes into embryos in various species. This is because IVM and ensuing development processes are vastly species-specific. For instance, cat oocytes can be activated by electric waves stimulation or ethanol, or calcium ionophore in the presence of protein synthesis inhibitors, protein phosphorylation, or histone kinases (Gómez et al. 2003; Kitiyanant et al. 2003). Exogenous hyaluronic acid enhances parthenogenetic development of porcine embryo in vitro possibly mediated by cellular response via receptor CD44 (Toyokawa et al. 2005).

Methods, such as providing electrical stimulus (Kim et al. 1996), ultrasound (Sato et al. 2005), or magnetic field (Grabiec et al. 2007), or including calcium ionophore (Funahashi et al. 1994), $CaCl_2$ (Macháty et al. 1996), EDTA saturated with Ca^{2+}, i.e., Ca-EDTA (Lee et al. 2007), cycloheximide (Nussbaum and Prather 1995; Grabiec et al. 2007), ethanol (Loi et al. 1998; Grabiec et al. 2007), butyric acid (Xiong et al. 2015), G protein stimulation (Macháty et al. 1996), ionomycin (Loi et al. 1998; Méo et al. 2007), protein kinase inhibitors (Mayes et al. 1995), 6-dimethylaminopurine (Méo et al. 2007), and strontium (Méo et al. 2004, 2007) are used either alone or in combination to activate oocytes and induce parthenogenesis.

The initial response in mature oocytes arrested at metaphase II involves an incremental change in intracellular Ca^{2+} concentration that is analogous to that induced by sperm during fertilization of oocytes. Thus, the sequence of events during parthenogenetic activation mimics the cascade of signal transduction elicited by sperm penetrating the oocyte.

Culture media and ingredients play a crucial role in development of activated oocytes to develop into embryos. A variety of embryo culture media and system have been suggested to support parthenogenetic development of oocytes. However, there are several major differences between sperm-induced and parthenogenetic activation of oocytes as revealed from the studies on porcine oocytes (Ruddock et al. 2001).

Occasionally, a co-culture system consisting of uterine or oviduct epithelial cells is used to support development of activated oocytes. Co-culture comprising of two- to four-cell stage surgically retrieved goat embryos in the presence of uterine or oviduct epithelial cells were found to be an effective system for improving their development (Prichard et al. 1992). On the contrary, a serum-free and cell-free culture medium could also support the development of bovine oocytes to the blastocyst stage. Compared to complex culture systems, the viability of the blastocysts was enhanced in a simpler medium (Shamsuddin et al. 1994).

22.7 Significance and Application of Parthenogenesis

Parthenogenesis is a simple, stable and primitive method of reproduction that serves to determine the sex in certain insects such as honeybees and wasps. It eliminates variations from populations, causes polyploidy in organisms, and encourages the development of mutations that might be beneficial. Parthenogenesis assists avoiding sterility in the races.

Parthenogenetic development is a necessary process in oocyte- or embryo-related technologies, such as ICSI and SCNT. During their in vitro development to the blastocyst stage, parthenotes are comparable to naturally produced embryos, and therefore may serve as useful tools for research intended to investigate embryo development, the impact of various agents and supplements, and other variables on embryos. Moreover, analysis of parthenogenetically activated human or murine blastocysts is an established way of studying novel imprinted genes that have a role in early stages of development (Ruddock et al. 2004). Development of a viable parthenogenetic mouse from oocyte having two haploid sets of maternal genes derived from non-growing and fully grown oocytes was proved to be possible by appropriate expression of *Igf2* and *H19* genes along with other imprinted genes. The parthenotes were able to produce offspring, indicating that paternal imprinting prevents parthenogenesis, and that paternal contribution is necessary in the process (Kono et al. 2004).

22.8 Development of Embryos for ESCs

Parthenogenesis has emerged as a tool to obtain haploid embryonic stem cells (haESCs). The technique was demonstrated during the early 1980s by using parthenogenetic embryos in mice (Kaufman et al. 1983). Androgenetic haploid blastocysts and parthenogenetic haploid blastocysts are convenient sources of haploid embryonic stem cells (haESCs). In addition, parthenogenesis is used to derive pluripotent stem cells in experimental non-human primates (Cibelli et al. 2002).

Tetraploid embryos have limited developmental potency and die during post-implantation development. If offspring survive, they have deleterious forebrain and eyes, vertebral axis and heart defects (Kaufman and Webb 1990). Nonetheless, tetraploid embryos can be used as complements for cloning offspring from ESCs (Eggan et al. 2001; Wang et al. 1997). In ESC-tetraploid aggregates, tetraploid cells are selected to form extra-embryonic membranes that contribute to form functional placenta (Everett and West 1998).

Obtaining haESCs from parthenogenetic embryos is an alternative way to obtain pluripotent stem cells for cell transplantation therapies or screening of drugs and pharmaceuticals (Revazova et al. 2007). Additionally, the parthenogenesis is an ideal solution to surmount ethical issues involved in the use of embryos to develop ESCs (Mai et al. 2007; Cheng 2008; Dighe et al. 2008).

Due to increasing interests in unraveling the underlying mechanisms of genomic imprinting, parthenogenetic and tetraploid development have gained the attention of molecular developmental biologists. Parthenogenetic embryos are defective in post-implantation development, in particular impaired or aberrant development of extra-embryonic membranes (Sturm et al. 1994; Newman-Smith and Werb 1995).

Oocyte activation allows cell cycle phase synchronization between the cytoplasm of the oocyte and the transferred nucleus, promoting nuclear reprograming and maintenance of correct ploidy (Campbell 1999). The parthenogenetically developed embryos are recognized as a suitable model for studying developmental mechanisms and obtaining haploid embryos and ESCs in experimental animals such as monkeys (*Macaca fascicularis*) (Yang et al. 2013).

22.9 Genetic Modifications in Animals

The haESCs are valuable tools for functional genetic studies and assisted reproduction. Both parthenogenetic and androgenetic (AG)-haESCs due to having one set of chromosomes are amenable for genetic analysis and alterations by modern gene-editing and knock-out methods (Bai et al. 2016).

Haploid stem cells are obtained from androgenetic embryos developed by injecting sperm into enucleated oocytes and then activating the injected oocytes, or by removing female pronucleus from the zygote. The genome of AG-haESCs was edited, and the cells were used as donor nuclei to produce semi-cloned gene-modified mice (Zhong and Li 2017).

The efficiency of offspring produced and their health condition is variable. In general, the mice pups born from haESCs had retarded growth and short life span because of the aberrant expression of imprinted genes. Improved imprinting profile, can however minimize the genetic errors and produce healthy offspring (Zhang et al. 2015). We foresee that the futuristic improvements in gene imprinting and reprograming of genome will lead to the birth of healthy animals.

22.10 Development of Cloned and Parthenogenetic Embryos

The parthenogenetic activation of oocytes has emerged as a legitimate strategy to investigate the comparative roles of paternal and maternal genomes in controlling early embryo development. Furthermore, parthenogenetic activation is relevant to cloning research, because artificial activation of oocytes is an essential component of nuclear transfer protocols (Kim et al. 1996). An optimized activation protocol may enhance better or complete reprograming of the reconstructed embryo, which might in turn increase the success of NT cloning. A study was carried out to evaluate the efficacy of various methods such as activation by spermatozoa and ICSI and mechanical methods to activate oocytes. It was inferred that chemical methods involving ionomycin plus 6-(Dimethylamino) purine (DMP) are capable of activating the oocytes and producing parthenogenetic cat (*Felis domesticus*) embryos (Kochan et al. 2018).

22.11 Parthenogenetic Stem Cells as Regenerative Medicine

haESCs obtained by parthenogenesis have been examined for their suitability in regenerative medicine. Most parthenogenetic stem cells are haploidentical for the major histocompatibility complex (MHC); hence, they are considered as potential sources of allogenic grafts having minimal immunogenic complexity (Johannsen et al. 2017). Besides, the parthenogenetic haESCs can be differentiated into other types of cells. Developmental chimeras were generated by injecting parthenogenetic ESCs into blastocysts of fumarylacetoacetate hydrolase (Fah)-deficient model mouse. It was found that injected ESCs could differentiate into hepatocytes that were functionally normal and enhanced the life span of chimera mice (Espejel et al. 2014).

In another report, parthenogenetic stem cells having stem cell properties such as pluripotency, self-renewal, and clonogenicity could be directed to differentiate into parthenogenetic mesenchymal cells (pMSCs). The pMSCs could differentiate into tenocytes for regenerating the tendon (Hirabayashi et al. 2017).

22.12 Outlook and Challenges

Parthenogenesis is observed in some invertebrates as well as a few vertebrates.

Parthenogenetic development requires additional stimuli to circumvent the effect of normal halving of ploidy. The study of parthenogenesis has explained a lot about underlying mechanisms of evolution of sex, the ubiquity of sex, and genetic recombination. Parthenogenetically developed embryos are essential alternative sources of ICM cells that are essential for establishing

ESCs; therefore, overcoming the political and ethical issues involved in damage to embryos during isolation of ESCs.

Therapeutic use of parthenogenetic stem cells is still in infancy. Using parthenogenesis in the treatment of fertility and regenerative medicine is a less explored area in veterinary sciences, hence necessitates further research in these important areas. Several hurdles have to be resolved for utilizing the potential of parthenogenesis in assisted reproduction, genetic manipulation, and regenerative medicine. The mechanisms regulating parthenogenesis are not clearly elucidated in most species. The process has certain disadvantages. For instance, the offspring resulting from parthenogenesis are not much successful in the endeavor of their survival because of the lack of recombination of genetic materials.

22.13 Conclusions

Parthenogenesis, the embryonic development in the unfertilized egg is mostly abortive in nature. The uniparental pluripotent stem cells obtained from parthenogenetic embryos are valuable alternatives to ESCs that can be used to derive more immunocompatible cells for regenerative medicine. The cells obtained from parthenogenesis are already tested for developing models to study diseases and using them for therapeutic or biomedical applications. Future research in this area will increase the impact of parthenogenetic stem cells and animals to evolve strategies to cure human and animal diseases.

References

Bai M, Wu Y, Li J (2016) Generation and application of mammalian haploid embryonic stem cells. J Int Med 280(3):236–245. https://doi.org/10.1111/joim.12503 (Epub 2016 May 3)

Beatty RA (1957) Parthenogenesis and polyploidy in mammalian development. Cambridge University Press, London

Campbell KH (1999) Nuclear equivalence, nuclear transfer, and the cell cycle. Cloning 1(1):3–15. Review

Cheng L (2008) More new lines of human parthenogenetic embryonic stem cells. Cell Res 18(2):215–217. https://doi.org/10.1038/cr.2008.19

Cibelli JB, Grant KA, Chapman KB, Cunniff K, Worst T, Green HL, Walker SJ, Gutin PH, Vilner L, Tabar V, Dominko T, Kane J, Wettstein PJ, Lanza RP, Studer L, Vrana KE, West MD (2002) Parthenogenetic stem cells in nonhuman primates. Science 295 (5556):819

Dev K, Giri SK, Kumar A, Yadav A, Singh B, Gautam SK (2012) Expression of transcriptional factor genes (Oct-4, Nanog, and Sox-2) and embryonic stem cell-like characters in placental membrane of Buffalo (Bubalus bubalis). J Membr Biol 245(4):177–183. https://doi.org/10.1007/s00232-012-9427-5

Dighe V, Clepper L, Pedersen D, Byrne J, Ferguson B, Gokhale S, Penedo MC, Wolf D, Mitalipov S (2008) Heterozygous embryonic stem cell lines derived from nonhuman primate parthenotes. Stem Cells 26(3):756–766. https://doi.org/10.1634/stemcells.2007-0869

Eggan K, Akutsu H, Loring J, Jackson-Grusby L, Klemm M, Rideout WM 3rd, Yanagimachi R, Jaenisch R (2001) Hybrid vigor, fetal overgrowth, and viability of mice derived by nuclear cloning and tetraploid embryo complementation. Proc Natl Acad Sci U S A 98(11):6209–6214 (Epub 2001 May 1)

Espejel S, Eckardt S, Harbell J, Roll GR, McLaughlin KJ, Willenbring H (2014) Brief report: parthenogenetic embryonic stem cells are an effective cell source for therapeutic liver repopulation. Stem Cells 32(7):1983–1988. https://doi.org/10.1002/stem.1726

Everett CA, West JD (1998) Evidence for selection against tetraploid cells in tetraploid <->diploid mouse chimaeras before the late blastocyst stage. Genet Res 72(3):225–228

Fukui Y, Sakuma Y (1980) Maturation of bovine oocytes cultured in vitro: relation to ovarian activity, follicular size and the presence or absence of cumulus cells. Biol Reprod 22(3):669–673. https://doi.org/10.1093/biolreprod/22.3.669

Funahashi H, Cantley TC, Stumpf TT, Terlouw SL, Day BN (1994) In vitro development of in vitro-matured porcine oocytes following chemical activation or in vitro fertilization. Biol Reprod 50 (5):1072–1077

Gómez MC, Jenkins JA, Giraldo A, Harris RF, King A, Dresser BL, Pope CE (2003) Nuclear transfer of synchronized african wild cat somatic cells into enucleated domestic cat oocytes. Biol Reprod 69 (3):1032–1041 (Epub 2003 May 28)

Grabiec A, Max A, Tischner M (2007) Parthenogenetic activation of domestic cat oocytes using ethanol, calcium ionophore, cycloheximide and a magnetic field. Theriogenology 67(4):795–800

Hagemann LJ, Hillery-Weinhold FL, Leibfried Rutledge ML, First NL (1995) Activation of murine oocytes with Ca^{2+} ionophore and cycloheximide. J Exp Zool 271(1):57–61

Hirabayashi M, Hara H, Goto T, Takizawa A, Dwinell MR, Yamanaka T, Hochi S, Nakauchi H (2017) Haploid embryonic stem cell lines derived from androgenetic and parthenogenetic rat blastocysts. J Reprod Dev 63(6):611–616. https://doi.org/10.1262/jrd.2017-074

Johannsen H, Muppala V, Gröschel C, Monecke S, Elsner L, Didié M, Zimmermann WH, Dressel R (2017) Immunological properties of murine parthenogenetic stem cells and their differentiation products. Front Immunol 8:924. https://doi.org/10.3389/fimmu.2017.00924 (eCollection 2017)

Kaufman MH, Webb S (1990) Postimplantation development of tetraploid mouse embryos produced by electrofusion. Development 110:1121–1132

Kaufman MH, Robertson EJ, Handyside AH, Evans MJ (1983) Establishment of pluripotential cell lines from haploid mouse embryos. J Embryol Exp Morphol 73:249–261

Kim NH, Simerly C, Funahashi H, Schatten G, Day BN (1996) Microtubule organization in porcine oocytes during fertilization and parthenogenesis. Biol Reprod 54(6):1397–1404

King WA, Xu KP, Sirard MA, Greve T, Leclerc P, Lambert RD, Jacques P (1988) Cytogenetic study of parthenogenetically activated bovine oocytes matured in vivo and in vitro. Gamete Res 20(3):265–274

Kitiyanant Y, Saikhun J, Pavasuthipaisit K (2003) Somatic cell nuclear transfer in domestic cat oocytes treated with IGF-I for in vitro maturation. Theriogenology 59(8):1775–1786

Kochan J, Nowak A, Niżański W, Prochowska S, Migdał A, Młodawska W, Partyka A, Witkowski M (2018) Developmental competence of cat (*Felis domesticus*) oocytes and embryos after parthenogeneticstimulation using different methods. Zygote 1–8. https://doi.org/10.1017/s0967199418000011

Kono T, Obata Y, Wu Q, Niwa K, Ono Y, Yamamoto Y, Park ES, Seo JS, Ogawa H (2004) Birth of parthenogenetic mice that can develop to adulthood. Nature 428(6985):860–864

Lee SR, Kim JW, Kim BS, Kim MO, Kim SH, Yoo DH, Shin MJ, Lee S, Park YS, Park YB, Ha JH, Ryoo ZY (2007) The parthenogenetic activation of canine oocytes with Ca-EDTA by various culture periods and concentrations. Theriogenology 67(4):698–703 (Epub 2006 Nov 27)

Loi P, Ledda S, Fulka J Jr, Cappai P, Moor RM (1998) Development of parthenogenetic and cloned ovine embryos: effect of activation protocols. Biol Reprod 58(5):1177–1187

Macháty Z, Funahashi H, Mayes MA, Day BN, Prather RS (1996) Effects of injecting calcium chloride into in vitro-matured porcine oocytes. Biol Reprod 54(2):316–322

Madan ML, Singla SK, Chauhan MS, Manik RS (1994) In vitro production and transfer of embryos in buffaloes. Theriogenology 41:139–143

Mai Q, Yu Y, Li T, Wang L, Chen MJ, Huang SZ, Zhou C, Zhou Q (2007) Derivation of human embryonic stem cell lines from parthenogenetic blastocysts. Cell Res 17(12):1008–1019

Mayes MA, Stogsdill PL, Prather RS (1995) Parthenogenic activation of pig oocytes by protein kinase inhibition. Biol Reprod 53(2) 270–275. https://doi.org/10.1095/biolreprod53.2.270

Méo SC, Leal CL, Garcia JM (2004) Activation and early parthenogenesis of bovine oocytes treated with ethanol and strontium. Anim Reprod Sci 81(1–2): 35–46

Méo SC, Yamazaki W, Ferreira CR, Perecin F, Saraiva NZ, Leal CL, Garcia JM (2007) Parthenogenetic activation of bovine oocytes using single and combined strontium, ionomycin and 6-dimethylaminopurine treatments. Zygote 15(4): 295–306

Nandi S, Chauhan MS, Plata P (2000) Effect of a corpus luteum in the recovery and developmental potential of buffalo oocytes. Vet Rec 147:580–581

Newman-Smith ED, Werb Z (1995) Stem cell defects in parthenogenetic peri-implantation embryos. Development 121(7):2069–2077

Nussbaum DJ, Prather RS (1995) Differential effects of protein synthesis inhibitors on porcine oocyte activation. Mol Reprod Dev 41(1):70–75

Pincus G, Enzmann EV (1936) The comparative behavior of mammalian eggs in vivo and in vitro. II. The activation of tubal eggs of the rabbit. J Exp Zool 73:195–208

Prichard JF, Thibodeaux JK, Pool SH, Blakewood EG, Menezo Y, Godke RA (1992) In-vitro co-culture of early stage caprine embryos with oviduct and uterine epithelial cells. Hum Reprod 7(4):553–557

Ramachandran R, Nascimento Dos Santos M, Parker HM, McDaniel CD (2018) Parental sex effect of parthenogenesis on progeny production and performance of Chinese Painted Quail (Coturnix chinensis). Theriogenology 118:96–102. https://doi.org/10.1016/j.theriogenology.2018.05.027 (Epub 2018 Jun 1)

Revazova ES, Turovets NA, Kochetkova OD, Kindarova LB, Kuzmichev LN, Janus JD, Pryzhkova MV (2007) Patient-specific stem cell lines derived from human parthenogenetic blastocysts. Cloning Stem Cells 9(3):432–449

Ruddock NT, Macháty Z, Cabot RA, Prather RS (2001) Porcine oocyte activation: differing roles of calcium and pH. Mol Reprod Dev 59(2):227–234

Ruddock NT, Wilson KJ, Cooney MA, Korfiatis NA, Tecirlioglu RT, French AJ (2004) Analysis of imprinted messenger RNA expression during bovine

preimplantation development. Biol Reprod 70 (4):1131–1135 (Epub 2003 Dec 10)
Sato K, Yoshida M, Miyoshi K (2005) Utility of ultrasound stimulation for activation of pig oocytes matured in vitro. Mol Reprod Dev 72(3):396–403
Savage TF, Harper JA (1986) Parthenogenesis in medium white turkeys selected for low and high semen ejaculate volumes. Poult Sci 65(2):401–402
Shamsuddin M, Larsson B, Gustafsson H, Rodriguez-Martinez H (1994) A serum-free, cell-free culture system for development of bovine one-cell embryos up to blastocyst stage with improved viability. Theriogenology 41(5):1033–1043
Singh B, Gautam SK, Verma V, Singla SK. 2012. Derivation of pluripotent stem cell-like cells from nuclear transferred cloned bubaline (Bubalus bubalis) embryos. Reprod Domest Anim 47(Suppl. 5): 108–109. (Abstract)
Sturm KS, Flannery ML, Pedersen RA (1994) Abnormal development of embryonic and extraembryonic cell lineages in parthenogenetic mouse embryos. Dev Dyn 201(1):11–28
Suomalainen E (1950) Pathenogenesis in animals. Adv Genet 3:193–253
Tanaka M, Daimon T (2018) First molecular genetic evidence for automictic parthenogenesis in cockroaches. Insect Sci. https://doi.org/10.1111/1744-7917.12572
Totey SM, Singh G, Taneja M, Pawshe CH, Talwar GP (1992) In vitro maturation, fertilization and development of follicular oocytes from buffalo (*Bubalus bubalis*). J Reprod Fertil 95(2):597–607
Toyokawa K, Harayama H, Miyake M (2005) Exogenous hyaluronic acid enhances porcine parthenogenetic embryo development in vitro possibly mediated by CD44. Theriogenology 64(2):378–392 (Epub 2005 Jan 11)
Verma V, Gautam SK, Singh B, Manik RS, Palta P, Singla SK, Goswami SL, Chauhan MS (2007) Isolation and characterization of embryonic stem cell-like cells from in vitro-produced buffalo (*Bubalus bubalis*) embryos. Mol Reprod Dev 74(4):520–529
Vogt G (2018) Annotated bibliography of the parthenogenetic marbled crayfish Procambarus virginalis, a new research model, potent invader and popular pet. Zootaxa 4418(4):301–352. https://doi.org/10.11646/zootaxa.4418.4.1
Wang ZQ, Kiefer F, Urbánek P, Wagner EF (1997) Generation of completely embryonic stem cell-derived mutant mice using tetraploid blastocyst injection. Mech Dev 62(2):137–145
White MJ, Contreras N, Chency J, Webb GC (1977) Cytogenetics of the parthenogenetic grasshopper Warramaba (formerly Moraba) virgo and its bisexual relatives. II. Hybridization studies. Chromosoma 61 (2):127–148
Witkowska A (1973) Parthenogenetic development of mouse embryos in vivo. II. Postimplantation development. J Embryol Exp Morphol 30(3):547–560 (No abstract available)
Xiong XR, Lan DL, Li J, Wang Y, Zhong JC (2015) Sodium butyrate improves the cloned yak embryo viability and corrects gene expression patterns. Zygote 23(1):19–26. https://doi.org/10.1017/S0967199413000245 (Epub 2013 Jun 12)
Yadav EN, Karche SD, Goel AK, Jindal SK, Johri DK (2007) Comparative efficacy of different techniques for oocytes recovery from prepubertal goat ovaries. Indian J Anim Sci 77:988–990
Yang H, Liu Z, Ma Y, Zhong C, Yin Q, Zhou C, Shi L, Cai Y, Zhao H, Wang H, Tang F, Wang Y, Zhang C, Liu XY, Lai D, Jin Y, Sun Q, Li J (2013) Generation of haploid embryonic stem cells from Macaca fascicularis monkey parthenotes. Cell Res 23(10):1187–1200. https://doi.org/10.1038/cr.2013.93 (Epub 2013 Jul 16)
Zhang M, Liu Y, Liu G, Li X, Jia Y, Sun L, Wang L, Zhou Q, Huang Y (2015) Rapidly generating knockout mice from H19-Igf2 engineered androgenetic haploid embryonic stem cells. Cell Discov 1:15031. https://doi.org/10.1038/celldisc.2015.31 (eCollection 2015)
Zhong C, Li J (2017) Efficient generation of gene-modified mice by haploid embryonic stem cell-mediated semi-cloned technology. Methods Mol Biol 1498:121–133

Transgenesis and Genetically Engineered Livestock as Live Bioreactors

23

Abstract

Transgenic animals are important components of modern biomedical and veterinary sciences. They produce recombinant proteins, enzymes and antibodies for treating diseases and serve as models for disease research, diagnosis, and prevention. A number of genetically engineered or recombinant products are currently obtained from animals, and some have been approved for humans use. With advances in precise genetic engineering and genome editing and producing cloned gene-edited animals, it is possible to produce larger quantities of recombinant proteins and biologics from transgenic animals.

Highlights

- The art of producing transgenic or genome-edited livestock has metamorphosed during recent two decades
- Animals contribute to human health by producing recombinant biomolecules and serving as model animals.

Keywords

Transgenic animals · Bioreactors · Recombinant therapeutics · Biopharming · Gene-editing

23.1 Introduction

Genetic selection is a scientific process adopted to improve fitness and production performance of crops as well as animals. Over the course of past three decades, researchers have evolved methods of genomic selection based on molecular markers as indicators of productivity as well as the well-being of animals. With advances in reproduction techniques, the livestock traits are improved more efficiently within minimum timescale.

Manipulating the genetic makeup of organisms aimed to develop transgenic animals is a fast method of improving productivity, expressing proteins with therapeutic import or developing model animals.

Introducing exogenous genes into pronucleus of zygote was the first attempt to produce transgenic embryos and animals (Pursel et al. 1989). Since then, numerous advances are achieved in biotechnology processes to produce recombinant proteins, and immunocompatible cells for clinical use. The commercial companies have adopted the molecular markers to commercialize livestock rearing and production. Development of transgenic animals as live bioreactors is attributed to genome sequencing, in vitro gene synthesis, and genome manipulation techniques.

B. Singh et al., *Advances in Animal Biotechnology*,
https://doi.org/10.1007/978-3-030-21309-1_23

Biologically active recombinant proteins and monoclonal antibodies (mAbs) and their derivatives are in greater demand as therapeutic agents in a variety of human and veterinary applications. Notably, the bacteria such as lactic acid bacteria (LAB) and *Escherichia coli* strains have discernible limitations such as inability of bacterial translation system to synthesize complex eukaryotic proteins (e.g., monoclonal antibodies or blood coagulation factors), which must be matured by posttranslational modifications (glycosylation, carboxylation, phosphorylation, and sulfatation, etc.) to be active or stable in vivo. Alternatively, yeasts (*Saccharomyces cerevisiae* and *Pichia pastoris*), insects, animal cells, and plants (e.g., *Arabidopsis thaliana*) are used to synthesize recombinant biomolecules (Kumar et al. 2016; Singh et al. 2017). Despite being eukaryotic cells, the yeasts are unable to synthesize large amounts of posttranslationally modified proteins. In addition, yeasts are unable to incorporate necessary multiple disulfide (S-S) bridges to proteins and inept glycosylation of human recombinant proteins by insect cells restrict their use as desirable hosts.

Use of mammalian cell lines, process control, and media formulations have increased therapeutic product titers by nearly 100-fold (Wurm 2004). However, major drawback of mammalian cell culture system is that these cells require a sterile setting, continued supply and substitution of culture media, and fail to produce higher titers of recombinant proteins (Niemann and Kues 2007; Houdbine 2009). Also, the narrow life span of mammalian cells in vitro and the problems associated with the acquisition of materials from normal and pathological tissues make them less suitable as producers of recombinant proteins.

23.2 Recombinant Drugs Approval for Humans

Till date only three drugs produced by transgenic animals are granted approval for human use (Woodfint et al. 2018). US FDA has for the first time granted approval for antithrombin produced in the milk of transgenic goats (Kling 2009). The second drug is Ruconest, a plasma-free C1-esterase inhibitor (C1-INH), used for treating hereditary angioedema (HAE), produced through milk of transgenic rabbit (Nature Biotechnology 2014). The "farmaceutical," a drug produced from transgenic chickens, was the third one produced in recombinant chickens (Sheridan 2016).

23.3 Biopharming Through Animal Transgenesis

The modern animal agriculture is being transformed by rapid advances in post-genomic era of genomics and biotechnology. Biopharming is new generation technology in which transgenic organisms are generated for cheaper and bulk production of pharmaceuticals (Fig. 23.1). If animals are tailored genetically to express and synthesize exogenous or recombinant proteins, the process is called "*animal pharming.*"

Since its inception (Hammer et al. 1985), the animal pharming industry has promoted transgenic animals as a cost-effective method of biopharmaceutical production. At present, numerous recombinant proteins, enzymes, and antibodies are produced using transgenic animals (Houdbine 2009). Transgenic fish, mice, pig, and birds (poultry) are the target species to produce recombinant therapeutics. We believe that ungulates, such as sheep, goats, buffaloes, antelopes, oxen, horses, donkeys, mule, deer, elk, caribou, buffaloes, camels, and llama, could also be used to produce recombinant proteins.

For animals, the term "transgenic" originally refers to the process of stable introduction of exogenous genetic material into their genome and germ line. This definition of transgenic animals has been extended for two reasons. First, the genetic engineering enables not only the additive gene transfer (gain of function), but also permits developing animals with deleted gene transfer (knockout, loss of function), replacement gene transfer (knock-in, exchange of gene function), and spatial–temporal gene transfer (conditional knockout) (Clark and Whitelaw 2003;

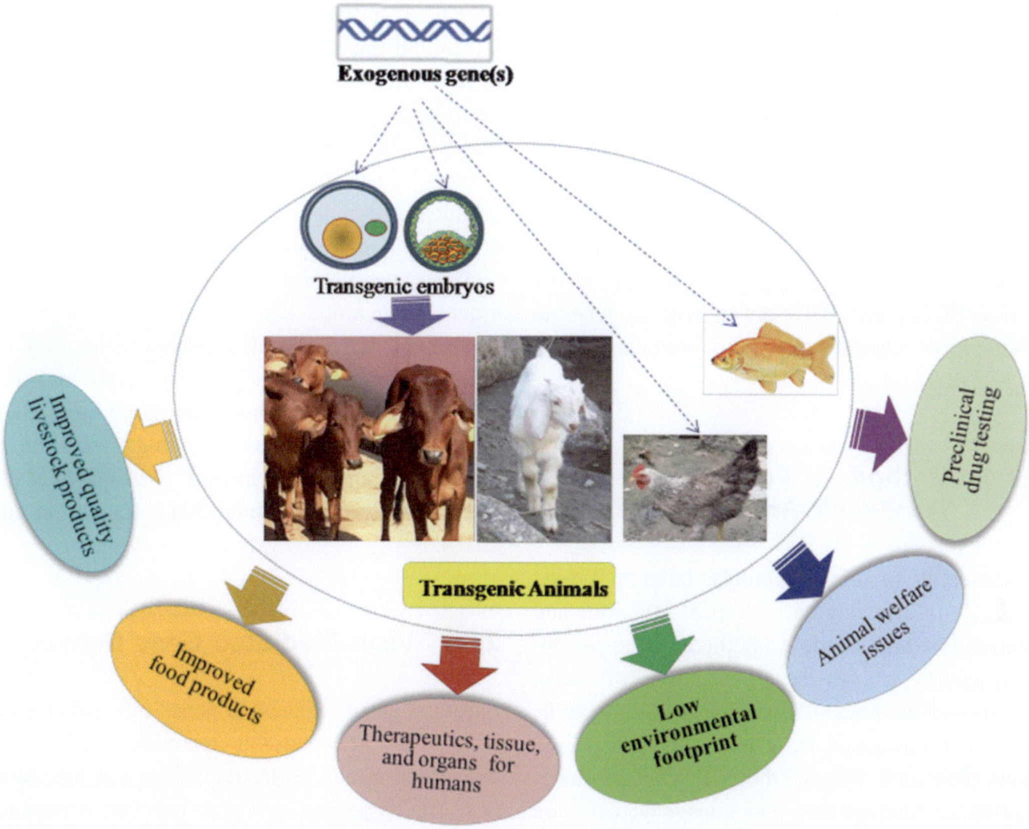

Fig. 23.1 Use of milch animals (such as cow, sheep, and goats), pig, chicken, and fish as bioreactors to produce biomolecules of pharmaceutical value, quality food, and address the environmental concerns such as minimizing GHG emissions from enteric fermentation, release of undigested nutrients (e.g., phytate), and saccharides into environment. The animals shown are merely for illustration

Ristevski 2005). Second, besides integrating transgenes into germline of animals, the normal gene transfer may result in (mostly transient) gene expression during the lifetime of transgenic animals. Diverse types of transgenic animals, knockout and conditional knockout animals, knock-in animals, and humanized animals models are now available for validating or discovering the function of genes of interest. Whereas traditional molecular genetics and rDNA technology have enabled to generate transgenic animals, the recent techniques such as engineered pluripotent stem cells, genome editing, and nanotechnology have contributed to create transgenic animals with more precision and rapidity.

Originally, the transgenic animals were created as basic research tools to study gene function. Hammer et al. (1985) for the first time reported the births of transgenic rabbits, sheep, and pigs using microinjection of foreign genes into pronuclei of the zygotes. However, the technique could not be exploited at large scale due to constraints such as random insertion of genes into genome of host species, uncontrolled copy number of the genes incorporated, and low survival rate of the embryos after microinjection. Another hitch was that expression of recombinant proteins may have adverse effects on the health of the host. The biologically active exogenous recombinant proteins accumulating in the blood may possibly alter the physiology and overall health status of the host and may not remain stable in the host.

When product yield is low due to random integration or targeted integration via homologous recombination, the chimera or transgenic animals are preferred. The technology is improved astonishingly and may have applications for both agriculture and biomedical purposes. Recombinant or genetically engineered ESCs, EGCs, and iPSCs are utilized to generate chimera or cloned transgenic animals (Singh et al. 2009).

23.4 Methods of Producing Transgenic Animals

Genetically modified animals have provided valuable information on biological functions and human diseases. Various methods, such as microinjection of genes, transfection of cells, transposon-mediated delivery of genes into cells, and then transgenic cells are transplanted into preimplantation stages embryos. This produces chimeras. Alternatively, the transgenic cells are used to develop transgenic SCNT embryos and live animals. However, dearth of efficient gene-delivery systems into mammalian cell or embryo is still an important area of research. In addition, the success of techniques of transgenesis (Table 23.1) depends much on target species (Fig. 23.2) and other factors.

23.5 Microinjection

Microinjection involves injecting microquantities of exogenous genes DNAs (gene constructs) (Fig. 23.3), RNA interferences (RNAi), peptides and drugs into ordinary somatic cells, stem cells, and pronucleus of the zygote (Zhang and Lu 2008). The resulting transgenic or recombinant cells, as the case is, are used as donor nuclei to produce cloned transgenic animals. Peptides and drugs are injected into cells for therapeutic rationale. In view of the ease of injection, but inherent limitations associated with microinjection, the technique should be improved. The advanced method of microinjection such as ultrasound image-guided microinjection includes micro-high-frequency ultrasound technology, custom imaging software, and a choice of platforms for image-guided microinjections. Table 23.2 summarizes the animals developed to produce different recombinant proteins using domestic animal species.

Despite availability of new and proven techniques of genome editing, pronuclear microinjection (Fig. 23.4) is still a proven technology to generate transgenic embryos and cells in commercial settings when producing large transgenes (>10 kb), such as those encoding synthesis of monoclonal antibodies in milk (Nakagawa et al. 2015; Delerue and Ittner 2017; Gavin et al. 2018).

23.6 Virus-Mediated Gene Delivery

Introduction of exogenous genes into eukaryotic cells is an indispensable technique in genetic manipulation and molecular biological laboratories. The virus-mediated gene delivery is used to transfect oocyte or early-stage embryos. Variety of recombinant gene-delivery vectors is available commercially. Understanding life cycle of virus, gene functions and their guided regulation have encouraged molecular biologists to choose the vectors to shuttle exogenous genes into cells of choice.

Vectors based on retro-, adeno-, flavi-, and parvoviruses are used for transfecting insect, fish, crustaceans, molluscan, avian, and mammalian cells. Virus-based vectors allow efficient integration and transient expression of the exogenous genes into host cell genome.

Modified baculovirus-based vectors serve as a new type of safe gene-delivery vehicle for transgene expression in mammalian cells. The virus-based vectors enter cells, but do not replicate there. With mammalian expression promoters, recombinant baculoviruses provide high transduction efficiencies. Hence, the innovations in mechanisms of intracellular delivery and integration of foreign genes into host genome by means of baculoviral vectors (Lufino et al. 2008;

Table 23.1 Summary of various techniques used in animal transgenesis, their salient features, and limitations

Methods	Salient features	Major limitations
Microinjection	Ease of injecting genes or drugs into host species	Low efficiency of integration of genes, random insertion of genes into genome of host, recurrence of cell or zygote mosaicism, inefficiency of microinjection in ruminants, the process requires expensive instruments and technical skills
SCNT	Production of complete cloned transgenic organism	The efficiency of the SCNT is low; rate of genetic alteration of donor nuclei is low, the nuclear transfer cloning needs expensive instruments and specialized laboratory, and animal health hazards are involved
Plasmids/ cosmids	Easy to prepare, and handle, wide range of commercial availability, safer to maintain	Gene-carrying capacity is low, hence not suitable for cloning large human/mammalian genes
Liposomes	Long-time gene expression, ranging for months	Low efficiency, but variable depending on cell type
Electroporation	Remains the mainstay of producing of gene-targeted cell lines	May cause transitory and reversible damage to cell membrane
Virus-mediated gene-transfer	Ease of transfection, high capacity (e.g., lentiviruses) to carry and introduce genes into oocytes or embryos	Applications hazards, threat of escape from laboratory, and integration of virus genes into humans or animals
Transposon-mediated gene transfer	Efficient and safer, faithful expression of resistant factors, e.g., siRNAs against specific viral, bacterial and prion diseases in large animals	Requires technical expertise to carry out the experimentation
SMGT/ICSI	Ease of use	Low efficiency, low fertilizing capacity of sperm carrying exogenous genes, requires expensive equipments and skilled personnel
Nanotechnology	Efficient delivery larger genes	The technique is still in initial phases Safety aspects are not aptly discovered
Enzymatic genome editing	Precision and high efficiency	Requirement of specialized instruments, microinjection is needed to inject edited genes into cells, the genome-edited cells are used as donor nuclei to produce cloned transgenic or gene-edited animals

Guijarro-Pardo et al. 2017; Ono et al. 2018) are an attractive choice to advance the current concept of *biopharming* by using insect and mammalian cells.

The mammary gland and the species such as cow, buffalo, goats, and sheep are more attractive candidates to express exogenous proteins in milk.

23.7 Formation of Chimera Animals

In Greek mythology, the chimera is described as a magnificent monster. In biological sciences, the chimera or hybrid or mosaic animal refers to a single animal comprising of two or genetically

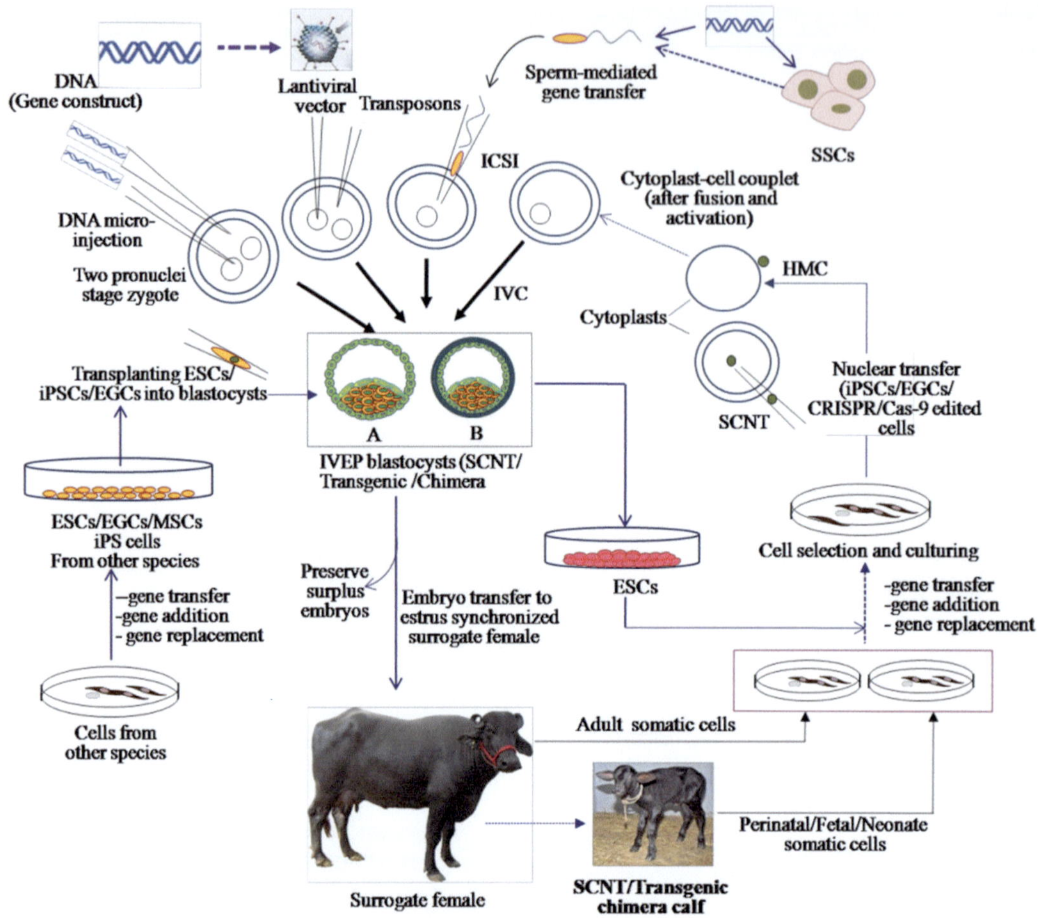

Fig. 23.2 Strategies to produce transgenic animals. Cloned animals such as pigs, goats, cattle, and sheep are already in use to produce recombinant proteins. Pigs are recommended to produce organs for xenotransplantation in humans. *Abbreviations* IVC—In vitro culture of embryos, ESCs—Embryonic stem cells, and SSCs—Spermatogonial stem cells

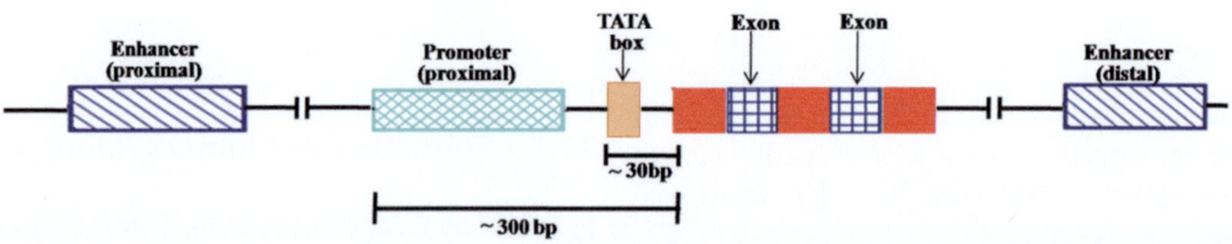

Fig. 23.3 Symbolic presentation of gene construct developed for introducing into cell nucleus by microinjection. A gene construct contains exogenous gene and cell-specific promoters. The gene construct is introduced into nucleus of cell to develop transgenic cell or into the nucleus of zygote to develop transgenic embryos

Table 23.2 A summary of chronology of milestone events in mammalian transgenesis

Achievement	Strategies and inferences
Transgenic mice	Birth of first transgenic mice through microinjection of purified DNA into oocytes (Gordon et al. 1980)
	Effect of *GH* genes was elucidated using transgenic technology (Palmiter et al. 1982), expression of ovine β-lactoglobulin in the milk of transgenic mice (Simons et al. 1987)
	Birth of CRISPR/Cas9-mediated knockout mice by microinjection of CRISPR/Cas9-edited genes (Nakagawa et al. 2015), generation of genetically tailored mice (Delerue and Ittner 2017), generating transgenic mice by SMGT (Wang et al. 2017b), generating CRISPR/Cas9-mediated obese and diabetic mice models (Roh et al. 2018)
Transgenic sheep	Microinjection of exogenous genes into the pronucleus of a zygote (Hammer et al. 1985), expression of human anti-hemophilic factor IX in the milk of transgenic sheep (Clark et al. 1989), high-level expression of active human α-1-antitrypsin in sheep milk (Wright et al. 1991), insulin production, and improved wool quality in transgenic sheep (Damak et al. 1996a, b), random integration of the genes producing human factor IX in sheep produced by NT cloning (Schnieke et al. 1997), first transgenic sheep produced through gene-replacement, and NT (McCreath et al. 2000)
	Birth of GM sheep expressing ovine Toll-like receptor 4 (TLR-4) (Deng et al. 2013); generating transgenic sheep by injecting lentiviral vectors into zygotes, the study reveals safety of HIV-1 based vectors (Cornetta et al. 2013), birth of rams over-expressing TLR-4 with no deleterious effects of TLR-4 on male reproduction traits *viz*., sperm quality, fertility and transgene transmission to offspring (Yao et al. 2017)
Transgenic goats	Birth of cloned transgenic goats from fibroblasts form transgenic goats (Reggio et al. 2001; Behboodi et al. 2004), production, processing and purification of rhLZ in the milk of transgenic goats (Maga et al. 2006), induction of repeat superovulation, non-surgical recovery of embryos, embryo transfer and kid births in transgenic dairy goats (Melican and Gavin 2008), high production of recombinant rhLZ in the milk of transgenic cloned goats produced from transgenic cells containing optimized lysozyme expression cassettes (β-casein/hLZ and β-lactoglobulin/hLZ) (Yu et al. 2013), development of gene-targeted goats containing β-lactoglobulin-modified, less-allergenic milk than the wild type milk, implying that gene-targeted transgenic animals could serve as effective tools for minimizing allergenic reactions and improving human nutrition (Zhu et al. 2016), generation of cloned transgenic cashmere goat kid expressing EGFP form GM fetal fibroblasts as nuclear donors, demonstrating usefulness of piggyBac transposition system to generate cashmere goats (Bai et al. 2012, 2017)
	Studies on properties and potential of rhLZ produced in the milk of transgenic goats, showing the antibacterial and nutraceutical potential (Carneiro et al. 2018); production of transgenic goats expressing a novel human plasminogen activator (He et al. 2018)
Transgenic cattle	The cloned cattle developed from fetal fibroblasts, random integration of the gene construct (Chan et al. 1998), birth of transgenic calf by transfection of oocytes with helper viruses (Chan et al. 1998)
	Transgenic bovine chimeric offspring produced from somatic cell-derived stem like cells (Cibelli et al. 1998), use of artificial chromosome to produce transgenic cattle (Golovan et al. 2001; Kuroiwa et al. 2002), production of transgenic calves expressing short hairpin RNA (shRNA) targeting myostatin (Tessanne et al. 2012), production of cloned transgenic cow expressing omega-3 fatty acids (Wu et al. 2012), transgenic cattle expressing human serum albumin (Luo et al. 2015) transgenic cattle with improved milk composition (Laible et al. 2016), birth of transgenic cattle producing a lipolytic recombinant human digestive enzyme bile salt-stimulated lipase (rhBSSL), and rhBSSL expression levels and characterization of enzyme activities (Wang et al. 2017a), expression of human β-defensin 3 as a strategy to reduce susceptibility to infectious *Mycobacterium bovis* (Su et al. 2016), transgenic cows expressing recombinant human lactoferrin N-glycans in milk (Parc et al. 2017)
	Expression of a novel glycosylated anti-CD20 (Rituxan) monoclonal antibodies in the milk of transgenic cattle (Zhang et al. 2018a, b), validation of ICSI-SMGT in cattle with sperm treated with streptolysin-O (Sánchez-Villalba et al. 2018)

(continued)

Table 23.2 (continued)

Achievement	Strategies and inferences
Transgenic buffaloes	Effects of various culture media and supplements on production of cloned transgenic embryos (Huang et al. 2008; Verma et al. 2008; Wadhwa et al. 2009), birth of first transgenic buffalo calves expressing EGFP (Meng et al. 2015)
Transgenic pigs	Attempts to produce influenza-resistant transgenic pigs (Müller et al. 1992), production of transgenic pig with eco-friendly exogenous genes (phytase) addressing environmental issues (McCreath et al. 2000), birth of heterozygous knockout (Lai et al. 2002), and homozygous knockout pigs (Phelps et al. 2003), development of strategy for gene transfer into mammalian zygotes via lentiviruses (Hofmann et al. 2003), pronucleus DNA microinjection in pigs (Kues et al. 2006), transgenic piglets produced using ICSI-SMGT in combination with recombinase RecA (Garcia-Vazquez et al. 2010), birth of Yorkshire pigs with phytase expressed in salivary glands (Meidinger et al. 2013), birth of pigs carrying human follistatin-344, showing role of follistatin-344 in increasing muscle mass in pig (Chang et al. 2017)
	CRISPR/Cas9-mediated generation of somatic cell-reprogramming free transgenic or myostatin-mutant pigs (Tanihara et al. 2016)
	Development of novel strains of transgenic pigs with single-copy quad-cistronic transgene, simultaneous expression of three microbial enzymes, namely β-glucanase, xylanase, and phytase in the salivary glands, thus reducing the escape of nutrients into environment (Zhang et al. 2018a, b), CRISPR/Cas9-mediated generation of cloned transgenic insulin-deficient piglets (Cho et al. 2018), production of transgenic pig as central nervous system disease model (Hwang et al. 2018)

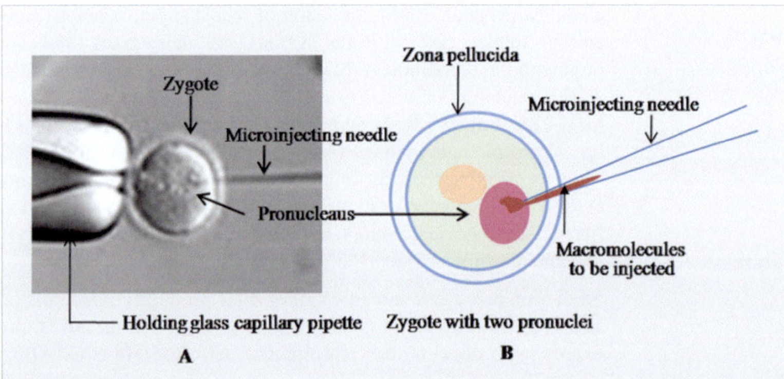

Fig. 23.4 Microinjection of exogenous genetic material into pronucleus of the zygote

different populations originating from different zygotes. Different cells originating from single zygote constitutes an organism, called mosaic. Chimeras are developed by incorporating normal or genetically engineered somatic or stem cell into early-stage developing embryos. The animals produced by this method represent tissues from different animals.

The chimera exists naturally and can be created artificially as well. Interspecies chimera models are important for basic and translational studies and developing patient-specific transplantable organs (Suchy and Nakauchi 2017; Wu et al. 2017), mammalian development, and assessing stem cell pluripotency. Human–animal chimeras, though raise some ethical concerns, are challenging resource for investigating the potential of human pluripotent stem cells and progenitor cells, creating transplantable organs (Masaki and Nakauchi 2017), and achieving regenerative medicine goals (Mascetti and Pedersen 2016).

23.8 Sperm-Mediated Gene Transfer

Sperm-mediated gene transfer (SMGT) is essentially based on the capacity of sperm to carry exogenous genes and transfer them into oocyte to produce transgenic offspring via development of transgenic zygote and embryo. Transplanting transgenic spermatogonial stem cells (SSCs) grafted into testes give rise to transgenic sperm.

There are many other ways to create transgenic sperm. Genes can also be introduced into sperm by transfection using suitable viral vectors. Recombinant spermatozoa or sperm carrying transgenes can be used for fertilizing IVM oocytes in vitro, or transferring the transgenic sperm head into oocyte by ICSI. Alternatively, the transgenic male can be used to produce transgenic offspring via sperm-mediated gene transfer in sexual reproduction.

The SMGT is a slow process, and success is low in large domestic animals. However, modified SMGT is still a method of choice to produce transgenic mice (Wang et al. 2017a), bovines (Arias et al. 2017), pig (Zaniboni et al. 2016) and sheep (Kumar Pramod et al. 2016).

23.9 Use of Nanotechnology

Preferably, a gene-delivery system should be stable and biocompatible to a living system and capable to transfer exogenous genes, antisense oligonucleotides, short interfering RNAs (siRNAs) into a cell, and integrate them at desired sites in the genome of host cells.

Micro- and nanosystems offer advantages over conventional systems of delivering genes, drugs, or therapeutics into cells. Nanocarriers enhance the delivery, extend the bioactivity of the drug by protecting them from environmental effects in biological media, show minimal side effects, demonstrate high-performance characteristics, and are more economical as very less of expensive drugs are required.

According to a report, the annual market value for successful gene-delivery technology may exceed US$ 30 billion, which urges researchers to study associated aspects of development and applications of organic nanoparticles as therapeutic gene-delivery system (Wong et al. 2017).

It is envisaged that nanotechnology will improve ARTs, stem cell engineering, regenerative, and reproductive medicine. Nanotechnology-based gene- or drug-carriers will facilitate site-specific drug/gene delivery into cells (Grześkowiak et al. 2016; Wang et al. 2017a). Polymer-coated viral vectors (Rajagopal et al. 2018) and non-viral nanomaterial-based vectors have been fabricated to transmit the genes into a cell and even into specific subcellular organelles such as mitochondria. The novel materials and systems are combined to reconstruct organs that would improve the quality of life of patients suffering from organ diseases and failure. Stem cells and nanotechnology are viewed as a newly emerging field to confer health benefits (Wang et al. 2009; Tabassum et al. 2018). As genetic material does not pass passively across cell wall or cell membranes, a carrier or gene-delivery system is required. Nanotechnology, as a tool to enhance reproductive medicine, may overcome some of the key limits of gene delivery into cells to create genetically engineered cells.

23.10 CRISPR/Cas 9 Systems—Gene and Genome Editing

Pronuclear microinjection was the first phase of developing transgenic animals in the 1980s. Meanwhile, SMGT, transplanting SSCs into testes, powerful gene-transfer vectors such as lentivirus system, transposons, RNAi, and site-specific recombinases continued to bolster the success of transgenesis. Nuclear transfer cloning or SCNT established in 1996 and induced genome reprogramming in 2006 by ectopic expression of genetic factors into somatic cells, heralded second era of manipulating cell fate, and producing transgenic cloned animals form engineered cells or stem cells.

The present era is regarded as third wave of genetic engineering technologies wherein endonucleases CRISPR/Cas 9 are used to precisely edit and create new genome. CRISPR/Cas 9 genome editing is developed to induce changes in the genes at desired nucleotide sequences. Among currently used genome-editing tools include zinc-finger nucleases (ZFNs), transcription activator-like endonucleases (TALENs), and CRISPR/Cas9, which are successfully utilized in animals including zebra fish, mice, rats, monkeys, pigs, cattle, sheep, and goats (Ruan et al. 2017; Cho et al. 2018), and insects (Dong et al. 2019; Ma et al. 2018). The modified versions of CRISPR/Cas 9 are also developed. More species with edited genome are expected in the near future. A novel diabetic pig models has been produced by CRISPR/Cas9 editing. The insulin-deficient pig might be useful in diabetes research to test the efficacy and safety of new drugs and will serve as recipient for islet transplantation to investigate cell transplantation therapies to cure diabetes (Cho et al. 2018).

23.11 Choice of the Animal Species and Tissue to Produce Recombinant Biomolecules

The animals that produce milk containing high protein contents, having normal reproductive health, being non-carriers of human diseases, and resistant to prion diseases are deemed as ideal candidates for producing recombinant proteins. Apart from laboratory animals, such as zebra fish and mice, pig is the most studied animal model. The history of mammals as transgenic animals began with success of stable integration of metallothionein-I (MT) promoter/regulator region fused to either rat or human growth hormone (hGH) structural genes into genome of pigs, sheep, and rabbit (Hammer et al. 1985). Since then, pigs have been the prime mammalian species for biomedicine and agriculture applications and as donors of organs for xenotransplantation.

Blood, milk (Houdbile 2002; Yu et al. 2013), urine, and seminal plasma, silk gland (Royer et al. 2005), egg white (Lillico et al. 2007; Yang et al. 2008), and insect larvae hemolymph (Markaki et al. 2007; Dong et al. 2018) are the possible systems to produce recombinant proteins. Sheep and goats, due to their small body size, small gestation period, and ability to produce milk, are utilized to produce recombinant proteins.

Mammary gland is an interesting tissue for producing pharmaceutical proteins due to a high level of protein expression, ability to perform posttranslational modifications, and releasing modified proteins into milk. The recombinant proteins excreted into milk can be easily purified. Numerous proteins have been produced in large amounts in the mammary glands of transgenic sheep, goat, cattle, pig, and rabbit (Kues and Niemann 2004; Monzani et al. 2016).

Early example of this technology was the production of transgenic sheep expressing human blood-clotting factor IX that is required for hemophilia patients. The human factor IX gene was expressed under the control of sheep genome that normally turns on the β-lactoglobulin gene in the mammary tissue. Combined with tissue-specific functional promoter elements, non-mammalian genes can also be expressed in farm animals to modify intermediary metabolism.

In theory, large quantities of human proteins can be produced in the milk of animals (Yu et al. 2013). Consumption of human lysozyme (hLZ)-containing milk from transgenic goats could protect the piglets, that are used as human models of childhood diarrhea, and enterotoxigenic *Escherichia coli*, signifying that hLZ-containing goat milk could be used as nutraceutical for effective treatment of *E. coli*-induced diarrhea (Cooper et al. 2013).

A 5–10 times higher levels of recombinant human lysozyme (rhLZ), being up to 6.2 g/L of rhLZ, were noted in the milk of transgenic goat. The study underscores the suitability of mammary gland for producing transgenic proteins. The hLZ-containing milk from transgenic goats could protect piglets against enterotoxigenic *E. coli* infections (Garas et al. 2017).

Also, the milk is a suitable medium to excrete monoclonal antibodies and enzymes. The appropriately glycosylated recombinant anti-CD20 mAb (Rituxan) (at a yield of up to ∼6.8 mg/mL), with ∼80% recovery rate, and

>99% purity were detected in the milk of transgenic cattle (Zhang et al. 2018a, b). Transgenic cattle, producing a lipolytic recombinant human digestive enzyme bile salt-stimulated lipase (rhBSSL) has been produced. The cows expressed higher levels (9.8 mg/ml) of rhBSSL having normal composition (N-terminal amino acid sequence, amino acid composition, and pI) and physicochemical and biological characteristics comparable to human native BSSL (Wang et al. 2017a).

However, some biological barriers present a challenge for efficient gene delivery, and as a result, the inefficient or low-efficiency integration of exogenous genetic material into genome of host remains a challenge in producing transgenic animals. Genetic engineering along with advanced cell culture technology may particularly be useful when used together, and the achievement of the goal seems to be possible.

It is evident that selection of genes, as opposed to phenotype, promises to be more accurate, efficient, and economically beneficial even before production traits are measured. Such selection will continue to be very successful in breeding farm animals and taking care of their welfare (Yadav et al. 2012).

23.12 Limitations of Large Mammal Transgenesis

Transgenic mice are most widely and routinely used mammalian models. One of the major limitations of the mouse as model organism is that certain human pathologies cannot be suitably modeled in transgenic mice because of differences in physiology between two species (reviewed by Garrels et al. 2012).

Active transgenesis is an evolving process for precision genetic engineering of domesticated ungulates. Domestic species as transgenic models are gaining importance for basic and applied biomedical research. Domestic animals, especially the pigs, goats and sheep serve as biomedical models and producers for exogenous proteins.

Mammary gland, in view of secreting large amounts of posttranslationally modified proteins, is the promising organ to express exogenous molecules. Cows, as producers of large quantities of milk over a long period of lactation, are considered ideal candidates for recombinant proteins in milk. However, there are some limitations, such as long gestation period, limited number of calves produced per gestation, insufficiency of transgenesis, long period for transgene detection, and expression of recombinant proteins, that hinder the speculated benefits (Monzani et al. 2016). Humanized pig would be a hope to manufacture tissues and organs for xenotransplantation.

23.13 Outlook and Challenges

Research in the fields of life science, pharmacology, and biomedical engineering has created demand for transgenic animals for producing experimental model animals, recombinant proteins of biomedical interest and organs for xenotransplantation.

Genetic modification of larger animals is a difficult task. However, it is tricky to foresee the large herds of transgenic livestock until the efficiency of producing transgenic animals is improved noticeably.

The recombinant biomolecules obtained from animals have to undergo rigorous testing for safety and efficacy they are recommended for humans.

Because of insertional mutagenesis, integration of transgenes into host genome, and concomitant expression, adverse effects of recombinant products on host health cannot be ruled out. This raises the concern about the health of transgenic animals.

Advanced CRISPR/Cas-9 gene editing provides new opportunities not only for improving growth performance, development of transgenic models. The drawback of genome editors is that they depend on pronuclear injection based transgenic and nuclear transfer cloning techniques. These techniques require expensive micromanipulators and highly skilled equipment operators. To overcome the problems, a new concept, named genome editing via oviductal nucleic acids delivery (GONAD) that directly delivers nucleic acids into zygotes within the oviducts, has been introduced (Sato et al. 2018). The technology avoids pronuclear injection of genome-editing tools and ex vivo handling of embryos.

23.14 Conclusions

The demand for recombinant proteins has increased in various areas of biological sciences. Transgenic technology represents potent biotechnological approaches that allow development of genetically modified animals for basic research for biomedical, veterinary, and agricultural applications. Milk appears to be the most mature biological system to produce recombinant biomolecules.

Animal-derived recombinant proteins and their derivatives are in demand as futuristic therapeutic agents. As the technique involves a high cost in large animals, most work is done in mice, pigs, and sheep. Genetic engineering along with "omics" to create transgenic animals is likely to prove practically viable.

References

Arias ME, Sánchez-Villalba E, Delgado A, Felmer R (2017) Effect of transfection and co-incubation of bovine sperm with exogenous DNA on sperm quality and functional parameters for its use in sperm-mediated gene transfer. Zygote 25(1):85–97. https://doi.org/10.1017/s096719941600037x (Epub 2016 Dec 8)

Bai DP, Yang MM, Chen YL (2012) PiggyBac transposon-mediated gene transfer in Cashmere goat fetal fibroblast cells. Biosci Biotechnol Biochem 76 (5):933–937 (Epub 2012 May 7)

Bai DP, Yang MM, Qu L, Chen YL (2017) Generation of a transgenic cashmere goat using the piggyBac transposition system. Theriogenology 15(93):1–6. https://doi.org/10.1016/j.theriogenology.2017.01.036 (Epub 2017 Jan 23)

Behboodi E, Memili E, Melican DT, Destrempes MM, Overton SA, Williams JL, Flanagan PA, Butler RE, Liem H, Chen LH, Meade HM, Gavin WG, Echelard Y (2004) Viable transgenic goats derived from skin cells. Transgenic Res 13(3):215–224

Carneiro IS, Menezes JNR, Maia JA, Miranda AM, Oliveira VBS, Murray JD, Maga EA, Bertolini M, Bertolini LR (2018) Milk from transgenic goat expressing human lysozyme for recovery and treatment of gastrointestinal pathogens. Eur J Pharm Sci 15 (112):79–86. https://doi.org/10.1016/j.ejps.2017.11.005 (Epub 2017 Nov 8)

Chan AW, Homan EJ, Ballou LU, Burns JC, Bremel RD (1998) Transgenic cattle produced by reverse-transcribed gene transfer in oocytes. Proc Natl Acad Sci USA 95:14028–14033

Chang F, Fang R, Wang M, Zhao X, Chang W, Zhang Z, Li N, Meng Q (2017) The transgenic expression of human follistatin-344 increases skeletal muscle mass in pigs. Transgenic Res 26(1):25–36. https://doi.org/10.1007/s11248-016-9985-x (Epub 2016 Oct 27)

Cho B, Kim SJ, Lee EJ, Ahn SM, Lee JS, Ji DY, Lee K, Kang JT (2018) Generation of insulin-deficient piglets by disrupting INS gene using CRISPR/Cas9 system. Transgenic Res 27(3):289–300. https://doi.org/10.1007/s11248-018-0074-1 (Epub 2018 Apr 24)

Cibelli JB, Stice SL, Golueke PL, Kane JJ, Jerry J, Blackwell C, Ponce de Leon FA, Robl JM (1998) Transgenic bovine chimeric offspring produced from somatic cell-derived stem like cells. Nat Biotechnol 16:642–646

Clark AJ, Whitelaw B (2003) A future for transgenic livestock. Nat Rev Genet 4:825–833

Clark AJ, Bessos H, Bishop JO, Brown P, Harris S, Lathe R, McClenaghan M, Prowse C, Simons JP, Whitelaw CB, Wilmut I (1989) Expression of human anti-hemophilic factor IX in the milk of transgenic sheep. Biotechnology 7:487–492

Cooper CA, Garas Klobas LC, Maga EA, Murray JD (2013) Consuming transgenic goats' milk containing the antimicrobial protein lysozyme helps resolve diarrhea in young pigs. PLoS One. 8(3):e58409. https://doi.org/10.1371/journal.pone.0058409 (Epub 2013 Mar 13)

Cornetta K, Tessanne K, Long C, Yao J, Satterfield C, Westhusin M (2013) Transgenic sheep generated by lentiviral vectors: safety and integration analysis of surrogates and their offspring. Transgenic Res 22 (4):737–745. https://doi.org/10.1007/s11248-012-9674-3 (Epub 2012 Nov 23)

Damak S, Su H, Jay NP, Bullock DW (1996a) Improved wool production in transgenic sheep expressing insulin-like growth factor 1. Biotechnology (N Y). 14(2):185–188

Damak S, Jay NP, Barrell GK, Bullock DW (1996b) Targeting gene expression to the wool follicle in transgenic sheep. Biotechnology (N Y). 14(2):181–184

Delerue F, Ittner LM (2017) Generation of genetically modified mice through the microinjection of oocytes. J Vis Exp. 124. https://doi.org/10.3791/55765

Deng S, Li G, Zhang J, Zhang X, Cui M, Guo Y, Liu G, Li G, Feng J, Lian Z (2013) Transgenic cloned sheep overexpressing ovine toll-like receptor 4.

Theriogenology 80(1):50–57. https://doi.org/10.1016/j.theriogenology.2013.03.008 (Epub 2013 Apr 25)

Dong Z, Hu Z, Qin Q, Dong F, Huang L, Long J, Chen P, Lu C, Pan M (2019) CRISPR/Cas9-mediated disruption of the immediate early-0 and 2 as a therapeutic approach to Bombyx mori nucleopolyhedrovirus in transgenic silkworm. Insect Mol Biol 28(1):112–122. https://doi.org/10.1111/imb.12529 (Epub 2018 Oct 8)

Garas LC, Cooper CA, Dawson MW, Wang JL, Murray JD, Maga EA (2017) Young pigs consuming lysozyme transgenic goat milk are protected from clinical symptoms of enterotoxigenic *Escherichia coli* infection. J Nutr 147(11):2050–2059. https://doi.org/10.3945/jn.117.251322 (Epub 2017 Sep 27)

Garcia-Vazquez FA, Ruiz S, Matas C, Izquierdo-Rico MJ, Grullon LA, De Ondiz A, Vieira L, Aviles-Lopez K, Gutierrez-Adán A, Gadea J (2010) Production of transgenic piglets using ICSI-sperm-mediated gene transfer in combination with recombinase RecA. Reproduction 140:259–272

Garrels W, Ivics Z, Kues WA (2012) Precision genetic engineering in large mammals. Trends Biotechnol 30:386–393

Gavin W, Blash S, Buzzell N, Pollock D, Chen L, Hawkins N, Howe J, Miner K, Pollock J, Porter C, Schofield M, Echelard Y, Meade H (2018) Generation of transgenic goats by pronuclear microinjection: a retrospective analysis of a commercial operation (1995–2012). Transgenic Res 27(1):115–122. https://doi.org/10.1007/s11248-017-0050-1 (Epub 2017 Dec 16)

Gordon JW, Scangos GA, Plotkin DJ, Barbosa JA, Ruddle FH (1980) Genetic transformation of mouse embryos by microinjection of purified DNA. Proc Natl Acad Sci USA 77:7380–7384

Golovan SP, Meidinger RG, Ajakaiye A, Cottrill, M, Weiderkehr MZ, Barney DJ, Plante C, Pollard JW, Fan MZ, Hayes MA, Laursen J, Hjorth, JP, Hacker RR, Phillips JP and Forsberg CW (2001) Pigs expressing salivary phytase produce low-phosphorus manure. Nat Biotechnol 19:741–745. Erratum in: Nat Biotechnol 19:979

Grześkowiak BF, Hryhorowicz M, Tuśnio K, Grzeszkowiak M, Załęski K, Lipiński D, Zeyland J, Mykhaylyk O, Słomski R, Jurga S, Woźniak A (2016) Generation of transgenic porcine fibroblast cell lines using nanomagnetic gene delivery vectors. Mol Biotechnol 58(5):351–361. https://doi.org/10.1007/s12033-016-9934-1

Guijarro-Pardo E, Gómez-Sebastián S, Escribano JM (2017) In vivo production of recombinant proteins using occluded recombinant AcMNPV-derived baculovirus vectors J Virol Methods 250:17–24. https://doi.org/10.1016/j.jviromet.2017.09.017 (Epub 2017 Sep 22)

Hammer RE, Pursel VG, Rexroad CE Jr, Wall RJ, Bolt DJ, Ebert KM, Palmiter RD, Brinster RL (1985) Production of transgenic rabbits, sheep and pigs by microinjection. Nature 315(6021):680–683

He Z, Lu R, Zhang T, Jiang L, Zhou M, Wu D, Cheng Y (2018) A novel recombinant human plasminogen activator: Efficient expression and hereditary stability in transgenic goats and in vitro thrombolytic bioactivity in the milk of transgenic goats. PLoS One 13(8): e0201788. https://doi.org/10.1371/journal.pone.0201788 (eCollection 2018)

Hofmann A, Kessler B, Ewerling S, Weppert M, Vogg B, Ludwig H, Stojkovic M, Boelhauve M, Brem G, Wolf E, Pfeifer A (2003) Efficient transgenesis in farm animals by lentiviral vectors. EMBO Rep 4(11):1054–1060 (Epub 2003 Oct 17)

Houdbile LM (2002) The methods to generate transgenic animals and transgene expression. J Biotechnol 98:145–160

Houdbine LM (2009) Production of pharmaceutical proteins by transgenic animals. Comp Immunol Microbiol Infect Dis 32:107–121

Huang B, Cui K, Li T, Wang XL, Lu F, Liu Q, da Silva FM, Shi D (2008) Generation of buffalo (*Bubalus bubalis*) transgenic chimeric and nuclear transfer embryos using embryonic germ-like cells expressing enhanced green fluorescent protein. Reprod Domest Anim 45:103–108

Hwang SU, Eun K, Yoon JD, Kim H, Hyun SH (2018) Production of transgenic pigs using a *pGFAP-CreER*$_{T2}$*/EGFP* $_{LoxP}$ inducible system for central nervous system disease models. J Vet Sci 19 (3):434–445. https://doi.org/10.4142/jvs.2018.19.3.434

Kling J (2009) First US approval for a transgenic animal drug. Nat Biotechnol 27(4):302–304. https://doi.org/10.1038/nbt0409-302 (No abstract available)

Kues WA, Niemann H (2004) The contribution of farm animals to human health. Review. Trends Biotechnol 22:286–294

Kues WA, Schwinzer R, Wirth D, Verhoeyen E, Lemme E, Herrmann D, Barg-Kues B, Hauser H, Wonigeit H, Niemann H (2006) Epigenetic silencing and tissue independent expression of a novel tetracycline inducible system in double-transgenic pigs. FASEB J 20:E1–E10

Kumar Pramod R, Kumar R, Mitra A (2016) Transgenic expression of green fluorescent protein in caprine embryos produced through electroporation-aided sperm-mediated gene transfer. Gene 576(1 Pt 3):505–511. https://doi.org/10.1016/j.gene.2015.10.066 (Epub 2015 Oct 30)

Kumar M, Yadav AK, Verma V, Singh B, Mal G, Nagpal R, Hemalatha R (2016) Bioengineered probiotics as a new hope for health and diseases: an overview of potential and prospects. Future Microbiol 11(4):585–600. https://doi.org/10.2217/fmb.16.4 (Epub 2016 Apr 12. Review)

Kuroiwa Y, Kasinathan P, Choi YJ, Naeem R, Tomizuka K, Sullivan EJ, Knott JG, Duteau A, Goldsby RA, Osborne BA, Ishida I, Robl JM (2002) Cloned transchromosomic calves producing human immunoglobulin. Nat Biotechnol 20:889–894

Lai L, Kolber-Simonds D, Park KW, Cheong HT, Greenstein JL, Im GS, Samuel M, Bonk A, Rieke A, Day BN, Murphy CN, Carter DB, Hawley RJ, Prather RS (2002) Production of alpha-1,3-galactosyltransferase knockout pigs by nuclear transfer cloning. Science 295:1089–1092

Laible G, Smolenski G, Wheeler T, Brophy B (2016) Increased gene dosage for β- and κ-casein in transgenic cattle improves milk composition through complex effects. Sci Rep 23(6):37607. https://doi.org/10.1038/srep37607

Lillico SG, Sherman A, McGrew MJ, Robertson CD, Smith J, Haslam C, Barnard P, Radcliffe PA, Mitrophanous KA, Elliot EA, Sang HM (2007) Oviduct-specific expression of two therapeutic proteins in transgenic hens. Proc Natl Acad Sci USA 104:1771–1776

Lufino MM, Esder PA, Wade-Martins R (2008) Advances in high-capacity extrachromosomal vector technology: episomal maintenance, vector delivery, and transgene expression. Mol Ther 16:1525–1538

Luo Y, Wang Y, Liu J, Lan H, Shao M, Yu Y, Quan F, Zhang Y (2015) Production of transgenic cattle highly expressing human serum albumin in milk by phiC31 integrase-mediated gene delivery. Transgenic Res 24 (5):875–883. https://doi.org/10.1007/s11248-015-9898-0 (Epub 2015 Jul 22)

Ma SY, Smagghe G, Xia QY (2018) Genome editing in Bombyx mori: new opportunities for silkworm functional genomics and the sericulture industry. Insect Sci. https://doi.org/10.1111/1744-7917.12609 (Epub ahead of print. Review)

Maga EA, Shoemaker CF, Rowe JD, Bondurant RH, Anderson GB, Murray JD (2006) Production and processing of milk from transgenic goats expressing human lysozyme in the mammary gland. J Dairy Sci 89(2):518–524

Markaki M, Drabek D, Livadaras I, Craig RK, Grosveld F, Savakis C (2007) Stable expression of human growth hormone over 50 generations in transgenic insect larvae. Transgenic Res 16:99–107

Masaki H, Nakauchi H (2017) Interspecies chimeras for human stem cell research. Development 144 (14):2544–2547. https://doi.org/10.1242/dev.151183

Mascetti VL, Pedersen RA (2016) Contributions of mammalian chimeras to pluripotent stem cell research. Cell Stem Cell 19(2):163–175. https://doi.org/10.1016/j.stem.2016.07.018

McCreath KJ, Howcroft J, Campbell KH, Colman A, Schnieke AE, Kind AJ (2000) Production of gene-targeted sheep by nuclear transfer from cultured somatic cells. Nature 405(6790):1066–1069. Erratum in Nature (2000) 408(6808):120

Meidinger RG, Ajakaiye A, Fan MZ, Zhang J, Phillips JP, Forsberg CW (2013) Digestive utilization of phosphorus from plant-based diets in the Cassie line of transgenic Yorkshire pigs that secrete phytase in the saliva. J Anim Sci 91(3):1307–1320. https://doi.org/10.2527/jas.2012-5575 (Epub 2013 Jan 7)

Melican D, Gavin W (2008) Repeat superovulation, non-surgical embryo recovery, and surgical embryo transfer in transgenic dairy goats. Theriogenology 69 (2):197–203 (Epub 2007 Oct 29)

Meng F, Li H, Wang X, Qin G, Oback B, Shi D (2015) Optimized production of transgenic buffalo embryos and offspring by cytoplasmic zygote injection. J Anim Sci Biotechnol. 6:44. https://doi.org/10.1186/s40104-015-0044-x (eCollection 2015)

Monzani PS, Adona PR, Ohashi OM, Meirelles FV, Wheeler MB (2016) Transgenic bovine as bioreactors: challenges and perspectives. Bioengineered 7(3):123–131. https://doi.org/10.1080/21655979.2016.1171429 (Epub 2016 May 11. Review)

Müller M, Brenig B, Winnacker EL, Brem G (1992) Transgenic pigs carrying cDNA copies encoding the murine Mx1 protein which confers resistance to influenza virus infection. Gene 121(2):263–270

Nakagawa Y, Sakuma T, Sakamoto T, Ohmuraya M, Nakagata N, Yamamoto T (2015) Production of knockout mice by DNA microinjection of various CRISPR/Cas9 vectors into freeze-thawed fertilized oocytes. BMC Biotechnol 15:33. https://doi.org/10.1186/s12896-015-0144-x

Nature Biotechnology (2014) Rabbit milk Ruconest for hereditary angioedema. 32:849. https://doi.org/10.1038/nbt0914-849d

Niemann H, Kues W (2007). Transgenic farm animals. Reprod Fertil Dev 19:762–770

Ono C, Okamoto T, Abe T, Matsuura Y (2018) Baculovirus as a tool for gene delivery and gene therapy. Viruses 10(9):pii: E510. https://doi.org/10.3390/v10090510 (Review)

Palmiter RD, Brinster RL, Hammer RE, Trumbauer ME, Rosenfeld MG, Birnberg NC, Evans RM (1982). Dramatic growth of mice that develop from eggs microinjected with metallothionein-growth hormone fusion genes. Nature 300:611–615

Parc AL, Karav S, Rouquié C, Maga EA, Bunyatratchata A, Barile D (2017) Characterization of recombinant human lactoferrin N-glycans expressed in the milk of transgenic cows. PLoS One 12(2): e0171477. https://doi.org/10.1371/journal.pone.0171477 (eCollection 2017)

Phelps CJ, Koike C, Vaught TD, Boone J, Wells KD, Chen SH, Ball S, Specht SM, Polejaeva IA, Monahan JA, Jobst PM, Sharma SB, Lamborn AE, Garst AS, Moore M, Demetris AJ, Rudert WA, Bottino R, Bertera S, Trucco M, Starzl TE, Dai Y, Ayares DL (2003) Production of alpha 1,3-galactosyltransferase-deficient pigs. Science 299 (5605):411–414 (Epub 2002 Dec 19)

Pursel VG, Pinkert CA, Miller KF, Volt DJ, Campbell RG, Palmiter RD, Brinster RL and Hammer RF (1989) Genetic engineering of livestock. Science 244:1281–1288

Rajagopal P, Duraiswamy S, Sethuraman S, Giridhara Rao J, Krishnan UM (2018) Polymer-coated viral vectors: hybrid nanosystems for gene therapy. J Gene

Med 20(4):e3011. https://doi.org/10.1002/jgm.3011 (Epub 2018 Mar 25. Review)

Reggio BC, James AN, Green HL, Gavin WG, Behboodi E, Echelard Y, Godke RA (2001) Cloned transgenic offspring resulting from somatic cell nuclear transfer in the goat: oocytes derived from both follicle-stimulating hormone-stimulated and non-stimulated abattoir-derived ovaries. Biol Reprod 65 (5):1528–1533

Ristevski S (2005) Making better transgenic models: conditional, temporal, and spatial approaches. Review. Mol Biotechnol 29:153–163

Roh JI, Lee J, Park SU, Kang YS, Lee J, Oh AR, Choi DJ, Cha JY, Lee HW (2018) CRISPR-Cas9-mediated generation of obese and diabetic mouse models. Exp Anim 67(2):229–237. https://doi.org/10.1538/expanim.17-0123 (Epub 2018 Jan 16)

Royer C, Jalabert A, Da Rocha M, Grenier AM, Mauchamp B, Couble P, Chavancy G (2005) Biosynthesis and cocoon-export of a recombinant globular protein in transgenic silkworms. Transgenic Res 14:463–472

Ruan J, Xu J, Chen-Tsai RY, Li K (2017) Genome editing in livestock: Are we ready for a revolution in animal breeding industry? Transgenic Res 26(6):715–726. https://doi.org/10.1007/s11248-017-0049-7 (Epub 2017 Nov 1. Review)

Sánchez-Villalba E, Arias ME, Loren P, Fuentes F, Pereyra-Bonnet F, Salamone D, Felmer R (2018) Improved expression of green fluorescent protein in cattle embryos produced by ICSI-mediated gene transfer with spermatozoa treated with streptolysin-O. Anim Reprod Sci 196:130–137. https://doi.org/10.1016/j.anireprosci.2018.07.005 (Epub 2018 Jul 17)

Sato M, Ohtsuka M, Nakamura S, Sakurai T, Watanabe S, Gurumurthy CB (2018) In vivo genome editing targeted towards the female reproductive system. Arch Pharm Res 41(9):898–910. https://doi.org/10.1007/s12272-018-1053-z (Epub 2018 Jul 4. Review)

Schnieke AE, Kind AJ, Ritchie WA, Mycock K, Scott AR, Ritchie M, Wilmut I, Colman A, Campbell KH (1997) Human factor IX transgenic sheep produced by transfer of nuclei from transfected fetal fibroblasts. Science 278:2130–2133

Sheridan C (2016) FDA approves 'farmaceutical' drug from transgenic chickens. Nat Biotechnol 34(2):117–119. https://doi.org/10.1038/nbt0216-117 (No abstract available)

Simons JP, McClenaghan M, Clark AJ (1987) Alteration of the quality of the milk by expression of sheep beta-lactoglobulin in transgenic mice. Nature 328:530–532

Singh B, Chauhan MS, Singla SK, Gautam SK, Verma V, Manik RS, Singh AK, Sodhi M, Mukesh M (2009) Reproductive biotechniques in buffalo (*Bubalus bubalis*): status, prospects and chllanges. Review. Reprod Fertil Dev 21:499–510

Singh B, Mal G, Marotta F (2017) Designer probiotics: paving the way to living therapeutics. Trends Biotechnol 35(8):679–682. https://doi.org/10.1016/j.tibtech.2017.04.001(Epub 2017 May 5)

Su F, Wang Y, Liu G, Ru K, Liu X, Yu Y, Liu J, Wu Y, Quan F, Guo Z, Zhang Y (2016) Generation of transgenic cattle expressing human β-defensin 3 as an approach to reducing susceptibility to Mycobacterium bovis infection. FEBS J 283(5):776–790. https://doi.org/10.1111/febs.13641 (Epub 2016 Jan 30)

Suchy F, Nakauchi H (2017) Lessons from interspecies mammalian chimeras. Annu Rev Cell Dev Biol 33:203–217. https://doi.org/10.1146/annurev-cellbio-100616-060654 (Epub 2017 Aug 14. Review)

Tabassum N, Verma V, Kumar M, Kumar A, Singh B (2018) Nanomedicine in cancer stem cell therapy: from fringe to forefront. Cell Tissue Res 374 (3):427–438. https://doi.org/10.1007/s00441-018-2928-5

Tanihara F, Takemoto T, Kitagawa E, Rao S, Do LT, Onishi A, Yamashita Y, Kosugi C, Suzuki H, Sembon S, Suzuki S, Nakai M, Hashimoto M, Yasue A, Matsuhisa M, Noji S, Fujimura T, Fuchimoto D, Otoi T (2016) Somatic cell reprogramming-free generation of genetically modified pigs. Sci Adv 2(9): e1600803. https://doi.org/10.1126/sciadv.1600803 (eCollection 2016 Sep)

Tessanne K, Golding MC, Long CR, Peoples MD, Hannon G, Westhusin ME (2012) Production of transgenic calves expressing an shRNA targeting myostatin. Mol Reprod Dev 79(3):176–185. 10.1002/mrd.22007 (Epub 2011 Dec 2)

Verma V, Gautam SK, Palta P, Manik RS, Singla SK, Chauhan MS (2008) Development of a pronuclear DNA microinjection technique for production of green fluorescent protein-expressing bubaline (*Bubalus bubalis*) embryos. Theriogenology 69(6):655–665. https://doi.org/10.1016/j.theriogenology.2007.09.035 (Epub 2008 Feb 12)

Wadhwa N, Kunj N, Tiwari S, Saraiya M, Majumdar SS (2009) Optimization of embryo culture conditions for increasing efficiency of cloning in buffalo (*Bubalus bubalis*) and generation of transgenic embryos via cloning. Cloning Stem Cells 11(3):387–395. https://doi.org/10.1089/clo.2009.0003

Wang Z, Ruan J, Cui D (2009) Advances and prospect of nanotechnology in stem cells. Nanoscale Res Lett 4 (7):593–605. https://doi.org/10.1007/s11671-009-9292-z

Wang Y, Ding F, Wang T, Liu W, Lindquist S, Hernell O, Wang J, Li J, Li L, Zhao Y, Dai Y, Li N (2017a) Purification and characterization of recombinant human bile salt-stimulated lipase expressed in milk of transgenic cloned cows. PLoS ONE 12(5): e0176864

Wang Y, Zhao X, Du W, Liu J, Chen W, Sun C, Cui B, Zeng Z, Shen Y, Gao F, Wang A, Liu G, Cui H (2017b) Production of transgenic mice through sperm-mediated gene transfer using magnetic nano-carriers. J Biomed Nanotechnol 13:1673–1681. https://doi.org/10.1166/jbn.2017.2456

Wong JKL, Mohseni R, Hamidieh AA, MacLaren RE, Habib N, Seifalian AM (2017) Will nanotechnology bring new hope for gene delivery? Trends Biotechnol 35(5):434–451. https://doi.org/10.1016/j.tibtech.2016.12.009 (Epub 2017 Jan 18. Review)

Woodfint RM, Hamlin E, Lee K (2018) Avian bioreactor systems: a review. Mol Biotechnol https://doi.org/10.1007/s12033-018-0128-x (Epub ahead of print. Review)

Wright G, Carver A, Cottom D, Reeves D, Scott A, Simons P, Wilmut I, Garner I, Colman A (1991) High level expression of active human alpha-1-antitrypsin in the milk of transgenic sheep. Biotechnology (N Y) 9(9):830–834

Wu X, Ouyang H, Duan B, Pang D, Zhang L, Yuan T, Xue L, Ni D, Cheng L, Dong S, Wei Z, Li L, Yu M, Sun QY, Chen DY, Lai L, Dai Y, Li GP (2012) Production of cloned transgenic cow expressing omega-3 fatty acids. Transgenic Res 21(3):537–543. https://doi.org/10.1007/s11248-011-9554-2 (Epub 2011 Sep 15)

Wu J, Platero-Luengo A, Sakurai M, Sugawara A, Gil MA, Yamauchi T, Suzuki K, Bogliotti YS, Cuello C, Morales Valencia M, Okumura D, Luo J, Vilariño M, Parrilla I, Soto DA, Martinez CA, Hishida T, Sánchez-Bautista S, Martinez-Martinez ML, Wang H, Nohalez A, Aizawa E, Martinez-Redondo P, Ocampo A, Reddy P, Roca J, Maga EA, Esteban CR, Berggren WT, Nuñez Delicado E, Lajara J, Guillen I, Guillen P, Campistol JM, Martinez EA, Ross PJ, Izpisua Belmonte JC (2017) Interspecies chimerism with mammalian pluripotent stem cells. Cell 168(3):473–486.e15. https://doi.org/10.1016/j.cell.2016.12.036

Wurm FM (2004) Production of recombinant protein therapeutics in cultivated mammalian cells. Nat Biotechnol 22:1393–1398

Yadav PS, Singh RK, Singh B (2012) Fetal stem cells in farm animals: applications in health and production. Agric Res 1:67–77

Yang P, Wang J, Gong G, Sun X, Zhang R, Du Z, Liu Y, Li R, Ding F, Tang B, Dai Y, Li N (2008) Cattle mammary bioreactor generated by a novel procedure of transgenic cloning for large-scale production of functional human lectoferrin. PLoS 3:e3453

Yao YC, Han HB, Song XT, Deng SL, Liu YF, Lu MH, Zhang YH, Qi MY, He HJ, Wang SM, Liu GS, Li W, Lian ZX (2017) Growth performance, reproductive traits and offspring survivability of genetically modified rams overexpressing toll-like receptor 4. Theriogenology 1(96):103–110. https://doi.org/10.1016/j.theriogenology.2017.04.006 (Epub 2017 Apr 5)

Yu H, Chen J, Liu S, Zhang A, Xu X, Wang X, Lu P, Cheng G (2013) Large-scale production of functional human lysozyme in transgenic cloned goats. J Biotechnol pii:S0168–1656(13)00456-2. https://doi.org/10.1016/j.jbiotec.2013.10.023 (Epub ahead of print)

Zaniboni A, Spinaci M, Zannoni A, Bernardini C, Forni M, Bacci ML (2016) X and Y chromosome-bearing spermatozoa are equally able to uptake and internalize exogenous DNA by sperm-mediated gene transfer in swine. Res Vet Sci 104:1–3. https://doi.org/10.1016/j.rvsc.2015.11.001 (Epub 2015 Nov 12)

Zhang Y, Lu L (2008) Microinjection as a tool of mechanical delivery. Curr Opin Microbiotechnol 19:506–510

Zhang R, Tang C, Guo H, Tang B, Hou S, Zhao L, Wang J, Ding F, Zhao J, Wang H, Chen Z, Dai Y, Li N (2018a) A novel glycosylated anti-CD20 monoclonal antibody from transgenic cattle. Sci Rep 8(1):13208. https://doi.org/10.1038/s41598-018-31417-2

Zhang X, Li Z, Yang H, Liu D, Cai G, Li G, Mo J, Wang D, Zhong C, Wang H, Sun Y, Shi J, Zheng E, Meng F, Zhang M, He X, Zhou R, Zhang J, Huang M, Zhang R, Li N, Fan M, Yang J, Wu Z (2018b) Novel transgenic pigs with enhanced growth and reduced environmental impact. Elife. 7. pii: e34286. https://doi.org/10.7554/elife.34286

Zhu H, Hu L, Liu J, Chen H, Cui C, Song Y, Jin Y, Zhang Y (2016) Generation of β-lactoglobulin-modified transgenic goats by homologous recombination. FEBS J 283(24):4600–4613. https://doi.org/10.1111/febs.13950 (Epub 2016 Dec 5)

Animal Stem Cells—A Perspective on Their Use in Human Health

24

Abstract

Stem cells have sparked a revolution in biomedical and veterinary medicine. The past two decades have witnessed astounding innovations in pursuit of stem cell applications in livestock production and health. Stem cells are reported from various domestic animals. The stem cells in livestock species are important candidates for genomic testing, selection, genome engineering, and developing model animals for investigating human diseases. Mesenchymal stem cells, due to the ease of attainment, pluripotency, and better proliferation activity have emerged as clinically important cells for treating injuries in pet and companion animals. Improved cell culture techniques, culture media, and supplements, insights into gene-environmental interactions may solve current bottlenecks associated with segregation, description, and applications of stem cells in livestock.

Highlights

- Stem cell technology is an important branch of animal reproduction and health sciences
- Animal stem cells serve to enhance reproduction engineering and cell-based therapies.

Keywords

Stem cell types · Embryonic stem cells · Mesenchymal stem cells · Livestock species · Stem cell therapeutics

24.1 Introduction

Stem cells, by definition, are the totipotent or pluripotent undifferentiated cells with remarkable self-renewing capacity, and aptitude to differentiate into one or more specialized tissues. The stem cells residing in specific enable body niches to assist in recovery from ordinary "wear and tear" and healing the general injuries.

Science of stem cell biology began in the early 1960s, when it was noted that infusions of cells collected from bone marrow could salvage mice whose hematopoietic cells were inactivated by irradiation. However, factual stem cells' research started around two decades later when true stem cells were isolated from inner cell mass (ICM) cells of developing embryos (Evans and Kaufman 1981; Martin 1981). The cells were named as embryonic stem cells (ESCs).

Now, it is evident that stem cells can differentiate into multiple cell lineages, therefore gives rise to various types of tissues and organs during

B. Singh et al., *Advances in Animal Biotechnology*,
https://doi.org/10.1007/978-3-030-21309-1_24

the course of development. Multipotent stem cells differentiate into specific cell types upon transplantation in vivo or in vitro.

Surge in the research on stem cells in domestic animals is attributed to their prospects in regenerative and reproduction medicine, as already observed in small experimental animals and humans. Notably, stem cells technology has metamorphosed amazingly in domestic animals and provides opportunities to improve livestock production, health, studying cell lineage commitment and developmental biological phenomena. Current innovations in cell culturing and preservation, genome, transcriptome and proteome sequencing, cellular metabolomics, nanobiotechnology, and high-throughput screening will boost the impact of stem cells in animal health and production. Stem cells, viz ESCs, perinatal stem cells (amniotic fluid stem cells and cord blood stem cells), and mesenchymal stem cells (MSCs) with a varying range of stemness features and therapeutic relevance are reported from mammalian livestock, wild animals, poultry, and fish.

Breakthroughs in creating stem cells from normal cells by insertional mutagenesis of Yamanaka factors (Takahashi and Yamanaka 2006), or non-genetic factors, maintaining stem cells by innovative cell culture techniques and stabilizing molecules such as Rho-associated kinase (ROCK) inhibitor (Watanabe et al. 2007), have ushered the golden era of stem cell technology. The modus operandi is suitable to improve stem cell techniques in animals. Advanced platform of gene therapy, stem cell therapy, and gene modification offers possibilities for evolving therapies and cure animal diseases. For reasons that are unclear, production of ESCs from ICM of blastocysts and the epiblast of slightly older embryos is still elusive in ungulates. Attempts to establish true stem cell lines of domestic animals are unsuccessful in part due to lack of suitable culture protocols for concerned species.

24.2 Main Types of Animal Stem Cells

24.2.1 Embryonic Stem Cells

As the name indicates, ESCs cells are pluripotent cells derived from ICM cells of pre-implantation embryo. ESCs can be propagated in vitro, express pluripotency markers, and differentiate into multiple cell lineages in vitro or in vivo. However, it is practically intricate to differentiate a pluripotent stem cell into all cell types in vitro. ESCs can differentiate into embryoid bodies (EBs) representing all the three germ layers. ICM cells of developing blastocyst (Fig. 24.1) are the key sources of ESCs which serve as an important model of mammalian development and genetic manipulation.

Subcutaneous transplantation of ESCs in immune-deficient mouse develops teratoma, an in vivo indicator of their pluripotency.

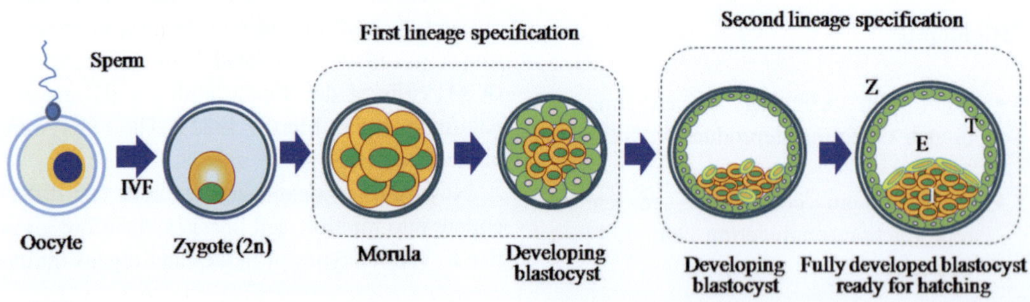

Fig. 24.1 Cellular differentiation during pre-implantation embryo development. ICM cells are the source of ESCs. Shortly after blastocyst stage, the ICM segregates into two layers: the hypoblast and the epiblast. The progenitor cells do exist in all organs and serve to replace the senescent or dead cells and maintain the functionality of the organs. True stem cells are reported only a few large animals

The cells are separated by mechanical separation, immunosurgery, and laser-assisted isolation. Stem cells exhibit a variable pattern of growth and differentiation. For instance, the mice ESCs grow faster in compact colonial groups or "nests" of cells that have a convex 3D shape and distinct and glistening edges. The ESCs' cells generally grow on top of feeder cells or in between feeder cells. If left undisturbed, they differentiate spontaneously at the periphery of colony forming a flat and irregularly cuboidal visceral endoderm.

ESCs may be totipotent or pluripotent as is evident from their ability to differentiate into a fetus and extraembryonic placental tissue. Stem cell lines can be propagated clonally in culture *ad infinitum*. These features make them promising candidates for genetic engineering, functional mammalian genetics, developmental biology, and creating animal models of human diseases.

A complex network of genetic and epigenetic interaction and signaling pathways are involved in maintaining pluripotency and self-renewal in stem cells. Scientists are piecing the cellular and molecular cascades that direct and maintain the stemness.

Nevertheless, the key aspects of animal stem cells, specifically the species-specific molecular signatures are not yet confirmed. It is pointed out that currently used human or murine stemness molecular markers are not fully relevant for livestock stem cells; hence, studies should focus on defining species-specific stemness markers.

24.2.2 Embryonic Stem Cell Research in Livestock Species

Ever since the first report on the establishment of putative ESC lines in pig (Evans et al. 1990), the stem cell technology has progressed in livestock species. There are multiple interests in obtaining pluripotent cells in animals with main expectations in genetic engineering and producing cloned transgenic animals (Table 24.1). Scientists and academic centers for stem cell research are using MSCs and ESCs to develop therapies for cancer, stroke, spinal cord injury (SCI), autoimmune diseases, and regeneration of bone, cartilage, and other organs.

Field of adult stem cell research in livestock species is increasing exponentially, and it encompasses objectives ranging from understanding the mechanisms of cellular development, creating organs for testing and evaluating drugs, and using stem cells for healing the injuries. A clear understanding of the immunology of allotransplantation as well as autotransplantation for malignancies and immunologically mediated diseases is a fundamental requirement of using stem cells as therapeutic candidates.

Isolating homogeneous cells from ICM cells is one of the major technological constraints in establishing the ESCs in most species, though innovative approaches such as "separate and seed" are suggested to improve the isolation of homogenous population of ESCs (Cao et al. 2009). Bovine ESC lines reported till date vary in morphology and marker expression, such as alkaline phosphatase (AP), stage-specific embryonic antigen-4 (SSEA-4), Nanog, and octamer-binding transcription factor (Oct-4). The bovine ESCs can grow in large, multicellular colonies resembling mice ESCs, and EGCs as well as human EGCs.

The long-lasting ESCs could be more suitable for genome editing by cutting edge techniques such as CRISPR/Cas9. Induced pluripotent bovine ESCs (bESCs) with stable morphology, karyotype, transcriptome, expression of pluripotency marker gene expression, and population doubling time are reported recently. bESCs could serve as competent donor cells to produce cloned embryos, thereby showing that bESCs could contribute to genomic selection and genome engineering (Bogliotti et al. 2018).

Not all of the stem cell research conducted in animals and using animal product is purely anthropocentric; some experiments are aimed to use them in health applications. When used as donor cells, the transgenic or genome-edited ESCs can enhance the efficiency of producing cloned animals.

Table 24.1 A summary of research achievements on stem cells in livestock species

Buffaloes	First reports on stem cell-like cells from IVF-derived embryos (Verma et al. 2007; Huang et al. 2010), development of multilineage ESCs from parthenogentic embryos (Sritanaudomchai et al. 2007), isolation of embryonic germ cell-like stem cells (Huang et al. 2007), in vitro culturing of spermatogonial stem cells (Xie et al. 2010), isolation and prolonged culturing of stem cell like AFSCs (Yadav et al. 2008a, b), expression of pluripotency-related genes in AFSCs (Mann et al. 2013; Yadav et al. 2011, 2012; Dev et al. 2012a, b), production of iPSCs (Deng et al. 2012)
Cattle	ESC-like cells in cattle (Stice et al. 1996), generation of cloned calves and transgenic chimeric embryos from bovine ESC-like cells (Cibelli et al. 1998; Iwasaki et al. 2000; Saito et al. 2003), elucidation of instability of Oct-4 expression in ESCs (Kurosaka et al. 2004; Yadav et al. 2005), comparative evaluation of expression of pluripotency-related genes during cattle ICM explants culture (Pant and Keefer 2009), isolation and culture of primary ESC by a novel method (Cao et al. 2009), culture of mammary epithelial stem cells (Li et al. 2009), isolation of bovine parthenogenetic ESCs from in vitro produced parthenotes (Pashaiasl et al. 2010), induction of pluripotency in quiescent cells by virus-free poly promoter under *chemically defin*ed conditions (Huang et al. 2011), derivation of germ cells from fetal testes (Fujihara et al. 2011), the studies revealed that compared to murine and swine, additional requirements are must for maintaining pluripotency in bovine ESCs (Maruotti et al. 2012) Identification of novel pluripotent SSEA (+) cells from a heterogenous population of fetal fibroblasts for enhancing efficiency of SCNT to(Pan et al. 2015), developing induced trophoblast cell lines from amnion-derived cells by doxycycline-inducible piggyback vectors (Kawaguchi et al. 2016) Establishing CDX2 knockdown ESC lines (Wu et al. 2016), derivation of ESCs with stable genome, karyotype, expression of pluripotency-regulating genes, and suitability as donor nuclei for producing SCNT cloned embryos, indicating that these cells might be promising for genome engineering (Bogliotti et al. 2018) Isolation of (bMSCs) from milk expressing pluripotency markers (CD90, CD73, and CD105), SOX2 and OCT-4, and differentiation into various cell lineages. The bMSCs cells might have role in regenerative medicine (Pipino et al. 2018)
Goat	Production of goat chimera by injecting EGSCs into a blastocyst (Jia et al. 2008), isolation of goat ESC-like cells (Yan et al. 2008; Pawar et al. 2009), development of enrichment protocols for muscle stem cells (Tripathi et al. 2010), identification of novel SSEA3 (+) stem cell-like cells from adult skin fibroblasts indicating that adult multipotent stem cells are present in goat skin that can be used to generate transgenic embryos (Yang et al. 2013), generating iPSCs from fibroblasts (Song et al. 2013), production of caprine iPSCs, and transgenic cloned dairy goat fibroblasts from fibroblasts (Song et al. 2016)
Sheep	Identification of embryonic cell lines (Meinecke-Tillmann and Meinecke 1996), ESC-like cells from in vitro-produced embryos, recommendation of Oct-4 as marker of stemness in sheep (Sanna et al. 2009), sheep placenta cotyledons revealed as abundant sources of non-invasive and pain- and risk-free source of MSCs (Ribitsch et al. 2017), studies proved the efficacy and applications of MSCs to heal superficial injuries and deep lesions (Martinello et al. 2018), reports on multipotent bone marrow-derived MSCs with clinical importance (Vivas et al. 2018)
Pig	First report on primitive ESCs (Evans et al. 1990), derivation of ESCs from culture of ICM cells (Piedrahita et al. 1990), isolation of pluripotent stem cells from cultured porcine primordial germ cells (Shim et al. 1997), isolation of pluripotent cell lines from embryos of diverse origins (Li et al. 2003), isolation of MSCs from blood and their use to produce cloned embryos (Faast et al. 2006), validation of therapeutic applications of MSCs in a porcine model (Dutton et al. 2010); derivation and characterization of iPSCs (Wu et al. 2009), stem Leydig cells reported from testicular interstitium (Yu et al. 2017), isolation and description of tendon-derived stem cells (TDSCs) with possible applications in tendon injuries (Yang et al. 2018)
Companion and pet animals	
Horse	First reports on ESC-like cells (Saito et al. 2002), isolation of homozygous stem cells (Lin et al. 2003), ESCs from ICM (Li et al. 2006), treating experimentally-induced tendinitis by adipose-derived nucleated fractions (Nixon et al. 2008), induction of pluripotency in adult equine somatic cells (Khodadadi et al. 2012), developing iPSCs by reprogramming equine fibroblasts using viral vectors coding for murine Oct-4, Sox2, c-Myc, and Klf4 sequences (Breton et al. 2013), isolation of iPSCs

(continued)

Table 24.1 (continued)

	from fibroblasts (Nagy et al. 2011), Producing equine-iPSCs by reprogramming equine adipose stem cells (e-ADSCs) by ectopic expression of Oct-4, Sox2, Klf4, and c-Myc by using a polycistronic lentiviral vector, and studying their ability to regenerate injured muscles (Lee et al. 2016), application of normoxic- or hypoxic-preconditioned umbilical cord-derived MSCs to heal wounds, inferring that MSC therapy could be promising in healing the distal extremity wounds (Textor et al. 2018)
Camel	Studies carries out to search for stem cells in camels which led to conclusion that cumulus cells of antral follicles are multipotent stem cell-like cells (Saadeldin et al. 2018)
Canines (Dog)	Induction of myogenesis and angiogenesis, and enhanced cardiac performance by combined cellular therapy to treat severe ischemic cardiomyopathy (Memon et al. 2005), description of adipose tissue-derived MSCs in serum-free culture medium (Liu et al. 2018), evaluating clinical potential of BMSCs, showing that these cells could serve as therapeutic candidates in cell-based therapies in disabling diseases such as SCI (Bhat et al. 2018), isolation of MSCs with higher chondrogenic potential from synovium, and their possible applications in canine cartilage regeneration for treating osteoarthritis (Sasaki et al. 2018)
Feline (Cat)	First report on bone marrow-derived (Martin et al. 2002), and adipose tissue-derived MSCs in cats (Webb et al. 2012), examining therapeutic traits of adipose-derived MSCs, and shoed that therapeutic potent is comparable to human MACs (Clark et al. 2017), comparative studies on clinical efficacy of fresh allogeneic versus fresh autologous adipose-derived MSCs for treating severe refractory feline chronic gingivostomatitis. The fresh autologous adipose-derived MSCs found to be more effective (Arzi et al. 2017)

24.2.3 Perinatal Stem Cells—Cord Blood, Amniotic and Placental Stem Cells

Fetus and placental tissue are inexhaustible alternative sources of multipurpose stem cells, also called as perinatal stem cells (Fig. 24.2). Fetal adnexa, entailing amniotic fluid (AF), umbilical cord blood, placenta and placental membrane, and Wharton's jelly long considered as waste products, offer significant and noncontroversial source of MSCs.

Important characteristics such as the similarity between term and mid-trimester amniotic fluid-derived stem cells, their multipotent behavior, and withstand cryostorage make perinatal stem cells candidates of choice for veterinary applications. Both ectodermal and mesenchymal cells are isolated from cells derived from full-term AF (Dolin et al. 2018).

The UCB is harvested, frozen, and stored as cord blood banks either as an individual resource (donor-specific source) or as a general resource for clinical applications. Advantages of cord blood include its availability, ease of harvest, and the reduced risk of graft-versus-host disease (GVHD). In addition, cord blood HSCs have a greater proliferative capacity than adult HSCs. Ex vivo expansion in tissue culture, to which cord blood cells are more amenable than adult cells, is another approach under active investigation. Umbilical cord matrix is a good source of stem cells. For this, umbilical cord (UC) is collected from neonates and processed for deriving hematopoietic cells from umbilical cord vein, and stem cell-like cells from Wharton's jelly (Yadav et al. 2012).

The AFSCs exhibit pluripotency and in vitro cell division (Dev et al. 2012a, b), and low immunogenicity. In livestock, perinatal stem cells are reported to be useful in cell-based therapies, SCNT, and genetic engineering. The initial studies on the characterization of the fetal stem cell in large herbivores and pet animals have raised the hopes of deriving somatic cells with clinical tissue engineering and therapeutic relevance from fetal stem cells (Fig. 24.3).

The bovine AFSC (bAFSCs) collected from abattoir-derived bovine fetuses exhibited fibroblast-like morphology, and expression of β-integrin, CD44, CD73, CD106, and Oct-4 but were negative for CD34 and CD45. The bAFSCs

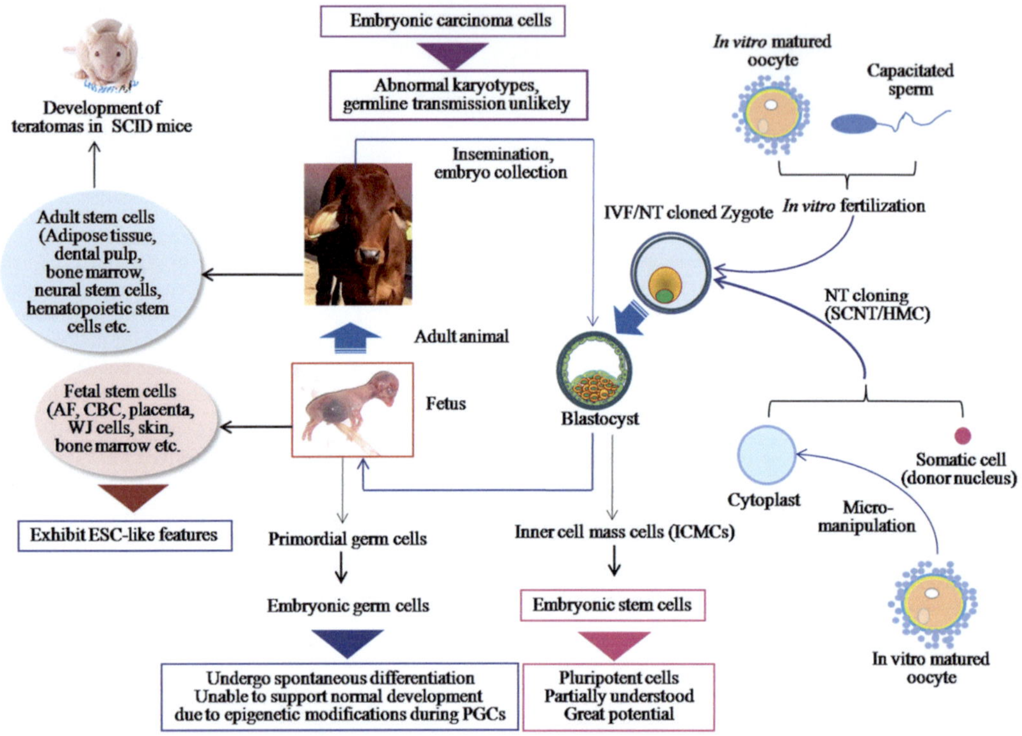

Fig. 24.2 Types, origins, and characteristics of pluripotent perinatal and adult MSCs in livestock species

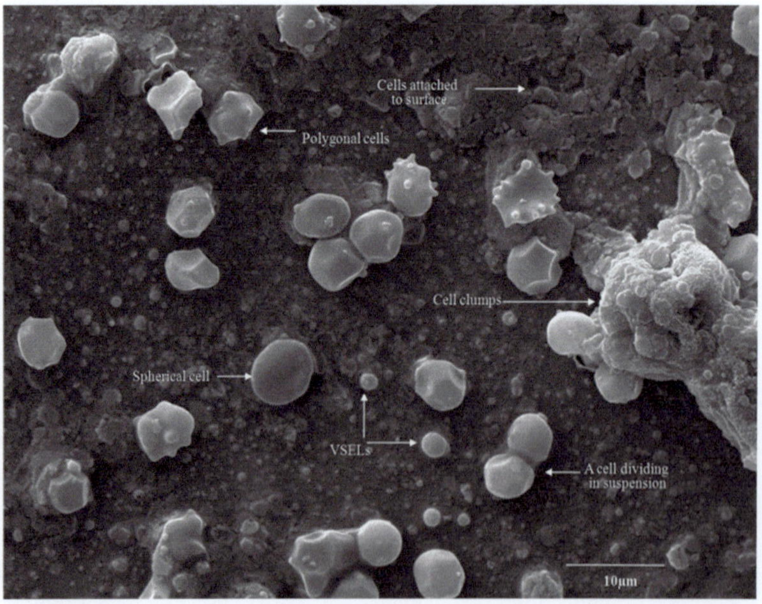

Fig. 24.3 Scanning electron micrograph of VSELs in buffaloes. Further studies are warranted to explore their properties, origin, and role in various livestock species

could alleviate bovine bilateral ovarian dystrophy and restore fertility in dairy cows (Chang et al. 2018).

24.2.4 Mesenchymal Stem Cells

Postnatal pluripotent MSCs are found in bone marrow, blood, skeletal muscles, adipose tissue, and skin tissue groups. These cells renew themselves and differentiate into specialized cell type in the concerned niches. Bone hematopoietic stem cells (HSCs) differentiate into blood cells, while others contribute to the development of skeleton or other types of cells, a phenomenon known as trans-differentiation or plasticity. For instance, spermatogonial stem cells originate from the primordial germ cells (PGCs), the progenitor cells which, in turn, are derived from the epiblast. PGCs are the single cells that under certain culture conditions form colonies of cells, that resemble undifferentiated ESCs. Spermatogonial stem cells (SSCs) are the only adult stem cells that contribute genetic material to the progeny. The spermatogonial stem cells have been extracted, cultured, and used for IVF or ICSI in many species including wild animal. SSCs are used to produce transgenic animals, albeit some limitations are there.

24.2.5 Leydig Stem Cells

Leydig cells located in testes produce testosterone. The adult Leydig cell population ultimately develops from undifferentiated mesenchymal-like stem. Leydig cells present in the interstitial compartment of early postnatal testis. Progenitor Leydig cells and immature Leydig cells serve as intermediate cells to produce fully developed or mature Leydig cells. Four distinct stages of adult Leydig cell development are: stem Leydig cells, progenitor Leydig cells, immature Leydig cells, and mature Leydig cells.

Stem Leydig cells renew themselves, differentiate into above cell types to replenish the Leydig cell niche. Progenitor Leydig cells give rise to immature Leydig cells which are spherical, contain larger quantities of smooth endoplasmic reticulum, and synthesize testosterone and testosterone-associated metabolites. Adult Leydig cells, which are distinct from the fetal Leydig cells, develop during puberty and supply testosterone for the onset of spermatogenesis (Chen et al. 2009). Studies on stem Leydig cells, the resulting cells from these cells have basic as well as clinical implications. Models have been developed based on Leydig cells for studying male infertility, associated issues and evolving therapies to cure male infertility (Arato et al. 2018).

24.2.6 Trophoblast Stem Cells

Trophoblast stem cells (TSCs), viewed as the developmental counterpart of ESCs, are precursors of the major differentiated cell types of the placenta (Tanaka et al. 1998). TSCs can be derived from outgrowths of either blastocyst polar trophectoderm (TE) or extraembryonic ectoderm (ExE), which originate from polar TE after implantation in uterine wall. The mouse TSCs niche appears to be located within the ExE adjacent to the epiblast, on which it depends for growth factors.

TSCs retain their indefinite self-renewal and differentiation repertoire into all trophoblast subtypes of the placenta. This is revealed from the production of animal chimeras by transplanting into ICM cells or grafting TSCs into pre-implantation embryos in mice, sheep, and goats (Ruffing et al. 1993; Latos and Hemberger 2016). Like other pluripotent stem cells, the TSCs serve as good model systems for studying trophoblast growth and differentiation (Rielland et al. 2008), CRISPR/Cas9-mediated genome editing, biochemical analysis, investigating transcriptional networks, characterizing epigenetic signatures, and examining the protein–protein interactions (Latos and Hemberger 2016).

Besides humans, TSCs, and cell lines are reported from a few model experimental species including mice and rhesus monkeys. Among ungulates, except cattle (Huang et al. 2014), the information is scarce on TSCs and their biology and applications. TSCs can be maintained as self-renewing stem cell populations in culture, and they retain their full differentiation repertoire in vitro or in vivo. Finally, differentiated somatic cells such as fibroblasts, amnion-derived bovine cells are reprogrammed into TSCs which new avenues for isolating TSCs from humans and other mammalian species (Kawaguchi et al. 2016).

It could be anticipated that pluripotent stem cells such as TSCs, ESCs, and artificial embryo-like entities and organoid cultures in combination with multiplex CRISPR/Cas9-based genome editing could provide a valuable platform to unravel the role of gap junction proteins such as connexins during lineage decision, differentiation, and morphogenesis in a cell culture model for mouse and human development.

24.2.7 Epiblast Stem Cells

Another category of stem cells, the epiblast stem cells (EpiSCs) are different from ESCs in terms of their origin and expression profile of markers of stemness and requirement of in vitro culture media and substrate for expansion and self-renewal. EpiSCs are derived from late epiblast of post-implantation embryo stage. It is noted that ESCs and EpiSCs share different signal transduction pathways, though ESCs and EpiSCs can be extracted with similar efficiency from IVF-derived and NT-cloned embryos.

Mice NT-EpiSCs are transcriptionally and epigenetically different from normal FT-EpiSCs, an observation that is in contrast to the findings observed with FT-ESCs versus NT-ESCs and suggests that pluripotent cell lines from NT cloned embryos at advanced developmental stages do retain the memory of epigenetic alterations that come into play during in vivo establishment of the epigenetic barrier in the epiblast during implantation (Maruotti et al. 2010). EpiSCs from murine post-implantation epiblasts are in a "primed" pluripotent state and can serve as a source for deriving other cells such as neural stem cells (Jang et al. 2014).

A novel class of epi-stem cells (epiSCs) was reported by (Brons et al. (2007) and Tesar et al. (2007). There are observable characteristics that distinguish this new stem cell type from germ cells. For example, EpiSCs did not express AP, or blimp1 and *stella*, demonstrating that EpiSCs are not derived from primordial germ layers. Under the present status of research, the information is scarce on EpiSCs in livestock species. It is expected that it might be easier to derive EpiSCs than true ESCs and discover the factors that regulate their pluripotency (Galli and Lazzari 2008).

24.2.8 Mammary Stem Cells

Milk plays vital roles in helping neonates to fulfill their genetic destiny. Besides, milk is the source of beneficial bacteria (probiotics), prebiotics and somatic cells. Majority of the cells in milk are leucocytes.

Indeed, the somatic cell count is an indicator of the quality of milk. Mammary stem cells (MaSCs) are indispensable for the growth of mammary tissue, renewal, and turnover of mammary epithelial cells. MaSCs are multipotent cells, which give rise to other cells with a specific function. These cells produce a lineage of daughter cells with a unidirectional terminal differentiation process.

MaSCs are potential targets for strategies to boost milk production in milch animals. Appropriate manipulation of MaSCs can potentially benefit milk production, dry-period management and used to revitalize the mammary tissue (Capuco et al. 2012). In addition, the MaSCs are of considerable interest to developmental biologists and cancer researchers. Due to the resemblance in cyto-architecture of animal mammary cells and human breasts, the MaSCs could serve as model for research on human breast cancer.

However, MaSCs are slow-cycling cells. Once the stem cell commits to differentiation, it enters a brief period of rapid proliferation. Asymmetrical cell division generates one daughter stem cell and one progenitor cell. This process allows for the maintenance of a stem cell pool for future growth while simultaneously generating a differentiated cell population (Stingl 2009). Possibilities that mammary stem cells can influence milk production, might be an important subject of scientific inputs for qualitative and quantitative alterations in milk production.

24.2.9 Very Small Embryonic Stem Cell-like Cells

In addition to conventional ESCs of ICM cells of blastocysts, bone marrow, cord blood, and AFSCs, a relatively new population of small-sized pluripotent stem cell-like cells, also called as very small embryonic stem cell-like cells (VSELs) are also noted in some species. Though reports are scarce on these type cells, they might be of significant concern.

We propose that these cells need further studies. In buffalo amnion, we have noted this population of cells (Fig. 24.4). Properties of VSELs, such as formation of teratoma, contribution to chimera formation are not yet reported. In mice also, similar cells are described as a rare population of small cells which over express pluripotency factors (Oct-4, Nanog, and Sox2) and maintain a demethylated Oct-4 promoter. A recent report delineates the characteristics and the possible role of VSELs as a backup pool for adult stem cells and host homeostasis, ***and that deficiency of*** VSELs ***may result in some cancer types*** (Bhartiya et al. 2016; Bhartiya and James 2017).

Further, like PGCs, the VSELs are quiescent cells, they do not readily multiply in culture, nor produce teratoma or integrate into developing embryos. It is thought that VSELs that survive oncotherapy can be targeted to induce endogenous regeneration of non-functional gonads (Bhartiya et al. 2016).

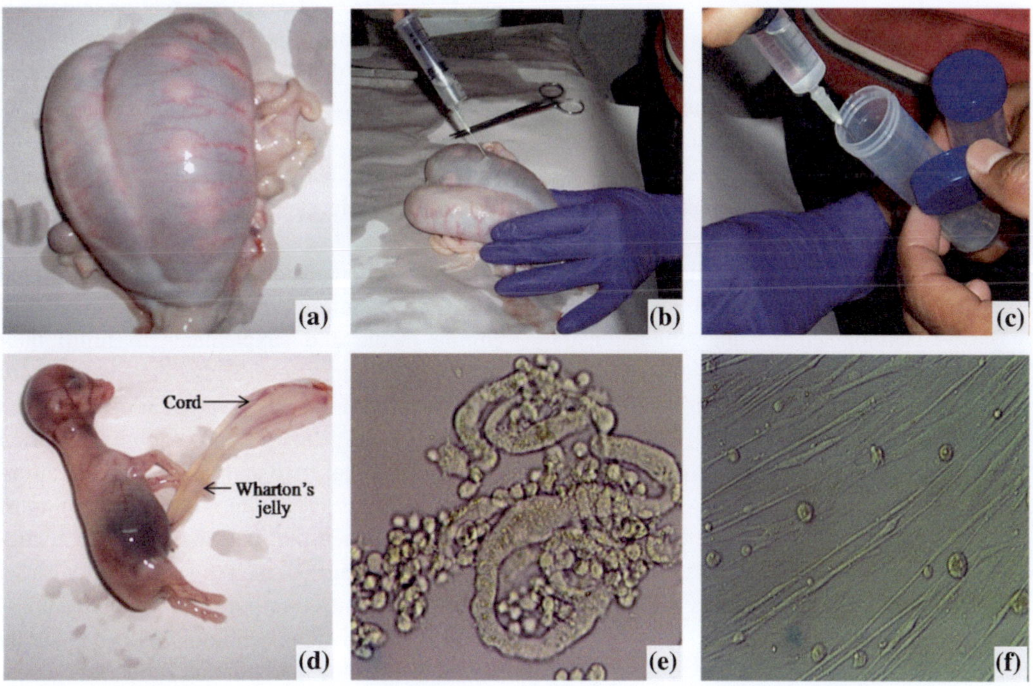

Fig. 24.4 Establishing perinatal stem cells. **a** caprine fetus; **b** and **c** collection of AF to culture AF cells; **d** young fetus with naval cord to collect cord blood stem cells and WJ cells; **e** partially trypsinized fetal stem cell monolayer showing various types of cells; **f** a cell monolayer after repeated passaging. Some very small embryonic stem cell-like cells are visible in cell monolayer (**f**)

24.2.10 Induced Pluripotent Stem Cells

A new era of stem cell technology started with the induced pluripotency through ectopic expression of genetic factors into murine skin fibroblasts (Takahashi and Yamanaka 2006). The method was termed "direct reprogramming" as it relied on forced reprogramming of differentiated somatic cells. The resulting cells were dubbed as induced pluripotent stem cells (iPSCs).

Technology has immensely progressed stem cell biology in ruminants, companion, and pet animals. Future improvements may overcome the problems in establishing pluripotent stem cells in livestock species. iPSCs have been reported from various livestock species including cattle, goats, sheep, pigs, and buffaloes (Table 24.1).

Box 1. Prospects of Stem Cell Science in Livestock

Validating cure of diseases in model animals.
Evolving stem cell-based therapies to treat tissue and organ injuries and damage in equines and camels.
Treating male fertility through spermatogonial stem cell transplantation.
Producing cloned transgenic animals, and/or transgenic chimera and animals.
Basic developmental and biomedical investigations.
In vitro meat production from the muscle stem cells.
Regeneration of damaged ligaments in species like equine or camels.
Using stem cells for testing of drugs and therapeutics intended for human use.
Using inferences from animal studies for human welfare.

24.3 Stem Cell Research in Livestock

A number of robust techniques are employed for isolation, culturing, and characterization of stem cells or stem cell-like cells in animals. Field of adult MSCs research is increasing exponentially and encompasses a wide range of topics—from deepening our understanding of cellular development to applying these findings to repair and create the organs.

Stem cell research in farm animals such as buffaloes, goats, and cattle is of interest for similar reasons as shown for mice and humans. Clinicians use adult stem cells to develop therapies for cancer, stroke, SCI, autoimmune diseases, and regeneration of bone, cartilage, and other tissues. A clear understanding of immunology for allotransplantation and autotransplantation for malignancies and immune-mediated diseases is essentially required.

Potential improvements in stem cells biology would come in the form of a reduction in number of animals required for research and development of cell-based therapies, generating transgenic animals or developing disease-resistance livestock herds.

Unlike mice and humans, true stem cell lines are not available in livestock species. Obtaining pure stem cell lines is tricky in animals due to generous differences in genetic makeup, and proteome, therefore differential requirements for culture media and supplements to fulfill their metabolic requirements. Yadav et al. (2005), for example, found that the expression of Oct-4 was optimal in the primary cultures of bovine ICM cells, which reduced to undetectable levels during subsequent passages. It implies that the utility of Oct-4 as stem cell-pluripotency marker is restricted to certain stages of their growth in vitro. Rapid loss of Oct-4 and pluripotency in cultured blastocysts and derivative cell lines are also recorded in rodents, bovine, and porcine epiblasts.

Nevertheless, the key aspects of animals ESCs specifically the identification of species-specific ESC-markers should be worked out. It has been pointed out that currently used human or murine ESC-specific molecular markers are not specific for bovine ESCs; hence, further research is warranted on the identification of valid stem cell markers in animals (Forest 1983). Despite inevitable hurdles, significant progress has been made in establishing stem cells from bovine, sheep and porcine, pet and companion animals (Table 24.1).

24.4 Stem Cell Therapy and Veterinary Clinical Applications

Stem cell therapy refers to the use of stem cell (e.g., bone marrow MSCs, adipose-origin MSCs, or umbilical stem cells) to treat disorders, diseases or disabilities. Stem cell therapy has important applications in veterinary medicine. Compared to humans, the progress in stem cell therapy and availability of commercial stem cells are slow in veterinary sciences.

Mostly rodents and primates are used as model animals for investigating human diseases and pathologies. Research using large animals could be promising as domestic animals and humans share many basal cellular metabolism rates, longer life span, comparable organ size, physiology and maintaining health style in response to environmental factors.

Among stem cells, the MSCs are extensively investigated for their clinical applications in veterinary medicine (Gugjoo et al. 2018; Bhat et al. 2018; Sultana et al. 2018). Treating chronic disorders and acute injuries are the important areas of stem cell applications in pets and companion animals such as horse. Application of normoxic- or hypoxic-preconditioned umbilical cord-derived MSCs to heal wounds shows that MSC therapy could be a promising approach in healing distal extremity wounds in equines (Yadav et al. 2012; Textor et al. 2018).

Pluripotent canine bone marrow-derived stem cells (BMSCs) were tested for their in vivo tracking in mice, and clinical evaluation in canine clinical paraplegia cases. When injected around the wound, the BMSCs were found to migrate and concentrate predominantly toward the center of the wound. The study reveals that BMSCs could serve as valuable therapeutic candidates for cell-based therapies such as spinal injuries in pets (Bhat et al. 2018).

Similarly, stem cell therapy is of interest in feline medicine. Specifically, the MSCs from bone marrow or adipose tissue were found to be useful for pet animals (Webb and Webb 2015; Quimby and Borjesson 2018).

Comparative studies on clinical efficacy of fresh allogeneic versus fresh autologous adipose-derived MSCs for treating severe refractory feline chronic gingivostomatitis showed that fresh autologous adipose-derived MSCs were more effective (Arzi et al. 2017). Allogeneic adipose-derived MSCs had been investigated for acute kidney injury (AKI) in ischemic kidney feline model. Time of treatment, but not treatment, had a significant effect on renal function. The results, however, did not support the use of allogeneic MSCs in AKI in this case. Kind of renal injury, stem cell dosage, MSC allogenicity, duration, and route or timing of administration could influence their efficacy (Rosselli et al. 2016).

Stem cell therapies are developed as baseline treatments for injuries such as tendon and ligament damage, orthopedic diseases, myocardial infarction, stroke, osteoarthritis, osteochondrosis, and muscular dystrophy (Yadav et al. 2012). Stem cells ***supported*** by certain transcription factors, growth factors (e.g., cytokines that promote cellular proliferation, and suppress inflammation), could have a role in functional regeneration of damaged tendon in equine (Conrad et al. 2018).

Equine-iPSCs (e-iPSCs) were generated by reprogramming skin fibroblasts (Breton et al. 2013), and equine adipose stem cells (e-ADSCs) by ectopic expression of Oct-4, Sox2, Klf4, and c-Myc using a polycistronic lentiviral vector. e-iPSCs were able to regenerate injured muscles (Lee et al. 2016). The iPSCs in equine is an important step toward understanding pluripotency and establishes iPSC technology as a tool for research and clinical applications in veterinary biomedicine.

In addition, the stem cells have evoked excitement in promoting cell-based therapies in pet animal health care for which current therapeutic strategies have minimal effectiveness. Combined autologous cellular therapy could induce myogenesis and angiogenesis and enhance cardiac performance, therefore suggesting that stem cells could have promised to treat severe ischemic cardiomyopathy in dogs (Memon et al. 2005).

Stem cells from pig, cattle, goat, and sheep might be useful for improving animal health and facilitate the development of animal models for cell-based therapies. Some salient advances and clinical achievements using stem cells in veterinary health are summarized in Table 24.1.

24.5 Stem Cells and in Vitro Derivation of Gametes

Stem cell research has opened new perspectives for regenerative and reproductive medicine. In vitro derivation of germ cells and viable gametes is a recently emerging field and a promising area of mammalian-assisted reproduction.

Research is going on to identify the precise mechanisms by which the differentiation of pluripotent stem cells into oocytes or sperm could be monitored. While evidences are there that ESCs can produce PGCs and germ cells, there remains a question as to whether the germ cells can be derived from adult stem cells outside the gonads. Germ cells are already reported from bone marrow (Shirazi et al. 2017), and fetal skin cells in some model animas (Dyce et al. 2006). It is anticipated that differentiation of ESCs into PGCs and then to oogonia and oocyte could herald a new hope in the area of animal-assisted reproduction and conservation of endangered animals. Also, the technology may offer ample opportunities for deriving oocytes in vitro for procedural refinement of genetic engineering where scarcity of oocytes is a limiting factor.

Differentiation of PGCs and germ cells from pluripotent stem cells has potential of becoming a reliable source of oocytes for research applications. Success was achieved for the first time in reprogramming mice ESCs and iPSCs into fully functional oocytes in vitro. Fully potent matured mouse oocytes were generated from ESCs, and iPSCs derived from fibroblasts (Hikabe et al. 2016). Process entailing in vitro differentiation (IVDi), in vitro growth (IVG), and in vitro maturation (IVM), which in total took around five weeks. Oocytes derived from both the sources produced embryos and healthy pups (Hayashi et al. 2017). However, till date, the information is scarce on oocytes or sperm derived from ESCs or iPSCs in large ruminants.

24.6 Animal Stem Cells in Human Health

Stem cell therapy is a promising advanced scientific topic and has contributed to an unprecedented revolution of science and medicine. Different types of human stem cells are developed for therapeutic applications. Stem cell treatment is going to be available for spinal cord injury in Japan (Cyranoski 2019).

Currently, there is a range of stem cells and their products for use as regenerative medicine (Harding et al. 2013). Animals are used as sources of stem cells for validating therapeutic applications of stem cells and screening drugs and pharmaceuticals. Some animals, such as pigs share physiological and anatomical similarities with humans and bridges the gap between rodent studies and human trials. This makes porcine iPSCs (piPSCs) very attractive tools for modeling human cell therapies, and testing the safety of iPSCs.

It is most important to understand full spectrum safety and therapeutic efficacy of stem cells intended for clinical applications. However, most of the studies are carried out using rodents as extant model animals. However, rodents as model animals for humans have certain limitations such as body size and physiology relative to humans. Larger animals such as rabbits, dogs, pigs, goats, sheep, and non-human primates are comparatively better predictors of responses.

24.7 Porcine Stem Cells in Human and Biomedical Sciences

A number of reports support the global use of pig as model animal for developmental biological and clinical studies, and bone tissue engineering (Cui et al. 2018). Wang et al. (2018) and Wu et al. (2019) achieved orthotopic whole tooth regeneration using allogenic cell re-association approach in swine model, supporting the potential of stem cells for regeneration teeth in situ.

MSCs from bronchoalveolar lavage fluid (BAL) of commercial pigs attenuated lipopolysaccharide (LPS)-induced acute lung injury (ALI) in a pig model. The study suggests that BAL-MSCs may be used in clinical settings to treat ALI in humans (Khatri and Richardson 2018).

Pig colonic stem cells (ASCL-2 and BM-1) sharing resemblance with human colonic stem cells can aid in the developing preventive strategies against gut bacterial dysbiosis and inflammation-promoted diseases, such as colon cancer (Charepalli et al. 2017). Adipose-derived porcine stem cells have potential as model cells in human clinical studies (Arrizabalaga and Nollert 2017).

24.8 Outlook and Challenges

There is considerable interest in pursuing stem cell research in domestic animals with economic value. Derivation and stable propagation of pluripotent ESCs in large ungulates are not fully successful despite the fact that the landscape of stem cell biology has radically changed during the past two decades.

Deriving and characterizing stem cells, originally quite finicky, is rather efficient and easier now. Today, the scientists know how to coax natural pluripotent stem cells or iPSCs into becoming specialized cells such as neurons, insulin-producing pancreatic cells, blood-producing hematopoietic cells or germ cells. Using induced pluripotency, it is possible to generate cells that are immunocompatible. Therefore, irrespective of their origins, the stem cells might have therapeutic potential, but it is necessary that associated pros and cons should be evaluated utterly.

It is likely that culture conditions developed for a particular species may not support the growth of stem cells in other species, which differ in factors and cellular signaling cascade indispensable for survival and maintenance of stem cell pluripotency. Therefore, it is necessary to focus on nutrient–gene interactions and accordingly evolve culture media for stem cells in different species.

Molecular markers for species specificity and stem cell pluripotency are not amply identified. Currently, methods are available to use small molecules and non-genetic elements to reprogram the somatic cells iPSCs. Hence, compared with MSCs and ESCs, the iPSCs seem to be promising as they will be immunocompatible to the recipients. Besides, there are gaps in the literature concerning ethics of using animal stem cells in human disease and injuries.

The potential improvements in stem cell biology would come in the form of reduction in number of animals used in drug testing, nutritional and immunological studies. In addition, the ESCs and perinatal stem cells may be useful for generating transgenic animals or developing disease-resistant cloned livestock herds.

24.9 Conclusions

The stem cells are found in all multicellular organisms and offer unparalleled information on early development of an individual. Remarkable feats are achieved in animal stem cells biology and engineering. The stem cells represent an attractive platform for examining basic human developmental biological processes and disease pathophysiology. The pig, as a large animal bridges the gap between rodent studies and human trials.

Whereas ESCs provide unparalleled information on early development, the perinatal stem cells and EpiSCs are important in vitro model for studying pluripotency in post-implantation epiblast tissues. It could be envisaged that adult MSCs, iPSCs, and fetal stem cells would provide potential sources of cells for regenerative treatment of cartilage and tendon repair. The optimisms is high for stem cell-based therapies that amend or minimize inflammatory cycle of disease, regenerate damaged tissue, or ideally both.

References

Arato I, Luca G, Mancuso F, Bellucci C, Lilli C, Calvitti M, Hansen BC, Milardi D, Grande G, Calafiore R (2018) An in vitro prototype of a porcine biomimetic testis-like cell culture system: a novel tool for the study of reassembled Sertoli and Leydig cells. Asian J Androl 20(2):160–165. https://doi.org/10.4103/aja.ja_47_17

Arrizabalaga JH, Nollert MU (2017) Properties of porcine adipose-derived stem cells and their applications in preclinical models. Adipocyte 6(3):217–223. https://doi.org/10.1080/21623945.2017.1312040 (Epub 2017 Mar 30. Review)

Arzi B, Clark KC, Sundaram A, Spriet M, Verstraete FJM, Walker NJ, Loscar MR, Fazel N, Murphy WJ, Vapniarsky N, Borjesson DL (2017) Therapeutic efficacy of fresh, allogeneic mesenchymal stem cells for severe refractory feline chronic gingivostomatitis. Stem Cells Transl Med 6(8):1710–1722. https://doi.org/10.1002/sctm.17-0035 (Epub 2017 Jun 15)

Bhartiya D, James K (2017) Very small embryonic-like stem cells (VSELs) in adult mouse uterine perimetrium and myometrium. J Ovarian Res 10(1):29. https://doi.org/10.1186/s13048-017-0324-5

Bhartiya D, Shaikh A, Anand S, Patel H, Kapoor S, Sriraman K, Parte S, Unni S (2016) Endogenous, very small embryonic-like stem cells: critical review, therapeutic potential and a look ahead. Hum Reprod Update 23(1):41–76 (Epub 2016 Sep 10. Review)

Bhat IA, T B S, Somal A, Pandey S, Bharti MK, Panda BSK, B I, Verma M, J A, Sonwane A, Kumar GS, Amarpal, Chandra V, Sharma GT (2018) An allogenic therapeutic strategy for canine spinal cord injury using mesenchymal stem cells. J Cell Physiol. https://doi.org/10.1002/jcp.27086

Bogliotti YS, Wu J, Vilarino M, Okamura D, Soto DA, Zhong C, Sakurai M, Sampaio RV, Suzuki K, Izpisua Belmonte JC, Ross PJ (2018) Efficient derivation of stable primed pluripotent embryonic stem cells from bovine blastocysts. Proc Natl Acad Sci U S A 115 (9):2090–2095. https://doi.org/10.1073/pnas.1716161115 (Epub 2018 Feb 9)

Breton A, Sharma R, Diaz AC, Parham AG, Graham A, Neil C, Whitelaw CB, Milne E, Donadeu FX (2013) Derivation and characterization of induced pluripotent stem cells from equine fibroblasts. Stem Cells Dev 22(4):611–621. https://doi.org/10.1089/scd.2012.0052 (Epub 2012 Sep 28)

Brons IG, Smithers LE, Trotter MW, Rugg-Gunn P, Sun B, de Sousa Chuva, Lopes SM, Howlett SK, Clarkson A, Ahrlund-Richter L, Pedersen RA, Vallier L (2007) Derivation of pluripotent epiblast stem cells from mammalian embryos. Nature 448 (7150):191–195 (Epub 2007 Jun 27)

Cao S, Wang F, Chen Z, Liu Z, Mei C, Wu H, Huang J, Li C, Zhou L, Liu L (2009) Isolation and culture of primary bovine embryonic stem cell colonies by novel method. J Exp Zool A Ecol Genet Physiol 311:368–376

Capuco AV, Choudhary RK, Daniels KM, Li RW, Evock-Clover CM (2012) Bovine mammary stem cells: cell biology meets production agriculture. Animal 6(3):382–393. https://doi.org/10.1017/S1751731111002369 (Review)

Chang LB, Peng SY, Chou CJ, Chen YJ, Shiu JS, Tu PA, Gao SX, Chen YC, Lin TK, Wu SC (2018) Therapeutic potential of amniotic fluid stem cells to treat bilateral ovarian dystrophy in dairy cows in a subtropical region. Reprod Domest Anim 53(2):433–441. https://doi.org/10.1111/rda.13123 (Epub 2017 Dec 14. Erratum in: Reprod Domest Anim. 2018 Aug; 53(4):1024)

Charepalli V, Reddivari L, Radhakrishnan S, Eriksson E, Xiao X, Kim SW, Shen F, Vijay-Kumar M, Li Q, Bhat VB, Knight R, Vanamala JKP (2017) Pigs, unlike mice, have two distinct colonic stem cell populations similar to humans that respond to high-calorie diet prior to insulin resistance. Cancer Prev Res (Phila) 10(8):442–450. https://doi.org/10.1158/1940-6207.CAPR-17-0010 (Epub 2017 Jun 2)

Chen H, Ge RS, Zirkin BR (2009) Leydig cells: from stem cells to aging. Mol Cell Endocrinol 306(1–2):9–16. https://doi.org/10.1016/j.mce.2009.01.023 (Epub 2009 Feb 7. Review)

Cibelli JB, Stice SL, Golueke PJ, Kane JJ, Jerry J, Blackwell C, Ponce de León FA, Robl JM (1998) Transgenic bovine chimeric offspring produced from somatic cell-derived stem-like cells. Nat Biotechnol 16 (7):642–646

Clark KC, Fierro FA, Ko EM, Walker NJ, Arzi B, Tepper CG, Dahlenburg H, Cicchetto A, Kol A, Marsh L, Murphy WJ, Fazel N, Borjesson DL (2017) Human and feline adipose-derived mesenchymal stem cells have comparable phenotype, immunomodulatory functions, and transcriptome. Stem Cell Res Ther 8(1):69. https://doi.org/10.1186/s13287-017-0528-z

Conrad S, Weber K, Walliser U, Geburek F, Skutella T (2018) Stem cell therapy for tendon regeneration: current status and future directions. Adv Exp Med Biol. https://doi.org/10.1007/5584_2018_194 (Epub ahead of print)

Cui Z, Li C, Jiang N, Zhang C, Wang Y, Gao H, Zhou Y (2018) Isolation and characterization of minipig perivascular stem cells for bone tissue engineering. Mol Med Rep 18(4):3555–3562. https://doi.org/10.3892/mmr.2018.9410 (Epub 2018 Aug 21)

Cyranoski D (2019) Japan's approval of stem-cell treatment for spinal-cord injury concerns scientists. Nature 565(7741):544–545. https://doi.org/10.1038/d41586-019-00178-x (No abstract available)

Deng Y, Liu Q, Luo C, Chen S, Li X, Wang C, Liu Z, Lei X, Zhang H, Sun H, Lu F, Jiang J, Shi D (2012) Generation of induced pluripotent stem cells from buffalo (*Bubalus bubalis*) fetal fibroblasts with buffalo defined factors. Stem Cells Dev 21:2484–2495

Dev K, Giri SK, Kumar A, Yadav A, Singh B, Gautam SK (2012a) Derivation, characterization and differentiation of buffalo (*Bubalus bubalis*) amniotic fluid derived stem cells. Reprod Domest Anim 47:704–711. https://doi.org/10.1111/j.1439-0531.2011.01947.x

Dev K, Giri SK, Kumar A, Yadav A, Singh B, Gautam SK (2012b) Expression of transcriptional factor genes (*Oct-4*, *Nanog*, and *Sox-2*) and embryonic stem cell-like characters in placental membrane of buffalo (*Bubalus bubalis*). J Membr Biol 245(4):177–183. https://doi.org/10.1007/s00232-012-9427-5 (Epub 2012 Apr 22)

Dolin CD, Chan MK, Basch RS, Young BK (2018) Human term amniotic fluid: a novel source of stem cells for regenerative medicine. Am J Obstet Gynecol 219(3):308–309. https://doi.org/10.1016/j.ajog.2018.05.035 (Epub 2018 Jun 2. No abstract available)

Dutton AQ, Choong PF, Goh JC, Lee EH, Hui JH (2010) Enhancement of miniscale repair in the avascular zone using mesenchymal stem cells in a porcine model. J Bone Joint Surg Br 92:169–175

Dyce PW, Wen L, Li J (2006) In vitro germline potential of stem cells derived from foetal porcine skin. Nat Cell Biol 8:384–390

Evans MJ, Kaufman MH (1981) Establishment in culture of pluripotent cells from mouse embryos. Nature 292:154–156

Evans MJ, Notarianni E, Laurie S, Moor RM (1990) Derivation and preliminary characterization of pluripotent cell lines from porcine and bovine blastocysts. Theriogenology 33:125–128

Faast R, Harrison SJ, Beebe LF, McIlfatrick SM, Ashman RJ, Nottle MB (2006) Use of adult mesenchymal stem cells isolated from bone marrow and blood for somatic cell nuclear transfer in pigs. Cloning Stem Cells 8(3):166–173

Forest MG (1983) Role of androgens in foetal and pubertal development. Horm Res 18:69–83

Fujihara M, Kim SM, Minami N, Yamada M, Imai H (2011) Characterization and in vitro culture of male germ cells from developing bovine testis. J Reprod Dev 57:355–364

Galli C, Lazzari G (2008) The manipulation of gametes and embryos in farm animals. Reprod Domest Anim 43(Suppl 2):1–7

Gugjoo MB, Amarpal, Chandra V, Wani MY, Dhama K, Sharma GT (2018) Mesenchymal stem cell research in veterinary medicine. Curr Stem Cell Res Ther. https://doi.org/10.2174/1574888x13666180517074444

Harding J, Roberts RM, Mirochnitchenko O (2013) Large animal models for stem cell therapy. Stem Cell Res Ther 4(2):23. https://doi.org/10.1186/scrt171 (Review)

Hayashi K, Hikabe O, Obata Y, Hirao Y (2017) Reconstitution of mouse oogenesis in a dish from pluripotent stem cells. Nat Protoc 12(9):1733–1744. https://doi.org/10.1038/nprot.2017.070 (Epub 2017 Aug 10)

Hikabe O, Hamazaki N, Nagamatsu G, Obata Y, Hirao Y, Hamada N, Shimamoto S, Imamura T, Nakashima K, Saitou M, Hayashi K (2016) Reconstitution in vitro of the entire cycle of the mouse female germ line. Nature 539(7628):299–303. https://doi.org/10.1038/nature20104 (Epub 2016 Oct 17)

Huang B, Xie TS, Shi DS, Li T, Wang XL, Mo Y, Wang ZQ, Li MM (2007) Isolation and characterization of EG-like cells from Chinese swamp buffalo (*Bubalus bubalis*). Cell Biol Int 31(10):1079–1088 (Epub 2007 Mar 19)

Huang B, Li T, Wang XL, Xie TS, Lu YQ, da Silva FM, Shi DS (2010) Generation and characterization of embryonic stem-like cell lines derived from in vitro fertilization buffalo (*Bubalus bubalis*) embryos. Reprod Domest Anim 45(1):122–128. https://doi.org/10.1111/j.1439-0531.2008.01268.x (Epub 2008 Dec 22)

Huang B, Li T, Alonso-Gonzalez L, Gorre R, Keatley S, Green A, Turner P, Kallingappa PK, Verma V, Oback B (2011) A virus-free poly-promoter vector induces pluripotency in quiescent bovine cells under chemically defined conditions of dual kinase inhibition. PLoS One 6(9):e24501. https://doi.org/10.1371/journal.pone.0024501 (Epub 2011 Sep 2)

Huang X, Han X, Uyunbilig B, Zhang M, Duo S, Zuo Y, Zhao Y, Yun T, Tai D, Wang C, Li J, Li X, Li R (2014) Establishment of bovine trophoblast stem-like cells from in vitro-produced blastocyst-stage embryos using two inhibitors. Stem Cells Dev 23(13):1501–1514. https://doi.org/10.1089/scd.2013.0329 (Epub 2014 Apr 10)

Iwasaki S, Campbell KH, Galli C, Akiyama K (2000) Production of live calves derived from embryonic stem-like cells aggregated with tetraploid embryos. Biol Reprod 62(2):470–475

Jang HJ, Kim JS, Choi HW, Jeon I, Choi S, Kim MJ, Song J, Do JT (2014) Neural stem cells derived from epiblast stem cells display distinctive properties. Stem Cell Res 12(2):506–516. https://doi.org/10.1016/j.scr.2013.12.012

Jia W, Yang W, Lei A, Gao Z, Yang C, Hua J, Huang W, Ma X, Wang H, Dou Z (2008) A caprine chimera produced by injection of embryonic germ cells into a blastocyst. Theriogenology 69(3):340–348 (Epub 2007 Nov 14)

Kawaguchi T, Cho D, Hayashi M, Tsukiyama T, Kimura K, Matsuyama S, Minami N, Yamada M, Imai H (2016) Derivation of Induced Trophoblast Cell Lines in Cattle by Doxycycline-Inducible piggyBac Vectors. PLoS One. 11(12):e0167550. https://doi.org/10.1371/journal.pone.0167550. eCollection 2016

Khatri M, Richardson LA (2018) Therapeutic potential of porcine bronchoalveolar fluid-derived mesenchymal stromal cells in a pig model of LPS-induced ALI. J Cell Physiol 233(7):5447–5457. https://doi.org/10.1002/jcp.26397

Khodadadi K, Sumer H, Pashaiasl M, Lim S, Williamson M, Verma PJ (2012) Induction of pluripotency in adult equine fibroblasts without c-MYC. Stem Cells Int 2012:429160

Kurosaka S, Sigrid E, McLaughlin KJ (2004) Pluripotent lineage definition in bovine embryos by *Oct-4* transcript localization. Biol Reprod 71:1578–1582

Latos PA, Hemberger M (2016) From the stem of the placental tree: trophoblast stem cells and their progeny. Development 143(20):3650–3660 (Review)

Lee EM, Kim AY, Lee EJ, Park JK, Park SI, Cho SG, Kim HK, Kim SY, Jeong KS (2016) Generation of equine-induced pluripotent stem cells and analysis of their therapeutic potential for muscle injuries. Cell Transplant 25(11):2003–2016. https://doi.org/10.3727/096368916X691691

Li M, Zhang D, Hou Y, Jiao L, Zheng X, Wang WH (2003) Isolation and culture of embryonic stem cells from porcine blastocysts. Mol Reprod Dev 65(4):429–434

Li X, Zhou SG, Imreh MP, Ahrlund-Richter L, Allen WR (2006) Horse embryonic stem cell lines from the proliferation of inner cell mass cells. Stem Cells Dev 15(4):523–531

Li JX, Zhang Y, Ma LB, Sun JH, Yin BY (2009) Isolation and culture of bovine mammary epithelial stem cells. J Vet Med Sci 71(1):15–19

Lin H, Lei J, Wininger D, Nguyen MT, Khanna R, Hartmann C, Yan WL, Huang SC (2003) Multilineage potential of homozygous stem cells derived from metaphase II oocytes. Stem Cells. 21(2):152–161

Liu Z, Screven R, Boxer L, Myers MJ, Devireddy LR (2018) Characterization of canine adipose-derived mesenchymal stromal/stem cells in serum-free medium. Tissue Eng Part C Methods 24(7):399–411. https://doi.org/10.1089/ten.TEC.2017.0409 (Epub 2018 Jun 20)

Mann A, Yadav RP, Singh J, Kumar D, Singh B, Yadav PS (2013) Culture, characterization and differentiation of cells from buffalo (*Bubalus bubalis*) amnion. Cytotechnology 65(1):23–30. https://doi.org/10.1007/s10616-012-9464-z (Epub 2012 Jul 21)

Martin G (1981) Isolation of a pluripotent cell line from early mouse embryos cultured in medium conditioned by teratocarcinoma stem cells. Proc Natl Acad Sci U S A 78:7634–7638

Martin DR, Cox NR, Hathcock TL, Niemeyer GP, Baker HJ (2002) Isolation and characterization of multipotential mesenchymal stem cells from feline bone marrow. Exp Hematol 30(8):879–886

Martinello T, Gomiero C, Perazzi A, Iacopetti I, Gemignani F, DeBenedictis GM, Ferro S, Zuin M, Martines E, Brun P, Maccatrozzo L, Chiers K, Spaas JH, Patruno M (2018) Allogeneic mesenchymal stem cells improve the wound healing process of sheep skin. BMC Vet Res 14(1):202. https://doi.org/10.1186/s12917-018-1527-8

Maruotti J, Dai XP, Brochard V, Jouneau L, Liu J, Bonnet-Garnier A, Jammes H, Vallier L, Brons IG, Pedersen R, Renard JP, Zhou Q, Jouneau A (2010) Nuclear transfer-derived epiblast stem cells are transcriptionally and epigenetically distinguishable from their fertilized-derived counterparts. Stem Cells 28 (4):743–752. https://doi.org/10.1002/stem.400

Maruotti J, Muñoz M, Degrelle SA, Gómez E, Louet C, Díez C, de Longchamp PH, Brochard V, Hue I, Caamaño JN, Jouneau A (2012) Efficient derivation of bovine embryonic stem cells needs more than active core pluripotency factors. Mol Reprod Dev. 79 (7):461–477. https://doi.org/10.1002/mrd.22051 (Epub 2012 May 31. Erratum in: Mol Reprod Dev. 2012 Dec; 79(12):888. Monforte, Carmen Díez [corrected to Díez, Carmen])

Meinecke-Tillmann S, Meinecke B (1996) Isolation of ES-like cell lines from ovine and caprine preimplantation embryos. J Anim Breed Genet 113:413–426

Memon IA, Sawa Y, Miyagawa S, Taketani S, Matsuda H (2005) Combined autologous cellular cardiomyoplasty with skeletal myoblasts and bone marrow cells in canine hearts for ischemic cardiomyopathy. J Thorac Cardiovasc Surg 130(3):646–653

Nagy K, Sung HK, Zhang P, Laflamme S, Vincent P, Agha-Mohammadi S, Woltjen K, Monetti C, Michael IP, Smith LC, Nagy A (2011) Induced pluripotent stem cell lines derived from equine fibroblasts. Stem Cell Rev 7(3):693–702. https://doi.org/10.1007/s12015-011-9239-5 (Erratum. In: Stem Cell Rev. 2012 Jun; 8(2):546)

Nixon AJ, Dahlgren LA, Haupt JL, Yeager AE, Ward DL (2008) Effect of adipose-derived nucleated cell fractions on tendon repair in horses with collagenase-induced tendinitis. Am J Vet Res 69(7): 928–937. https://doi.org/10.2460/ajvr.69.7.928

Pan S, Chen W, Liu X, Xiao J, Wang Y, Liu J, Du Y, Wang Y, Zhang Y (2015) Application of a novel population of multipotent stem cells derived from skin fibroblasts as donor cells in bovine SCNT. PLoS ONE. 10(1):e0114423. https://doi.org/10.1371/journal.pone.0114423 (eCollection 2015)

Pant D, Keefer CL (2009) Expression of pluripotency-related genes during bovine inner cell mass explant culture. Cloning Stem Cells 11(3): 355–365. https://doi.org/10.1089/clo.2008.0078

Pashaiasl M, Khodadadi K, Holland MK, Verma PJ (2010) The efficient generation of cell lines from bovine parthenotes. Cell Reprogram 12(5):571–579. https://doi.org/10.1089/cell.2009.0118

Pawar SS, Malakar D, De AK, Akshey YS (2009) Stem cell-like outgrowths from in vitro fertilized goat blastocysts. Indian J Exp Biol 47(8):635–642

Piedrahita JA, Anderson GB, Bondurant RH (1990) On the isolation of embryonic stem cells: comparative behavior of murine, porcine and ovine embryos. Theriogenology 34(5):879–901

Pipino C, Mandatori D, Buccella F, Lanuti P, Preziuso A, Castellani F, Grotta L, Di Tomo P, Marchetti S, Di Pietro N, Cichelli A, Pandolfi A, Martino G (2018) Identification and characterization of a stem cell-like population in bovine milk: a potential new source for regenerative medicine in veterinary. Stem Cells Dev. https://doi.org/10.1089/scd.2018.0114 [Epub ahead of print]

Quimby JM, Borjesson DL (2018) Mesenchymal stem cell therapy in cats: Current knowledge and future potential. J Feline Med Surg 20(3):208–216. https://doi.org/10.1177/1098612X18758590

Ribitsch I, Chang-Rodriguez S, Egerbacher M, Gabner S, Gueltekin S, Huber J, Schuster T, Jenner F (2017) Sheep placenta cotyledons: a noninvasive source of ovine mesenchymal stem cells. Tissue Eng Part C Methods 23(5):298–310. https://doi.org/10.1089/ten.tec.2017.0067

Rielland M, Hue I, Renard JP, Alice J (2008) Trophoblast stem cell derivation, cross-species comparison and use of nuclear transfer: new tools to study trophoblast growth and differentiation. Dev Biol 322(1):1–10. https://doi.org/10.1016/j.ydbio.2008.07.017 (Epub 2008 Jul 22. Review)

Rosselli DD, Mumaw JL, Dickerson V, Brown CA, Brown SA, Schmiedt CW (2016) Efficacy of allogeneic mesenchymal stem cell administration in a model of acute ischemic kidney injury in cats. Res Vet Sci 108:18–24. https://doi.org/10.1016/j.rvsc.2016.07.003 (Epub 2016 Jul 5)

Ruffing NA, Anderson GB, Bondurant RH, Currie WB, Pashen RL (1993) Effects of chimerism in sheep-goat concepti that developed from blastomere-aggregation embryos. Biol Reprod 48(4):889–904

Saadeldin IM, Swelum AA, Elsafadi M, Mahmood A, Alfayez M, Alowaimer AN (2018) Cumulus cells of camel (*Camelus dromedarius*) antral follicles are multipotent stem cells. Theriogenology 15(118):233–242. https://doi.org/10.1016/j.theriogenology.2018.06.009 (Epub 2018 Jun 19)

Saito S, Ugai H, Sawai K, Yamamoto Y, Minamihashi A, Kurosaka K, Kobayashi Y, Murata T, Obata Y, Yokoyama K (2002) Isolation of embryonic stem-like cells from equine blastocysts and their differentiation in vitro. FEBS Lett 531(3):389–396

Saito S, Sawai K, Ugai H, Moriyasu S, Minamihashi A, Yamamoto Y, Hirayama H, Kageyama S, Pan J, Murata T, Kobayashi Y, Obata Y, Yokoyama KK (2003) Generation of cloned calves and transgenic chimeric embryos from bovine embryonic stem-like cells. Biochem Biophys Res Commun 309(1):104–113

Sanna D, Sanna A, Mara L, Pilichi S, Mastinu A, Chessa F, Pani L, Dattena M (2009) *Oct-4* expression in in-vitro-produced sheep blastocysts and embryonic-stem-like cells. Cell Biol Int 34(1):53–60. https://doi.org/10.1042/CBI20090008

Sasaki A, Mizuno M, Ozeki N, Katano H, Otabe K, Tsuji K, Koga H, Mochizuki M, Sekiya I (2018) Canine mesenchymal stem cells from synovium have a higher chondrogenic potential than those from infrapatellar fat pad, adipose tissue, and bone marrow. PLoS ONE 13(8):e0202922. https://doi.org/10.1371/journal.pone.0202922 (eCollection 2018)

Shim H, Gutiérrez-Adán A, Chen LR, BonDurant RH, Behboodi E, Anderson GB (1997) Isolation of pluripotent stem cells from cultured porcine primordial germ cells. Biol Reprod 57(5):1089–1095

Shirazi R, Zarnani AH, Soleimani M, Nayernia K, Ragerdi Kashani I (2017) Differentiation of bone marrow-derived stage-specific embryonic antigen 1 positive pluripotent stem cells into male germ cells. Microsc Res Tech 80(4):430–440. https://doi.org/10.1002/jemt.22812 (Epub 2016 Dec 19)

Song H, Li H, Huang M, Xu D, Gu C, Wang Z, Dong F, Wang F (2013) Induced pluripotent stem cells from goat fibroblasts. Mol Reprod Dev 80(12):1009–1017. https://doi.org/10.1002/mrd.22266 (Epub 2013 Nov 27)

Song H, Li H, Huang M, Xu D, Wang Z, Wang F (2016) Big animal cloning using transgenic induced pluripotent stem cells: a case study of goat transgenic induced pluripotent stem cells. Cell Reprogram 18(1):37–47. https://doi.org/10.1089/cell.2015.0035

Sritanaudomchai H, Pavasuthipaisit K, Kitiyanant Y, Kupradinun P, Mitalipov S, Kusamran T (2007) Characterization and multilineage differentiation of embryonic stem cells derived from a buffalo parthenogenetic embryo. Mol Reprod Dev 74(10):1295–1302

Stice SL, Strelchenko NS, Keefer CL, Matthews L (1996) Pluripotent bovine embryonic cell lines direct embryonic development following nuclear transfer. Biol Reprod 54(1):100–110

Stingl J (2009) Detection and analysis of mammary gland stem cells. J Pathol 217(2):229–241. https://doi.org/10.1002/path.2457 (Review)

Sultana T, Lee S, Yoon HY, Lee JI (2018) Current status of canine umbilical cord blood-derived mesenchymal stem cells in veterinarymedicine. Stem Cells Int 15 (2018):8329174. https://doi.org/10.1155/2018/8329174 (eCollection 2018. Review)

Takahashi K, Yamanaka S (2006) Induction of pluripotent stem cells from mouse embryonic and adult fibroblast cultures by definedfactors. Cell 126(4):663–676 (Epub 2006 Aug 10)

Tanaka S, Kunath T, Hadjantonakis AK, Nagy A, Rossant J (1998) Promotion of trophoblast stem cell proliferation by FGF4. Science 282(5396):2072–2075

Tesar PJ, Chenoweth JG, Brook FA, Davies TJ, Evans EP, Mack DL, Gardner RL, McKay RD (2007) New cell lines from mouse epiblast share defining features with human embryonic stem cells. Nature 448(7150):196–199 (Epub 2007 Jun 27)

Textor JA, Clark KC, Walker NJ, Aristizobal FA, Kol A, LeJeune SS, Bledsoe A, Davidyan A, Gray SN, Bohannon-Worsley LK, Woolard KD, Borjesson DL (2018) Allogeneic stem cells alter gene expression and improve healing of distal limb wounds in horses. Stem Cells Transl Med 7(1):98–108. https://doi.org/10.1002/sctm.17-0071 (Epub 2017 Oct 24)

Tripathi AK, Ramani UV, Ahir VB, Rank DN, Joshi CG (2010) A modified enrichment protocol for adult caprine skeletal muscle stem cell. Cytotechnology 62 (6):483–488. https://doi.org/10.1007/s10616-010-9306-9 (Epub 2010 Sep 24)

Verma V, Gautam SK, Singh B, Manik RS, Palta P, Singla SK, Goswami SL, Chauhan MS (2007) Isolation and characterization of embryonic stem cell-like cells from in vitro-produced buffalo (*Bubalus bubalis*) embryos. Mol Reprod Dev 74(4):520–529

Vivas D, Caminal M, Oliver-Vila I, Vives J (2018) Derivation of multipotent mesenchymal stromal cells from ovine bone marrow. Curr Protoc Stem Cell Biol 44:2B.9.1–2B.9.22. https://doi.org/10.1002/cpsc.43

Wang F, Wu Z, Fan Z, Wu T, Wang J, Zhang C, Wang S (2018) The cell re-association-based whole-tooth regeneration strategies in large animal, Sus scrofa. Cell Prolif 51(4):e12479. https://doi.org/10.1111/cpr.12479 (Epub 2018 Jul 20)

Watanabe K, Ueno M, Kamiya D, Nishiyama A, Matsumura M, Wataya T, Takahashi JB, Nishikawa S, Nishikawa S, Muguruma K, Sasai Y (2007) A ROCK inhibitor permits survival of dissociated human embryonic stem cells. Nat Biotechnol 25(6):681–686 (Epub 2007 May 27)

Webb TL, Webb CB (2015) Stem cell therapy in cats with chronic enteropathy: a proof-of-concept study. J Feline Med Surg 17(10):901–908. https://doi.org/10.1177/1098612X14561105 (Epub 2014 Dec 5)

Webb TL, Quimby JM, Dow SW (2012) In vitro comparison of feline bone marrow-derived and adipose tissue-derived mesenchymal stem cells. J Feline Med Surg 14(2):165–168. https://doi.org/10.1177/1098612X11429224

Wu Z, Chen J, Ren J, Bao L, Liao J, Cui C, Rao L, Li H, Gu Y, Dai H, Zhu H, Teng X, Cheng L, Xiao L (2009) Generation of pig induced pluripotent stem cells with a drug-inducible system. J Mol Cell Biol 1(1):46–54. https://doi.org/10.1093/jmcb/mjp003 (Epub 2009 Jun 3)

Wu X, Song M, Yang X, Liu X, Liu K, Jiao C, Wang J, Bai C, Su G, Liu X, Li G (2016) Establishment of bovine embryonic stem cells after knockdown of CDX2. Sci Rep 20(6):28343. https://doi.org/10.1038/srep28343

Wu Z, Wang F, Fan Z, Wu T, He J, Wang J, Zhang C, Wang S (2019) Whole tooth regeneration by allogeneic cell reassociation in pig jawbone. Tissue Eng Part A. https://doi.org/10.1089/ten.tea.2018.0243 (Epub ahead of print)

Xie B, Qin Z, Huang B, Xie T, Yao H, Wei Y, Yang X, Shi D, Jiang H (2010) In vitro culture and differentiation and of buffalo (*Bubalus bubalis*) spermatogonia. Reprod Domest Anim 45:275–282

Yadav PS, Kues WA, Herrmann D, Carnwath JW, Niemann H (2005) Bovine ICM derived cells express the *Oct-4* ortholog. Mol Reprod Dev 72(2):182–190

Yadav RP, Yadav PS, Nanda T, Singh I (2008a) Isolation and culture of stem cells like cells from buffalo amnion. In: First international stem cell submit-2008, held at IIT, Chennai, 14–16 Nov 2008

Yadav PS, Tokas J, Sharma RK, Singh I, Sethi RK (2008b) Buffalo amniotic fluid, umbilical cord matrix, and early foetal explants as possible source of adult stem cells. In: IX annual conference of Indian society of animal genetics and breeding, held at NASC Complex, Delhi, July 3–4, 2008

Yadav PS, Mann A, Singh V, Yashveer S, Sharma RK, Singh I (2011) Expression of pluripotency genes in buffalo (*Bubalus bubalis*) amniotic fluid cells. Reprod Domest Anim 46(4):705–711. https://doi.org/10.1111/j.1439-0531.2010.01733.x (Epub 2010 Dec 30)

Yadav PS, Singh RK, Singh B (2012) Animal fetal stem cells-potential health applications. Agric Res 1: 67–77

Yan L, Lei L, Yang C, Gao Z, Lei A, Ma X, Dou Z (2008) Isolation and cultivation of goat embryo stem cells. Sheng Wu Gong Cheng Xue Bao 24:1670–1676

Yang Z, Liu J, Liu H, Qiu M, Liu Q, Zheng L, Pang M, Quan F, Zhang Y (2013) Isolation and characterization of SSEA3(+) stem cells derived from goat skin fibroblasts. Cell Reprogram 15(3):195–205. https://doi.org/10.1089/cell.2012.0080 (Epub 2013 May 13)

Yang J, Zhao Q, Wang K, Ma C, Liu H, Liu Y, Guan W (2018) Isolation, culture and biological characteristics of multipotent porcine tendon-derived stem cells. Int J Mol Med 41(6):3611–3619. https://doi.org/10.3892/ijmm.2018.3545 (Epub 2018 Mar 7)

Yu S, Zhang P, Dong W, Zeng W, Pan C (2017) Identification of stem leydig cells derived from pig testicular interstitium. Stem Cells Int 2017:2740272. https://doi.org/10.1155/2017/2740272 (Epub 2017 Jan 24)

Transgenesis and Poultry as Bioreactors

25

Abstract

The birds as live bioreactors have unique advantages. Short generation time, appropriate post-translational modification of newly synthesized proteins as opposed to prokaryotes and yeasts, higher titers of recombinant proteins produces compared to cultured animal cells, and secretion of transgenic biomolecules into sterile egg white or ovalbumin are the features that make poultry as a preferred system for producing transgenic therapeutic proteins.

Highlights

- Poultry birds are fine hosts for producing therapeutic proteins
- Remarkable progress has been made in developing transgenic birds and the therapeutic proteins from them.

Keywords

Avian transgenesis · Egg ovalbumin · Genome editing · Therapeutics · Germ cells

25.1 Introduction

Poultry is an important component of the human food chain, nutrition, and livelihood of marginal or landless families. Chickens, ducks, geese, pheasants or quails, turkeys, guinea fowl, and emus (Fig. 25.1) are reared for meat and egg production and are considered as poultry species of national economic interest.

The chicken believed to originate from Red Jungle Fowl over 8000 years ago, is now a main food animals all over the world. The poultry industry, at present, is one of the largest animal-based entrepreneurs, and has vast scopes of expansion and improvement in near future (Park et al. 2013, 2015).

With the discovery of the map of genomics, whole genome sequencing and a high-density SNP map in chickens, the avian biodiversity can be monitored and recognized rapidly and efficiently. This will facilitate poultry breeding programs based on the presence of candidate genes or molecular markers associated with production performance and quality parameters. Commercial companies have come up to develop high-yielding broiler chickens and laying hens using genomics and assisted reproduction techniques. Stem cells of various origins with potential and properties comparable to those obtained from mice, pig, and ruminants are reported from avian species of commercial interest (Ma et al. 2018; Houdebine 2018).

The current acquaintance with human and animal development is greatly advanced and aided by insights obtained from chick ESCs as model systems (Nishijima and Iijima 2013).

Due to unique advantages, and ease of management, the birds are preferred as the source of eggs, meat, and as host for producing transgenic

B. Singh et al., *Advances in Animal Biotechnology*,
https://doi.org/10.1007/978-3-030-21309-1_25

Fig. 25.1 Some important dual purpose poultry birds. **a** Domestic chicken; **b** Emu; **c** ducks; and **d** guinea fowl. Some other species such as quail are also used as meat producing animals

proteins. There are several advantages (Box 1) associated with the use of avian embryos, stem cells, and poultry birds as bioreactors and model animals. The birds, compared to bacteria, yeasts and mammalian cells are better suited for producing recombinant proteins. The recombinant proteins are released into egg white, an almost germ-free medium which makes it easy to purify them.

Box 1. Adaptive Features of Poultry Birds Making them as Appropriate Live Bioreactors

- Shorter generation interval, low cost management, and fecundity.
- Low age at maturity and short reproductive cycle.
- Large number of eggs obtained per bird.
- High protein contents in eggs.
- External embryo development and shorter incubation time to reach the hatching stage.
- Less time between generating transgenic strains and production of recombinant proteins.
- Ease of purifying recombinant proteins or enzymes from ovalbumin.

25.2 Transgenesis and Transgenic Chicken

The genetic makeup of farm animals, including chickens, is altered by selection and breeding for desirable traits. Chicken is one of the few animal species that is approved for producing recombinant proteins. The first transgenic chicken was developed by incorporating foreign genes into early chicken germ cell line through wild-type and avian leukosis viruses. Some of the viremic male chicks, designated as G0 or Generation-0, were mosaic, with the transmission frequency of proviral DNA ranging from 1 to 11% (Salter et al. 1987).

Since mid-2000, many transgenic avian varieties are generated using retroviral vectors and broad range promoters. The proteins produced using poultry species include human parathyroid hormone, human erythropoietin (hEPO), a green fluorescent protein (GFP), tumor necrosis factor (TNF) receptor/Fc fusion protein, and chimeric monoclonal antibodies (reviewed in Woodfint et al. 2018). Human lysosomal acid lipase (Kanuma$^{(R)}$) (sebelipase α) produced in chicken egg white has been approved for use by the US FDA (Sheridan 2016). This means, more recombinant therapeutic proteins have to be produced through transgenic chicken for human and veterinary applications.

Transgenic birds are cost-effective systems to produce pure, high-quality, biological active therapeutic cytokines (Herron et al. 2018). Transgenic broilers expressing enhanced green fluorescent protein (EGFP) were produced by microinjection of Tol2 plasmid-liposome complex into early embryonic dorsal aorta. The study reveals that appropriate egg windowing and administration of genes can improve the production of transgenic birds (Wang et al. 2018).

25.3 Methods of Generating Transgenic Cell Lines and Birds

Poultry birds are useful organisms for food and research. Transgenic techniques are aimed to improve the genetic make-up of existing strains or varieties for resistance to pathogens and parasitic infestation and lowering cholesterol levels in meat and eggs. Notably, in addition to the domestic chicken, some other avian species, such as quails (Sato and Lansford 2013), are produced and used to derive therapeutic proteins (Table 25.1).

In some cases, it is necessary to monitor embryo growth or introduce modified cells into embryo during development. This is achieved by a process called egg windowing. As chick develops inside the hard eggshell, it is necessary to open eggshell carefully at the broader end and reclose it to avoid perturbation in the growth of embryos due to contamination or infection.

Primordial germ cells (PGCs) obtained from embryos can be cultured in vitro and manipulated to express recombinant proteins. When reintroduced into other embryo, the transgenic PGCs contribute to germ line. The GSCs, can therefore contribute to the formation of germline chimera or transgenic birds (Collarini et al. 2019).

The techniques such as microinjection of gene or gene constructs, electroporation, virus vector-based gene transfer, spermatogonial stem cell (SSC)-mediated gene transfer, enzyme-based genome editing, and stem cell-based gene transfer (Fig. 25.2) are used to generate transgenic birds.

25.4 Viral-Vector Mediated Gene Transfer

Viruses have developed mechanisms to introduce their genetic material into host cells and utilize host machinery for their replication. In addition, it is possible to manipulate virus genes and pack the modified genes into the virus head or capsid. Hence, viruses are developed as vectors to transfer or introduce genes of interest into host cells or eggs.

Appreciable progress has been made in generating transgenic avian cells and birds using virus vectors. Vick et al. (1993), had for the first time reported chimera bird by use of PGCs obtained from the germinal crescent region and transfected with defective retroviruses. The birds could grow to sexual maturity and produce offspring. Retroviral vectors are among the most widely used vectors (Scott and Lois 2005; Sato and Lansford 2013; Ahn et al. 2015). Alternatively, genetically modified PGCs can be used to generate chimera or transgenic birds.

For this, PGCs are isolated from the developing embryos, purified, and acclimatized to

Table 25.1 Summary of the salient achievements in transgenic birds, and the therapeutics obtained from them

Year	Salient achievements and inferences (References)
1987	Transgenic chicken developed by insertion of genes into early chicken germ line through wild-type and avian leukosis viruses (Salter et al. 1987)
1993	First report on generation of chimera bird by using PGCs obtained from germinal crescent region and transfected with defective retroviruses. The birds could grow to sexual maturity, and produced offspring (Vick et al. 1993)
2003	Expression of biologically active glycosylated human interferon α-2b (hIFN) in transgenic hens, the study emphasizes opportunities of producing biologically active proteins using hens as bioreactors (Rapp et al. 2003)
2008	Production of chimeric chicken expressing human erythropoietin (hEPO) (Kodama et al. 2008), production of bioactive human granulocyte-colony stimulation factor (Kwon et al. 2008), TNF receptor/Fc fusion protein (Kyogoku et al. 2008) in transgenic birds, generating transgenic chicken expressing GFP in cock TTSSCs (Li et al. 2008)
2009	Tetracycline-dependent production of monoclonal antibodies by chimeric poultry (Kamihira et al. 2009)
2010	Expression of erythropoietin gene (Koo et al. 2010) and recombinant human erythropoietin/Fc fusion protein in transgenic chicken (Penno et al. 2010), description of stem cell-specific markers in SSCs, and production of EGFP-transgenic sperm by electroporation, liposome-mediated transfer and calcium acid phosphate precipitation (Yu et al. 2010)
2011	Expression of tetracycline-inducible GFP in transgenic chicken (Kwon et al. 2011)
2012	Genetic modification of chicken PGCs by piggyBac and Tol2 transposons inferring that transposable elements can serve as efficientvectors for genetic variation of chicken PGCs (Macdonald et al. 2012)
2013	Production of human urokinase-type plasminogen activator (Lee et al. 2013), development of Lipofectamine 2000 complexed with Tol2 transposon and transposase plasmids as a new method to stably transform avian PGCs in vivo, and using the transformed cells to generate germ line transgenic chicken (Tyack et al. 2013), description of immunoglobulin knockout chickens via homologous recombination in PGCs (Schusser et al. 2013), establishing a quail model, named Tg(tie1:H2B-eYFP), to study structure and blood vascular system in higher vertebrates (Sato and Lansford 2013)
2016	Utilizing CRISPR/Cas9 as feasible system for gene-targeting in chicken to produce ovomucoid chickens (Oishi et al. 2016), applications of CRISPR/Cas9 for targeted mutagenesis in chicken cells (Oishi et al. 2016), United States FDA approval for human lysosomal acid lipase (Kanuma$^{(R)}$) (sebelipase α) produced in the egg white (Sheridan 2016)
2018	Using CRISPR/Cas9 protocol to produce human interferon β (hIFN-β) in ovlabumin (Oishi et al. 2018), production of transgenic broilers expressing EGFP, by microinjection of Tol2 plasmid-liposome complex into the early embryonic dorsal aorta (Wang et al. 2018), production of transgenic chicken expressing IFIT5 to ameliorate pathogenic avian influenza, and velogenic Newcastle disease viruses (Rohaim et al. 2018), development of divalent vaccine against NDV and infectious bronchitis virus (IBV) (Shirvani et al. 2018)

grow in vitro. The culturable PGCs are transfected by viral vectors equipped with desired genes or by other means (Vick et al. 1993), and then transplanted into the developing embryos or injected into testes of chicks. Various vectors are used to transfer transgenes into PGCs (Schusser et al. 2013). Transgenic birds were produced by PGCs transformed by piggyback and Tol2 (Park and Han 2012; Macdonald et al. 2012; Tyack et al. 2013).

Some researchers have used lentiviral vectors to generate tissue-specific transgenic birds (which birds) (Scott and Lois 2005). Lipofectamine 2000 complexed with Tol2 transposon and transposase plasmids as a new method to stably transform avian PGCs in vivo and using the transformed PGCs to generate germ line transgenic chicken (Tyack et al. 2013). The transgenic hens were produced that have high concentrations (4810–6600 IU/ml, equivalent to

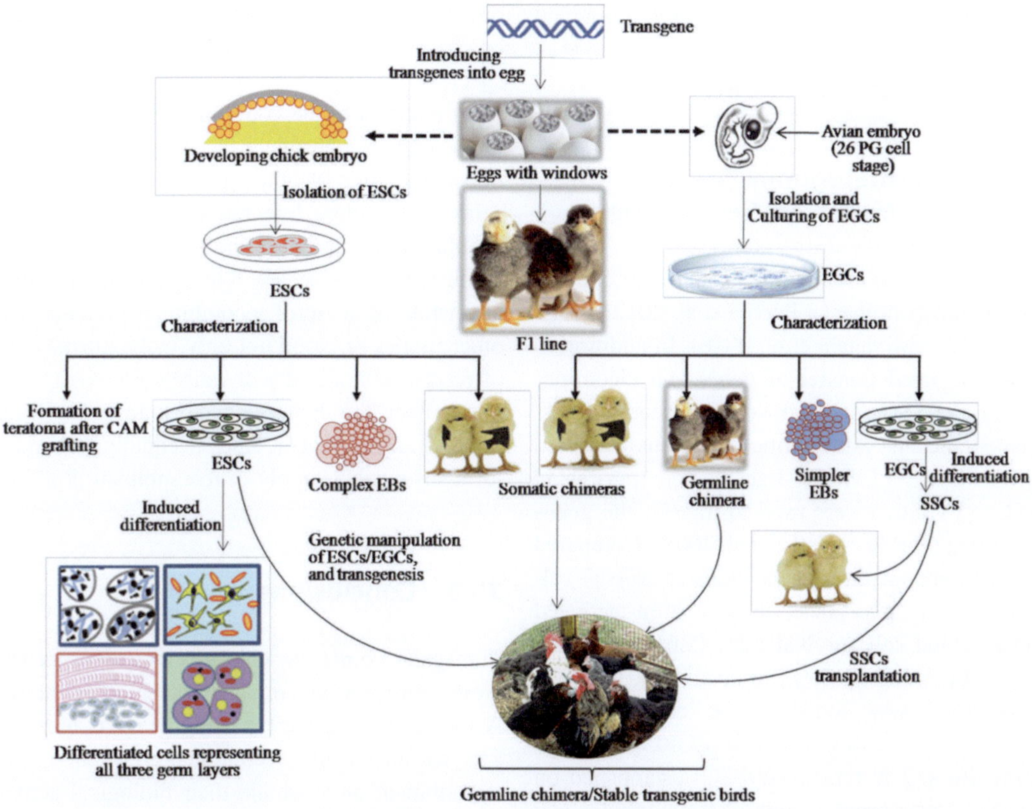

Fig. 25.2 Diagrammatic illustration of producing transgenic birds. The gene constructs are introduced by microinjection, electroporation, or virus-mediated gene transfection. SSC-mediated-gene transfer is another option to transmit transgenes. The egg has to be opened at one end to introduce transgenic cells or genes into the egg. *Abbreviations* EBs—embryoid bodies, ESCs—embryonic stem cells, EGCs—embryonic germ stem cells

40.1–55.0 μg/ml) of human erythropoietin (hEPO) in egg ovalbumin. Biological activities of transgenic hEPO and commercially available hEPO were similar (Kwon et al. 2018).

25.5 Genome Editing and Avian Transgenesis

Researchers have started using dominant genome and gene-editing tools such as CRISPR/Cas9 to produce genetically modified avian cell lines, germ cell lines, and chicken and quails. CRISPR/Cas9 protocol was used to produce hIFNβ-knock-in germline chimeric roosters by transplanting genetically modified cells into recipient embryos (Oishi et al. 2018). Two egg white genes, namely ovalbumin and ovomucoid in PGCs were mutagenized by transfection of a plasmid encoding Cas9, a single guide RNA, and a gene responsible for drug resistance. The CRISPR-induced mutated PGCs were transplanted into a chicken embryo, and finally a total of three germline chimeric roosters (G0) were obtained and were used to generate donor-derived offspring, named G1. Crossing G1 mutants provided a pure homozygous mutant offspring, named G2 mutants (Oishi et al. 2018). CRISPR/Cas9 was used to insert genes of interest, i.e., eGFP under the control of tissue-specific promoters (GAPDH) in PGCs aimed to produce transgenic birds. The effectiveness of

CRISPR/Cas9-mediated homology directed repair was increased up to 90% via G418 enrichment (Antonova et al. 2018).

25.6 Spermatogonial Stem Cell-Mediated Gene Transfer

The PGCs have multiple applications in animals and avians (Chojnacka-Puchta et al. 2012; Tyack et al. 2013; Miyahara et al. 2016). In addition to virus-mediated transfer of genes and electroporation, the SSCs are also used to generate transgenic chicken. Also termed as testis-mediated gene transfer (TMGT), transplanted transfected SSCs (TTSSCs) were used to transfer the genes encoding GFP in cock (Li et al. 2008). Compared to liposome-mediated gene transfer, versus calcium acid phosphate precipitation-mediated transfer and cell survival rates (5.61%; 69.86% vs. 65.00% and 51.16%, respectively), the electroporation was found to be more efficient (20.52% vs. 9.75%) to transfer EGFP genes into fowl SSCs. The transgenic SSCs established on the basal membrane of seminiferous tubules and could differentiate into sperm cells (Yu et al. 2010).

Recombinant Newcastle disease viruses (rNDV) expressing S1, S2, and S proteins have been developed. The rNDV expressing S protein of infectious bronchitis virus (IBV) conferred better protection in chicks than rNDV expressing S1 or S2 proteins of IBV, implying that S protein is more protective against threats posed by NDV and IBV. Hence, S protein can be used as an effective bivalent vaccine against NDV and IBV (Shirvani et al. 2018).

25.7 Outlook and Challenges

Birds have unique features that make them suitable host for inducing genetic alterations and obtaining pharmaceutical proteins and nutraceuticals. Much progress has been made during the past two decades to generate transgenic birds and proteins of therapeutic significance. The transgenesis or genetic engineering is a dynamic process. However, progress in avian genetic engineering is slow compared to mammalian species.

In addition to the common poultry birds, it is interesting to investigate the usefulness of other transgenic birds of other species for biologically active molecules. Integration of foreign genes should be tissue- and site-specific so that recombinant proteins are released explicitly into egg white. Release of recombinant proteins into other tissues or blood not only make purification of protein difficult, it will cause cytotoxicity and may affect the health of the transgenic bird. Maintaining transgenic lines is equally important for a sustainable supply of recombinant proteins.

25.8 Conclusions

Transgenic birds or chicken is a prospective bioreactor for producing pharmaceutical proteins. A lot has been achieved using birds as bioreactors for producing recombinant proteins. The recombinant proteins and their biological activities being comparable to those existing under natural physiological conditions imply the aptness of transgenic avian species to produce recombinant therapeutic proteins.

References

Ahn J, Shin S, Suh Y, Park JY, Hwang S, Lee K (2015) Identification of the avian RBP7 gene as a new adipose-specific gene and RBP7 promoter-driven GFP expression in adipose tissue of transgenic quail. PLoS ONE 10(4):e0124768. https://doi.org/10.1371/journal.pone.0124768 (eCollection 2015)

Antonova E, Glazova O, Gaponova A, Eremyan A, Zvereva S, Grebenkina N, Volkova N, Volchkov P (2018) Successful CRISPR/Cas9 mediated homologous recombination in a chicken cell line. Version 2. F1000Res [revised 2018 Jan 1]; 7:238. https://doi.org/10.12688/f1000research.13457.2 (eCollection 2018)

Chojnacka-Puchta L, Kasperczyk K, Płucienniczak G, Sawicka D, Bednarczyk M (2012) Primordial germ cells (PGCs) as a tool for creating transgenic chickens. Pol J Vet Sci. 15(1):181–188

Collarini EJ, Leighton PA, Van de Lavoir MC (2019) Production of transgenic chickens using cultured primordial germ cells and gonocytes. Methods Mol

Biol 1874:403–430. https://doi.org/10.1007/978-1-4939-8831-0_24

Herron LR, Pridans C, Turnbull ML, Smith N, Lillico S, Sherman A, Gilhooley HJ, Wear M, Kurian D, Papadakos G, Digard P, Hume DA, Gill AC, Sang HM (2018) A chicken bioreactor for efficient production of functional cytokines. BMC Biotechnol 18(1):82. https://doi.org/10.1186/s12896-018-0495-1

Houdebine LM (2018) Production of pharmaceutical proteins by transgenic animals. Rev Sci Tech 37 (1):131–139. https://doi.org/10.20506/rst.37.1.2746

Kamihira M, Kawabe Y, Shindo T, Ono K, Esaka K, Yamashita T, Nishijima K, Iijima S (2009) Production of chimeric monoclonal antibodies by genetically manipulated chickens. J Biotechnol 141(1–2):18–25. https://doi.org/10.1016/j.jbiotec.2009.02.022 (Epub 2009 Mar 17)

Kodama D, Nishimiya D, Iwata K, Yamaguchi K, Yoshida K, Kawabe Y, Motono M, Watanabe H, Yamashita T, Nishijima K, Kamihira M, Iijima S (2008) Production of human erythropoietin by chimeric chickens. Biochem Biophys Res Commun 367 (4):834–839. https://doi.org/10.1016/j.bbrc.2008.01.020 (Epub 2008 Jan 15)

Koo BC, Kwon MS, Lee H, Kim M, Kim D, Roh JY, Park YY, Cui XS, Kim NH, Byun SJ, Kim T (2010) Tetracycline-dependent expression of the human erythropoietin gene in transgenic chickens. Transgenic Res 19(3):437–447. https://doi.org/10.1007/s11248-009-9327-3 (Epub 2009 Oct 1)

Kwon MS, Koo BC, Choi BR, Park YY, Lee YM, Suh HS, Park YS, Lee HT, Kim JH, Roh JY, Kim NH, Kim T (2008) Generation of transgenic chickens that produce bioactive human granulocyte-colony stimulating factor. Mol Reprod Dev 75(7):1120–1126. https://doi.org/10.1002/mrd.20860

Kwon MS, Koo BC, Roh JY, Kim M, Kim JH, Kim T (2011) Production of transgenic chickens expressing a tetracycline-inducible GFP gene. Biochem Biophys Res Commun 410(4):890–894. https://doi.org/10.1016/j.bbrc.2011.06.088

Kwon MS, Koo BC, Kim D, Nam YH, Cui XS, Kim NH, Kim T (2018) Generation of transgenic chickens expressing the human erythropoietin (hEPO) gene in an oviduct-specific manner: Production of transgenic chicken eggs containing human erythropoietin in egg whites. PLoS ONE 13(5):e0194721. https://doi.org/10.1371/journal.pone.0194721 (eCollection 2018)

Kyogoku K, Yoshida K, Watanabe H, Yamashita T, Kawabe Y, Motono M, Nishijima K, Kamihira M, Iijima S (2008) Production of recombinant tumor necrosis factor receptor/Fc fusion protein by genetically manipulated chickens. J Biosci Bioeng 105 (5):454–459. https://doi.org/10.1263/jbb.105.454

Lee SH, Gupta MK, Ho YT, Kim T, Lee HT (2013) Transgenic chickens expressing human urokinase-type plasminogen activator. Poult Sci 92(9):2396–2403. https://doi.org/10.3382/ps.2013-03223

Li B, Sun G, Sun H, Xu Q, Gao B, Zhou G, Zhao W, Wu X, Bao W, Yu F, Wang K, Chen G (2008) Efficient generation of transgenic chickens using the spermatogonial stem cells in vivo and ex vivo transfection. Sci China C Life Sci. 51(8):734–742. https://doi.org/10.1007/s11427-008-0100-2 (Epub 2008 Aug 3)

Ma C, Wang K, Ji H, Wang H, Guo L, Wang Z, Ren H, Wang X, Guan W (2018) Multilineage potential research of Beijing duck amniotic mesenchymal stem cells. Cell Tissue Bank. https://doi.org/10.1007/s10561-018-9701-6 (Epub ahead of print)

Macdonald J, Taylor L, Sherman A, Kawakami K, Takahashi Y, Sang HM, McGrew MJ (2012) Efficient genetic modification and germ-line transmission of primordial germ cells using piggyBac and Tol2 transposons. Proc Natl Acad Sci U S A 109(23): E1466–E1472. https://doi.org/10.1073/pnas.1118715109 (Epub 2012 May 14)

Miyahara D, Oishi I, Makino R, Kurumisawa N, Nakaya R, Ono T, Kagami H, Tagami T (2016) Chicken stem cell factor enhances primordial germ cell proliferation cooperatively with fibroblast growth factor 2. J Reprod Dev 62(2):143–149. https://doi.org/10.1262/jrd.2015-128 (Epub 2015 Dec 28)

Nishijima K, Iijima S (2013) Transgenic chickens. Dev Growth Differ 55(1):207–216. https://doi.org/10.1111/dgd.12032 (Epub 2012 Dec 26. Review)

Oishi I, Yoshii K, Miyahara D, Kagami H, Tagami T (2016) Targeted mutagenesis in chicken using CRISPR/Cas9 system. Sci Rep 6(6):23980. https://doi.org/10.1038/srep23980

Oishi I, Yoshii K, Miyahara D, Tagami T (2018) Efficient production of human interferon beta in the white of eggs from ovalbumin gene-targeted hens. Sci Rep 8(1): 10203. https://doi.org/10.1038/s41598-018-28438-2

Park TS, Han JY (2012) piggyBac transposition into primordial germ cells is an efficient tool for transgenesis in chickens. Proc Natl Acad Sci U S A 109(24): 9337–9341. https://doi.org/10.1073/pnas.1203823109

Park TS, Kang KS, Han JY (2013) Current genomic editing approaches in avian transgenesis. Gen Comp Endocrinol 1(190):144–148. https://doi.org/10.1016/j.ygcen.2012.11.020 (Epub 2012 Dec 14)

Park TS, Lee HG, Moon JK, Lee HJ, Yoon JW, Yun BN, Kang SC, Kim J, Kim H, Han JY, Han BK (2015) Deposition of bioactive human epidermal growth factor in the egg white of transgenic hens using an oviduct-specific minisynthetic promoter. FASEB J 29 (6):2386–2396. https://doi.org/10.1096/fj.14-264739

Penno CA, Kawabe Y, Ito A, Kamihira M (2010) Production of recombinant human erythropoietin/Fc fusion protein by genetically manipulated chickens. Transgenic Res 19(2):187–195. https://doi.org/10.1007/s11248-009-9310-z (Epub 2009 Aug 4)

Rapp JC, Harvey AJ, Speksnijder GL, Hu W, Ivarie R (2003) Biologically active human interferon alpha-2b produced in the egg white of transgenic hens. Transgenic Res 12(5):569–575

Rohaim MA, Santhakumar D, Naggar RFE, Iqbal M, Hussein HA, Munir M (2018) Chickens expressing IFIT5 ameliorate clinical outcome and pathology of highly pathogenic avian influenza and velogenic newcastle disease viruses. Front Immunol 9:2025. https://doi.org/10.3389/fimmu.2018.02025 (eCollection 2018)

Salter DW, Smith EJ, Hughes SH, Wright SE, Crittenden LB (1987) Transgenic chickens: insertion of retroviral genes into the chicken germ line. Virology 157(1):236–240

Sato Y, Lansford R (2013) Transgenesis and imaging in birds, and available transgenic reporter lines. Dev Growth Differ 55(4):406–421. https://doi.org/10.1111/dgd.12058 (Epub 2013 Apr 29. Review)

Schusser B, Collarini EJ, Yi H, Izquierdo SM, Fesler J, Pedersen D, Klasing KC, Kaspers B, Harriman WD, van de Lavoir MC, Etches RJ, Leighton PA (2013) Immunoglobulin knockout chickens via efficient homologous recombination in primordial germ cells. Proc Natl Acad Sci U S A 110(50):20170–20175. https://doi.org/10.1073/pnas.1317106110 (Epub 2013 Nov 26)

Scott BB, Lois C (2005) Generation of tissue-specific transgenic birds with lentiviral vectors. Proc Natl Acad Sci U S A 102(45):16443–16447 (Epub 2005 Oct 31)

Sheridan C (2016) FDA approves'farmaceutical' drug from transgenic chickens. Nat Biotechnol 34(2):117–119. https://doi.org/10.1038/nbt0216-117

Shirvani E, Paldurai A, Manoharan VK, Varghese BP, Samal SK (2018) A recombinant newcastle disease virus (NDV) expressing S protein of infectious bronchitis virus (IBV) protects chickens against IBV and NDV. Sci Rep 8(1):11951. https://doi.org/10.1038/s41598-018-30356-2

Tyack SG, Jenkins KA, O'Neil TE, Wise TG, Morris KR, Bruce MP, McLeod S, Wade AJ, McKay J, Moore RJ, Schat KA, Lowenthal JW, Doran TJ (2013) A new method for producing transgenic birds via direct in vivo transfection of primordial germ cells. Transgenic Res 22(6):1257–1264. https://doi.org/10.1007/s11248-013-9727-2 (Epub 2013 Jun 27)

Vick L, Li Y, Simkiss K (1993) Transgenic birds from transformed primordial germ cells. Proc Biol Sci. 251 (1332):179–182

Wang ZB, Du ZQ, Na W, Jing JH, Li YM, Leng L, Luan P, Wu CY, Zhang K, Wang YX, Liu WL, Yuan H, Liu ZH, Mu YS, Meng QW, Wang N, Yang CX, Li H (2018) Production of transgenic broilers by non-viral vectors via optimizing egg windowing and screening transgenic roosters. Poult Sci. https://doi.org/10.3382/ps/pey321 (Epub ahead of print)

Woodfint RM, Hamlin E, Lee K (2018) Avian bioreactor systems: a review. Mol Biotechnol 60(12):975–983. https://doi.org/10.1007/s12033-018-0128-x (Review)

Yu F, Ding LJ, Sun GB, Sun PX, He XH, Ni LG, Li BC (2010) Transgenic sperm produced by electrotransfection and allogeneic transplantation of chicken fetal spermatogonial stem cells. Mol Reprod Dev 77 (4):340–347. https://doi.org/10.1002/mrd.21147

Transgenic Fish

26

Abstract

Fish is an important component of revenue and nutrition to millions of people. In current scenario, when demand and cost of producing land-based animal protein are increasing, the fish can complement the requirement of proteins. Reducing operating costs, minimal ecological pollution, and generating income are the prime objectives of fish production in controlled environment. Genetically engineered fish offers not only increased growth, but also serve as system to produce therapeutics and nutraceuticals.

Key Points

- Fish is a valued source of essential nutrients, viz. proteins, vitamins, and minerals
- The genetic engineering tools developed for other animals are used to develop transgenic fish to boost aquaculture production.

Keywords

Fish · Genetic engineering · Transgenesis · Therapeutics · Nutraceuticals

26.1 Introduction

Fishery has important role in alleviating hunger, providing essential nutrients, viz. proteins, long-chain fatty acids, vitamins A and D, and minerals such as iodine and calcium. Ranging from its role in human nutrition, to basic biological and advanced biomedical sciences, and the characteristic attributes (Box 1), the fish has proved to be a valuable species (Zhu et al. 2018). Some species, such as zebrafish, due to optically clear embryos and rapid development are considered as important models for developmental and neuronal studies (Chitramuthu and Bennett 2018).

The prime concern of developing transgenic fish is to enhance growth efficiency, induce resistance to pathogens and parasites, generate ornamental aquarium fish, and produce therapeutic biomolecules. In addition, efforts have been made to induce additional genetic determinants to increase adaption of fish to adverse environment such as salinity and low temperature (Fig. 26.1). As growth and gain in body weight are slow under natural conditions, the very basic

B. Singh et al., *Advances in Animal Biotechnology*,
https://doi.org/10.1007/978-3-030-21309-1_26

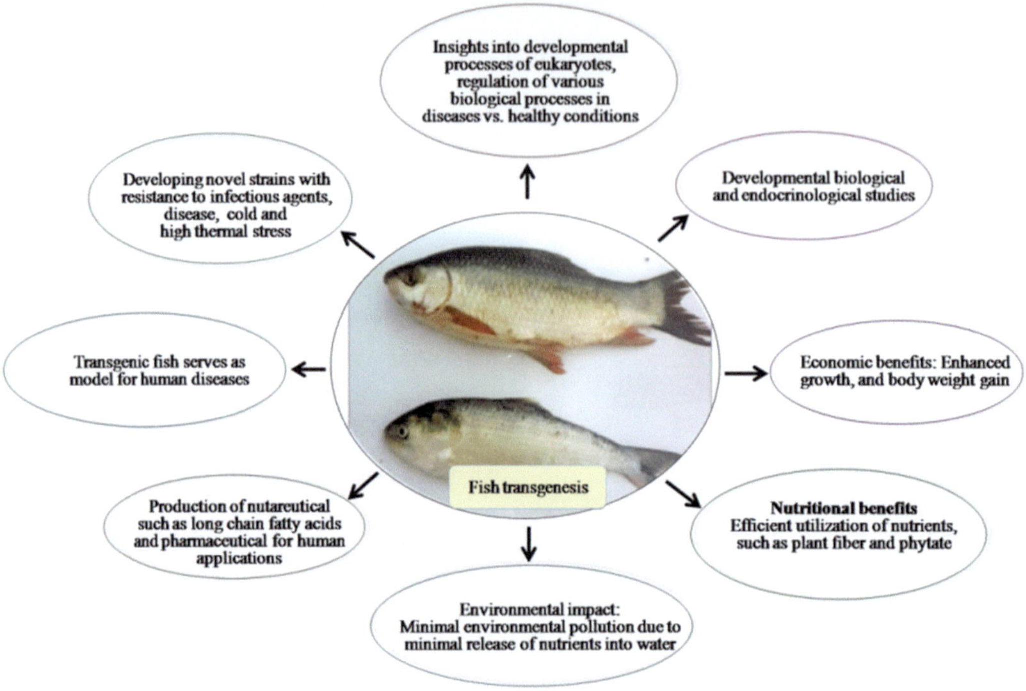

Fig. 26.1 Benefits of bioengineering or transgenic fish in human nutrition and biomedical sciences

purpose of introducing exogenous genes into them is to enhance their growth rate.

Box 1. Features of Fish for Genetic Manipulation or Transgenesis

- Wide distribution including freshwater, marine or brackish water, and cold water.
- Ease of maintenance in laboratory, ponds, and cages in water.
- High genetic diversity, therefore more opportunities of genetic and phenotypic selection.
- Compared to birds and mammals, the number of eggs and embryos is high in fish.
- Shorter life span if some species, e.g., *Oryzias latipes*, and *Brachydanio rerio*.
- Ability to produce eggs non-breeding season in some species, external mode of fertilization and embryo development, thus making oocytes and embryos available for in vitro studies.
- Ease of micromanipulation in oocytes and embryos.
- Optically transparent and fast-growing embryos zebrafish facilitate several studies.

26.2 Motivating Factors Behind Fish Genome Engineering and Transgenesis

Transgenic technologies are the promising means of producing genetically superior broodstock for aquaculture by introducing desirable traits into cultured fish species (Yaskowiak et al. 2006).

Several species of fish such as channel fish, Japanese medaka (*O. latipes*), salmon, rainbow trout, tilapia, and zebrafish (*B. rerio*) are used as experimental models to produce transgenic fish. Rohu (*Labeo rohita*), a member of farmed Indian major carp (IMC), is manipulated for better growth and increased body weight.

The art of producing transgenic fish began in mid-1980s with microinjection of exogenous genes of growth hormones (GH) into eggs of experimental and food fish (Chourrout 1986; Dunham et al. 1987). Zhu et al. (1985) published the first report of transgenes microinjected into fertilized goldfish eggs. For most fish studies, the focus was on enhancing growth rate and resistance to stress (Gong and Hew 1995). Fluorescent zebrafish was the first transgenic fish marketed (Table 26.1).

(i) **Faster growth rate and body weight gain**

The prime concern of transgenesis in fish is promoting growth and biomass production. Transgenic fish was generated by microinjecting GH gene into fertilized and non-activated Atlantic salmon eggs (Du et al. 1992). The fish possessing transgenic GH had higher body weight compared to control or normal fish.

Fish of several species with transgenic for GH genes are larger than wild strains. Some aspects of molecular basis of gain in body size are described in coho salmon, *Oncorhynchus kisutch* (Hill et al. 2000). Information is scarce on proteomic basis of enhanced transgene-phenotypes in GH transgenic fish. A recent study based on skeletal muscle proteomics in *Oncorhynchus kisutch* revealed many astoundingly significant proteomic divergences between GH transgenic and selectively bred *O. kisutch* strains (Causey et al. 2018).

(ii) **Resistance against biotic (pathogens and parasites) and abiotic stress**

Diseases are the utmost threat to aquatic animals and therefore accounts for extensive economic loss. This is because fish are exposed to assorted microbiota including pathogens (bacteria, yeasts, and fungi), protozoa (e.g., *Amyloodinium ocellatum*, *Cryptobia salmositica,* and *Ichthyophthirius multifiliis*), and eukaryotic parasites originating from aquatic and terrestrial animals and humans.

In addition, several species of trematodes, cestodes, nematodes, acanthocephalans, arthropods, and crustaceans (e.g., *Lernaea* sp.) invade fish and live as external or internal parasites.

Disease outbreak occurs more frequently in cage cultured fishes (Woo 2007). Pathogenic bacteria (e.g., *Aeromonas hydrophila*, *Aeromonas salmonicida*, *Listonella* (*Vibrio*) *anguillarum*, *Moritella viscose*, *Piscirickettsia salmonis*, *Renibacterium salmonis*, *Vibrio salmonicida*, *Yersinia ruckeri*, etc.) are of prime concern to fish health. In addition, many species of streptococci, e.g., *Streptococcus pneumoniae*, zoonotic *S. iniae*, and *S. agalactiae* affect fish.

Antimicrobial peptides (AMPs) are reported in many fish species. Transgenic fish has been developed with enhanced resistance to pathogenic bacteria. Hepcidin is a hepatic AMP that has important role in mammalian iron homeostasis. A transgenic zebrafish containing tilapia hepcidin (TH)2-3 as a transgene could effectively resist *Vibrio vulnificus* (Hsieh et al. 2010). Transgenic zebrafish overexpressing a tilapia piscidin 3 (TP3), named as TP3/DsRed transgenic zebrafish, was resistant to *S. agalactiae* infection (Su et al. 2018). Similarly, *Spironucleus* sp. and *Hezamita* sp. protozoa inhabit lumen of fish intestine. Heavy infestation of parasitic protozoa is associated with necrosis and sloughing of intestinal epithelium which cause inappetence, poor growth, and occasionally mortality. The coccidial infections of GI tract and kidney caused by *Eimeria* sp. and *Goussia* sp. reduce general health and production of fish.

Virus infections are recorded in fish. Therefore, it is necessary to utilize fish immune system and develop vaccines to protect them against pathogens. Information is scarce on transgenic fish showing immunity toward virus infections. DNA vaccines are developed to curtail virus infection in aquaculture, but modes of action of DNA vaccines, including mechanisms of immune responses elicited by DNA vaccine and their longevity are not thoroughly investigated (Collins et al. 2018).

Creating fish with resistance to infectious disease and parasites is of top concern. A recombinant immobilizing-antigen vaccine was developed against ichthyophthiriasis caused by

Table 26.1 Summary of chronology and salient achievements in developing transgenic fish

Year	Salient achievements
1986	Development of chromosome manipulation techniques in rainbow trout (Chourrout 1986)
1989	Expression of chloramphenicol acetyltransferase (*CAT*) reporter genes using a plasmid vector instilled into fertilized eggs of *O. latipes*, indicating that larger genes can be introduced into fish eggs for studying gene functions (Chong and Vielkind 1989)
1990	Electroporation-mediated transfer of recombinant plasmid, pMV-GH, containing rainbow trout GH cDNA fused to mouse metallothionein I promoter, into *Oryzias latipes* eggs (Inoue et al. 1990), microinjection and expression of recombinant plasmid containing Rous sarcoma virus long terminal repeat (RSV-LTR) promoter (pRSV)-rainbow trout GH genes into fertilized *Cyprinus carpio* eggs (Zhang et al. 1990)
1992	Expression of a transgene containing *lacZ* in neurons of zebrafish (Bayer and Campos-Ortega 1992), electroporation-mediated introduction of rainbow trout (*Oncorhynchus mykiss*) GH genes into fertilized eggs of various fish species viz., into fertilized eggs of *B. rerio*, *I. punctatus*, and *C. carpio* (Powers et al. 1992)
1994	High growth in transgenic salmon (Devliņ et al. 1994)
1995	Development of transgenic *C. auratus* carrying AFP genes from *M. americanus*, exploring the feasibility of that *AFP* genes might confer protection against cold stress (Wang et al. 1995)
2000	Development of transgenic coho salmon (*O. kisutch*) exhibiting higher rate of muscle hyperplasia (Hill et al. 2000)
2005	Development of transgenic Japanese medaka (*Oryzias latipes*) expressing phytase genes of *Aspergillus niger*. The fish could overcome anti-nutritional effects of phytate by means of phytase (Hostetler et al. 2005)
2006	Developing a stable line of GH transgenic *Salmo salar* using an "all fish" gene construct) (opAFP-GHc2) containing a GH cDNA from chinook salmon (Yaskowiak et al. 2006)
2007	Detection of altered hepatic gene expression related to iron metabolism, innate immunity, and growth in GH transgenic fish in response to expression of growth hormone (Mori et al. 2007)
2008	Demonstration of Tol1 germline transgenesis in zebrafish by means of Tol 1 transposon (Koga et al. 2008), formation of two lines of transgenic Atlantic salmon (*Salmo salar*) using gene constructs derived from ocean pout gene (t-OP5a-AFP), and second containing GH transgene (EO-1α) consisting of chinhook salmon GH cDNA almost identical to t-OP5a-AFP. The transgenic strains had faster growth (Hobbs and Fletcher 2008)
2010	Development of a transgenic zebrafish containing tilapia hepcidin (TH)2-3 as a transgene that could inhibit *Vibrio vulnificus* infection (Heish et al. 2010)
2012	Generating transgenic zebrafish with hepatic-specific expression of EGFP-Lc3, a useful model for studying generation of autophagosomes during autophagy (Cui et al. 2012), development of a blue fluorescent protein-expressing transgenic fish, called PhOTO (that allows photoconvertible optical tracking of nuclear and membrane), to study tissue regeneration mechanisms in vivo (Dempsey et al. 2012)
2016	Applications of TALEN and CRISPR/Cas9 demonstrated as highly efficient to modify *sp7* and *myostatin* gene of common carp (Zhong et al. 2016)
2018	Proteomic analysis of growth phenotypes in transgenic coho salmon (*O. kisutch*) revealed a significant different molecular mechanism of growth-enhanced phenotypes in GH transgenic fish (Causey et al. 2018; Alzaid et al. 2018) Development of a stable transgenic albino *Oryzias latipes* strain, by editing of *tyr* gene. The albino *O. latipes* might be a valuable model to study pigmentation in animals (Fang et al. 2018), generation of transgenic zebrafish exhibiting resistance to *Streptococcus agalactiae* (Su et al. 2018) Development of CRISPR/Cas9-mediated genome engineering method via non-homologous end joining (NHEJ) to produce knock-in transgenic *O. latipes* (Watakabe et al. 2018), production of gene-edited catfish (Dunham et al. 2018)

Ichthyophthirius multifiliis. The live *Cryptobia* vaccine protected salmonids from infections, while DNA-vaccine stimulated synthesis of antibody to neutralize the disease-causing metalloprotease in cryptobiosis (Woo 2007).

(iii) **Improved nutrient utilization**

Like poultry and pig, the fish is monogastric and hence cannot effectively utilize dietary ingredients such as plant fiber, oxalates, and phytic acid (Dersjant-Li et al. 2015). The unutilized phytic acid is released into environment which creates phosphorus pollution and causes water eutrophication. General and recombinant microorganisms, developed as probiotics or feed supplements, may enhance nutrient utilization (Kumar et al. 2016; Singh et al. 2017), confer immunity against infections and diseases (Qin et al. 2018), and improve reproduction efficiency (Vílchez et al. 2015).

Alternatively, genes encoding enzymes required to utilize certain nutrients are introduced into genome of fish, to produce recombinant proteins in specific tissue or glands to improve nutrient utilization. Japanese medaka (*O. latipes*) transgenic for *Aspergillus niger* phytase gene had shown improved utilization of phytate (Hostetler et al. 2005).

(iv) **Resistance to climate and salinity**

Global warming has posed threat to aquatic diversity. Water salinity has increased as a consequence of global warming. Enhancing tolerance to salinity or temperature variation and producing stable lines of freeze-resistance or salinity-resistance by modifications in their genetic makeup would be realistic approaches.

The fish is sensitive to alterations in temperature. For instance, the salmonids die if they are exposed to ice or subzero temperature. In an attempt to increase their tolerance to cold, Hew et al. (1992) introduced antifreeze protein (AFP) genes from winter flounder (*Pseudopleuronectes americanus*) into Atlantic salmon. The transgenic flounders (F0) were germline mosaics. Low levels of AFP precursors were noted in blood of all transgenic offspring (F1), and that 50% of the progeny resulting from cross between F1 and wild strains, possessed AFP. The study demonstrated that it was possible to produce germline transformed Atlantic salmonids (Hew et al. 1992).

Development of transgenic goldfish (*Carassius auratus*) carrying AFP genes from ocean pout (*Macrozoarces americanus*) implies that *AFP* genes might confer protection against cold stress in fish that were earlier susceptive to low temperature (Wang et al. 1995).

Similarly, there is need to enhance tolerance against detrimental effects of environment in other economically fish species such as Tilapia and IMCs.

(v) **Transgenic fish as model to study human diseases**

Application of transgenic fish is rather a recent development to understand genetic mechanisms, and developmental processes, improving aquaculture, and producing pharmaceuticals (Lee et al. 2018). It is evident that like many other animals, the fish also serves as valuable models for studying human disease. Transgenic fish strains serve as ideal candidates for cell development, traumatic injury and regeneration, cancer progression, and stem cell activities (Dempsey et al. 2012; Zhu et al. 2018).

A transgenic zebrafish line expressing hepatic EGFP-Lc3, a widely utilized receptor of autophagy, is reported to be a useful model to understand the mechanisms of hepatic autophagy in liver diseases (Cui et al. 2012). In addition, the embryonic zebrafish is nearly ideal model system for studying development of vertebrate nervous system.

26.3 Techniques Used to Produce Transgenic Fishes

Various methods are used to inculcate exogenous genes into target cells such as eggs, zygote, and embryos. Prior to 1990s, transgenic fish were developed using virus vectors, microinjection, electroporation, microprojectiles, and liposome-mediated genetic transformation. Genetic engineering and genome modification have progressed substantially by pursuit of enzyme-based genome-editing methods summarized herein.

(i) **Microinjection**

Microinjection (Palmiter et al. 1982) has the same principal and mechanism to introduce genetic materials into target cells or eggs or embryos. For fish, microinjection is the preferred technique, with a success rate of about 10% (meaning that out of every 100 eggs injected, about 10 will integrate and express the exogenes introduced), though inheritance of the transgenes is much lower being approximately 1%. However, production of transgenic female fish is particularly attractive for genetic modifications and molecular developmental biological studies.

Microinjection of a gene expression cassette directly into oocytes or early embryos is the most antique technique. It is still a method of choice to introduce functional genes, proteins (Elaswad et al. 2018), enzyme inhibitors, therapeutics, or cryoprotective agents into cell (Alam et al. 2018). The advantages associated with microinjection include optimum quantity of genes to be introduced, thereby increasing chances of integrative transformation, and precise delivery into cells or zygotes. Due to difficulty in locating pronuclei in fish zygote, the microinjection is a preferred technique to inoculate transgenes into cytoplasm.

The microinjection has some limitations. The chances of insertion of genes at desired locus or site are less. It is likely that this technique may give rise to chimera mosaic fish. The technique is slow and random; it requires technical skills and advanced micromanipulators and microinjectors. Microinjection may also result in cell or egg mortality. The alternative strategies, such as electroporation, retrovirus-mediated transfer of genes, liposomal-reverse-phase-evaporation, sperm-mediated gene transfer, and high-velocity microprojectile bombardments, are used to convey genes and drug biomolecules into cells.

(ii) **Electroporation**

Electroporation involves generating transient, nanometer-sized pores in the cell membrane exposed a brief electric pulse. The technical steps may vary depending on type of cells used and the genes to be introduced. Wong and Neumann (1982) used this technique to launch a recombinant plasmid into murine fibroblasts.

Electroporation was used to introduce recombinant plasmid, pMV-GH, containing rainbow trout GH cDNA fused to mouse metallothionein I promoter, into *O. latipes* eggs. The transgenes were inherited to F1 offspring when transgenic male was allowed to mate with normal female. Transgenes were detected in F1, and F2 (80%) offspring generated by mating transgenic F1 fishes (Inoue et al. 1990).

Electroporation is used in many studies to transfer transgenic genes into gametes and cells of various fish species. Electroporation-mediated introduction of rainbow trout (*Oncorhynchus mykiss*) GH1 (rtGH1) or rtGH2 cDNA into fertilized eggs of various fish species, viz. *Brachydanio rerio*, *Ictalurus punctatus*, and *Cyprinus carpio* (Powers et al. 1992), and Drosophila β-actin promoter coupled to a β-galactosidase gene cassette into fertilized eggs of red abalone (*Haliotis rufescens*) (Powers et al. 1995), revealed the aptness of electroporation as method of choice to transfer genes.

(iii) **Spermatogonial stem cell mediated gene transfer**

Germ cell transplantation has opened new avenues for reproductive biotechnology and genetic manipulation of fish. Spermatogonial stem cells (SSCs) are the specialized germ cells, unipotent in terms of producing sperms only, self-renewal and generating specialized cells committed to form sperm. Studies have shown that SSCs can be reprogrammed to pluripotent stem cells with or without feeder cells (Tonelli et al. 2017; Lee et al. 2018).

Isolation and description and biotechnological applications of SSCs in aquaculture and fish are limited primarily because of lack of specific molecular markers. Some highly conserved molecular markers such as *Gfra1* and *Oct4/Pou5f1* serve as important candidate genes for describing SSCs in fish. However, unlike mammalian SSCs, the fish SSCs and oogonial stem cells exhibit high sexual plasticity that depends much on somatic microenvironment (Lacerda et al. 2014).

SSCs express some pluripotency markers. For instance, *Sox 2* (Patra et al. 2015) and *Nanog* genes have been identified in SSCs of rohu (*Labeo rohita*) (Patra et al. 2018). Fertile sperm were obtained from SSCs of catfish, the *Clarias batrachus* (Nayak et al. 2016), and zebrafish when SSCs transferred into Sertoli feeder cells (Kawasaki et al. 2016). The transgenic SSCs can transfer the transgenes to offspring. Offspring were obtained from cultured SSCs in zebrafish (Kawasaki et al. 2012).

(iv) **Genome editing and transgenic fish**

It is imperative to know difference between gene-delivery methods and transgenesis or genetic engineering techniques. Gene-delivery methods intend to introduce genes or genetic materials into cytoplasm or nucleus of target cells. Genes introduced may or may not integrate into genome of target cell. The plasmid, cosmid, or virus vector-mediated gene delivery, and genome editing ensure integration of genes into genome of host cells.

Zinc-finger nuclease (ZFN), transcription activator-like effector nuclease (TALEN), and clustered regularly interspaced short palindromic repeats associated with Cas9 (CRISPR/Cas9) have emerged as promising tools for precise genome editing (Zhu et al. 2018).

CRISPR/Cas9 has just made entry into genome manipulation in fish. TALEN combined with CRISPR/Cas9 was confirmed as highly efficient method to modify genome of common carp to generate more muscular cells (Zhong et al. 2016). CRISPR/Cas9 was used to edit genome of medaka, i.e., *O. latipes tyr* gene to generate a stable albino strain that could be used to study the mechanisms of regulation of pigmentation (Fang et al. 2018). Watakabe et al. (2018) have used CRISPR/Cas9-mediated genome engineering via non-homologous end joining (NHEJ) to produce mutant or knock-in transgenic *O. latipes*.

26.4 Problems Associated with Fish Transgenesis

In fish, larger males behave as dominant and avail high mating opportunities. The transgenic males being large may invade natural fish populations and replace normal shorter males. Also, it is probable that transgenes may imbalance the mating balance due to altered phenotypes and behavioral traits. Though transgenes are rare, if transgenic fish are released into wild populations, they may propagate undesirable effects such as loss of genetic diversity and extinction of native genetic resources. Hence, risks should be thoroughly evaluated before transgenic fish are released into natural genetic stock (Muir and Howard 1999).

Description of complex cell activities (cell divisions, movement, morphological changes, etc.) underlying embryonic development and adult tissue regeneration requires efficient means to examine and monitor the cells with high fidelity in space and time. To satisfy these criteria, Dempsey et al. (2012) developed a set of transgenic zebrafish line, named PhOTO (photoconvertible optical tracking of), that allows photoconvertible optical tracking of nuclear and membrane dynamics in vivo. These lines combine the benefits of global and sparse imaging approaches for lineage analysis (Dempsey et al. 2012).

26.5 Outlook and Challenges

The transgenic fish technology will play an important role in nutrition, health, and medicine. The transgenic or GM fish were originally developed to enhance growth, resistance to stress and disease, and altered color to promote ornamental fish business. Transgenic fish for GH could be more competitive in utilizing feed under natural conditions. Transgenic fish carrying resistance to biotic and abiotic stress would increase profitability and producing colored or ornamental fishes.

The ongoing tweaks in bioengineering of fish will improve reproduction, aquaculture performance, and production of nutraceuticals or pharmaceuticals for human and veterinary applications. The efficiency of conventional approaches to create transgenic fish is poor due to low rates of integration of exogenous genes into genome of host cell and embryo. Therefore, it is necessary to opt for site-directed integration of a transgene into genome of the host or target cells. New technologies, such as TALEN and CRISPR/Cas9, offer encouraging opportunities to modify the fishes.

However, gene-transfer technology in fish has lagged behind that of mammals, largely due to lack of suitable regulatory elements to control transgene expression. A growing concern to prevent farmed fish (both GM and non-GM) from escaping and competing with wild strains calls for propagating transgenic sterile fish that fail to reproduce and interbreed with wild populations in natural habitats. The increased vulnerability to predators, reduced swimming ability, lack of increased growth when foraging, and unchanged spawning percentage of transgenic fish examples indicate that some transgenic fish may not compete well under natural conditions or may cause main ecological or environmental damage. There is an urgent need to develop data on consequences of releasing the transgenic fish into natural habitats, and the safety of transgenic fish and their products intended for human use.

26.6 Conclusions

Fish is among the largest group of vertebrates having values for food, income generation, and biomedical investigations. It is important to develop transgenic fish to meet human nutrition and economic needs and maintain a balance between commercial fish production and conservation of native fishes. This calls for international research collaboration at the regional and local level which can develop appropriate expertise for implementing the new technologies.

References

Alam MA, Rahman SM, Yamamoto Y, Hattori RS, Suzuki T, Watanabe M, Strüssmann CA (2018) Optimization of protocols for microinjection-based delivery of cryoprotective agents into Japanese whiting Sillago japonica embryos. Cryobiology. pii: S0011-2240(18)30202-5. https://doi.org/10.1016/j.cryobiol.2018.10.007 (Epub ahead of print)

Alzaid A, Kim JH, Devlin RH, Martin SAM, Macqueen DJ (2018) Growth hormone transgenesis in coho salmon disrupts muscle immune function impacting cross-talk with growth systems. J Exp Biol 221(Pt 13). pii: jeb173146. https://doi.org/10.1242/jeb.173146

Bayer TA, Campos-Ortega JA (1992) A transgene containing lacZ is expressed in primary sensory neurons in zebrafish. Development. 115(2):421–426

Causey DR, Kim JH, Stead DA, Martin SAM, Devlin RH, Macqueen DJ (2018) Proteomic comparison of selective breeding and growth hormone transgenesis in fish: unique pathways to enhanced growth. J Proteomics. pii: S1874-3919(18)30325-7. https://doi.org/10.1016/j.jprot.2018.08.013 (Epub ahead of print)

Chitramuthu BP, Bennett HPJ (2018) Application of zebrafish and knockdown technology to define progranulin neuronal function. Methods Mol Biol 1806:207–231. https://doi.org/10.1007/978-1-4939-8559-3_15

Chong SS, Vielkind JR (1989) Expression and fate of CAT reporter gene microinjected into fertilized medaka (*Oryzias latipes*) eggs in the form of plasmid DNA, recombinant phage particles and its DNA. Theor Appl Genet 78(3):369–380. https://doi.org/10.1007/BF00265299

Chourrout D (1986) Techniques of chromosome manipulation in rainbow trout: a new evaluation with karyology. Theor Appl Genet 72(5):627–632. https://doi.org/10.1007/BF00289000

Collins C, Lorenzen N, Collet B (2018) DNA vaccination for finfish aquaculture. Fish Shellfish Immunol. pii: S1050-4648(18)30410-8. https://doi.org/10.1016/j.fsi.2018.07.012 (Epub ahead of print)

Cui J, Sim TH, Gong Z, Shen HM (2012) Generation of transgenic zebrafish with liver-specific expression of EGFP-Lc3: a new in vivo model for investigation of liver autophagy. Biochem Biophys Res Commun 422 (2):268–273. https://doi.org/10.1016/j.bbrc.2012.04.145 (Epub 2012 May 3)

Dempsey WP, Fraser SE, Pantazis P (2012) PhOTO zebrafish: a transgenic resource for in vivo lineage tracing during development and regeneration. PLoS ONE 7(3):e32888. https://doi.org/10.1371/journal.pone.0032888 (Epub 2012 Mar 14)

Dersjant-Li Y, Awati A, Schulze H, Partridge G (2015) Phytase in non-ruminant animal nutrition: a critical review on phytase activities in the gastrointestinal tract and influencing factors. J Sci Food Agric 95(5):878–896. https://doi.org/10.1002/jsfa.6998 (Epub 2014 Dec 19. Review)

Devlin RH, Yesaki TY, Biagi CA, Donaldson EM, Swanson P, Chan W-K (1994) Extraordinary salmon growth. Nature 371:209–210

Du SJ, Gong ZY, Fletcher GL, Shears MA, King MJ, Idler DR, Hew CL (1992) Growth enhancement in transgenic Atlantic salmon by the use of an "all fish" chimeric growth hormone gene construct. Biotechnology (N Y) 10(2):176–181

Dunham RA, Eash J, Askins J, Townes TM (1987) Transfer of the metallothioneinhuman growth hormone fusion gene into channel catfish. Trans Am Fish Soc 116:87–91

Dunham RA, Elaswad A, Qin Z (2018) Gene editing in channel catfish via double electroporation of zinc-finger nucleases. Methods Mol Biol 1867:201–214. https://doi.org/10.1007/978-1-4939-8799-3_15

Elaswad A, Khalil K, Cline D, Page-McCaw P, Chen W, Michel M, Cone R, Dunham R (2018) Microinjection of CRISPR/Cas9 protein into channel catfish, *Ictalurus punctatus*, embryos for gene editing. J Vis Exp 131. https://doi.org/10.3791/56275

Fang J, Chen T, Pan Q, Wang Q (2018) Generation of albino medaka (*Oryzias latipes*) by CRISPR/Cas9. J Exp Zool B Mol Dev Evol 330(4):242–246. https://doi.org/10.1002/jez.b.22808 (Epub 2018 Jun 5)

Gong Z, Hew CL (1995) Transgenic fish in aquaculture and developmental biology. Curr Top Dev Biol 30:177–214 (Review. No abstract available)

Hew CL, Davies PL, Fletcher G (1992) Antifreeze protein gene transfer in Atlantic salmon. Mol Mar Biol Biotechnol 1(4–5):309–317

Hill JA, Kiessling A, Devlin RH (2000) Coho salmon (*Oncorhynchus kisutch*) transgenic for a growth hormone gene construct exhibit increased rates of muscle hyperplasia and detectable levels of differential gene expression. Can J Fish Aquat Sci 57:939–950

Hobbs RS, Fletcher GL (2008) Tissue specific expression of antifreeze protein and growth hormone transgenes driven by the ocean pout (*Macrozoarces americanus*) antifreeze protein OP5a gene promoter in Atlantic salmon (*Salmo salar*). Transgenic Res 17(1):33–45 (Epub 2007 Sep 2)

Hostetler HA, Collodi P, Devlin RH, Muir WM (2005) Improved phytate phosphorus utilization by Japanese medaka transgenic for the Aspergillus niger phytase gene. Zebrafish 2(1):19–31. https://doi.org/10.1089/zeb.2005.2.19

Hsieh JC, Pan CY, Chen JY (2010) Tilapia hepcidin (TH) 2-3 as a transgene in transgenic fish enhances resistance to Vibrio vulnificus infection and causes variations in immune-related genes after infection by different bacterial species. Fish Shellfish Immunol 29 (3):430–439. https://doi.org/10.1016/j.fsi.2010.05.001 (Epub 2010 May 12)

Inoue K, Yamashita S, Hata J, Kabeno S, Asada S, Nagahisa E (1990) Fujita T electroporation as a new technique for producing transgenic fish. Cell Differ Dev 29(2):123–128

Kawasaki T, Saito K, Sakai C, Shinya M, Sakai N (2012) Production of zebrafish offspring from cultured spermatogonial stem cells. Genes Cells 17(4):316–325. https://doi.org/10.1111/j.1365-2443.2012.01589.x (Epub 2012 Mar 5)

Kawasaki T, Siegfried KR, Sakai N (2016) Differentiation of zebrafish spermatogonial stem cells to functional sperm in culture. Development. 143(4):566–574. https://doi.org/10.1242/dev.129643 (Epub 2015 Dec 30)

Koga A, Cheah FS, Hamaguchi S, Yeo GH, Chong SS (2008) Germline transgenesis of zebrafish using the medaka Tol1 transposon system. Dev Dyn 237 (9):2466–2474. https://doi.org/10.1002/dvdy.21688

Kumar M, Yadav AK, Verma V, Singh B, Mal G, Nagpal R, Hemalatha R (2016) Bioengineered probiotics as a new hope for health and diseases: an overview of potential and prospects. Future Microbiol 11(4):585–600. https://doi.org/10.2217/fmb.16.4 (Epub 2016 Apr 12. Review)

Lacerda SM, Costa GM, de França LR (2014) Biology and identity of fish spermatogonial stem cell. Gen Comp Endocrinol 1(207):56–65. https://doi.org/10.1016/j.ygcen.2014.06.018 (Epub 2014 Jun 23)

Lee SW, Wu G, Choi NY, Lee HJ, Bang JS, Lee Y, Lee M, Ko K, Schöler HR, Ko K (2018) Self-reprogramming of spermatogonial stem cells into pluripotent stem cells without microenvironment of feeder cells. Mol Cells 41(7):631–638. https://doi.org/10.14348/molcells.2018.2294 (Epub 2018 Jul 10)

Mori T, Hiraka I, Kurata Y, Kawachi H, Mano N, Devlin RH, Nagoya H, Araki K (2007) Changes in hepatic gene expression related to innate immunity, growth and iron metabolism in GH-transgenic amago salmon (*Oncorhynchus masou*) by cDNA subtraction and microarray analysis, and serum lysozyme activity. Gen Comp Endocrinol 151(1):42–54 (Epub 2007 Jan 12)

Muir WM, Howard RD (1999) Possible ecological risks of transgenic organism release when transgenes affect mating success: sexual selection and the Trojan gene hypothesis. Proc Natl Acad Sci U S A 96(24):13853–13856

Nayak S, Ferosekhan S, Sahoo SK, Sundaray JK, Jayasankar P, Barman HK (2016) Production of fertile sperm from in vitro propagating enriched spermatogonial stem cells of farmed catfish. Clarias batrachus. Zygote 24(6):814–824 (Epub 2016 Jul 15)

Palmiter RD, Brinster RL, Hammer RE, Trumbauer ME, Rosenfeld MG, Birnberg NC, Evans RM (1982) Dramatic growth of mice that develop from eggs microinjected with metallothionein-growth hormone fusion genes. Nature 300(5893):611–615

Patra SK, Chakrapani V, Panda RP, Mohapatra C, Jayasankar P, Barman HK (2015) First evidence of molecular characterization of rohu carp Sox2 gene being expressed in proliferating spermatogonial cells. Theriogenology 84(2):268–276.e1. https://doi.org/10.1016/j.theriogenology.2015.03.017 (Epub 2015 Mar 25)

Patra SK, Vemulawada C, Soren MM, Sundaray JK, Panda MK, Barman HK (2018) Molecular characterization and expression patterns of *Nanog* gene validating its involvement in the embryonic development and maintenance of spermatogonial stem cells of farmed carp, *Labeo rohita*. J Anim Sci Biotechnol 9:45. https://doi.org/10.1186/s40104-018-0260-2 (eCollection 2018)

Powers DA, Hereford L, Cole T, Chen TT, Lin CM, Kight K, Creech K, Dunham R (1992) Electroporation: a method for transferring genes into the gametes of zebrafish (*Brachydanio rerio*), channel catfish (*Ictalurus punctatus*), and common carp (*Cyprinus carpio*). Mol Mar Biol Biotechnol 1(4–5):301–308

Powers DA, Kirby VL, Cole T, Hereford L (1995) Electroporation as an effective means of introducing DNA into abalone (*Haliotis rufescens*) embryos. Mol Mar Biol Biotechnol 4(4):369–375

Qin C, Xie Y, Wang Y, Li S, Ran C, He S, Zhou Z (2018) Impact of *Lactobacillus casei* BL23 on the host transcriptome, growth and disease resistance in larval zebrafish. Front Physiol 9:1245. https://doi.org/10.3389/fphys.2018.01245 (eCollection 2018)

Singh B, Mal G, Marotta F (2017) Designer probiotics: paving the way to living therapeutics. Trends Biotechnol 35(8):679–682. https://doi.org/10.1016/j.tibtech.2017.04.001 (Epub 2017 May 5)

Su BC, Lai YW, Chen JY, Pan CY (2018) Transgenic expression of tilapia piscidin 3 (TP3) in zebrafish confers resistance to Streptococcus agalactiae. Fish Shellfish Immunol 74:235–241. https://doi.org/10.1016/j.fsi.2018.01.001 (Epub 2018 Jan 6)

Tonelli FMP, Lacerda SMSN, Tonelli FCP, Costa GMJ, de França LR, Resende RR (2017) Progress and biotechnological prospects in fish transgenesis. Biotechnol Adv. 35(6):832–844. https://doi.org/10.1016/j.biotechadv.2017.06.002 (Epub 2017 Jun 8. Review)

Vílchez MC, Santangeli S, Maradonna F, Gioacchini G, Verdenelli C, Gallego V, Peñaranda DS, Tveiten H, Pérez L, Carnevali O, Asturiano JF (2015) Effect of the probiotic Lactobacillus rhamnosus on the expression of genes involved in European eel spermatogenesis. Theriogenology 84(8):1321–1331. https://doi.org/10.1016/j.theriogenology.2015.07.011 (Epub 2015 Jul 17)

Wang R, Zhang P, Gong Z, Hew CL (1995) Expression of the antifreeze protein gene in transgenic goldfish (*Carassius auratus*) and its implication in cold adaptation. Mol Mar Biol Biotechnol 4(1):20–26

Watakabe I, Hashimoto H, Kimura Y, Yokoi S, Naruse K, Higashijima SI (2018) Highly efficient generation of knock-in transgenic medaka by CRISPR/Cas9-mediated genome engineering. Zoological Lett 4:3. https://doi.org/10.1186/s40851-017-0086-3 (eCollection 2018)

Wong TK, Neumann E (1982) Electric field mediated gene transfer. Biochem Biophys Res Commun 107 (2):584–587 (No abstract available)

Woo PT (2007) Protective immunity in fish against protozoan diseases. Parassitologia 49(3):185–191 (Review)

Yaskowiak ES, Shears MA, Agarwal-Mawal A, Fletcher GL (2006) Characterization and multi-generational stability of the growth hormone transgene (EO-1alpha) responsible for enhanced growth rates in Atlantic salmon. Transgenic Res 15(4):465–480 (Erratum in Transgenic Res. 2007 Apr; 16(2):253–259)

Zhang PJ, Hayat M, Joyce C, Gonzalez-Villaseñor LI, Lin CM, Dunham RA, Chen TT, Powers DA (1990) Gene transfer, expression and inheritance of pRSV-rainbow trout-GH cDNA in the common carp, *Cyprinus carpio* (Linnaeus). Mol Reprod Dev 25(1):3–13

Zhong Z, Niu P, Wang M, Huang G, Xu S, Sun Y, Xu X, Hou Y, Sun X, Yan Y, Wang H (2016) Targeted disruption of sp7 and myostatin with CRISPR-Cas9 results in severe bone defects and more muscular cells in common carp. Sci Rep 15(6):22953. https://doi.org/10.1038/srep22953

Zhu Z, Li G, He L, Chen S (1985) Novel gene transfer into the fertilized eggs of goldfish (*Carassius auratus* L. 1758). Z Angew Ichthyol 1:31–34

Zhu XY, Wu SQ, Guo SY, Yang H, Xia B, Li P, Li CQ (2018) A zebrafish heart failure model for assessing therapeutic agents. Zebrafish 15(3):243–253. https://doi.org/10.1089/zeb.2017.1546 (Epub 2018 Mar 20)

Reproduction Biotechnology in Goats 27

Abstract

Goat is an important small ruminant livestock species with ubiquitous distribution in most countries. Goats are known for their unique browsing habits, and quality milk, meat, and skin production. A number of transgenic goats are developed for use as bioreactors to produce recombinant proteins of therapeutic importance. Reproduction biotechniques, namely embryo cryopreservation, sperm sexing and cryopreservation, nuclear transfer cloning, and genome editing, have important contribution goat production.

Key Points

- Goats are important multipurpose livestock species reared all over the world
- Due to peculiar characteristics, the goats are preferred for producing recombinant proteins.

Keywords

Reproduction biotechnology · Transgenic goats · Recombinant therapeutics · Biomedical applications

27.1 Introduction

Herding of goats has evolved about 10,000 years ago in the mountains of Iran, indicating that goats are one of the oldest animals domesticated by humans (Haenlein and Ramirez 2007). Goat milk was venerated in ancient Egypt with some Pharaoh supposedly placing these foods among other treasures in the burial tombs (Smith 2006). As a milch animal, the goat plays an important role in human nutrition in the areas acknowledged as cradles of modern civilization (Hatziminaoglou and Boyazoglu 2004). At least ten countries depend on goats and sheep for 30–70% of milk demand (Haenlein 2001).

Asia and Africa are the hubs of goats and goat genetic resources (Figs. 27.1 and 27.2). It is estimated that over 80% of world's goat population is located in Asia and Africa (Morand-Fehr et al. 2004). What makes goats so popular is their ability to provide high-quality food (meat and milk) under adverse and diverse climatic conditions and resilience to impulsive environments (Silanikove 2000).

27.2 Goats as Valuable Asset

Research in food science and nutrition has exponentially grown recently, and changing the way food is considered. It provides components with specific functions and nutritional properties which include potential benefits as well as possible detrimental effects on health. Functional compounds such as flavonoids, phenolic acids, vitamins, ω3-fatty acids, proteins, and peptides are current topics of studies with reference to

B. Singh et al., *Advances in Animal Biotechnology*,
https://doi.org/10.1007/978-3-030-21309-1_27

Fig. 27.1 Migratory goats reared in cold deserts of Northwest Himalayan Region (NWHR). The *Chegu* and *Changthangi* goats are the multipurpose goats reared for fiber (Cashmere), meat, and milk. The milk yield is low in these goats

their health benefits. Some of these metabolites are obtained from goats.

Goat milk is a natural source of bioactive peptides with antioxidant and immune-modulating activities. It is rich in various physiologically functional biomolecules including proteins, vitamins, flavonoids, and carotenoids with antioxidant properties. The goats can transform carotenoid pigments into vitamin A, so goat milk contains a higher amount of vitamin A compared to cattle milk. In addition, the goat milk is easily digestible, less allergenic, naturally homogenized, easier to digest, and resembles closely to human milk (https://mtcapra.com/2010/08/20/benefits-of-goat-milk-vs-cow-milk/, accessed on Dec 1, 2018).

Goat meat or mutton, being low in calories, saturated fats and cholesterol compared to other red meats, is among most commonly utilized used meat. The mutton improves blood cholesterol levels, eases inflammatory responses, and stabilizes heart rhythms. Selenium and choline in goat meat assist in preventing from cancer.

27.3 Reproduction Biotechniques in Goats

Starting from natural selection of animals, to genotyping of superior goats, and embryo production and transfer, the reproduction biotechnology has contributed to goat production. Besides milk and meat, the goats are used as model species in human studies.

27.4 Sperm Sorting and Artificial Insemination

Generally, natural mating is practiced in goats. The nomadic goat owners keep a naturally selected fertile breeder bucks in their flocks.

Bucks are used as semen donors for commercial dissemination of quality germplasm. Semen is collected by small artificial vagina, transrectal ultrasound-guided massage of accessory sex glands (TUMASG), and electroejaculation (EE).

Fig. 27.2 Milk yielding pure breed or crossbreed goats in plane tropical regions. The goats are used primarily for milk. Some are dual purpose, while breeds such as Boer are primarily meat producing goats

Though quality of semen collected by EE is superior, TUMASG is preferred due to being less stressful (Abril-Sánchez et al. 2017).

The goat buck seminal plasma contains proteins that impede sperm quality. The protocols for cryopreservation of goat semen are improved by incorporating cholesterol-loaded cyclodextrin (CLC) and eliminating seminal plasma (Salmon et al. 2017). Does are inseminated with freshly collected or frozen-thawed semen.

There are reports on sexing goat buck sperm (Parrilla et al. 2004) and use of sexed sperm to establish pregnancies and produce kids. Sex-sorted sperm frozen in pellets displayed high post-thaw motility than the sperm cryopreserved using plastic straws. The Saanen goat semen was sorted by a modified flow cytometer, MoFlo SX®, cryopreserved, and frozen-thawed for inseminating the goats. Thirty-eight percent (5/18) of the inseminated goats kidded with 83% (3/5) of kids of anticipated sex. The study shows the feasibility of sperm sexing, cryopreservation, and using sexed sperm for producing presexed kids (Bathgate et al. 2013).

The sperm is subjected to oxidative stress during flow cytometric sorting. Supplementing ascorbic acid-2-glucoside (AA-2G) as antioxidant during sorting of sperm and deep uterine insemination using sex-sorted sperm resulted in pregnancy rates comparable to those obtained by using normal sperm in cashmere goats (Qin et al. 2018).

27.5 Oocyte and Embryo Banking

Enough ovaries are available from abattoir as goats are slaughtered routinely all over the world. Oocytes are aspirated from the ovaries and used for in vitro production of embryos by IVF, parthenogenetic activation or nuclear transfer cloning.

Compared to other animals, fewer efforts are made to cryopreservation of goat oocytes and embryos. However, depending on situation, extra oocytes or embryos are subjected to cryopreservation so that they can be transported to other laboratories or goat farms as and when needed. Slow-freezing and vitrification are used for cryopreservation of oocytes and embryos.

The cryopreserved goat embryos have higher survival than sheep embryos subjected to similar cryopreservation and thawing protocols (Traldi et al. 1999). Embryos cryopreserved at expanding stage show higher survival after cryopreservation. Studies have been carried out to evaluate embryo survival and kidding rate from transferring vitrified-warmed embryos, and vitrified-warmed and CPAs-free embryos. Kidding rates tended to be unaffected by both methods, but embryo survival increased when 0.4 M sucrose was included in the vitrifying solution, and vitrified-warmed embryos were directly transferred to females (Guignot et al. 2006).

Conventional cryopreservation and vitrification using DMSO have similar efficacies for cryopreserving caprine IVF embryos (Ferreira-Silva et al. 2017). It is likely that minor differences may exist in seminal proteins and sensitivity of oocytes and embryos to cryopreservation in different caprine breeds. This should be investigated while cryopreserving the goat oocytes and embryos.

27.6 Somatic Cell Nuclear Transfer Cloning

After successful SCNT cloning of sheep, studies were focused to clone other species of animals. Goats, in view of their role as model animals and as producer of transgenic or recombinant proteins, were one of the prime focus of SCNT cloning.

Baguisi et al. (1999) succeeded in producing transgenic cloned goats by nuclear transfer of genetically altered fetal fibroblasts. The goats were found to produce a high concentration of human antithrombin-III in their milk. Keefer et al. (2002) evaluated suitability of various somatic cells, viz. fibroblasts and granulosa cells, and concluded that adult fibroblasts could be reprogrammed to support embryo development and birth of cloned offspring. Subsequent studies focused on optimizing various parameters such as use of different somatic cells as donor nuclei vis-a-vis their developmental competence (Mao et al. 2018), retrieving oocytes from different sources, preparing cytoplast either by micromanipulation or bisectioning of zona-removed IVM oocytes, and developing culture media to enhance development efficiency of cloned embryos. The inferences drawn from these studies were used to generate cloned transgenic animals. SCNT was used to improving milk production and improving quality of milk or producing therapeutic proteins in mammary gland of goats (Table 27.1).

As goat is an important animal for meat, SCNT was used to produce transgenic myostatin-targeted goats for enhancing meat production (Zhou et al. 2013).

27.7 Stem Cells in Goats

There are plentiful reports on isolation of stem cells from various origins, viz. embryonic stem cells, fetal or perinatal stem cells (Kumar et al. 2016), hair follicle stem cells (He et al. 2016), primordial germ cells, and induced pluripotent stem cells (iPSCs). The stem cells are shown to express variable markers many of which are noted in other species also.

Mammary epithelial cells (MECs) were isolated from early lactation milk, and expression of stem cell marker stage-specific embryogenic antigen-4 (SSEA-4) and *Capra hircus* β-casein after induced expression were observed. Isolation

Table 27.1 Summary of milestone achievements in ARTs in goats

Year	Achievements, inferences, and references
1999	Birth of SCNT cloned transgenic goats, producing a high concentration of human antithrombin-III in their milk (Baguisi et al. 1999)
2001	Birth of cloned transgenic goats from fibroblasts (Reggio et al. 2001)
2004	Studies on flow cytometric identification of X- and Y-chromosome bearing sperm (Parrilla et al. 2004) and birth of cloned transgenic goats from fibroblasts (Behboodi et al. 2004)
2006	Production, processing, and purification of recombinant hLZ lysozyme in the milk of transgenic goats (Maga et al. 2006)
2008	Induction of repeat superovulation, non-surgical recovery of embryos, embryo transfer, and kid births in transgenic dairy goats (Melican and Gavin 2008)
2013	First report on sorting of goat sperm and use of sex-sorted sperm to produce kids of desired sex (Bathgate et al. 2013) High production of recombinant rhLZ in the milk of transgenic cloned goats using transgenic donor cells containing optimized lysozyme expression cassettes (β-casein/hLZ and β-lactoglobulin/hLZ) (Yu et al. 2013), expression of human coagulation factor IX in milk gland of goat (Amiri Yekta et al. 2013)
2015	Using CRISPR/Cas9 editing to edit target cells for *MSTN* or *FGF5* genes at several places, thus showing the feasibility of editing target cells at multiple sites in large animals (Wang et al. 2015) Formation goat iPSCs by forced expression of reprogramming factors, namely POU5F1, SOX2, MYC, KLF4, LIN-28, and NANOG, in combination with a MIR302/367 cluster, delivered by lentiviral vectors (Sandmaier et al. 2015)
2016	Isolation of hair follicle stem cells and description of stem cell like features, namely Oct4, Nanog, Sox2, AKP, and TERT in the cells (He et al. 2016), development of gene-targeted goats containing β-lactoglobulin-modified, less-allergenic milk than the wild-type milk, implying that gene-targeted transgenic animals could serve as effective tools for minimizing allergenic reactions and improving human nutrition (Zhu et al. 2016), TALENs-mediated modification of myostatin genes to produce site-modified goats (Yu et al. 2016)
2017	Reprogramming goat fibroblasts by means of "Yamanaka factors," and differentiation of the reprogrammed cells into oocyte-like cells (Guo et al. 2017), induced reprogramming of goat embryonic fibroblasts (GEFs) to iPSCs, transfection of the cells by Oct4, Sox2, Klf4, and c-Myc mRNAs (Chen et al. 2017), birth of cloned transgenic cashmere goat kid showing EGFP (Bai et al. 2017), birth of *β-lactoglobulin* (*BLG*) gene knockout goat chimera kids by means of CRISPR/Cas9 editing (Zhou et al. 2017) Induced differentiation of diploid SSCs into functional haploid sperm in Saanen dairy goats (Deng et al. 2017)
2018	Using CRISPR/Cas9 genome editing to develop cashmere goats carrying an EDAR gene mutant (Hao et al. 2018), studies on properties and potential of recombinant human hLZ lysozyme produced in the milk of transgenic goats, revealing the antibacterial and nutraceutical potential of recombinant lysozyme (Carneiro et al. 2018); production of transgenic goats expressing a novel human plasminogen activator (He et al. 2018)

of MECs is a step forward for animal welfare and replacement of the animal used for studying mammary gland functions (Saipin et al. 2018).

iPSCs were established by enforced expression of reprogramming factors, namely POU5F1, SOX2, MYC, KLF4, LIN-28, and NANOG, in combination with a MIR302/367 cluster, delivered into cells by lentiviral vectors. The cells showed morphology typical of human and mice ESCs, expression of alkaline phosphatase and formation of teratoma in vivo (Sandmaier et al. 2015).

Chen et al. (2017) produced goat iPSCs by induced reprogramming of goat embryonic fibroblasts (GEFs) when GEFs were transfected by Oct4, Sox2, Klf4, and c-Myc mRNAs. The study is shown to be a safer and efficient approach to produce goat iPSCs.

Goat MSCs have been studies with reference to their use as regenerative medicine. MSCs isolated from goat amniotic fluid and bone marrow improved experimentally created dermal wounds. The AF-derived MSCs had higher healing properties by improving epithelialization, neovascularization, and collagen development (Pratheesh et al. 2017).

27.8 Genetic Manipulation, Transgenesis, and Recombinant Protein Production

Among transgenic animals, the goats are among most widely used species to produce recombinant proteins. In addition, the genetically modified goats are also used as model species in developmental biological studies.

Like other species, various strategies, such as microinjection of genes, viral vector-mediated transfer, electroporation, primordial germ cell-mediated gene transfer, or spermatogonial stem cell transplantations, are used to introduce produce recombinant metabolites.

Microinjection has been used in various programs to introduce genes of interest into multiple stages of embryo development. Despite the availability of potent gene transfer methods and gene editing tools, the microinjection is a method chosen to produce transgenic cells and animals (Gavin et al. 2018).

Enhancing quality of milk is one of the prime objectives of bioengineering of animals. Recently evolved methods such as CRISPR/Cas9 genome editing is used to generate transgenic caprine cells for generating cloned transgenic goats. Some examples of genome and gene editing are cited here. The goats having modified myostatin (MSTN) or fibroblast growth factor 5 (FGF5), or both interrupted, were produced by co-injection of single-cell-stage embryo with Cas9 mRNA and sgRNA targeting these genes. The study shows that it is possible to edit target cells at more than one different sites (Wang et al. 2015). CRISPR/Cas9 was used to generate goat chimera having interrupted BLG (Zhou et al. 2017). Wang et al. (2018) reported applications of CRISPR/Cas9 to edit the goat genome through microinjection of casmRNA and sgRNAs targeting MSTN and FGF5 in caprine embryos. Hao et al. (2018) generated a goat carrying an ectodysplasin-A receptor (*EDAR*) gene mutant.

27.9 Outlook and Challenges

The goat is an adapted livestock species and an important contributor of meat, milk, fiber, and therapeutic products. Because of easy maintenance, adaptation to varied climates and large range of forages, low age at maturity, and shorter gestation period, the goat is preferred for experimental studies. In particular, the goat is used as transgenic animal to produce recombinant therapeutic biomolecules.

One of the major challenges in goat embryos techniques is the small size of the animal. Surgical methods are used to collect oocytes, and transfer embryos into surrogate does. Performing these actions requires anesthesia which includes fasting and drugs to secure the animals. In addition, the protocols need expert veterinarians to perform the action.

Some laboratories have developed non-surgical methods of embryo transfer in goats. These protocols should be simplified further to simplify embryo transfer.

27.10 Conclusions

The goats are multipurpose livestock used for milk, meat, manure, and livelihood. Goats have an important contribution to nutrition, livelihood and economy of farmers, and biomedical sciences as transgenic animal.

References

Abril-Sánchez S, Freitas-de-Melo A, Beracochea F, Damián JP, Giriboni J, Santiago-Moreno J, Ungerfeld R (2017) Sperm collection by transrectal ultrasound-guided massage of the accessory sex glands is less stressful than electroejaculation without altering sperm characteristics in conscious goat bucks. Theriogenology 98:82–87. https://doi.org/10.1016/j.theriogenology.2017.05.006

Amiri Yekta A, Dalman A, Eftekhari-Yazdi P, Sanati MH, Shahverdi AH, Fakheri R, Vazirinasab H, Daneshzadeh MT, Vojgani M, Zomorodipour A, Fatemi N, Vahabi Z, Mirshahvaladi S, Ataei F, Bahraminejad E, Masoudi N, Rezazadeh Valojerdi M, Gourabi H (2013) Production of transgenic goats expressing human coagulation factor IX in the mammary glands after nuclear transfer using transfected fetal fibroblast cells. Transgenic Res 22(1):131–142. https://doi.org/10.1007/s11248-012-9634-y

Baguisi A, Behboodi E, Melican DT, Pollock JS, Destrempes MM, Cammuso C, Williams JL, Nims SD, Porter CA, Midura P, Palacios MJ, Ayres SL, Denniston RS, Hayes ML, Ziomek CA, Meade HM, Godke RA, Gavin WG, Overström EW, Echelard Y (1999) Production of goats by somatic cell nuclear transfer. Nat Biotechnol 17(5):456–461

Bai DP, Yang MM, Qu L, Chen YL (2017) Generation of a transgenic cashmere goat using the piggyBac transposition system. Theriogenology 15(93):1–6. https://doi.org/10.1016/j.theriogenology.2017.01.036

Bathgate R, Mace N, Heasman K, Evans G, Maxwell WM, de Graaf SP (2013) Birth of kids after artificial insemination with sex-sorted, frozen-thawed goat spermatozoa. Reprod Domest Anim 48(6):893–898. https://doi.org/10.1111/rda.12182

Behboodi E, Memili E, Melican DT, Destrempes MM, Overton SA, Williams JL, Flanagan PA, Butler RE, Liem H, Chen LH, Meade HM, Gavin WG, Echelard Y (2004) Viable transgenic goats derived from skin cells. Transgenic Res 13(3):215–224

Carneiro IS, Menezes JNR, Maia JA, Miranda AM, Oliveira VBS, Murray JD, Maga EA, Bertolini M, Bertolini LR (2018) Milk from transgenic goat expressing human lysozyme for recovery and treatment of gastrointestinal pathogens. Eur J Pharm Sci 15 (112):79–86. https://doi.org/10.1016/j.ejps.2017.11.005

Chen H, Zuo Q, Wang Y, Song J, Yang H, Zhang Y, Li B (2017) Inducing goat pluripotent stem cells with four transcription factor mRNAs that activate endogenous promoters. BMC Biotechnol 17(1):11. https://doi.org/10.1186/s12896-017-0336-7

Deng S, Wang X, Wang Z, Chen S, Wang Y, Hao X, Sun T, Zhang Y, Lian Z, Liu Y (2017) In vitro production of functional haploid sperm cells from male germ cells of Saanen dairy goat. Theriogenology 90:120–128. https://doi.org/10.1016/j.theriogenology.2016.12.002

Ferreira-Silva JC, Moura MT, Silva TD, Oliveira LRS, Chiamenti A, Figueirêdo Freitas VJ (2017) Oliveira MAL Full-term potential of goat in vitro produced embryos after different cryopreservation methods. Cryobiology 75:75–79. https://doi.org/10.1016/j.cryobiol.2017.01.009 (Epub 2017 Jan 27)

Gavin W, Blash S, Buzzell N, Pollock D, Chen L, Hawkins N, Howe J, Miner K, Pollock J, Porter C, Schofield M, Echelard Y, Meade H (2018) Generation of transgenic goats by pronuclear microinjection: a retrospective analysis of a commercial operation (1995–2012). Transgenic Res 27(1):115–122. https://doi.org/10.1007/s11248-017-0050-1

Guignot F, Bouttier A, Baril G, Salvetti P, Pignon P, Beckers JF, Touzé JL, Cognié J, Traldi AS, Cognié Y, Mermillod P (2006) Improved vitrification method allowing direct transfer of goat embryos. Theriogenology 66(4):1004–1011

Guo Y, Yu T, Lei L, Duan A, Ma X, Wang H (2017) Conversion of goat fibroblasts into lineage-specific cells using a direct reprogramming strategy. Anim Sci J 88(5):745–754. https://doi.org/10.1111/asj.12700

Haenlein GF (2001) Past, present, and future perspectives of small ruminant dairy research. J Dairy Sci 84 (9):2097–2115

Haenlein GF, Ramirez RG (2007) Potential mineral deficiencies on arid rangelands for small ruminants with special reference to Mexico. Small Rumin Res 68:35–41

Hao F, Yan W, Li X, Wang H, Wang Y, Hu X, Liu X, Liang H, Liu D (2018) Generation of cashmere goats carrying an *EDAR* gene mutant using CRISPR-Cas9-mediated genome editing. Int J Biol Sci 14(4):427–436. https://doi.org/10.7150/ijbs.23890 (eCollection 2018)

Hatziminaoglou Y, Boyazoglu J (2004) The goat in ancient civilisations: from the Fertile Crescent to the Aegean Sea. Small Rumin Res 21:123–129

He N, Dong Z, Zhu B, Nuo M, Bou S, Liu D (2016) Expression of pluripotency markers in Arbas Cashmere goat hair follicle stem cells. Vitro Cell Dev Biol Anim 52(7):782–788. https://doi.org/10.1007/s11626-016-0023-3

He Z, Lu R, Zhang T, Jiang L, Zhou M, Wu D, Cheng Y (2018) A novel recombinant human plasminogen activator: Efficient expression and hereditary stability in transgenic goats and in vitro thrombolytic bioactivity in the milk of transgenic goats. PLoS One 13(8): e0201788. https://doi.org/10.1371/journal.pone.0201788 (eCollection 2018)

Keefer CL, Keyston R, Lazaris A, Bhatia B, Begin I, Bilodeau AS, Zhou FJ, Kafidi N, Wang B, Baldassarre H, Karatzas CN (2002) Production of cloned goats after nuclear transfer using adult somatic cells. Biol Reprod 66(1):199–203

Kumar K, Agarwal P, Das K, Mili B, Madhusoodan AP, Kumar A, Bag S (2016) Isolation and characterization of mesenchymal stem cells from caprine umbilical cord tissue matrix. Tissue Cell 48(6):653–658. https://doi.org/10.1016/j.tice.2016.06.004

Maga EA, Shoemaker CF, Rowe JD, Bondurant RH, Anderson GB, Murray JD (2006) Production and processing of milk from transgenic goats expressing human lysozyme in the mammary gland. J Dairy Sci 89(2):518–524

Mao T, Han C, Deng R, Wei B, Meng P, Luo Y, Zhang Y (2018) Treating donor cells with 2-PCPA corrects aberrant histone H3K4 dimethylation and improves cloned goat embryo development. Syst Biol Reprod Med 64(3):174–182. https://doi.org/10.1080/19396368.2018.1446229

Melican D, Gavin W (2008) Repeat superovulation, non-surgical embryo recovery, and surgical embryo transfer in transgenic dairy goats. Theriogenology 69 (2):197–203

Morand-Fehr P, Boutonnet JP, Devendra C, Dubeuf JP, Haenlein GFW, Holst P, Mowlem L, Caote J (2004) Strategy for goat farming in the 21st century. Small Rumin Res 51(2):175–183. https://doi.org/10.1016/j.smallrumres.2003.08.013

Parrilla I, Vazquez JM, Roca J, Martinez EA (2004) Flow cytometry identification of X- and Y-chromosome-bearing goat spermatozoa. Reprod Domest Anim 39 (1):58–60

Pratheesh MD, Dubey PK, Gade NE, Nath A, Sivanarayanan TB, Madhu DN, Somal A, Baiju I, Sreekumar TR, Gleeja VL, Bhatt IA, Chandra V, Amarpal Sharma B, Saikumar G, Taru Sharma G (2017) Comparative study on characterization and wound healing potential of goat (*Capra hircus*) mesenchymal stem cells derived from fetal origin amniotic fluid and adult bone marrow. Res Vet Sci 112:81–88. https://doi.org/10.1016/j.rvsc.2016.12.009

Qin Y, Yang S, Xu J, Xia C, Li X, An L, Tian J (2018) Deep insemination with sex-sorted Cashmere goat sperm processed in the presence of antioxidants. Reprod Domest Anim 53(1):11–19. https://doi.org/10.1111/rda.13045

Reggio BC, James AN, Green HL, Gavin WG, Behboodi E, Echelard Y, Godke RA (2001) Cloned transgenic offspring resulting from somatic cell nuclear transfer in the goat: oocytes derived from both follicle-stimulating hormone-stimulated and non-stimulated abattoir-derived ovaries. Biol Reprod 65 (5):1528–1533

Saipin N, Noophun J, Chumyim P, Rungsiwiwut R (2018) Goat milk: non-invasive source for mammary epithelial cell isolation and in vitro culture. Anat Histol Embryol 47(3):187–194. https://doi.org/10.1111/ahe.12339

Salmon VM, Leclerc P, Bailey JL (2017) Novel technical strategies to optimize cryopreservation of goat semen using cholesterol-loaded cyclodextrin. Cryobiology 74:19–24. https://doi.org/10.1016/j.cryobiol.2016.12.010

Sandmaier SE, Nandal A, Powell A, Garrett W, Blomberg L, Donovan DM, Talbot N, Telugu BP (2015) Generation of induced pluripotent stem cells from domestic goats. Mol Reprod Dev 82(9):709–721. https://doi.org/10.1002/mrd.22512

Silanikove N (2000) Effects of heat stress on the welfare of extensively managed domestic ruminants. Livest Prod Sci 67:1–18

Smith V (2006) Food fit for the soul of a Pharaoh. The Mortuary temple's bakeries and breweries. Expedition 48:27–30

Traldi AS, Leboeuf B, Cognié Y, Poulin N, Mermillod P (1999) Comparative results of in vitro and in vivo survival of vitrified in vitro produced goat and sheep embryos. Theriogenology 51(1):175

Wang X, Yu H, Lei A, Zhou J, Zeng W, Zhu H, Dong Z, Niu Y, Shi B, Cai B, Liu J, Huang S, Yan H, Zhao X, Zhou G, He X, Chen X, Yang Y, Jiang Y, Shi L, Tian X, Wang Y, Ma B, Huang X, Qu L, Chen Y (2015) Generation of gene-modified goats targeting MSTN and FGF5 via zygote injection of CRISPR/Cas9 system. Sci Rep 10(5):13878. https://doi.org/10.1038/srep13878

Wang X, Niu Y, Zhou J, Zhu H, Ma B, Yu H, Yan H, Hua J, Huang X, Qu L, Chen Y (2018) CRISPR/Cas9-mediated MSTN disruption and heritable mutagenesis in goats causes increased body mass. Anim Genet 49(1):43–51. https://doi.org/10.1111/age.12626

Yu H, Chen J, Liu S, Zhang A, Xu X, Wang X, Lu P, Cheng G (2013) Large-scale production of functional human lysozyme in transgenic cloned goats. J Biotechnol. pii: S0168-1656(13)00456-2. https://doi.org/10.1016/j.jbiotec.2013.10.023

Yu B, Lu R, Yuan Y, Zhang T, Song S, Qi Z, Shao B, Zhu M, Mi F, Cheng Y (2016) Efficient TALEN-mediated myostatin gene editing in goats. BMC Dev Biol 16(1):26. https://doi.org/10.1186/s12861-016-0126-9

Zhou ZR, Zhong BS, Jia RX, Wan YJ, Zhang YL, Fan YX, Wang LZ, You JH, Wang ZY, Wang F (2013) Production of myostatin-targeted goat by nuclear transfer from cultured adult somatic cells. Theriogenology 79(2):225–233. https://doi.org/10.1016/j.theriogenology.2012.08.006

Zhou W, Wan Y, Guo R, Deng M, Deng K, Wang Z, Zhang Y, Wang F (2017) Generation of beta-lactoglobulin knock-out goats using CRISPR/Cas9. PLoS ONE 12(10):e0186056. https://doi.org/10.1371/journal.pone.0186056 (eCollection 2017)

Zhu H, Hu L, Liu J, Chen H, Cui C, Song Y, Jin Y, Zhang Y (2016) Generation of β-lactoglobulin-modified transgenic goats by homologous recombination. FEBS J 283(24):4600–4613. https://doi.org/10.1111/febs.13950

Part III

Livestock Genomics

Animal Genomics—A Current Perspective

28

Abstract

The livestock sector as an important subsector of agriculture plays an important role in national economy. In current era of molecular genetics, genomics and sequencing technologies applicable to genome, proteome and transriptome sequencing, and access to gigantic sequence data, a number of markers are identified and used to analyze genomic diversity. Molecular markers have emerged as amazing tools for selection of farm animals for beneficial traits, and detecting diseases prior to clinical symptoms. Researchers and scholars as beginners in livestock production and management must be familiar with types, advantages, and prospects of these valuable tools.

Key Points

- Genetic evaluation of livestock is indispensable for their conservation and utilization
- Molecular genetic markers have emerged as incredible tools for livestock selection and disease diagnosis.

Keywords

Livestock genomics · Genomic diversity · Diversity analysis · Microsattalite markers, RFLP · DNA bar coding · Microaarays

28.1 Introduction

Livestock is the major asset for marginal and resource-poor farmers, and contribute significantly to national economy. Apart from being the source of milk and meat, the livestock is an important source of draft power and traction, and recycling of nutrient that helps to sustain crop production. The animals not only provide nutritive food rich in high biological value protein, but also contribute toward family income and generating employment in the rural sector, particularly among landless, small, marginal farmers, and women. The share of working men and women in animal production (per 1000 persons) in a rural area is 10.2 and 88.3, respectively, whereas the same figures out to be 5.5 and 25.6 in the urban area.

Currently, India ranks first in total livestock population (512.05 million), milk production, cattle population (190.90 million), buffalo population (108.70 million), carabeef production, goat milk production, and total bovine population (299.99 million). Total milk, meat and egg production is 165.4 MT (million tons), 7.4 MT and 88.1 billion, respectively, whereas the total wool production in India is 43.5 million kg (Singh 2018; https://www.vetextension.com/current-livestock-animal-husbandry-statistics-india/, June 15, 2019). Total export of livestock and livestock products was 185,233,966 lakhs in the year 2016–17. In the year

B. Singh et al., *Advances in Animal Biotechnology*,
https://doi.org/10.1007/978-3-030-21309-1_28

2015–16, the share of agriculture and allied in gross value added (GVA)) was 17.5% while the same for livestock sector is 4.5% with GVA for total, agriculture and allied, and livestock sector being 12458642; 2175547, and 560613 crores, respectively (https://www.nddb.coop/information/stats/GDPcontrib: National Accounts Statistics 2016; Central Statistical Organization; GoI). These statistics clearly indicate the importance of the livestock sector for the economy and livelihood.

28.2 Genomics in Livestock

Genomics, a multidisciplinary approach (Fig. 28.1), is an effective method of prediction of breeding values, selecting breeds for ideal production (milk/meat/wool) traits, and developing nutritional strategies to support better physiology and production performance (Li et al. 2017). Genomics-based markers are used to predict mammary gland and peripheral tissue health, and overall disease status of prior to the onset of disease.

28.3 Animal Genomic Diversity and Genetic Resources

Various livestock genetic resources have originated from respective wild ancestor species and evolved according to diverse environments including domestication, forage resources, and agro-climatic conditions. Animals were domesticated in different parts of the world at different time periods and eventually spread over the

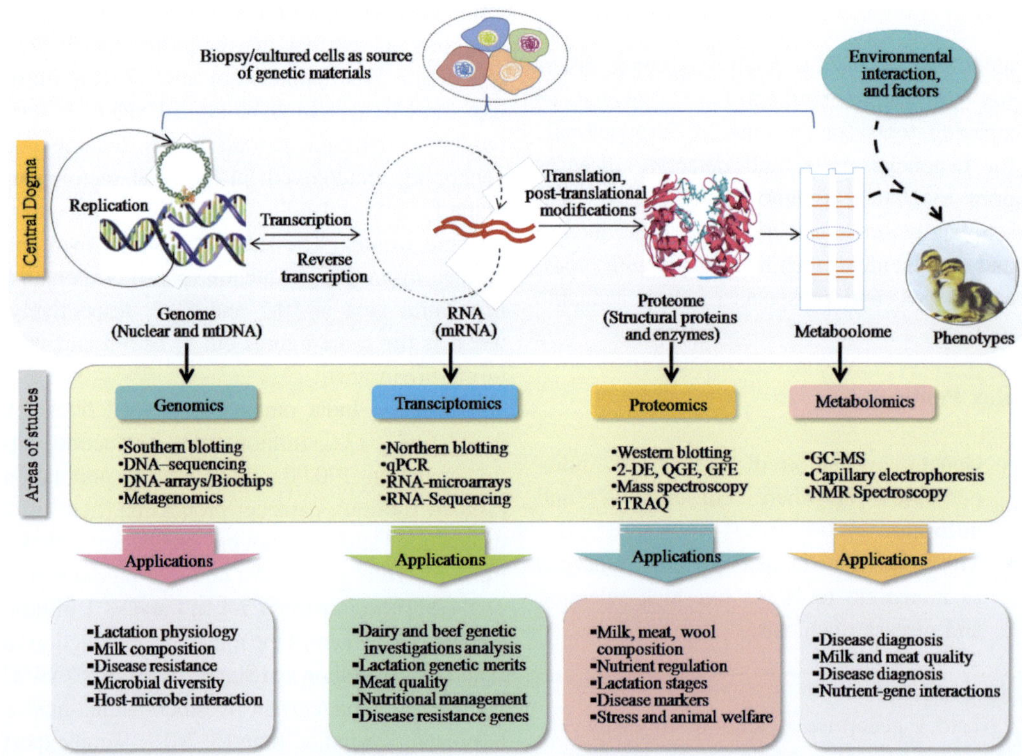

Fig. 28.1 Diagrammatic illustration of applications of various 'omics' tools in livestock production. NGS is faster and has overcome the limitations associated with conventional sequencing of nucleic acids. *Abbreviations* 2-DE—2-dimensional gel electrophoresis; GC-MS—gas chromatography–mass spectrometry; iTRAQ—isobaric tags for absolute and relative quantification; NMR—nuclear magnetic resonance spectrometry; qPCR—quantitative PCR; QGE—quantitative gel electrophoresis; GC-MS; mRNA—messenger RNA

inhabited regions of the globe along with human migrants. Inhabitation across diverse climate conditions required adaptation as per the climate, available vegetation, and animal rearing practices. This resulted in phenotypic as well as genotypic diversity in different livestock species. These variations have been shaped through the processes of mutations, genetic drift, and natural and artificial selection. The unique genetic makeup of different livestock species evolved through these processes defines variations in their production and adaptive capabilities.

However, during the last few years, there has been a net loss of diversity of animal genetic resources because of the increased rate of disappearance of native or locally domesticated breeds of livestock species (FAO 2007c). Intensive livestock development programs are increasingly promoting the widespread use of a few preferred "improved" breeds. This has resulted in a reduction in population numbers for many of the indigenous breeds as well as the genetic variability within species. Availability of efficient methods of breeding and emerging techniques used for accurately estimating breeding values is further reducing the genetic variability within breeds and effective population size. This has further been accelerated due to habitat change or destruction, biological invasions, global expansion, and global warming. The loss of commercial relevance of the breeds is also one of the reasons that important breeds are facing extinction now.

It is estimated that in the twentieth century, about 600 out of 3000 breeds of the seven major mammalian species, viz. cattle, pig, sheep, goat, horse, donkey, and buffalo are already lost (Ruane 1999). This situation is even more alarming as per FAO estimates, where it is suspected that every week at least one traditional breed of livestock becomes extinct. Scenario for India and other developing countries would also be of major concern, as many of the breeds would be lost without ever having been adequately characterized or studied (Köhler-Rollefson 1997; FAO 2001).

Loss of genetic diversity in developing countries is mainly due to the rapid intensification of livestock production, failure to characterize local breeds and their importance, and inappropriate breed replacement or crossbreeding facilitated by the availability of high-performing breeds (FAO 2007b). Use of indiscriminate crossbreeding strategies to infuse native breeds with exotic germplasm is a major cause in dilution or loss of important gene pool. Although crossbreeding may offer benefits, such as producing high yielding animals, it could also result in the vanishing of valuable traits developed over years of natural selection and adaptation. For instance, due to mass level cattle, crossbreeding several native breeds have become extinct or mixed with foreign breeds.

Conservation of livestock is a global issue and to preserve the maximum amount of genetic diversity and further set the conservation priorities, several efforts are being undertaken worldwide. The objectives of Food and Agricultural Organization (FAO) Global Plan of Action for Animal Genetic Resources (FAO 2007a) and the Convention on Biological Diversity (CBD 1992), clearly reflect the global concern to safeguard the genetic diversity in livestock breeds. However, to preserve the maximum amount of genetic diversity and further set conservation priorities, it is important to characterize different breeds so as to know how unique or different a breed is from other populations. Although it is difficult to confirm the difference between the breeds in terms of agriculturally important genes, general genetic variability are most suitable criteria to identify the breeds for genetic uniqueness, the important criterion of selecting breeds for conservation. The underlying assumption is that the taxonomically distinct breeds are most likely to have special adaptation and gene combinations that are missing in other breeds (Hall and Bradley 1995). Therefore, genetic characterization is the first step to answer questions on taxonomy, evolution, domestication processes, management of genetic resources and setting conservation plans for their effective utilization. By selecting for conservation those populations with unique evolutionary histories a maximum amount of diversity could be preserved.

In addition, with new innovations in computational technology, novel strategies such as whole-genome SNP chips and DNA Barcoding have also emerged. The expansion in DNA information will facilitate the study of genome-wide diversity and such information is much more precise for the assessment of genetic diversity than previous markers.

28.4 Methods for Assessing Animal Genetic Diversity

28.4.1 Morphological Markers

Traditionally, phenotypic characterization based on morphological features, physical body measurements, production traits, reproductive traits, and adaptive traits were used in the description of breeds. Morphological markers are the visible external features such as coat color, body shape, hair, and skin structure. They are decided based on direct visual observation and measurement. Being easily visible and comparable within the population, morphological markers are the primary factors of phenotypic characterization (Yang et al. 2013).

Reports are available wherein assignments of indigenous breeds were based on phenotypic/subjective data and information generated from the local sources. However, an animal's phenotype is determined by its genetic background and the environment it experiences. Characterization of animal genetic resources using morphological markers is based on subjective judgments and descriptions, and the conclusions reached are often not wholly perfect. These markers have limited applications in evaluation of qualitative traits as it is not easy to eliminate the effects of environmental factors on phenotypes of an organism. Furthermore, it is not always feasible to describe a breed on the basis of phenotypic characteristics alone because of the difficulty of combining the various measures used, and interbreed phenotypic comparison is difficult or may not be reliable. This led to the development of concept of using alternative markers for genetic characterization.

28.4.2 Cytological and Biochemical Markers

The switch was toward cytological and biochemical markers. Many researchers have used cytological markers including chromosome karyotypes, bandings, repeats, deletions, translocations, and inversions for characterization of livestock diversity as well as for assessment of their evolution from wild ancestors (Popescu et al. 1976; Bitgood and Shoffner 1990). As biochemical markers (either proteins or isozymes) are products of gene expression, they have been used for investigating the genetic variation within species and phylogenetic relationships between species. These biochemical markers were used extensively, but it was difficult to predict or measure genetic divergence and gene flow among closely related populations with these classical markers because of their low discriminating power (Goldstein and Schlötterer 1999), attributed to extremely low mutation rate due to which the occurrence of any new mutation and subsequent fixation in a population was minimal. Also, the biochemical markers are vulnerable to environmental impacts, individual growth, sex-limited, and age-dependent effects. However, they provide more detailed representation of polymorphisms than morphological or cytological markers, and are rapid, economic, and straightforward tools, hence still widely used in elucidating the origin and classification of some species (Jonker et al. 1982).

28.5 Molecular Tools for Diversity Analysis

With recent advances in genome studies and genotyping techniques, much attention is focused on ultimate level of variability, i.e., at the DNA level. "Molecular markers" or "Genomic markers" capable of detecting genetic variation both within coding as well as non-coding sequences of DNA have been developed. It is now possible to uncover genetic variations/polymorphism and use them as markers for understanding the genetic basis of observed phenotypic variability. Use of DNA polymorphism as molecular markers has opened many vistas in genetic characterization, conservation, improvement, and molecular evolution studies in livestock species. Adequate maternally, paternally, and bi-parentally inherited genetic markers are available for particular species, and can be used to estimate evolutionary history, population subdivision, dispersal, gene flow, effective population size, extended pedigrees, levels of relatedness, breeding structure, etc. (Goldstein and Schlötterer 1999). Broadly, based on their functionality, genetic markers have been grouped into two types (O'Brien 1994): Type I markers are gene-based and possess a relatively low degree of polymorphism, but have extensive evolutionary conservation. Type II markers usually have no identifiable biological function, but they are highly polymorphic, and not well conserved between species. Molecular markers include microsatellite markers (simple tandem repeat, STR), single-nucleotide polymorphism (SNP), variable number tandem repeats (VNTRs), random amplified polymorphic DNA (RAPD), single-strand conformation polymorphisms (SSCPs), amplified fragment length polymorphisms (AFLPs), and restriction fragment length polymorphisms (RFLPs). All these markers reflect differences in DNA sequences usually with a trade-off between precision and convenience (Table 28.1).

28.6 Restriction Fragment Length Polymorphism

Restriction fragment length polymorphism (RFLP) was the first DNA-based marker to detect polymorphisms among different individuals (Botstein et al. 1980). Herein, total cellular DNA is digested with restriction endonucleases (REs) that cleave DNA at specific recognition sites of varying length and sequence, to reduce the genome to a large pool of restriction fragments of different sizes. HindIII, EcoRI, and BamHI are the most preferred REs as they generally provide the best size distribution of DNA fragments, and are inexpensive. As REs recognition sites are present both in coding and non-coding regions, all the variations in the genome are trapped. Restriction fragments thus obtained are separated by their size on an agarose gel by electrophoresis. If two individuals have differences in their DNA sequences at particular restriction sites, then the RE will cut DNA into fragments of different lengths; therefore, differences in the number of DNA fragments will be observed among two or more individuals. Fractionated DNA is then transferred to a nylon membrane by a process called Southern blotting. Specific DNA fragments are visualized by hybridizing the DNA fragments bound to the nylon membrane with a radioactively-labeled or chemiluminescent homologous probes that have sequences complementary to the DNA of interest. These are processed by an X-ray film and the different fragments are visible by autoradiography. Before hybridization, the DNA fragments on the nylon membrane as well as the probe are denatured to get single stranded sequences enabling them to pair with their complementary DNA sequence.

RFLP analysis was among the early methods of characterization of genetic diversity or breeding patterns in animal populations. It detects DNA polymorphisms among different individuals based on nucleotide base substitutions, insertions, deletions, duplications, and inversions within whole genome that either eliminates or generates new restriction sites. Disadvantages of RFLP markers include detection of relatively low polymorphisms as RFLP provides information related to specific mutations at restriction enzyme cut sites; being labor-intensive and time-consuming. In addition, detection of polymorphism is based on radioisotope-based methods which are hazardous. However, being Co-dominant markers, these can distinguish heterozygotes from homozygotes. Also these have high reliability, reproducibility, and selective neutrality refers that is different alleles of a certain gene confer equal fitness.

Table 28.1 A summarized comparative overview of different molecular markers used in animal health and production

Marker	Advantages	Limitations
SNPs	Low mutation rate High abundance Easy to type New analytical approaches are available Cross-study comparisons are possible as data repositories are available	Substantial rate heterogeneity among sites Expensive to isolate Ascertainment bias Low information content of a single SNP
RAPDs	Low cost Produces large number of bands which can act as single locus marker	Low reproducibility Mainly dominant Difficult to analyze and automate Difficult cross comparison
Microsatellites	Highly informative (large number of alleles, high heterozygosity Low ascertainment bias Easy to isolate Easy to automate	High mutation rate Complex mutation behavior Less abundance Cross-study comparisons are possible, if data repositories are available
DNA sequencing	Highest level of resolution Unbiased; Time effective High precision; Easy cross comparison	Higher cost

28.6.1 Random Amplified Polymorphic DNA

Random amplified polymorphic DNA (RAPD) markers have been widely used molecular marker type in all species for genetic characterizations. They are the first of PCR-based markers developed independently by Welsh and McClelland (1990) and Williams et al. (1990). Main reason for popularity of RAPD marker system is that unlike other PCR-based genetic marker systems prior DNA sequence information is not required for designing PCR primers. RAPD amplifies the target genomic DNA using a very small amount of template DNA (usually less than 10 nanograms) and short and arbitrary primers (commonly 10 bp) in a PCR reaction. It can be used to yield relatively elaborate DNA profiles for detecting augmented fragment length polymorphisms between individuals. The single RAPD primer anneals to many locations in the genome revealing multiple loci and thus it is possible to obtain a large number of RAPD genetic markers in a short span of time and at relatively low cost. Since the arbitrary primers match different parts of the genomic DNA, PCR products differ in number and size (polymorphism). RAPD-PCR fingerprints have been successfully used to deduce genetic diversity among different species including wild boar, pig, cattle, horse, buffalo, venison, dog, cat, rabbit by generate specific fingerprint patterns.

28.7 Amplified Fragment Length Polymorphism

Amplified fragment length polymorphism (AFLP) markers are comparatively fresh development (Vos et al. 1995). It is a combination of the RFLP and PCR techniques. These are like RAPDs as many markers, generally being dominant can be assayed swiftly by PCR. Like RFLPs, because they survey the genome for the presence of RFLPs, and therefore appear to be more repeatable than RAPDs. Here, the genomic DNA is digested with a REs followed by ligation of complementary double-stranded adaptors to the ends of the digested fragments. Fragments are then amplified with specified primers that are complementary to a selective sequence on adaptors. Subsequent separation of the amplified fragments is obtained by selective primers, and visualized using autoradiography or fluorescence methodologies (Blears et al. 1998).

AFLP technology has the capability to detect various polymorphisms in different genomic regions, simultaneously. AFLPs have an upper edge over RFLPs as these are comparatively quicker and require less labor. AFLP is used in population genetics to determine slight differences within populations, and in linkage studies to generate maps for quantitative trait loci (QTL) analysis (Ajmone-Marsan et al. 2002; Negrini et al. 2007). However, being dominant bi-allelic markers, AFLPs are unable to distinguish between dominant homozygous vs. dominant heterozygous individuals.

28.8 Microsatellite DNA Markers

Microsatellite markers were first developed for use in genetic mapping in humans (Litt and Luty 1989; Weber and May 1989). Microsatellites are also known as simple sequence repeats (SSR), short tandem repeats (STR) and sequence-tagged microsatellite repeats (STMR). These are densely and evenly distributed throughout the genome and often exhibit substantial variation/polymorphism due to site-specific length variation, as a consequence of the occurrence of different number of repeat units. Simple sequence motifs repeated in tandem, one to six nucleotides (mono-, di-, tri-, tetra-, penta-, and hexanucleotides tandem repeats) in length. For example,

mono-nucleotides, AAAAAAAAAAA would be referred to as $(A)_{11}$
di nucleotides, GTGTGTGTGTGT would be referred to as $(GT)_6$
tri nucleotides, CTGCTGCTGCTG would be referred to as $(CTG)_4$
tetra-nucleotides, ACTCACTCACTCACTC would be referred to as $(ACTC)_4$.

The difference in repeat number can be reliably distinguished, and the variants are inherited as alleles at each locus. For example, one allele might have 10 copies of the AC tandem repeat $(AC)_{10}$, whereas another allele would have 11 copies $(AC)_{11}$, another 12 copies $(AC)_{12}$, and so on. Polymorphic nature of this type of locus with variations many times more common than is non-repetitive sequence makes microsatellite ideal for examining genetic variation within a species. With the availability of high-throughput systems, the most frequently used markers in genetic diversity studies are the microsatellite markers.

Microsatellites occur at a frequency of 1 SSR per 10 kb DNA and numbering a total of about 50–100 thousand in the mammalian genome. Microsatellites occur very frequently and randomly in most eukaryotic DNA including plants. Human genomic DNA contains on an average one microsatellites every 6 bp (Beckman and Weber 1992). Their short lengths make them amenable to amplifications by PCR and subsequent separation by polyacrylamide gels with the resolution of alleles differing by as low as a single base. Additionally, with the automation in sequencing and genotyping technologies, it has now become much easier genotype microsatellite loci in large number samples. The FAO has formulated an integrated program for global management of genetic resources of various livestock species using species-specific lists of microsatellite loci (about 30 per species) for cattle, chicken, sheep, swine, and buffalo diversity studies.

An advisory group of the International Society for Animal Genetics (ISAG) in collaboration with FAO (FAO MoDAD project) has established, for each species of interest, a set of microsatellite markers to be used as the standard set for the calculation of genetic distances. Adherence to such recommendations allows for reasonable comparison of parallel or overlapping studies and helps combine results in meta-analyses. To attain a certain precision for different levels of resolution or discrimination among breeds, it is recommended to sample at least 25 animals per breed (mainly blood samples, hair or tissues may be taken) and investigate 25 microsatellite loci with 4–10 alleles per locus. The primer sequences and map position of each of these markers can be obtained from Domestic Animal Diversity Information System (DAD-IS-MoDAD) and are also available at site http://dad.fao.org/dad-is/data/molecula/index.html.

To select appropriate microsatellite, the working group of MoDAD project has issued the following criteria.

- The microsatellite marker should be in public domain
- Wherever possible, microsatellite loci that have been identified in mapping studies be used and should preferably be known to be unlinked
- The microsatellite variants should be shown to exhibit Mendelian inheritance
- Each microsatellite locus should exhibit at least four alleles
- There should be information on the microsatellite loci in a published report
- The microsatellite loci suitable for several related species (heterologous markers) should be preferred
- Microsatellite markers used should be suitable for multiplexing with automated DNA sequencer.

These criteria were agreed in a meeting of the EU-AIR concerted action group on "Analysis of genetic diversity in cattle to preserve future breeding option," held in Dublin in 1995. List of microsatellite markers was compiled as per above recommendations for universal use in molecular genetic characterization of animal breeds so that joint analysis of future data from different laboratories would be possible for prioritizing breeds for conservation in terms of genetic uniqueness.

28.9 Isolation of Microsatellite Markers

Development of simple sequence repeat genetic markers requires considerable efforts. Genomic DNA libraries rich in microsatellite sequences needs to be created and screened for clones containing SSR sequences (Ostrander et al. 1992). Further, there is an absolute requirement to sequence these clones to know about unique sequence regions flanking the SSR so as to design the PCR primers for amplification of microsatellites in individual animals. After developing the PCR primers, the regions should be amplified to see the polymorphism.

Broadly, two strategies are used for the isolation of microsatellite markers.

(A) **Cosmid-derived microsatellite markers**

Herein, the genomic DNA, after digestion with REs is cloned into suitable vectors mostly cosmids, thus forming a cosmid genomic library. Cosmids are then screened with a labeled $(CA)_n$ or $(GT)_n$ polynucleotide probe. Clones that hybridize to the probes are detected by autoradiography. The positive clones are isolated and the insert (microsatellite) which they harbor is sequenced and characterized. Appropriate primers are designed from the flanking regions.

(B) **Microdissected chromosome-derived microsatellite markers**

In this strategy, a chromosome spread is obtained from a blood culture, and the chromosome of interest is identified under microscope. This chromosome is dissected using a micromanipulator. Microdissected chromosomal fragments are then used to construct genomic DNA library which is screened with radiolabeled $(CA)_n$ or $(GT)_n$ probes. The positive clones are isolated and subjected to PCR amplification. PCR products are sequenced and the sequences checked for uniqueness to develop PCR primers. A modification of this method involves amplification of the microdissected chromosomal fragments by PCR using degenerate oligonucleotide primers. Biotinylated $(CA)_n$ probes are added to amplified products. After denaturation and annealing, the annealed DNA is added to streptavidin paramagnetic particles and incubated to capture DNA fragments hybridized to biotinylated $(CA)_n$ probes. The bound DNA is eluted and amplified using appropriate primers. The amplified products are purified and sequenced for use as markers.

28.10 Evolution of Microsatellites

It is believed that when DNA is being replicated, errors occur in the process and extra sets of these repeated sequences are added to the strand. Although a clear understanding of the origin and evolution of microsatellites is lacking, the number of repeats increases or decreases by a single repeat unit, though sometimes more. Simple repeats might have generated mostly by slipped-strand mispairing (Moxon and Wills 1999) (summarized below), or by insertions or substitutions (Zhu et al. 2000).

28.11 Slipped-Strand Mispairing

In this process, the number of microsatellite repeats increases or decreases during DNA replication. An increase in the number of microsatellite repeats develops when slippage occurs on newly synthesized strand during its binding to the template strand. DNA polymerase adds the nucleotides to fill in the gap, thereby increasing the strand by one repeat. The decrease in the repeat number occurs when old or template strand slips, therefore, resulting in the repair enzymes deleting a repeat. DNA polymerase has a very high rate of slippage or templates containing simple repeats in vivo, but most of these errors are corrected by cellular mismatch repair systems. The instability of simple repeats observed for some human diseases may be a consequence of either an increased rate of DNA polymerase slippage, or a decreased efficiency of mismatch repair (Strand et al. 1993).

28.12 Insertions and Substitutions

Slippage of DNA polymerase depends on mispairing of tandem repeats during DNA replication, so it may not occur when there are few tandem repeats. Studies of slippage mutations show that they are more common in loci with longer repeats. Loci with fewer than five repeats are rarely polymorphic. For slippage to occur on longer repeats, some mechanism other than slippage must occur on shorter repeats from which longer repeats have evolved. Microsatellite sequences are exceptionally vulnerable to spontaneous insertion or deletion mutations. Non-triplet microsatellites when located in coding sequences are expected to introduce frameshift mutations at high frequency. Substitutions are much more common than insertions and they are the dominant sources of new two-repeat loci. Microsatellites may mutate at the rate of 10^3 to 10^5 mutations per gamete.

28.13 Theoretical Models of Microsatellite Mutations

Theoretical mutation models have been derived to explain the evolutionary processes of microsatellites from which genetic distances and population differentiation are estimated. The infinite allele model (IAM) was given by Kimura and Crow (1964). According to this model, a mutation can involve any number of tandem repeats and always results in a new allele state not previously existing in the population. But this model does not confer with the slipped-strand mispairing mechanism responsible for microsatellite length variation. This mechanism leads to small changes in the repeat numbers and alleles may mutate toward allele states that are already present in the population. In order to explain the discrepancies in the mutational processes, the step-wise mutation model (SMM) was introduced in the 1970s. The model assumes that the entire sequence of allelic states can be expressed as integers and mutation results in a change in one repeat unit either by insertion or deletion (Kimura and Ohta 1978). In addition to this model, Di Rienzo et al. (1994) described the two-phase model (TPM), where a limited proportion of mutations involve several repeats.

28.14 Limitations

28.14.1 Null Alleles

Failure of amplification of some alleles due to mutations in the binding regions results in reduction or loss of PCR products. These are termed as null alleles, and may lead to serious underestimation of heterozygosity. In a heterozygote of two different alleles, if one allele fails to amplify due to failure of primer annealing, then phenotype will appear as a single-banded homozygote. The problem may be overcome by designing new primers though it is a tedious task.

28.14.2 Slippage

This problem is due to the activity of the *Taq* polymerase used in the PCR. During PCR amplification, the thermo-polymerase tends to "slip" leading to the production of differently sized products. These products are less intense and are also referred to as shadow bands. Further, the *Taq* polymerase has a tendency to add an additional ATP at the 3'end of the amplified PCR products. This can also lead to difficulties in scoring the bands.

28.14.3 Homoplasy

Homoplasy can be defined as the co-occurrence of alleles that are identical by descent. If two alleles are inherited without any mutation from the same ancestral allele, they are identical by descent. But, two alleles may have the same structure and even the same sequence, but may not have been inherited from the same ancestral allele. Such alleles are identical in the state (Jarne and Lagoda 1996).

28.15 Advantages of Microsatellite as Markers for Genetic Diversity Studies

Microsatellites are considered as the most powerful genetic markers as they overcome many of the difficulties associated with other types of marker. Major advantages of these highly polymorphic microsatellites are their locus specificity, abundance and random distribution over the genome, co-dominant inheritance, ease and speed of application, and suitable for automated analysis. Their short length makes them amenable to amplification by PCR and subsequent separation either by polyacrylamide gel electrophoresis (PAGE) with the resolution of alleles differing by as little as single base, or genotyping by automated sequencer. All these features make them valuable for linkage analysis, genome mapping, parentage control, and phylogenetic analysis. Due to their remarkable features (Box 1), the microsatellite DNA markers are the preferred tool for analysis of genetic variations in closely related populations of organisms.

Box 1. Advantages or Merits of Microsatellite Markers

- Microsatellite markers have essentially replaced other markers because of their high polymorphic nature.
- The whole procedure of microsatellite genotyping using PCR can be automated and as many as 10 microsatellite loci can be analyzed simultaneously using multiplex.
- In contrast to RAPD and other multilocus genetic fingerprinting, where results may be difficult to interpret genetically and may be difficult to compare and replicate in different laboratories, result with microsatellite analysis is easily comparable between different laboratories.
- Ease to standardize, and are reproducible. The genotyping can be done on

most tissues and cell types with the same ease.
- For most livestock species there are now many microsatellite markers to choose from, far more than is necessary to compare allele frequencies between breeds.

28.16 Application of Microsatellite Markers in Diversity Analysis and Population Structure

The microsatellite allelic frequency data have successfully been exploited in numerous studies to assess genome evolution, population dynamics, breed demarcation, individual assignments, and phyogenetic structuring in different species. International comparison tests under ISAG to establish an international standard also exists. High level of polymorphism coupled with the ease of analysis has made this type of marker being one of the most widely used for genetic analysis. The usefulness of microsatellites for estimation of genetic distances among closely related population has been documented by numerous studies. Different breeds of livestock may have similar phenotypic characters, and sometimes the animals of same breeds may look very different. Breeds, animals sharing the same alleles at similar frequencies, are genetically closely related, whereas those having the same alleles at different frequencies or carrying many different alleles are genetically distant.

Genetic distance is a reliable measure of differences among breeds and can be estimated from the differences in the frequencies of different genetic variants (alleles) at a number of microsatellite loci. From the patterns of within-population genetic variation at microsatellite loci, it is possible to deduce the demographic factors important for conservation of domestic cattle diversity.

28.17 Biodiversity Analysis

It is possible to predict overall magnitude of genetic diversity within breed by analyzing microsatellite profiles for each individual across different loci. The priority breeds for conservation should be the ones with large within-breed diversity. Large number of projects studying genetic variation/characterization of domestic animals species has been performed with the aim of providing information for breed prioritization for conservation based on DNA markers (Kantanen et al. 2000; Bjørnstad et al. 2000; Bjørnstad and Røed 2001; Saitbekova et al. 2001). About 87 projects from 93 countries have been completed, mostly on ruminants using microsatellite loci (Baumung et al. 2004).

The microsatellites, therefore, are the molecular markers that could help in assessing the genetic variation within and between breeds and provide a rational basis for ranking the breeds in terms of genetic uniqueness. Also, microsatellite can help in solving the problems ranging from individual specificity, such as the questions of relatedness and parentage, the genetic structure of populations, the comparison among breeds/populations/species to the linkage analysis, and gene mapping.

As the importance for characterization and conservation of livestock genetic resources was realized, almost all the descript breeds of major livestock (cattle, buffalo, sheep, goat, horse, and pig) and poultry have also been characterized using microsatellite markers. Initial attempts were made primarily to assess the genetic diversity of indigenous animal genetic resources. Further, concerted efforts were put into address the different issues related to molecular genetic characterization. Species-wise attempts were made to evaluate the genetic diversity and relationship among breeds using species-specific microsatellite markers. Efforts were also made to characterize buffalo (Mishra et al. 2009a, b), cattle (Sodhi et al. 2011a, b; Mukesh et al. 2009), goat genetic resources (Fatima et al. 2008; Dixit et al. 2010), pig and horse (Koringa et al. 2008), and poultry (Pandey et al. 2002).

Besides, the microsatellite markers have also been helpful to identify genetic uniqueness of the breeds. Microsatellite-based genetic diversity analysis has revealed Banni buffaloes of Kutch, Gujarat (India), as genetically unique population among other buffaloes of Western India, and thus indicated as a separate buffalo breed (Mishra et al. 2010).

Microsatellites are thus most suitable to determine the relationships, expressed as genetic distances (Takezaki and Nei 1996) among breeds, possible levels of inbreeding in each breed, gene flow in livestock populations, most diverse and distinctive, i.e., "genetically unique" breeds/populations for higher priority in conservation programs, and relative contribution of each breed to the total (species) genetic diversity. These markers have been successfully used for differentiation of closely related breeds and assignment of individuals to specific breeds.

Across the globe, microsatellites have been successfully used to assess the genetic variation and relationship between various cattle breeds (Martín-Burriel et al. 1999; Kantanen et al. 2000). The relationship among breeds of species other than cattle have also been estimated, viz. goats (Saitbekova et al. 1999), horse (Bjørnstad et al. 2000), and donkey (Jordana et al. 2001).

28.18 Admixture Analysis

Microsatellites have the potential to assess the admixture in the breeds and assignment of individuals to their specific population. This is of importance as in developing countries, there is a practice of unrecorded breeding of animal breeds, and consequently, a large population of such interbred non-descript breeds has been evolved. So, it is pertinent to assess the genetic structure of economically important breeds of livestock. In one of such studies, genetic structure of important indigenous milch cattle breeds: Rathi, Tharparkar, Gir and other breeds Kankrej, and Mewati and Nagori in close proximity was analyzed (Sodhi et al. 2011a, b). Correspondence analysis plotted Rathi, Tharparkar, and Gir individuals into three separate areas of multivariate space, whereas Kankrej, Mewati, and Nagori cattle showed low breed-specific clustering. This reflected the existence of discrete genetic structure for Tharparkar, Rathi, and Gir, the prominent dairy cattle breeds of the region, while admixture was observed for Kankrej, Mewati, and Nagori individuals.

Microsatellite markers thus can prove invaluable for measuring the genetic structure of populations and determining the phylogenetic status of potentially endangered livestock species/breeds. In future, increasing use of such information will be of paramount importance to animal breeders and evolutionary biologists in effective prioritization of breeds on the basis of genetic uniqueness. Ideally, all available conservation options should be used to preserve the existing farm animal biodiversity. However, the limited financial resources and given the large numbers of diverse animal breeds that exist in our country, we may need to prioritize the breeds using the molecular genetic data and other associated factors, that are most important for conservation purposes.

28.19 Parentage Testing

Microsatellite typing can be used as a tool for identity or paternity testing by detection of hypervariable sequences. Identity testing and parentage determination are useful in artificial insemination and progeny testing programs and also in paternity related disputes. DNA analysis allows a far greater accuracy of parent identification through comparison of microsatellite sequences of an individual and its candidate parents. A DNA-based technique can be used to identify parentage in situations with multiple-sire mating. In addition, these molecular markers also serve as a useful tool for animal identification, particularly for verification of the semen used for artificial insemination. ISAG has recommended panels of microsatellite markers for parentage verification in horse, dog, cattle, sheep, goat, and pig (www.isag.us/Docs/consignmentforms/02_PVpanels_LPCGH.doc).

28.20 Population Bottleneck

A population bottleneck is a drastic reduction in the size of a population that may be caused by natural calamities, habitat destruction, or endemic disease. The decrease in population number directly impacts the genetic diversity which also decreases. When populations are under strong natural selection or artificial selection, only a subset of individuals in the population will reproduce; therefore, relatively few individuals contribute alleles to subsequent generations. Alleles for gene-regions that are not under selection are present in the post-selection population as a random subset of the original allelic diversity. The probability of an allele being present in subsequent generations is equivalent to its frequency in the original population therefore high-frequency alleles have a greater probability of being present in the post-selection population than low-frequency alleles (Luikart et al. 1998a, b). If selection pressure lasts for many generations, rare alleles will be lost simply by chance resulting in a post-selection population with fewer alleles and lower heterozygosity than the original population. Unfortunately, it is often difficult to identify losses of variability because levels of genetic variability prior to a population decline are generally unknown (Spencer et al. 2000) A number of statistical methods now make it possible to investigate a population's history without the need for information on past population sizes (Spencer et al. 2000). These tests typically quantify deviations from expected patterns in allele sizes, allele numbers, heterozygosity levels, or allele distributions, often using microsatellites data as these molecular markers are important modern tools for estimating the level of genetic diversity in endangered populations (O'Brien 1994).

28.21 Identification of Disease Carriers

Many incurable diseases result from defects or mutations in genome of an organism. DNA polymorphism occurring within a gene helps to understand the molecular mechanism and genetic control of several genes and metabolic disorders and allows the identification of heterozygous carrier animals. Identification of carrier animals of weaver disease (progressive degenerative myeloencephalopathy) in cattle has been accomplished using TGLA116 microsatellite marker. Georges et al. (1993) performed an extensive linkage study in a bovine pedigree segregating for the weaver condition and identified a microsatellite locus closely linked to the weaver gene and by extension, the weaver locus was assigned to bovine synteny group 13. Microsatellite TGLA116 can be used to identify weaver carriers, to select against this genetic defect.

28.22 Mapping of QTL

The most important application of microsatellites includes mapping of QTL by linkage. Such mapping information if available for genes of economic importance can be used in breeding programmes of either within breed's manipulations like marker-assisted selection of young sires or between breed's introgression programs. Microsatellites have been adopted widely for use in heritage mapping studies of the farm animals to the point that they are now the favored polymorphic marker for this purpose. Microsatellite marker D21S4 has shown significant association with effects on milk and protein yields in cattle. The presence of QTL for milk production on five chromosomes (namely chromosome no. 1, 6, 9, 10 and 20) has also been demonstrated in 14 US Holstein half-sib families using 159 microsatellites. Significant association of microsatellite markers with somatic cell score (SCS, an indicator for susceptibility to mastitis), productive herd life, and milk production traits has also been established. Potential QTL for SCS, fat yield, fat percentage, and protein percentage have also been identified using microsatellite (Ron et al.1994; Ashwell et al. 1996, 1997). Characterization of QTL for economically important traits using microsatellite markers will help in formulating more efficient breeding programmes using MAS. The map would also help in identification, isolation, and manipulation of animals with predetermined phenotype by modifying the candidate genes.

28.23 Candidate Gene-Based Diversity Analysis

The typing of allelic variations of the genes of importance might be a promising strategy for trait specific gene characterization, genetic differentiation, and population structure of native cattle breeds. Several class—I (gene specific) polymorphisms have been reported in different exotic cattle breeds, but their status (gene frequencies and gene diversity, differences between breeds) was unknown in the Indian breeds. In the regime of patenting and IPR related to animal agriculture, it becomes important to identify the unique gene/gene combination of Indian cattle breeds which are known to harbor important traits viz. milk composition and yield, disease resistance, and most adaptive traits. No systematic genomic initiative has been taken so far in India, though in the developed world genomic programmes have reached advance stages. Considering the fact that some of the countries such as Brazil, Australia, and Egypt already have germplasm of some of Indian cattle breeds like Ongole, Gir, Sahiwal, etc., identification and characterization of major genes associated with production, reproduction, and disease resistance become the focal research objective for the immediate future for becoming globally competitive.

Recently, studies have attempted to characterize the genes of economic significance in several indigenous cattle breeds. However, most of these efforts were isolated and mainly concentrated to identify polymorphism and compare the frequency distribution of allele types/genotypes of few important genes viz. α-lactalbumin, β-lacglobulin, insulin-like growth factors, BoLA-DRB3.2, Complement C5, and interleukin 12 genes (Kumar et al. 2004).

At ICAR-NBAGR Karnal (India), much systematic initiatives were undertaken to genotype variations at important gene loci known to influence milk production traits so as to understand the molecular basis of performance traits in different cattle breeds. The allele frequency spectra at specific candidate gene loci (Butyrophilin1,3, β-Lactoglobulin, prolactin, pituitary-specific transcription factor, kappa-casein, β-casein, αS1- casein, αS2-casein, Bovine growth hormone, α-lactalbumin, diacylglycerol acyltransferase, etc., was generated across diversified Indian native cattle breeds. The results pointed toward the existence of Indian native cattle breed-specific allelic profile that is quite distinct from *Bos taurus* breeds. The study highlighted the maintenance of indicine characteristics (e.g., A allele at kappa-casein, MspI-allele at GH, K allele at DGAT I, A2 allele at β-CN, A allele at BTN1, and C allele at αS2-CN had frequency >0.80), and the near absence of taurine influence/introgression in these Indian zebu cattle on the studied genic region (Mukesh et al. 2009). The near fixation observed at these candidate genes strongly argues for the involvement of additional alleles/genes influencing milk traits and thus indicates the necessity of identification of new gene markers for marker-phenotype association studies in Indian zebuine cattle.

Some of the studies have also been extended to estimate the association of a particular genotypic variant with traits of importance like birth weight, growth rate, milk fat percentage, milk yield, etc. These association studies of polymorphic candidate gene with economic traits were aimed to help the breeders to search some genetic marker for economic traits which might be used as an aid to the selection of bulls at an early stage. In one such study, Biswas et al. (2003) attempted to reveal polymorphism in growth hormone gene by PCR-RFLP approach and the genotype effect on birth weight in cattle and buffalo. Similarly, Pal et al. (2004) found significant effect of growth hormone gene genotype on birth weight, three-month body weight, and average body weight gain in Karan Fries

Table 28.2 List of different available databases

Databases	Link/sources
DNA sequence databases	GenBank: http://www.ncbi.nlm.nih.gov/
	European Molecular Biology Lab (EMBL): http://www.ebi.ac.uk/embl/index.html
	DNA Data Bank of Japan (DDBJ): http://www.ddbj.nig.ac.jp
Protein databases	SWISS-PROT: http://www.expasy.ch/sprot/sprottop.html
	Protein Information Resource (PIR): http://pir.georgetown.edu/pirwww/
	Protein Data Bank (PDB): http://www.rcsb.org/pdb/
Gene identification Portals	MOLBIOL: http://www.molbiol.net/
	Pedro's biomolecular research tools: http://www.biophys.uni-duesseldorf.de/BioNet/Pedro/research_tools.htm
	ExPASy molecular biology server: http://www.expasy.ch/

bulls. In this study, LL homozygotes showed to favor the increased growth traits (average birth weight, average three-month body weight, and daily body weight gains) in comparison to LV heterozygotes in Karan Fries cattle. In another report on polymorphism and frequency distribution of β-lactoglobulin gene, Badola et al. (2004) found no significant differences of allele frequency and association of polymorphism with milk traits (milk fat percentage) in *Bos indicus*, *Bos taurus*, and Indicine x Taurine crossbred cattle. Nucleotide sequencing has become the much-preferred tool to identify sequence variations/SNPs associated with traits of interest in indigenous farm animals. Following this approach, attempts have been made to sequence characterize the casein cluster gene. The studies (Sodhi et al. 2011a, b) have revealed the variations specific to Indian zebu cattle in coding as well-regulatory regions of these important genes. With the ongoing trends, the analysis of functionally relevant variants would become possible in the near future by comparing the candidate gene sequence information across many individuals to predict the different haplotypes and further classification of these haplotypes into functionally related categories. Considering the limitations of suitable resource population in Indian zebu cattle and lack of accurate population data in the present scenario, it seems logical to employ candidate gene approach to understand the genetics of important traits and generate SNP markers. Further, with the availability of electronic databases (Table 28.2), information for cross comparison is easily accessible. Advances in bioinformatics has made possible to combine information from different sources and generate new knowledge from existing data. This has helped to characterize and compare the genetic level difference within and between breeds.

Recent developments in molecular genetics have provided new powerful tools, called molecular markers, to assess the evolutionary and demographic history of livestock species, domestication events, and geographic distribution of their diversity. DNA-based marker methods are commonly used in ecological, evolutionary, and genetic approaches to analyze efficiently genetic structure in both animal and plant species (Tarnita et al. 2009). These markers have helped in identification of the wild ancestors of modern livestock and the nature of livestock expansion in past millennia. Such information tells us about the history and the way in which extraordinary biological diversity has been shaped in a relatively short period of time. With development of molecular technologies, DNA-based polymorphisms became the markers of choice for molecular-based survey of genetic variation. Different genetic markers provide different levels of genetic diversity information among which mtDNA sequences are the markers of choice for significant insights into the domestication and past migration history of livestock species. Use of mtDNA has broadened the perspective on the origin and evolution of domesticated cattle (Maji et al. 2009). Further, one of the persistent challenges in the analysis of population genetic data is to account for the spatial arrangement (non-random distribution of genetic variation among individuals within populations) of samples and populations. mtDNA data have been extensively used to understand the spatial distribution of genetic lineages within species allowing the historical factor with the highest effect on the lineages spatial patterns. mtDNA has been used for the identification of maternal and paternal lineages as well as test hypothesis related to past genetic history and evolution of different species (Hebsgaard et al. 2007). mtDNA can also tell about the recent demographic processes affecting a population, for example, whether a population has undergone a recent demographic expansion or has a more complex history. The recognition of mtDNA molecule as a genetic marker in population and evolutionary biology derives in part from the relative ease with which clearly homologous sequences can be isolated and compared. Simple sequence organization, maternal inheritance, and absence of recombination make mtDNA an ideal marker for tracing maternal genealogies.

28.24 Mitochondrial DNA Vis-a-Vis Genomic DNA

Mitochondrial DNA is cytoplasmic and maternal (uniparental) in inheritance and in contrast to nuclear DNA, it is present in abundance per cell as each cell may contain several hundred of mitochondria and each mitochondrion contains multiple copies of its own duplex circular genome. The presence of mtDNA in abundance facilitates its extraction from even forensic/ancient cell or tissue and provides many insights on the genetic diversity, origin, and taxonomy of breeds, besides estimating the time depth of domestication history. The mutation rate of mtDNA is higher than that of nuclear DNA, making it useful for addressing genetic relationships within and between populations and the presence of population bottlenecks (Berggren et al. 2005).

Cytoplasmically inherited mtDNA evolves five to ten times more rapidly than nuclear genome (Brown et al. 1979) due to the presence of a hypervariable displacement-loop (D-loop) and tend to enhance recent demographic events. In general, the non-coding D-loop region exhibit elevated levels of variation relative to coding sequences such as the *cytochrome-b* gene, presumably due to reduced functional constraints and relaxed selection pressure (Brown et al. 1993). The D-loop is also known as a control region since it is the site of transcriptional and replicational control (Anderson et al. 1982) as it contains the two major transcriptional promoters (P_H and P_L) and the origin of replication (Ori_H) within it. Overall, the mtDNA structure is highly conserved in higher animals. mtDNA has many advantages over genomic DNA in that it has been well characterized at both population and molecular levels and yields data readily amenable to analysis. A rapid rate of sequence divergence (at least in vertebrates) allows discrimination of recently diverged lineages. Studies of mtDNA from a diversity of animal groups have revealed significant variation among taxa in mtDNA sequence dynamics, gene order and genome size.

28.25 Mitochondrial DNA Genetic Markers

To address the impact of population historic events, markers must meet certain criterion: (i) the markers must be evolutionarily conserved to allow the identification of the wild taxon or population from which the species descends, (ii) variable and structured enough across the geographical range of the species so that the approximate locality of domestication can be identified, and (iii) the markers should evolve rapidly but at a constant rate to date the polymorphism. The mtDNA presents all these characteristics. Further, mtDNA is also almost exclusively maternally inherited, is effectively haploid, and is free from genetic recombination implying that each individual has a single haplotype and that phylogenetic analyses are relatively straight forward to interpret. Because of these features mtDNA has been the predominant molecule used to determine maternal-based phylogenies unobscured by genetic. This property of unipaternal inheritance makes mtDNA distinct from the highly polymorphic microsatellite markers, which are co-dominantly inherited. Also, the presence of D-loop that evolves much rapidly provides an upper edge over information based on Y-chromosome DNA sequences, which are less variable within species (Bruford et al. 2003) which makes its routine use difficult for phylogenetic analyses. Analysis of the Y-chromosome also lacks the power of multiple band profile. mtDNA is highly variable within species, such that in humans for just one highly variable section of mtDNA control region, over 500 distinct haplotypes have been recorded. Thus, due to combination of genetic characteristics such as uniparental mode of inheritance, lack of recombination, and presence of hypervariable region mtDNA diversity have been the primary focus for maternal-based genetic studies (MacHugh and Bradley 2001).

28.26 SNP Array—An Opportunity in Post-genomic Era

With the advances in our understanding of the genome and accompanying technological innovations, new possibilities are opening up for characterization, direct identification, and selection of animals carrying the best genes: selection on genotype. A recent technological advancement in selection breeding program called genomic selection (GS), using the DNA markers distributed throughout the entire genome, is revolutionizing livestock breeding. The recent sequencing of the genome of several livestock species has led to the availability of hundreds of thousands of single-nucleotide polymorphisms (SNP). Taking the advantage of the availability of sequence draft (chicken, 2004; cattle, 2005; rabbit, 2006; pig, 2007; sheep, 2009; camel, 2010), and recent technologies (next generation sequencing, deep sequencing, reduced reductional libraries), SNP (single-nucleotide polymorphism) arrays have already been developed for chicken (Illumina 18 K iSelect chip), cattle (Bovine 50 K Illumina™ iSelect chip), pig (Illumina iSelect pig DNA chip, 60 K) and sheep (Illumina's OvineSNP50) and it is under development for other livestock species. These dramatic developments have made the SNP genotyping very cost-effective leading to the successful implementation of whole-genome selection (WGS).

SNP chip provides a platform where hundreds or thousands of markers are genotyped in a cost-effective manner. As compared to current DNA tests based on dozens to hundreds of SNP, whole-genome selection is costlier; however, it is efficient and effective as the SNP in the test span the entire genome and hence same set of SNP could be used for all traits. The genomic abundance and amenability to high-throughput genotyping has made the SNPs as the most preferred class of genetic marker in genome wide association studies, genomic selection and the dissection of quantitative traits. Additionally, the whole-genome SNP panel in different breeds of cattle identified several levels of population substructure with the greatest level of genetic differentiation detected between *Bos taurus* and *Bos indicus* breeds (McKay et al. 2008). SNP array also offers the solution to know the extent of genetic diversity existing in the breeding stocks so as to take care of future needs. SNP chip is also useful for estimation of a pattern of linkage disequilibrium (LD). Estimation of LD in cattle from Europe and Africa indicated that LD decreases with the increasing physical distance across breeds revealing footprints of ancestral LD at short distances. With the availability of SNP chip, it is now possible to undertake the whole-genome scan and thus genomic selection approach can be accomplished to correlate phenotype and genotype across the whole genome simultaneously rather than 1 locus at a time. Thus, SNP-based platform is available to predict the molecular genetic value of individual animal for specific traits viz. milk production in dairy cattle, marbling and tenderness in beef cattle, and high fecundity in sheep.

28.27 DNA Barcoding Markers

The concept of DNA barcoding was suggested by Herbert et al. (2003). A DNA barcode is a short DNA sequence segment selected from a standardized part of genome. DNA barcode is used to identify a particular species, screening of genes, or reference genes to identify new species, or assigning unknown individuals to a species. The DNA barcoding being highly accurate is currently used in diverse biological, aquaculture, and veterinary applications such as describing diversity of animal parasites and evolving strategies for their control (Zhang et al. 2017), and developing dietary management plans to investigate and conserve wildlife possessing similar dietary habits (Gebremedhin et al. 2016; Mallott et al. 2017). The DNA barcoding is expected to increase the pace the discovery of new species, identification, and determining genetic diversity of animals, insects, and plants.

28.28 Outlook and Challenges

Molecular characterization is an important tool to unravel the genetic diversity, distinctiveness, and population structure of animal genetic resources. It also serves as an aid in the genetic management of small populations, to avoid excessive inbreeding. A number of investigations have described within- and between-population diversity. However, these studies are fragmented and are difficult to compare and integrate. It is required to have a comprehensive study facilitating global comparison. Viewing the past and present status of animal genomics scenario, and pace of characterization of genetic diversity, identification and commercial utilization of economically important genes in several exotic species, it is pertinent to further strengthen studies on exploring the genetic architecture of livestock in a speedy way by employing the latest technologies available for genome analysis. Such strategy will provide the basic information on genetic makeup of native breeds that will further be helpful in their improvement and sustainable utilization in a better way. Also, it will help in sharing the scientific advances in the field of molecular characterization across the globe, thereby contributing to an improved understanding, utilization, and conservation of the animal genetic resources for the good of present and future human generations.

The diversity of species and genes affects the ability of ecological communities to resist or recover from disturbances or face the unseen challenges in terms of diseases or climate change. Conservation of native genetic group should be of a priority since the loss of adaptive traits will restrict the options available to meet future unknown requirements of mankind.

28.29 Conclusions

In order to facilitate and rationalize the maintenance of domestic animal diversity, it is essential that simple assays be quickly developed taking advantage of molecular genetic tools. At present, DNA array-based techniques to type polymorphic loci for detecting diversity at the DNA level are available and are being exploited globally to construct genetic profile for different populations, breeds, and strains of farm animals.

References

Ajmone-Marsan P, Negrini R, Milanesi E, Bozzi R, Nijman IJ, Buntjer JB, Valentini A, Lenstra JA (2002) Genetic distances within and across cattle breeds as indicated by biallelic AFLP markers. Anim Genet 33 (4):280–286. https://doi.org/10.1046/j.1365-2052.2002.00865.x

Anderson S, de Bruijn MH, Coulson AR, Eperon IC, Sanger F, Young IG (1982) Complete sequence of bovine mitochondrial DNA. Conserved features of the mammalian mitochondrial genome. J Mol Biol 156 (4):683–717 (No abstract available)

Ashwell MS, Rexroad CE Jr, Miller RH, VanRaden PM (1996) Mapping economic trait loci for somatic cell score in Holstein cattle using microsatellite markers and selective genotyping. Anim Genet 27(4):235–242

Ashwell MS, Rexroad CE Jr, Miller RH, VanRaden PM, Da Y (1997) Detection of loci affecting milk production and health traits in an elite US Holstein population using microsatellite markers. Anim Genet 28:216–222

Beckman JS, Weber JL (1992) Survey of human and rat microsatellites. Genomics 12:627–631

Badola S, Bhattacharya TK, Biswas TK, Shivakumar BM, Kumar P, Sharma A (2004) A comparison on polymorphism of Beta-lactoglobulin gene in *Bos indicus*, *Bos taurus* and Indicine × Taurine crossbred cattle. Asian Aust J Anim Sci 17(6):733–736. https://doi.org/10.5713/ajas.2004.733

Berggren KT, Ellegren H, Hewitt GM, Seddon JM (2005) Understanding the phylogeographic patterns of European hedgehogs, Erinaceus concolor and E. europaeus using the MHC. Heredity (Edinb) 95(1):84–90

Baumung R, Simianer H, Hoffmann I (2004) Genetic diversity studies in farm animals—a survey Studien zur genetischen Diversität landwirtschaftlicher Nutztiere- ein globaler Überblick. J Anim Breed Genet 121 (6):361–373

Biswas TK, Bhattacharya TK, Narayan AD, Badola S, Kumar P, Sharma A (2003) Growth hormone gene polymorphism and its effect on birth weight in cattle and buffalo. Asian Aust J Anim Sci 16(4):494–497. https://doi.org/10.5713/ajas.2003.494

Bitgood JJ, Shoffner RN (1990) Cytology and cytogenetics. Poultry breeding. Genetics 22:401–427

Bjørnstad G, Røed KH (2001) Breed demarcation and potential for breed allocation of horses assessed by microsatellite markers. Anim Genet 32(2):59–65

Bjørnstad G, Gunby E, Røed KH (2000) Genetic structure of Norwegian horse breeds. J Anim Breed Genet 117:307–317

Blears MJ, De Grandis SA, Lee H, Trevors JT (1998) Amplified fragment length polymorphism (AFLP): a review of the procedure and its applications. J Ind Microbiol Biotech 21:99–114. https://doi.org/10.1038/sj.jim.2900537

Botstein D, White RL, Skolnick M, Davis RW (1980) Construction of a genetic linkage map in man using restriction fragment length polymorphisms. Am J Hum Genet 32(3):314–331 (Review)

Brown JR, Beckenbach AT, Smith MJ (1993) Intraspecific DNA sequence variation of the mitochondrial control region of white sturgeon (*Acipenser transmontanus*). Mol Biol Evol 10(2):326–341

Brown WM, George M Jr, Wilson AC (1979) Rapid evolution of animal mitochondrial DNA. Proc Natl Acad Sci U S A 76(4):1967–1971

Bruford MW, Bradley DG, Luikart G (2003) DNA markers reveal the complexity of livestock domestication. Nat Rev Genet 4(11):900–910 (Review)

Convention on Biological Diversity (CBD) (1992) https://en.wikipedia.org/wiki/Conventionon_Biological_Diversity

Di Rienzo A, Peterson AC, Garza JC, Valdes AM, Slatkin M, Freimer NB (1994) Mutational processes of simple-sequence repeat loci in human populations. Proc Natl Acad Sci U S A 91:3166–3170. https://doi.org/10.1073/pnas.91.8.3166

Dixit SP, Verma NK, Aggarwal RAK, Vyas MK, Rana J, SharmaA Tyagia P, Arya P, Ulmek BR (2010) Genetic diversity and relationship among southern Indian goat breeds based onmicrosatellite markers. Small Rumin Res 91:153–159

FAO (2001) The states of the world's animal genetic resource food and agriculture FAO. Rome, Italy

FAO (2007a) In: Rischkowsky B, Pilling D (eds) The state of the world's animal genetic resources for food and agriculture. Rome. http://www.fao.org/docrep/010/a1250e/a1250e00.htm

FAO (2007b) Global plan of action for animal genetic resources and the interlakendeclaration. Rome. http://www.fao.org/docrep/010/a1404e/a1404e00.htm

FAO (2007c) In: In: Rischkowsky B, Pilling D (eds) The state of the world's animal genetic resources for food and agriculture. Rome

Fatima S, Bhonga CD, Ranka DN, Joshi CG (2008) Genetic variability and bottleneck studies in Zalawadi, Gohilwadi and Surti goat breeds of Gujarat (India) usingmicrosatellites. Small Rumin Res 77:58–64

Gebremedhin B, Flagstad Ø, Bekele A, Chala D, Bakkestuen V, Boessenkool S, Popp M, Gussarova G, Schrøder-Nielsen A, Nemomissa S, Brochmann C, Stenseth NC, Epp LS (2016) DNA metabarcoding reveals diet overlap between the endangered Walia Ibex and domestic goats—implications for conservation. PLoS ONE 11(7):e0159133. https://doi.org/10.1371/journal.pone.0159133 (eCollection 2016)

Georges M, Dietz AB, Mishra A, Nielsen D, Sargeant LS, Sorensen A, Steele MR, Zhao X, Leipold H, Womack JE, Lathrop M (1993) Microsatellite mapping of the gene causing weaver disease in cattle will allow the study of an associated quantitative trait locus. Proc Natl Acad Sci U S A 90(3):1058–1062

Goldstein DB, Schlötterer C (1999) Microsatellites: evolution and applications. Oxford University Press, New York, 368p

Hall SJ, Bradley DG (1995) Conserving livestock breed biodiversity. Trends Ecol Evol 10(7):267–270

Hebert PD, Cywinska A, Ball SL, deWaard JR (2003) Biological identifications through DNA barcodes. Proc Biol Sci 270(1512):313–321

Hebsgaard MB, Wiuf C, Gilbert MT, Glenner H, Willerslev E (2007) Evaluating Neanderthal genetics and phylogeny. J Mol Evol 64(1):50–60 (Epub 2006 Dec 1)

Jarne P, Lagoda PJ (1996) Microsatellites, from molecules to populations and back. Trends Ecol Evol 11 (10):424–429

Jonker M, Van Meurs G, Balner H (1982) Typing for RhLA-D in rhesus monkeys: II. Genetics of the D antigens and their association with DR antigens in a population of unrelated animals. Tissue Antigens 19:69–78. https://doi.org/10.1111/j.1399-0039.1982.tb01417.x

Jordana J, Folch P, Aranguren JA (2001) Microsatellite analysis of genetic diversity in the Catalonian donkey breed. J Anim Breed Genet 118:57–63

Kantanen J, Olsaker I, Holm LE, Lien S, Vilkki J, Brusgaard K, Eythorsdottir E, Danell B, Adalsteinsson S (2000) Genetic diversity and population structure of 20 North European cattle breeds. J Hered 91(6):446–457

Kimura M, Crow JF (1964) The number of alleles that can be maintained in a finite population. Genetics 49:725–738 (No abstract available. PMID: 14156929)

Kimura M, Ohta T (1978) Stepwise mutation model and distribution of allelic frequencies in a finite population. ProcNatlAcadSci U S A 75(6):2868–2872

Köhler-Rollefson I (1997) Indigenous practices of animal genetic resource management and their relevance for the conservation of domestic animal diversity in developing countries. J. Anim Breed Genet 114:231–238. https://doi.org/10.1111/j.1439-0388.1997.tb00509.x

Kumar KG, Ponsuksili S, Schellander K, Wimmers K (2004) Molecular cloning and sequencing of porcine C5 gene and its association with immunological traits. Immunogenetics 55(12):811–817 (Epub 2004 Feb 10)

Koringa PG, Joshi CG, Solanki JV, Rank DN (2008) Genetic characterization and bottleneck studies in Kathiawari horse breed of India. Haryana Vet 47:77–83

Li S, Wang Q, Lin X, Jin X, Liu L, Wang C, Chen Q, Liu J, Liu H (2017) The use of "omics" in lactation research in dairy cows. Int J Mol Sci 18(5). pii: E983. https://doi.org/10.3390/ijms18050983 (Review)

Litt M, Luty JA (1989) A hypervariable microsatellite revealed by in vitro amplification of a dinucleotide repeat within the cardiac muscle actin gene. Am J Hum Genet 44(3):397–401

Luikart G, Allendorf FW, Cornuet JM, Sherwin WB (1998a) Distortion of allele frequency distributions provides a test for recent population bottlenecks. J Hered 89(3):238–247

Luikart G, Sherwin WB, Steele BM, Allendorf FW (1998b) Usefulness of molecular markers for detecting population bottlenecks via monitoring genetic change. Mol Ecol. 7(8):963–974

MacHugh DE, Bradley DG (2001) Livestock genetic origins: goats buck the trend. Proc Natl Acad Sci U S A 98(10):5382–5384 (No abstract available)

Maji S, Krithika S, Vasulu TS (2009) Phylogeographic distribution of mitochondrial DNA macrohaplogroup M in India. J Genet 88(1):127–139

Mallott EK, Garber PA, Malhi RS (2017) Integrating feeding behavior, ecological data, and DNA barcoding to identify developmental differences in invertebrate foraging strategies in wild white-faced capuchins (*Cebus capucinus*). Am J Phys Anthropol 162(2):241–254. https://doi.org/10.1002/ajpa.23113 (Epub 2016 Oct 5)

Martín-Burriel I, García-Muro E, Zaragoza P (1999) Genetic diversity analysis of six Spanish native cattle breeds using microsatellites. Anim Genet 30(3):177–182

McKay SD, Schnabel RD, Murdoch BM, Matukumalli LK, Aerts J, Coppieters W, Crews D, Dias Neto E, Gill CA, Gao C, Mannen H, Wang Z, Van Tassell CP, Williams JL, Taylor JF, Moore SS (2008) An assessment of population structure in eight breeds of cattle using a whole genome SNP panel. BMC Genet 20(9):37. https://doi.org/10.1186/1471-2156-9-37

Mishra BP, Kataria RS, Bulandi SS, Prakash B, Kathiravan P, Mukesh M, Sadana DK (2009a) Riverine status and genetic structure of Chilika buffalo of eastern India as inferred from cytogenetic and molecular marker-based analysis. J Anim Breed Genet 126(1):69–79. https://doi.org/10.1111/j.1439-0388.2008.00759.x

Mishra BP, Kataria RS, Kathiravan P, Bulandi SS, Singh KP, Sadana DK (2009b) Evaluation of genetic variability and mutation drift equilibrium of Banni buffalo using multi locus microsatellite markers. Trop Anim Health Prod 41(7):1203–1211. https://doi.org/10.1007/s11250-008-9301-8 (Epub 2009 Jan 9)

Mishra BP, Kataria RS, Kathiravan P, Singh KP, Sadana DK, Joshi BK (2010) Microsatellite based genetic structuring reveals unique identity of Banni among river buffaloes of Western India. Livestock Sci 127:257–261

Moxon ER, Wills C (1999) DNA microsatellites: agents of evolution? Sci Am 280(1):94–99 (No abstract available)

Mukesh M, Sodhi M, Kataria RS, Mishra BP (2009) Use of microsatellite multilocus genotypic data for individual assignment assay in six native cattle breeds from North West Region of India. Livestock Sci 121:72–77

Negrini R, Nijman IJ, Milanesi E, Moazami-Goudarzi K, Williams JL, Erhardt G, Dunner S, Rodellar C, Valentini A, Bradley DG, Olsaker I, Kantanen J, Ajmone-Marsan P, Lenstra JA (2007) European cattle genetic diversity consortium. Differentiation of European cattle by AFLP fingerprinting. Anim Genet 38(1):60–66

O'Brien SJ (1994) A role for molecular genetics in biological conservation. Proc Natl Acad Sci U S A 91(13):5748–5755 (Review)

Ostrander EA, Jong PM, Rine J, Duyk G (1992) Construction of small-insert genomic DNA libraries highly enriched for microsatellite repeat sequences. Proc Natl Acad Sci U S A 89(8):3419–3423

Pal A, Chakravarty AK, Bhattacharya TK, Joshi BK, Sharma A (2004) Detection of polymorphism of growth hormone gene for the analysis of relationship between allele type and growth traits in Karan Fries cattle. Asian Aust J Anim Sci 17:1334–1337

Pandey AK, Tantia MS, Kumar D, Mishra B, Chaudhury P, Vijh RK (2002) Microsatellite analysis of three poultry breeds of India. Asian Aus J Anim Sci 15(11):1536–1542. https://doi.org/10.5713/ajas.2002.1536 (Published online January 1, 2002)

Popescu NC, Evans CH, DiPaolo JA (1976) Chromosome patterns (G and C bands) of in vitro chemical carcinogen-transformed guinea pig cells. Cancer Res 36(4):1404–1413

Ron M, Band M, Yanai A, Weller JI (1994) Mapping quantitative trait loci with DNA microsatellites in a commercial dairy cattle population. Anim Genet 25 (4):259–264

Ruane J (1999) A critical review of the value of genetic distance studies in conservation of animal genetic resources. J Anim Breed Genet 116:317–323. https://doi.org/10.1046/j.1439-0388.1999.00205.x

Saitbekova N, Gaillard C, Obexer-Ruff G, Dolf G (1999) Genetic diversity in Swiss goat breeds based on microsatellite analysis. Anim Genet 30(1):36–41

Saitbekova N, Schlapfer J, Dolf G, Gaillard C (2001) Geneticrelationships in Swiss sheep breeds based on microsatellite analysis. J Anim Breed Genet 118: 379–387

Singh (2018). https://www.vetextension.com/current-livestock-animal-husbandry-statistics-india/ (June 15, 2019)

Sodhi M, Mukesh M, Mishra BP, Ahlawat SP, Prakash B, Sobti RC (2011a) Microsatellite analysis of genetic population structure of zebu cattle (*Bos indicus*) breeds from north-western region of India. Anim Biotechnol 22(1):16–29. https://doi.org/10.1080/10495398.2011.536091

Sodhi M, Mukesh M, Mishra BP, Parvesh K, Joshi BK (2011b) Analysis of genetic variation at the prolactin-RsaI (PRL-RsaI) locus in Indian native cattle breeds (*Bos indicus*). Biochem Genet 49(1–2):39–45. https://doi.org/10.1007/s10528-010-9383-7

Spencer CC, Neigel JE, Leberg PL (2000) Experimental evaluation of the usefulness of microsatellite DNA for detecting demographic bottlenecks. Mol Ecol 9 (10):1517–1528

Strand M, Prolla TA, Liskay RM, Petes TD (1993) Destabilization of tracts of simple repetitive DNA in yeast by mutations affecting DNA mismatch repair. Nature. 365(6443):274–276 (Erratum in: Nature 1994 Apr 7; 368(6471):569)

Takezaki N, Nei M (1996) Genetic distances and reconstruction of phylogenetic trees from microsatellite DNA. Genetics 144(1):389–399

Tarnita CE, Antal T, Ohtsuki H, Nowak MA (2009) Evolutionary dynamics in set structured populations. Proc Natl Acad Sci U S A 106(21):8601–8604. https://doi.org/10.1073/pnas.0903019106 (Epub 2009 May 11)

Vos P, Hogers R, Bleeker M, Reijans M, van de Lee T, Hornes M, Frijters A, Pot J, Peleman J, Kuiper M, Zabeau M (1995) AFLP: a new technique for DNA fingerprinting. Nucleic Acids Res 23(21):4407–4414

Weber JL, May PE (1989) Abundant class of human DNA polymorphisms which can be typed using the polymerase chain reaction. Am J Hum Genet 44 (3):388–396

Welsh J, McClelland M (1990) Fingerprinting genomes using PCR with arbitrary primers. Nucleic Acids Res 18(24):7213–7218

Williams JG, Kubelik AR, Livak KJ, Rafalski JA, Tingey SV (1990) DNA polymorphisms amplified by arbitrary primers are useful as genetic markers. Nucleic Acids Res 18(22):6531–6535

Yang W, Kang X, Yang Q, Lin Y, Fang MY (2013) Review on the development of genotyping methods for assessing farm animal diversity. J Anim Sci Biotechnol 4:2. https://doi.org/10.1186/2049-1891-4-2

Zhang R, Zhao A, Wang X, Zhang Z (2017) Diversity of tick species on domestic animals in Shandong Province, China, using DNA barcoding. Exp Appl Acarol 73(1):79–89. https://doi.org/10.1007/s10493-017-0161-7

Zhu Y, Strassmann JE, Queller DC (2000) Insertions, substitutions, and the origin of microsatellites. Genet Res 76(3):227 236

Genome Mapping and Analysis

29

Abstract

Genome map is the milestone of genes on genome. The landmarks include genes, short DNA or nucleotide sequences, and regulatory sites that regulate genes and their expression. The objective of genome mapping is to analyze genomes and determine the position of genes, and find out the distance between genes on a chromosome. Genome mapping is a highly valuable tool to utilize genetic merit of animals.

Highlights

- Genome mapping assists in identifying genes and their arrangement on genome
- Genome mapping is not a final product, but the work is in progress.

Keywords

Genome mapping · Livestock applications · Physical genome mapping · Optical mapping

29.1 Introduction

Genome mapping refers to identify the location of genes and the distance between genes on a chromosome. Genome mapping has provided a critical starting point for Human Genome Project. The principles and tools are used in genome sequencing and mapping in other species.

Mapping genome provides a guide for sequencing experiments by revealing the location of genes and other distinctive transmissible features on chromosome. High-density genome maps have greatly facilitated the genome sequencing in eukaryotic genome. The whole genome is sequenced with the help of high-throughput sequencing.

Genomes of several domestic animals cattle, pig, sheep, horse, buffalo, and chicken are partially or completely sequenced (Amaral et al. 2008; Michelizzi et al. 2010; Bai et al. 2018). Genome maps have been developed for these animals.

29.2 DNA Markers for Genome Mapping

Though genes themselves could be useful markers, they are not the ideal markers. This is because a genome map based on genes is not descriptive, and in eukaryotes, the genes are spread over large gaps and widely spaced out. To overcome the limitation, other types of markers are used.

DNA sequences that may not be the genes indeed are used as DNA markers. DNA marker should have at least two alternative alleles. Restriction fragment length polymorphisms (RFLPs), simple sequence length polymorphisms (SSLPs), and single nucleotide polymorphisms (SNPs) are main types of DNA markers employed in various species of plants and animals.

B. Singh et al., *Advances in Animal Biotechnology*,
https://doi.org/10.1007/978-3-030-21309-1_29

29.3 Restriction Fragment Length Polymorphisms (RFLP)

RFLP is scored by two methods, viz. classical Southern hybridization and by PCR analysis. In Southern hybridization, DNA is extracted from cells or biological samples and purified. Purified DNA is digested using suitable restriction endonucleases (REs). Resulting DNA fragments are separated using agarose gel electrophoresis and transferred onto a specialized nylon membrane for probing. Fluorescently labeled or radioactive probes are used to detect desired DNA sequences. Presence of two restriction fragments confirms the desired segment.

Detection of RFLP by PCR amplification and analysis involves annealing primers to polymorphic restriction sites. After PCR amplification, the amplicons are digested with REs. DNA fragments are analyzed using agarose gel electrophoresis. Presence of desirable site is confirmed by two DNA bands, one band shows the absence of a particular site.

RFLPs are highly robust, accurate, and sensitive and can detect single nucleotide mutations, insertions, or deletions in genes, animals, and plants.

29.4 Methods of Genome Mapping

Different methods and markers are used to map genes on a genome map. Two general types, namely genetic mapping and physical mapping are summarized herein.

29.5 Genetic Mapping

Genetic mapping looks at how genes are shuffled between chromosomes or different regions in the same chromosome during recombination or crossing over of meiotic division.

Map of a chromosome of *Drosophila melanogaster* was the first animal genome map constructed in 1913. It was shown that genes were arranged on chromosome in a linear way, and genes responsible for specific traits are located at defined regions. Some dogmas, of gene inheritance, such as frequency of recombination during crossing over, linkage distance, and likelihood of transfer of separate inheritance of genes located far apart on a chromosome, and the genes located adjacent (smaller linkage distance) to each other are inherited together, and could be explained based on chromosome map https://www.yourgenome.org/facts/how-do-you-map-a-genome, accessed on Feb. 28, 2019).

A map showing linkage or relation of a gene with other genes on a chromosome is known as linkage map. Position of genes on chromosome is determined based on the exact frequency of genetic recombination. Genetic mapping relies on genetic techniques such as breeding experiments and pedigree analysis.

29.6 Physical Genome Mapping

Physical mapping describes or determines the physical distance between genes. Various techniques used in physical genome mapping include restriction mapping, fluorescent in situ hybridization (FISH) mapping, and sequence-tagged sites (STS) mapping. Physical genome maps enhance the quality of genome assemblies, improve gene annotation, and serve as frameworks for population genomics (Sharakhova et al. 2019). DNA FISH is developed and be applied for gene mapping and detecting genetic aberrations. RNA FISH provides information about transcriptome or gene expression in cells and tissues (Magaraki et al. 2018).

Genome maps are precious resources to correlate phenotypic variation with underlying genomic variations, and are essential tools to understand the biology underpinning the animals so closely associated with humans. Previously described genetic and radiation hybrid maps of cattle genome are used to identify genomic regions and genes that affect particular traits.

A physical map (Barendse et al. 1997; Snelling et al. 2007), linkage maps (Martín-Burriel et al. 1997; Ihara et al. 2004; Snelling et al. 2005), and radiation hybrid maps (Kurar et al. 2003) have been reported cattle genome. Further refinement of the maps and greater integration into genome assembly process may contribute to a high-quality assembly (Snelling et al. 2007).

29.7 Restriction Mapping

Here, isolated genomic DNA is digested or cut with the help of REs. A physical map is generated by aligning different restriction maps along chromosome. Two types or restriction maps, viz. optical and fingerprint maps are documented.

In fingerprinting mapping (Marra et al. 1997), the genomic DNA is purified from any nucleated cells and cloned using suitable host species such *Escherichia coli*. The resulting clones are digested using REs and analyzed using agarose gel electrophoresis.

A map is constructed by comparing restriction profiles from all the DNA fragments to find the areas of similarity. Nucleotide sequences having similar patterns are grouped to constitute a map. Restriction enzyme fingerprinting was used to sequence genomes of human, mice, zebrafish, and pig.

29.8 Optical Mapping

Optical mapping is a unique system based on single nucleotides stretched and held in place on a slide. DNA is purified and digested with the help of REs leaving behind the gaps in DNA molecule. RE-digested fragments are stained with fluorescent dyes and visualized under fluorescent microscope. Optical maps of single DNA molecule are constructed based on the intensity of fluorescence. These are combined and overlapped to generate a global profile of the genome.

The basic principle of an optical chart based on RE profile of genomic DNA is based on the order the DNA fragments are generated by restriction digestion. Profiles of several molecules are compiled to develop a genome map. An intact high molecular weight genomic DNA is required to generate the accurate optical map. Single-cell optical mapping performed after whole-genome amplification of DNA extracted from a cell on-chip is suggested to be less complex than classical NGS sequencing (Marie et al. 2018).

Optical mapping produces high-resolution, high-throughput genome map data that provide information about the physical organization of the genome. Optical mapping is used to scaffolding contigs and for assembly validation of large-scale sequencing projects. The technique has been used in the sequencing of goat genomes. Error correcting methods such as cOMet improve assembly of Rmap data (Mukherjee et al. 2018).

29.9 Fluorescent in Situ Hybridization

Fluorescent in situ hybridization (FISH) mapping, also called as cytogenetic mapping, uses fluorescent probes to detect DNA sequences. Conceptually, FISH essentially consists of hybridizing a DNA probe to its complementary nucleotide sequences on chromosomes already fixed on a solid surface or glass slides. Hybridized target DNA-probe complex is visualized by microscope. Basic FISH protocols have been modified to make them suitable for a specific task (Volpi and Bridger Volpi and Bridger 2008).

FISH is a paradigm-changing technique in cytogenetic studies to address queries related to structure, mutations, and evolution of individual chromosome as well as entire genome (Jiang 2019). It is achieved with the help of synthetic probes based on oligonucleotides (oligos).

FISH has several applications in livestock genomics and breeding management such as detecting male-specific DNA sequence in bovine embryos (Kobayashi et al. 1998); detecting sexed embryos (Kobayashi et al. 2004), and identifying freemartinism (Sohn et al. 2007). FISH is a proficient technique to detect and identify specific bacteria in meat (Zadernowska et al. 2014), detecting specific microorganisms in tissues in natural and experimentally infected samples. FISH can detect pathogens in fixed deparaffinized tissues mounted on glass slides (Jensen et al. 2015).

Use of FISH has been expanded to detect the differential distribution and prevalence of pathogens genitourinary tract. FISH examination of biopsies obtained from different sections of uterine tissue of Holstein cows revealed a differential distribution of *Fusobacterium necrophorum* and *Porphyromonas levii* in postpartum uterine disease of cattle, which shows that tissue invasiveness is an important step in infectious genitourinary pathogenesis (Karstrup et al. 2017).

A modification of basic FISH, the ViewRNA ISH method, aimed to detect distribution of classical swine fever virus (CSFV) RNA in PK15 porcine cells revealed that ViewRNA ISH was highly specific for CSFV. The assay facilitates studying CSFV RNA life cycle (Zhang et al. 2017).

29.10 Sequence-Tagged Site (STS) Mapping

STS (Olson et al. 1989) maps the position of short DNA sequences (200–500 bp) which exist only once in genome, and can be easily recognizable. STS is detected by isolating genomic DNA and digesting the same using REs. Sequences are cloned using suitable bacteria as host to create a library of DNA clones. STSs are detected by specialized primers developed from known DNA sequences, and analyzed by PCR amplification. STSs serve as landmarks for constructing a physical map of large genome.

29.11 Outlook and Challenges

From early domestication to modern breeding practice, humans have shaped the genomes of animals. Genetic diversity in livestock is a valuable reserve for developing and improving breed characteristics for improved productivity, disease-resistance, and adaptation to diverse environments and stresses.

Various methods are used to describe genetic maps. Genetic and radiation hybrid maps of bovine genome were used to identify genomic regions and genes responsible for specific traits.

The goal of gene mapping is to investigate regulation and expression of genes and to use the inferences to utilize and manipulate the genes. Gene mapping is like preparing a roadmap wherein markers are considered as indicators of particular trait. Animal genome sequences provide important resources for further improvements in animal rearing through scientific methods of nutrition, health and reproduction. A variety of methods have been developed during post-genomic era to describe and the genome maps. Many genes of biomedical and veterinary significances are found in the genome databases.

29.12 Conclusions

Genome sequencing has progressed significantly in recent years. Draft genome sequences of several livestock species have been assembled. Genetic mapping and physical mapping are two main methods of gene mapping. By mapping the genomes, agriculture researchers perform genetic comparisons among different species and utilize the traits of economic interest. Each method has characteristic features and limitations associated with it.

References

Amaral ME, Grant JR, Riggs PK, Stafuzza NB, Filho EA, Goldammer T, Weikard R, Brunner RM, Kochan KJ, Greco AJ, Jeong J, Cai Z, Lin G, Prasad A, Kumar S, Saradhi GP, Mathew B, Kumar MA, Miziara MN, Mariani P, Caetano AR, Galvão SR, Tantia MS, Vijh RK, Mishra B, Kumar ST, Pelai VA, Santana AM, Fornitano LC, Jones BC, Tonhati H, Moore S, Stothard P, Womack JE (2008) A first generation whole genome RH map of the river buffalo with comparison to domestic cattle. BMC Genom 24(9):631. https://doi.org/10.1186/1471-2164-9-631

Bai H, Sun Y, Liu N, Liu Y, Xue F, Li Y, Xu S, Ni A, Ye J, Chen Y, Chen J (2018) Genome-wide detection of CNVs associated with beak deformity in chickens using high-density 600 K SNP arrays. Anim

Genet 49(3):226–236. https://doi.org/10.1111/age.12652 (Epub 2018 Apr 11)

Barendse W, Vaiman D, Kemp SJ, Sugimoto Y, Armitage SM, Williams JL, Sun HS, Eggen A, Agaba M, Aleyasin SA, Band M, Bishop MD, Buitkamp J, Byrne K, Collins F, Cooper L, Coppettiers W, Denys B, Drinkwater RD, Easterday K, Elduque C, Ennis S, Erhardt G, Li L, et al (1997) A medium-density genetic linkage map of the bovine genome. Mamm Genome.8(1):21–8. Erratum in: Mamm Genome 8(10):798. Lil, L [corrected to Li, L]

Ihara N, Takasuga A, Mizoshita K, Takeda H, Sugimoto M, Mizoguchi Y, Hirano T, Itoh T, Watanabe T, Reed KM, Snelling WM, Kappes SM, Beattie CW, Bennett GL, Sugimoto Y (2004) A comprehensive genetic map of the cattle genome based on 3802 microsatellites. Genome Res 14(10A):1987–1998

Jensen HE, Jensen LK, Barington K, Pors SE, Bjarnsholt T, Boye M (2015) Fluorescence in situ hybridization for the tissue detection of bacterial pathogens associated with porcine infections. Methods Mol Biol 1247:219–234. https://doi.org/10.1007/978-1-4939-2004-4_17

Jiang J (2019) Fluorescence in situ hybridization in plants: recent developments and future applications. Chromosome Res. https://doi.org/10.1007/s10577-019-09607-z (Epub ahead of print) Review

Karstrup CC, Agerholm JS, Jensen TK, Swaro LRV, Klitgaard K, Rasmussen EL, Krogh KM, Pedersen HG (2017a) Presence and localization of bacteria in the bovine endometrium postpartum using fluorescence in situ hybridization. Theriogenology 1(92):167–175. https://doi.org/10.1016/j.theriogenology.2017.01.026 (Epub 2017 Jan 17)

Kobayashi J, Sekimoto A, Uchida H, Wada T, Sasaki K, Sasada H, Umezu M, Sato E (1998) Rapid detection of male-specific DNA sequence in bovine embryos using fluorescence in situ hybridization. Mol Reprod Dev 51 (4):390–394

Kobayashi J, Nagayama H, Uchida H, Oikawa T, Numabe T, Takada N, Sasada H, Sato E (2004) Selection of sexed bovine embryos using rapid fluorescence in situ hybridisation. Vet Rec 154(25):789–91. No abstract available

Karstrup CC, Agerholm JS, Jensen TK, Swaro LRV, Klitgaard K, Rasmussen EL, Krogh KM, Pedersen HG (2017b) Presence and localization of bacteria in the bovine endometrium postpartum using fluorescence in situ hybridization. Theriogenology 1(92):167–175. https://doi.org/10.1016/j.theriogenology.2017.01.026 (Epub 2017 Jan 17)

Kurar E, Womack JE, Kirkpatrick BW (2003) A radiation hybrid map of bovine chromosome 24 and comparative mapping with human chromosome 18. Anim Genet 34(3):198–204

Magaraki A, Loda A, Gribnau J, Baarends WM (2018) Simultaneous RNA-DNA FISH in mouse preimplantation embryos. Methods Mol Biol 1861:131–147. https://doi.org/10.1007/978-1-4939-8766-5_11

Marie R, Pedersen JN, Bærlocher L, Koprowska K, Pødenphant M, Sabatel C, Zalkovskij M, Mironov A, Bilenberg B, Ashley N, Flyvbjerg H, Bodmer WF, Kristensen A, Mir KU (2018) Single-molecule DNA-mapping and whole-genome sequencing of individual cells. Proc Natl Acad Sci U S A 115 (44):11192–11197. https://doi.org/10.1073/pnas.1804194115 (Epub 2018 Oct 15)

Marra MA, Kucaba TA, Dietrich NL, Green ED, Brownstein B, Wilson RK, McDonald KM, Hillier LW, McPherson JD, Waterston RH (1997) High throughput fingerprint analysis of large-insert clones. Genome Res 7(11):1072–1084

Martín-Burriel I, Osta R, Barendse W, Zaragoza P (1997) New polymorphism and linkage mapping of the bovine lactotransferrin gene. Mamm Genome. 8 (9):704–705. No abstract available

Michelizzi VN, Dodson MV, Pan Z, Amaral ME, Michal JJ, McLean DJ, Womack JE, Jiang Z (2010) Water buffalo genome science comes of age. Int J Biol Sci 6(4):333–349. Review

Mukherjee K, Washimkar D, Muggli MD, Salmela L, Boucher C (2018) Error correcting optical mapping data. Gigascience 7(6). https://doi.org/10.1093/gigascience/giy061

Olson M, Hood L, Cantor C, Botstein D (1989) A common language for physical mapping of the human genome. Science 245(4925):1434–1435

Sharakhova MV, Artemov GN, Timoshevskiy VA, Sharakhov IV (2019) Physical genome mapping using fluorescence in situ hybridization with mosquito chromosomes. Methods Mol Biol 1858:177–194. https://doi.org/10.1007/978-1-4939-8775-7_13

Snelling WM, Casas E, Stone RT, Keele JW, Harhay GP, Bennett GL, Smith TP (2005) Linkage mapping bovine EST-based SNP. BMC Genom 19(6):74

Snelling WM, Chiu R, Schein JE, Hobbs M, Abbey CA, Adelson DL, Aerts J, Bennett GL, Bosdet IE, Boussaha M, Brauning R, Caetano AR, Costa MM, Crawford AM, Dalrymple BP, Eggen A, Everts-van der Wind A, Floriot S, Gautier M, Gill CA, Green RD, Holt R, Jann O, Jones SJ, Kappes SM, Keele JW, de Jong PJ, Larkin DM, Lewin HA, McEwan JC, McKay S, Marra MA, Mathewson CA, Matukumalli LK, Moore SS, Murdoch B, Nicholas FW, Osoegawa K, Roy A, Salih H, Schibler L, Schnabel RD, Silveri L, Skow LC, Smith TP, Sonstegard TS, Taylor JF, Tellam R, Van Tassell CP, Williams JL, Womack JE, Wye NH, Yang G, Zhao S (2007) International Bovine BAC Mapping Consortium. A physical map of the bovine genome. Genome Biol. 8(8):R165

Sohn SH, Cho EJ, Son WJ, Lee CY (2007) Diagnosis of bovine freemartinism by fluorescence in situ hybridization on interphase nuclei using a bovine Y chromosome-specific DNA probe. Theriogenology 68 (7):1003–1011

Volpi EV, Bridger JM (2008) FISH glossary: an overview of the fluorescence in situ hybridization technique.

Biotechniques 45(4):385–386, 388, 390 passim. https://doi.org/10.2144/000112811. Review

Zadernowska A, Chajęcka-Wierzchowska W, Kłębukowska L (2014) Vidas UP-enzyme-linked fluorescent immunoassay based on recombinant phage protein and fluorescence in situ hybridization as alternative methods for detection of Salmonella enterica serovars in meat. Foodborne Pathog Dis 11(9):747–752. https://doi.org/10.1089/fpd.2014.1738

Zhang Q, Xu L, Zhang Y, Wang T, Zou X, Zhu Y, Zhao Y, Li C, Chen K, Sun Y, Sun J, Zhao Q, Wang Q (2017) A novel ViewRNA in situ hybridization method for the detection of the dynamic distribution of Classical Swine Fever Virus RNA in PK15 cells. Virol J 14(1):81. https://doi.org/10.1186/s12985-017-0734-4

30 Genome Sequencing Technologies in Livestock Health System

Abstract

Sequencing technologies are vital components of biological sciences. The multiple "omics" technologies have revolutionized animal healthcare management by improving speed, specificity, and sensitivity of diagnostic assays, and decreasing the probability of false-positive assays. It is possible to diagnose diseases much before the appearance of clinical symptoms.

Highlights

- Sequencing is now an integral component of scientific livestock health and production
- The bioinformatics methods are obligatory to analyze and infer the sequencing data.

Keywords

Genome sequencing · Next-generation sequencing · Genome analysis · Drug targets

30.1 Introduction

The next-generation sequencing (NGS) is a hot topic of research in biological sciences including scientific animal farming. The NGS technologies including genomic, transcriptomic, and proteomic analysis have transformed the livestock production from a small scale to a commercial business enterprise. At present, genomes of various livestock species such as cattle, buffalo, sheep, horse, swine, and chicken are partly or completely sequenced (Table 30.1). A number of sequences are available in public domains to access and analysis. Significance of the genome sequencing, transcriptomics, and metagenomics to unravel genetic diversity or organisms are highlighted in several studies (Miller et al. 2013; Bayliss et al. 2017).

NGS technologies have revolutionized the genomics research through analysis of whole-genome genotyping, transcriptome or RNA sequencing, de novo assembling of genomes, genome-wide structural variations, detection of mutations, and complex diseases (Greenwood et al. 2016; Anis et al. 2018).

The genome sequence is crucial for biological research, and it involves different techniques for handling the sequence data. NGS technologies produce 100 times more data compared to first-generation nucleic acid sequencing by Sanger and Maxam–Gilbert sequencing methods. Hence, the massive data generated by NGS can be managed, interpreted, and utilized by means of advanced bioinformatics software and in silico methods (Zhang et al. 2011; Sharma et al. 2019).

The genome sequencing of the livestock species has helped in understanding how genes govern quantitative traits at cellular and molecular levels. Applications of NGS have broadened from basic to applied research, uncovering

B. Singh et al., *Advances in Animal Biotechnology*,
https://doi.org/10.1007/978-3-030-21309-1_30

Table 30.1 Various livestock genomes sequenced with their genome size, GC content, sequencing technology used, genome coverage, and sequencing submitter information

Sl. No.	*Animal* sp.	Genome size (Mb)	GC content (%)	Sequencing technology	Genome coverage	Sequencing submitter	Total SRA experiments[a]
1	*Bos taurus*	2715.85	41.94	Illumina and Pacific Bioscience s	80X	United States Department of Agriculture, Agricultural Research Service	32,418
2	*Sus scrofa*	2501.91	41.97	Pacific Bioscience s	65X	The Swine Genome Sequencing Consortium	19,123
3	*Ovis aries*	2615.52	41.98	Illumina and Pacific Bioscience s	166X	International Sheep Genome Consortium	7208
4	*Equus caballus*	2506.97	41.53	Illumina and Pacific Bioscience s	88X	University of Louisville	5552
5	*Gallus gallus*	1065.37	42.30	Pacific Bioscience s	82X	Washington University Genome Sequencing Center	12,732
6	*Bubalus bubalis*	2655.78	41.81	Pacific Bioscience s	69X	Italian Buffalo Genome Consortium	1021
7	*Anas platyrhynchos*	1105.05	41.20	Solexa	60X	China Agricultural University	3655
8	*Canis lupus familiaris*	2410.98	41.30	Sanger	7X	Dog Genome Sequence Consortium	10,250

[a]SRA (Sequence Read Archive) contains raw sequence data from high-throughput next-generation sequencing platforms with in NCBI

DNA–protein interactions, insight into epigenetics modification, and gene expression networks.

30.2 NGS Technologies in Animal Sciences

The sequencing technologies have made progressive improvements in overall performance of animals. These platforms have enabled the scientists to realize biological world of a broader and useful perspective (Zhang et al. 2011). As high-throughput technologies rely on different procedural steps, substrates and equipment, their outputs are also different. The individual technologies have their own advantages and capabilities (Loman et al. 2012; Buermans and Den Dunnen 2014).

The sequencing platforms were initially restricted to genome sequencing, but later on NGS was used in different biological studies including transcriptomics, proteomics, animal breeding, unraveling host-pathogen interactions, and diagnosis of non-infectious diseases such as cancer, (Zhang et al. 2011; Bai et al. 2012; Anis et al. 2018). The 454 Sequencer system by Life

Sciences was the first, and Solexa 1G by Illumina was the second commercial platform for NGS (Morozova and Marra 2008). The 454 Platform is based on the principle of pyrosequencing, Illumina on sequencing by synthesis, and ABI-SOLiD is based on sequencing by ligation.

Different platforms and technologies are available. The platform based on single-molecule real-time sequencing (SMRT) offers advantages in terms of simplicity, low cost, and long read lengths. As SMRT is not involved in clonal amplification of DNA, the chances of errors are associated with clonal amplification are minimum. The PacBio (Pacific Biosciences) is based on SMRT sequencing technology which sequences long single DNA molecules in real time (Nakano et al. 2017). Other sequencing technologies, such as Oxford nanopore, are different from NGS platforms and do not involve amplification of DNA, but uses electric signal to detect the nucleotides (Buermans and Den Dunnen 2014; Senol et al. 2018).

30.3 NGS in Livestock Genomics

In early era of genomics, when NGS technologies were not available, the animal breeding programs primarily relied on molecular marker-assisted selection (Williams 2005). The genome sequencing indicates the genetic makeup of an organism, and RNA sequencing displays the sequences that are actively expressed in the cell. The transcriptome or RNA sequencing (RNA-seq) is a genomic reduction method used to study ecological genomics, adaptation, and single-nucleotide polymorphism (SNP) in different species (McCormack et al. 2013).

The progress in the sequencing technologies over the past years benefited the animal research in multiple ways (Tellam et al. 2009; Williams et al. 2017). The NGS techniques have generated massive data at very low cost. Livestock genome sequencing information is helpful in identification of genetic markers used in animal breeding programs and detecting diseases (Ramos et al. 2009; Aslam et al. 2012; Diaz-Sanchez et al. 2013). Among various livestock species, the dog was the first species whose genome was sequenced in 2005 (Lindblad-Toh et al. 2005). Cattle (*Bos taurus*) genome was sequenced shortly thereafter, followed by *Sus scrofa* (Table 30.1).

NGS has opened new approaches to investigate the relationship between phenotypic and genetic diversity. Whole-genome sequences and draft genome sequences of livestock form different species are available under various public domains, and various gene sequencing projects are underway. The data generated by NGS is used to identify molecular markers in the whole genome (Zhang et al. 2011).

The advent of NGS has pioneered a new area of research and identification of SNPs in livestock species including cattle, sheep, and pig (Djari et al. 2013). The interpretation of NGS data is more advantageous over SNP array, as it provides better information on genomics and genomic prediction.

30.4 NGS in Infectious Diseases

It is important that livestock should be free of pathogens so that diseases do not affect livestock produce and the international trade involving livestock products. However, conditions are not always favorable for living beings. The climatic stress renders animals susceptible to pathogens, parasites, and pests (Gibbs 2005). The infectious diseases pose a serious threat to livestock and economic gains from livestock (Brooks-Pollock et al. 2015). Infectious diseases cause huge economic loss due to massive culling of livestock affected from zoonotic diseases (Tomley and Shirley 2009; Lefrançois and Pineau 2014). In addition, there is a serious concern of the presence of antibiotic residues in livestock and agriculture products meant for export to other countries.

NGS has allowed improved understanding of livestock genome, transcriptome, and epigenome. NGS technologies are used for monitoring and diagnosis of infectious diseases including host–pathogen interactions.

Diagnosis of epidemics and infectious diseases caused by multiple and mixed pathogens is very challenging. Strategies have been developed to detect infectious agents utilizing NGS (Anis et al. 2018). The situation is aggravated as all the pathogens are unable to grow in isolation in vitro. To overcome the problem, metagenomic approaches are employed to amplify and discover all the infectious agents in biological samples (Miller et al. 2013). However, the metagenomic data is enormous and needs processing of millions of reads to conclude identification of suspected pathogens or pathogenic genetic markers.

The targeted NGS is another technique that involves selective amplification of defined genomic regions of interest (Mertes et al. 2011). Compared to metagenomics, the targeted NGS has limited prospects for discovering new pathogens. Therefore, metagenomic sequencing is the important tool which is progressively being used in disease diagnosis and management by detecting most pathogens.

NGS is presently the leading technology in molecular genetics research. The development of NGS technologies has changed the vision of researchers to analyze the genome, transcriptome, and proteome of an organism. It has provided the biological insight and showed the potential of NGS technologies which directly have an impact to control animal infectious disease (Deurenberg et al. 2017).

NGS is used in biology of pathogen for screening transposon libraries and identifying pathways and genes that enhance their survival in different environments (Van Opijnen et al. 2009; Langridge et al. 2009). RNA-Seq has revolutionized the bacterial research by providing the extensive perspective of pathogen transcriptomes (Sorek and Cossart 2010). RNA-Seq has been applied to decipher various bacterial transcriptomes (Martin et al. 2010; Camarena et al. 2010; Sharma et al. 2010). NGS can be used to unravel the transcriptome of pathogen present in tissue biopsies of patients or host (Azhikina et al. 2010). Nowadays, high-throughput sequencing is combined with translational research. The high-throughput sequencing of pathogen genome has helped to identify the immunogenic proteins, based on which diagnostic immunoassays have been developed (Greub et al. 2009).

NGS is used to study the evolution of the infectious agents. Different studies have been performed to get insight into evolution of pathogenic bacteria by using NGS (Srivatsan et al. 2008; Bryant et al. 2012). Mutations leading to resistance in bacteria and in viral genomes are efficiently detected by genomics (Wang et al. 2007; Feng et al. 2009; Urbaniak et al. 2018).

The NGS facilitates sequencing of microorganisms without prior culturing. It helps in identification of novel metabolites, enzymes, microorganisms, and antimicrobial peptides or bacteriocins (Singh et al. 2008; Challis 2014). The genomic sequencing of bacterial genomes by NGS helps in identifying genes and pathways involved in synthesis of new compounds or pharmaceuticals which are of utmost economic importance (Gomez-Escribano et al. 2015).

Earlier, various genes were identified from gastrointestinal ecosystem of herbivores (Singh et al. 2012; Sharma et al. 2017). Novel genes can be identified which can be used to improve nutrition and health (Kim et al. 2017).

30.5 Development of Drug Targets from Sequence Data

Genome, transcriptome, and proteome database are valuable resources. The unique metabolic pathways of pathogens harbor considerable amount of information. In the future, the microbial ecologists, epidemiologist, and clinicians will depend more on reliable microbiological diagnostics and rapid molecular testing of the pathogen. Clinicians need to interpret the sequencing data generated by the next-generation sequencing (NGS) technologies which is different from the conventional biological methods (Sharma et al. 2019).

In silico tools are used to identify and characterize targets by molecular dynamic simulations. As these targets are crucial for survival and multiplication of a pathogen, they can be used as potential targets to develop drugs to curtail the pathogen. In silico analysis of 46 proteins in *Orientia tsutsugamushi* (Ott) showed that 25 proteins were probable drug targets possessing druggable characteristics. The study provides insight to understand the mechanism of pathogenesis of Ott and suggest feasibility of developing an effective treatment against this pathogen (Sharma et al. 2019).

30.6 Bioinformatics and Web-Based Tools for Genome Sequence Analysis

High-throughput sequence data has revolutionized the molecular biology and genomics. Almost all areas of modern life sciences including biomedical and veterinary science are becoming more and more data intensive. Whole-genome shotgun-based NGS of transcriptome and metagenome generate massive 100–1000 GB sequence data from tens of thousands of different genes or microbial species (Shi et al. 2019). Making sense of high-throughput data and assembly of datasets requires bioinformatics software and computational methods (Blankenberg et al. 2010;)

This poses a significant challenge for most researchers and data analysts and shifts the bottleneck of scientific discovery from data generation to data analysis. Hence, the progress in genomics and genome research is impeded due to inadequate know-how of bioinformatics methods. Powerful and easy-to-use genome analysis tools (Table 30.2) have been developed to address these issues, enabling the biologists to perform complex bioinformatics analyses. Many of the tools are available in public domain and are used by system biologists, bench scientists, computer scientists, and bioinfomaticians (Mi et al. 2019).

Table 30.2 Summary of some platforms, bioinformatics, and Web-based tools used for genome sequence analysis

Software	Salient features and applications (References)
AlignGraph	AlignGraph is an algorithm for extending and joining de novo-assembled contigs or scaffolds guided by closely related reference genomes. It aligns paired-end (PE) reads and preassembled contigs or scaffolds to a close reference. From the obtained alignments, it builds a novel data structure, called the PE multipositional De Bruijn graph. The AlignGraph is accessible at: https://github.com/baoe/AlignGraph (Bao et al. 2014)
EpiGRAPH	EpiGRAPH is a Web service (http://epigraph.mpi-inf.mpg.de/) that enables us to discover hidden associations in genome and epigenome datasets of vertebrates. EpiGRAPH is used to test multiple attributes such as DNA sequence, chromatic structure, epigenome modification, evolutionary conservation attributes and monoallelic gene expression (Bock et al. 2009)
Galaxy	Galaxy (http://galaxyproject.org) is a Web-based genome sequence analysis software system. Galaxy provides this support through a framework that offers the researcher's simple interfaces to powerful tools, while automatically managing the computational details with just a Web browser (Blankenberg et al. 2010)
	EpiGRAPH (http://epigraph.mpi-inf.mpg.de/) and Galaxy (http://galaxyproject.org/) were used jointly is to identify epigenetic modifications that are characteristic of highly polymorphic SNO-rich promoters. This article describes a step-by-step guide to use the software (Bock et al. 2010)
GFinisher	GFinisher is a new bioinformatics approach, based on GC skew to helpfinish prokaryotic microbial genome sequences. The software along with the source code is accessible at http://gfinisher.sourceforge.net/ (Guizelini et al. 2016)
Helmsman	Helmsman is computationally efficient software to evaluate mutations is massive sequencing datasets. It is offered at https://github.com/carjed/helmsman (Carlson et al. 2018)
PAGIT	Post-assembly genome-improvement toolkit (PAGIT) is developed toobtain annotated genomes from contigs. The PAGIT improves the quality of draft genomes to improve scaffolding and generating annotations (Swain et al. 2012)

(continued)

Table 30.2 (continued)

Software	Salient features and applications (References)
PANTHER	Protein analysis through evolutionary relationship (PANTHER) is a comprehensive classification system for inferring the functions of genes in a large curated biological database of gene or protein families and their functionally active related subfamilies. The project was initiated in 1998 at Molecular Application Group. Several improved versions of PANTHER are developed since 2005. It is available at (http://www.pantherdb.org)
PanWeb	PanWeb is a Web interface for pan-genomic data analysis. It is freely accessible at http://www.computationalbiology.ufpa.br/panweb (Pantoja et al. 2017)
PhageWeb	It is a computational tool with graphical interface and Web service that is used to identify phage regions through homology search and gene clustering. It is accessible at http://computationalbiology.ufpa.br/phageweb (de Sousa et al. 2018)
qPortal	qPortal is an important feature-enriched platform for publically available data-driven biomedical sciences. It provides users with an intuitive way to manage and analyze quantitative biological data and allows users to carryout digital project management, and performing the analysis (Mohr et al. 2018)
RAAT	Rapid Annotation Transfer Tool (RAAT) is used to transfer annotations from a high-quality reference to a new genome on the basis of conserved synteny. RATT is freely accessible at http://ratt.sourceforge.net (Otto et al. 2011)
Rapture	RAD Capture (Rapture) is an improved restriction site associated DNA (RAD) sequencing protocol that combines the benefits of RAD and a sequence capture, i.e., rapid library preparations. Rapture is expected tominimize the cost involved in gene sequencing (Ali et al. 2016)
RGAAT	The Reference-based Genome Assembly and Annotation Tool (RGAAT) is a new flexible toolkit for resequencing-based consensus building and annotation update. RGAAT is able to detect sequence variants with comparable precision, specificity, and sensitivity to genome analysis toolkit. RGAAT displays better performance for annotating transfer between different genome assemblies, species, and strains, genome modification, and genome comparison. The software is available at https://sourceforge.net/projects/rgaat/ and https://github.com/wushyer/RGAAT_v2 (Liu et al. 2018)
SpaRC	SparkReadClust (SpaRC) is an Apache Spark-based sequence clustering application. SpaRC partitions read based on their molecules of origin and enable downstream assembly optimization, produced high clustering performance on transcriptomes and metagenomes generated from long-read and short-read sequencing. It is a cost-effective approach. The SpaRC is available at: https://bitbucket.org/berkeleylab/jgi-sparc (Shi et al. 2019)
svist4get	svist4get is a command-line tool that visualizes signal tracks at a given genomic location and aggregates data from various tracks along with transcriptome annotation. It is implemented in Python 3, runs on Linus, and is accessible at https://bitbucket.org/artegorov/svist4get (Egorov et al. 2019)
Tavaxy	Tavaxy, the combination of Taverna and Galaxy workflows, is a stand-alone system for creating and executing workflows based on using an extensible set of reusable workflow patterns. It offers a new set of features to simplify the sequence analysis. Tavaxy is available at http://www.tavaxy.org. It can be accessed through a cloud-enabled Web interface or downloaded and installed to analyze the data within the user's local environment (Abouelhoda et al. 2012)

30.7 Outlook and Challenges

The present time is the era of genomic revolution. NGS technologies have revolutionized the research in genomics and transcriptomics epidemiology. High-quality genomic and transcriptomic data generated NGS platforms will help in diagnosis of infectious and non-infectious diseases, and developing drugs against them.

The NGS technologies are progressively used in the study of genomics, transcriptomics, evolution, etiology, host–pathogen interactions, and epidemiology of livestock infectious diseases. Genome-wide data provides opportunities to identify novel genes and metabolic pathways of therapeutic and economic interest. The genome sequence data should be analyzed for predicting new genes and proteins as targets for therapeutic interventions.

The microbial genome sequences act as contaminants in whole-genome sequence of an organism. It is important to remove the false positives or contaminating sequences of bacteria, archaea, and viruses in NGS diagnostic samples (Lu and Salzberg 2018).

The global market for animal health products and treatments is a multibillion dollar business. The cost of animal disease outbreak is likely to increase with urbanization and a growing demand for animal products.

30.8 Conclusions

NGS technologies have facilitated the livestock genomics studies. Sequencing of genome has already changed the study of infectious agents. NGS can be applied to the infectious agents isolated from the pure culture which can lead to our better understanding of epidemiological infection and pathogenesis caused by the pathogen.

References

Abouelhoda M, Issa SA, Ghanem M (2012) Tavaxy: integrating taverna and galaxy workflows with cloud computing support. BMC Bioinformatics 4(13):77. https://doi.org/10.1186/1471-2105-13-77

Ali OA, O'Rourke SM, Amish SJ, Meek MH, Luikart G, Jeffres C, Miller MR (2016) RAD capture (Rapture): flexible and efficient sequence-based genotyping. Genetics 202(2):389–400. https://doi.org/10.1534/genetics.115.183665 (Epub 2015 Dec 29)

Anis E, Hawkins IK, Ilha MRS, Woldemeskel MW, Saliki JT, Wilkes RP (2018) Evaluation of targeted next-generation sequencing for detection of bovine pathogens in clinical samples. J Clin Microbiol 56(7). pii: e00399-18. https://doi.org/10.1128/jcm.00399-18. Print 2018 Jul

Aslam ML, Bastiaansen JW, Elferink MG, Megens HJ, Crooijmans RP, le Blomberg A, Fleischer RC, Van Tassell CP, Sonstegard TS, Schroeder SG, Groenen MA, Long JA (2012) Whole genome SNP discovery and analysis of genetic diversity in Turkey (Meleagris gallopavo). BMC Genom 14(13):391. https://doi.org/10.1186/1471-2164-13-391

Azhikina TL, Skvortsov TA, Radaeva TV, Mardanov AV, Ravin NV, Apt AS, Sverdlov ED (2010) A new technique for obtaining whole pathogen transcriptomes from infected host tissues. Biotechniques 48(2):139–144

Bai Y, Sartor M, Cavalcoli J (2012) Current status and future perspectives for sequencing livestock genomes. J Anim Sci Biotechnol 3(1):8

Bao E, Jiang T, Girke T (2014) AlignGraph: algorithm for secondary de novo genome assembly guided by closely related references. Bioinformatics 30(12): i319–i328. https://doi.org/10.1093/bioinformatics/btu291

Bayliss SC, Verner-Jeffreys DW, Bartie KL, Aanensen DM, Sheppard SK, Adams A, Feil EJ (2017) The promise of whole genome pathogen sequencing for the molecular epidemiology of emerging aquaculture pathogens. Front Microbiol 3(8):121. https://doi.org/10.3389/fmicb.2017.00121 (eCollection 2017). Review

Blankenberg D, Von Kuster G, Coraor N, Ananda G, Lazarus R, Mangan M, Nekrutenko A, Taylor J (2010) Galaxy: a web-based genome analysis tool for experimentalists. Curr Protoc Mol Biol Chapter 19:Unit 19.10.1-21. https://doi.org/10.1002/0471142727.mb1910s89

Bock C, Halachev K, Büch J, Lengauer T (2009) EpiGRAPH: user-friendly software for statistical analysis and prediction of (epi)genomic data. Genome Biol 10(2):R14. https://doi.org/10.1186/gb-2009-10-2-r14

Bock C, Von Kuster G, Halachev K, Taylor J, Nekrutenko A, Lengauer T (2010) Web-based analysis of (Epi-) genome data using EpiGRAPH and Galaxy. Methods Mol Biol 628:275–296. https://doi.org/10.1007/978-1-60327-367-1_15

Brooks-Pollock E, de Jong MC, Keeling MJ, Klinkenberg D, Wood JL (2015) Eight challenges in modelling infectious livestock diseases. Epidemics 10:1–5. https://doi.org/10.1016/j.epidem.2014.08.005 (Epub 2014 Aug 26)

Bryant J, Chewapreecha C, Bentley SD (2012) Developing insights into the mechanisms of evolution of bacterial pathogens from whole-genomesequences. Future Microbiol 7(11):1283–1296. https://doi.org/10.2217/fmb.12.108 Review

Buermans HP, den Dunnen JT (2014) Next generation sequencing technology: advances and applications. Biochim Biophys Acta. 1842(10):1932–1941. https://doi.org/10.1016/j.bbadis.2014.06.015 (Epub 2014 Jul 1). Review

Camarena L, Bruno V, Euskirchen G, Poggio S, Snyder M (2010) Molecular mechanisms of ethanol-induced pathogenesis revealed by RNA-sequencing. PLoS Pathog 6(4):e1000834. https://doi.org/10.1371/journal.ppat.1000834

Carlson J, Li JZ, Zöllner S (2018) Helmsman: fast and efficient mutation signature analysis for massive sequencing datasets. BMC Genom 19(1):845. https://doi.org/10.1186/s12864-018-5264-y

Challis GL (2014) Exploitation of the Streptomyces coelicolor A3(2) genome sequence for discovery of new natural products and biosynthetic pathways. J Ind Microbiol Biotechnol 41(2):219–232. https://doi.org/10.1007/s10295-013-1383-2 (Epub 2013 Dec 10). Review

de Sousa AL, Maués D, Lobato A, Franco EF, Pinheiro K, Araújo F, Pantoja Y, da Costa da Silva AL, Morais J, Ramos RTJ (2018) PhageWeb—Web interface for rapid identification and characterization of prophages in bacterial genomes. Front Genet 9:644. https://doi.org/10.3389/fgene.2018.00644 (eCollection 2018)

Deurenberg RH, Bathoorn E, Chlebowicz MA, Couto N, Ferdous M, García-Cobos S, Kooistra-Smid AM, Raangs EC, Rosema S, Veloo AC, Zhou K, Friedrich AW, Rossen JW (2017) Application of next generation sequencing in clinical microbiology and infection prevention. J Biotechnol 243:16–24. https://doi.org/10.1016/j.jbiotec.2016.12.022 (Epub 2016 Dec 29). Review

Diaz-Sanchez S, Hanning I, Pendleton S, D'Souza D (2013) Next-generation sequencing: the future of molecular genetics in poultry production and food safety. Poult Sci 92(2):562–572. https://doi.org/10.3382/ps.2012-02741 Review

Djari A, Esquerré D, Weiss B, Martins F, Meersseman C, Boussaha M, Klopp C, Rocha D (2013) Gene-based single nucleotide polymorphism discovery in bovine muscle using next-generation transcriptomic sequencing. BMC Genom 7(14):307. https://doi.org/10.1186/1471-2164-14-307

Egorov AA, Sakharova EA, Anisimova AS, Dmitriev SE, Gladyshev VN, Kulakovskiy IV (2019) svist4get: a simple visualization tool for genomic tracks from sequencing experiments. BMC Bioinform 20(1):113. https://doi.org/10.1186/s12859-019-2706-8

Feng J, Lupien A, Gingras H, Wasserscheid J, Dewar K, Légaré D, Ouellette M (2009) Genome sequencing of linezolid-resistant Streptococcus pneumoniae mutants reveals novel mechanisms of resistance. Genome Res 19(7):1214–1223. https://doi.org/10.1101/gr.089342.108 (Epub 2009 Apr 6)

Gibbs EP (2005) Emerging zoonotic epidemics in the interconnected global community. Vet Rec 157(22):673–679

Gomez-Escribano JP, Castro JF, Razmilic V, Chandra G, Andrews B, Asenjo JA, Bibb MJ (2015) The streptomyces leeuwenhoekii genome: de novo sequencing and assembly in single contigs of the chromosome, circular plasmid pSLE1 and linear plasmid pSLE2. BMC Genom 30(16):485. https://doi.org/10.1186/s12864-015-1652-8

Greenwood JM, Ezquerra AL, Behrens S, Branca A, Mallet L (2016) Current analysis of host-parasite interactions with a focus on next generation sequencing data. Zoology (Jena) 119(4):298–306. https://doi.org/10.1016/j.zool.2016.06.010 (Epub 2016 Jun 30)

Greub G, Kebbi-Beghdadi C, Bertelli C, Collyn F, Riederer BM, Yersin C, Croxatto A, Raoult D (2009) High throughput sequencing and proteomics to identify immunogenic proteins of a new pathogen: the dirty genome approach. PLoS ONE 4(12):e8423. https://doi.org/10.1371/journal.pone.0008423

Guizelini D, Raittz RT, Cruz LM, Souza EM, Steffens MB, Pedrosa FO (2016) GFinisher: a new strategy to refine and finish bacterial genome assemblies. Sci Rep 10(6):34963. https://doi.org/10.1038/srep34963

Kim M, Park T, Yu Z (2017) Metagenomic investigation of gastrointestinal microbiome in cattle. Asian-Australas J Animal Sci 30(11):1515

Langridge GC, Phan MD, Turner DJ, Perkins TT, Parts L, Haase J, Charles I, Maskell DJ, Peters SE, Dougan G, Wain J, Parkhill J, Turner AK (2009) Simultaneous assay of every Salmonella Typhi gene using one million transposon mutants. Genome Res 19(12):2308–2316. https://doi.org/10.1101/gr.097097.109 (Epub 2009 Oct 13)

Lefrançois T, Pineau T (2014) Public health and livestock: emerging diseases in food animals. Anim Front 4(1):4–6

Lindblad-Toh K, Wade CM, Mikkelsen TS, Karlsson EK, Jaffe DB, Kamal M, Clamp M, Chang JL, Kulbokas EJ 3rd, Zody MC, Mauceli E, Xie X, Breen M, Wayne RK, Ostrander EA, Ponting CP, Galibert F, Smith DR, DeJong PJ, Kirkness E, Alvarez P, Biagi T, Brockman W, Butler J, Chin CW, Cook A, Cuff J, Daly MJ, DeCaprio D, Gnerre S, Grabherr M, Kellis M, Kleber M, Bardeleben C, Goodstadt L, Heger A, Hitte C, Kim L, Koepfli KP, Parker HG, Pollinger JP, Searle SM, Sutter NB, Thomas R, Webber C, Baldwin J, Abebe A, Abouelleil A, Aftuck L, Ait-Zahra M, Aldredge T, Allen N, An P, Anderson S, Antoine C, Arachchi H, Aslam A, Ayotte L, Bachantsang P, Barry A, Bayul T, Benamara M, Berlin A, Bessette D, Blitshteyn B, Bloom T, Blye J, Boguslavskiy L, Bonnet C, Boukhgalter B, Brown A, Cahill P, Calixte N, Camarata J, Cheshatsang Y, Chu J, Citroen M, Collymore A, Cooke P, Dawoe T, Daza R, Decktor K, DeGray S, Dhargay N, Dooley K, Dooley K, Dorje P, Dorjee K, Dorris L, Duffey N, Dupes A, Egbiremolen O, Elong R, Falk J, Farina A, Faro S, Ferguson D, Ferreira P, Fisher S, FitzGerald M, Foley K, Foley C, Franke A, Friedrich D, Gage D, Garber M, Gearin G, Giannoukos G, Goode T, Goyette A, Graham J, Grandbois E, Gyaltsen K, Hafez N, Hagopian D, Hagos B, Hall J, Healy C, Hegarty R, Honan T, Horn A, Houde N, Hughes L, Hunnicutt L, Husby M, Jester B, Jones C, Kamat A, Kanga B, Kells C, Khazanovich D, Kieu AC, Kisner P, Kumar M, Lance K, Landers T, Lara M, Lee W, Leger JP, Lennon N, Leuper L, LeVine S, Liu J, Liu X, Lokyitsang Y, Lokyitsang T, Lui A, Macdonald J, Major J, Marabella R, Maru K, Matthews C, McDonough S, Mehta T, Meldrim J, Melnikov A, Meneus L, Mihalev A, Mihova T, Miller K, Mittelman R, Mlenga V, Mulrain L, Munson G, Navidi A, Naylor J, Nguyen T, Nguyen N, Nguyen C, Nguyen T, Nicol R, Norbu N, Norbu C, Novod N, Nyima T, Olandt P, O'Neill B, O'Neill K,

Osman S, Oyono L, Patti C, Perrin D, Phunkhang P, Pierre F, Priest M, Rachupka A, Raghuraman S, Rameau R, Ray V, Raymond C, Rege F, Rise C, Rogers J, Rogov P, Sahalie J, Settipalli S, Sharpe T, Shea T, Sheehan M, Sherpa N, Shi J, Shih D, Sloan J, Smith C, Sparrow T, Stalker J, Stange-Thomann N, Stavropoulos S, Stone C, Stone S, Sykes S, Tchuinga P, Tenzing P, Tesfaye S, Thoulutsang D, Thoulutsang Y, Topham K, Topping I, Tsamla T, Vassiliev H, Venkataraman V, Vo A, Wangchuk T, Wangdi T, Weiand M, Wilkinson J, Wilson A, Yadav S, Yang S, Yang X, Young G, Yu Q, Zainoun J, Zembek L, Zimmer A, Lander ES (2005) Genome sequence, comparative analysis and haplotype structure of the domestic dog. Nature 438(7069):803–819

Liu W, Wu S, Lin Q, Gao S, Ding F, Zhang X, Aljohi HA, Yu J, Hu S (2018) RGAAT: a reference-based genome assembly and annotation tool for new genomes and upgrade of known genomes. Genomics Proteomics Bioinform 16(5):373–381. https://doi.org/10.1016/j.gpb.2018.03.006 (Epub 2018 Dec 21)

Loman NJ, Misra RV, Dallman TJ, Constantinidou C, Gharbia SE, Wain J, Pallen MJ (2012) Performance comparison of benchtop high-throughput sequencing platforms. Nat Biotechnol 30(5):434–439. https://doi.org/10.1038/nbt.2198 Erratum. In: Nat Biotechnol. 2012 Jun;30(6):562

Lu J, Salzberg SL (2018) Removing contaminants from databases of draft genomes. PLoS Comput Biol 14(6): e1006277. https://doi.org/10.1371/journal.pcbi.1006277 (eCollection 2018 Jun)

Martin J, Zhu W, Passalacqua KD, Bergman N, Borodovsky M (2010) Bacillus anthracis genome organization in light of whole transcriptome sequencing. BMC Bioinformatics 11(Suppl 3):S10. https://doi.org/10.1186/1471-2105-11-S3-S10

McCormack JE, Hird SM, Zellmer AJ, Carstens BC, Brumfield RT (2013) Applications of next-generation sequencing to phylogeography and phylogenetics. Mol Phylogenet Evol 66(2):526–538. https://doi.org/10.1016/j.ympev.2011.12.007 (Epub 2011 Dec 14). Review

Mertes F, Elsharawy A, Sauer S, van Helvoort JM, van der Zaag PJ, Franke A, Nilsson M, Lehrach H, Brookes AJ (2011) Targeted enrichment of genomic DNA regions for next-generation sequencing. Brief Funct Genomics 10(6):374–386. https://doi.org/10.1093/bfgp/elr033 (Epub 2011 Nov 26)

Mi H, Muruganujan A, Huang X, Ebert D, Mills C, Guo X, Thomas PD (2019) Protocol Update for large-scale genome and gene function analysis with the PANTHER classification system (v.14.0). Nat Protoc 14(3):703–721. https://doi.org/10.1038/s41596-019-0128-8 (Epub 2019 Feb 25)

Miller RR, Montoya V, Gardy JL, Patrick DM, Tang P (2013) Metagenomics for pathogen detection in public health. Genome Med 5(9):81. https://doi.org/10.1186/gm485 (eCollection 2013). Review

Mohr C, Friedrich A, Wojnar D, Kenar E, Polatkan AC, Codrea MC, Czemmel S, Kohlbacher O, Nahnsen S (2018) qPortal: a platform for data-driven biomedical research. PLoS One 13(1):e0191603. https://doi.org/10.1371/journal.pone.0191603 (eCollection 2018)

Morozova O, Marra MA (2008) Applications of next-generation sequencing technologies in functional genomics. Genomics 92(5):255–264. https://doi.org/10.1016/j.ygeno.2008.07.001 (Epub 2008 Aug 24). Review

Nakano K, Shiroma A, Shimoji M, Tamotsu H, Ashimine N, Ohki S, Shinzato M, Minami M, Nakanishi T, Teruya K, Satou K, Hirano T (2017) Advantages of genome sequencing by long-read sequencer using SMRT technology in medical area. Hum Cell. 30(3):149–161. https://doi.org/10.1007/s13577-017-0168-8 (Epub 2017 Mar 31). Review

Otto TD, Dillon GP, Degrave WS, Berriman M (2011) RATT: rapid annotation transfer tool. Nucleic Acids Res 39(9):e57. https://doi.org/10.1093/nar/gkq1268 (Epub 2011 Feb 8)

Pantoja Y, Pinheiro K, Veras A, Araújo F, Lopes de Sousa A, Guimarães LC, Silva A (2017) Ramos RTJ. PanWeb: a web interface for pan-genomic analysis. PLoS One. 12(5):e0178154. https://doi.org/10.1371/journal.pone.0178154 (eCollection 2017)

Ramos AM, Crooijmans RP, Affara NA, Amaral AJ, Archibald AL, Beever JE, Bendixen C, Churcher C, Clark R, Dehais P, Hansen MS, Hedegaard J, Hu ZL, Kerstens HH, Law AS, Megens HJ, Milan D, Nonneman DJ, Rohrer GA, Rothschild MF, Smith TP, Schnabel RD, Van Tassell CP, Taylor JF, Wiedmann RT, Schook LB, Groenen MA (2009) Design of a high density SNP genotyping assay in the pig using SNPs identified and characterized by next generation sequencing technology. PLoS ONE 4(8):e6524. https://doi.org/10.1371/journal.pone.0006524

Senol Cali D, Kim JS, Ghose S, Alkan C, Mutlu O (2018) Nanopore sequencing technology and tools for genome assembly: computational analysis of the current state, bottlenecks and future directions. Brief Bioinform. https://doi.org/10.1093/bib/bby017 (Epub ahead of print)

Sharma CM, Hoffmann S, Darfeuille F, Reignier J, Findeiss S, Sittka A, Chabas S, Reiche K, Hackermüller J, Reinhardt R, Stadler PF, Vogel J (2010) The primary transcriptome of the major human pathogen Helicobacter pylori. Nature 464(7286):250–255. https://doi.org/10.1038/nature08756 (Epub 2010 Feb 17)

Sharma D, Mal G, Kannan A, Bhar R, Sharma R, Singh B (2017) Degradation of euptox a by tannase-producing rumen bacteria from migratory goats. J Appl Microbiol 123(5):1194–1202

Sharma D, Sharma A, Verma SK, Singh B (2019) Targeting metabolic pathways proteins of Orientia tsutsugamushi using combined hierarchical approach to combat scrub typhus. J Mol Recognit 32(4):e2766. https://doi.org/10.1002/jmr.2766 (Epub 2018 Oct 21)

Shi L, Meng X, Tseng E, Mascagni M, Wang Z (2019) SpaRC: scalable sequence clustering using Apache Spark. Bioinformatics 35(5):760–768. https://doi.org/10.1093/bioinformatics/bty733

Singh B, Gautam SK, Verma V, Kumar M, Singh B (2008) Metagenomics in animal gastrointestinal ecosystem: potential biotechnological prospects. Anaerobe 14(3):138–144

Singh B, Bhat TK, Sharma OP, Kanwar SS, Rahi P, Gulati A (2012) Isolation of tannase-producing Enterobacter ludwigii GRT-1 from the rumen of migratory goats. Small Rumin Res 102(2–3):172–176

Sorek R, Cossart P (2010) Prokaryotic transcriptomics: a new view on regulation, physiology and pathogenicity. Nat Rev Genet 11(1):9–16. https://doi.org/10.1038/nrg2695 (Epub 2009 Nov 24). Review

Srivatsan A, Han Y, Peng J, Tehranchi AK, Gibbs R, Wang JD, Chen R (2008) High-precision, whole-genome sequencing of laboratory strains facilitates genetic studies. PLoS Genet 4(8):e1000139. https://doi.org/10.1371/journal.pgen.1000139

Swain MT, Tsai IJ, Assefa SA, Newbold C, Berriman M, Otto TD (2012) A post-assembly genome-improvement toolkit (PAGIT) to obtain annotated genomes from contigs. Nat Protoc 7(7):1260–1284. https://doi.org/10.1038/nprot.2012.068

Tellam RL, Lemay DG, Van Tassell CP, Lewin HA, Worley KC, Elsik CG (2009) Unlocking the bovine genome. BMC Genom 24(10):193. https://doi.org/10.1186/1471-2164-10-193

Tomley FM, Shirley MW (2009) Livestock infectious diseases and zoonoses. Philos Trans R Soc Lond B Biol Sci 364(1530):2637–2642. https://doi.org/10.1098/rstb.2009.0133

Urbaniak C, Sielaff AC, Frey KG, Allen JE, Singh N, Jaing C, Wheeler K, Venkateswaran K (2018) Detection of antimicrobial resistance genes associated with the international space station environmental surfaces. Sci Rep 8(1):814. https://doi.org/10.1038/s41598-017-18506-4

van Opijnen T, Bodi KL, Camilli A (2009) Tn-seq: high-throughput parallel sequencing for fitness and genetic interaction studies in microorganisms. Nat Methods 6(10):767–772. https://doi.org/10.1038/nmeth.1377 (Epub 2009 Sep 20)

Wang C, Mitsuya Y, Gharizadeh B, Ronaghi M, Shafer RW (2007) Characterization of mutation spectra with ultra-deep pyrosequencing: application to HIV-1 drug resistance. Genome Res 17(8):1195–1201 (Epub 2007 Jun 28)

Williams JL (2005) The use of marker-assisted selection in animal breeding and biotechnology. Rev Sci Tech 24(1):379–391. Review

Williams JL, Iamartino D, Pruitt KD, Sonstegard T, Smith TPL, Low WY, Biagini T, Bomba L, Capomaccio S, Castiglioni B, Coletta A, Corrado F, Ferré F, Iannuzzi L, Lawley C, Macciotta N, McClure M, Mancini G, Matassino D, Mazza R, Milanesi M, Moioli B, Morandi N, Ramunno L, Peretti V, Pilla F, Ramelli P, Schroeder S, Strozzi F, Thibaud-Nissen F, Zicarelli L, Ajmone-Marsan P, Valentini A, Chillemi G, Zimin A (2017) Genome assembly and transcriptome resource for river buffalo, Bubalus bubalis (2n = 50). Gigascience 6(10):1–6. https://doi.org/10.1093/gigascience/gix088

Zhang J, Chiodini R, Badr A, Zhang G (2011) The impact of next-generation sequencing on genomics. J Genet Genomics 38(3):95–109. https://doi.org/10.1016/j.jgg.2011.02.003 (Epub 2011 Mar 15). Review

Whole-Genome Selection in Livestock 31

Abstract

Selective breeding is a traditional method of improving livestock, but molecular genetic revolution in the last decade of the twentieth century has initiated the modern era of genomics. Molecular genetics has influenced the breeding strategies in a big way by providing genetic maps, individual genes, and quantitative trait loci (QTL) related to performance traits in livestock species. QTL detection in animals led the shift from conventional selective breeding to marker-assisted selection (MAS) and SNPs related to performance traits. Advancements in genomics have motivated animal breeder to formulate high-density SNP chips comprising of lakhs of SNPs covering the whole genome of a species. Selection on the basis of whole-genome markers could make selection of genetically superior animals at very early age and at the same time with the accuracy of 0.8 in predicting their breeding value. The SNP chip analysis has been very popular in livestock in predicting breeding potential at early age, but some traits, i.e., traits involving nonadditive effects and epigenetic effects, are still out of the reach of genomic selection.

Highlights

- Continuously increasing demand for livestock-based products puts pressure to increase livestock productivity
- Livestock can be better characterized by genomics approaches and selected for production.

Keywords

Quantitative trait loci · Marker-assisted selection · Marker-assisted tests · SNP selection

31.1 Introduction

Selection for improvement of livestock is in practice since the evolution of civilization, only the phases have changed with the advancement of technologies. Though the accuracy of traditional methods of selection is high, selection for certain traits is limited by the fact that they can only be measured in one sex or are difficult to measure. Also, majority of the traits of interest in livestock species are quantitative in nature as they are controlled by many genes with few genes of large effect or many genes of small effects. For such traits, genomic tools are expected to provide major impetus. Traditionally, livestock improvement programs have utilized selection based on phenotypic information on individuals and their relatives. The genetic gain cannot be increased using these methods as all the milk and economic traits have low heritability. The molecular markers can enhance the accuracy of selection of sires to increase the

B. Singh et al., *Advances in Animal Biotechnology*,
https://doi.org/10.1007/978-3-030-21309-1_31

genetic gain two- to threefold. It can provide an impetus to livestock improvement programs by reducing the cost of progeny testing and selection of bulls based on reliable parameters. With the genomic advancements, the genetic component was also incorporated in the selection scheme. Selective breeding has produced enormous improvement in farm animals resulting in annual genetic gain of about 1–3% of the mean performance of various economically important traits in different livestock species. Such progress has largely been achieved through selection on phenotype: the identification of genetically elite animals through their own performance and physical characteristics and those of their relatives. With the use of DNA markers in the breeding scheme, i.e., MAS, the efficiency of selective programs has further increased. With the recent advances in our understanding of the genome and accompanying technological innovations, new possibilities are opening up for direct identification and selection of animals possessing best genes (Dekkers 2004; Hayes et al. 2009; Dong et al. 2013). A recent technological advancement in selection breeding program, called genomic selection (GS), using the DNA markers distributed throughout the entire genome, is revolutionizing livestock breeding. In this article, we will focus on the different approaches used in the selective breeding programs.

31.2 Marker-Assisted Selection

The MAS involves selection of animals carrying genomic regions that are involved in expressing the traits of economic interest. Research related to MAS initiated in 1990 when mapping projects to discover genes that were responsible for variation in growth and meat quality traits of beef cattle started. Under these research initiatives, chromosomes harboring numerous QTL were identified. Different approaches were followed throughout the world for detecting QTL in livestock depending upon the availability of data on phenotypes of interest and information on molecular markers (Khatkar et al. 2004). The possibility of QTL identification is a function of the size of the gene effect, the family or population structure under study, and the density of informative DNA marker. Hence, whole-genome/specific chromosome scans were carried out to identify the QTL. In dairy cattle, several QTL associated with protein and fat percentage and milk, protein, and fat yield have been identified on bovine chromosome 6-BTA6 (Freyer et al. 2003; Ashwell et al. 2004; Olsen et al. 2004; Szyda et al. 2005; Chen et al. 2006; Kucerová et al. 2006), BTA7 (Ron et al. 2004; Weller et al. 2008), BTA11 (Ashwell et al. 2004; Kucerová et al. 2006), BTA14 (Weller et al. 2003; Ashwell et al. 2004; Schrooten et al. 2004; Kučerová et al. 2006), and BTA23 (Mosig et al. 2001; Schrooten et al. 2004). Similarly, existence of QTL for resistance to mastitis in cattle has been reported on almost all chromosomes (Rupp and Foucras 2010).

Till now, with the technical as well as theoretical developments, QTL mapping in various livestock species has been very successful, and to date, QTL for thousands of traits have been reported (http://www.animalgenome.org/QTLdb/).

With the development and availability of an array of molecular markers and dense molecular genetic maps in animals, MAS has become possible for traits governed by both major genes and QTL. Once associations between genetic markers and performance have been detected, they can be harnessed in a breeding program. Thus, MAS has the potential to increase genetic gain and is expected to be more effective than traditional selection systems (Abdel-Azim and Freeman 2002). Selection for a marker allele known to be associated with a beneficial QTL allele will increase the frequency of that allele and hence increase the production performance. One of the potential major benefits of selection based on marker information is that marker genotypes can be determined based on easily collected samples (e.g., hair and blood) that can be taken from an animal as soon as it is born. Thus, marker information can be used to predict genotype of an animal before it has records for the trait or even for animals which may never express the trait.

The basic principle of MAS is to exploit linkage disequilibrium (LD) between markers and QTLs, i.e., the fact that marker alleles are not randomly associated with QTL alleles. If there is strong enough LD, then one can select (directly) on the markers in order to (indirectly) increase the frequencies of linked QTL alleles of interest. Additionally, MAS can be based on DNA marker in linkage equilibrium with a quantitative trait locus (LE-MAS) or based on selection of the actual mutation causing a QTL effect (gene MAS or GAS). All three types of MAS are currently used in the livestock industries (Dekkers 2004). Out of these three types, the LE-MAS is most difficult to implement because marker-QTL phase within each family must be established before an increase in selection response can be realized. Gene MAS is successful, but requires enormous amount of work and resources.

Compared to classical (phenotypic) selection, MAS is: (i). easier (e.g., when phenotypes are difficult to record); (ii). faster (when phenotyping is long, e.g., progeny testing; also, MAS permits selection at an early stage of development, ability to reduce the generation interval by selecting MAS and breeding at an earlier age); (iii). cheaper (if the cost of genotyping is lower than the cost of phenotyping); (iv). more efficient (because 'heritability' at the marker itself (not the linked QTL) is 1(one) if there is no genotyping error); and v). gains are larger where traditional selection is most difficult, e.g., traits displayed only in females.

Though the advantages of MAS are not really in question and mostly depend on biological or technical particularities of the species and/or breeding scheme considered, the increased 'efficiency' (not heritability) is not truly implemented, because the ultimate goal of selection is obviously not the marker itself, but rather the linked QTL or even the better trait. Thus, a good efficiency of MAS presupposes that LD between markers and QTLs does not vanish, and that QTL effects are well estimated and sustained (e.g., over time or environmental conditions). However, if the marker is in linkage equilibrium with the QTL, all QTL alleles in founder animals are considered to be different, and hence, the number of QTL alleles whose effects are to be estimated is further increased.

A number of genetic tests (Table 31.1) are available to predict the potential performance of an animal. However, these efforts require exhaustive screening. The QTL search requires screening of half sib family sizes of not less than 300 daughters per sire with phenotypic records, molecular markers, and their linkage map for the detection of QTL markers with significant effect. In case of granddaughter design, maintenance of pedigree records for minimum two to three generation is required (depending on the model used).

Also, through MAS, only a limited proportion of the total genetic variance can be captured as most of the traits in livestock species are polygenic in nature and governed by large number of loci, MAS generally results in proportionally small genetic gains. In addition, inability to identify the causal mutations underlying QTLs has made it impossible to implement MAS in commercial populations, since allele phase relationships for linked markers are generally unknown.

31.3 Genomic Selection

Meuwissen et al. (2001) proposed a variant of MAS, called genomic selection (GS), to overcome the limitations associated with MAS. In this approach, markers covering the whole genome are used so that potentially all the genetic variance is explained by the markers; the markers are assumed to be in LD with the QTL so that the number of effects per QTL to be estimated is small. The advantage of this approach is that DNA markers covering the whole genome are taken into consideration and effects of all genes or chromosomal positions are taken care of simultaneously. In simpler terms, it is the simultaneous selection of several thousands of genetic markers that cover the entire genome. Traditional MAS focuses only on those regions which are relatively certain to influence the trait of interest and leaves most of the genome and much of the genetic variation unaccounted.

Table 31.1 List of available marker-assisted tests

Breed	Available test	Source
Cattle	Tibial hemimelia (TH), Pulmonary Hypoplasia with Anasarca (PHA), black/red coat color (CC), dilution (DL), idiopathic epilepsy (IE), arthrogryposis multiplex (AM), or curly calf syndrome analysis	AgriGenomics, Inc.
	Parentage, freemartin, coat color, leptin, meat quality, BSE resistance, Johne's disease, bovine viral diarrhea (BVD)	Biogenetic Services
	Arthrogryposis multiplex (AM), parentage, coat color, Seek-Black, Seek-Tender, bovine viral diarrhea (BVD-PI), identity tracking, 50 K SNP CHIP genotyping	GeneSeek
	Coat color, prolactin (CMP), BLAD, citrullinemia, DUMPS, kappa-casein, βeta-lactoglobulin, complex vertebral malformation (CVM), Calpain 316/530, freemartin	Genetic Visions
	Arthrogryposis multiplex (AM) or curly calf syndrome analysis, neuropathic hydrocephalus (NH), idiopathic epilepsy (IE), osteopetrosis (OS), Pulmonary Hypoplasia with Anasarca (PHA), and tibial hemimelia (TH), Igenity profile analysis (tenderness, marbling, quality grade, fat thickness, ribeye area, hot carcass weight, yield grade, heifer pregnancy rate, stayability, calving ease, weaning weight, docility, residual feed intake, and average daily gain), DoubleBLACK coat color, dilution (DL), horned–polled, bovine viral diarrhea (BVD-PI), identity tracking, parentage, myostatin	Igenity
	Arthrogryposis multiplex (AM) or curly calf syndrome analysis, neuropathic hydrocephalus (NH), osteopetrosis (OS), parentage, Tru-Marbling™, Tru-Tenderness™, MMIG Homozygous Black, polled/horned	MMI Genomics
	Arthrogryposis multiplex (AM) or curly calf syndrome analysis, neuropathic hydrocephalus (NH), osteopetrosis (OS), GeneSTAR® MVP™ (feed efficiency, marbling, tenderness), GeneSTAR® Elite Tender, GeneSTAR® Quality Grade, GeneSTAR® Tenderness 2, GeneSTAR® Feed Efficiency, GeneSTAR® BLACK, parentage, identity tracking	Pfizer Animal Genetics (previously Bovigen)
	Leptin	Quantum genetics
	Fertility-associated antigen (FAA)	Repro Tec Inc.
	Parentage, freemartin, coat color, Dexter cattle (Dexter Dun, extension (red/black), bulldog dwarfism (Chondrodysplasia), freemartin, karyotyping)	Veterinary Genetics Laboratory (UC Davis)
	Breed identification (AnguSure™)	Viagen
Sheep	Scrapie, spider lamb syndrome	Biogenetic Services
	Scrapie	GeneSeek
Fish	Whirling disease	Biogenetic Services
Swine	Parentage, RN gene, porcine stress syndrome, breed purity	GeneSeek

Conversely genomic selection puts the greatest emphasis on those regions with the largest effects, while still accounting appropriately for the more ambiguous genetic variation in the remainder of the genome. Further, prediction of genomic breeding value (GEBV) through GS is based on identical by state (IBS) rather than identical by descent (IBD) followed in traditional estimates of breeding value, wherein knowledge of the genotypes of ancestors is essential. The comparison between GS and QTL mapping is depicted in Table 31.2.

This approach holds importance as in livestock genes affecting most economically important traits are distributed throughout the genome and there are relatively few that have large effects

Table 31.2 Differences between GS and QTL mapping

QTL mapping	Genomic selection
Significant effect for an evaluated locus is required	All effects are estimated simultaneously
Estimated QTL effect may be biased because only one QTL is fitted at the time	Average (across SNPs) bias may be limited.
Most of the QTL mapping studies so far used have used linkage analysis and sparse marker maps	Genomic selection depends on LD between marker-QTL, persistent across population and dense marker maps

with many more genes with progressively smaller effects (Hayes and Goddard 2001; Sanna et al. 2008, VanRaden et al. 2009).

The advent of next-generation sequencing (NGS) technologies, advancement in genotyping tools, and low-cost genotyping strategies have resulted in implementation of genomic selection in livestock in a big way, especially in North America and European countries. Due to this, in last few years, the landscape of livestock genomic research has been dramatically changed as genomic selection has revolutionized the animal breeding, in particular the dairy cattle breeding in recent years by increasing the accuracy of estimated breeding value (EBV).

31.4 Recent Advances in Genomic Selection

In genomic selection, the animals from reference population with recorded phenotypes are genotyped for thousands of markers (single-nucleotide polymorphism—SNP) densely distributed throughout the genome. This genotypic information forms the basis to calculate the marker effects and predict the genomic breeding value of the animal as the sum of effects of all markers. Several studies have indicated that genomic selection strategies could result in about 50% increase in annual genetic gain in dairy cattle (Schaeffer 2006; Pyrce et al. 2010). These studies have established the fact that breeding program based on genomic selection has much higher efficiency in genetic improvement in comparison with traditional breeding programs. Using the approach of GS, genomic breeding value of an individual animal can be estimated based on previously estimated marker effect in a reference population without using the phenotype data. Through GEBV, breeding decision can be made at very young stage of the animals or even at embryo stage and that too with higher accuracies in comparison with breeding value based on a pedigree index (Pryce and Daetwyler 2012).

Genomic selection exploits the concept of linkage disequilibrium with the assumption that the effects of the chromosome segments are same across the populations because the markers are in LD with the QTL that they bracket. The implementation of genomic selection requires a large number of markers distributed throughout the entire genome ensuring that all QTL are in LD with a marker or haplotype of markers.

Identification of the large set of markers and their genotyping has not been cost-effective till date. With the latest technologies of sequencing, sequence draft of most of the livestock species (Table 31.3) is available. The first drafts of the sequencing of the chicken and cattle genomes were completed in 2005 (http://www.genome.gov/12512874). In 2006, the sequencing of horse and pig genomes was also begun. The sequencing of livestock genome further led to the availability of hundreds of thousands of single-nucleotide polymorphisms. Taking the advantage of the availability of sequence draft and recent technologies (NGS, deep sequencing, and reduced reduction libraries), SNP arrays could be developed for different livestock species. The draft DNA sequence provides a template for identifying genetic polymorphisms, and the approach of reduced representation libraries (RRLs) helps in the production of a large number of SNPs. Herein, DNA isolated from genetically

Table 31.3 Details of the livestock genome sequence and available SNP chips (2009)

Cow (*Bos taurus*)	Draft assembly (7X)	Completed	Bovine 50K Illumina™ iSelect chip Illumina High-density Bovine BeadChip Array Affymetrix Axiom genome-wide BOS1 array
Chicken (*Gallus gallus*)	Draft assembly (6, 6X)	Completed	Illumina 18K iSelect chip
Dog (*Canis lupus familiaris*)	Draft assembly	Completed	–
Horse (*Equus caballus*)	Draft assembly (7X)	Completed	Equine Illumina 60K SNP chip; Illumina Equine SNP50 BeadChip
Pig (*Sus scrofa*)	Draft assembly (BAC to BAC)	Completed	Illumina iSelect pig DNA chip, 60K
Buffalo (*Bubalus bubalis*)	Draft assembly	Completed	–
Sheep (*Ovis aries*)	Draft assembly	Completed	Illumina's OvineSNP50
Goat (*Capra hircus*)	Draft assembly	Completed	IlluminaGoatSNP50 BeadChip

diverse individuals is mixed together, restriction digested and purified by agarose gel electrophoresis. Plasmid libraries with reduced representation of genomes are prepared using subset of restriction fragments (Fig. 31.1).

Size selection of fragments produced by complete restriction endonuclease (RE) digestion reduces the complexity of the pools of DNA from multiple individuals. After restriction digestion, Sanger sequencing (two–fivefold) of the libraries is performed, and the sequences are aligned and screened for polymorphism. Through this approach, it is possible to reduce the fraction of genome in the reduced representational library by 1-2 orders of magnitude. It also has the advantage that independently prepared libraries

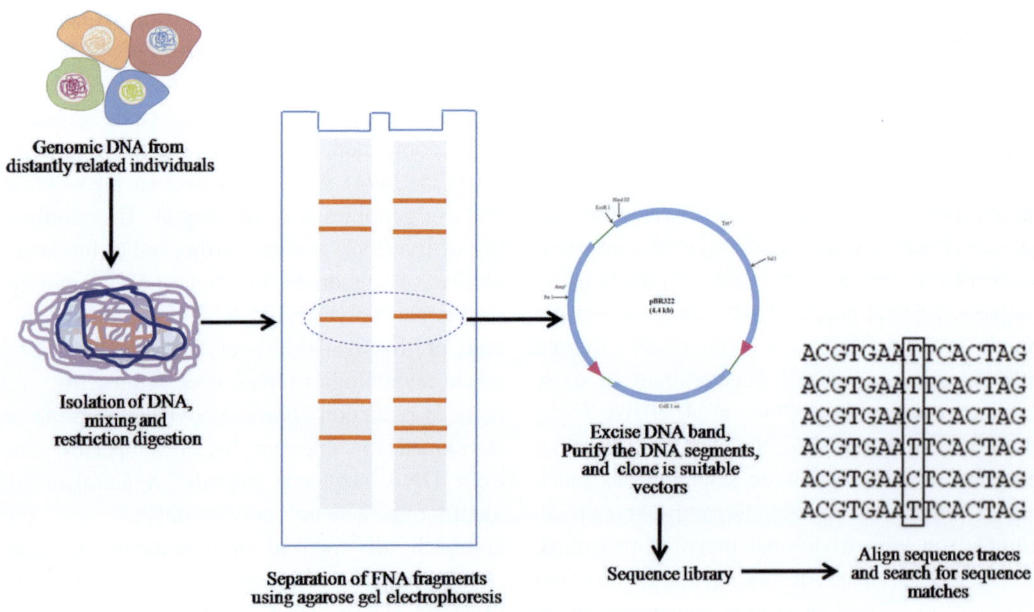

Fig. 31.1 Reduced representation shotgun (RRS) for SNP discovery

have identical fragment populations. The approach based on the deep sequencing of libraries of reduced complexity constructed from pooled DNA samples using NGS is cost-effective and efficient for concurrent SNP discovery, validation, and characterization. This approach of deep sequencing has been followed by USDA, Beltsville (USA), scientist to discover SNPs in cattle and develop high-density SNP chip (Matukumalli et al. 2009). In silico comparison with draft, bovine sequence assembly (Btau3.1) was performed to compare and characterize the fragments produced by complete RE digests of the whole genome. The BovineSNP50 BeadChip is a 12-sample genotyping product and is commercially available for detecting genetic variation in any breed of cattle. The chip features more than 54,000 SNPs and presents an average SNP spacing of 51.5 Kb across the entire genome. This Bovine 50 K Illumina™ iSelect SNP chip (51,386 polymorphic SNP markers) has been designed using a combination of publicly available SNPs along with highly informative novel SNPs discovered. The chip includes more than 23,800 novels, validated SNPs discovered from deeply sequencing of the *Bos taurus* genome; over 12,000 common SNPs derived from the Bovine HapMap Consortium; 15,000 SNPs from the bovine genome reference Btau3.1; and 116 parentage markers from the Clay Center.

High-density swine SNP chip has been developed as part of an integrated effort of several European and American institutions involved in swine genomics research. This pig SNP chip also includes already validated SNPs as well as SNPs identified de novo using second-generation sequencing on the Illumina 1 G analyzer (Solexa), and the Roche 454 sequencer. For the purpose, a total of 15 DNA libraries were prepared using pooled DNA samples from global wild boar (35 samples from Europe and Asia) and ten European, North American, and Asian domesticated breeds were digested with three REs (AluI, HaeIII, and MspI). Fragments in the size range of 150–200 bp were selected for sequencing using Solexa and 454 platforms. A bioinformatics platform was developed to identify SNPs with >0.05 minor allele frequencies covering the genome at spaced intervals. These SNPs were then used to design an Illumina iSelect pig DNA chip (60,000 SNPs) for defining wild boar and domesticated populations. The knowledge gained from the inclusion of wild boar and other outgroups would be of tremendous value for interpreting SNP variation in domesticated animals and making inferences of ancestral derived alleles and coalescent patterns across the genome.

The high-density SNP chips are an extremely valuable tool for the livestock genomics community for a variety of applications including haplotype sharing, signatures of selection, QTL and LD mapping. These chips are being employed widely to drive gene discovery, association analyses, and eventually whole-genome selection. Chip-based genomic selection approach can be accomplished to correlate phenotype and genotype across the whole genome simultaneously rather than one locus at a time.

The list of SNP arrays presently available for livestock and poultry species includes chicken, cattle, pig, and sheep (Illumina's OvineSNP50), and it is under development for other livestock species (Table 31.3). In these high-density oligonucleotide SNP arrays, thousands of probes are arrayed on a small chip, allowing for many SNPs to be interrogated simultaneously in a cost-effective manner leading to the successful implementation of whole-genome selection (WGS). As compared to current DNA tests based on dozens to hundreds of SNP, whole-genome selection is costlier.

These platforms have been extensively utilized in genotyping tens of thousands of animals for genome-wide association study (GWAS). However, to estimate the genetic merit these markers should be spaced less than 10 Kb in order to obtain consistent LD phase across population (de Roos et al. 2008). Keeping this in view, these SNP chips are upgraded continuously; e.g., the usefulness of Bovine SNP50 chip in estimating the genetic merit was observed to be limited (Hayes et al. 2009) due to lower density of SNP (spaced at 50 Kb intervals). With the availability of two high-density platforms from Illumina and Affymetrix, this limitation was resolved and chips with gap size of about 5.2 Kb

(Illumina high-density Bovine BeadChip Array with 777,962 SNPs) and 6.2 kb Affymetrix Axiom genome-wide BOS1 array with 648,874 SNP) have been obtained in Holstein and Jersey populations (Rincon et al. 2011). These chips are powerful tools for fine QTL mapping in cattle.

31.5 Population Structure of Livestock and WGS

The key feature of GS is that the markers (SNPs) are assumed to be in LD with the QTL so that the number of effects to be estimated per QTL is small. The number of SNPs required for GS depends on the pattern of LD in the population. Fortunately, the population structure of livestock species makes them particularly useful for GWS. The populations of livestock species resemble to some extent recombinant inbred lines, although most of them are not highly inbred.

Most livestock breeds have been formed from large population by dividing them into many smaller often closed populations. This has led to a reduced genetic diversity within breeds and therefore large haplotype blocks. Therefore, a modest number of SNPs are required for GS in livestock species. Furthermore, the same causal mutation is shared between populations of livestock breeds, e.g., IGF2 locus in pigs (Van Laere et al. 2003). This indicates that high-resolution mapping can be obtained using more than one breed, as shared haplotype blocks are substantially shorter between breeds than within breeds.

Conceptually, the implementation of genomic selection requires:

(1) Availability of reference population with genotype and phenotype records
(2) High-quality herd recording data for different traits of economic importance
(3) Availability of high-density SNP array platforms for genotyping.

Technological advancements during last five years have revolutionized the fields of genomics and proteomics. Now we can generate the required genomic resources for a species/population in very short period, but the phenomics resources (reference population with standard phenotypic records) need to be established to identify and validate molecular markers which could be extended to other test populations. The livestock industry is making a great progress in countries where pedigreed information on a large number of individuals with phenotypes on traits of interest is available. For example, the GS strategy has been a huge success, especially in North American and European countries due to availability of sufficient animal numbers with good phenotypes and high-density SNP chip platforms. With the availability of high-throughput resources and low-cost genotyping tools, remarkable advancement has been witnessed in GWAS for several traits in bovines since the inception of this technology (Table 31.4) (Sharma et al. 2015). In livestock species, GWAS has gained popularity in mapping QTL to several traits of economic importance like milk yield, fat and protein percentage, meat quality and quantity, calving ease, fertility traits, disease resistance traits, and egg production. If performed carefully, GWAS has proved to be an ideal strategy to identify genes associated with various phenotypes and to elucidate the mechanisms of complex traits.

In short span of time, GWAS has become successful in mapping QTL to several traits of economic importance like dairy traits, disease resistance, meat quality and quantity traits, and reproductive and egg production. It has the potential not only to identify genes associated with various phenotypes, but also to explain the mechanisms underlying complex traits. In bovines, a number of studies have been carried out to identify genes/SNPs associated with desirable traits in dairy and beef cattle (Sharma et al.

Table 31.4 Summary of genome-wide association studies in livestock

Species/breeds	SNP array	Trait	Chr. no.	Candidate genes	References
Cattle—Angus, Murray Grey, Shorthorn and Hereford, Brahman, Santa Gertrudis, and Belmont Red	BovineSNP50K, Parallele SNP10K chip	RFI, ADG, and mMWT	8	–	Bolormaa et al. (2011)
		RFI	5	HSD17B3, SHC3	
		RFI, ADG, and DFI	2	IGFBP2	
Cattle—Native Hanwoo	50K	Carcass weight	14	FAM110B, SCDBP, and TOX	Lee et al. (2013)
Cattle—Brazilian Nellore	777K	Birth weight and size	14		Utsunomiya et al. (2013)
Cattle—HF, Nordic Red and Jersey		Mastitis	6, 13, 16, 19, and 20	Vitamin D-binding protein precursor and neuropeptide FF receptor 2	Sahana et al. (2016)
Cattle—Angus, Charolais, Gelbvieh, Hereford, Limousin, Red Angus, and Simmental	50K	Growth	6	SPP1, NCAPG	Snelling et al. (2010)
Cattle—Holstein-Friesian	50K, 700K	Stature	3, 5, 11 and 12	–	Tan (2013)
Cattle—German Holstein	54K	General production and environmental sensitivity of milk	14	DGAT1, PPARGC1, and casein cluster	Streit et al. (2013)
Cattle— Holstein Friesian		Foot-and-mouth disease resistance	17, 22, and 15	Myosin XVIIIB and seizure related 6 homolog (mouse)-like	Lee et al. (2015)
Cattle—Holstein-Friesian	40,668	Fat yield and somatic cell score	20	GHR and PRLR	Meredith et al. (2012)
Cattle—lactating Holstein cows	39,759	Rectal temperature	24	U1, NCAD	Dikmen et al. (2013)
			16	SNORA19, RFWD2, and SCARNA3	
Goats—Yunnan Black	Illumina Goat60 KiSelect	Lambing number	1, 21, 28	SLC4A10, TBR1, hypothetical protein , WDFY4, TMEM26, and BICC1	Lan et al. (2015)
Goats—South African Boer goat	GoatSNP50 BeadChip	Pigmentation	17	EDNRA	Menzi et al. (2016)
Baluchi sheep	OvineSNP50 BeadChip	Greasy fleece weight	17, 20	Multiple genes	Gholizadeh et al. (2015); Pasandideh et al. (2018)

2015). For beef industry, the interest is for big stature of the animal as well as its better quality (tenderness, juiciness, and taste), while in dairy cattle focus is on milk yield, and fat and protein percentage. SNP arrays have helped to identify the regions associated with these desirable/valuable traits.

An et al. (2018) studied genome-wide association for heart, liver, spleen, lung, and kidney weight in Simmental cattle using Bovine HD 770 K. These traits are found to be linked to many valuable traits, such as growth, health, and productivity, and are of high value to industry and breeders. SNP array was performed as in beef cattle, and internal organs hold economic interest to breeders. Thirty-eight significant SNPs located within or near the *NDUFAF4, LCORL, BT.94996, SLIT2, FAM184B, LAP3, BBS12, MECOM, CD300LF, HSD17B3, TLR4, MXI1*, and *MB21D2* genes were identified for the internal organ weight traits. In addition, 4 haplotype blocks containing 18 significant SNPs on chr6 associated with spleen weight were also identified.

In similar lines with bovines, GWAS has been carried on in other livestock species and facilitated the identification of genomic regions with potential candidate genes. For example, in goat, genome-wide study using Illumina Caprine 50 K BeadChip has helped to identify genomic regions contributing to coat color and mohair characteristics in Iranian Markhoz goat (Nazari-Ghadikolaei et al. 2018). Regions within or near *ASIP, ITCH, AHCY, RALY* genes on chromosome 13 were significantly associated with black and brown coat color, whereas *KIT* and *PDGFRA* genes on chromosome 6 were found to be associated with white coat color. Potential candidate genes identified for various mohair characteristics included: *POU1F1/chr1* (mohair quality), *MREG/chr 2* (mohair volume), *DUOX1/chr 10* (yearling fleece weight), and *ADGRV1/chr 7* (grease percentage). Other traits for which associated putative genes have been identified through GWAS in goats include brown coat colour in Coppernecked goats (Becker et al. 2015); lambing number in Yunnan Black goats (Lan et al. 2015), limb development and outgrowth or congenital appendages, also called as wattles in goats (Reber et al. 2015), white spotting phenotype in South African Boer goat (Menzi et al. 2016); supernumerary teats in French Alpine and Saanen dairy goats (Martin et al. 2016), and cashmere fibre traits in cashmere goat population (Li et al. 2017).

Though progress made in GWAS is very rapid, there are few precautions to be taken to minimize the errors. The fundamental requirements for GWAS are: (i) The population under study must be genetically homogeneous; i.e., there should be no population stratification; (ii) all subjects in the samples must represent statistically independent units drawn from that population. If not taken care of, the tests of association may lead to spurious associations or may have inflated type I error rates. Another scenario is that the related individuals share both causal and non-causal alleles, and that linkage disequilibrium between these sites can lead to artifacts. A powerful method, i.e., mixed model that handles population structure by accounting for the amount of phenotypic covariance to deal with the artifacts, is practiced in the field of animal breeding. This model, if applied to GWAS, can markedly reduce the number of false-positive associations. Lack of statistical knowledge remains one of the major glitches in GWAS projects. Hence, GWAS leads to meaningful and valuable results only if performed carefully.

If these criteria are not satisfied while conducting tests of association, spurious associations may be observed or there might be inflated type I error rates. Further, in case the causal or non-causal alleles are shared between related individuals, artifacts may be observed due to linkage disequilibrium between these sites. In animal breeding programs, to deal with the artifacts mixed model that handles population structure by accounting for the amount of phenotypic covariance is used. Through this approach, GWAS can markedly reduce the number of false-positive associations. Another major difficulty in applying GWAS is proficiency in statistical knowledge. Slowly, with the advancement of knowledge these hurdles are being addressed and GWAS is being applied with meaningful and valuable results.

31.6 Estimation of Genomic Breeding Value

Practical utility of genome-wide selection lies on accurate determination of genomic breeding values (GEBVs). The GEBV is calculated as the sum of the effects of densely spaced genetic markers or their haplotypes spread across the whole genome. It potentially captures all the QTL contributing to variation in a trait. To calculate GEBV, a prediction equation is derived on the basis of SNP data. The entire genome is divided into small segments, and effect of all these segments is estimated in reference population (population wherein animals have both phenotypic and genotypic records). Herein, the effects of all loci contributing to genetic variation are taken into account, even if the effects or contribution of the individual loci is very small. To predict breeding value in subsequent generations, animals can be genotyped for the markers to determine the respective chromosome segments they carry and then estimated effects of these segments are summed across the whole genome (Fig. 31.2).

This breeding value is termed as GEBV. Meuwissen et al. (2001) demonstrated that it was possible to make very accurate selection decisions when breeding values were predicted from dense marker data alone. Using simulation, they have shown that the breeding value could be predicted with an accuracy of 0.85 from marker data alone (where accuracy is the correlation between true breeding value and EBV, and reliability is the square of this result).

Genome of animal X (markers A, B,..., J, possibly associated with QTL:

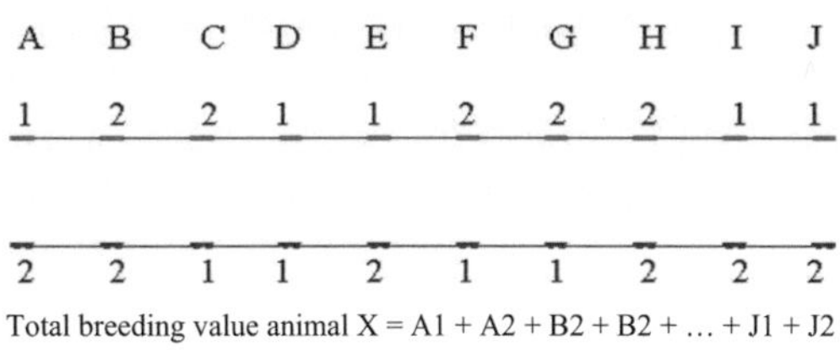

Fig. 31.2 Estimation of breeding value

31.7 Integration of Whole-Genome Selection in Breeding Program

The genomic selection can very well be executed/incorporated in the breeding program. The primary requirement is the 'reference population' (discovery dataset) with a moderate number of animals for which phenotypes are recorded for all the relevant traits. This dataset is genotyped with a large number of SNPs. The marker data is used as input for the prediction equation to estimate breeding value (EBV). The second requirement is validation sample consisting of a large number of animals with phenotypes recorded for the traits. The validation samples are genotyped with markers that are to be used commercially. The prediction equation is tested to assess its accuracy on this independent sample set. Then, selection candidates are genotyped for the markers and the prediction equation is estimated in the discovery data used to calculate GEBV. Phenotyping recording for the selection candidate is not required. The accuracy is assumed to be similar to that observed in validation samples. Practical implementation might be complex, but distinction between discovery, validation, and selection candidates is still useful. Using this information, estimation of QTL effects can be carried out on animals that are completely different from the selection candidates. Summing up, the QTL effects, inferred from either haplotypes or individual single-nucleotide polymorphism markers, are first estimated in a large reference population with phenotypic information and in subsequent generations, only marker information is required to calculate GEBV.

31.8 Potential of Genomic Selection

One of the most economically important aspects of GS is the possibility to use it for improving selective breeding. In near future, the DNA marker data as well as phenotypes and pedigrees on potential selection candidates will be available. All this data could be combined to estimate improved EBV. EBV obtained from this 'genome-wide' selection is of comparable robustness to a conventional EBV based on analysis of pedigree and phenotype data. Once the associations between chromosome segments and phenotypes are established, it should be possible to make selection without phenotypic information for a number of generations. Genomic selection approach is particularly attractive for expensive-to-measure traits. GEBV can be used to estimate traits not normally recorded, such as efficiency of nutrient utilization and various animal behaviors. It also can be more easily applied to traits where the heritability is low and genetic change is slow, and traits that are difficult to measure. This presupposes, however, that the data on the trait and the QTL or the genes it links to are known.

The most likely first application of this approach will be in dairy cattle where, in principle, bulls could be selected on the basis of genomic information alone and used for mating long before their progeny test results would be known considerably reducing generation interval and accelerating the genetic gain. In dairy cattle, GS could be used initially to select young bull calves for progeny testing. However, if it is successful, sires of sons and sires of replacement heifers will be selected based on genetic markers and formal progeny testing will disappear. The simulation results suggest that the accuracy of the GEBV for a bull calf can be as high as the accuracy of an EBV after a progeny test. Potentially, genomic selection could lead to a doubling of the rate of genetic gain through selection and breeding from bulls at 2 years of age rather than 5 years of age or later. By avoiding progeny testing, bull breeding companies could save up to 92% of their costs. The rate of selection gain can further be improved by extensive genotyping for selection intensities and thereby increase in the rates of genetic gain.

31.9 Limitations of SNP Selection

The major limitations of SNP selection are: (i) SNP analysis requires good phenotypic information from a large number of animals in the population of interest (Hayes et al. 2009). (ii) It cannot be applied to different populations unless they overlap; for example, the system developed for American Holsteins will likely be fairly accurate for European Holsteins because there is considerable overlap in their genetic makeup. The system developed for *Bos taurus* cannot be applied to *Bos indicus*. (iii) SNP analyses degrade very slightly with each new generation in the population due to the crossing-over that occurs during meiosis. If a crossing-over event occurs between a marker and the allele it is marking, one not only gets the wrong information, but it is exactly opposite, which equates to selection for the undesirable allele.

31.10 Implementation of GEBV

The reliability of GEBV has been evaluated in experiments in the USA, New Zealand, Australia, and the Netherlands (Harris et al. 2008; Hayes et al. 2009; VanRaden et al. 2009). The reference populations used were between 650 and 4500 progeny-tested Holstein-Friesian bulls, genotyped for approximately 50,000 genome-wide markers. GEBV reliability for the young bulls in the reference population was between 20 and 67%. The major factors that determined the reliability were: (i) heritability of the trait evaluated, (ii) the number of bulls in the reference population, (iii) the statistical method used to estimate the single-nucleotide polymorphism effects in the reference population, and (iv) method used to calculate the reliability. Another outcome of these studies is that polygenic effect (parent average breeding value) should be included in the GEBV calculations in order to capture any genetic variance not associated with the markers and to put some selection pressure on low-frequency QTL that may not be captured by the markers. All the studies in four different countries showed reliabilities achieved for GEBV were significantly greater than the reliability of parental average breeding values, the current criteria for selection of bull calves to enter progeny test teams. The increase in reliability is sufficiently high that at least two dairy breeding companies are already marketing bull teams for commercial use based on their GEBV only, at 2 years of age. This strategy should at least double the rate of genetic gain in the dairy industry. The similar strategy can also be applied to beef cattle; however till now it is not certain that how far it would be useful. It would be difficult to apply this strategy to traits that can be measured in both sexes, like weaning weights.

31.11 Outlook and Challenges

The ever-increasing population of the world will require 70–100% more food by mid of this century; hence, animal breeders and geneticists have to adopt latest interventions to provide improved livestock with minimal or no impact on environment and animal welfare. Genomics is a prospective area to augment livestock production through selecting livestock based on molecular markers of superior traits. As molecular markers can enhance the accuracy of selection, emphasis should be on identifying novel markers of genetic superiority in animals. Already, a surge has been noted in genetic gains, especially in dairy animals proliferating. In developed countries, the semen used for AI is from the sires that are selected based on molecular markers. Like developing countries, regular genomic prediction of sires is needed in developing countries.

Lack of quality animal breeding infrastructure, viz., lack of routine recording of reliable phenotypes and analytical methods for genotyping of animals, and timely feedback improve management and husbandry techniques (Mrode et al. 2019). Genomics has revolutionized the livestock production in large parts of the world. The livestock in developing tropical countries possess unique adaptive merits should be documented by molecular genetics methods and candidate genes be identified. Further advances in genomics will help in collecting more information about the animal genetics, but issues pertaining to ethical, environmental, animal welfare, and social values should also be looked into.

31.12 Conclusions

There are three practical ways of increasing genetic progress: (1). increase the accuracy of selection; (2). increase selection intensity; and (3). decrease the generation interval, which also results in more selection steps per unit time. WGS is one of the few tools that can affect all components affecting genetic progress. Accuracy is clearly increased with genomic selection using SNPs. Generation interval can be lowered easily because SNP evaluations of embryos are equally valid as evaluations for young or old animals. Because the technology can be applied broadly at relatively low cost (screening hundreds of embryos or calves), practical opportunities are provided for increasing selection intensity as well. The combination of these advantages, when added to pedigree and phenotypic information on each respective individual, becomes the most powerful, practical approach available for making genetic changes. Although there are many challenges associated with implementation of genomic selection including increasing the accuracy of GEBV, integrating genomic information into national genetic evaluations, and managing long-term genetic gain, this is the best-suited technology for livestock improvement.

References

Abdel-Azim G, Freeman AE (2002) Superiority of QTL-assisted selection in dairy cattle breeding schemes. J Dairy Sci 85(7):1869–1880

An B, Xia J, Chang T, Wang X, Miao J, Xu L, Zhang L, Gao X, Chen Y, Li J, Gao H (2018) Genome-wide association study identifies loci and candidate genes for internal organ weights in Simmental beef cattle. Physiol Genomics 50(7):523–531. https://doi.org/10.1152/physiolgenomics.00022.2018

Ashwell MS, Heyen DW, Sonstegard TS, Van Tassell CP, Da Y, VanRaden PM, Ron M, Weller JI, Lewin HA (2004) Detection of quantitative trait loci affecting milk production, health, and reproductive traits in Holstein cattle. J Dairy Sci 87(2):468–475

Becker D, Otto M, Ammann P, Keller I, Drögemüller C, Leeb T (2015) The brown coat colour of Copper-necked goats is associated with a non-synonymous variant at the TYRP1 locus on chromosome 8. Anim Genet 46(1):50–54. https://doi.org/10.1111/age.12240 (Epub 2014 Nov 13). Erratum in: Anim Genet. 2015 Aug;46(4):470

Bolormaa S, Hayes BJ, Savin K, Hawken R, Barendse W, Arthur PF, Herd RM, Goddard ME (2011) Genome-wide association studies for feedlot and growth traits in cattle. J Anim Sci 89(6):1684–1697. https://doi.org/10.2527/jas.2010-3079

Chen HY, Zhang Q, Yin CC, Wang CK, Gong WJ, Mei G (2006) Detection of quantitative trait loci affecting milk production traits on bovine chromosome 6 in a Chinese Holstein population by the daughter design. J Dairy Sci 89(2):782–790

de Roos AP, Hayes BJ, Spelman RJ, Goddard ME (2008) Linkage disequilibrium and persistence of phase in Holstein-Friesian, Jersey and Angus cattle. Genetics 179(3):1503–1512. https://doi.org/10.1534/genetics.107.084301 (Epub 2008 Jul 13)

Dekkers JC (2004) Commercial application of marker- and gene-assisted selection in livestock: strategies and lessons. J Anim Sci 82 E-Suppl:E313–328

Dikmen S, Cole JB, Null DJ, Hansen PJ (2013) Genome-wide association mapping for identification of quantitative trait loci for rectal temperature during heat stress in Holstein cattle. PLoS ONE 8(7):e69202. https://doi.org/10.1371/journal.pone.0069202 Print 2013

Dong Y, Xie M, Jiang Y, Xiao N, Du X, Zhang W, Tosser-Klopp G, Wang J, Yang S, Liang J, Chen W, Chen J, Zeng P, Hou Y, Bian C, Pan S, Li Y, Liu X, Wang W, Servin B, Sayre B, Zhu B, Sweeney D, Moore R, Nie W, Shen Y, Zhao R, Zhang G, Li J, Faraut T, Womack J, Zhang Y, Kijas J, Cockett N, Xu X, Zhao S, Wang J, Wang W (2013) Sequencing and automated whole-genome optical mapping of the genome of a domestic goat (*Capra hircus*). Nat Biotechnol 31:135–141. https://doi.org/10.1038/nbt.2478

Freyer G, Sørensen P, Kühn C, Weikard R, Hoeschele I (2003) Search for pleiotropic QTL on chromosome BTA6 affecting yield traits of milk production. J Dairy Sci 86:999–1008

Gholizadeh M, Rahimi-Mianji G, Nejati-Javaremi A (2015) Genomewide association study of body weight traits in Baluchi sheep. J Genet 94:143–6

Harris BL, Johnson DL, Spelman RJ (2008) Genomic selection in New Zealand and the implications for national genetic evaluation. Proc, Interbull Meeting, Niagara Falls, Canada

Hayes BJ, Bowman PJ, Chamberlain AJ, Goddard ME (2009) Invited review: genomic selection in dairy cattle: progress and challenges. J Dairy Sci 92(2):433–443. https://doi.org/10.3168/jds.2008-1646 Review. Erratum. In: J Dairy Sci. 2009 Mar;92(3):1313

Hayes BJ, Goddard ME (2001) The distribution of the effects of genes affecting quantitative traits in livestock. Genet Sel Evol 33:209–229

Khatkar MS, Thomson PC, Tammen I, Raadsma HW (2004) Quantitative trait loci mapping in dairy cattle: review and meta-analysis. Genet Sel Evol 36:163–190

Kucerová J, Lund MS, Sørensen P, Sahana G, Guldbrandtsen B, Nielsen VH, Thomsen B, Bendixen C (2006) Multitrait quantitative trait Loci mapping for milk production traits in danish Holstein cattle. J Dairy Sci 89(6):2245–2256

Lan R, Lan Zhu L, Yao X, Wang P, Shao Q, Hong Q (2015) Yunnan institute of animal science and veterinary, sheep breeding promotion center in Yunnan province. Chin J Anim Vet Sci 4:549–554

Lee SH, Choi BH, Lim D, Gondro C, Cho YM, Dang CG, Sharma A, Jang GW, Lee KT, Yoon D, Lee HK, Yeon SH, Yang BS, Kang HS, Hong SK (2013) Genome-wide association study identifies major loci for carcass weight on BTA14 in Hanwoo (Korean cattle). PLoS ONE 8(10):e74677. https://doi.org/10.1371/journal.pone.0074677 (eCollection 2013)

Sharma A, Lee, JS, Dang, CG, Sudrajad P, Kim HC, Yeon SH, Kang HS, Lee SH (2015) Stories and challenges of genome wide association studies in livestock-A review. Asian-Australas J Anim Sci 28:1371–1379. https://doi.org/10.5713/ajas.14.0715

Lee B-Y, Lee K-N, Lee T, Park J-H, Kim S-M, Lee H-S, Chung D-S, Shim H-S, Lee H-K, Kim H (2015) bovine genome-wide association study for genetic elements to resist the infection of foot-and-mouth disease in the field. Asian-Australas J Anim Sci 28 (2):166–170. https://doi.org/10.5713/ajas.14.0394

Li X, Su R, Wan W, Zhang W, Jiang H, Qiao X, Fan Y, Zhang Y, Wang R, Liu Z, Wang Z, Liu B, Ma Y, Zhang H, Zhao Q, Zhong T, Di R, Jiang Y, Chen W, Wang W, Dong Y, Li J (2017) Identification of selection signals by large-scale whole-genome resequencing of cashmere goats. Sci Rep 7(1):15142. https://doi.org/10.1038/s41598-017-15516-0

Martin P, Palhière I, Tosser-Klopp G, Rupp R (2016) Heritability and genome-wide association mapping for supernumerary teats in French Alpine and Saanen dairy goats. J Dairy Sci 99(11):8891–8900. https://doi.org/10.3168/jds.2016-11210 (Epub 2016 Aug 17)

Matukumalli LK, Lawley CT, Schnabel RD, Taylor JF, Allan MF, Heaton MP, O'Connell J, Moore SS, Smith TP, Sonstegard TS, Van Tassell CP (2009) Development and characterization of a high density SNP genotyping assay for cattle. PLoS One 4(4):e5350. https://doi.org/10.1371/journal.pone.0005350 (Epub 2009 Apr 24)

Menzi F, Keller I, Reber I, Beck J, Brenig B, Schütz E, Leeb T, Drögemüller C (2016) Genomic amplification of the caprine EDNRA locus might lead to a dose dependent loss of pigmentation. Sci Rep 22(6):28438. https://doi.org/10.1038/srep28438

Meredith BK, Kearney FJ, Finlay EK, Bradley DG, Fahey AG, Berry DP, Lynn DJ (2012) Genome-wide associations for milk production and somatic cell score in Holstein-Friesian cattle in Ireland. BMC Genet 26 (13):21. https://doi.org/10.1186/1471-2156-13-21

Meuwissen TH, Hayes BJ, Goddard ME (2001) Prediction of total genetic value using genome-wide dense marker maps. Genetics 157(4):1819–1829

Mosig MO, Lipkin E, Khutoreskaya G, Tchourzyna E, Soller M, Friedmann A (2001) A whole genome scan for quantitative trait loci affecting milk protein percentage in Israeli-Holstein cattle, by means of selective milk DNA pooling in a daughter design, using an adjusted false discovery rate criterion. Genetics 157(4):1683–1698

Mrode R, Ojango JMK, Okeyo AM, Mwacharo JM (2019) Genomic selection and use of molecular tools in breeding programs for indigenous and crossbred cattle in developing countries: current status and future prospects. Front Genet 9(9):694. https://doi.org/10.3389/fgene.2018.00694 (eCollection 2018). Review

Nazari-Ghadikolaei A, Mehrabani-Yeganeh H, Miarei-Aashtiani SR, Staiger EA, Rashidi A, Huson HJ (2018) Genome-wide association studies identify candidate genes for coat color and mohair traits in the Iranian Markhoz Goat. Front Genet 9:105. https://doi.org/10.3389/fgene.2018.00105. (eCollection 2018)

Olsen HG, Lien S, Svendsen M, Nilsen H, Roseth A, Aasland Opsal M, Meuwissen TH (2004) Fine mapping of milk production QTL on BTA6 by combined linkage and linkage disequilibrium analysis. J Dairy Sci 87(3):690–698

Pasandideh M, Rahimi-Mianji G, Gholizadeh M (2018) A genome scan for quantitative trait loci affecting average daily gain and Kleiber ratio in Baluchi Sheep. J Genet 97(2):493–503

Pryce JE, Bolormaa S, Chamberlain AJ, Bowman PJ, Savin K, Goddard ME, Hayes BJ (2010) A validated genome-wide association study in 2 dairy cattle breeds for milk production and fertility traits using variable length haplotypes. J Dairy Sci 93(7):3331–3345. https://doi.org/10.3168/jds.2009-2893

Pryce JE, Daetwyler HD (2012) Designing dairy cattle breeding schemes under genomic selection: a review of international research. Anim Prod Sci 52:107–114. https://doi.org/10.1071/AN11098

Reber I, Keller I, Becker D, Flury C, Welle M, Drögemüller C (2015) Wattles in goats are associated with the FMN1/GREM1 region on chromosome 10. Anim Genet 46:316–320. https://doi.org/10.1111/age.12279

Rincon G, Weber KL, Eenennaam AL, Golden BL, Medrano JF (2011) Hot topic: performance of bovine high-density genotyping platforms in Holsteins and Jerseys. J Dairy Sci 94(12):6116–6121. https://doi.org/10.3168/jds.2011-4764

Ron M, Feldmesser E, Golik M, Tager-Cohen I, Kliger D, Reiss V, Domochovsky R, Alus O, Seroussi E, Ezra E, Weller JI (2004) A complete genome scan of the Israeli Holstein population for quantitative trait loci by a daughter design. J Dairy Sci 87(2):476–490

Rupp R, Foucras G (2010) ©CAB International 2010. In: Bishop SC et al (eds) Breeding for disease resistance in farm animals, 3rd edn

Sahana G, Iso-Touru T, Wu X, Nielsen US, de Koning DJ, Lund MS, Vilkki J, Guldbrandtsen B (2016) A 0.5-Mbp deletion on bovine chromosome 23 is a strong candidate for stillbirth in Nordic Red cattle. Genet Sel Evol 48:35. https://doi.org/10.1186/s12711-016-0215-z

Sanna S, Jackson AU, Nagaraja R, Willer CJ, Chen WM, Bonnycastle LL, Shen H, Timpson N, Lettre G, Usala G, Chines PS, Stringham HM, Scott LJ, Dei M, Lai S, Albai G, Crisponi L, Naitza S, Doheny KF, Pugh EW, Ben-Shlomo Y, Ebrahim S, Lawlor DA, Bergman RN, Watanabe RM, Uda M, Tuomilehto J, Coresh J, Hirschhorn JN, Shuldiner AR, Schlessinger D, Collins FS, Davey Smith G, Boerwinkle E, Cao A, Boehnke M, Abecasis GR, Mohlke KL (2008) Common variants in the GDF5-UQCC region are associated with variation in human height. Nat Genet 40(2):198–203. https://doi.org/10.1038/ng.74 (Epub 2008 Jan 13)

Schaeffer LR (2006) Strategy for applying genome-wide selection in dairy cattle. J Anim Breed Genet 2006 (123):218–223

Schrooten C, Bink MC, Bovenhuis H (2004) Whole genome scan to detect chromosomal regions affecting multiple traits in dairy cattle. J Dairy Sci 87(10):3550–3560

Snelling WM, Allan MF, Keele JW, Kuehn LA, McDaneld T, Smith TP, Sonstegard TS, Thallman RM, Bennett GL (2010) Genome-wide association study of growth in crossbred beef cattle. J Anim Sci 88(3):837–848. https://doi.org/10.2527/jas.2009-2257 (Epub 2009 Dec 4)

Streit M, Reinhardt F, Thaller G, Bennewitz J (2013) Genome-wide association analysis to identify genotype × environment interaction for milk protein yield and level of somatic cell score as environmental descriptors in German Holsteins. J Dairy Sci 96 (11):7318–7324. https://doi.org/10.3168/jds.2013-7133 (Epub 2013 Sep 18)

Szyda J, Liu Z, Reinhardt F, Reents R (2005) Estimation of quantitative trait loci parameters for milk production traits in German Holstein dairy cattle population. J Dairy Sci 88(1):356–367

Tan ME (2013) Genome-wide association study for stature in New Zealand Dairy Cattle. M. Sc. Thesis. Massey University; Palmerston North, New Zealand

Utsunomiya YT, do Carmo AS, Carvalheiro R, Neves HH, Matos MC, Zavarez LB, Pérez O'Brien AM, Sölkner J, McEwan JC, Cole JB, Van Tassell CP, Schenkel FS, da Silva MV, Porto Neto LR, Sonstegard TS, Garcia JF (2013) Genome-wide association study for birth weight in Nellore cattle points to previously described orthologous genes affecting human and bovine height. BMC Genet 14:52. https://doi.org/10.1186/1471-2156-14-52

Van Laere AS, Nguyen M, Braunschweig M, Nezer C, Collette C, Moreau L, Archibald AL, Haley CS, Buys N, Tally M, Andersson G, Georges M, Andersson L (2003) A regulatory mutation in IGF2 causes a major QTL effect on muscle growth in the pig. Nature 425(6960):832–836

VanRaden PM, Van Tassell CP, Wiggans GR, Sonstegard TS, Schnabel RD, Taylor JF, Schenkel FS (2009) Invited review: reliability of genomic predictions for North American Holstein bulls. J Dairy Sci 92(1):16–24. https://doi.org/10.3168/jds.2008-1514

Weller JI, Golik M, Reikhav S, Domochovsky R, Seroussi E, Ron M (2008) Detection and analysis of quantitative trait loci affecting production and secondary traits on chromosome 7 in Israeli Holsteins. J Dairy Sci 91(2):802–813. https://doi.org/10.3168/jds.2007-0367

Weller JI, Golik M, Seroussi E, Ezra E, Ron M (2003) Population-wide analysis of a QTL affecting milk-fat production in the Israeli Holstein population. J Dairy Sci 86(6):2219–2227

32 Digital PCR

Abstract

Precise detection of mutations is important for a variety of clinical and research applications. The digital PCR offers precise quantification of nucleic acids, viz. DNA, cDNA and RNA, and mutations. Digital PCR, used in the past also by different terms, is now at the apex of progression and applications in biological sciences owing to developments in chemical sciences, instrumentation, and accessibility of genome sequence data.

Highlights

- Compared to qPCR, the digital PCR is more precise accurate for determining DNA and RNA quantities and mutations therein
- Digital PCR is used in a variety of nucleic acid analytical and quantification methods.

Keywords

Digital PCR · Types of digital PCR · Livestock applications · Adulterations detection

32.1 Introduction

The digital PCR (dPCR) is an approach used for detection and absolute quantification of nucleic acids including DNA, cDNA or RNA, and mutations therein. The term digital PCR (dPCR) was used for the first time is published by Vogelstein and Kinzler (1999) to determine the magnitude of *ras* mutations in PCR products. The method described was not new, as it was already developed in 1990 and 1991 by terms "single-molecule PCR" or "limiting dilution PCR." (Brisco et al. 1991; Sykes et al. 1992) (Table 32.1) (reviewed in Morley 2014).

The limiting dilution PCR was an improved version of previous techniques, such as competitive PCR, for quantifying the PCR targets. These methods were soon verified and validated in different experiments.

The classical or general PCR is used for qualitative or quantitative purposes, to unravel properties or a number of target molecules. The dPCR, on contrary, is an alternate to real-time quantitative PCR (qPCR), and more precise in terms of detecting rare alleles, point mutations, copy number variations, and absolute quantification (Table 32.2). The dPCR is used to infer the difference between two nucleic acid macromolecules to the level of single nucleotides. dPCR is regarded as third-generation PCR that is designed to achieve an absolute quantification of the nucleic acid.

dPCR due to its unique attributes, i.e., simplicity, reproducibility with no need of standard curve, calibration curve, and endogenous control for quantification of sample target (Box 1), has become popular in biological sciences. Similar to conventional PCR, dPCR also carries out a single

B. Singh et al., *Advances in Animal Biotechnology*,
https://doi.org/10.1007/978-3-030-21309-1_32

Table 32.1 Historical and chronological events in the advancements of dPCR

1988	Description of in vitro amplification of DNA (Saiki et al. 1988)
1990	First time publication on the use of methodology of "single molecule PCR" or "limiting dilution PCR" (Simmonds et al. 1990)
1991–92	The methods described for the first time under the terms "single-molecule PCR" or "limiting dilution PCR" or "single-molecule PCR" (Brisco et al. 1991; Sykes et al. 1992)
1999	The term dPCR coined by Vogelstein and Kinzler (1999)
2018	Development and evaluation of a droplet dPCR assay for detecting fowl adenovirus serotypes in attenuated vaccines (Dong et al. 2018), development of a smartphone-based mobile dPCR device for DNA quantitative assay (Gou et al. 2018) Using dPCR to noninvasive fetal sex diagnosis in plasma of early week pregnant female (D'Aversa et al. 2018), description of a novel droplet digital polymerase chain reaction for quantification of gut nematodes in sheep (Elmahalawy et al. 2018)

Table 32.2 Comparative overview of advantages and disadvantages of dPCR and classical qPCR

Advantages	Disadvantages
dPCR provides absolute quantification	Reaction volume is very limited
No need of standard curve	Smaller dynamic range and dropout
Improved inter-laboratory communicability	Less accuracy of large amplicons
Less affected by sample inhibitors, and poor	More complex to perform, needs amplification efficiency expensive instruments More precise, beetter detection of low copy number variants

reaction within a sample, but a major working difference in dPCR is that each sample is separated into several partitions, and each partition is used for setting up reaction individually. The concept behind this is that some of the partitioned samples may contain target nucleic acid (positive) while others may not (negative). In dPCR analysis, as samples are diluted and separated into a large number of partitions, the reactions provide higher sensitivity and accuracy.

Box 1. Unique Characteristics and Merits of dPCR

- High accuracy, sensitivity, and absolute quantification of DNA, cDNA, and RNA
- Increased precision by using more PCR replicates
- High tolerance to PCR inhibitory substances
- Ability to analyze complex biological samples and mixtures
- Ease of linear detection of small-fold alterations
- Direct display of results on touch screen of the instrument, and their viewing and analyses
- Enhancing the performance of the present TaqMan Assays, thus offering enhanced sensitivity, accuracy, and absolute quantification

The technique of qPCR is used as a routine practice in several research and biomedical laboratories. However, due to its limited counting resolution, this method does not meet the more rigorous quantification demand, especially when target concentration is relatively low or PCR inhibitors are present and perturb the exquisite assay. Moreover, as the techniques such as next-generation sequencing (NGS) and single-cell analyses continue to flourish, an interest in the quantification of nucleic acids is drawn to unprecedented single-molecule level. This has given rise to the prosperity of dPCR technologies.

As mentioned above, the uniqueness of dPCR technique is in the partitioning of sample reaction mix into a large number of individual wells following a Poisson distribution. This kind of splitting of samples in a large number of individual wells results in the presence of either zero or one target molecule per well. The end point PCR amplification allows absolute estimation of number of target molecules present in the sample based on fluorescent signals. The wells with fluorescent signals are scored as one (positive); while wells with background signal are scored as zero (negative). In order to determine the absolute concentration of target present in the initial sample, analysis based on Poisson statistics is generally employed. Additionally, the studies have shown that dPCR is more resistant to PCR inhibitors as compared to the real-time PCR (Dingle et al. 2013). Thus, dPCR may be less dependent on the purity of the DNA template. Therefore, the technique is better suited for various types of disease diagnosis applications. Bosman et al. (2015) showed that dPCR enables direct and accurate quantification of HIV DNA. The chip-based digital PCR exhibits superior technical qualities and promises to be a superior method for quantifying miRNA levels in the circulation, which may become a potential as a more accurate and reproducible method for directly quantifying miRNAs, particularly for use in large multi-center clinical trials (Robinson et al. 2018).

32.2 Types of Digital PCR

The dPCR extends capabilities of PCR applications. The different versions of dPCR are in use depending on the task to be performed. Various manufacturers supply the instruments to carry out dPCR with enhanced sensitivity, precision, and absolute quantification.

32.3 Droplet Digital PCR

Droplet digital PCR (ddPCR) technology partitions a single-PCR reaction into many nanoliter-sized droplets and amplifies DNA sequences in each droplet to provide an accurate copy number of a single gene without the use of an external reference standard and multiple replicates (Hindson et al. 2013). Technically, in ddPCR assay, >20,000 individual monodispersed droplets are generated from each DNA sample and automated droplet generation oil for probes followed by amplification of target template in each droplet utilizing sequence-specific primers and fluorescently labeled TaqMan probes (Yung et al. 2009).

32.4 CRISPR-Typing PCR (ctPCR)

It is a new Cas9-based DNA detection method for detecting and typing target DNA based on Cas9 nuclease or CRISPR-typing PCR (Wang et al. 2018). In this method, the first PCR allows the determination of target DNA (such as virus infection), while the second PCR determines genotypic characteristics of target DNA (such as virus subtypes). The advantage of this technique relies on the fact that it can detect and type target DNA easily, rapidly, specifically, and sensitively. This technique detects target DNA in three steps: (1) amplifying target DNA with PCR by using a pair of universal primers (PCR1); (2) treating PCR1 products with a process referred to as CAT, representing Cas9 cutting, A tailing and T adaptor ligation; (3) amplifying the CAT-treated DNA with PCR by using a pair of general-specific primers (gs-primers) (PCR2). This method was verified by detecting human papillomavirus HPV16 and HPV18 L1 gene in 13 different high-risk HPV subtypes. The method was verified by detecting L1 and E6–E7 genes of two high-risk HPVs (HPV16 and 18) in cervical carcinoma cells, and many clinical samples. Based on these proof-of-concept experiments, this study provides a new CRISPR/Cas9-based DNA detection and typing method.

32.5 Chip-Based Digital PCR

It is a novel detection method for quantifying microRNAs in biological samples. In this technique, a PCR sample is loaded on a partitioning the sample across 20,000 reaction wells, essentially allowing 20,000 separate PCR reactions to happen (Conte et al. 2015). The QuantStudio® system uses for Poisson statistical analysis of fluorescent signals from positive and negative wells allows for absolute quantification, without need for referencing to a standard control (Conte et al. 2015).

Among many other applications, researchers have used digital PCR to distinguish differential expression of alleles (Chen et al. 2012), to identify specific viruses infecting individual bacterial cells (Tadmor et al. 2011), to quantify cancer genes in patient specimens (Wang et al. 2010), and to detect fetal chromosomal aneuploidy in circulating blood (Lo et al. 2007). dPCR has also been widely used for the analysis of clinical samples, such as the detection of allelic discrimination (Hindson et al. 2011; Pinheiro et al. 2012), the determination of single-cell expression profile (Guo et al. 2010; Yung et al. 2009), detection of single-nucleotide polymorphisms (SNPs) (Ludlow et al. 2014, 2018), and detecting low-copy targets (Miotto et al. 2014).

32.6 Major Applications of Digital PCR in Livestock Sciences

Some of the areas where digital PCR has played an important role in livestock sector are summarized below.

32.7 Detecting Meat Adulterations

Meat adulteration is a worldwide menace today (Tang et al. 2016; O'Mahony 2013; Cawthorn et al. 2013), probably to earn more benefits. dPCR is extensively and routinely used to detect species and amount of its meat content in processed and canned forms. Tian et al. (2015) performed microfluidics dPCR system to detect bovine DNA in ovine meat. For instance, sheep and goat meats are often adulterated with chicken (Ren et al. 2017; Tang et al. 2016). Similarly, fox meat or pork is being added to beef or mutton to entrap consumers (Tian et al. 2015). dPCR-based DNA testing is quite useful in confirming if a product is authentic or not.

32.8 Detecting Contaminant in Cell Culture System

Droplet digital PCR (ddPCR) technique has been also used for absolute quantification of the bovine viral diarrhea virus genome copy numbers in cell culture supernatants used for virus propagation and raw materials (fetal bovine serum) employed for the maintenance of cells in culture (Flatschart et al. 2015).

32.9 Detecting Animal Pathogens

dPCR is used to detect pathogens in clinical microbiology (Kuypers and Jerome 2017). The droplet dPCR is used for accurate and absolute quantification of Shiga toxin-producing *Escherichia coli* (STEC) in cattle feces in comparison with conventional PCR (Verhaegen et al. 2016). Lai et al. (2017) reported increased expression of inflammation-related miRNAs; miR-21, miR-146a, miR-155, miR-222, and miR-383 in the bovine milk affected by mastitis using droplet digital PCR, suggesting the potential role of these miRNAs as biomarkers to bovine mastitis. Droplet dPCR has been reported to detect STEC in feces of western Canadian slaughter cattle (Paquette et al. 2018).

32.10 Detecting Contaminants in Foods and Food Materials

Milk and dairy products can harbor various microorganisms such as *Campylobacter* spp., *Salmonella* spp., *Listeria monocytogenes*, *verocytotoxin-producing Escherichia coli* arising from animal reservoirs, and which can become important sources of food-borne illness. The demand of milk products is very high in the market worldwide due to their superior nutritional properties (Quigley et al. 2013). Therefore, it is mandatory to establish the absence of food-borne pathogens or their toxins to ensure the quality and safety of food products. Digital droplet PCR has emerged as a method for high-throughput screening of food-borne pathogens in dairy microbiology due to its accuracy and precision. Cremonesi et al. (2016) have used this technique to identify food-borne pathogens in dairy products. This technique has also been adopted for routine analyses of genetically modified organisms in food and animal feed (Morisset et al. 2013; Gerdes et al. 2016).

32.11 Miscellaneous Uses

The droplet digital PCR (ddPCR) assay has been used to evaluate the absolute copy number variation of the starch-metabolizing gene amylase-2b, *AMY2B* in canines (Pendleton et al. 2018; Arendt et al. 2014). This technique has a potential to estimate the DNA copy numbers more precisely and resolve high copy number gene duplications more accurately which in particular is a limitation of conventional qPCR (Hindson et al. 2011; Pinheiro et al. 2012). ddPCR has also been used to identify differential expression of alleles imprinted in the genome of chicken (Frésard et al. 2014).

32.12 Outlook and Challenges

The dPCR circumventing external calibration and absolutely quantifying nucleic acids has popular applications in biological sciences. The dPCR has increasing role in animal sciences and food and feed industries. The modified versions of original dPCR are already being used in animal and veterinary sciences.

32.13 Conclusions

The dPCR circumvents external calibration and provides precise quantification of nucleic acids. The future will be both fascinating and fruitful for applications of dPCR in biological studies.

References

Arendt M, Fall T, Lindblad-Toh K, Axelsson E (2014) Amylase activity is associated with AMY2B copy numbers in dog: implications for dog domestication, diet and diabetes. Anim Genet 45(5):716–722. https://doi.org/10.1111/age.12179 (Epub 2014 Jun 28)

Bosman KJ, Nijhuis M, van Ham PM, Wensing AM, Vervisch K, Vandekerckhove L, De Spiegelaere W (2015) Comparison of digital PCR platforms and semi-nested qPCR as a tool to determine the size of the HIV reservoir. Sci Rep. 9(5):13811. https://doi.org/10.1038/srep13811

Brisco MJ, Condon J, Sykes PJ, Neoh SH, Morley AA (1991) Detection and quantitation of neoplastic cells in acute lymphoblastic leukaemia, by use of the polymerase chain reaction. Br J Haematol 79(2):211–217

Cawthorn D-M, Steinman HA, Hoffman LC (2013) A high incidence of species substitution and mislabelling detected in meat products sold in South Africa. Food Control 32:440–449

Chen R, Mias GI, Li-Pook-Than J, Jiang L, Lam HY, Chen R, Miriami E, Karczewski KJ, Hariharan M, Dewey FE, Cheng Y, Clark MJ, Im H, Habegger L, Balasubramanian S, O'Huallachain M, Dudley JT, Hillenmeyer S, Haraksingh R, Sharon D, Euskirchen G, Lacroute P, Bettinger K, Boyle AP, Kasowski M, Grubert F, Seki S, Garcia M, Whirl-Carrillo M, Gallardo M, Blasco MA, Greenberg PL, Snyder P, Klein TE, Altman RB, Butte AJ, Ashley EA, Gerstein M, Nadeau KC, Tang H, Snyder M (2012) Personal omics profiling reveals dynamic molecular and medical phenotypes. Cell 148(6):1293–1307. https://doi.org/10.1016/j.cell.2012.02.009

Conte D, Verri C, Borzi C, Suatoni P, Pastorino U, Sozzi G, Fortunato O (2015) Novel method to detect microRNAs using chip-based QuantStudio 3D digital PCR. BMC Genom 23(16):849. https://doi.org/10.1186/s12864-015-2097-9

Cremonesi P, Cortimiglia C, Picozzi C, Minozzi G, Malvisi M, Luini M, Castiglioni B (2016) Development of a droplet digital polymerase chain reaction for rapid and simultaneous identification of common foodborne pathogens in soft cheese. Front Microbiol 7:1725 (eCollection 2016)

D'Aversa E, Breveglieri G, Pellegatti P, Guerra G, Gambari R, Borgatti M (2018) Non-invasive fetal sex diagnosis in plasma of early weeks pregnants using droplet digital PCR. Mol Med 24(1):14. https://doi.org/10.1186/s10020-018-0016-7

Dingle TC, Sedlak RH, Cook L, Jerome KR (2013) Tolerance of droplet-digital PCR vs real-time quantitative PCR to inhibitory substances. Clin Chem 59 (11):1670–1672. https://doi.org/10.1373/clinchem.2013.211045 (Epub 2013 Sep 3)

Dong G, Meng F, Zhang Y, Cui Z, Lidan H, Chang S, Zhao P (2018) Development and evaluation of a droplet digital PCR assay for the detection of fowl adenovirus serotypes 4 and 10 in attenuated vaccines. J Virol Methods. pii: S0166-0934(18)30417-8. https://doi.org/10.1016/j.jviromet.2018.09.005 (Epub ahead of print)

Elmahalawy ST, Halvarsson P, Skarin M, Höglund J (2018) Droplet digital polymerase chain reaction (ddPCR) as a novel method for absolute quantification of major gastrointestinal nematodes in sheep. Vet Parasitol 15(261):1–8. https://doi.org/10.1016/j.vetpar.2018.07.008 (Epub 2018 Jul 18)

Flatschart RB, Almeida DO, Heinemann MB, Medeiros MN, Granjeiro JM, Folgueras-Flatschart AV (2015) Absolute quantification of Bovine Viral Diarrhea Virus (BVDV) RNA by the digital PCR technique. J Phys Conf Ser 575(1):012038 (IOP Publishing)

Frésard L, Leroux S, Servin B, Gourichon D, Dehais P, Cristobal MS, Marsaud N, Vignoles F, Bed'hom B, Coville JL, Hormozdiari F, Beaumont C, Zerjal T, Vignal A, Morisson M, Lagarrigue S, Pitel F (2014) Transcriptome-wide investigation of genomic imprinting in chicken. Nucleic Acids Res 42(6):3768–3782. https://doi.org/10.1093/nar/gkt1390 (Epub 2014 Jan 21)

Gerdes L, Iwobi A, Busch U, Pecoraro S (2016) Optimization of digital droplet polymerase chain reaction for quantification of genetically modified organisms. Biomol Detect Quantif 7:9–20. https://doi.org/10.1016/j.bdq.2015.12.003 (eCollection 2016 Mar)

Gou T, Hu J, Wu W, Ding X, Zhou S, Fang W, Mu Y (2018) Smartphone-based mobile digital PCR device for DNA quantitative analysis with high accuracy. Biosens Bioelectron 30(120):144–152. https://doi.org/10.1016/j.bios.2018.08.030 (Epub 2018 Aug 17)

Guo G, Huss M, Tong GQ, Wang C, Li Sun L, Clarke ND, Robson P (2010) Resolution of cell fate decisions revealed by single-cell gene expression analysis from zygote to blastocyst. Dev Cell 18 (4):675–685. https://doi.org/10.1016/j.devcel.2010.02.012

Hindson BJ, Ness KD, Masquelier DA, Belgrader P, Heredia NJ, Makarewicz AJ, Bright IJ, Lucero MY, Hiddessen AL, Legler TC, Kitano TK, Hodel MR, Petersen JF, Wyatt PW, Steenblock ER, Shah PH, Bousse LJ, Troup CB, Mellen JC, Wittmann DK, Erndt NG, Cauley TH, Koehler RT, So AP, Dube S, Rose KA, Montesclaros L, Wang S, Stumbo DP, Hodges SP, Romine S, Milanovich FP, White HE, Regan JF, Karlin-Neumann GA, Hindson CM, Saxonov S, Colston BW (2011) High-throughput droplet digital PCR system for absolute quantitation of DNA copy number. Anal Chem 83(22):8604–8610. https://doi.org/10.1021/ac202028g (Epub 2011 Oct 28)

Hindson CM, Chevillet JR, Briggs HA, Gallichotte EN, Ruf IK, Hindson BJ, Vessella RL, Tewari M (2013) Absolute quantification by droplet digital PCR versus analog real-time PCR. Nat Methods 10(10):1003–1005. https://doi.org/10.1038/nmeth.2633 (Epub 2013 Sep 1)

Kuypers J, Jerome KR (2017) Applications of digital PCR for clinical microbiology. J Clin Microbiol 55(6):1621–1628. https://doi.org/10.1128/jcm.00211-17 (Epub 2017 Mar 15. Review)

Lai YC, Fujikawa T, Maemura T, Ando T, Kitahara G, Endo Y, Yamato O, Koiwa M, Kubota C, Miura N (2017) Inflammation-related microRNA expression level in the bovine milk is affected by mastitis. PLoS ONE 12(5):e0177182. https://doi.org/10.1371/journal.pone.0177182 (eCollection 2017)

Lo YM, Lun FM, Chan KC, Tsui NB, Chong KC, Lau TK, Leung TY, Zee BC, Cantor CR, Chiu RW (2007) Digital PCR for the molecular detection of fetal chromosomal aneuploidy. Proc Natl Acad Sci U S A. 104(32):13116–13121

Ludlow AT, Robin JD, Sayed M, Litterst CM, Shelton DN, Shay JW, Wright WE (2014) Quantitative telomerase enzyme activity determination using droplet digital PCR with single cell resolution. Nucleic Acids Res 42(13):e104. https://doi.org/10.1093/nar/gku439

Ludlow AT, Shelton D, Wright WE, Shay JW (2018) ddTRAP: a method for sensitive and precise quantification of telomerase activity. Methods Mol Biol 1768:513–529. https://doi.org/10.1007/978-1-4939-7778-9_29

Miotto E, Saccenti E, Lupini L, Callegari E, Negrini M, Ferracin M (2014) Quantification of circulating miRNAs by droplet digital PCR: comparison of EvaGreen- and TaqMan-based chemistries. Cancer Epidemiol Biomarkers Prev 23(12):2638–2642. https://doi.org/10.1158/1055-9965.EPI-14-0503

Morley AA (2014) Digital PCR: a brief history. Biomol Detect Quantif 1(1):1–2 (eCollection 2014 Sep)

Morisset D, Štebih D, Milavec M, Gruden K, Žel J (2013) Quantitative analysis of food and feed samples with droplet digital PCR. PLoS ONE 8(5):e62583. https://doi.org/10.1371/journal.pone.0062583 (Print 2013)

O'Mahony PJ (2013) Finding horse meat in beef products—a global problem. QJM 106(6):595–597. https://doi.org/10.1093/qjmed/hct087

Paquette SJ, Stanford K, Thomas J, Reuter T (2018) Quantitative surveillance of shiga toxins 1 and 2, *Escherichia coli* O178 and O157 in feces of western-Canadian slaughter cattle enumerated by droplet digital PCR with a focus on seasonality and slaughterhouse location. PLoS One 13(4):e0195880.

https://doi.org/10.1371/journal.pone.0195880 (eCollection 2018)

Pendleton AL, Shen F, Taravella AM, Emery S, Veeramah KR, Boyko AR, Kidd JM (2018) Comparison of village dog and wolf genomes highlights the role of the neural crest in dog domestication. BMC Biol 16(1):64. https://doi.org/10.1186/s12915-018-0535-2

Pinheiro LB, Coleman VA, Hindson CM, Herrmann J, Hindson BJ, Bhat S, Emslie KR (2012) Evaluation of a droplet digital polymerase chain reaction format for DNA copy number quantification. Anal Chem 84 (2):1003–1011. https://doi.org/10.1021/ac202578x (Epub 2011 Dec 21)

Quigley L, O'Sullivan O, Stanton C, Beresford TP, Ross RP, Fitzgerald GF, Cotter PD (2013) The complex microbiota of raw milk. FEMS Microbiol Rev 37(5):664–698. https://doi.org/10.1111/1574-6976.12030

Ren J, Deng T, Huang W, Chen Y, Ge Y (2017) A digital PCR method for identifying and quantifying adulteration of meat species in raw and processed food. PLoS One 12(3):e0173567. https://doi.org/10.1371/journal.pone.0173567 (eCollection 2017)

Robinson S, Follo M, Haenel D, Mauler M, Stallmann D, Heger LA, Helbing T, Duerschmied D, Peter K, Bode C, Ahrens I, Hortmann M (2018) Chip-based digital PCR as a novel detection method for quantifying microRNAs in acute myocardial infarction patients. Acta Pharmacol Sin 39(7):1217–1227. https://doi.org/10.1038/aps.2017.136

Saiki RK, Gelfand DH, Stoffel S, Scharf SJ, Higuchi R, Horn GT, Mullis KB, Erlich HA (1988) Primer-directed enzymatic amplification of DNA with a thermostable DNA polymerase. Science 239(4839): 487–491

Simmonds P, Balfe P, Peutherer JF, Ludlam CA, Bishop JO, Brown AJ (1990) Human immunodeficiency virus-infected individuals contain provirus in small numbers of peripheral mononuclear cells and at low copy numbers. J Virol 64(2):864–872

Sykes PJ, Neoh SH, Brisco MJ, Hughes E, Condon J, Morley AA (1992) Quantitation of targets for PCR by use of limiting dilution. Biotechniques 13(3):444–449

Tadmor AD, Ottesen EA, Leadbetter JR, Phillips R (2011) Probing individual environmental bacteria for viruses by using microfluidic digital PCR. Science 333 (6038):58–62. https://doi.org/10.1126/science.1200758

Tang SP, Zhang Y, Huang JH (2016) Analysis of beef and lamp products adulteration in Guangdong province. J Food Saf Qual 7(5):1882–1886 (ref. 15)

Tian Q, Mu Y, Xu Y, Song Q, Yu B, Ma C, Jin W, Jin Q (2015) An integrated microfluidic system for bovine DNA purification and digital PCR detection. Anal Biochem 15(491):55–57. https://doi.org/10.1016/j.ab.2015.08.030

Verhaegen B, De Reu K, De Zutter L, Verstraete K, Heyndrickx M, Van Coillie E (2016) Comparison of droplet digital PCR and qPCR for the quantification of Shiga toxin-producing *Escherichia coli* in bovine feces. Toxins (Basel) 8(5). pii: E157. https://doi.org/10.3390/toxins8050157

Vogelstein B, Kinzler KW (1999) Digital PCR. Proc Natl Acad Sci U S A. 96(16):9236–9241

Wang J, Ramakrishnan R, Tang Z, Fan W, Kluge A, Dowlati A, Jones RC, Ma PC (2010) Quantifying EGFR alterations in the lung cancer genome with nanofluidic digital PCR arrays. Clin Chem 56(4):623–632. https://doi.org/10.1373/clinchem.2009.134973

Wang Q, Zhang B, Xu X, Long F, Wang J (2018) CRISPR-typing PCR (ctPCR), a new Cas9-based DNA detection method. Sci Rep. 8(1):14126. https://doi.org/10.1038/s41598-018-32329-x

Yung TK, Chan KC, Mok TS, Tong J, To KF, Lo YM (2009) Single-molecule detection of epidermal growth factor receptor mutations in plasma by microfluidics digital PCR in non-small cell lung cancer patients. Clin Cancer Res 15(6):2076–2084. https://doi.org/10.1158/1078-0432.CCR-08-2622

33 Transcriptomics: Genome-Wide Expression Analysis in Livestock Research

Abstract

Transcriptomics or genome-wide expression analysis has always played a central role in the field of functional genomics. Simultaneous quantification of thousands of genes at genome-wide scale has revolutionized the research in livestock. It has helped in unraveling the complexity of gene regulation and providing insights into gene networks and molecular pathways relevant to functional traits.

Highlights

- Transcriptomics describes and estimates the genes expressed in cells or organisms.
- Several candidate genes are identified by transcriptomic analysis and suggested to understand diseases, stress, and genetic merits of livestock.

Keywords

Transcriptomics · RNA-Seq · Expression microarrays · Microarray platforms · Adaptive traits · Disease diagnosis · Lactation physiology

33.1 Introduction

Developments in bioinformatics, computational, and molecular biology have provided tools to investigate and quantify the expression level of genes in microorganisms, animals, and plants. The methods used to assess gene expression vary from conventional Northern blotting to highly advanced techniques such as subtractive hybridization (SH), expressed sequence tags (ESTs), serial analysis of gene expression (SAGE), microarrays, and RNA sequencing (RNA-Seq).

Gene expression microarrays and RNA-Seq are the genotyping methods that have unraveled expression of thousands of genes (Schneider and Niemeyer 2018). The high-throughput techniques can ultimately link entire genome expression and whole organism function by allowing for the study of the expression of a huge number of genes under a range of investigational situations. The advancements have allowed concurrent quantification of mRNA expression of the entire genome in cells and tissues. The swiftly increasing esteem of these technologies in livestock species is evident from the increasing number of publications in recent years.

The microarray and RNA-Seq have transformed functional genomic research in microorganisms, plants, animals, and humans by improving the quantification of thousands of genes together. The objective of high-throughput research is to investigate the genes which are differentially up- or down-regulated in cells between, a control group, and the cells that have undergone some treatment, or between cells of animals of different genetic background (e.g. control models *vis-à-vis* genetically modified animals), or between cells in healthy and

B. Singh et al., *Advances in Animal Biotechnology*,
https://doi.org/10.1007/978-3-030-21309-1_33

diseased tissues, or between cells at different stages of development.

33.2 Expression Microarrays

The use of cDNA microarrays for large-scale expression measurement was first reported by Patrick Brown's group at Stanford University Medical Centre, CA, USA (Schena et al. 1995). Microarrays consist of a glass slide or membrane spotted (arrayed) with DNA fragments or oligonucleotides that represent specific coding regions of genes obtained from complementary DNA (cDNA). Gene expression microarrays have largely been used for genome-wide differential expression analysis, expression-based quantitative trait loci (eQTL) mapping, and metabolic pathway analysis in biological systems.

The microarray technique is based on the phenomenon of preferential complementary base pairing known as hybridization, which generates signals by parallel hybridization of labeled targets to specific probes that are immobilized on a solid surface in an ordered array. The core principle behind microarrays is the hybridization between two DNA strands the property of complementary nucleic acid sequences to specifically pair with each other through hydrogen bonding. Thus, microarrays are an orderly array of DNA probe material immobilized onto a solid substrate. The substrate may be a specialized glass or other solid support.

The most important requirements of a typical microarray platform are system reproducibility that offers the means for high confidence experiments, and precise comparison across multiple samples; and high sensitivity, for detecting significant gene expression changes across multiple gene sets. All components of microarray workflow, such as probe design, printing process, RNA sample quality, labeling, microarray processing, scanning of the images and feature extraction algorithms, collectively determine the quality of the data generated.

33.3 Types of Microarray Platforms

There are two principal microarray methods based upon the nature of the "target"-arrayed DNA material (cDNA or oligonucleotide microarrays) and the method of spotting DNA (mechanical microspotting or photolithography). The number of "target" genes that make up an array can range from a small number of specific well-characterized genes to a pool of thousands of genes that may encompass entire genome.

For model organisms, such as yeast, mouse, and human, both cDNA and oligonucleotide arrays are commercially available and are used for various applications, such as medical diagnostics and drug discovery. Custom arrays can also be constructed for non-model species from a number of different "target" DNA sources including cDNAs clones obtained from normalized libraries, ESTs, oligonucleotides, genomic clones or genomic DNA. At present, oligonucleotide arrays and whole-genome arrays are preferred over cDNA arrays, due to more reliability, spot uniformity, and avoiding certain pitfalls of cDNA arrays.

There are various types of microarray platforms that are commercially available for different species. Arrays can also be tissue specific such as mammary and immune response genes specific or whole genome specific, representing full repertoire of the genome. Among two main requirements of a microarray platform include system reproducibility that is must for high confidence experiments and accurate comparison across multiple samples and high sensitivity to detect even minor changes across multiple gene sets.

Various commercial companies manufacture microarray chips. For instance, Agilent whole-genome bovine 44 K chip harboring 60-mer oligos is a popular platform for detecting differential expression of genes in bovines. Bovine whole-genome platforms from Affymetrix have shorter oligos (25–35 mer) developed by photolithographic masks. Besides, the microarray platforms from Illumina are also available for bovine and other livestock species. The bead chip from Illumina consists of 50-mer oligos attached to beads randomly.

33.4 Strategies to Utilize Gene Expression Microarrays

Microarrays are used to perform direct comparison of the expression profile of the "target" genes spotted on an array between biological samples collected or obtained under two conditions or treatments. Different fluorophores are used to label cDNA prepared from total RNA or mRNA, typically representing the control and treatment or experimental conditions. Many types of fluorescent dyes are used to label cDNA for microarray experiments, but Cy3 and Cy5 are more commonly used.

Fluorescently labeled cDNAs are mixed, and hybridized to 'target DNA already spotted on the arrays. The two dyes have nonlinear sample labeling and hybridization kinetics, which means that they do not provide equal sensitivity across the whole range of transcripts in a sample. More specifically, they have differential labeling and scanning efficiencies and also exhibit gene-specific bias. To combat this, the dyes are often exchanged and the procedures of hybridization and scanning repeated, known as a *dye-swap*, means exchanging the dye labels across the samples. Taking a suitable average of both dye-swap pair ratios removes dye-bias, giving more reliable results. Recently, because of the availability of high-quality microarrays and robust workflow, several groups have started utilizing one-color (intensity-based) microarrays that are simpler to perform than the traditionally used two-color (ratio-based) microarrays. In one-color intensity-based microarrays, it is possible to simply hybridize a sample on one microarray. Therefore, a color microarray makes it feasible to compare the measured gene expression output of a microarray directly across other microarrays.

Hence, one-color microarrays differ from two-color microarray approach, where all gene expression ratios are generated only from two samples. For one-color microarray experiments, mostly Cy3 is used as due to being more resistant to degradation by ozone, pH, and organic solvents. Automated processes are used to calculate a comparative quantity of the genes expressed. The overall expression pattern of all genes also called as "expression profile", wherein genes that are up-regulated or down-regulated can be monitored. DNA microarrays, their features, and protocols to use them are available from commercial sources.

33.5 Analysis of Microarray Gene Expression Data

With the generation of the large amounts of microarray data, it is imperative to analyze the quality of data. The major concern of data quality control is to detect problematic raw probe-level data to facilitate the decision of whether to remove this array from further analysis. Data quality control is followed by two preprocessing steps. The first step is "data normalization'. It is a fundamental step which is used to remove systematic bias and noise variability developed from technical and experimental artifacts. The normalization procedure is carried out only on the background corrected values for each spot. The subsequent step, also called as "data filtering", aims to discard the probe sets with very low expression across the samples in order to reduce noise in data.

Computational data analysis, such as data mining which includes clustering and classification, are used to extract valuable information from microarray data. Clustering is one of the unsupervised approaches to classify data into groups of genes or samples with similar patterns that are characteristic to the group. Clustering is used in gene expression data matrix analysis and finding co-regulated and functionally related groups. Clustering methods can be hierarchical (grouping objects into clusters and specifying relationships among objects in a cluster, resembling a phylogenetic tree) or non-hierarchical (grouping into clusters without specifying relationships between objects in a cluster).

Classification is also known as class prediction, discriminant analysis, or supervised learning.

In addition, differential gene expression is the most prominent feature of typical microarray analysis relating gene expression data with other biological information; provides information on transcription factor binding site analysis, pathway analysis, and protein–protein interaction network analysis. Differentially expressed genes are the genes whose expression levels are significantly different between two groups of experiments. In addition, relating expression data with other biological information further helps in identification of transcription factor binding sites, pathways and protein–protein interaction networks.

Several univariate statistical methods are used to measure either the expression or the relative expression of a gene from normalized microarray data, including t-tests, modified t-test known as SAM, two-sample t-tests, F-statistic and Bayesian models. In some instance, Analysis of Variance (ANOVA) technique is also used.

Owing to a large number of genes represented on a microarray, it is possible that false-positive calls or "Type 1 error" may be detected. This is solved by the concept of the false discovery rate (FDR). Factors determining FDR are the proportion of truly differentially expressed genes, the distribution of the true differences, measurement variability, and sample size. Various approaches, such as protein-level FDR (Wu et al. 2018), multiple testing procedures such as Bon-EV (Li et al. 2017a, b), are used to detect FDRs.

Identification of functional elements such as transcription-factor binding sites (TFBS) on a whole-genome level can be one of the challenging tasks for gene regulation studies. Transcription factors play a prominent role in transcription regulation; identifying and characterizing their binding sites are central to annotate genomic regulatory regions and understanding the gene regulatory networks. Protein–protein interactions (PPI) are also useful to investigate the cellular functions of genes. It is a core of the entire interactomics system of any living cell.

Databases, such as the Biomolecular Interaction Database (BIND), Database of Interacting Proteins (DIP), IntAct, STRING and the Molecular Interaction Database (MINT), have been developed to score the protein interactions. Combining co-expressed as well as interacting genes in the same cluster several meaningful predictions related to gene functions, evolutionary relationships, and pathways can be made.

The next promising method for analyzing microarray data is pathway analysis as it involves the cascade of network interactions. Analyzing the microarray data in a pathway perspective could lead to a higher level of understanding of the system. This integrates the normalized array data and their annotations, such as metabolic pathways and gene ontology and functional classifications. Metabolic pathway analysis can identify more subtle changes in expression than the gene lists that originate from univariate statistical analysis. Gene Set Enrichment Analysis (GSEA) is a computational method that determines whether a set of genes shows statistically significant and concordant differences between two biological states. The gene sets are defined based on prior biological knowledge, e.g., published information about biochemical pathways, located in the same cytogenetic band, sharing the common Gene Ontology category or any user-defined set.

33.6 Microarray-Based Transcriptomic Applications

With recent developments in sequencing of genomes of different livestock species, the availability of species-specific microarray platform has enabled the researcher to discover genes and address the underlying genetic mechanisms of normal physiological processes such as cell differentiation, pregnancy, and lactation. Cellular and tissue models are developed to investigate genetic regulation of traits of economic importance in livestock species. Inferences from global transcriptomics studies have unfolded critical aspects in bovine health and normal physiology.

33.7 Detection of Diseases

Vertebrates, including ruminants and non-ruminants, have a strong association with microorganisms colonizing their external as well as internal parts. The microbiota is associated with health, onset of diseases, and production performance of animals. Disease state and biological conditions are marked by characteristic patterns of gene expression. Microarray-based studies have contributed a lot to decipher host–microbe symbiosis and host–pathogen interaction to better understand the immune functions and regulation of genes controlling immunity traits. RNA-Seq-based transcriptome (discussed ahead) data of peripheral blood leukocytes of cattle was generated to gain a deeper knowledge of the host transcriptional response to *Mycobacterium bovis* infection (McLoughlin et al. 2014).

Progress in RNA-Seq with NGS has increased the understanding of molecular mechanisms of pathogenesis of various cancers such as ovarian (Wang et al. 2014), pancreatic (Lin et al. 2019), pulmonary (Xue et al. 2018), and other types of cancers. RNA-Seq principles are used to detect invasive urothelial carcinoma and identifying potential therapeutic targets in canines (Maeda et al. 2018).

33.8 Transcriptomic Studies in Milk Production

The "omics" technologies and data analysis via bioinformatics serve as a key technical tools to study and develop nutritional, genetic and management policies to improve milk production and milk quality in dairy animals (Li et al. 2017a, b).

With a goal to better understand bovine mammary gland biology, Suchyta et al. (2003) examined and compared the gene expression profiles of lactating bovine mammary gland against non-lactating tissues (adipose, hepatic, adrenal lymph, thymus, spleen, gut, and developing mammary gland) using BMAM microarray. Gene expression analysis confirmed that many novel and interesting genes including oncogenes (VAV3, C-myc), mediators of apoptosis (Caspase 8), and cell cycle regulators (LASP1) are expressed explicitly in lactating mammary tissue.

High-yielding dairy cows need a special care so that they are able to continue milk production without adverse effects on their health. Mammary liver and adipose organs allow dairy animals to adapt to physiological environment to synthesize milk for neonate during transition from pregnancy to lactation. This period is characterized by general decrease in feed intake. Particularly, in high-yielding dairy cattle, the increase in milk production and concomitant reduction in feed intake is the main cause of marked negative energy balance (NEB), i.e., an imbalance between energy intake and the energy output (Grummer et al. 2004).

The significant progress has been made in understanding the physiology and mammary gland and hepatic tissue genomic responses of high-yielding milch animals during stressful periparturient stage, metabolic disorders like ketosis (Loor et al. 2007; Moyes et al. 2010), and associated infectious diseases such as mastitis (Kosciuczuk et al. 2017). Using a bovine microarray chip, the changes in key metabolic and signaling network have been described during nutrition-induced ketosis and liver lipidosis in periparturient Holstein Friesian cows. Several previously unknown alterations were detected in genes playing key roles in hepatic metabolism adaptations to NEB and changing physiological state near the time of parturition (Loor et al. 2007).

To decipher the complexity that underlies mammary gland development and function, microarray expression data was generated that provided insights into genetic mechanisms regulating mammary gland functioning such as lactose synthesis, lipid metabolism, protein synthesis, angiogenesis, epigenetic regulation, and immune functions. They were found to be coordinated in specified manner throughout puberty, pregnancy, lactation, and involution (Bionaz et al. 2012; Gao et al. 2013).

Elucidating the signaling mechanisms underlying the functional development of mammary gland and regulation of milk fat and protein synthesis by generating whole-genome expression pattern coupled with metabolic/hormonal pathways are the high-priority study areas in animal genomics. This can generate valuable information about mechanisms of adaptations in response to a particular physiological stage of the animal. Such inputs can relate functional development of mammary gland of dairy animals with coordinated changes in global expression pattern to understand the basic biology of development of mammary gland.

33.9 Transcriptomics in Adaptive Traits

In order to assess the impact of thermal stress in bubaline (*Bubalus bubalis*) mammary epithelial cells (MECs), Agilent 44 K bovine oligonucleotide array was used to understand the transcriptional response of MECs exposed to thermal stress. Several genes belonging to heat shock family, and functional classes, viz. chaperons, immune responsive, cell proliferation and metabolism related, and pathways associated with heat stress response have been identified (Kapila et al. 2016). Transcriptome signatures of indigenous cattle from high-altitude arid and temperate Ladakh (India) vs. Sahiwal (*Bos indicus*) cattle from tropical region were examined using Agilent 44 K microarray chip. Both types of cattle were found to be genetically distinct and had a differential expression of genes related to adaptation of Ladakh cattle to high altitude (Verma et al. 2018a). Hypoxia-inducing factor-1 (HIF-1), and its associated genes, viz. glucose transporter 1 (GLUT1), vascular endothelial growth factor (VEGF), and hexokinase 2 (HK2) were found to accumulate in cattle adapted to high-altitude hypoxic conditions. These genes serve as candidate genes for homeostatic response to hypoxia at high altitude (Verma et al. 2018b).

33.10 Meat Quality

Pig, cattle, buffaloes, goats, and sheep are the prominent meat-producing animals. To fulfill the increasing demand of meat, it is necessary to improve the quality of the meat. It is imperative to develop molecular markers to assess the quality of meat. Meat quality is affected by various factors including breed, nutrition, production management, and post-slaughter handling.

Insights into bovine muscle biology have been obtained by profiling the cattle muscle transcriptome by microarray techniques. For instance, Byrne et al. (2005) carried out gene expression profiling of muscle tissue in Brahmen steers to decipher the cellular and molecular processes associated with remodeling of muscles in response to nutritional stress. Gene expression profiling was also conducted in different muscle types to better understand the muscle characteristics to evaluate meat quality traits. A microarray-based comparison of the *longissimus dorsi* (LM) muscle from Japanese Black and Holstein cattle over an extended intensive feeding period was made to identify the genes that might be involved in determining the unique ability of Japanese Black cattle to accumulate intramuscular fats (Wang et al. 2005). Markers of meat tenderness and muscle growth have been identified based on inferences from transcriptomic analysis of bovine muscles (Sudre et al. 2005; Reecy et al. 2006). Microarrays-based transcriptome analysis of longissimus thoracis muscle from 25 Charolais bull calves revealed a subsistence of a marker, named DNAJA1, that constituted a new marker of beef sensory quality (Bernard et al. 2007).

To look at underlying mechanism responsible for the differences of meat quality between different pork breeds or cuts, the LM and the *biceps femoris* (BF) muscle from Min and Large White pigs were investigated. A total of 1371 genes were differentially expressed between two breeds, and an expression of 114 genes was different between LM and BF. It was concluded that differential expression of genes (DEGs) is responsible for difference in fatty acid metabolism, intramuscular fat deposition and skeletal muscle growth in two types of breeds. The study shows its import to improve meat quality through molecular breeding (Liu et al. 2017).

33.11 Insights into Embryonic Developmental Biology

Microarray technology is broadly used to unravel key insights of reproductive biology. Caetano et al. (2004) identified differentially expressed genes in ovaries and ovarian follicles of pigs selected for increased ovulation rate to search for insights into ovarian physiology and the quantitative genetic control of reproduction traits in swine.

Ushizawa et al. (2004) conducted a custom designed utero-placental cDNA microarray analysis of bovine embryo gene expression profiles to search new molecules and genes involved in embryonic and extra-embryonic membrane development. Novel genes and molecules were detected with possible roles in trophoblast differentiation and embryonic development.

Follicular development in ovaries is controlled by multiples genes, hormones, and signal pathways. Hayashi et al. (2010) carried out differential microarray and quantitative real-time PCR genome-wide gene expression profiling of largest (F1) and second-largest bovine ovarian follicles (F2) to identify the genes associated with growth of dominant follicles. It was inferred that F1 and F2 follicles had a differential gene expression profile, and the expression of stage-specific gene might be associated with ovarian follicular growth or atresia. Hence, it is essential that strategies should be developed to promote candidate genes responsible for ovarian follicular development (Hayashi et al. 2010). Microarray analysis or serine protease inhibitors (SERPINs), a class of proteases having role in fibrinolysis, coagulation, inflammation, cell mobility, cellular differentiation, and apoptosis, revealed a characteristic expression profile in healthy versus atretic bovine ovarian follicles (Hayashi et al. 2011).

33.12 RNA Sequencing (RNA-Seq)

High-throughput sequencing technologies (next- or second-generation sequencing) have introduced a new alternative to microarrays, called RNA-Seq. After years of extensive investigations based on the characterization of genome-wide gene expression through oligonucleotide-based array technologies, transcriptomics has gained new momentum. The RNA-Seq, also called whole-transcriptome shotgun sequencing (WTSS), refers to the use of high-throughput sequencing technologies for characterizing the RNA content and composition of RNA in a biological sample at a given time. RNA-Seq quantifies continuously changing cellular gene expression by sequencing short strands of cDNA, aligning sequences obtained back to the genome or transcriptome, and counting the aligned reads for each gene.

Until the advent of RNA-Seq, the microarrays were the standard tool for gene expression quantification. But, with the development of new sequencing technologies and bioinformatic tools, RNA-Seq has emerged as an appealing alternative to classical microarrays in measuring global genomic expressions profile (Wolf et al. 2010). RNA-Seq technology, unlike microarray, does not depend on the prerequisite knowledge of the reference transcriptome. Compared to microarray technology, the RNA-Seq data contains very low background signal, a higher dynamic range of expression levels, and also relatively small amount of total RNA required for quantification. Therefore, gene detection in RNA-Seq, unlike microarray, does not depend on probe design; rather, it relies on short nucleotide reads mapping which can attain exceedingly high resolution.

Although both RNA-Seq and microarrays are good when it comes to relative gene expression quantification (Nookaew et al. 2012), RNA-Seq has additional advantages as it offers sufficient coverage and captures a wider range of gene expression values. RNA-Seq can identify transcripts that are not previously annotated, and it can quantify very low transcripts (unlike microarrays where there is background noise interference), as well as very high transcripts. As a digital measure (count data), it scales linearly even at extreme values, whereas microarrays show saturation of analog-type fluorescent signals (Marioni et al. 2008). RNA-Seq further provides information on RNA splice events that cannot be easily detected by standard microarrays (Mortazavi et al. 2008).

RNA-Seq transcriptomics has become an important tool in many research studies in farm animals, including bovines, buffalo, pig, chicken, sheep, and goat. The RNA-Seq has been used to study potential candidate genes associated with muscle growth, meat quality, lactation, reproduction efficiency, and response to diseases, and stresses.

33.13 RNA-Seq Based Transcriptomics in Livestock Science

RNA-Seq technology is comparatively a new tool to explore various new frontiers in the field of animal genetic structuring, muscle biology, assisted reproduction (Gòdia et al. 2018;, Yang et al. 2018), gut ecosystem dynamics and animal nutrition (Elekwachi et al. 2017; Gruninger et al. 2017; Denman et al. 2018; Iannaccone et al. 2018), disease investigations (Zhao et al. 2017), mammary development and lactation physiology (Paten et al. 2015), and adaptation to stress (Park et al. 2019).

Whole-transcriptome sequencing was performed to reveal the genetic structure in six Chinese cattle populations (Wang et al. 2018a, b). RNA-Seq dataset on LM from 80 beef cattle revealed several genes and pathways involved in feed intake, feed efficiency, and metabolism (Keel et al. 2018).

Feed conversion efficiency depends on various factors such as type of digestive system of animals, genetic system involved in regulating feed efficiency. Selective breeding of animals with high-feed conversion efficiency is an important goal of dairy and meat animal producers. Salleh et al. (2017) generated bovine RNA-Seq expression data from liver biopsies of 19 dairy cows from high- and low-feed efficiency categories to identify pathways and biomarkers of feed efficiency. The DEGs identified were CYP's (Cytochrome P450, Family polypeptides), and GIMAP genes for the Holstein and Jersey cows, respectively. Notably, these gene families are responsible for primary immunodeficiency pathway and play a major role in feed utilization and the metabolism of lipids, sugars, and proteins. In another recent study, RNA-Seq was applied to identify potential long noncoding RNAs (lncRNAs) in transcriptome data of jejunum tissue of calves of German Holstein cows fed two different milk diets. It was found that differentially expressed lncRNAs might play potential regulatory roles in modulating biological processes affecting the barrier function of intestinal epithelial cells of calves in response to different feeding regimens during preweaning period (Weikard et al. 2018).

RNA-Seq has emerged as a major transcriptome profiling tool and has contributed a lot to understand lactation biology and the genes involved in lactogenesis in dairy animals. As obtaining mammary tissue is difficult and ethically undesirable, mammary epithelial cells are used as model cells to understand lactation biology. Cui et al. (2014) used RNA-Seq to investigate the complexity of mammary gland of lactating Holstein cows. Integrated analysis of differential gene expression and the already reported QTL and genome-wide association study (GWAS) data indicates that *TRIB3*, *SAA* (*SAA1*, *SAA3*, and *M-SAA3.2*), *VEGFA*, *PTHLH*, and *RPL23A* were the important candidate genes that affect milk composition traits, i.e., protein, and fat quality.

The RNA-Seq data was used to identify genes with differential expression pattern in milk fat globule membrane (MFGM)-enriched fractions from Holstein, Jersey, yak, buffalo, goat, horse, camel, and human. The purpose was to identify some promising candidate genes for milk production traits in dairy cattle. A total of 520 proteins were identified in MFGM fractions across the species (Yang et al. 2015).

Another comprehensive study by Wickramasinghe et al. (2012) resulted in identification of stage-specific transcripts/genes in milk somatic cells of Holstein cows expressed during early, peak, and late stages of lactation. In addition, a few genes had a role in each lactation stage indicating that milk somatic cells adapt to different settings as per need. RNA-Seq analysis of milk transcriptome of Spanish *Churra* and *Assaf* sheep breeds, which differ in their milk production traits, showed a total of 573 DEGs across lactation points (Suárez-Vega et al. 2015)

RNA-Seq was used to generate global profile of bovine embryonic transcriptomes in order to understand pre-implantation embryonic development at a fine scale (Huang and Khatib 2010).

A comprehensive catalogue of transcripts was generated in germinal vesicle, metaphase II oocytes, and IVF embryos at the 4 cell, 8 cell, 16 cell, and blastocyst stage embryos produced by IVFs. This study provides a leading transcriptome data and the specific genes activated during maturation of oocytes and early embryonic development (Graf ct al. 2014).

It is evident that the major embryonic genome activation (EGA) in bovine commences during 8- to 16-cell stages of embryo development. Lavagi et al. (2018) have reported RNA sequencing of single cell heterogeneity of transcriptome profiles among single cells in bovine 2- and 3-day embryos in order to understand major embryonic genome activation, suggesting asynchronous embryonic development during the developmental phase of bovine EGA.

Huang et al. (2017) applied RNA-Seq technology to identify the transcriptomic difference of subcutaneous adipose tissues between Wagyu and Holstein cattle to explore the molecular mechanisms of fat deposition. The study has provided a basis for better understanding of molecular mechanism of fat metabolism and deposition in beef cattle. Additionally, RNA-Seq has been used in adipose tissue of sheep (Miao et al. 2015), pig (Sodhi et al. 2014), and cattle (Sheng et al. 2014) to generate genome-wide transcriptome data for identifying genes associated with fat metabolism. In order to identify genes and pathways associated with heat stress response, RNA-Seq-based transcriptome analysis was carried out in Holstein calves (Srikanth et al. 2017).

33.14 Outlook and Challenges

The development of bioinformatics and functional genomic approaches for improved annotation combined with new data analysis tools that enable cross-species comparisons will greatly enhance the extraction of biological information from species-specific microarrays/RNA-seq and advance our understanding of livestock biology. From the economic point of view, the importance and impact of genome-wide tools in modern livestock production are likely to increase in the future. Over the longer term, these high-throughput technologies would reshape the livestock biology in terms of functional annotation and discovery of new gene regulating trait of economic importance, complete description, and understanding of cellular pathways (e.g., metabolism, proliferation, cell–cell interaction), understanding genomic-environment interaction (e.g., developmental pathways, abiotic stress, nutritional genomics and infectious diseases). This would further help in identification of target molecules for improvement and selection of better performing livestock species to ensure food security meeting the challenges of increasing global population.

The genes identified in transcriptomics studies could be used as candidate genes for improving production performance and enhancing thermal tolerance through breeding by marker-assisted selection.

33.15 Conclusions

Transcriptome profiling or transcriptomics is a valuable tool to in various aspects of livestock production including detecting differential gene expression in during embryo development, somatic cells of mammary gland at various stages of lactation, adipose tissue, and muscle cells. Detecting transcriptomics by different methods leads to identification of genes that could be used as candidate markers to promote embryo development, augmenting lactone to enhance milk yield in lactating animals, and improving meat quality, or enhancing tolerance of animals to thermal stress.

References

Bernard C, Cassar-Malek I, Le Cunff M, Dubroeucq H, Renand G, Hocquette JF (2007) New indicators of beef sensory quality revealed by expression of specific genes. J Agric Food Chem 55(13):5229–5237 Epub 2007 Jun 5

Bionaz M, Periasamy K, Rodriguez-Zas SL, Everts RE, Lewin HA, Hurley WL, Loor JJ (2012) Old and new stories: revelations from functional analysis of the bovine mammary transcriptome during the lactation cycle. PLoS One 7(3):e33268. https://doi.org/10.1371/journal.pone.0033268 (Epub 2012 Mar 12)

Byrne KA, Wang YH, Lehnert SA, Harper GS, McWilliam SM, Bruce HL, Reverter A (2005) Gene expression profiling of muscle tissue in Brahman steers during nutritional restriction. J Anim Sci 83 (1):1–12

Caetano AR, Johnson RK, Ford JJ, Pomp D (2004) Microarray profiling for differential gene expression in ovaries and ovarian follicles of pigs selected for increased ovulation rate. Genetics 168(3):1529–1537

Cui X, Hou Y, Yang S, Xie Y, Zhang S, Zhang Y, Zhang Q, Lu X, Liu GE, Sun D (2014) Transcriptional profiling of mammary gland in Holstein cows with extremely different milk protein and fat percentage using RNA sequencing. BMC Genom 24(15):226. https://doi.org/10.1186/1471-2164-15-226

Denman SE, Morgavi DP, McSweeney CS (2018) Review: the application of omics to rumen microbiota function. Animal 12(s2):s233–s245. https://doi.org/10.1017/s175173111800229x (Epub 2018 Sep 28 Review)

Elekwachi CO, Wang Z, Wu X, Rabee A, Forster RJ (2017) Total rRNA-seq analysis gives insight into bacterial, fungal, protozoal and archaeal communities in the rumen using an optimized RNA isolation method. Front Microbiol 8:1814. https://doi.org/10.3389/fmicb.2017.01814 (eCollection 2017)

Gao Y, Lin X, Shi K, Yan Z, Wang Z (2013) Bovine mammary gene expression profiling during the onset of lactation. PLoS One 8(8):e70393. https://doi.org/10.1371/journal.pone.0070393 (eCollection 2013)

Graf A, Krebs S, Zakhartchenko V, Schwalb B, Blum H, Wolf E (2014) Fine mapping of genome activation in bovine embryos by RNA sequencing. Proc Natl Acad Sci U S A. 111(11):4139–4144. https://doi.org/10.1073/pnas.1321569111 (Epub 2014 Mar 3)

Gòdia M, Mayer FQ, Nafissi J, Castelló A, Rodríguez-Gil JE, Sánchez A, Clop A (2018) A technical assessment of the porcine ejaculated spermatozoa for a sperm-specific RNA-seqanalysis. Syst Biol Reprod Med. 64(4):291–303. https://doi.org/10.1080/19396368.2018.1464610 (Epub 2018 Apr 26)

Grummer RR, Mashek DG, Hayirli A (2004) Dry matter intake and energy balance in the transition period. Vet Clin North Am Food Anim Pract 20(3):447–470 (Review)

Gruninger RJ, Reid I, Forster RJ, Tsang A, McAllister TA (2017) Identification of genes involved in the degradation of lignocellulose using comparative transcriptomics. Methods Mol Biol 1588:279–298. https://doi.org/10.1007/978-1-4939-6899-2_21

Hayashi KG, Ushizawa K, Hosoe M, Takahashi T (2010) Differential genome-wide gene expression profiling of bovine largest and second-largest follicles: identification of genes associated with growth of dominant follicles. Reprod Biol Endocrinol 5(8):11. https://doi.org/10.1186/1477-7827-8-11

Huang W, Guo Y, Du W, Zhang X, Li A, Miao X (2017) Global transcriptome analysis identifies differentially expressed genes related to lipid metabolism in Wagyu and Holstein cattle. Sci Rep. 7(1):5278. https://doi.org/10.1038/s41598-017-05702-5

Huang W, Khatib H (2010) Comparison of transcriptomic landscapes of bovine embryos using RNA-Seq. BMC Genom 17(11):711. https://doi.org/10.1186/1471-2164-11-711

Hayashi KG, Ushizawa K, Hosoe M, Takahashi T (2011) Differential gene expression of serine protease inhibitors in bovine ovarian follicle: possible involvement in follicular growth and atresia. Reprod Biol Endocrinol. 27(9):72. https://doi.org/10.1186/1477-7827-9-72

Iannaccone M, Elgendy R, Giantin M, Martino C, Giansante D, Ianni A, Dacasto M, Martino G (2018) RNA sequencing-based whole-transcriptome analysis of friesian cattle fed with grape pomace-supplemented diet. Animals (Basel) 8(11). pii: E188. https://doi.org/10.3390/ani8110188

Kapila N, Sharma A, Kishore A, Sodhi M, Tripathi PK, Mohanty AK, Mukesh M (2016) Impact of heat stress on cellular and transcriptional adaptation of mammary epithelial cells in riverine buffalo (*Bubalus Bubalis*). PLoS ONE 11(9):e0157237. https://doi.org/10.1371/journal.pone.0157237 (eCollection 2016. Erratum. In: PLoS One. 2018 Jan 11;13(1):e0191380)

Keel BN, Zarek CM, Keele JW, Kuehn LA, Snelling WM, Oliver WT, Freetly HC, Lindholm-Perry AK (2018) RNA-Seq meta-analysis identifies genes in skeletal muscle associated with gain and intake across a multi-season study of crossbred beef steers. BMC Genom 19(1):430. https://doi.org/10.1186/s12864-018-4769-8

Kosciuczuk EM, Lisowski P, Jarczak J, Majewska A, Rzewuska M, Zwierzchowski L, Bagnicka E (2017) Transcriptome profiling of Staphylococci-infected cow mammary gland parenchyma. BMC Vet Res. 13 (1):161. https://doi.org/10.1186/s12917-017-1088-2

Lavagi I, Krebs S, Simmet K, Beck A, Zakhartchenko V, Wolf E, Blum H (2018) Single-cell RNA sequencing reveals developmental heterogeneity of blastomeres during major genome activation in bovine embryos. Sci Rep. 8(1):4071. https://doi.org/10.1038/s41598-018-22248-2

Li D, Xie Z, Zand M, Fogg T, Dye T (2017a) Bon-EV: an improved multiple testing procedure for controlling false discovery rates. BMC Bioinformatics 18 (1):1. https://doi.org/10.1186/s12859-016-1414-x

Li S, Wang Q, Lin X, Jin X, Liu L, Wang C, Chen Q, Liu J, Liu H (2017b) The use of "Omics" in lactation research in dairy cows. Int J Mol Sci 18(5). pii: E983. https://doi.org/10.3390/ijms18050983 (Review)

Lin J, Wu YJ, Liang X, Ji M, Ying HM, Wang XY, Sun X, Shao CH, Zhan LX, Zhang Y (2019) Network-based integration of mRNA and miRNA profiles reveals new target genes involved in pancreatic cancer. Mol Carcinog 58(2):206–218. https://doi.org/10.1002/mc.22920 (Epub 2018 Oct 29)

Liu Y, Yang X, Jing X, He X, Wang L, Liu Y, Liu D (2017) Transcriptomics analysis on excellent meat quality traits of skeletal muscles of the chinese indigenous min pig compared with the large white breed. Int J Mol Sci 19(1). pii: E21. https://doi.org/10.3390/ijms19010021

Loor JJ, Everts RE, Bionaz M, Dann HM, Morin DE, Oliveira R, Rodriguez-Zas SL, Drackley JK, Lewin HA (2007) Nutrition-induced ketosis alters metabolic and signaling gene networks in liver of periparturient dairy cows. Physiol Genomics 32 (1):105–116 (Epub 2007 Oct 9)

Maeda S, Tomiyasu H, Tsuboi M, Inoue A, Ishihara G, Uchikai T, Chambers JK, Uchida K, Yonezawa T, Matsuki N (2018) Comprehensive gene expression analysis of canine invasive urothelial bladder carcinoma by RNA-Seq. BMC Cancer 18(1):472. https://doi.org/10.1186/s12885-018-4409-3

Marioni JC, Mason CE, Mane SM, Stephens M, Gilad Y (2008) RNA-seq: an assessment of technical reproducibility and comparison with gene expression arrays. Genome Res 18(9):1509–1517. https://doi.org/10.1101/gr.079558.108 (Epub 2008 Jun 11)

McLoughlin KE, Nalpas NC, Rue-Albrecht K, Browne JA, Magee DA, Killick KE, Park SD, Hokamp K, Meade KG, O'Farrelly C, Gormley E, Gordon SV, MacHugh DE (2014) RNA-seq transcriptional profiling of peripheral blood leukocytes from cattle infected with mycobacterium bovis. Front Immunol 5:396. https://doi.org/10.3389/fimmu.2014.00396 (eCollection 2014)

Miao X, Luo Q, Qin X, Guo Y, Zhao H (2015) Genome-wide mRNA-seq profiling reveals predominant down-regulation of lipid metabolic processes in adipose tissues of Small Tail Han than Dorset sheep. Biochem Biophys Res Commun 467(2):413–420. https://doi.org/10.1016/j.bbrc.2015.09.129 (Epub 2015 Sep 28)

Mortazavi A, Williams BA, McCue K, Schaeffer L, Wold B (2008) Mapping and quantifying mammalian transcriptomes by RNA-Seq. Nat Methods 5(7):621–628. https://doi.org/10.1038/nmeth.1226 (Epub 2008 May 30)

Moyes KM, Drackley JK, Morin DE, Rodriguez-Zas SL, Everts RE, Lewin HA, Loor JJ (2010) Mammary gene expression profiles during an intramammary challenge reveal potential mechanisms linking negative energy balance with impaired immune response. Physiol Genomics 41(2):161–170. https://doi.org/10.1152/physiolgenomics.00197.2009 (Epub 2010 Jan 26)

Nookaew I, Papini M, Pornputtapong N, Scalcinati G, Fagerberg L, Uhlén M, Nielsen J (2012) A comprehensive comparison of RNA-Seq-based transcriptome analysis from reads to differential gene expression and cross comparison with microarrays: a case study in Saccharomyces cerevisiae. Nucleic Acids Res 40 (20):10084–10097. https://doi.org/10.1093/nar/gks804 (Epub 2012 Sep 10)

Park W, Srikanth K, Lim D, Park M, Hur T, Kemp S, Dessie T, Kim MS, Lee SR, Te Pas MFW, Kim JM, Park JE (2019) Comparative transcriptome analysis of Ethiopian indigenous chickens from low and high altitudes under heat stress condition reveals differential immune response. Anim Genet 50(1):42–53. https://doi.org/10.1111/age.12740 (Epub 2018 Nov 25)

Paten AM, Duncan EJ, Pain SJ, Peterson SW, Kenyon PR, Blair HT, Dearden PK (2015) Functional development of the adult ovine mammary gland–insights from gene expression profiling. BMC Genom 5(16):748. https://doi.org/10.1186/s12864-015-1947-9

Reecy JM, Spurlock DM, Stahl CH (2006) Gene expression profiling: insights into skeletal muscle growth and development. J Anim Sci 84 Suppl:E150-4 (Review)

Salleh MS, Mazzoni G, Höglund JK, Olijhoek DW, Lund P, Løvendahl P, Kadarmideen HN (2017) RNA-Seq transcriptomics and pathway analyses reveal potential regulatory genes and molecular mechanisms in high- and low-residual feed intake in Nordic dairy cattle. BMC Genom 18(1):258. https://doi.org/10.1186/s12864-017-3622-9

Schena M, Shalon D, Davis RW, Brown PO (1995) Quantitative monitoring of gene expression patterns with a complementary DNA microarray. Science 270 (5235):467–470

Schneider AK, Niemeyer CM (2018) DNA surface technology: from gene sensors to integrated systems for life and materials sciences. Angew Chem Int Ed Engl 57(52):16959–16967. https://doi.org/10.1002/anie.201811713

Sheng X, Ni H, Liu Y, Li J, Zhang L, Guo Y (2014) RNA-seq analysis of bovine intramuscular, subcutaneous and perirenal adipose tissues. Mol Biol Rep 41 (3):1631–1637. https://doi.org/10.1007/s11033-013-3010-8 (Epub 2014 Jan 8)

Sodhi SS, Park WC, Ghosh M, Kim JN, Sharma N, Shin KY, Cho IC, Ryu YC, Oh SJ, Kim SH, Song KD, Hong SP, Cho SA, Kim HB, Jeong DK (2014) Comparative transcriptomic analysis to identify differentially expressed genes in fat tissue of adult Berkshire and Jeju Native Pig using RNA-seq. Mol Biol Rep 41(9):6305–6315. https://doi.org/10.1007/s11033-014-3513-y (Epub 2014 Jul 11)

Srikanth K, Kwon A, Lee E, Chung H (2017) Characterization of genes and pathways that respond to heat stress in Holstein calves through transcriptome analysis. Cell Stress Chaperones 22(1):29–42. https://doi.org/10.1007/s12192-016-0739-8 (Epub 2016 Nov 16)

Suárez-Vega A, Gutiérrez-Gil B, Klopp C, Robert-Granie C, Tosser-Klopp G, Arranz JJ (2015) Characterization and comparative analysis of the milk transcriptome in two dairy sheep breeds using RNA sequencing. Sci Rep 18(5):18399. https://doi.org/10.1038/srep18399

Suchyta SP, Sipkovsky S, Halgren RG, Kruska R, Elftman M, Weber-Nielsen M, Vandehaar MJ, Xiao L, Tempelman RJ, Coussens PM (2003) Bovine mammary gene expression profiling using a cDNA microarray enhanced for mammary-specific transcripts. Physiol Genomics 16(1):8–18

Sudre K, Cassar-Malek I, Listrat A, Ueda Y, Leroux C, Jurie C, Auffray C, Renand G, Martin P, Hocquette JF (2005) Biochemical and transcriptomic analyses of two bovine skeletal muscles in Charolais bulls divergently selected for muscle growth. Meat Sci 70 (2):267–277. https://doi.org/10.1016/j.meatsci.2005.01.012 (Epub 2005 Mar 16)

Ushizawa K, Herath CB, Kaneyama K, Shiojima S, Hirasawa A, Takahashi T, Imai K, Ochiai K, Tokunaga T, Tsunoda Y, Tsujimoto G, Hashizume K (2004) cDNA microarray analysis of bovine embryo gene expression profiles during the pre-implantation period. Reprod Biol Endocrinol 24(2):77

Verma P, Sharma A, Sodhi M, Thakur K, Kataria RS, Niranjan SK, Bharti VK, Kumar P, Giri A, Kalia S, Mukesh M (2018a) Transcriptome analysis of circulating PBMCs to understand mechanism of high altitude adaptation in native cattle of Ladakh Region. Sci Rep 8(1):7681. https://doi.org/10.1038/s41598-018-25736-7

Verma P, Sharma A, Sodhi M, Thakur K, Bharti VK, Kumar P, Giri A, Kalia S, Swami SK, Mukesh M (2018b) Overexpression of genes associated with hypoxia in cattle adapted to Trans Himalayan region of Ladakh. Cell Biol Int. https://doi.org/10.1002/cbin.10981 (In press)

Wang J, Dean DC, Hornicek FJ, Shi H, Duan Z (2018a) RNA sequencing (RNA-Seq) and its application in ovarian cancer. Gynecol Oncol 152(1):194–201. https://doi.org/10.1016/j.ygyno.2018.10.002 (Epub 2018 Oct 5. Review)

Wang W, Wang H, Tang H, Gan J, Shi C, Lu Q, Fang D, Yi J, Fu M (2018b) Genetic structure of six cattle populations revealed by transcriptome-wide SNPs and gene expression. Genes Genomics. 40(7):715–724. https://doi.org/10.1007/s13258-018-0677-1 (Epub 2018 Mar 26)

Wang X, Zhou G, Xu X, Geng R, Zhou J, Yang Y, Yang Z, Chen Y (2014) Transcriptome profile analysis of adipose tissues from fat and short-tailed sheep. Gene 549(2):252–257. https://doi.org/10.1016/j.gene.2014.07.072 (Epub 2014 Aug 1)

Wang YH, Byrne KA, Reverter A, Harper GS, Taniguchi M, McWilliam SM, Mannen H, Oyama K, Lehnert SA (2005) Transcriptional profiling of skeletal muscle tissue from two breeds of cattle. Mamm Genome 16(3):201–210

Weikard R, Hadlich F, Hammon HM, Frieten D, Gerbert C, Koch C, Dusel G, Kuehn C (2018) Long noncoding RNAs are associated with metabolic and cellular processes in the jejunum mucosa of pre-weaning calves in response to different diets. Oncotarget. 9(30):21052–21069. https://doi.org/10.18632/oncotarget.24898 (eCollection 2018 Apr 20)

Wickramasinghe S, Rincon G, Islas-Trejo A, Medrano JF (2012) Transcriptional profiling of bovine milk using RNA sequencing. BMC Genom 25(13):45. https://doi.org/10.1186/1471-2164-13-45

Wolf JB, Bayer T, Haubold B, Schilhabel M, Rosenstiel P, Tautz D (2010) Nucleotide divergence vs. gene expression differentiation: comparative transcriptome sequencing in natural isolates from the carrion crow and its hybrid zone with the hooded crow. Mol Ecol 19 Suppl 1:162–175. https://doi.org/10.1111/j.1365-294x.2009.04471.x

Wu G, Wan X, Xu B (2018) A new estimation of protein-level false discovery rate. BMC Genom 19

(Suppl 6):567. https://doi.org/10.1186/s12864-018-4923-3

Xue L, Xie L, Song X, Song X (2018) Identification of potential tumor-educated platelets RNA biomarkers in non-small-cell lung cancer by integrated bioinformatical analysis. J Clin Lab Anal 32(7):e22450. https://doi.org/10.1002/jcla.22450 (Epub 2018 Apr 17)

Yang H, Wang F, Li F, Ren C, Pang J, Wan Y, Wang Z, Feng X, Zhang Y (2018) Comprehensive analysis of long noncoding RNA and mRNA expression patterns in sheep testicular maturation. Biol Reprod 99(3):650–661. https://doi.org/10.1093/biolre/ioy088

Yang Y, Zheng N, Zhao X, Zhang Y, Han R, Ma L, Zhao S, Li S, Guo T, Wang J (2015) Proteomic characterization and comparison of mammalian milk fat globule proteomes by iTRAQ analysis. J Proteomics 26(116):34–43. https://doi.org/10.1016/j.jprot.2014.12.017 (Epub 2015 Jan 8)

Zhao FR, Xie YL, Liu ZZ, Shao JJ, Li SF, Zhang YG, Chang HY (2017) Transcriptomic analysis of porcine PBMCs in response to FMDV infection. Acta Trop 173:69–75. https://doi.org/10.1016/j.actatropica.2017.05.009 (Epub 2017 May 8)

Proteomics: Applications in Livestock 34

Abstract

"Proteomics" is the large-scale study of proteins, particularly their composition, structure, function, and interactions with each other, and directing the activities of a cell. Techniques, such as 2D gel electrophoresis, MALDI-TOF/MS, X-ray crystallography, NMR, protein microarrays, two-hybrid screening, and western blotting are used in proteomic analysis. The goal of proteomics is to identify new and potentially unexpected changes in protein expression, interaction, or modifications. Generation of large proteomic datasets is expected to demonstrate the interdependence of cellular processes important for normal cell growth or a cell's response to abnormal or disease conditions. Proteomic approach enables an investigator to step back and view the whole picture of cellular functions instead of one particular action of one protein. The scope of proteomics in livestock is broad and includes the characterization and monitoring of changes that take place during the growth, development, and other physiological processes.

Highlights

- Proteomics depicts the whole picture of cellular functions
- Proteomics applications in livestock system are diverse and highly useful.

Keywords

Proteomics · Biomarkers · Farm animals · Aquaculture · Animal health · Disease detection · Livestock product quality

34.1 Introduction

Proteins are one of the most important classes of biomolecules which are responsible for a wide range of cellular activities including cellular organization, biochemical reactions, metabolic activities, cell signaling, and cell-to-cell interactions. Proteins are made up of a single or several chains of amino acid monomeric units. A typical amino acid has the amino, carboxyl, and "R" group (also called as side chain). There are a total of 20 amino acids, and the nature of R-group varies from one amino acid to other. Protein chains are held together by peptide bonds, which are simply amide linkages between alpha amino and carboxylic group of next amino acids.

All the cells of multicellular organisms have the same set of genes, but the proteins produced are different in different tissues. The complete set of proteins of a cell, tissue, organ, or organism at a certain time encoded by a particular genome is known as "proteome" (Blackstock and Weir 1999; Reinhardt et al. 2012). It is defined as the study of the entire set of proteins produced by a cell type at a particular physiological state in order to understand its structure and function

B. Singh et al., *Advances in Animal Biotechnology*,
https://doi.org/10.1007/978-3-030-21309-1_34

(Chandrasekhar et al. 2014; Jayasri et al. 2014). The expression profile of total protein content in a particular cell and tissue depends on time, micro- and macro-environmental conditions.

The term "proteomics" was first coined in 1997. Proteome analysis depends on four major components, i.e., protein separation, identification, characterization, and quantification. The advances in proteomics include quantification and identification of differentially expressed proteins across distinct conditions, identification and characterization of protein–protein interactions, and characterization of protein post-translational modifications (Mann and Jensen 2003). The proteomics deals with the problems which could not be resolved by DNA analysis. Thus, it gives a much better understanding of biological pathways and structural organization of an organism.

Proteome can be studied using the knowledge of genome because genes code for mRNAs and the mRNAs encode proteins. But, the proteome is more complex than the genome or the transcriptome. This is because RNAs can be alternatively spliced and many proteins can be modified after translation by processes such as proteolytic cleavage, phosphorylation, glycosylation, and ubiquitination. The protein–protein interactions also complicate the study of proteomes (Mann and Jensen 2003).

34.2 Types of Proteomics

34.2.1 Structural Proteomics

Structural proteomics is the large-scale analysis of protein structures that help to understand the three-dimensional shape and structural complexities of functional proteins present in a specific cellular organelle. It is also known as "cell map" (Blackstock and Weir 1999; Graves and Haystead 2002a, b). It gives detailed information about the structure of proteins and protein complexes and further helps to figure out their interactions with each other. The comparison of protein structures also helps to identify the functions of newly discovered genes (homology modeling). The structural analysis of proteins is achieved by using X-ray crystallography and NMR spectroscopy technologies.

34.2.2 Expression Proteomics

The qualitative and quantitative expression of total proteins under two different conditions can be studied using expression proteomics. It is a large-scale analysis of protein expression. Proteins that express due to a stress or disease can be compared to proteins in a normal state to detect the protein that is responsible for the stress or diseased state (Graves and Haystead 2002a, b). A protein present only in a diseased state may represent a useful therapeutic target or diagnostic marker. The technologies used to study the differential expression of proteins mainly are 2D-PAGE and mass spectrometry.

34.2.3 Functional Proteomics

Functional proteomics determines the function of proteins that help in unrevealing the molecular mechanisms within the cell. In other words, the characterization of protein–protein interactions helps to determine protein functions and biological pathways (Almeida et al. 2015). Thus, it is the large-scale analysis of protein interactions. It shows how proteins assemble in larger complexes and participate in the cellular signaling pathways. The technologies such as affinity purification, mass spectrometry, and the yeast-two-hybrid system are particularly used to elucidate the role of proteins in cellular processes.

34.3 Proteomic Strategies for the Identification and Analysis of Proteins

One of the rate-limiting steps in any proteomic analysis study is obtaining, and then handling, sufficient quantities of target protein (s) from its original biological source. The classical method of quantitative and qualitative expression proteomics combines protein separation by high-resolution 2D gel electrophoresis with mass spectrometer (MS) or MS/MS identification of selected protein spots. Because, even the best 2D gels can routinely separate no more than 2000 proteins, this technique is limited to the most abundant proteins if a crude protein mixture (whole-cell lysate) is used. 2D gel electrophoresis is limited by the amount of material that can be applied to the first-dimension immobilized pH gradient gel (~150 ug to low milligram quantities). Hence, 2D gels have limited "scale-up" capability. For this reason, it is often desirable to "trace enrich" for a particular subclass of proteins. By analyzing proteins in a cellular compartment or organelle, it is possible to reduce the complexity and differences in the abundance of a subset of proteins within a cell. There have been several advances in the area of proteomics since the development of 2D gel electrophoresis and the incorporation of mass spectrometry for the identification of proteins. The three-dimensional structure of a protein or protein complex can be easily detected with X-ray crystallography and NMR. Protein microarrays further helped to resolve the protein–protein interactions.

34.4 Two-Dimensional Gel Electrophoresis

In two-dimensional (abbreviated as 2DE or 2D) gel electrophoresis, proteins are separated based on two distinct properties. Two-step electrophoresis is performed to resolve the protein samples in two dimensions. The samples are separated on the basis of their net charge in the first dimension, and on the basis of molecular weight in second dimension (Rabilloud and Lelong 2011). This results in an image of thousands of small spots, each spot represents a protein. It can resolve approximately 1000–2000 protein spots that appear as dots in a gel after staining (Chandrasekhar et al. 2014). The primary application of 2D gel electrophoresis is mainly protein expression profiling to qualitatively and quantitatively compare two similar samples to find the specific protein differences. It is also used to map proteins from cellular organelles and protein complexes. The amalgamation of these two techniques produces resolution much better than one-dimensional electrophoresis (Graves and Haystead 2002a, b). It also has a remarkable property to resolve proteins that have undergone some sort of post-translational modification (Mann and Jensen 2003). Despite its several uses, it has some limitations as it is a labor-intensive and time-consuming process. 2D is unable to detect low-copy proteins in the total cell lysate.

The technique is used in several analytical strategies in livestock such as developing a reference map of bull seminal plasma (Sarsaifi et al. 2015), understanding proteomic basis of meat tenderness (Carvalho et al. 2014), differentiating beef of different cattle breeds (Rodrigues et al. 2017), and identifying proteins as markers of bull fertility (Soggiu et al. 2013).

34.5 Mass Spectrometry

Mass spectrometry is an analytical technique that is useful to detect the size of a protein or protein complex (Cottrell 2011; Soares et al. 2012). It is also used to identify and determine the characteristics of protein molecule (Soares et al. 2012; Chandrasekhar et al. 2014). It works by ionizing chemical compounds to generate charged molecules. Then, an electric field is applied to accelerate the ionized molecules. Lighter ions accelerate faster and detected first. These differences in the acceleration of the ions are used to measure their unique mass-to-charge ratio (Soares et al. 2012). If the mass spectrum of ions is measured with precision, then the composition of a protein molecule can be identified and the sequence of the protein can be identified. The advances in mass spectrometry have permitted the analysis of very small quantities of proteins which, otherwise, were difficult to detect in the presence of abundant proteins. It is more sensitive and is amenable to high-throughput operations. It has replaced Edman sequencing as the protein identification tool of choice (Graves and Haystead 2002a, b). The most useful technique for protein identification is matrix-assisted laser desorption/ionization–time of flight (MALDI-TOF) mass spectrometry. Current mass spectrometers can detect and identify peptides in the femtomole (10^{-15}) to attomole (10^{-18}) range. There are many types of mass spectrometers that can be used for proteomic studies, and each accomplishes the task of peptide identification in a slightly different way. However, the basic process of identifying a protein using a mass spectrometer is consistent between the various types. After initial protein digestion typically with trypsin, peptides must be ionized to enter the mass spectrometer. Peptides are then detected, isolated, and finally, fragmented and sequenced by the mass spectrometer. Ionization of peptides is the first step in mass spectrometry of proteomes. The two frequently used ionization methods are electrospray ionization (ESI) and matrix-assisted laser desorption ionization. One advantage of ESI is that this method of ionization allows for the direct linkage between liquid chromatography and mass spectrometry because of the volatility of the HPLC solutions. Charged gas phase peptides are generated by ESI when the acidic HPLC solution containing peptides is sprayed from a tip, and the solution evaporates. The MALDI requires mixing of the peptide with a UV-absorbing molecule and the formation of crystals. When a laser strikes the crystalline structure, the results are the sublimation of the matrix and ionization and release of the associated peptides. The peptides are then analyzed by the mass spectrometer and the peptide mass determined. The peptide's mass is typically expressed as a ratio of mass divided by the charge of the peptide (m/z). Both ESI and MALDI cause peptides to gain protons. The same peptide population may gain a different number of protons. Therefore, a peptide with a mass of 1000 Da will be detected on a mass spectrometer with an m/z of 1001 if it gained 1 proton, 501 if it gained 2 protons, and 334.3 if it gained 3 protons during ionization.

There are two basic mass spectrometry methods for the identification of proteins. The first-called peptide fingerprinting is often associated with 2D-PAGE protein separation scheme. Individual spots from the 2D-PAGE gel are isolated and the proteins digested with trypsin. These proteins are typically analyzed with a MALDI-TOF time of flight mass spectrometer. The mass spectrometer will record all the peptide m/z detected in the gel spot. Identification of a protein is based on the measurement of multiple peptides that can come from that protein. For example, after a mass spectrometer has determined the m/z for the peptides from a gel spot, this information will be matched to a protein database. A successful protein match will be based on the number of peptides matched to the protein and the accuracy of the matches. The second method to identify proteins involves the use of tandem mass spectrometers that allow for the sequencing of individual peptides.

34.6 Differential Display Proteomics

A fundamental aspect of proteomic research is the determination of protein-expression levels between two different states of a biological system (e.g., relative quantification of protein levels), such as that encountered between normal and diseased cells or tissues. This is referred to as differential display or comparative proteomics. This can be done in different ways such as running and comparing samples in 2D-SDS-PAGE, with LC-MS and Isotope tags or isobaric tags for relative and absolute quantitation *(iTRAQ)*.

For differences in the protein-expression profiling, 2D gels from two different samples for differences in the occurrence or intensity of protein spots were compared. This approach provides a useful means of comparing proteomes. However, the identification of protein is cumbersome and difficult by this procedure. Application of peptide mass fingerprinting and

LC-MS-MS analysis now makes it possible to identify essentially any protein one can detect by staining the gel. Therefore, the critical task in comparative proteomics with 2D gels is identifying the features that differ between the gels.

The LC-MS approach to proteome comparison is conceptually opposite to the 2D gel approach. Whereas, the 2D gel approach separates proteins and begins with an image comparison, the LC-MS approach separates peptides and ends with data mining to assess differences between samples. Two protein samples are treated with reagents to "tag" them. The tags are chemically identical, except that one contains heavy isotopes (e.g., ^{2}H, ^{15}N, ^{13}C, ^{18}O, etc.) and the other contains light isotopes. The samples are digested and the peptides are analyzed by LC-MS-MS. Analysis of the MS-MS data allows the identification of the protein present. Examination of the full-scan spectra corresponding to each MS-MS scan then allows measurement of the ratio of the light- and heavy-isotope tagged peptides. This ratio corresponds to the ratio of that protein in two samples. This approach provides a relative quantification of the level of a particular protein in two samples.

34.7 Isobaric Tags for Relative and Absolute Quantitation (ITRAQ) for Biomarker Discovery

Isobaric tags for relative and absolute quantitation (iTRAQ) is a non-gel-based multiplexed protein quantitation technique that provides relative and absolute measurements of all peptides from different samples/treatments. iTRAQ is ideally suited for comparing normal, diseased, and treated samples, time course studies, biological replicates, and relative quantitation. Quantitation of differences in protein amount between different samples can be done by iTRAQ isotope tagging of peptides and 2D-LC separations (up to 8 separate samples with 600–1200 proteins identified and quantitated in a single experiment). Applied BiosystemsiTRAQ isobaric affinity labels allow for multiplexing up to 4 or 8 samples (4-Plex or 8-Plex) in a single experiment.

The labeling reagent consists of a quantification group (N-methylpiperazine), a balanced group (carbonyl), and a hydroxyl succinimide ester group that reacts with the N-terminal amino groups of peptides and the amino groups of lysine. Furthermore, the isobaric tag consists of a balanced group and a reporter group. There are eight kinds of iTRAQ® reagents, each having stable isotopes that are uniquely distributed between the balanced and reporter groups. Identical peptides obtained from different samples can be labeled with different iTRAQ® reagents to be used as iTRAQ®-modified peptides having the same mass. When iTRAQ®-modified peptides are analyzed by MS/MS, the ratios of peptide quantities among different samples are expressed as signal intensity ratios of the reporter groups (m/z: 113, 114, 115, 116, 117, 118, 119, and 121). After 1) denaturation, reduction and alkylation, 2) enzyme digestion, and 3) iTRAQ® modification, multiple peptides are labeled, including peptides with post-translational modifications. Each sample will be labeled individually, and they will be pooled and fractionated into up to 25 or 30 fractions by SCX. NanoLC-MS/MS using latest HCD technology in OrbiTrap or Q-TOF are normally performed on each fraction. Identical peptides derived from different samples have the same mass. These peptides share similar chromatographic properties, allowing both peptide identification and quantification to be derived from the same MS/MS spectrum. iTRAQ chemistry offers more reliable quantification than other shotgun approaches that require highly reproducible chromatography runs. In MS/MS analysis, the signal intensity ratios of the reporter groups indicate the ratios of the peptide quantities and can be used to determine the relative quantities of the peptides. The MS/MS spectra of the individual peptides show signals reflecting amino acid sequences and also show reporter ions reflecting the protein contents of the samples. A database search is then performed using fragmentation data to identify the labeled peptides and hence the corresponding proteins while the iTRAQ mass reporter ion is used to relatively quantify the peptides. Quantitation of protein from multiple samples can be achieved in the same run.

34.8 X-ray Crystallography

The three-dimensional structure of a protein or protein complex can be determined using X-ray crystallography (Okada et al. 2000). It enables researchers to determine the structure of a protein crystal at atomic resolution. The protein is first crystallized to amplify the signals, as the scattering of X-rays from a single protein molecule is not enough to allow the detection (Manjasetty et al. 2012). X-rays with low wavelengths range are used since they are of the same order of magnitude as the distance of atoms within proteins ($\sim$0.1 nm). Crystallographers aim X-rays at a tiny crystal containing trillions of identical molecules (Higashiura et al. 2013). The crystal scatters the X-rays onto an electronic detector which converts into a three-dimensional digital image of the molecule with the help of computers.

34.9 Nuclear Magnetic Resonance

Nuclear magnetic resonance (NMR), another protein imaging technique, uses the magnetic properties of atoms to determine the three-dimensional structure of proteins (Yee et al. 2002). If a highly concentrated solution of protein is available, then it is a unique technique to reveal the atomic structure of proteins in the solution (Rehm et al. 2002). This technique depends on the principle that certain atomic nuclei are intrinsically magnetic and the chemical shift of nuclei depends on their local environment (Shin et al. 2008). NMR has an advantage that it does not require crystals. However, the structure determination is limited by size constraints, lengthy data collection, and analysis times. NMR-based structural proteomics together with X-ray crystallography provide a comprehensive structural database to predict the basic biological functions of hypothetical proteins identified by the genome projects.

34.10 Protein Microarrays and Two-Hybrid Screening

After determining the nature, size, and three-dimensional structure of proteins, it is a prerequisite to determine the interactions between proteins to comprehend the precise biological functions of proteins (Brückner et al. 2009). This could be easily studied with the help of yeast-two-hybrid systems and protein microarray technique. The protein microarray is a large-scale version of the basic two-hybrid screen (Zhu and Qian 2012). Generally, the most eukaryotic transcription factors have two domains: an activation domain (AD) and a DNA-binding domain (BD). The idea behind the two-hybrid screen is that transcription can still activate even when these domains split into two separate fragments, as long as the fragments are brought within close proximity to each other. Thus, in the two-hybrid screen, one protein of interest is genetically fused to the BD and another protein is fused to the AD (Suter et al. 2008; Rajagopala et al. 2012). The reporter gene will get activated if both the proteins of interest interact with each other. This is because due to the interaction of the two-hybrid proteins, the BD and AD domains will also come closer to each other (Causier 2004; Brückner et al. 2009). The protein microarray technology provides a versatile platform for the characterization of hundreds or thousands of proteins in a highly parallel and high-throughput manner.

34.11 Application in Livestock

The applications of proteomics in livestock are extensive and include monitoring of proteome changes in the tissue and body fluids to deduce the physiological processes during growth, development, and production. This technology is also applied for detection/management of diseases and drug/vaccine production (Soares et al. 2012; Jayasri et al. 2014). However, the diversity of species (cattle, sheep, goats, equines, chickens, and fishes) as well as the diversity of biological samples (milk, serum, plasma, tear, saliva, urine, semen, etc.) complicate the analysis and interpretation of proteome data.

34.12 Biomarkers

The protein biomarkers could be easily identified in biological fluids such as milk, plasma, serum, saliva, tear, respiratory exudates, urine, semen, genital secretions, egg yolk, and egg white (Roncada et al. 2012; Oskoueian et al. 2016). These biomarkers are useful in monitoring the health of animals, detecting animal pathogens, elucidating disease mechanisms, vaccine development and assessing pharmacologic response to therapeutics (Oskoueian et al. 2016). They are also crucial for the genetic selection and breeding of livestock. The biomarkers that can help to optimize a sustainable balance between productivity, product quality, and animal welfare are in high demand these days (Roncada et al. 2012; Kurpińska et al. 2014). For instance, bovine mastitis is one of the major diseases that reduce the milk production and results in severe economic losses to the dairy industry. The early detection and treatment of bovine mastitis provide a quick recovery with no significant decline in the milk production (Reinhardt et al. 2012). However, detection of mastitis in a late stage decreases the milk production escorted with reproductive disorders in dairy cows. The advantage of protein biomarkers is that the samples used are from readily available biological fluids which are non-invasive or minimally invasive sources of sample collection. They are also real-time and cost-effective sources for diagnosis purposes such as monitoring therapeutic response and development of vaccines against various diseases.

34.13 Quality Characteristic of the Dairy Product

Farm animal products, meat, and milk provide the basic source of protein in food for human consumption and contribute to a balanced diet for the majority of the population. The type of proteins present in an animal product determines the nutritional value and edibility of the food (Jayasri et al. 2014). The quality traits are influenced both by genetics, environmental factors, and processing conditions (Hollung et al. 2007). Hence, understanding the biological traits that impact yield, quality, and nutritional value of these products is of prime importance for the dairy industry. For instance, both the SDS-PAGE and 2DE analyses of milk from six Indian goats showed remarkable variability as well as the similarity in protein forms and suggests this variability exist in all the individual milk samples (Kumar et al. 2013). This complexity is the consequence of post-translational modifications and due to the presence of numerous genetic variants of proteins and bioactive peptides in milk. Therefore, proteomics has several applications to ensure the quality of milk, meat, and other dairy products.

34.14 Early Diagnosis of Disease

The main aim of the research is to prevent or cure disease at its onset. Proteomics analysis can enable the development of diagnostic methods to reveal pathological changes before any of clinical symptoms appears (Kurpińska et al. 2014). For instance, the majority of proteomic studies in bovines have been carried out to elucidate pathogenic mechanisms of mastitis and endometritis (Almeida et al. 2015). Glutathione S-transferase is diagnosed as a significant component of bovine nasal secretion in bovine respiratory disease. Additionally, g-Glutamyl transferase and alkaline phosphatase have also been identified in nasal secretion as a part of active host defense mechanisms. Proteomics is also used to identify markers of stress and plant–insect interactions to monitor animal welfare in increasingly intensified farm management practices.

34.15 Outlook and Challenges

Unlike other methodologies that analyze a few proteins at a time, proteomics can analyze thousands of proteins in a single experiment. This ability to analyze thousands of proteins gives the field of proteomics a unique capability to demonstrate how cells can dynamically respond to changes in their environment.

Unlike in human and model species, the studies on proteomics have been relatively neglected in farm animal research. Now, the tide is turning in the application of proteomics to livestock with the growing awareness to deeply understand the phenotype, physiology, pathophysiology, and productivity of farm animals. On the other hand, variations in protein expression in diseased states, however, can be difficult to discern. Proteins are naturally unstable molecules, which makes proteomic analysis much more difficult than genomic analysis.

34.16 Conclusions

With the recent advancement in technologies, the speed and ease by which farm animal proteomes can be studied will soon be comparable to that of classical model organisms like rodents, zebrafish, and fruit fly.

References

Almeida AM, Bassols A, Bendixen E, Bhide M, Ceciliani F, Cristobal S, Eckersall PD, Hollung K, Lisacek F, Mazzucchelli G, McLaughlin M, Miller I, Nally JE, Plowman J, Renaut J, Rodrigues P, Roncada P, Staric J, Turk R (2015) Animal board invited review: advances in proteomics for animal and food sciences. Animal 9(1):1–17. https://doi.org/10.1017/s1751731114002602 (Epub 2014 Oct 31 (Review))

Blackstock WP, Weir MP (1999) Proteomics: quantitative and physical mapping of cellular proteins. Trends Biotechnol 17(3):121–127 (Review)

Brückner A, Polge C, Lentze N, Auerbach D, Schlattner U (2009) Yeast two-hybrid, a powerful tool for systems biology. Int J Mol Sci 10(6):2763–2788. https://doi.org/10.3390/ijms10062763 (Review)

Carvalho ME, Gasparin G, Poleti MD, Rosa AF, Balieiro JC, Labate CA, Nassu RT, Tullio RR, Regitano LC, Mourão GB, Coutinho LL (2014) Heat shock and structural proteins associated with meat tenderness in Nellore beef cattle, a Bos indicus breed. Meat Sci 96(3):1318–1324. https://doi.org/10.1016/j.meatsci.2013.11.014 (Epub 2013 Nov 22)

Causier B (2004) Studying the interactome with the yeast two-hybrid system and mass spectrometry. Mass Spectrom Rev 23(5):350–367 (Review)

Chandrasekhar K, Dileep A, Ester Lebonah D, Kumari JP (2014) A short review on proteomics and its applications. Int Lett Nat Sci 17:77–84

Cottrell JS (2011) Protein identification using MS/MS data. J Proteomics. 74(10):1842–1851. https://doi.org/10.1016/j.jprot.2011.05.014 (Epub 2011 May 15)

Graves PR, Haystead TAJ (2002a) Molecular biologist's guide to proteomics. Microbiol Mol Biol Rev 66 (1):39–63

Graves PR, Haystead TA (2002) Molecular biologist's guide to proteomics. Microbiol Mol Biol Rev 66(1): 39–63 (table of contents (Review))

Higashiura A, Ohta K, Masaki M, Sato M, Inaka K, Tanaka H, Nakagawa A (2013) High-resolution X-ray crystal structure of bovine H-protein using the high-pressure cryocooling method. J Synchrotron Radiat 20(Pt 6):989–993. https://doi.org/10.1107/S090904951302373X (Epub 2013 Oct 5)

Hollung K, Veiseth E, Jia X, Faergestad EM, Hildrum KL (2007) Application of proteomics to understand the molecular mechanisms behind meat quality. Meat Sci 77:97–104

Jayasri K, Padmaja K, Prasad PE (2014) Proteomics in animal health and production. IOSR J Agri Vet Sci 7 (4):50–56

Kumar A, Rout PK, Mohanty BP (2013) Identification of Milk Protein Polymorphism in Indian Goats by 2D Gel Electrophoresis. J Proteomics Bioinform 6(1): 001–004

Kurpińska AJ, Skrzypczak WF (2014) Proteomic studies in pregnant and lactating cows. A review. J Anim Feed Sci 23(3):203–211

Manjasetty BA, Bussow K, Panjikar S, Turnbull AP (2012) Current methods in structural proteomics and its applications in biological sciences. 3 Biotech 2:89–113

Mann M, Jensen ON (2003) Proteomic analysis of post-translational modifications. Nat Biotechnol 21 (3):255–261 (Review)

Okada T, Le Trong I, Fox BA, Behnke CA, Stenkamp RE, Palczewski K (2000) X-Ray diffraction analysis of three-dimensional crystals of bovine rhodopsin obtained from mixed micelles. J Struct Biol 130(1):73–80

Oskoueian E, Eckersall PD, Bencurova E, Dandekar T (2016) Application of proteomic biomarkers in livestock disease management. In: Salekdeh GH (ed) Agricultural proteomics, vol 2. Environmental stresses. Springer International Publishing, Switzerland, pp 299–310

Rabilloud T, Lelong C (2011) Two-dimensional gel electrophoresis in proteomics: a tutorial. J Proteomics. 74(10):1829–1841. https://doi.org/10.1016/j.jprot.2011.05.040 (Epub 2011 Jun 12)

Rajagopala SV, Sikorski P, Caufield JH, Tovchigrechko A, Uetz P (2012) Studying protein complexes by the yeast two-hybrid system. Methods 58(4):392–399

Rehm T, Huber R, Holak TA (2002) Application of NMR in structural proteomics: screening for proteins amenable to structural analysis. Structure 10(12):1613–1618

Reinhardt TA, Lippolis JD, Nonnecke BJ, Sacco RE (2012) Bovine milk exosome proteome. J Proteomics. 75(5):1486–1492. https://doi.org/10.1016/j.jprot.2011.11.017 (Epub 2011 Nov 23)

Rodrigues RT, Chizzotti ML, Vital CE, Baracat-Pereira MC, Barros E, Busato KC, Gomes RA, Ladeira MM, Martins TD (2017) Differences in beef quality between angus (Bos taurus taurus) and Nellore (Bos taurus indicus) cattle through a proteomic and phosphoproteomic approach. PLoS One 12(1):e0170294. https://doi.org/10.1371/journal.pone.0170294 (eCollection 2017)

Roncada P, Piras C, Soggiu A, Turk R, Urbani A, Bonizzi L (2012) Farm animal milk proteomics. J Proteomics 75 (14):4259–4274. https://doi.org/10.1016/j.jprot.2012.05.028 (Epub 2012 May 26. Review)

Sarsaifi K, Haron AW, Vejayan J, Yusoff R, Hani H, Omar MA, Hong LW, Yimer N, Ju TY, Othman AM (2015) Two-dimensional polyacrylamide gel electrophoresis of Bali bull (Bos javanicus) seminal plasma proteins and their relationship with semen quality. Theriogenology 84(6):956–968. https://doi.org/10.1016/j.theriogenology.2015.05.035 (Epub 2015 Jun 6)

Shin J, Lee W, Lee W (2008) Structural proteomics by NMR spectroscopy. Expert Rev Proteomics. 5(4):589–601. https://doi.org/10.1586/14789450.5.4.589 (Review)

Soares R, Franco C, Pires E, Ventosa M, Palhinhas R, Koci K, Martinho de Almeida A, Varela Coelho A (2012) Mass spectrometry and animal science: protein identification strategies and particularities of farm animal species. J Proteomics 75(14):4190–4206. https://doi.org/10.1016/j.jprot.2012.04.009

Soggiu A, Piras C, Hussein HA, De Canio M, Gaviraghi A, Galli A, Urbani A, Bonizzi L, Roncada P (2013) Unravelling the bull fertility proteome. Mol BioSyst 9(6):1188–1195. https://doi.org/10.1039/c3mb25494a (Epub 2013 Feb 7)

Suter B, Kittanakom S, Stagljar I (2008) Two-hybrid technologies in proteomics research. Curr Opin Biotechnol 19(4):316–323. https://doi.org/10.1016/j.copbio.2008.06.005 (Epub 2008 Jul 23. Review)

Yee A, Chang X, Pineda-Lucena A, Wu B, Semesi A, Le B, Ramelot T, Lee GM, Bhattacharyya S, Gutierrez P, Denisov A, Lee CH, Cort JR, Kozlov G, Liao J, Finak G, Chen L, Wishart D, Lee W, McIntosh LP, Gehring K, Kennedy MA, Edwards AM, Arrowsmith CH (2002) An NMR approach to structural proteomics. Proc Natl Acad Sci U S A 99(4):1825–1830

Zhu H, Qian J (2012) Applications of functional protein microarrays in basic and clinical research. Adv Genet 79:123–155. https://doi.org/10.1016/B978-0-12-394395 8.00004-9 (Review)

Metabolomics in Livestock Sciences 35

Abstract

Metabolomics is the systematic study of all the metabolites present in a cell, tissue, or organ. It is as an important functional genomics tool which provides a complete picture of a living organism. The amalgamation of metabolomics with other omic-based approaches is of growing interest, as it leads to a better comprehensive understanding of physiology of an organism. The metabolome is the end product of all the biological processes; thus, it can also explain the link between genotype and phenotype. This expanding technology can aid the disease research, pharmaceutical drug development and development of healthier and safer foods, and routinely used personal care products.

Highlights

- Metabolomics is the study of entire range of metabolites including small molecules in cells, tissue, and biofluids
- The technique is based on a number of advanced analytical instrumentation and bioinformatics tools
- Metabolomics allows researchers to focus on measuring end products, and use the inferences to detect diseases and livestock monitoring.

Keywords

Metabolomics · Metabolite profiling · Biomarker discovery · Metabolome database · Disease diagnosis · Livestock health

35.1 Introduction

Metabolomics is the comprehensive, qualitative, and quantitative study of all the metabolites in an organism which are influenced by both genetic and environmental factors. Metabolites are the small molecules whose mass ranges from 50 to 1500 daltons (Da), such as sugars, lipids, amino acids, and fatty acids. In other words, metabolites are all the substrates and products of metabolic pathways in an organism. They are the end products of complex interactions occurring inside the cell (the genome) and outside the cell or organism (the environment). Hence, metabolomics is the study that permits a highly sensitive and complete description of the phenotype.

In 1998, another term "metabolome" was coined and used to describe the metabolites of living tissues (Oliver et al. 1998). Thus, "metabolome" is defined as the complete set of metabolites within a cell, tissue, or biological sample at a given point of time. It is noteworthy that metabolites exclude polymers of amino acids

B. Singh et al., *Advances in Animal Biotechnology*,
https://doi.org/10.1007/978-3-030-21309-1_35

and sugars. It just includes the intermediary substrates and products used to form the macromolecular structures. It also comprises the other small molecules which participate in important metabolic functions and has critical roles in cellular activities such as metabolic intermediates, hormones, signaling molecules, or secondary metabolites.

Metabolites of foreign origin such as drugs are known as "xenometabolites." Thus, metabolomics is useful to assess responses to environmental stresses, nutrition, comparing mutants, toxicology, diabetes, cancer, drug discovery, studying global effects of genetic manipulation, comparing different growth stages, and natural product discovery (Zhang et al. 2012).

Metabolomics can be studied through two approaches: untargeted and targeted. Untargeted approach requires an unbiased detection and quantification of all of the metabolites of a biological sample. On the other hand, targeted approach involves the identification of specific metabolites which are already known and have some biological function. However, the analysis of metabolites is not very straightforward as they are generally labile species and chemically very diverse and often present in a wide dynamic range. This implies that the study of metabolomics is very complex and vast, but when used in combination with genomics, transcriptomics, or proteomics, this could lead to a deeper understanding about an organism's organization and function. Thus, metabolomics provides a snapshot of the metabolic state of an organism at a given time.

Metabolomics studies generate copious amount of data and various databases are available online to store and retrieve this data such as Livestock Metabolome Database (LMDB), Bovine Metabolome Database (BMDB), Human Metabolome Database (HMDB), and Kyoto Encyclopedia of Genes and Genomes (KEGG). The LMDB and BMDB are freely available electronic databases which contain the detailed information about metabolites present in milk. The LMDB contains 1070 metabolite entries including both lipid soluble and water-soluble metabolites. It also supports extensive text, sequence, chemical structure, and relational query searches (Goldansaz et al. 2017).

35.2 Types of Metabolomics

There are several ways to study metabolomics; mainly three approaches for the metabolome analysis are used: metabolite profiling, metabolic fingerprinting, and metabonomics.

35.3 Metabolite Profiling

The concept of "metabolic profiling" was first introduced in 1971 using mass spectrometry (Horning and Horning 1971). It is a hypothesis-driven approach that focuses on the analysis of a large group of metabolites either related to a specific metabolic pathway (example: glycolysis) or a class of compounds (such as lipids). It basically identifies and quantifies the metabolome (Bino et al. 2004). Therefore, it can provide a more comprehensive description of metabolic perturbations than information offered by a single biomarker. Consequently, the data obtained from "metabolite profiling" can be integrated with other "omics" data or pathway maps, which further enhance the biological understanding (Dettmer et al. 2007). Profiling of metabolites is often performed using Liquid chromatography (LC) or gas chromatography (GC) coupled with mass spectrometry (MS) (Schuhmacher et al. 2013).

35.4 Metabolic Fingerprinting

"Metabolite fingerprinting" is the rapid classification of samples to compare patterns or "fingerprints" of metabolites that change in a given biological system such as in response to disease, toxin exposure, environmental, or genetic alteration. But, it does not necessarily give specific metabolite information. The analytical methods used for metabolite fingerprinting are direct infusion MS (without prior separation of the sample constituents), NMR spectroscopy, and IR spectroscopy (Schuhmacher et al. 2013).

35.5 Metabonomics

Metabonomics is defined as "quantitative measurement of the dynamic multiparametric metabolic response of living systems to pathophysiological stimuli or genetic modification." It is the study of the interactions of metabolites over a period of time in a complex system. It can provide information on gene function, drug toxicology, and disease diagnosis (Faber et al. 2007). This approach is usually restricted to microbiological and non-botanical studies.

35.6 Methods of Metabolomics

There are several analytical methods to study metabolomics. The important criteria for selection of a method are dynamic range, accuracy, precision, selectivity, coverage, detection limit, and price per sample. The most popular methods are gas chromatography–mass spectrometry (GC-MS), liquid chromatography–mass spectrometry (LC-MS), and capillary electrophoresis–mass spectrometry (CE-MS). The recent development of a range of analytical platforms including MS, HPLC, UPLC, CE, and NMR spectroscopy enhance the separation, detection, characterization, and quantification of metabolites and related metabolic pathways. Still, there is not a single analytical platform that can be applied to detect all the metabolites in a biological sample due to the complexity of the metabolome and the diverse properties of metabolites. Till date, there is not a software tool that can comprehensively translate all the data into a biologically meaningful context. Thus, multiple approaches are required to increase the coverage of detected metabolites that cannot be achieved by single-analysis techniques. For example, the use of multiple metabolomics platforms and technologies, such as NMR spectroscopy, gas chromatography–mass spectrometry (GC-MS), inductively coupled plasma–mass spectroscopy (ICP-MS), direct flow injection (DFI) mass spectrometry, and lipidomics, substantially enhances the level of metabolomic coverage (Saleem et al. 2013).

35.7 Nuclear Magnetic Resonance Spectroscopy

Nuclear magnetic resonance (NMR) spectroscopy is the most widely used technique of metabolomics and provides direct information on chemical structure. It is a robust, non-destructive, relatively fast, highly reproducible high-throughput analytical platform that requires minimal sample preparation (Sotelo and Slupsky 2013). NMR spectroscopy is commonly applied for untargeted metabolomic studies and the absolute concentration of each metabolite can also be measured from an NMR spectrum. However, it suffers from relatively low sensitivity as compared to the mass spectrometry.

35.8 Mass Spectrometry

Mass spectrometry (MS) techniques are very sensitive and selective to characterize the metabolome of an organism with high throughput and depth of coverage (Aretz and Meierhofer 2016). It is also used to characterize the complex metabolic effects of foods or nutrients. It allows the identification of both targeted profiling (in which metabolites are known a priori) and fingerprinting (the identity of the metabolites of interest is established a posteriori) (Scalbert et al. 2009). This technique has considerable potential, but still lack well established and standardized procedures. There are also some difficulties in detection of the differentiating metabolites. Therefore, to better analyze the different types of metabolites in biological samples, proper metabolite separation is needed. For this, a chromatographic technique is performed before MS, such as gas chromatography–mass spectrometry (GC-MS), liquid chromatography–mass spectrometry (LC-MS), and ultra-performance liquid chromatography–mass spectrometry (UPLC-MS).

35.9 Gas Chromatography-Mass Spectrometry

Gas chromatography–mass spectrometry (GC-MS) is applied to metabolites which have a low boiling point and which can remain in the gas phase at the temperature range of 50-350°C. This approach is commonly applied in metabolomics and detect copious amount of metabolites in a biological sample. Universal electron ionization (EI) is most commonly used in GC-MS. The advantages of this technique are high separation efficiency, greater resolution and reproducible retention times, which can be compared between different laboratories via the retention index concept using retention time markers (Scalbert et al. 2009). The disadvantage of GC-MS is that some amino acids are not very stable and their derivatization can degrade during injection and separation.

35.10 Liquid Chromatography–Mass Spectrometry

Liquid chromatography–mass spectrometry (LC-MS) has several advantages over other techniques as in liquid chromatography no derivatization is required for the analysis of polar or high molecular weight metabolites. The specificity of LC-MS depends upon chromatographic retention time (RT) and mass-to-charge ratio (m/z). There are many types of LC-MS systems that limit the establishment of a single optimized analytical procedure. This method generally has poorer reproducibility for retention time compared to GC, thus it hampers the comparison of LC-MS chromatograms between laboratories (Moco et al. 2006).

35.11 Ultra-Performance Liquid Chromatography-Mass Spectrometry

In UPLC-MS, standard HPLC upgrades to ultra-performance liquid chromatography. UPLC systems can significantly increase the resolution sensitivity and peak capacity due to the reduced particle size, while decreasing sample volumes and mobile phases. UPLC can operate at high operating pressures and use sub-2-μm porous packing.

35.12 Capillary Electrophoresis–Mass Spectrometry

Capillary electrophoresis–mass spectrometry (CE-MS) is considered a relatively new analytical technique in the field of metabolomics for the profiling of highly polar and ionic metabolites in ultra-small complex biological samples. This technique is also known as capillary zone electrophoresis-mass spectrometry. In CE, metabolites are first separated based on charge and size in an electric field and, then, selectively detected using MS by screening ions over a large range of m/z values (Soga, 2007). All the metabolites should be charged to allow the mobility to occur. CE-MS became a complementary approach to LC-MS for analyzing cations, anions, and neutral particles in a single run. In contrast to chromatography-based methods, the separation efficiency of CE is very high as it does not involve mass transfer between phases. Unfortunately, the use of CE-MS is low for metabolomics studies due to relatively poor reproducibility and sensitivity (Zhang et al. 2017).

All the techniques used in metabolomic studies have their own advantages and disadvantages. They also have different mechanisms to separate and detect the different sets of metabolites. There is no single technique yet that can detect all the metabolites. Therefore, applying more than one analytical technique can only provide detection of a greater number of metabolites.

35.13 Applications of Metabolomics

Metabolomics has many applications in all the field of biological sciences. It is useful to assess the responses to environmental stresses, nutrition, comparing mutants, toxicology, diabetes, cancer, drug discovery, studying global effects of genetic manipulation, comparing different growth stages, and natural product discovery (Zhang et al. 2012). These days, it is increasingly used in livestock research and monitoring. Fundamental knowledge of animal metabolism provides an important foundation for health, nutrition, production, reproduction, physiology, and products from livestock such as milk, meat, and yogurt.

35.14 Biomarker Discovery

Biomarkers, in metabolomics, are small molecules that can be used to distinguish two groups of samples, typically a disease and control group. For example, several potential biomarkers have been discovered in the rumen, serum, urine, and milk from dairy cows aiming to discover for milk quality under two types of forage using GC-TOF/MS-based metabolomics (Sun et al. 2015). Furthermore, hippuric acid (HUA) and N-methyl-glutamic (NML-Glu) identified as potential biomarkers for discriminating alfalfa hay and corn stover diets and could be used for milk yield-related mechanistic investigations (Sun et al. 2016). Similarly, metabolomics analysis of follicular fluid of Nellore cows using high-resolution mass spectrometry has provided biomarkers that can differentiate between high and low fertility cows (Guerreiro et al. 2018). The use of metabolome as biomarkers is also gradually increasing in drug discovery and development and is leading to the discovery of new and improved therapeutic strategies for many serious and life-threatening diseases (Yeung, 2018).

35.15 Physiological and Metabolic Mechanisms

Metabolomics remains successful to access the physiological and metabolic state of livestock in different physiological conditions. It is also focused on measuring the end products of complex, hard-to-decipher genetic, epigenetic, and environmental interactions. A metabolomics study conducted to analyze the metabolomics differences in the milk between heat stress-free and heat stress dairy cows using 1H-NMR spectroscopy, and LC-MS has discovered several differential metabolites. These include the metabolites involved in pathways of amino acid, carbohydrate, lipid, and gut microbiome-derived metabolism (Tian et al. 2016). Metabolic pathway analysis of milk synthesis of Chinese Holstein, Jersey, buffalo, yak, goat, horse, and camel revealed that Chinese Holstein shares glycerophospholipid metabolism and biosynthesis of valine, leucine, and isoleucine with ruminant animals (Jersey, buffalo, yak, and goat). It also shares unsaturated fatty acids biosynthesis with the non-ruminant animals (camel and horse). Choline and succinic acids are two metabolites identified that can distinguish Holstein milk from other animals (Yang et al. 2016).

35.16 Early Disease Diagnosis and Improved Treatment Strategies

Studying the complete metabolome of an organism can lead to the better understanding of a disease state, its prevention and cure. For example, analysis of plasma metabolomics of type I ketotic, type II ketotic and healthy cows by 1H-nuclear magnetic resonance technology (1H NMR) revealed differential metabolic profile that could lead to early diagnosis and prevention of type I and type II ketosis in dairy cows (Xu et al. 2015a, b). NMR-based metabolomics is also successfully used to diagnose an etiologic agent of Caseous Lymphadenitis (*Corynebacterium pseudotuberculosis*) in sheep, which cause enormous damages such as reduction in milk, meat, and wool production or death of infected sheep (De Moraes Pontes et al. 2017). It is noticed that the metabolomic profile of milk is dramatically affected by udder health status and environmental temperature. Milk lactate and citrate are the detected markers for udder inflammation and heat stress, respectively (Love et al. 2016). Therefore, metabolomics offers the potential for quick identification of hundreds of metabolites in a given biological samples (biofluid, tissue, cells, etc.) enabling us to diagnose disease states much earlier.

35.17 Outlook and Challenges

Metabolomics has profound capacity to identify the potential biomarkers, drug discovery, disease diagnosis, quality control of dairy products and to identify the molecular basis of environmental stress. There are several approaches for metabolomic discovery and analysis such as NMR, GC-MS, LC-MS, and CE-MS. Although individually these different approaches may not provide interpretable outcome, all generate abundant useful information that can be merged to achieve the ultimate goal. Metabolic studies are mainly hampered by its diversity, variation of metabolite concentration by several orders of magnitude and biological data interpretation. Still, recent technical advances in the techniques could lead to better comprehensive knowledge of "Metabolomics."

References

Aretz I, Meierhofer D (2016) Advantages and pitfalls of mass spectrometry based metabolome profiling in systems biology. Int J Mol Sci 17(5). pii: E632. https://doi.org/10.3390/ijms17050632 (Review)

Bino RJ, Hall RD, Fiehn O, Kopka J, Saito K, Draper J, Nikolau BJ, Mendes P, Roessner-Tunali U, Beale MH, Trethewey RN, Lange BM, Wurtele ES, Sumner LW (2004) Potential of metabolomics as a functional genomics tool. Trends Plant Sci 9(9):418–425 (No abstract available)

De Moraes Pontes JG, De Santana FB, Portela RW, Azevedo V, Poppi RJ, Tasic L (2017) Biomarkers of the *Caseous Lymphadenitis* in sheep by NMR-based metabolomics. Metabolomics 7(2):1–7

Dettmer K, Aronov PA, Hammock BD (2007) Mass spectrometry-based metabolomics. Mass Spectrom Rev 26(1):51–78 (Review)

Faber JH, Malmodin D, Toft H, Maher AD, Crockford D, Holmes E, Nicholson JK, Dumas ME, Baunsgaard D (2007) Metabonomics in diabetes research. J Diabetes Sci Technol. 1(4):549–557

Goldansaz SA, Guo AC, Sajed T, Steele MA, Plastow GS, Wishart DS (2017) Livestock metabolomics and the livestock metabolome: A systematic review. PLoS ONE 12(5):e0177675. https://doi.org/10.1371/journal.pone.0177675

Guerreiro TM, Gonçalves RF, Melo CFOR, de Oliveira DN, Lima EO, Visintin JA, de Achilles MA, Catharino RR (2018) A metabolomic overview of follicular fluid in cows. Front Vet Sci 5:10. https://doi.org/10.3389/fvets.2018.00010 (eCollection 2018)

Horning EC, Horning MG (1971) Human metabolic profiles obtained by GC and GC/MS. J Chromatogr Sci 9(3):129–140

Love S, Salama A, Mehaba N, Caja G (2016) Milk metabolomics of dairy goats with mammary inflammation under heat stress conditions. J Anim Sci 94 (5):616

Moco S, Bino RJ, Vorst O, Verhoeven HA, de Groot J, van Beek TA, Vervoort J, de Vos CH (2006) A liquid chromatography-mass spectrometry-based metabolome database for tomato. Plant Physiol 141 (4):1205–1218

Oliver SG, Winson MK, Kell DB, Baganz F (1998) Systematic functional analysis of the yeast genome. Trends Biotechnol 16(9):373–378 (Review. Erratum in: Trends Biotechnol 1998 Oct;16(10):447)

Saleem F, Bouatra S, Guo AC, Psychogios N, Mandal R, Dunn SM, Ametaj BN, Wishart DS (2013) The bovine ruminal fluid metabolome. Metabolomics 9(2):360–378

Scalbert A, Brennan L, Fiehn O, Hankemeier T, Kristal BS, van Ommen B, Pujos-Guillot E, Verheij E, Wishart D, Wopereis S (2009) Mass-spectrometry-based metabolomics: limitations and recommendations for future progress with particular focus on nutrition research. Metabolomics 5:435–458

Schuhmacher R, Krska R, Weckwerth W, Goodacre R (2013) Metabolomics and metabolite profiling. Anal Bioanal Chem 405(15):5003–5004. https://doi.org/10.1007/s00216-013-6939-5 (No abstract available)

Soga T (2007) Capillary electrophoresis-mass spectrometry for metabolomics. In: Weckwerth W (eds) Metabolomics. Methods in molecular biology, vol 358, pp 129–137. Humana Press

Sotelo J, Slupsky CM (2013) Metabolomics using nuclear magnetic resonance (NMR). In: Metabolomics in food and nutrition, pp 29–43. University of California, Davis, USA. Woodhead Publishing Limited

Sun HZ, Wang DM, Wang B, Wang JK, Liu HY, le Guan L, Liu JX (2015) Metabolomics of four biofluids from dairy cows: potential biomarkers for milk production and quality. J Proteome Res 14(2):1287–1298. https://doi.org/10.1021/pr501305g (Epub 2015 Jan 28)

Sun H, Wang B, Wang J, Liu H, Liu J (2016) Biomarker and pathway analyses of urine metabolomics in dairy cows when corn stover replaces alfalfa hay. J Anim Sci Biotechnol. 7(1):49. https://doi.org/10.1186/s40104-016-0107-7 (eCollection 2016)

Tian H, Zheng N, Wang W, Cheng J, Li S, Zhang Y, Wang J (2016) Integrated metabolomics study of the milk of heat-stressed lactating dairy cows. Sci Rep 6 (6):24208. https://doi.org/10.1038/srep24208

Xu C, Li Y, Xia C, Zhang HY, Sun LW, Xu CC (2015a) 1H NMR-based plasma metabolic profiling of dairy cows with Type I and Type II ketosis. Pharmaceutica Analytica Acta 6(2):328

Xu C, Shu S, Xia C, Wang P, Sun Y, Xu C, Li C (2015b) Mass spectral analysis of urine proteomic profiles of dairy cows suffering from clinical ketosis. Vet Q. 35(3):133–141. https://doi.org/10.1080/01652176.2015.1055352 (Epub 2015 Jun 18)

Yang Y, Zheng N, Zhao X, Zhang Y, Han R, Yang J, Zhao S, Li S, Guo T, Zang C, Wang J (2016) Metabolomic biomarkers identify differences in milk produced by Holstein cows and other minor dairy animals. J Proteomics. 16(136):174–182. https://doi.org/10.1016/j.jprot.2015.12.031 (Epub 2016 Jan 11)

Yeung PK (2018) Metabolomics and biomarkers for drug discovery. Metabolites 8:11

Zhang A, Sun H, Wang P, Han Y, Wang X. Modern analytical techniques in metabolomics analysis. Analyst. 2012 Jan 21;137(2):293–300. https://doi.org/10.1039/c1an15605e (Epub 2011 Nov 21. Review)

Zhang W, Hankemeier T, Ramautar R (2017) Next-generation capillary electrophoresis-mass spectrometry approaches in metabolomics. Curr Opin Biotechnol 43:1–7. https://doi.org/10.1016/j.copbio.2016.07.002 (Epub 2016 Jul 22. Review)

Synthetic Biology 36

Abstract

Developments in sequencing, "OMICS" and genetic engineering have given rise to several innovative concepts and research in biological sciences. The field of synthetic biology has made rapid progress from a concept to an established discipline with immense industrial and biotechnological applications besides its role in environmental, agriculture and veterinary sciences.

Highlights

- Synthetic biology is capable of altering natural genomes with precision
- The approach as multiple benefits in animal health, dairy, and food production.

Keywords

Synthetic biology · Artificial cells · Artificial microbes · Genome engineering · Microbial engineering · Novel therapeutics

36.1 Introduction

Synthetic biology is an emerging collaborative and trans-disciplinary field at the intersection of engineering and basic sciences such as bioinformatics, metabolic engineering, nanotechnology, computer science, and chemical engineering. Just as synthetic organic chemistry had once revolutionized the synthesis of new molecule, that under no circumstances existed naturally, the cell-free synthetic biology has provided the means to improve toolbox and processes to not only harness and expand the biological processes (Moreno 2012; Harris and Jewett 2012).

Synthetic biology offers development of new biological systems and devices and reshapes existing biological systems for more useful purposes. Synthetic or artificial biology tools are used to develop novel biological systems ranging from simpler proteins to cell organelles and cells. The purpose of synthetic biology is to make biological processes competitive and proficient for human and animal welfare. Researchers and experts of the convergent fields of chemistry, biology, computer science, and engineers share a common platform to lay foundations in the field of medicine, environment, and energy.

During past few decades, the synthetic biology has progressed remarkably (Cameron et al. 2014). These early achievements in genetic engineering paved the means for modern synthetic biology. Indeed, advent of recombinant DNA technology has spurred the concept of developing synthetic minimal cells (Glass et al. 2017), by assembling minimum possible cell components that are necessary for survival of a cell (Fig. 36.1) and recombinant genome to produce desirable substances.

In 2016, the researchers at J. Craig Venter Research Institute (JCVI), Maryland and California, designed a new meaning of life, i.e., the

B. Singh et al., *Advances in Animal Biotechnology*,
https://doi.org/10.1007/978-3-030-21309-1_36

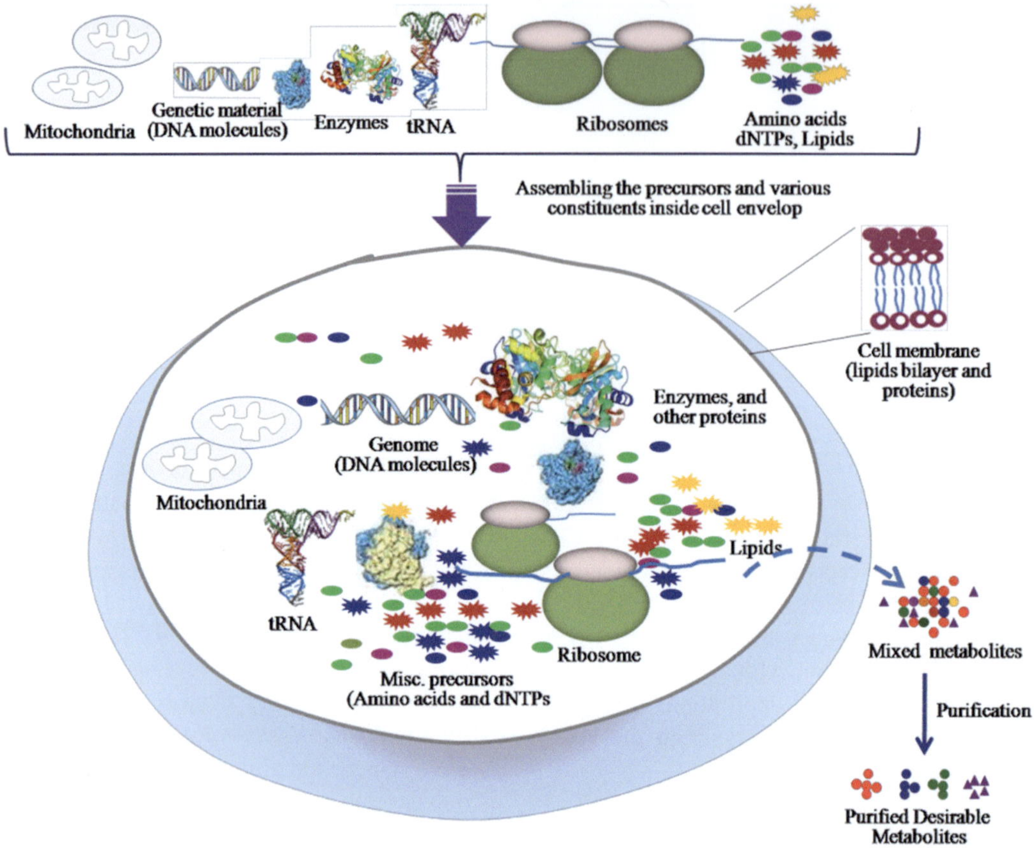

Fig. 36.1 A diagrammatic illustration of a synthetic cell possessing minimal essential components. Understanding the functions of genes and economic production of desirable metabolites are among prime objectives of synthetic cells

minimal number of genes required for bacterial cell to survive. The work proved the principle for producing cells based on computer-designed genome sequences. Whole-genome design and complete chemical synthesis approaches were used to minimize 1079-kilobase pair synthetic genome of *Mycoplasma mycoides* JCVI-syn1.0. The newly developed strain (JCVI-syn3.0 (531 kilobase pairs, 473 genes) with much smaller genome had almost all the essential components needed for sustenance (Hutchison 2016).

The approach of synthetic biology provides how new cells that are rationally engineered to act as specialized chassis. These re-engineered organisms will change lives over the coming years leading to production of therapeutics at low cost, and compounds meant for use as fuels and targeted therapies against superbugs and diseases, viz. cancer metabolic and cognitive disorders. The de novo bioengineered genetic circuits, biological modules, and synthetic pathways have already made their impact in the field of biomedical and clinical sciences (Box 1).

Box 1. Summary of Some Remarkable Advances in Synthetic Biology

- Precise transcription control of transgenes that forms the functional basis for designing synthetic gene networks and heterologous transcription control circuits,
- Creating transgene expression circuits that could be monitored by endogenous cell metabolites,

- Development of electricity-inducible gene circuits (Webber et al. 2008),
- Development of designer bacteria with knack to target and destroy or locate tumors (Singh et al. 2017),
- Evolution of new synthetic microbial species with a minimal set of known genes,
- Developing novel gene circuits for biological applications (Xie and Fussenegger 2018),
- Development of non-biological polymers with new structure and properties, and functions,
- Producing synthetic probiotic-like microbes against *Vibrio cholera* (Bugaj and Schaffer 2012),
- Developing of highly proficient photosynthetic microbial or cyanobacterial cell factories aimed to generate biofuel (Masukawa et al. 2012),
- Design and development of bacteria with minimal synthetic genome (Hutchison et al. 2016),
- Metabolic engineering of microorganisms for producing biomolecules for cell engineering and treating medical disorders (Katz et al. 2018).

36.2 Experimental Approaches

Synthetic biology has been divided into two approaches that utilize metabolic engineering and genetic techniques to impart new functions in living cells (in vivo), and the second one is based on creating new biological systems by combining together the non-living components (in vitro).

36.3 In Vivo Approach

In this approach, cellular system is the subject of engineering. Microorganisms are engineered for large-scale production or biotransformation systems. Efforts are made to reprogram or redesign mammalian cells to produce complex proteins. Two approaches, namely "top-down approach" and "bottom-up approach," are used. In "top-down approach," the cells are simplified by removing certain components from them. It helps to understand minimum number of genetic instructions and genes essential for an organism. Using minimal genome knowledge, it is possible to design new cell factories.

In "bottom-up approach," synthetic biological components are introduced into a cell for creating essentially minimal genome. New genetic information is introduced in order to modify the system so that it can carry the instructions for making types of protein unknown in nature.

36.4 In Vitro Approach

In this approach, a non-cellular biological system is the subject of engineering. The synthesized biological components or structures are assembled into a functioning entity. Systems created using this approach are based on biological building blocks such as amino acids and lipids or similar molecules. The large number of possible building blocks offers combinations for engineering non-cellular biological system. The disadvantage of these systems is that they are unable to replicate.

36.5 Role of Synthetic Biology

The biological systems are unparalleled in their abilities to produce complex biomolecules and polymers naturally by utilizing simpler precursors. This ability of biological system is exploited by manipulating the fundamental machinery of living systems to produce biomolecules with altered architecture, bioavailability and activities. Genetic engineering and synthetic biology are two major tools that have revolutionized manufacture and availability of novel metabolites such as drugs, vaccines, antibiotics, and stem cells.

36.6 Development of Novel Therapeutic

Dearth of progress in development of new antibiotics, and continually emerging resistance toward existing antibiotics, is the main challenge in context of current medication and necessitates the search for alternative therapeutic interventions to combat microbial pathogens and infectious diseases (Singh et al. 2017; Guzmán-Trampe et al. 2017). Further, low specificity and toxicity of some of available antimicrobial compounds have made imperative the search for new antibiotics with enhanced bioactivity.

Synthetic mammalian gene circuits (Weber et al. 2008) and living cells are used as platforms for applications of synthetic biology to develop novel antimicrobial pathways, antimicrobial peptides (AMPs), non-traditional antimicrobials (Zakeri and Lu 2013), redesigning RNA and DNA molecules to create new synthetic systems, and engineer antibiotic clusters for overexpression of antibiotics or improving the functions of existing antimicrobial genetic system (Guzmán-Trampe et al. 2017).

Synthetic biology and bioengineering (Fig. 36.2) have been used to develop recombinant biotherapeutics or improve the efficacy of probiotic microorganisms targeting human clinical applications (Ozdemir et al. 2018; Duo and Bennet 2018).

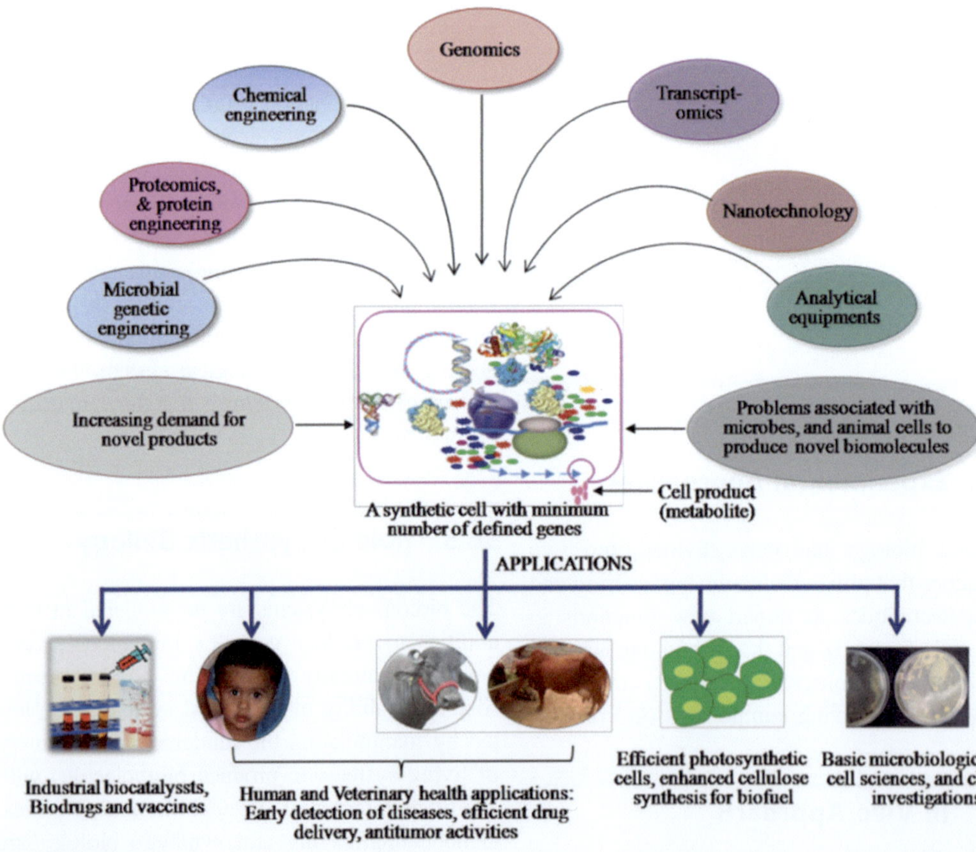

Fig. 36.2 Factors contributing to development and advancement of synthetic biology and its applications in biomedical, veterinary sciences

36.7 Biofuels

Plants, algae, and cyanobacteria are designed to enhance their ability to trap and accumulate hydrocarbons in them for use as ingredient to generate biofuel. Efforts are being made to develop microbial systems equipped with artificial or de novo biosynthetic pathways that make them highly proficient in breakdown of abundant renewable energy source—the cellulose biomass to produce biofuels that are less expensive and environmentally sustainable (Okano et al. 2018).

Cyanobacteria and microalgae are used to produce numerous industrial and biotechnological products. Genetic engineering and genome precision technique are applied to develop strains with superior properties such as enhanced photosynthesis through increased RuBisCO activity and truncation of light-harvesting antennae (Gomaa et al. 2016). Certain algae produce oil. However, harvesting oil from these tiny species is difficult and expensive. With the help of synthetic biology, it is possible to construct genetic circuits to enhance genetic capabilities of algae such as *Nannochloropsis gaditana* to increase oil production.

36.8 Environment

Contaminants and pollutants are major threats to environment. Synthetic biology is expected to get rid of the problem and improve environmental health. New microbes can be designed with enhanced ability to consume toxic chemicals from water or soil that would not otherwise decompose. In addition, development of new eco-friendly products such as fatty acid-based biodiesel (Steen et al. 2010), gasoline (Choi and Lee 2013), and bio-plastics (Yim et al. 2011) is the prospective areas.

36.9 Agriculture

Efficient utilization of agricultural waste and plant residues for producing biofuel would lessen the load on crops used for the purpose. However, development of third-generation or "advanced bio-ethanol" from algae, rather than traditional land-grown crops, is desirable. Also, conversion to agriculture waste to quality compost will improve health of soil. Genetic engineering and synthetic biology might help in building biological systems that enhance the ability of plant-growth promoting phytomicrobiomes and plants to fix atmospheric nitrogen. This will lessen the dependence of farmers on chemical fertilizers (Martínez-Hidalgo et al. 2018). Synthetic biology is also used to redesign plants for producing novel food additives. Microbes are also engineered for artificial photosynthesis process.

36.10 Livestock

If milk was produced by a genetically engineered cow, or honey from a genetically engineered bee, the same holds for ingredients that are produced via synthetically engineered yeast or algae—especially in cases when they are intended for use as animal feed. The modern synthetic biology is logical extension of recombinant DNA technology, genome-editing tools supported by DNA computational tools, modern genome sequencing, proteomics, and nanotechnology (Xie and Fussenegger 2018).

Synthetic biology approaches in livestock genetic and breeding may deliver long-term genetic gains. For instance, horns are not important to animals managed under controlled farms and safe settings. The aggressive bulls pose threat to each other and caretakers at farms. One of the important envisaged uses of synthetic biology in livestock is producing hornless livestock to enhance safety of workers and animal welfare.

36.11 Conservation of Ecology and Biodiversity

Human activities have affected planet's natural system and have intensified during recent decades, causing disruption and transformation of natural systems including atmosphere, oceans, and terrestrial ecosystems. The changes have posed threat to human life and well-being (2018), and crops and food production systems (Myers et al. 2017).

A number of natural wild flora and fauna are endangered, and many have become extinct. Efforts such as in situ conservation and reproductive biotechniques are applied to conserve or repopulate wild animal species. Synthetic biology and conservation communities can fulfill equivalent goals and motives.

It is therefore imperative to protect natural systems to protect the health of future generations. Synthetic biology could slow down the adverse impact of anthropological activities overall biodiversity loss. Many alternative forms of life are being created and patented (Redford et al. 2013, 2014). For example, minimizing dependence on pesticides and herbicides is one of the most important and desirable contributions of synthetic biology.

Green biotechnology, genetic engineering, and synthetic biology driven by techniques such as CRISPR/Cas9, and cheap DNA synthesis may offer solutions to increasing demand of food, water, energy, and medicine (Mortimer 2018). Collaborative efforts of conservation communities and practitioners of synthetic biology are urgently needed to avoid or minimize deleterious ecological loss (Piaggio et al. 2017).

36.12 Manufacturing Animal Proteins

Synthetic biological approaches can enable us fulfil meat and dairy needs in sustainable manners. Animal proteins such as casein, β-lactoglobulin, and lacto-albumin can be synthesized using yeasts or other suitable hosts.

Producing meat without sacrificing animals is another important area where synthetic biology can contribute a lot. Also called as clean meat, the synthetic meat is a form of cellular meat, produced by growing the animal cells in vitro, and assemble them artificially in the form of the meat.

The plan for lab-grown burgers has won support from animal welfare and vegetarian groups, who feel that this approach addresses their concerns about animal suffering. For this, muscle stem cells are isolated from animals (cattle, buffalo, sheep, or goat) and allowed to grow in vitro. After few weeks, the cell mass is harvested and used as meat. This meat is free from bones and adipose tissue.

Clean meat is free from residual growth hormone, antibiotics, microbial contaminant such as *Escherichia coli*, *Salmonella* sp., and *Listeria monocytogenes*, above all avoid slaughter of animals. If scientists are successful in commercializing this technology in near future, it will be a big relief to environment, and methane emission is the major problem associated with meat producing ruminant animals. However, many issues are still to be resolved before artificial meat is brought to commercial production and using for human consumption.

36.13 Modern Meadow (Bio Leather)

Biofabrication industries are modern meadow which offers alternatives to leather of animals. Bacterial cellulose is a potential alternative to footwear industry (García and Prieto 2018). Genetic engineering and genome-editing tools can enhance capacity of microorganism to produce metabolites that can substitute for animal skin.

36.14 Personalized Medicine

Development of concept of personalized medicine, the prescription of therapeutics best befitting to an individual, is an important contribution of synthetic biology. Personalized medicine aims to improve health care based on early detection of disease, preventive medicine, drug discovery, and monitoring of treatment. Synthetic biology, by reprogramming patient-specific cell to an iPSCs state, will open new avenues in regenerative medicines. This will open new opportunities to utilize synthetic biology to advance personalized medicines (Xie and Fussenegger 2018).

Synthetic biology can be used to alter the genetics of parasitic protozoa such as *Theileria parva* that causes East coast fever in African cattle and *Mycoplasma* that causes contagious bovine pleuropneumonia (CBPP).

36.15 Outlook and Challenges

Synthetic biology along with synthetic chemistry has advanced bioproducts beyond their natural limitations. Synthetic biology has matured into a true engineering discipline to develop model microorganisms for industrial and biomedical applications. Immediate challenges include methods of improving ability of ribosomes to incorporate non-genetically encoded amino acids and enhancing the number of available codons for engineering in extract-based systems (Harris and Jewett 2012).

Synthetic biology should be utilized to develop microorganisms-based therapies to curtail the menace of antibiotic resistance in human, veterinary and zoonotic pathogens. Especially, the efforts should focus on personalized medicine by using cells as therapeutics.

Combining synthetic biology with traditional processes of drug developments will enhance the speed of discovery of new drug molecules, process design, and manufacturing. Some of the synthetic systems are in clinical trial stages, proving the proficiency of synthetic biology in terms of pharmacological activities, reduced cost of production of pharmaceuticals and their biosafety.

36.16 Conclusions

Cell-free synthetic biology as a remarkable discipline of engineering-governed biological cell development, ectopic expression of transgenes of biological origins, or synthetic genes has emerged as a powerful approach to understand, harness, and expand the potential of biological systems. Applications of synthetic biology in regenerative medicine, blood transfusion therapies, biofuel production, and biofertilizers have already proven to be a reality. Role of synthetic biology in animal production system has already begun, but should be expedited to improve health management by means of novel antibiotics, vaccines, and antibodies.

References

Bugaj LJ, Schaffer DV (2012) Bringing next-generation therapeutics to the clinic through synthetic biology. Curr Opin Chem Biol 16(3–4):355–361. https://doi.org/10.1016/j.cbpa.2012.04.009

Cameron DE, Bashor CJ, Collins JJ (2014) A brief history of synthetic biology. Nat Rev Microbiol 12(5):381–390. https://doi.org/10.1038/nrmicro3239

Choi YJ, Lee SY (2013) Microbial production of short-chain alkanes. Nature 502(7472):571–574. https://doi.org/10.1038/nature12536

Dou J, Bennett MR (2018) Synthetic biology and the gut microbiome. Biotechnol J 13(5):e1700159. https://doi.org/10.1002/biot.201700159 (Epub 2017 Oct 27)

García C, Prieto MA (2018) Bacterial cellulose as a potential bioleather substitute for the footwear industry. Microb Biotechnol. https://doi.org/10.1111/1751-7915.13306

Glass JI, Merryman C, Wise KS, Hutchison CA 3rd, Smith HO (2017) Minimal cells-real and imagined. Cold Spring Harb Perspect Biol 9(12). pii: a023861. https://doi.org/10.1101/cshperspect.a023861 (Review)

Gomaa MA, Al-Haj L, Abed RM (2016) Metabolic engineering of cyanobacteria and microalgae for enhanced production of biofuels and high-value products. J Appl Microbiol 121(4):919–931. https://doi.org/10.1111/jam.13232 Epub 2016 Aug 28

Guzmán-Trampe S, Ceapa CD, Manzo-Ruiz M, Sánchez S (2017) Synthetic biology era: improving antibiotic's world. Biochem Pharmacol 15(134):99–113. https://doi.org/10.1016/j.bcp.2017.01.015 (Epub 2017 Jan 31)

Harris DC, Jewett MC (2012) Cell-free biology: exploiting the interface between synthetic biology and synthetic chemistry. Curr Opin Biotechnol 23(5):672–678. https://doi.org/10.1016/j.copbio.2012.02.002 (Epub 2012 Apr 4. Review)

Hutchison CA 3rd, Chuang RY, Noskov VN, Assad-Garcia N, Deerinck TJ, Ellisman MH, Gill J, Kannan K, Karas BJ, Ma L, Pelletier JF, Qi ZQ, Richter RA, Strychalski EA, Sun L, Suzuki Y, Tsvetanova B, Wise KS, Smith HO, Glass JI, Merryman C, Gibson DG, Venter JC (2016) Design and synthesis of a minimal bacterial genome. Science 351(6280):aad6253. https://doi.org/10.1126/science.aad6253 (Erratum in: ACS Chem Biol. 2016 May 20;11(5):1463)

Katz L, Chen YY, Gonzalez R, Peterson TC, Zhao H, Baltz RH (2018) Synthetic biology advances and applications in the biotechnology industry: a perspective. J Ind Microbiol Biotechnol. https://doi.org/10.1007/s10295-018-2056-y

Martínez-Hidalgo P, Maymon M, Pule-Meulenberg F, Hirsch AM (2018) Engineering root microbiomes for healthier crops and soils using beneficial, environmentally safe bacteria. Can J Microbiol. https://doi.org/10.1139/cjm-2018-0315

Masukawa H, Kitashima M, Inoue K, Sakurai H, Hausinger RP (2012) Genetic engineering of cyanobacteria to enhance biohydrogen production from sunlight and water. Ambio 41(Suppl 2):169–173. https://doi.org/10.1007/s13280-012-0275-4

Moreno E (2012) Design and construction of "synthetic species". PLoS One 7(7):e39054. https://doi.org/10.1371/journal.pone.0039054 (Epub 2012 Jul 25. Review)

Mortimer JC (2018) Plant synthetic biology could drive a revolution in biofuels and medicine. Exp Biol Med (Maywood). 24:1535370218793890. https://doi.org/10.1177/1535370218793890

Myers SS (2018) Planetary health: protecting human health on a rapidly changing planet. Lancet 390 (10114):2860–2868. https://doi.org/10.1016/S0140-6736(17)32846-5 (Epub 2017 Nov 13)

Myers SS, Smith MR, Guth S, Golden CD, Vaitla B, Mueller ND, Dangour AD, Huybers P (2017) Climate change and global food systems: potential impacts on food security and undernutrition. Annu Rev Public Health 38:259–277. https://doi.org/10.1146/annurev-publhealth-031816-044356 (Epub 2017 Jan 6. Review)

Okano K, Honda K, Taniguchi H, Kondo A (2018) De novo design of biosynthetic pathways for bacterial production of bulk chemicals and biofuels. FEMS Microbiol Lett 365(20). https://doi.org/10.1093/femsle/fny215

Ozdemir T, Fedorec AJH, Danino T, Barnes CP (2018) Synthetic biology and engineered live biotherapeutics: toward increasing system complexity. Cell Syst. 7(1):5–16. https://doi.org/10.1016/j.cels.2018.06.008 (Review)

Piaggio AJ, Segelbacher G, Seddon PJ, Alphey L, Bennett EL, Carlson RH, Friedman RM, Kanavy D, Phelan R, Redford KH, Rosales M, Slobodian L, Wheeler K (2017) Is it time for synthetic biodiversity conservation? Trends Ecol Evol 32(2):97–107. https://doi.org/10.1016/j.tree.2016.10.016

Redford KH, Adams W, Mace GM (2013) Synthetic biology and conservation of nature: wicked problems and wicked solutions. PLoS Biol 11(4):e1001530. https://doi.org/10.1371/journal.pbio.1001530 (Epub 2013 Apr 2)

Redford K, Adams W, Carlson R, Mace G, Ceccarelli B (2014) Synthetic biology and the conservation of biodiversity. Oryx 48(03):330–336

Singh B, Mal G, Marotta F (2017) Designer probiotics: paving the way to living therapeutics. Trends Biotechnol 35(8):679–682. https://doi.org/10.1016/j.tibtech.2017.04.001 (Epub 2017 May 5)

Steen EJ, Kang Y, Bokinsky G, Hu Z, Schirmer A, McClure A, Del Cardayre SB, Keasling JD (2010) Microbial production of fatty-acid-derived fuels and chemicals from plant biomass. Nature 463(7280):559–562. https://doi.org/10.1038/nature08721

Weber W, Schoenmakers R, Keller B, Gitzinger M, Grau T, Daoud-El Baba M, Sander P, Fussenegger M (2008) A synthetic mammalian gene circuit reveals antituberculosis compounds. Proc Natl Acad Sci U S A. 105(29):9994–9998. https://doi.org/10.1073/pnas.0800663105 (Epub 2008 Jul 9)

Xie M, Fussenegger M (2018) Designing cell function: assembly of synthetic gene circuits for cell biology applications. Nat Rev Mol Cell Biol 19(8):507–525. https://doi.org/10.1038/s41580-018-0024-z

Yim H, Haselbeck R, Niu W, Pujol-Baxley C, Burgard A, Boldt J, Khandurina J, Trawick JD, Osterhout RE, Stephen R, Estadilla J, Teisan S, Schreyer HB, Andrae S, Yang TH, Lee SY, Burk MJ, Van Dien S (2011) Metabolic engineering of Escherichia coli for direct production of 1,4-butanediol. Nat Chem Biol 7 (7):445–452. https://doi.org/10.1038/nchembio.580 (Comment in Metabolic engineering: from retrofitting to green field. [Nat Chem Biol. 2011])

Zakeri B, Lu TK (2013) Synthetic biology of antimicrobial discovery. ACS Synth Biol. 2(7):358–372. https://doi.org/10.1021/sb300101g (Epub 2012 Dec 4)

Part IV
Health Biotechnology

Animal Biotechnology in Human Health

37

Abstract

The eventual purpose of animal biotechnology is using the animals and animal cells to improve human health through nutrition, novel recombinant therapeutics and producing organs for transplantation. Utilizing the curative potential of microbiota is another motive of the pursuit of biotechnology into animal gut ecosystem.

Highlights

- Human–animal interaction is associated with health and well-being
- Various livestock species are important agriculturally and relevant as model organisms
- Transgenic animals produce recombinant drugs, vaccines, and monoclonal antibodies.

Keywords

Animal biotechnology · Recombinant therapeutics · Nutraceuticals · Nutritional supplements · Health security · Model organisms · Animal health

37.1 Introduction

Ever since the civilization, the animals have remained as integral components of human evolution. Humans, pets, companion animals, and domestic livestock sharing common niches and climate influence each other. The relationship is a give-and-take principle: Pets and livestock rely on humans for survival, and in return, they bestow health benefits on humans (Friedman and Krause-Parello 2018). Pets and companion animals boost human health by alleviating stress and promoting the immunity (Gupta 2017).

Healthy livestock is important not only for producing quality products, but also for the health of livestock owners, and consumers. It is clear that animals have been benefited from therapeutics developed for the welfare of humans, and vice versa. Antibiotics and vaccines, such as naked DNA vaccines, viral vector vaccines, have eliminated several deadly infectious diseases of animals. Vaccines are also developed to eradicate non-communicable diseases such as cancer.

37.2 Contribution of Animals to Human Health

Many of current serious diseases such as cancer, diabetes, cardiovascular diseases, atherosclerosis, AIDS, Alzheimer's diseases are scantily investigated. Much remains to understand about these and other emerging health issues. Historically, animals have contributed to our knowledge of endocrine function, fertilization, reproduction, developmental studies. Animals have played crucial role to understand various phenomena of genetics, selection and evolution. Surgical

B. Singh et al., *Advances in Animal Biotechnology*,
https://doi.org/10.1007/978-3-030-21309-1_37

methods, pain-relieving drugs, treatment of shock trauma and blood diseases, cancer cure, gene therapy, insulin, pacemakers, nutrition supplements, tissue and organ transplantation, stem cell therapies, etc., all have been originally tested in experimental animals before permitting their use in humans.

37.3 Recombinant Therapeutics

Animal biotechnology has developed the concept of commercial pharming. Pharming refers to production of recombinant pharmaceuticals using live animals. Ever since the first report of birth of "super mice," (Palmiter et al. 1982), and development of transgenic mice to produce a human drug, called tissue plasminogen activator (tPA) to treat blood clots, animals are engineered to produce recombinant drugs. Already a variety of transgenic animals, viz. insects (Chen et al. 2018; Qian et al. 2018), pig (Chung et al. 2018; Kim et al. 2019), cattle (Monzani et al. 2016), goats (Feng et al. 2015; An et al. 2019), sheep (Menchaca et al. 2016), and chicken (Woodfint et al. 2018), have been generated to produce functionally stable recombinant biomolecules.

Recombinant nutraceuticals and therapeutics produced from transgenic animals include omega-3 fatty acids (Wu et al. 2012), human serum albumin (Luo et al. 2015), recombinant human bile salt-stimulated lipase (rhBSSL) (Wang et al. 2017), recombinant human lactoferrin N-glycans (Parc et al. 2017), novel glycosylated anti-CD20 (Rituxan) monoclonal antibodies (Zhang et al. 2018a), and plasminogen activator (He et al. 2018). US FDA has already approved human lysosomal acid lipase (Kanuma[(R)]) (sebelipase α) produced in chicken egg white for therapeutic applications (Sheridan 2016).

Commercial companies will utilize the technology to develop more transgenic animals in the future. This is because transgenic animals serving as bioreactors to produce complex human or eukaryotic therapeutic proteins prove to be economical than producing these molecules using bacteria, yeast, or cultured animal cells.

37.4 Tissues and Organs for Humans

Biotechnology has offered new prospects in human regenerative medicine. Five broader areas are of significant concern. These include in vitro production of mammalian cells and tissues, producing organs using large animals such as pig, three-dimensional (3D) bioprinting using biocompatible materials, ex vivo decellularization, and xenotransplantation.

There is a worldwide dearth of organs for transplantation into patients who are at end-stage organ failures. Organ cryopreservation has certain physical and biological limitations (Finger and Bischof 2018). Hence, the demand for human tissues and organs for transplantation far surpasses the supply of organs from deceased donors. To address the problem, transplant surgeons have suggested the use of xenografts, i.e., transplanting animal tissues or organs into humans. Genetically engineered pigs could provide tissues and cells for treating health conditions such as diabetes, renal failure, Parkinson's disease, and corneal blindness (Hryhorowicz et al. 2017; Cooper et al. 2018).

Notably, there are ethical concerns in use of organs from non-human primates. As an alternative, pig is preferred as donor of organs because of its large litter size (up to 8 offspring at a time), short gestation period (around 120 days), and anatomical and physiological resemblance with humans. Replacement tissues and heart valves from pig are more suitable for therapies.

Structural degeneration or immunological barriers impede the success of transplanted organs. The problem can be surmounted by the use of genome editors such as ZFNs, TALENs, and CRISPR/Cas9. Thus, the transgenic animals should be immunocompatible to humans. Evidences suggest that bioprosthetic valves obtained from genetically modified pigs lacking xenogeneic antigens (namely Gal, Neu5Gc, and Sd[a]), also termed triple-knockout pigs, would function significantly longer in human recipients than those obtained from ordinary or genetically unmodified animals (Smood et al. 2019). A multifactorial approach of inclusion of some human genes into pig genome appears to bestow superior xenograft outcomes. This kind of genetic engineering enhances the prospects of long-term survival of cardiac xenograft and consequent clinical applications (Chan et al. 2017).

Though pig organs (heart and kidney) are suitable for xenotransplantation, odds of endogenous retroviruses in pig genome raise the specter of zoonotic viral infections in humans. It is suggested that problem can be overcome by generating gene-edited animals free of replication-competent virus (Denner 2017). Transgenic strategies have been developed to minimize the potential risk of infectious viruses in pigs (Kwon et al. 2017; Pan et al., 2019).

37.5 Nutritional and Environmental Security

During past three decades, the human living standard, income status, and food habits have changed a lot. Biotechnology has contributed to fulfill the needs of human nutrition. As noticed by Kishore and Shewmaker (1999), the crop biotechnology and agriculture had initially focussed on meeting the nutritional needs of humans. However, the future challenges are far beyond simply addressing the needs of ever-growing global human population.

At the same time, care should be taken to fulfil nutritional demand of livestock intended for producing food. This should be achieved by improving the quality of forages and feed.

Biotechnology can, and is, enhance the quantity and quality of food obtained from animals. Commercial livestock entrepreneurs opt to select their animal breeding stock based on molecular markers associated with enhanced growth, superior milk and meat traits and resistance to diseases. Genetic engineered and genome editing is recommended to produce pigs, cattle, sheep, and goats that are resistant to infectious diseases, or have high growth rates compared to conventionally domesticated animals (Xie et al. 2018; Petersen 2018).

Dietary nitrogen, and phosphorous released from monogastric animals (pig, poultry and fish) pollute water. Genetically modified pigs that utilize dietary fiber, nitrogen, and phosphorous efficiently could reduce environmental pollution (Petersen 2018; Zhang et al. 2018b).

Intensive animal production has an adverse impact on environment. Ruminants raised for milk and meat require lot of fodder, grains, and water and emit greenhouse gases such as CO_2 and CH_4. Alternative sources of animal proteins, such as insects, cultured or in vitro produced meat, are suggested to lessen dependence on animals for food. Animal biotechnology will be an important component of the holistic system to ensure supply of proteins and energy.

37.6 Therapeutics from Gut Microbiome

The animals are important reservoirs of microorganisms that have substantial health potential. Anaerobes are valuable sources of fascinating bioactive compounds (Mamo 2016). The gastrointestinal microbiome has drawn attention of microbial ecologists and clinicians over the past decades. Bacteriocins and bacteriophages are of concern in view of their interaction and targeted killing of pathogenic bacteria. In addition, free fatty acids, amino acids, vitamins, quorum-sensing autoinducers, and miscellaneous metabolites produced by microorganisms are of clinical importance (Li et al. 2018).

Bacteriocins, antimicrobial peptides (AMPs), and other antagonistic compounds of gut microbiome may serve as novel alternatives to antibiotic therapy (Kumar et al. 2013; Garcia-Gutierrez et al. 2019). The gut microbes of herbivores are potent degraders of toxic phytometabolites such as oxalates, non-protein amino acids, phytates, and tannin-polyphenols. We foresee the use of such microorganisms and their enzymes to alleviate toxic effects of some dietary phytometabolites in humans.

37.7 Outlook and Challenges

Biotechnology promises to fulfill the nutritional and health demands of humans. At present, the animal biotechnology is a well-established area of specialization aimed to improve human health and livelihood. Transgenic milch animals are admirable systems for producing nutritional and therapeutic human recombinant proteins. Research has begun to tackle some complicated human medical problems such as stress, stress-induced depression, and cancers. Animal trials are indispensable components of such types of studies.

Pig-to-human tissue and organ transplantation is a doable substitute to overcome the shortage of tissues and organs. Transgenic pigs produced by knockout, knock-in, or gene editing can serve as donors of organs for human patients. The key problems, such as immunological tolerance of host against transplanted organs, deleterious effects of the transgenes on donor animals, are not properly investigated.

Similarly, the focus should be to generate animals that cause minimum environmental pollution. Some genetically modified pigs are already produced that degrade dietary fiber and phytate more efficiently. It is imperative to increase the efficiency of process of transgenesis, and lowering the cost involved in producing transgenic animals. Acceptance of milk and meat from transgenic animals is a social issue that should be addressed based on scientific research trials on the safety of products obtained from transgenic animals.

37.8 Conclusions

Animals have been and will continue to be essential models to combat human illness. Indeed, scientific and medical research is the endowment of the experimental trials conducted in animals. Animal biotechnology research will play more important role in the future. Whole organism or animal studies are needed to resolve many complex diseases that are still inadequately understood.

References

An L, Yang L, Huang Y, Cheng Y, Du F (2019) Generating goat mammary gland bioreactors for producing recombinant proteins by gene targeting. Methods Mol Biol 1874:391–401. https://doi.org/10.1007/978-1-4939-8831-0_23

Chan JL, Singh AK, Corcoran PC, Thomas ML, Lewis BG, Ayares DL, Vaught T, Horvath KA, Mohiuddin MM (2017) Encouraging experience using multi-transgenic xenografts in a pig-to-baboon cardiac xenotransplantation model. Xenotransplantation 24(6). https://doi.org/10.1111/xen.12330 (Epub 2017 Sep 22)

Chen W, Wang F, Tian C, Wang Y, Xu S, Wang R, Hou K, Zhao P, Yu L, Lu Z, Xia Q. (2018) Transgenic silkworm-based silk gland bioreactor for large scale production of bioactive human platelet-derived growth factor (pdgf-bb) in silk cocoons. Int J Mol Sci 19(9). pii: E2533. https://doi.org/10.3390/ijms19092533

Chung HJ, Park HJ, Baek SY, Park JK, Lee WY, Kim KW, Jo YM, Hochi S, Kim YM, Choi TJ, Cho ES, Cho KH (2018) Production of human tissue-type plasminogen activator (htPA) using in vitro cultured transgenic pig mammary gland cells. Anim Biotechnol 1–6. https://doi.org/10.1080/10495398.2018.1521824

Cooper DKC, Gaston R, Eckhoff D, Ladowski J, Yamamoto T, Wang L, Iwase H, Hara H, Tector M, Tector AJ (2018) Xenotransplantation-the current status and prospects. Br Med Bull 125(1):5–14. https://doi.org/10.1093/bmb/ldx043

Denner J (2017) Paving the path toward porcine organs for transplantation. N Engl J Med 377(19):1891–1893. https://doi.org/10.1056/NEJMcibr1710853 (No abstract available)

Feng X, Cao S, Wang H, Meng C, Li J, Jiang J, Qian Y, Su L, He Q, Zhang Q (2015) Production of transgenic dairy goat expressing human α-lactalbumin by somatic cell nuclear transfer. Transgenic Res 24(1):73–85. https://doi.org/10.1007/s11248-014-9818-8 (Epub 2014 Aug 20)

Finger EB, Bischof JC (2018) Cryopreservation by vitrification: a promising approach for transplant organ banking. Curr Opin Organ Transplant 23(3):353–360. https://doi.org/10.1097/MOT.0000000000000534

Friedman E, Krause-Parello CA (2018) Companion animals and human health: benefits, challenges, and the road ahead for human-animal interaction. Rev Sci Tech 37(1):71–82. https://doi.org/10.20506/rst.37.1.2741

Garcia-Gutierrez E, Mayer MJ, Cotter PD, Narbad A (2019) Gut microbiota as a source of novel antimicrobials. Gut Microbes 10(1):1–21. https://doi.org/10.1080/19490976.2018.1455790 (Epub 2018 May 22)

Gupta S (2017) Puppy power. Nature 543:S48–S49

He Z, Lu R, Zhang T, Jiang L, Zhou M, Wu D, Cheng Y (2018) A novel recombinant human plasminogen activator: efficient expression and hereditary stability in transgenic goats and in vitro thrombolytic bioactivity in the milk of transgenic goats. PLoS ONE 13

(8):e0201788. https://doi.org/10.1371/journal.pone.0201788 (eCollection 2018)

Hryhorowicz M, Zeyland J, Słomski R, Lipiński D (2017) Genetically modified pigs as organ donors for xenotransplantation. Mol Biotechnol 59(9–10):435–444. https://doi.org/10.1007/s12033-017-0024-9 (Review)

Kim GA, Lee EM, Cho B, Alam Z, Kim SJ, Lee S, Oh HJ, Hwang JI, Ahn C, Lee BC (2019) Generation by somatic cell nuclear transfer of GGTA1 knockout pigs expressing soluble human TNFRI-Fc and human HO-1. Transgenic Res 28(1):91–102. https://doi.org/10.1007/s11248-018-0103-0 (Epub 2018 Dec 14)

Kishore GM, Shewmaker C (1999) Biotechnology: enhancing human nutrition in developing and developed worlds. Proc Natl Acad Sci U S A 96(11):5968–5972

Kumar M, Nagpal R, Verma V, Kumar A, Kaur N, Hemalatha R, Gautam SK, Singh B. (2013) Probiotic metabolites as epigenetic targets in the prevention of colon cancer. Nutr Rev 71(1):23–34. https://doi.org/10.1111/j.1753-4887.2012.00542.x (Epub 2012 Nov 9. Review)

Kwon DJ, Kim DH, Hwang IS, Kim DE, Kim HJ, Kim JS, Lee K, Im GS, Lee JW, Hwang S (2017) Generation of α-1,3-galactosyltransferase knocked-out transgenic cloned pigs with knocked-in five human genes. Transgenic Res 26(1):153–163. https://doi.org/10.1007/s11248-016-9979-8

Li Z, Quan G, Jiang X, Yang Y, Ding X, Zhang D, Wang X, Hardwidge PR, Ren W, Zhu G (2018) Effects of metabolites derived from gut microbiota and hosts on pathogens. Front Cell Infect Microbiol 8:314. https://doi.org/10.3389/fcimb.2018.00314 (eCollection 2018. Review)

Luo Y, Wang Y, Liu J, Lan H, Shao M, Yu Y, Quan F, Zhang Y (2015) Production of transgenic cattle highly expressing human serum albumin in milk by phiC31 integrase-mediated gene delivery. Transgenic Res 24 (5):875–883. https://doi.org/10.1007/s11248-015-9898-0 (Epub 2015 Jul 22)

Mamo G (2016) Anaerobes as sources of bioactive compounds and health promoting tools. Adv Biochem Eng Biotechnol 156:433–464 (Review)

Menchaca A, Anegon I, Whitelaw CB, Baldassarre H, Crispo M (2016) New insights and current tools for genetically engineered (GE) sheep and goats. Theriogenology 86(1):160–169. https://doi.org/10.1016/j.theriogenology.2016.04.028

Monzani PS, Adona PR, Ohashi OM, Meirelles FV, Wheeler MB (2016) Transgenic bovine as bioreactors: challenges and perspectives. Bioengineered 7(3):123–131. https://doi.org/10.1080/21655979.2016.1171429 (Epub 2016 May 11)

Palmiter RD, Brinster RL, Hammer RE, Trumbauer ME, Rosenfeld MG, Birnberg NC, Evans RM (1982) Dramatic growth of mice that develop from eggs microinjected with metallothionein-growth hormone fusion genes. Nature 300:611–615

Pan D, Liu T, Lei T, Zhu H, Wang Y, Deng S (2019) Progress in multiple genetically modified minipigs for xenotransplantation in China. Xenotransplantation 26 (1):e12492. https://doi.org/10.1111/xen.12492 (Review)

Parc AL, Karav S, Rouquié C, Maga EA, Bunyatratchata A, Barile D (2017) Characterization of recombinant human lactoferrin N-glycans expressed in the milk of transgenic cows. PLoS ONE 12(2): e0171477. https://doi.org/10.1371/journal.pone.0171477 (eCollection 2017)

Petersen B (2018) Transgenic pigs to the rescue. Elife 7. pii: e37641. https://doi.org/10.7554/elife.37641

Qian Q, You Z, Ye L, Che J, Wang Y, Wang S, Zhong B (2018) High-efficiency production of human serum albumin in the posterior silk glands of transgenicsilkworms, Bombyx mori L. PLoS ONE 13(1):e0191507. https://doi.org/10.1371/journal.pone.0191507 (eCollection 2018)

Sheridan C (2016) FDA approves 'farmaceutical' drug from transgenic chickens. Nat Biotechnol 34(2):117–119. https://doi.org/10.1038/nbt0216-117

Smood B, Hara H, Cleveland D, Cooper DKC (2019) In search of the ideal valve: optimizing genetic modifications to prevent bioprosthetic degeneration. Ann Thorac Surg. pii: S0003-4975(19)30251-6. https://doi.org/10.1016/j.athoracsur.2019.01.054 (Epub ahead of print Review)

Wang Y, Ding F, Wang T, Liu W, Lindquist S, Hernell O, Wang J, Li J, Li L, Zhao Y, Dai Y, Li N (2017) Purification and characterization of recombinant human bile salt-stimulated lipase expressed in milk of transgenic cloned cows. PLoS ONE 12(5):e0176864

Woodfint RM, Hamlin E, Lee K (2018) Avian bioreactor systems: a review. Mol Biotechnol 60(12):975–983. https://doi.org/10.1007/s12033-018-0128-x

Wu X, Ouyang H, Duan B, Pang D, Zhang L, Yuan T, Xue L, Ni D, Cheng L, Dong S, Wei Z, Li L, Yu M, Sun QY, Chen DY, Lai L, Dai Y, Li GP (2012) Production of cloned transgenic cow expressing omega-3 fatty acids. Transgenic Res 21(3):537–543. https://doi.org/10.1007/s11248-011-9554-2 (Epub 2011 Sep 15)

Xie Z, Pang D, Yuan H, Jiao H, Lu C, Wang K, Yang Q, Li M, Chen X, Yu T, Chen X, Dai Z, Peng Y, Tang X, Li Z, Wang T, Guo H, Li L, Tu C, Lai L, Ouyang H (2018) Genetically modified pigs are protected from classical swine fever virus. PLoS Pathog 14(12): e1007193. https://doi.org/10.1371/journal.ppat.1007193 (eCollection 2018 Dec)

Zhang R, Tang C, Guo H, Tang B, Hou S, Zhao L, Wang J, Ding F, Zhao J, Wang H, Chen Z, Dai Y, Li N (2018a) A novel glycosylated anti-CD20 monoclonal antibody from transgenic cattle. Sci Rep 8(1):13208. https://doi.org/10.1038/s41598-018-31417-2

Zhang X, Li Z, Yang H, Liu D, Cai G, Li G, Mo J, Wang D, Zhong C, Wang H, Sun Y, Shi J, Zheng E, Meng F, Zhang M, He X, Zhou R, Zhang J, Huang M, Zhang R, Li N, Fan M, Yang J, Wu Z (2018b) Novel transgenic pigs with enhanced growth and reduced environmental impact. Elife 7. pii: e34286. https://doi.org/10.7554/elife.34286

Designer Milk

38

Abstract

Milk is an important component of nutrition of human and animal neonates and has attracted interest of food technologists, clinicians, and biochemists. Composition of milk can be modified by dietary manipulations of milch animals and altering the genetic make-up of milk-producing species.

Highlights

- Milk can be modified by dietary and genetic manipulation
- Milk having modified constituents has applications to improve health.

Keywords

Milk · Milk derived therapeutics · Transgenic animals · Milk composition

38.1 Introduction

Milk is a source of essential nutritive elements such as biologically active peptides, immunoglobulins, immunoprotective agents such as lactoferrin and microbiota (LeMay-Nedjelski et al. 2018; Metzger et al. 2018) that have pro-health role in infants. Food-origin peptides are the potential health-promoting nutraceuticals. While bioactive peptides can be generated from various foods, milk peptides are fundamental constituents of several commercially available functional food products. Based on their source or origin and physiological effect on the body, the milk bioactive peptides can be antihypertensive, antithrombotic, opioid, casein phosphopeptides (CPPs), antimicrobial, cytomodulatory, immunomodulatory, and miscellaneous peptides (Hayes et al. 2007; Mal et al. 2018).

Whey proteins, such as lactoferrin, lactoperoxidase, lysozyme, and immunoglobulins, are antimicrobial in nature. Other minor bioactive substances present in milk include oligosaccharides, fucosylated oligosaccharides, hormones, growth factors, mucin, gangliosides, and endogenous peptides.

38.2 Milk-Producing Livestock

Milk is an essential as well as commercial livestock product consumed, and produced all around the world. One-third of the global milk production is from a few countries including India, USA, and China (Whitelaw et al. 2016). Various livestock species are reared for producing milk depending on adaptation of animal species to prevailing agroclimatic situations. Cows, buffaloes, and goat are the prime milk-producing livestock species. In addition, the less known milk-producing species include

B. Singh et al., *Advances in Animal Biotechnology*,
https://doi.org/10.1007/978-3-030-21309-1_38

camel, yak, mithun, donkey, sheep, and reindeer. Their contribution to gross milk production is less, but they fulfill the milk demand of native populations. The milk from these animals is used locally for human consumption besides use of surplus milk in ethnic sweets and dairy products.

38.3 Concept and Benefits of Designer Milk

Organic milk is already available, but there is need and demand to develop designer milk that contains additional components with additional health benefits (Table 38.1). Dairy and food industry will be benefited immensely from commercial scale production of designer milk. It was realized that animals yielding good quality milk, easy to manage under farm or laboratory conditions could serve as bioreactors to produce drugs in their milk. For nutrition and human health measures, the actions that would be beneficial include (i) generate a greater proportion of unsaturated fatty acids (USFA) in milk fat, (ii) reduce lactose content so that milk can be used by persons suffer from lactose intolerance, and (iii). eliminate or minimize β-LG from milk (Fig. 38.1). From a technological point of view, there exist vast opportunities in altering primary structure of casein to improve technological properties of milk, producing high-protein milk for dairy industry, modify milk composition for cheese manufacturing that leads to accelerated curd clotting time, and design milk that has enhanced nutraceuticals (Box 1). All this could be possible after years of research on rDNA technology, reproduction biotechniques, high-throughput sequencing and bioinformatics.

38.4 Designer Milk-Historical Background

Consumption of milk and dairy products is increasing in Western as well as developing countries. The scopes of biotechnological intervention in dairy industry are enormous. Animals that produce milk containing recombinant therapeutic agents such as insulin, human plasma proteins, drugs, and vaccines, are already available in some countries.

Ever since the inception of genetic manipulation techniques to produce mice expressing β-lactoglobulin (BLG) (Simons et al. 1987), the milk has been the target for genetic engineering. Efforts were aimed to develop milch animals as bioreactors for producing transgenic proteins, enzymes, and antibodies (Clark et al. 1989).

Table 38.1 Summary of pro-health attributes and effects of designer milk

Parameters	Advantages
Lactose levels	Overcoming problems of lactose intolerance by lowering lactose contents, and using low-lactose milk to prepare milk-based infants diets
Altered milk fats (increasing ω fatty acids)	Lowering risk of cardiovascular disease, autoimmune disorders, allergies, obesity, diabetes, dementia, and certain cognitive disorders
Minimized saturated fats	Lowering the incidences of obesity, cholesterol levels, and cardiovascular diseases
Altered protein contents	Using proteins for nutrition and beverages such as instant drinks and infant formulas, and dairy products
	Altering primary structure of casein to improve technological properties, increasing casein to obtain cheese yield, lowering BLG levels in milk for alleviating milk-induced allergy
Humanized milk	Developing infants diets based on animal milk containing human proteins

Box 1. Potential targets for use of designer milk

- Development of value-added dairy products with enhanced shelf life
- Producing less-allergenic milk with low lactose and BLG
- Enhancing unsaturated fatty acids in milk
- Producing high-proteins, engineering milk for accelerated curd clotting
- Increasing the milk yield and manufacturing milk-based products having nutraceuticals
- Developing milk food formulations for health augmentation, and protecting lambs, kids, piglets, and calves against GI infectious agents.

Using gene promoters from milk protein genes, human proteins are produced in mammary cells of various livestock species. The gene expression cassette including necessary elements to integrate the transgenes into genome of animals was microinjected into preimplantation stage embryos.

Development of techniques to clone animals by SCNT was a new era of enhancing production of complete transgenic cloned animals. Using SCNT, a number of cloned animals have been produced. The recombinant technologies with enhanced gene-delivery systems and precision of integration of recombinant genes into the genome of host have improved the genetic engineering processes of livestock species for large-scale production of novel products and processes, and improving genetics of animals (Menchaca et al. 2016).

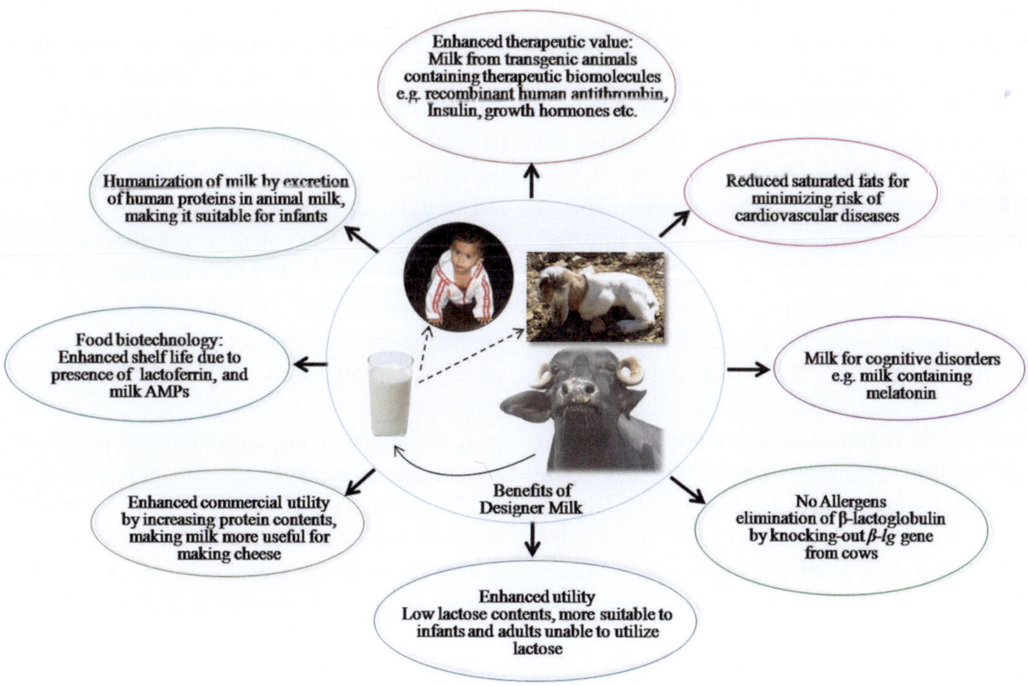

Fig. 38.1 Benefits and motives of producing designer milk. Bioactive milk peptides are valuable nutraceuticals with various pro-health attributes for preventing diarrhea, hypertension, thrombosis, and immunodeficiency. Besides industrial applications, the designer milk is aimed to improve health of human and animal neonates

38.5 Strategies to Alter Composition of Milk

The milk components can be altered at various levels. In ruminants, the rumen microbiota is the origin of bioactive fatty acids that are incorporated into milk and meat of the animals.

38.6 Dietary Interventions to Produce Designer Milk

Less amounts of saturated fats, increased conjugated *linoleic acid (CLA)* and omega-fatty acids, more proteins, low lactose, and minimal or nil BLG, are the desirable attributes of milk.

Dietary ingredients have impact on composition of milk. It is feasible to alter composition of milk by dietary interventions (Henno et al. 2018). Supplementing extruded linseed, rich in α-linolenic acid, in the diet of lactating cows increased the proportion of potentially pro-health milk fatty acids viz., oleic acid, vaccenic acid, rumenic acid, α-linolenic acid, total polyunsaturated fatty acids (PUFA) (Meignan et al. 2017).

Notably, the rumen ciliate protozoa do not participate in biohydrogenation of lipids, yet protozoal lipids contain higher amounts of CLAs than the lipids found in rumen bacteria. This is possibly because rumen protozoa gulp tiny forage fragments and plant organelles, such as chloroplasts, which are partially metabolized by bacteria that are ingested along with feed particles (Lourenço et al. 2010). The ruminant diet generally has polyunsaturated fatty acids (PUFA), but ruminant products such as meat or milk contain saturated fatty acids and some amounts of conjugated linoleic acids (CLAs). This is due to microbial enzymatic lipolysis and biohydrogenation of PUFA in the rumen (Lourenço et al. 2010). *Butyrivibrio fibrisolvens* are the major biohydrogenating bacteria cultured from the rumen.

Maneuvring rumen ecosystem offers opportunities to modify lipid profile of meat and milk by altering nature of fatty acids (FA) for uptake by intramuscular and mammary tissue (Toral et al. 2018). Probiotics or microbial feed supplements favorably modify lipid metabolism, though information is scarce on use of microbial feed supplements to change milk composition. Phytometabolites such as tannins, polyphenol oxidase, essential oils, oxygenated FA, and saponins have varying effects on milk composition (Jochum et al. 2017; Toral et al. 2018).

Hence, varying rumen microbiome or microbial metabolites can change the composition of milk. As grazing regime changes fatty composition of rumen microbiota, it can be used to alter rumen microbial communities, thereby shifting the fatty acids profile of milk (Bainbridge et al. 2018).

38.7 Genome Engineering and Genetic Manipulation

Dietary manipulation is a transitory method of changing the composition of milk. Applications of genomics and identification of genes responsible for various components of milk allow researchers to breed animals with selective traits. Success is achieved in producing genetically tailored farm animals that secrete human lactoferrin, lysozyme, and lipase in the milk. Similarly, milk allergenicity can be reduced by eliminating the BLG genetic determinants. It is argued that transgenic cattle platform offers an efficient, safe, and cost-effective method for producing large amounts of biopharmaceuticals, which is essential to develop innovative healthcare products.

It is comparatively convenient to harvest recombinant proteins from milk (Wang et al. 2017). Generating mammary gland bioreactors via gene targeting methods is an asset for producing recombinant proteins in the milk of lactating species such as goats (Lu et al. 2018; An et al. 2019), sheep (Wang et al. 2018), and cattle (Wei et al. 2018) (Table 38.2). Regulatory authorities and commercial entrepreneurs are also involved in producing transgenic animals for producing recombinant proteins through milk. US FDA has granted approval for antithrombin

Table 38.2 Examples of some human transgenic proteins and enzymes produced in milk

Proteins	Expressing species	Promoters
Bovine β- and K-casein	Cow	Bovine β casein (Brophy et al. 2003)
CuZn-SOD, EC-SOD	Bitransgenic goats	Goat β-casein 5′ and 3′ regulatory elements (Lu et al. 2018)
Lysostaphin	Cow	Ovine β-lactoglobulin (BLG) (Wall et al. 2005)
Lysozyme	Goat	Bovine αs1-casein (Maga et al. 2006)
hIGF-1	Pig	Bovine α-lactalbumin (Monaco et al. 2005)
hIGF-1	Rabbit	Bovine αS1-casein (Wolf et al. 1997)
hα-LA	Pig	Bovine α lactalbumin (Bleck et al. 1998)
hα-LA	Goat (GT)	Human (*α-LA* gene) integrated into BLG locus (An et al. 2019)
hLF	Cow	Bovine αS1-casein (Krimpenfort et al. 1991)
hLF	Mice	Bovine αS1-casein (Platenburg et al. 1994)

Abbreviations CuZn-SOD—human copper/zinc superoxide dismutase; EC-SOD human extracellular superoxide dismutase; hα-LA—human α-lactalbumin; hLF—human lactoferrin; hIGF-1—human insulin-like growth factor 1; GT—gene-targeted goats

produced in the milk of transgenic goats (Kling 2009), and Ruconest, a plasma-free C1-esterase inhibitor (C1-INH), used for treating hereditary angioedema (HAE) produced in milk of transgenic rabbit (Nature Biotechnology 2014).

Genome editing is used to eliminate specific genes form milk or meat producing animals to improve the quality of animal-derived products. Zygote-mediated genome editing for BLG led to production of knockout cattle that produced hypoallergenic milk (Wei et al. 2018).

38.8 Outlook and Challenges

The use of recombinant therapeutic, enzymes, proteins, and antibodies has gradually increased during the past two decades. The FAO, in 2000, had predicted that dairy protein products consumption will reach 700 m tonnes annually by 2020. The envisaged dairy production is already reached (reviewed in Whitelaw et al. 2016). Further increase seems to be knotty merely by means of conventional methods of animal husbandry. Hence, genetic engineering and transgenic animal production remain to be viable options.

There is renewed enthusiasm in new technologies of genetic manipulation of livestock and crops. By using enzyme directed genome-editing methods such as CRISPR/Cas9, it is possible to remove or add genes of interest into an organism. With these methods, it is possible to produce humanized cattle milk, i.e., the protein containing human milk proteins. There are hopes to enhance the production of therapeutic proteins and minimize their cost of production.

It is not practically feasible to depend solely on genetic manipulation to boost livestock production. Management tactics, such as genetic selection, optimal nutrition management, housing and health of animals, should be taken into account to utilize production merits of high yielding animals.

The ultimate acceptability of the "designer" products will depend on ethical issues such as animal welfare and safety, besides better health benefits and increased profitability of products manufactured by the novel techniques. While genetic engineering strategies may provide some solutions, bringing GM organisms to market is still a difficult task owing to legislative regulations and pressure from consumers.

38.9 Conclusions

The food industry is in continuous transition, developing steadily from its traditional raw material-driven basis to innovative and market-orientated position. This transformation is accompanied by an increasing demand for high value-added products that meet consumers' demands of superior taste and convenience, health and well-being, as well as safety. Biotechnological innovations are likely to assist in industrializing the production of indigenous fermented foods. Pursuit of genomics and genetic engineering tools will lead to production of fermented foods of improved quality, safety, and consistency.

References

An L, Yang L, Huang Y, Cheng Y, Du F (2019) Generating goat mammary gland bioreactors for producing recombinant proteins by gene targeting. Methods Mol Biol 1874:391–401. https://doi.org/10.1007/978-1-4939-8831-0_23

Bainbridge ML, Saldinger LK, Barlow JW, Alvez JP, Roman J, Kraft J (2018) Alteration of rumen bacteria and protozoa through grazing regime as a tool to enhance the bioactive fatty acid content of bovine milk. Front Microbiol 9:904. https://doi.org/10.3389/fmicb.2018.00904 (eCollection)

Bleck GT, White BR, Miller DJ, Wheeler MB (1998) Production of bovine alpha-lactalbumin in the milk of transgenic pigs. J Anim Sci 76(12):3072–3078

Brophy B, Smolenski G, Wheeler T, Wells D, L'Huillier P, Laible G (2003) Cloned transgenic cattle produce milk with higher levels of beta-casein and kappa-casein. Nat Biotechnol 21(2):157–162

Clark AJ, Ali S, Archibald AL, Bessos H, Brown P, Harris S, McClenaghan M, Prowse C, Simons JP, Whitelaw CBA, Wilmut I (1989) The molecular manipulation of milk composition. Genome 31 (2):950–955 (Review)

Hayes M, Stanton C, Fitzgerald GF, Ross RP (2007) Putting microbes to work: dairy fermentation, cell factories and bioactive peptides. Part II: bioactive peptide functions. Biotechnol J 2(4):435–449 (Review)

Henno M, Ariko T, Kaart T, Kuusik S, Ling K, Kass M, Jaakson H, Leming R, Givens DI, Sterna V, Ots M (2018) The fatty acid composition of Estonian and Latvian retail milk; implications for human nutrition compared with a designer milk. J Dairy Res 85 (2):247–250. https://doi.org/10.1017/S0022029918000183

Jochum F, Alteheld B, Meinardus P, Dahlinger N, Nomayo A, Stehle P (2017) Mothers' consumption of soy drink but not black tea increases the flavonoid content of term breast milk: a pilot randomized, controlled intervention study. Ann Nutr Metab 70 (2):147–153. https://doi.org/10.1159/000471857 (Epub 2017 Apr 8)

Kling J (2009) First US approval for a transgenic animal drug. Nat Biotechnol 27(4):302–304. https://doi.org/10.1038/nbt0409-302

Krimpenfort P, Rademakers A, Eyestone W, van der Schans A, van den Broek S, Kooiman P, Kootwijk E, Platenburg G, Pieper F, Strijker R et al (1991) Generation of transgenic dairy cattle using 'in vitro' embryo production. Biotechnology (NY) 9(9):844–847

LeMay-Nedjelski L, Copeland J, Wang PW, Butcher J, Unger S, Stintzi A, O'Connor DL (2018) Methods and strategies to examine the human breastmilk microbiome. Methods Mol Biol 1849:63–86. https://doi.org/10.1007/978-1-4939-8728-3_5

Lourenço M, Ramos-Morales E, Wallace RJ (2010) The role of microbes in rumen lipolysis and biohydrogenation and their manipulation. Animal 4(7):1008–1023. https://doi.org/10.1017/S175173111000042X

Lu R, Zhang T, Wu D, He Z, Jiang L, Zhou M, Cheng Y (2018) Production of functional human CuZn-SOD and EC-SOD in bitransgenic cloned goat milk. Transgenic Res https://doi.org/10.1007/s11248-018-0080-3

Maga EA, Cullor JS, Smith W, Anderson GB, Murray JD (2006) Human lysozyme expressed in the mammary gland of transgenic dairy goats can inhibit the growth of bacteria that cause mastitis and the cold-spoilage of milk. Foodborne Pathog Dis 3(4):384–392. https://doi.org/10.1089/fpd.2006.3.384

Mal G, Singh B, Mane BG, Sharma V, Sharma R, Bhar R, Dhar JB (2018) Milk composition, antioxidant activities and protein profile of *Gaddi* goat milk. J Food Biochem (in press). https://doi.org/10.1111/jfbc.12660

Meignan T, Lechartier C, Chesneau G, Bareille N (2017) Effects of feeding extruded linseed on production performance and milk fatty acid profile in dairy cows: a meta-analysis. J Dairy Sci 100(6):4394–4408. https://doi.org/10.3168/jds.2016-11850

Menchaca A, Anegon I, Whitelaw CB, Baldassarre H, Crispo M (2016) New insights and current tools for genetically engineered (GE) sheep and goats. Theriogenology 86(1):160–169. https://doi.org/10.1016/j.theriogenology.2016.04.028 (Epub. Review)

Metzger SA, Hernandez LL, Suen G, Ruegg PL (2018) Understanding the milk microbiota. Vet Clin North Am Food Anim Pract 34(3):427–438. https://doi.org/10.1016/j.cvfa.2018.06.003

Monaco MH, Gronlund DE, Bleck GT, Hurley WL, Wheeler MB, Donovan SM (2005) Mammary specific transgenic over-expression of insulin-like growth factor-I (IGF-I) increases pigmilk IGF-I and IGF binding proteins, with no effect on milk composition or yield. Transgenic Res 14(5):761–773

Nature Biotechnology (2014) Rabbit milk Ruconest for hereditary angioedema. 32:849. https://doi.org/10.1038/nbt0914-849d

Platenburg GJ, Kootwijk EP, Kooiman PM, Woloshuk SL, Nuijens JH, Krimpenfort PJ, Pieper FR, de Boer HA, Strijker R (1994) Expression of human lactoferrin in milk of transgenic mice. Transgenic Res 3(2):99–108

Simons JP, McClenaghan M, Clark AJ (1987) Alteration of the quality of milk by expression of sheep beta-lactoglobulin in transgenic mice. Nature 328 (6130):530–532

Toral PG, Monahan FJ, Hervás G, Frutos P, Moloney AP (2018) Review: modulating ruminal lipid metabolism to improve the fatty acid composition of meat and milk. Challenges and opportunities. Animal. 1–10. https://doi.org/10.1017/s1751731118001994

Wall RJ, Powell AM, Paape MJ, Kerr DE, Bannerman DD, Pursel VG, Wells KD, Talbot N, Hawk HW (2005) Genetically enhanced cows resist intramammary Staphylococcus aureus infection. Nat Biotechnol 23(4):445–451 (Epub 2005 Apr 3. Erratum in: Nat Biotechnol. 23(7):897)

Wang Y, Ding F, Wang T, Liu W, Lindquist S, Hernell O, Wang J, Li J, Li L, Zhao Y, Dai Y, Li N (2017) Purification and characterization of recombinant human bile salt-stimulated lipase expressed in milk of transgenic cloned cows. PLoS One 12(5):e0176864

Wang S, Deng S, Cao Y, Zhang R, Wang Z, Jiang X, Wang J, Zhang X, Zhang J, Liu G, Lian Z (2018) Overexpression of toll-like receptor 4 contributes to phagocytosis of salmonella enterica serovar typhimurium via phosphoinositide 3-kinase signaling in sheep. Cell Physiol Biochem 49(2):662–677. https://doi.org/10.1159/000493032 (Epub 2018 Aug 30)

Wei J, Wagner S, Maclean P, Brophy B, Cole S, Smolenski G, Carlson DF, Fahrenkrug SC, Wells DN, Laible G (2018) Cattle with a precise, zygote-mediated deletion safely eliminate the major milk allergen beta-lactoglobulin. Sci Rep 8(1):7661. https://doi.org/10.1038/s41598-018-25654-8

Whitelaw CB, Joshi A, Kumar S, Lillico SG, Proudfoot C (2016) Genetically engineering milk. J Dairy Res 83 (1):3–11. https://doi.org/10.1017/S0022029916000017 (Review)

Wolf E, Jehle PM, Weber MM, Sauerwein H, Daxenberger A, Breier BH, Besenfelder U, Frenyo L, Brem G (1997) Human insulin-like growth factor I (IGF-I) produced in the mammary glands of transgenic rabbits: yield, receptor binding, mitogenic activity, and effects on IGF-binding proteins. Endocrinology 138(1):307–313

Marine Bioresources—Animals and Veterinary Applications

39

Abstract

The marine life constituting around one-half of the global biodiversity serves as a rich source of bioactive ingredients for health and industrial applications. Indeed, the vast majority of the marine microorganisms is only partly explored.

Highlights

- Marine ecosystem comprises of a vast category of microorganisms, majority of which is poorly understood
- The marine microorganisms are treasure of valuable microbes and microbial metabolites
- It is high time to harness marine biodiversity and bioresources.

Keywords

Marine biome · Marine microorganisms · Metagenomics · Bioactive molecules · Bryozoan · Molluscs

39.1 Introduction

The aquatic part constitutes around 71 percent of the surface area of earth. Presence of over 95% of the biosphere as organisms, *viz*. microorganisms, plants, and animals throughout the water column represents living organisms adapted to extreme temperature, salinity, light, darkness, and pressure. Such a vast living world provides myriad of prokaryotic microorganisms multicellular sedentary (sponges and coelenterates) and moving organisms and their living parts with unprecedented curative and industrial biotechnological potential.

There are two main reasons to look into marine bioresources, exhibiting adaptations ranging from cold polar seas at −2 °C to the high temperature and pressure at ocean floor, where hydrothermal waves continuously spew forth. First, the marine biota represents a major share of the earth's biological resources. Second, the marine biota possesses unique genomes, metabolic pathways and sensory and defense mechanisms, probably because they have to survive in extreme environments. Hence, the marine biota has developed unique adaptive features through precise modifications in genetic makeup to synthesize metabolites that are quite different than those found in their terrestrial counterparts.

The marine extremophiles are important organisms in view of valuable enzymes and pharmaceuticals and therapeutics for fish, veterinary, and human applications (Fig. 39.1). The pace of discovery of new species and products that are potentially useful to pharmacology is high for marine species than for terrestrial organisms (Li et al. 2017; Giordano et al. 2018).

Marine biotechnology encompasses the efforts that involve the marine resources of the world, either as the source or as the target of

B. Singh et al., *Advances in Animal Biotechnology*,
https://doi.org/10.1007/978-3-030-21309-1_39

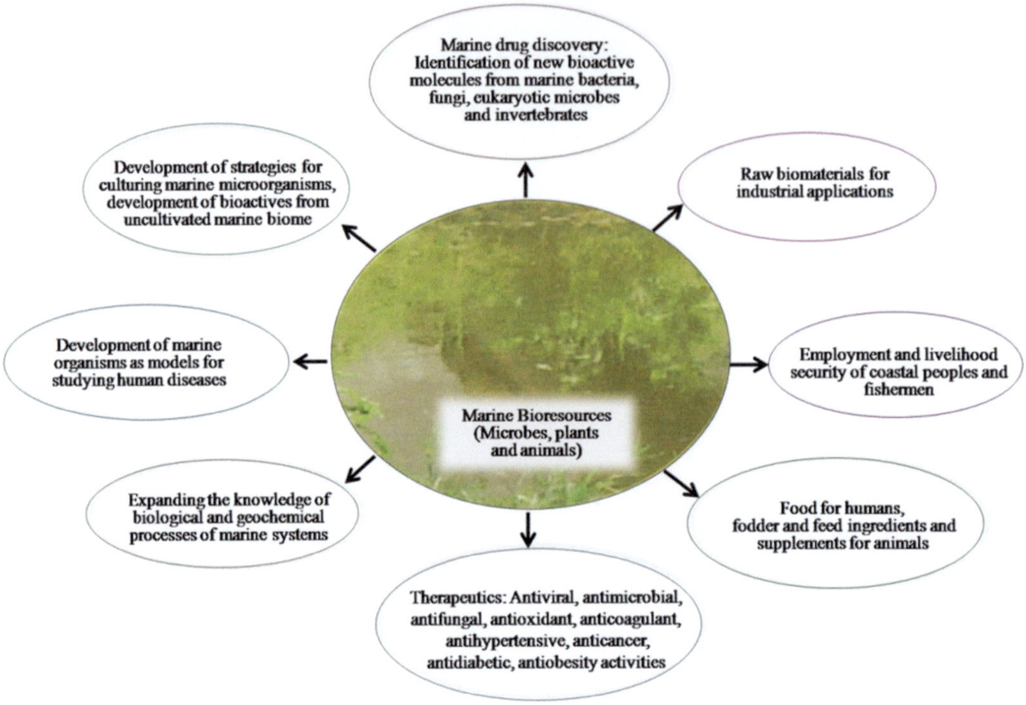

Fig. 39.1 Marine biome as a massive source of bioactive metabolites. Among various ecosystems, marine biome and metagenomics have vast potential to serve and promote health, pharmaceutical, and bio-business sector

biotechnology applications. Many marine cyanobacterial natural products are used as anti-proliferative or anticancer agents. Some metabolites from marine microbes and higher animals are found to have anti-inflammatory activities. Anti-infective compounds obtained from marine animals include viridamides A and B, gallinamide A, dragonamide E, and the almiramides (Nunnery et al. 2010a, b). Comparative genomic and metabolomic analysis of marine bioresources discloses the variety of biosynthetic gene clusters (BGCs) coding for non-ribosomal peptides (NRPs) in them, antimicrobial molecules, and the potential for biosynthesis of novel specialized metabolites (Amiri Moghaddam et al. 2018).

Commercial applications of marine biotechnology include bioprospecting, improving production of marine organisms, extraction of novel products, particularly food and feed products and diagnostics (Correia-da-Silva et al. 2017; Schiller et al. 2018). Marine-derived drugs, such as analgesic (Prialt®) obtained from *Conus magus*, a fish-eating snail, and a cancer drug (Yondelis®) are obtained from a mangrove tunicate. The enzymes such as Vent-DNA polymerase are obtained from hydrothermal vent microbes and used in polymerase chain reaction. Ziconotide is a powerful analgesic drug used to get rid of severe and chronic pain. It is derived from *Conus magus*.

39.2 Bioprospecting Marine Resources

Indeed, the marine organisms have emerged as wealthy sources of bioactive natural products. Bioprospecting includes screening and identifying bioactive molecules, and evaluating them for efficacy and safety testing. The process involves multiple experimental steps. The bioprospecting strategies are generally analogous for all animal and plant phyla from slow or deep seabed ecosystems.

Research into pharmacological properties of marine natural products has led to the discovery of many compounds considered worthy of biomedical and veterinary applications.

Great potential in bioprospecting from marine resources has just started to bloom.

Ara-C and Ara-A drugs were derived from sponges in the early 1950s. There are hopes of obtaining drugs, foodstuffs, nutraceuticals, adhesives, paints, cosmetics, and environmental remediation from marine bioresources.

39.3 Therapeutics from Marine Resources

The marine sponges and associated microbes produce several therapeutically important metabolites. For instance, the Class Demospongiae and the orders Halichondrida, Poecilosclerida, and Dictyoceratida are among the rich sources of therapeutic biomolecules. Similarly, marine actinobacteria and Ascomycota are the sources of therapeutic metabolites. The intense competition for survival in tough environment such as higher levels of halogenations and salinity, limited availability of resources have introduced development of distinct chemical defense systems in marine organisms (Williams et al. 2010; Mayer et al. 2010, 2017).

This offers great potential for obtaining new medicinal compounds and enzymes from marine organisms. Sea hares (order Opisthobranchia, subclass Gastropoda) have attracted attention of researchers who are interested in unraveling chemical defense mechanisms.

39.4 Therapeutics Form Marine Actinobacteria

Marine actinobacteria are among the prevalent microorganisms in marine habitats and serve as prolific producers of valuable metabolites such as antimicrobial, anti-proliferative, antitumor, and anticancer biomolecules. Marine actinobacteria have attracted the attention of microbial ecologists in view of unique features of secondary metabolites identified and obtained from them. They are used for synthesizing new drugs to combat pathogens and treat diseases like cancers.

39.5 Therapeutics Form Marine Cyanobacteria

To ward off predation by diverse types of macrograzers, the marine cyanobacteria have acquired genetic capability to produce several bioactive molecules as self-defense mechanisms (Nagle and Paul 1999). Several marine cyanobacterial natural products due to their unique intriguing structures and anti-proliferative and pronounced anti-inflammatory activities are focus of research. In addition, anti-infective compounds such as viridamides A and B, gallinamide A, dragonamide E, and almiramides are of considerable health interest (Nunnery et al. 2010a, b).

Abundantly available marine prokaryotic microorganisms such as cyanobacteria, favorable environmental conditions such as shallow tropical waters, have become prime targets of natural product chemists and biochemical analysts. Combined with an increasing knowledge of how these molecules are made at the biochemical and genetic levels, they are deemed as potential sources for biotechnological applications (Jones et al. 2010; Sammet et al. 2010).

Enzymes-derived marine extremophiles offer advantages over traditional enzymes used in food processing due to their ability to function at extremes (high as well as low) of temperature and pH. Polysaccharides, namely algins, carrageenans, and agar, are used for as thickeners and stabilizers in a variety of foods. Besides, ω-3 polyunsaturated fatty acids (PUFA) and photosynthetic pigments are the important nutraceuticals.

39.6 Cryptophycins

Cryptophycins were initially reported from *Nostoc* sp. The Cryptophycins have antifungal bioactivity. Cryptophycin is an inhibitor of tubulin assembly in animal cells. Recently identified and synthesized cryptophycin derivatives and analogues show high biological activity in cytotoxic assays (Sammet et al. 2010).

39.7 Thiocoraline

Thiocoraline, a potential candidate for clinical trials. Belonging to the chemical family known as the thiodepsipeptides, thiocoraline was first isolated from *Micromonospora marina*, an actinomycete bacterium. The Spanish marine drug company PharmaMar has reported that the compound is active against several standard drug screens, including breast cancer, colon cancer, renal cancer, and melanoma. Thiocoraline induces profound perturbations of cell cycle arrest, and is under investigations for safety and preclinical assessment.

39.8 Bioactive Metabolites from Tropical Marine Sponges and Tunicates

The sponges belong to phylum Porifera. The marine sponges are sessile filter feeders, some of which have developed symbiosis with other microorganisms, while others have developed unique mechanisms of self-defense against invading bacteria, and eukaryotes. The marine sponges are regarded as valuable sources of biologically active metabolites. Indeed, many compounds have been derived from marine sponges.

Xestospongia muta, a largest marine sponge found in coral reefs of the Caribbean Sea, exhibits association with microorganisms belonging to microbial phyla such as Cyanobacteria, Firmicutes, Actinobacteria, and Spirochete (Villegas-Plazas et al. 2018).

Similarly, the microorganisms, such as bacteria, actinobacteria, and cyanobacteria, and marine fungi are associated with Ascidians, the sessile filter feeding invertebrates (subphylum Tunicata). Hence the ascidians are important sources of natural products.

The bioactive chemicals obtained from marine resources are categorized into six classes including terpenes, polyketides, peptides, shikimates, alkaloids, and saccharides. More than 1000 natural products have been identified and characterized that exhibit anti-inflammatory, antibacterial, and antifungal activities.

39.9 Bryozoan-Derived Therapeutics

Bryozoa, also known as Polyzoa, moss animals or Octapoda are sessile aquatic invertebrates in some reef frameworks. They are typically 0.5 long, filter feeders, feeding on small cellular algae, protozoa, etc. Most bryozoa live in colonies and depend on each others; each individual is called as zooids.

The marine bryozoa are of interest from drug discovery perspective as they produce bioactive natural metabolites (Ruan and Zhu 2012; Tian et al. 2018). The bryozoa of family Flustridae are rich sources of important secondary metabolites (Hansen et al. 2017).

39.10 Bryostatins

Bryostatin 1 is a macrocyclic lactone, originally isolated from *Bugula neritina*. The bryostatin is a powerful protein kinase C (PKC) agonists with anti-neoplastic, synergistic, and chemotherapeutic activities and as therapy to prevent Alzheimer's disease. In addition, bryostatin is beneficial in promoting cognitive health (Ruan and Zhu 2012). Other bryozoa such as *Bugula simplex* also have bryostatin-like metabolites. Some bryostatins with anti-tumor activities are purified from bryozoan found in Japanese (Ueno et al. 2012).

Suppression of matrix metalloproteinases (MMPs), and blood–brain barrier (BBB) damage owing to anti-inflammatory and antioxidant activities, makes Bryostatin-1 a potent candidate to treat multiple sclerosis (MS). In addition, the Bryostatin-1 is proposed to be effective cure in multiple sclerosis due to its neuroprotective effects, promoting neurogenesis and inducing differentiation of oligodendrocytes progenitor stem cells (Safaeinejad et al. 2018).

Tian et al., (2018) have reviewed structural and functional attributes of 164 new compounds with therapeutic potential, originating from 24 marine bryozoans.

Sacuramine derivatives from Arctic bryozoa (*Securiflustra securifrons*) of Flustridae family displayed cytotoxicity against human skin, colon, and breast cancer cell lines, and non-malignant lung fibroblasts (Hansen et al. 2017) (Table 39.1).

39.11 Compounds Derived from Mollusks

Mollusca are the second largest phylum among invertebrates representing more than 85,000 extant species. The marine mollusks constitute about 23% of total mollusks and are a diverse group of invertebrates with multiple roles in human nutrition, health, and economy. The marine mollusks are divided into five subcategories. The phylum mollusca include snails, slugs, gastropods, clams and other bivalves, squids, and associated cephalopods.

Table 39.1 Therapeutically relevant metabolites from marine sources

Metabolites	Origin	Biological activities
Actinomycetes and others		
Abyssomicins (novel polycyclic Polyketides)	*Verrucosispora* sp. strain AB-18-032	Antibacterial activities against multi-resistant clinical origin *Staphylococcus aureus* (Riedlinger et al. 2004)
Angucycline glycosides	*Streptomyces* sp. (from intertidal sediments)	Cytotoxic activity against hepatoma carcinoma cell lines (Peng et al. 2018)
Bonactin (1)	*Streptomyces* sp. (from marine sediment)	Antifungal and antibacterial activity against Gram-positive as well as Gram-negative bacteria (Schumacher et al. 2003)
Ecteinamycin	*Actinomadura* sp.	Antimicrobial activities against Clostridium difficile via potassium transport dysregulation mechanisms (Wyche et al. 2017)
Marinopyrroles (A and B)	*Streptomyces* sp.	Antibacterial activities against methicillin-resistant *Staphylococcus aureus* (Hughes et al. 2008)
Marinopyrroles (A–F)	*Streptomyces* sp.	Antibacterial activities against methicillin-resistant *Staphylococcus aureus* (Hughes et al. 2010)
Thiocoraline	*Verrucosispora* sp. (from sponge *Chondrilla caribensis* f. *caribensis*)	Thiocoraline is an analogue of thiodepsipeptide thiocoraline A, expresses potent cytotoxic activity against human cancer cell lines (A549) (Wyche et al. 2011)
Bryozoans		
Bryostatin	*Bugula neritina B. simplex*	Bryostatins are anti-neoplastic, synergistic, and chemotherapeutic activities, beneficial in beneficial in promoting cognitive health (Ruan and Zhu 2012)
		Decrease in matrix metalloproteinases (MMPs) activity, and blood–brain barrier (BBB) damage owing to anti-inflammatory and antioxidant activities. In addition, the Bryostatin-1 is proposed to be an effective treatment in multiple sclerosis due to its neuroprotective effects, promote neurogenesis and differentiation of oligodendrocytes progenitor stem cells (Safaeinejad et al. 2018)

(continued)

Table 39.1 (continued)

Metabolites	Origin	Biological activities
Sacuramine	*Securiflustra securifrons*	Cytotoxicity against human non-malignant lung fibroblasts, skin, colon, and breast cancer cell lines (Hansen et al. 2017)
Mollusks		
Dolastatins	*Dollabella auricularia*	Dolastatin 10 and Dolastatin 15 are potent antitumor agents (Yokosaka et al. 2018)
Ziconotide	*Conus magus*	Prescribed for managing chronic refractory pain, but may have cognitive side effects (Wallace et al. 2008; Burdge et al. 2018)
Marine sponges		
Dictyostatin	*Spongia* sp.	Stabilization of microtubules (MTs), inhibition of expansion of certain cancer cell lines and cells that have become resistant to Taxol, hence emerging as potent anticancer drug (Raccor et al. 2008)
Girolline	*Cymbastela cantharella*	Cytotoxic, anticancer, and antiplasmodial activities (Benoit-Vical et al. 2008)

Mollusk-derived metabolites (terpenes, steroids, peptides, polyketides, or nitro compounds) including those produced by mollusks themselves as well as originating from their dietary sources have various anticancer properties. The anti-proliferative activities are identified based on characteristic properties of mollusk metabolites to selectively inhibit cancer cells, but not normal cells, and circumventing multidrug resistant phenotypes (Ciavatta et al. 2017).

There is demand for bio-derived or biosynthetically produced polymers. Naturally occurring proteins of marine squid proteins can be used as next generation biomaterials in medicine, electronics, energy, security, and defense sector (Pena-Francesch and Demirel 2019).

39.12 Dolastatins

Dolastatins (Dolastatin 10 and Dolastatin 15) are the natural marine products obtained from Indian sea hare, *Dolabella auricularia.* Dolastatins are potent antitumor agents with activities against mammary and hepatic cancers, solid tumors, and leukemias (https://adcreview.com/adc-university/adcs-101/cytotoxic-agents/dolastatins-dolastatin-10-and-dolastatin-15/, accesses on Nov. 24, 2018). The dolastatins are microtubule inhibitors (Matthew et al. 2007). Dolastatin 10 is a linear peptide having N,N-dimethyl Val-OH, l-valine, (3R,4S,5S)-dolaisoleucine, (2R,3R,4S)-dolaproine, and (S)-dolaphenine (Zhou et al. 2017). Analogs of Dolastatin 10 are used as payloads in antibody drug conjugates. Dolastatin 10 analogs having different functional groups exhibited potency against tumor cells (Yokosaka et al. 2018). Some cyanobacteria, e.g., *Caldora penicillata* gen.nov., comb. nov. (Engene et al. 2015) forming part of diet of mollusks, also produce dolastatins.

39.13 Ziconotide (Prialt)

Ziconotide or SNX-111 is non-opid analgesic drug obtained from venom of *Conus geographus* and *Conus magus*, the fish predating cone snails. This drug is a member of newly described chemical family, called the conopeptides, and prescribed for managing chronic refractory pain. The fact that ziconotide and other conotoxins are short (usually 20–30 amino acids) peptides means that it is easy to synthesize their derivatives.

The contraindications associated with the use of ziconotide include induction of psychosis or cognitive impairment or changes in mood depending on case history of patients (Wallace et al. 2008; Burdge et al. 2018).

39.14 Sponge-Derived Compounds

Marine sponges, the sessile, filter-feeder multicellular aquatic animals existing for millions of years, belong to phylum Porifera (Metazoa). They are among best sources of natural products with health and industrial importance. As sponges are sessile, they are threatened from predation by other animals such as turtles, fish, and mobile mollusks. To defend themselves, the sponges have developed chemical means of their defense. Hence, sponges serve as huge diversity of structurally and functionally active secondary metabolites that are discovered over the past 50 years (Hertiani et al. 2010; Moitinho-Silva et al. 2017).

The bioactive metabolites derived from sponges include antibacterial compounds, such as cyclic peptides like Discodermins B, C and D (Matsunaga et al. 1985), alkaloid such as 6-hydroxymanzamine E (Rao et al. 2004), nitrogen heterocyclic Cribrostatin 3 and Cribrostatin (Pettit et al. 2004), anti-viral alkaloids such as Dragmacidin F (Cutignano et al. 2000), anti-malarial alkaloids such as Manzamine A (D'Ambrosio et al. 1998), are already identified from marine sponges.

39.15 Dictyostatin

Dictyostatin is an anticancer agent, originally isolated from a marine sponge belonging to Genus *Spongia*. Later, the dictyostatin was isolated from a sponge belonging to family Corallistidae. Dictyostatin inhibits the expansion of certain cancer cell lines and cells that have become resistant to Taxol.

The microtubule stabilizers are potent anticancer agents and pioneer a prudent path to anticancer drug discovery (Cao et al. 2018). The naturally occurring compound (-)-dictyostatin binds to MTs, causes cell cycle arrest in G(2)/M at nanomolar concentrations, and retains anti-proliferative activity in paclitaxel-resistant cell lines, making dictyostatin an attractive anti-neoplastic agent (Raccor et al. 2008). The dictyostatin shares structural similarity with discodermolide and has affinity taxpod binding site on cellular tubulin.

One of the important features of dictyostatin is that, it crosses blood–brain barrier (BBB), as illustrated in experimental mice, and shows extended brain retention and stabilizes MTs in brain. This implies that dictyostatin is a promising tauopathy curative substance (Brunden et al. 2013).

39.16 Girolline

Girolline is a 2-aminoimidazole derivative extracted from *Cymbastela cantharella* (a New-Caledonian sponge) (Benoit-Vical et al. 2008). The Girolline possesses cytotoxic and antitumor activity. It was shown to inhibit protein synthesis at the termination steps of the translation process rather than at initiation or chain elongation steps. The simple structure of the compound facilitates the synthesis of alternative synthetic molecules with anticancer properties.

The Girolline shows anti-malarial activity by inhibiting *Plasmodium falciparum* and was 100% inhibitory to *P. falciparum* (Benoit-Vical et al. 2008). Bromotyrosine alkaloid derivatives from *Hyattella* sp. an Australian marine sponge were inhibitory to *P. falciparum* (Yang et al. 2010).

39.17 Therapeutics from Marine Fungi

The fungi and bacteria inhabiting marine or estuarine milieu in ocean depths, seawater, and coastal water are of paramount economic interest (Wiese and Imhoff 2018). Around 444 species of marine fungi, including seven genera and ten species of basidiomycetes, and 177 genera and 360 species of ascomycetes, are documented so far. In addition, some marine fungi are chytrids and mitosporic or asexual fungi (https://en.wikipedia.org/wiki/Marine_fungi, accessed on Dec. 2018).

The list of bioactive molecules obtained from marine fungi and bacteria is enormous. For instance, the compounds with neuroprotective activities are obtained from a marine-derived fungus *Penicillium* sp. (Yurchenko et al. 2018). Two new glucosidase inhibitors are identified from an endophytic fungus, Aspergillus flavus (Wu et al. 2018). Aspergixanthones I¯K obtained from marine-derived Aspergillus sp. ZA-01 have shown strong antagonistic activity against *Vibrio parahaemolyticus, V. anguillarum, and V. alginolyticus* (Zhu et al. 2018).

39.18 Products from Marine Vertebrates

39.18.1 Squalamine

Squalamine is a small aminosterol molecule isolated from the stomach and liver of the spiny dogfish, *Squalus acanthias*. It has a novel intracellular mechanism of action. Squalamine shows a broad range of antibiotic activity.

Squalamine prevents aberrant neovascularization by inhibiting multiple protein growth factors such as vascular endothelial growth factor (VEGF), platelet-derived growth factor (PDGF), and basic fibroblast growth factor (bFGF), with high potency at nanomolar concentrations. In addition, squalamine shows anti-angiogenic activity under certain conditions.

39.19 Outlook and Challenges

It is evident that marine biodiversity is a mammoth source of natural products with interesting chemistry, biosynthetic pathways, and enzymes. There is vast potential to discover new biologicals and compounds with therapeutic and industrial applications from marine organisms (Box 2).

Despite vast possibilities for the use of marine organisms in the food industry, further biotechnological interventions are required to improve recovery of unique bioactive metabolites. Similarly, proteins from some marine fishes are found to operate at relatively low temperatures and can be used in heat-sensitive processes such as gelling and clarifying.

Besides the development of novel marine models to study the molecular basis of poorly understood human disease processes, the development of biochemical sensors to monitor human and environmental health indicators in the marine environment should be another prioritized area of interest.

In near future, marine bioprospecting efforts should focus not only on natural products from ocean life, but also on the potential for biotechnology to exploit the genetic wealth of these organisms. Increasingly, new projects are expected to target the unidentified biomolecular potential of vast, and almost entirely unknown marine microbial community. These research efforts could be instrumental in developing next generation of pharmaceuticals from marine ecosystem for human health, as well as contributing to improved livestock and agricultural production.

However, there are challenges ahead. Overutilization of marine bioresources poses threat to diversity of these organisms. Already some habitats are threatened due to human interventions. Alteration in diversity of coral reefs and mangroves will result in the loss of beneficial marine organisms. With rapid depletion of natural marine resources, the "omocs" technologies might offer hopes to efficiently produce novel peptides with unique pharmacological activities against pathogens. Non-conventional collaborations and interdisciplinary approaches in marine biotechnology research and development are likely to result in innovative and transformative technologies in the future.

Contamination of water bodies by industrial effluents containing halogenated organic contaminants, heavy metal toxicity, and accidental oil spillages is among major threats to marine biome.

Box 2: The Research Priorities in Marine Biotechnology

- Exploring marine microbiome by means of genomic and metagenomic approaches and evolving strategies to culture precious microbiota
- Applying metabolomics to analyze complex metabolome of marine organisms and identify the prospective biomolecules
- Separation and purification of desirable metabolites by advanced analytical methods
- Insights into microbial physiology, genetics, biochemistry, and ecology in order to provide model systems for research and production systems for commerce
- Elaborating the use of biotechnology to improve health, reproduction, development, growth, and overall well-being of cultivated aquatic organisms promoting the interdisciplinary development of environmentally sensitive, sustainable systems that will enable significant commercialization of aquaculture.

39.20 Conclusions

The marine ecosystem has emerged as precious treasure of raw materials, enzymes, and therapeutic agents for treating cancer, diabetes, infectious diseases mediated by bacteria, fungi, and parasites, inflammation and cancer. Development of these products and services, as well as the fundamental research, from which they are derived, will be enhanced by collaborative and interdisciplinary sciences such as pharmacology, chemical ecology, molecular biology, genomics, metagenomics, computational and combinatorial chemistry and biology.

References

Amiri Moghaddam J, Crüsemann M, Alanjary M, Harms H, Dávila-Céspedes A, Blom J, Poehlein A, Ziemert N, König GM, Schäberle TF (2018) Analysis of the genome and metabolome of marine myxobacteria reveals high potential for biosynthesis of novel specialized metabolites. Sci Rep 8(1):16600. https://doi.org/10.1038/s41598-018-34954-y

Benoit-Vical F, Saléry M, Soh PN, Ahond A, Poupat C (2008) Girolline: a potential lead structure for antiplasmodial drug research. Planta Med 74(4):438–444. https://doi.org/10.1055/s-2008-1034348

Brunden KR, Gardner NM, James MJ, Yao Y, Trojanowski JQ, Lee VM, Paterson I, Ballatore C, Smith AB 3rd (2013) MT-stabilizer, dictyostatin, exhibits prolonged brain retention and activity: potential therapeutic implications. ACS Med Chem Lett 4 (9):886–889. https://doi.org/10.1021/ml400233e (eCollection)

Burdge G, Leach H, Walsh K (2018) Ziconotide-induced psychosis: A case report and literature review. Ment Health Clin 8(5):242–246. https://doi.org/10.9740/mhc.2018.09.242 (eCollection)

Cao YN, Zheng LL, Wang D, Liang XX, Gao F, Zhou XL (2018) Recent advances in microtubule-stabilizing agents. Eur J Med Chem. 143:806–828. https://doi.org/10.1016/j.ejmech.2017.11.062 (Epub 2017 Nov 24. Review)

Ciavatta ML, Lefranc F, Carbone M, Mollo E, Gavagnin M, Betancourt T, Dasari R, Kornienko A, Kiss R (2017) marine mollusk-derived agents with antiproliferative activity as promising anticancer agents to overcome chemotherapy resistance. Med Res Rev. 37(4):702–801. https://doi.org/10.1002/med.21423 (Epub 2016 Dec 7. Review)

Correia-da-Silva M, Sousa E, Pinto MMM, Kijjoa A (2017) Anticancer and cancer preventive compounds from edible marine organisms. Semin Cancer Biol 46:55–64. https://doi.org/10.1016/j.semcancer.2017.03.011 (Epub 2017 Apr 7. Review)

Cutignano A, Bifulco G, Bruno I, Casapullo A, Gomez-Paloma L, Riccio R, Dragmacidin F (2000) A new antiviral bromoindole alkaloid from the mediterranean sponge *Halicortex* sp. Tetrahedron 56:3743–3748

D'Ambrosio M, Guerriero A, Deharo E, Debitus C, Munoz V, Pietra F (1998) New types of potentially antimalarial agents: epidioxy-substituted norditerpene and norsesterpenes from the marine sponge *Diacarnus levii*. Helv Chim Acta 81:1285–1292

Engene N, Tronholm A, Salvador-Reyes LA, Luesch H, Paul VJ (2015) *Caldora penicillata* gen. nov., comb. nov. (cyanobacteria), a pantropical marine species with biomedical relevance. J Phycol 51(4):670–681

Giordano D, Costantini M, Coppola D, Lauritano C, Núñez Pons L, Ruocco N, di Prisco G, Ianora A, Verde C (2018) Biotechnological applications of bioactive peptides from marine sources. Adv Microb Physiol 73:171–220. https://doi.org/10.1016/bs.ampbs.2018.05.002 (Epub 2018 Jun 20)

Hansen KØ, Isaksson J, Bayer A, Johansen JA, Andersen JH, Hansen E (2017) Securamine derivatives from the arctic bryozoan securiflustra securifrons. J Nat Prod 80(12):3276–3283. https://doi.org/10.1021/acs.jnatprod.7b00703

Hertiani T, Edrada-Ebel R, Ortlepp S, van Soest RW, de Voogd NJ, Wray V, Hentschel U, Kozytska S, Müller WE, Proksch P (2010) From anti-fouling to biofilm inhibition: new cytotoxic secondary metabolites from two Indonesian Agelas sponges. Bioorg Med Chem 18 (3):1297–1311. https://doi.org/10.1016/j.bmc.2009.12.028 (Epub 2010 Jan 12)

Hughes CC, Prieto-Davo A, Jensen PR, Fenical W (2008) The marinopyrroles, antibiotics of an unprecedented structure class from a marine *Streptomyces* sp. Org Lett 10(4):629–631. https://doi.org/10.1021/ol702952n (Epub 2008 Jan 19)

Hughes CC, Kauffman CA, Jensen PR, Fenical W (2010) Structures, reactivities, and antibiotic properties of the marinopyrroles A–F. J Org Chem 75(10):3240–3250. https://doi.org/10.1021/jo1002054

Jones AC, Monroe EA, Eisman EB, Gerwick L, Sherman DH, Gerwick WH (2010) The unique mechanistic transformations involved in the biosynthesis of modular natural products from marine cyanobacteria. Nat Prod Rep 27(7):1048–1065. https://doi.org/10.1039/c000535e (Epub 2010 May 4. Review)

Li T, Ding T, Li J (2017) Medicinal purposes: bioactive metabolites from marine-derived organisms. Mini Rev Med Chem. https://doi.org/10.2174/1389557517666170927113143

Matthew S, Ross C, Rocca JR, Paul VJ, Luesch H (2007) Lyngbyastatin 4, a dolastatin 13 analogue with elastase and chymotrypsin inhibitory activity from the marine cyanobacterium Lyngbya confervoides. J Nat Prod 70(1):124–127

Matsunaga S, Fusetani N, Konosu S (1985) Bioactive marine metabolites, IV. Isolation and the amino acid composition of discodermin A, an antimicrobial peptide, from the marine sponge Discodermia kiiensis. J Nat Prod 48(2):236–241

Mayer AM, Glaser KB, Cuevas C, Jacobs RS, Kem W, Little RD, McIntosh JM, Newman DJ, Potts BC, Shuster DE (2010) The odyssey of marine pharmaceuticals: a current pipeline perspective. Trends Pharmacol Sci 31(6):255–265. https://doi.org/10.1016/j.tips.2010.02.005

Mayer AMS, Rodríguez AD, Taglialatela-Scafati O, Fusetani N (2017) Marine pharmacology in 2012–2013: marine compounds with antibacterial, antidiabetic, antifungal, anti-inflammatory, antiprotozoal, antituberculosis, and antiviral activities; affecting the immune and nervous systems, and other miscellaneous mechanisms of action. Mar Drugs 15(9). pii: E273. https://doi.org/10.3390/md15090273

Moitinho-Silva L, Nielsen S, Amir A, Gonzalez A, Ackermann GL, Cerrano C, Astudillo-Garcia C, Easson C, Sipkema D, Liu F, Steinert G, Kotoulas G, McCormack GP, Feng G, Bell JJ, Vicente J, Björk JR, Montoya JM, Olson JB, Reveillaud J, Steindler L, Pineda MC, Marra MV, Ilan M, Taylor MW, Polymenakou P, Erwin PM, Schupp PJ, Simister RL, Knight R, Thacker RW, Costa R, Hill RT, Lopez-Legentil S, Dailianis T, Ravasi T, Hentschel U, Li Z, Webster NS, Thomas T (2017) The sponge microbiome project. Gigascience 6(10):1–7. https://doi.org/10.1093/gigascience/gix077

Nagle DG, Paul VJ (1999) Production of secondary metabolites by filamentous tropical marine cyanobacteria: ecological functions of the compounds. J Phycol 35:1412–1421

Nunnery JK, Mevers E, Gerwick WH (2010a) Biologically active secondary metabolites from marine cyanobacteria. Curr Opin Biotechnol 21(6):787–793. https://doi.org/10.1016/j.copbio.2010.09.019 (Epub 2010 Oct 26)

Nunnery JK, Mevers E, Gerwick WH (2010b) Biologically active secondary metabolites from marine cyanobacteria. Curr Opin Biotechnol 21(6):787–793. https://doi.org/10.1016/j.copbio.2010.09.019 (Epub 2010 Oct 26. Review)

Pena-Francesch A, Demirel MC (2019) Squid-inspired tandem repeat proteins: functional fibers and films. Front Chem (in press). https://doi.org/10.3389/fchem.2019.00069

Peng A, Qu X, Liu F, Li X, Li E, Xie W (2018) Angucycline glycosides from an intertidal sediments strain *Streptomyces* sp. and their cytotoxic activity against hepatoma carcinoma cells. Mar Drugs 16(12). pii: E470. https://doi.org/10.3390/md16120470

Pettit RK, Fakoury BR, Knight JC, Weber CA, Pettit GR, Cage GD, Pon S (2004) Antibacterial activity of the marine sponge constituent cribrostatin 6. J Med Microbiol 53(Pt 1):61–65

Raccor BS, Vogt A, Sikorski RP, Madiraju C, Balachandran R, Montgomery K, Shin Y, Fukui Y, Jung WH, Curran DP, Day BW (2008) Cell-based and biochemical structure-activity analyses of analogs of the microtubule stabilizer dictyostatin. Mol Pharmacol 73(3):718–726 (Epub 2007 Dec 11)

Rao VK, Kasanah N, Wahyuono S, Tekwani BL, Schinazi RF, Hamann MT (2004) Three new manzamine alkaloids from a common Indonesian sponge and their activity against infectious and tropical parasitic diseases. J Nat Prod 67:1314–1318

Riedlinger J, Reicke A, Zähner H, Krismer B, Bull AT, Maldonado LA, Ward AC, Goodfellow M, Bister B, Bischoff D, Süssmuth RD, Fiedler HP (2004) Abyssomicins, inhibitors of the para-aminobenzoic acid pathway produced by the marine Verrucosispora strain AB-18-032. J Antibiot (Tokyo) 57(4):271–279

Ruan BF, Zhu HL (2012) The chemistry and biology of the bryostatins: potential PKC inhibitors in clinical development. Curr Med Chem 19(16):2652–2664

Safaeinejad F, Bahrami S, Redl H, Niknejad H (2018) Inhibition of inflammation, suppression of matrix metalloproteinases, induction of neurogenesis, and antioxidant property make bryostatin-1 a therapeutic choice for multiple sclerosis. Front Pharmacol 9:625. https://doi.org/10.3389/fphar.2018.00625 (eCollection)

Sammet B, Bogner T, Nahrwold M, Weiss C, Sewald N (2010) Approaches for the synthesis of functionalized cryptophycins. J Org Chem 75(20):6953–6960. https://doi.org/10.1021/jo101563s

Schiller L, Bailey M, Jacquet J, Sala E (2018) High seas fisheries play a negligible role in addressing global food security. Sci Adv 4(8):eaat8351. https://doi.org/10.1126/sciadv.aat8351 (eCollection 2018 Aug)

Schumacher RW, Talmage SC, Miller SA, Sarris KE, Davidson BS, Goldberg A (2003) Isolation and structure determination of an antimicrobial ester from a marine sediment-derived bacterium. J Nat Prod 66 (9):1291–1293

Tian XR, Tang HF, Tian XL, Hu JJ, Huang LL, Gustafson KR (2018) Review of bioactive secondary metabolites from marine bryozoans in the progress of new drugs discovery. Future Med Chem 10(12):1497–1514. https://doi.org/10.4155/fmc-2018-0012 (Epub 2018 May 23)

Ueno S, Yanagita RC, Murakami K, Murakami A, Tokuda H, Suzuki N, Fujiwara T, Irie K (2012) Identification and biological activities of bryostatins from Japanese bryozoan. Biosci Biotechnol Biochem 76(5):1041–1043

Villegas-Plazas M, Wos-Oxley ML, Sanchez JA, Pieper DH, Thomas OP, Junca H (2018) Variations in microbial diversity and metabolite profiles of the tropical marine sponge xestospongia muta with season and depth. Microb Ecol. https://doi.org/10.1007/s00248-018-1285-y

Wallace MS, Kosek PS, Staats P, Fisher R, Schultz DM, Leong M (2008) Phase II, open-label, multicenter study of combined intrathecal morphine and ziconotide: addition of ziconotide in patients receiving intrathecal morphine for severe chronic pain. Pain Med 9(3):271–281. https://doi.org/10.1111/j.1526-4637.2007.00355.x

Wiese J, Imhoff JF (2018) Marine bacteria and fungi as promising source for new antibiotics. Drug Dev Res. https://doi.org/10.1002/ddr.21482

Williams P, Sorribas A, Liang Z (2010) New methods to explore marine resources for Alzheimer's therapeutics. Curr Alzheimer Res 7(3):210–213 (Review)

Wu Y, Chen Y, Huang X, Pan Y, Liu Z, Yan T, Cao W, She Z (2018) α-glucosidase inhibitors: diphenyl ethers and phenolic bisabolane sesquiterpenoids from the mangrove endophytic fungus *Aspergillus flavus* QQSG-3. Mar Drugs 16(9). pii: E307. https://doi.org/10.3390/md16090307

Wyche TP, Hou Y, Braun D, Cohen HC, Xiong MP, Bugni TS (2011) First natural analogs of the cytotoxic thiodepsipeptide thiocoraline A from a marine *Verrucosispora* sp. J Org Chem 76(16):6542–6547. https://doi.org/10.1021/jo200661n

Wyche TP, Alvarenga RFR, Piotrowski JS, Duster MN, Warrack SR, Cornilescu G, De Wolfe TJ, Hou Y, Braun DR, Ellis GA, Simpkins SW, Nelson J, Myers CL, Steele J, Mori H, Safdar N, Markley JL, Rajski SR, Bugni TS (2017) Chemical genomics, structure elucidation, and in vivo studies of the marine-derived anticlostridial ecteinamycin. ACS Chem Biol 12(9):2287–2295. https://doi.org/10.1021/acschembio.7b00388

Yang X, Davis RA, Buchanan MS, Duffy S, Avery VM, Camp D, Quinn RJ (2010) Antimalarial bromotyrosine derivatives from the Australian marine sponge *Hyattella* sp. J Nat Prod 73(5):985–987. https://doi.org/10.1021/np900834g

Yokosaka S, Izawa A, Sakai C, Sakurada E, Morita Y, Nishio Y (2018) Synthesis and evaluation of novel dolastatin 10 derivatives for versatile conjugations. Bioorg Med Chem 26(8):1643–1652. https://doi.org/10.1016/j.bmc.2018.02.011

Yurchenko EA, Menchinskaya ES, Pislyagin EA, Trinh PTH, Ivanets EV, Smetanina OF, Yurchenko AN (2018) Neuroprotective activity of some marine fungal metabolites in the 6-hydroxydopamin- and paraquat-induced Parkinson's disease models. Mar Drugs 16(11). pii: E457. https://doi.org/10.3390/md16110457

Zhou W, Nie XD, Zhang Y, Si CM, Zhou Z, Sun X, Wei BG (2017) A practical approach to asymmetric synthesis of dolastatin 10. Org Biomol Chem 15 (29):6119–6131. https://doi.org/10.1039/c7ob01395g

Zhu A, Zhang XW, Zhang M, Li W, Ma ZY, Zhu HJ, Cao F (2018) Aspergixanthones I⁻K, New anti-*vibrio* prenylxanthones from the marine-derived fungus *Aspergillus* sp. ZA-01. Mar Drugs 16(9). pii: E312. https://doi.org/10.3390/md16090312

RNA Interference: A Veterinary Health Perspective

40

Abstract

RNA interference (RNAi) is a naturally evolved and a conserved self-defensive mechanism triggered by double-stranded RNA (dsRNA) that limits transcript levels either by suppressing transcription or activating sequence-specific degradation of RNA. RNAi protects the host from endogenous and exogenous invading nucleic acids by regulating gene expression. RNAi is used as a promising tool in functional genomics and targeting microbial infections.

Highlights

- RNAi is a natural mechanism of self-defense in living organisms
- RNAi is used as a tool to prevent zoonotic infections and non-infectious diseases.

Keywords

RNAi · RNA isoforms · Virus · Bacteria · Vaccines · Gene suppression · CRISPR/Cas-9 · *Drosophila melanogaster*

40.1 Introduction

Living organisms have developed several mechanisms to protect themselves against invading infectious agents. Some of the protection mechanisms include restriction digestion of the DNA of pathogens by restriction endonucleases, modification and editing of pathogens' genome by clustered regularly interspaced short palindromic repeats (CRISPR) system and CRISPR-associated protein 9 (Cas9) (CRISPR/Cas-9), and gene silencing by means of double-stranded RNA (dsRNA).

40.2 RNA as a Versatile Molecule

RNA is versatile nucleic acid molecule not only in abundance, but also in the functions performed. There is a range of applications of RNAi such as suppressing or enhancing expression of genes through diverse mechanisms. Gene suppression is mediated through RNA interference (RNAi), microRNA (miRNA), small interfering RNA (siRNA), antisense RNA, and synthetic hairpin RNA (shRNA). Protein synthesis is altered by messenger RNA (mRNA) instilled into cells. The process is like gene therapy where genes introduced into cells produce desirable proteins to cure genetic disorder or exert beneficial effects (Balmayor and Evans 2019).

A class of small RNAs, especially miRNAs and short interference RNAs (SRNAs) are of paramount biomedical significance, and practical interest in gene-regulation mechanisms in plants and animals organisms (Carrington and Ambros 2003).

As a novel gene regulatory mechanism, the RNAi limits the transcription levels by suppressing transcription (transcription gene

B. Singh et al., *Advances in Animal Biotechnology*,
https://doi.org/10.1007/978-3-030-21309-1_40

silencing or TGS), or by activating RNA degradation in a sequence-specific manner (post-transcriptional gene silencing or PTGS) (Agrawal et al. 2003). Initially, the PTGS/RNAi were noted in plants, but later on RNAi-associated events were found in many other organisms including protozoa, nematodes, arthropods (insects and ticks), mammalian species, and cell lines (Agrawal et al. 2003).

The first evidence of RNAi came from the studies on *C. elegans*, followed by subsequent works in *Drosophila melanogaster* (Elbashiret al. 2001a), and mammalian cells (Elbashir et al. 2001b) that contributed to understand the nature of RNAi pathways.

RNAi is a potent means that requires only a few molecules of dsRNA per cell to mute gene expression. The process is triggered by various types of dsRNAs (Box 1). Under natural conditions, the dsRNA comes from viruses and transposon activities, and it can also be introduced into cells experimentally. RNAi is achieved by producing virus-derived siRNAs that when introduced into cell bind with perfect complementarity to target viral sequence (Fay and Langlois 2018).

Box 1. Various small RNA isoforms involved in RNAi

1. **Small interference RNAs**: They are 21–23 nucleotide long dsRNA molecules with 2–3 nucleotides overhangs at 3′ termini. They generate as a result of cleavage of long dsRNAs by the action of RNase III Dicer (Zamore et al. 2000). The 3′ termini are hydroxylated for siRNA-primed amplification, and 5′ termini are phosporylated by endogenous kinases to enter RNA-induced silencing complex (RISC). siRNA are globally conserved among various species (Denli et al. 2004).
2. **MicroRNAs (miRNAs)**: miRNAs are 19–25 nucleotide small RNA species generated by Dicer-mediated cleavage of endogenous 70-nucleotide non-coding stem-loop precursors. miRNAs allow mismatches and repress target mRNA translation in mammals. More than 2000 miRNAs are identified in animals and plants. They are associated with a large number of processes.
3. **Tiny non-coding RNAs (TncRNAs)**: They were first identified in *C. elegans*, and are similar to miRNAs in terms of their size, and being single-stranded structures, lack precise complimentarity to a given mRNA (Ambros et al. 2003). The developmental origin of TncRNAs is not fully understood.

40.3 Prospects of RNAi in Livestock Health

40.3.1 RNAi in Functional Genomics

Various methods have been used to SiRNA knockdown of genes in mammalian species. Much of the information concerning their role in functional genomics is obtained from *C. elegans*. In addition, RNAi is used in studies aimed to gene silencing in many species including model species such as *D. melanogaster* and mammalian species. The RNAi gene libraries have also been generated in humans (Hu et al. 2009).

40.4 Functional Genomics of Vectors and Parasites

A number of pests and parasites affect animal health and production performance. Flatworms, nematodes, annelids (leech), and arthropods (ticks, flies, and mosquitoes) affect animals in humid tropical climates, suck blood of domestic and wild animals, and spread diseases.

PTGS initiated by dsRNA is discovered in eukaryotes. Hence, RNAi is used as tool for functional genomics, and gene-silencing technique in ticks where alternative approaches of their prevention are less effective (Ramakrishnan et al. 2005; Zhou et al. 2006; Kocan et al. 2011). The injection of dsRNA is the promising strategy to study gene regulation and system biology, characterization of tick-pathogen interface, and tick protective antigens. RNAi may assist in developing vaccines to control tick infestation and transmission of tick-borne parasites and pathogens (Kocan et al. 2011).

The first tick genome published in 2016 has provided an invaluable tool for studying molecular biological basis of tick-pathogen interaction, metabolic pathways in ticks, and evolving alternative strategies to curtail the tick infestation. RNAi is a helpful tool for studying tick gene function, the characterization of the tick-pathogen interface, and screening and characterization of antigens protective against tick (de la Fuente et al. 2007; Jongejan et al. 2007; Cabezas-Cruz et al. 2019). Vaccines are expected to improve in future with novel and efficient molecular technologies aimed to discover antigens. There is an urgent need for effective strategies to control ticks and minimize the overuse of conventional acaricides, especially in regions where ticks have developed resistance against them.

Parasitic nematodes are common in animals, and share camaraderie in parasitic genes. For instance, filarial worms *Setaria digitata* novel protein (SDNP) having sequence homologies with those of *Wucheria bancrofti*, is nematode-specific and has sequence similarities with other nematodes such as *W. bancrofti*, *Brugia malayi*, *Loa loa*, and *Onchocerca volvulus*.

Microinjection of siRNAi aimed to knock down the adult *S. digitata* SDNP revealed a significant reduction in SDNP transcriptome levels. It was found that SDNP is involved in development and movement of nematode muscles, and could serve as potential drug target to prevent *S. digitata* (Somarathne et al. 2018).

Some neglected diseases such as fasciolosis caused by *Fasciola gigentica* in cattle, buffaloes, and sheep are often found in Asia and Africa. The parasite has developed resistance to commonly used anthelmintics. RNAi by microinjection of dsRNA targeting six genes, namely superoxide dismutase (*SOD*), σ class of glutathione-s-transferase (*GST*), and cathepsin (*Cat*) *L1-D*, *Cat B1*, *Cat B2,* and *Cat B3* showed strong transcriptional silencing of the targets following exposure of the newly excysted juveniles to 170–223 nucleotide dsRNA. The knockdown was concentration-dependent and opens prospects of developing vaccines and therapeutic targets against this type of neglected parasitic infections (Anandanarayanan et al. 2017).

40.5 Development of Therapeutics

The vaccine development is a complex process that needs public and private participation for development, testing, and regulation. At the end of nineteenth century, several vaccines such as smallpox, plague, rabies, typhoid, and cholera vaccines were developed. Simultaneous efforts were also made to develop vaccines against infections in livestock species.

With the development of molecular biological techniques, the technology of developing vaccines also changed. Live-attenuated vaccines are the effective way to establish robust, long-lasting immunity against viral infections. However, the possibility of reversion of live-attenuated vaccine including miRNA targeted vaccines to wild-type replication and associated pathogenicity are the possible threats associated with live-attenuated vaccines (Fay and Langlois 2018).

Gene silencing mediated by siRNA serves as an antiviral defense mechanisms in eukaryotes.

SiRNA duplexes targeted against Rift Valley fever virus (RVFV) nucleoprotein effectively inhibits replication of RVFV in human and African green monkey cell lines. The individual or complex siRNAs targeting the RVFV nucleoprotein gene completely inhibited viral protein expression and prevented the degradation of host innate antiviral factor, the protein kinase R (PKR). In addition, pretreatment of cells with the nucleoprotein-specific siRNAs markedly reduced the virus load. The effect of siRNAs was not attributable to interferon or the interferon response effector molecule, PKR. It was inferred that antiviral activity of RVFV nucleoprotein-specific siRNAs may help in developing novel therapeutic strategy against infectious RVFV (Faburay and Richt 2016).

RNAi can inhibit influenza virus. The shRNA targeting the non-coding region of the viral RNA of nucleoprotein might be a supporting tool in developing influenza-resistant chicks (Fujimoto et al. 2019).

40.6 Molecular Insights into Stem Cell Biology and Genetic Engineering

One of the outstanding features of ESCs is their pluripotency and self-renewal under in vitro as well as in vivo conditions. A thorough understanding of molecular mechanisms of biological properties of stem cells, and the mechanisms involved in reprogramming of the cells and their differentiation into other cell lineages is of significant interest.

To fully utilize their biomedical potential, it is necessary to understand the molecular mechanism involved in regulation and pluripotency of the ESCs. RNAi technology has contributed immensely to revolutionize the functional genetics in animal cells. The genome-wide screening of RNAi has provided comprehensive information about key scientific areas such as ESC biology, transcription factors, posttranslational modulators, and chromatin remodelers (Zheng and Hu 2014).

RNAi technology in combination with Sleeping Beauty transposon system constitutes a suitable method to generate transgenic cells which can be used to generate nuclear transfer cloned transgenic animals. Using these combinations, transgenic sheep resistant to foot-and-mouth disease were generated (Deng et al. 2017).

40.7 Outlook and Challenges

RNAi is a promising technology in understanding functional genomics. It is rapidly becoming a method of choice for analyzing gene functions and holds promise for developing therapeutic agents based on the phenomenon of gene silencing (Balmayor and Evans 2019). The progress in RNAi has shown promises for use in reverse genetic and therapy. RNAi is used to decipher basic regulatory mechanism of gene regulation and modulation.

Extraneous RNA modulates innate immune nucleic acid sensors, and may provoke inflammatory responses. One of the major challenges associated with RNAi is the mechanistic complexities that are only feebly understood till now. In addition, RNA is unstable and degraded by ribonucleases (RNases) that are prevalent in nature. It is difficult to deliver RNA across hydrophobic cell membranes.

In general, miRNA targeting is a proficient platform for developing safe, effective vaccines and provides better plasticity over traditional live-attenuated vaccine strategies. As siRNA do not integrate into host genome, the RNAi response from siRNA is transient. RNAi-based targeted gene silencing is a quick, cost-effective and reliable method to study gene expression, function of the genes in cell or organisms. RNAi technology combined with CRISPR/Cas9 provides the opportunity to produce genetically modified animals that are resistant to viral infection. This will minimize the economic losses to commercial livestock companies due to animal mortalities or culling of infected animals (Xie et al. 2018).

It is, therefore, necessary that transfection methods for applying RNAi should have minimal side effects.

40.8 Conclusions

Small non-coding RNAs such as miRNAs involved in post-transcriptional regulation of gene expression orchestrate a wide range of functional biological and pathological processes. RNAi causes disruption of gene expression in order to determine gene function or its effects on metabolic pathway. The process has been used to study stem cells and prevent animal health by controlling parasites, pests and pathogens.

References

Agrawal N, Dasaradhi PV, Mohmmed A, Malhotra P, Bhatnagar RK, Mukherjee SK (2003) RNA interference: biology, mechanism, and applications. Microbiol Mol Biol Rev. 67(4):657–685. Review

Ambros V, Lee RC, Lavanway A, Williams PT, Jewell D (2003) MicroRNAs and other tiny endogenous RNAs in C. elegans. Curr Biol 13(10):807–818

Anandanarayanan A, Raina OK, Lalrinkima H, Rialch A, Sankar M, Varghese A (2017) RNA interference in Fasciola gigantica: Establishing and optimization of experimental RNAi in the newly excysted juveniles of the fluke. PLoS Negl Trop Dis. 11(12): e0006109. https://doi.org/10.1371/journal.pntd.0006109 (eCollection 2017 Dec)

Balmayor ER, Evans CH (2019) RNA therapeutics for tissue engineering. Tissue Eng Part A 25(1–2):9–11. https://doi.org/10.1089/ten.TEA.2018.0315

Cabezas-Cruz A, Espinosa P, Alberdi P, de la Fuente J (2019) Tick-Pathogen Interactions: The Metabolic Perspective. Trends Parasitol pii: S1471-4922(19) 30018-2. https://doi.org/10.1016/j.pt.2019.01.006. (In press)

Carrington JC, Ambros V (2003) Role of microRNAs in plant and animal development. Science 301 (5631):336–338

de la Fuente J, Kocan KM, Almazán C, Blouin EF (2007) RNA interference for the study and genetic manipulation of ticks. Trends Parasitol 23(9):427–433 (Epub 2007 Jul 25). Review

Deng S, Li G, Yu K, Tian X, Wang F, Li W, Jiang W, Ji P, Han H, Fu J, Zhang X, Zhang J, Liu Y, Lian Z, Liu G (2017) RNAi combining Sleeping Beauty transposon system inhibits ex vivo expression of foot-and-mouth disease virus VP1 in transgenic sheep cells. Sci Rep 7(1):10065. https://doi.org/10.1038/s41598-017-09302-1

Denli AM, Tops BB, Plasterk RH, Ketting RF, Hannon GJ (2004) Processing of primary microRNAs by the microprocessor complex. Nature 432(7014):231–235 (Epub 2004 Nov 7)

Elbashir SM, Lendeckel W, Tuschl T (2001a) RNA interference is mediated by 21- and 22-nucleotide RNAs. Genes Dev 15(2):188–200

Elbashir SM, Harborth J, Lendeckel W, Yalcin A, Weber K, Tuschl T (2001b) Duplexes of 21-nucleotide RNAs mediate RNA interference in cultured mammalian cells. Nature 411(6836):494–498

Faburay B, Richt JA (2016) Short interfering RNA inhibits rift valley fever virus replication and degradation of protein kinase R in human cells. Front Microbiol 7:1889 (eCollection 2016)

Fay EJ, Langlois RA (2018) MicroRNA-attenuated virus vaccines. Noncoding RNA 4(4). pii: E25. https://doi.org/10.3390/ncrna4040025

Fujimoto Y, Kyogoku K, Takeda K, Ozaki K, Yamamoto S, Suyama H, Ono E (2019) Antiviral effects against influenza A virus infection by a short hairpin **RNA** targeting the non-coding terminal region of the viral nucleoprotein gene. J Vet Med Sci https://doi.org/10.1292/jvms.18-0436 (Epub ahead of print)

Hu G, Kim J, Xu Q, Leng Y, Orkin SH, Elledge SJ (2009) A genome-wide RNAi screen identifies a new transcriptional module required for self-renewal. Genes Dev 23(7):837–848. https://doi.org/10.1101/gad.1769609

Jongejan F, Nene V, de la Fuente J, Pain A, Willadsen P (2007) Advances in the genomics of ticks and tick-borne pathogens. Trends Parasitol 23(9):391–396

Kocan KM, Blouin E, de la Fuente J (2011) RNA interference in ticks. J Vis Exp (47). pii: 2474. https://doi.org/10.3791/2474

Ramakrishnan VG, Aljamali MN, Sauer JR, Essenberg RC (2005) Application of RNA interference in tick salivary gland research. J Biomol Tech 16(4):297–305. Review

Somarathne MBCL, Gunawardene YINS, Chandrasekharan NV, Dassanayake RS (2018) Development of siRNA mediated RNA interference and functional analysis of novel parasitic nematode-specific protein of Setaria digitata. Exp Parasitol 186:42–49. https://doi.org/10.1016/j.exppara.2018.02.001 (Epub 2018 Feb 12)

Xie Z, Pang D, Yuan H, Jiao H, Lu C, Wang K, Yang Q, Li M, Chen X, Yu T, Chen X, Dai Z, Peng Y, Tang X, Li Z, Wang T, Guo H, Li L, Tu C, Lai L, Ouyang H (2018) Genetically modified pigs are protected from classical swine fever virus. PLoS Pathog 14(12): e1007193. https://doi.org/10.1371/journal.ppat.1007193 (eCollection 2018 Dec)

Zamore PD, Tuschl T, Sharp PA, Bartel DP (2000) RNAi: double-stranded RNA directs the ATP-dependent cleavage of mRNA at 21–23 nucleotide intervals. Cell 101(1):25–33

Zheng X, Hu G (2014) Use of genome-wide RNAi screens to identify regulators of embryonic stem cell pluripotency and self-renewal. Methods Mol Biol 1150:163–173. https://doi.org/10.1007/978-1-4939-0512-6_10

Zhou D, He QS, Wang C, Zhang J, Wong-Staal F (2006) RNA interference and potential applications. Curr Top Med Chem 6(9):901–11. Review

Big from Small: MicroRNA in Relation to Veterinary Sciences

41

Abstract

Eukaryotic transcriptome sequence data reveals that majority of the genome is transcribed into distinct noncoding RNA species, called microRNAs (miRNA). The miRNA is subset of tiny noncoding RNAs, about 21–23 nucleotides long, that play a role in controlling cognate mRNA degradation by cleavage or inhibiting the translational process. miRNA is popularly referred to as "micromanager of gene regulation." In humans, miRNA disruption is correlated to diseases like cancers. In addition, miRNA has important implications in diseases and biological processes such as development, memory establishment, cell proliferation, apoptosis, and infections. Knowledge acquired from understanding miRNA pathway should be exploited to design artificial miRNAs with potential uses in therapeutics and livestock development.

Highlights

- miRNAs are abundantly present in many organisms
- The impact of miRNA-driven gene regulation is enormous; hence, it is a key field of study
- In view of their involvements in diseases and biological processes, miRNAs offer opportunities to develop curative interventions.

Keywords

miRNAs · Noncoding RNAs · miRNAs diversity · Biological functions · Developmental biology · Microbial inhibition

41.1 Introduction

The RNA in eukaryotes was thought to have main function of transferring (as tRNA and rRNA) and carrying (as mRNA) the information from DNA to synthesize the proteins. Analysis of transcriptome sequence data has revealed subsistence of a plethora of noncoding RNAs (ncRNAs) (Li et al. 2007).

The microRNAs are a class of noncoding RNAs (ncRNAs) encoded by the genome. miRNAs were first identified as a regulator of developmental timing in *Caenorhabditis elegans*. The discovery of these noncoding RNAs and subsequent revelation of their functional significance revamped the prevailing dogma that a large portion of genome is "junk" (Tannenbaum 2006; Xiong et al. 2012). It is now clear that most of the noncoding genome is transcribed into various classes of small RNAs that regulate several aspects of cellular physiology by fine-tuning of gene expression.

The miRNAs are small, endogenous, noncoding RNA molecules of 21–23 nucleotide (nt) length that are essential for regulation of

B. Singh et al., *Advances in Animal Biotechnology*,
https://doi.org/10.1007/978-3-030-21309-1_41

cellular processes through gene silencing (Lv et al. 2018). They function post-transcriptionally to regulate gene expression via cleavage of target mRNA or by inhibiting its translation, depending on the extent of complementarity between mature miRNA and target mRNA. The *lin-4,* i.e., the founder member of miRNA family, was found to encode a 61-nt-long, noncoding precursor that matured into 22 nucleotide regulatory RNA, which was partially complementary to 3′ UTR of *lin-14* mRNA.

41.2 Biological Roles of miRNA

miRNAs play roles in diverse cellular processes such as control of developmental timing (Reinhart et al. 2000), cell death (Chhabra et al. 2009), cell differentiation (Schickel et al. 2008), fat metabolism (Lin et al. 2009), and haematopoiesis (Bellon et al. 2009).

Numerous miRNAs are discovered in fungi, plants, animals, and viruses. With a rapid increase in the numbers of miRNAs, a uniform nomenclature system has been adopted (Griffiths-Jones et al. 2006), wherein first three to four letter prefixes are used to identify the species. For instance, "has" denotes miRNA in humans (*Homo sapiens*), "mmu" in mouse (*Mus musculus*), and "dme" in *Drosophila melanogaster*. Next three letters indicate the matured (miR) or precursor (mir) form. This is followed by a numerical, which is given by the miRBase Registry, in the order of its discovery. The numerical assignment provides information regarding homology. For instance, hsa-miR-101 is orthologue of mmu-miR-101.

41.3 Regulation of miRNA Gene Transcription

Experimental data suggests that expression of several miRNA genes is regulated temporally and spatially. Transcription of canonical miRNA genes is controlled by a process similar to protein-coding genes. Promoters of miRNA genes have specific features like TATA box, CpG islands, enhancers, initiation elements, transcription factor (TF) binding sequences (Cai et al. 2004; Ozsolak et al. 2008). TFs regulate transcription of specific miRNAs in a tissue-specific or developmental stage-specific manner.

Post-maturation, miRNAs require their integration into miRISC, a large multiprotein complex that has sequence-specific mRNA silencing activity. ARGONAUTE (AGO) family protein is a important component of miRISC. In *Drosophila*, other protein members of RISC include Vasa Intronic Gene (VIG), Fragile X mental retardation protein (dFXR), TUDOR-SN-an exonuclease, Ribosomal proteins L5 & L11, and putative helicase proteins such as P68 (*Drosophila*)/GEMIN3 (humans) (Nelson et al. 2004; Mourelatos et al. 2002).

Majority of animal miRNAs bind their target mRNAs with imperfect complementarity in the 3’ UTR region of mRNA and usually lead to translation repression, without affecting the levels of target mRNA (Reinhart et al. 2000; Stark et al. 2005). Several miRNAs that pair with target mRNA with imperfect complementarity are shown to affect the levels of target transcripts. Most likely, this affect is not due to cleavage of targets, as near complete complementarity is required for slicer activity.

41.4 Diversity of miRNAs

The miRNA registry has expanded enormously during recent years. The first two examples of miRNAs in *C. elegans*, *lin-4*, *let-7,* and *lys-6* were discovered in forward genetic screens, i.e., the phenotypes were already known and the mutations were characterized in the genes encoding the respective miRNAs (Lee et al. 1993; Reinhart et al. 2000). Isolation, cloning, and sequencing of libraries of small RNA prepared from various cell types/tissues/organs in different species have led to discovery of a large number of miRNAs.

The miRNAs can be detected by wet laboratory methods, such as high-speed capillary sieving electrophoresis (Wang et al. 2018), RT-qPCR (Xu et al. 2018). According to recent information, the miRBase is associated with cataloging, naming, and distribution of microRNA gene sequences. miRBase is publicly available and freely accessible at http://mirbase.org/. The latest miRBase (v22) has miRNA sequences from 271 organisms: 38,589 hairpin precursors and 48,860 mature miRNAs (Kozomara et al. 2018).

Another source, RNAcentral is comprehensive database of noncoding NRA (ncRNA) sequences, that collates information pertaining to ncRNA sequences of all types of organisms. This database has genomic locations of ncRNA sequences of 296 species, besides functional annotations such as tRNA secondary structures, gene ontology annotations, and miRNA-target interactions. The RNAcentral is freely accessible at https://rnacentral.org (The RNAcentral Constortium 2018).

41.5 Artificial miRNAs and miRNA Technology

Aberrant expression of miRNAs is associated with several human multigenic diseases and can serve as a biomarker for diagnostics. High-resolution expression profiling of miRNAs in cancer cell lines and human tumors suggest that miRNA signatures may be useful in categorizing, detecting, and predicting the course of human cancers (Gusev and Brackett 2007; Lu et al. 2005). miRNA expression profiles can be used to distinguish the lineage of solid tumors and classify poorly differentiated cancer specimens (Volinia et al. 2006; Swanton and Caldas 2009). Asuragen, Inc. (USA) has launched a qRT-PCR-based miRNA diagnostic test for differentiation between chronic pancreatitis and pancreatic cancer patients. A blood-based non-invasive miRNA signature assay has been developed for detecting malignant lung cancer. Rosetta Genomics, Inc. has launched a miRview meets, a miRNA-based diagnostic test for identification of primary origin of cancer of uncertain or unknown origin. This test offers a wide panel of 42 identifiable tumor origins for accurate classification of tumors.

Apart from its use in diagnostics, miRNAs could also prove to be novel therapeutic targets (Kole et al. 2012; Broderick and Zamore 2011). miRNA levels may be reduced by targeting them with artificial antisense RNAs called antagomirs. Stability of antagomirs is increased by modifications such as a 2′ *O*-methyl modification that renders them resistant to nucleases and provides high thermodynamic duplex stability. Antagomirs (antisense oligonucleotides; ASO) are sequence-specific, long-lasting, and have little or no short-term toxicity (Krützfeldt et al. 2005). Targeting miR-122 in the mouse liver using an anti-miR-122 antagomir results in complete degradation of miR-122 in a dose-dependent manner (Elmén et al. 2008). In many cases where miRNA expression has reduced and/or down regulating of the targets is desired, synthetic canonical miRNAs could be transfected in cells, or delivered using viral vectors. Artificial miRNAs or miRNA mimics could be used for down regulating the sought mRNAs. These artificial miRNAs could be designed as a mimic of natural miRNA, such that their binding to target is determined by 2–8 nucleotide seed region at 5′ end with bulges in other regions, due to mismatches (McBride et al. 2008; Broderick and Zamore 2011). Artificial miRNAs could be used to reduce the expression of oncogenes, which may slow down the process of development of tumors. Hairpins mimicking the precursor miRNAs could be designed and delivered through gene therapy. These hairpins could also be designed such that it is processed into more than one mature miRNAs thus giving the potential to simultaneously target several genes. In an interesting case involving huntington's disease (HD), a fatal neurodegenerative disease caused by a polyglutamine repeat expansion in HD gene, it was shown that artificial miRNAs have significant advantages over shRNAs. It was also shown that although down regulation of target was achieved using RNAi in HD murine model, active as well as control shRNAs induced significant neurotoxicity. However, when the same sequences were placed into artificial miRNA expression systems, molecular as well as neuropathological readouts of neurotoxicity were significantly reduced (McBride et al. 2008).

41.6 Functions of miRNAs in Animals

41.6.1 Role in Genomic Reprogramming

The miRNAs virtually affect and regulate all biological processes including cell self-renewal and gene silencing that act by targeting specific mRNAs for degradation or by suppressing their translation. The miRNA plays a vital role in cellular reprogramming. Ectopic introduction of ESC-specific miRNAs reverts a somatic cell back to pluripotent cell stage (Kuo and Ying 2012).

miRNAs and their varied effects on stem cells have provided a far better understanding of the molecular mechanisms that fine-tune the complex gene regulatory networks controlling the proliferation and the differentiation of stem cells. Expression of stem cell-specific miRNAs promotes cellular reprogramming. miRNA from miR-302 cluster has a role in promoting genome reprogramming. This method holds a distinct advantage over the four-factor- or Yamanaka factors-induced reprogramming.

Hence, based on the pivotal role of miRNAs in embryogenesis and somatic cell reprogramming, further studies in this area must continue to gain a better understanding of the role of miRNAs in regulating stemness and pluripotency.

41.7 Roles in Development

In general, miRNAs play an important role in development. Proteins involved in biogenesis or maturation of miRNAs are detected by mutational analysis. In *C. elegans*, mutation in *Dicer 1* causes defects in germline development (Knight and Bass 2001; Lagos-Quintana et al. 2001). Removal of maternal and zygotic *Dicer 1* causes embryonic mortality (Grishok et al. 2001). In *D. melanogaster,* removal of *Dicer 1* causes defects in germline development and sterility (Hatfield et al. 2005). Other components of siRNA/miRNA pathway such as *Dicer 2* and *armitage* are implicated in memory development in *Drosophila* (Ashraf et al. 2006). Zebrafish *Dicer 1* mutant embryos are unable to continue growth and die eventually (Wienholds et al. 2003).

Conditional mouse mutants, who are allowed to grow beyond early stages of embryonic development, show defects in morphogenesis of hair follicles, limb development, and maturation of immune cells (Harfe et al. 2005).

Deep sequencing of skeletal muscle cells reveals the role of miRNAs in myogenesis by regulation of two genes, namely bone morphogenetic protein 2 (*BMP2*) and mitogen-activated protein kinase 1 (*MAPK1*) (Hou et al. 2012). The miRNAs play important role in myogenesis, enhancing muscle mass, muscle fiber type, and muscle-related diseases. The muscle-specific miRNAs, namely miR-206, miR-1, and miR-133, are most widely studied with reference to their role in differentiation of skeletal muscles (Luo et al. 2013).

miRNA-1 targets histone deacetylase 4 (HDAC4) to promote differentiation of duck myoblasts, while miRNA-133 might affect SRF and transforming growth factor-beta (TGF-β) receptor type 1 (TGFBR1) expression that promotes proliferation, indicating that miRNA-1 and miRNA-133 have role in skeletal muscle development in some species of ducks (Wu et al. 2018).

The miRNAs are present in milk in extracellular vesicles. Several mammalian species share miRNAs. The milk miRNAs are implicated in immune-related functions, besides their role in cell growth and signal transduction, and regulate cell functions in neonates (van Herwijnen et al. 2018).

41.8 Roles in Infection and Disease

Deletion of chromosomal region 13q14, encoding hsa-miR-15a and hsa-miR-16-1, is associated with most cases of chronic lymphocytic leukemia. These miRNAs target *B cell lymphoma 2*

(*Bcl2*), an anti-apoptotic gene (Cimmino et al. 2005). Overexpression of miRNAs with "oncogenic" potential such as miR-17-92 accelerates *C-Myc* induced tumorigenesis by suppressing proapoptotic factor E2F1 (He et al. 2005). miR-372 and miR-373 target tumor suppressor LATS2 and induce tumorigenesis. These miRNAs are expressed specifically in testicular germ cell tumors (Voorhoeve et al. 2006). miR-155 overexpression is associated with B cell lymphomas.

Involvement of miRNAs in human disease is not limited to cancers. miRNAs are also involved in the interactions with pathogenic viruses. Virally encoded miRNAs interfere with normal cellular physiology such that viruses can continue their life cycle in host. miR-S1, a miRNA encoded by simian vacuolating virus 40 (simian virus 40 or SV40), helps to keep the infected cells veiled from the immune system. T antigen for degradation. Herpes simplex virus-1 (HSV1) encoded miRNA, miR-LAT, targets proapoptotic transforming growth factor β (TGFβ) and SMAD3, thereby inhibiting apoptosis of latently infected neurons.

Besides, the host cells also make use of miRNA to regulate replication of certain viruses. The host cell, for example, encodes miR-32 that inhibit replication of primate foamy virus type 1 (PFV1), while PFV1 encodes a protein Tas that suppresses microRNA-directed functions in mammalian cells (Lecellier et al. 2005).

We at the moment have not complete understanding of the mechanisms by which miRNAs regulate gene expression, nor do we know the complete range of mRNAs each miRNA controls. Nevertheless, developing effective strategies to block miRNA suggests that there is need to develop anti-miRNA drugs for clinical applications against diseases and pathologies caused by miRNAs (Broderick and Zamore 2011).

41.9 Outlook and Challenges

MicroRNAs (miRNAs) are among growing classes of small noncoding RNAs that negatively regulate gene expression by degrading target mRNAs by cleavage or inhibiting their translation. The association of miRNAs with diverse processes is evident from their widespread abundance in metazoan genomes. miRNAs play active role in a number of biological processes including development, memory, cellular proliferation, and apoptosis and prevention of infectious diseases.

Abnormalities in expression of miRNAs lead to impaired cellular functions and development of pathologies such as inflammation, cancer, and autoimmunity. Insights from biology of miRNAs have opened avenues to develop RNA-based therapies, bioengineering of miRNAs with potential veterinary clinical applications, and genetic engineering. Inducing pluripotency in somatic cells by means of miRNA is a promising area in stem cell engineering.

Despite plentiful information available on miRNA biogenesis, maturation process, and their modes of functions, we are only in beginning to understand the regulatory aspects of miRNAs. Furthermore, the mechanism by which miRNA mediates repression of genes and regulatory pathways needs to be deciphered. The subtle differences between classes of small RNA are at beginning. On the other hand, genetic screening is continually adding to list of miRNAs and their involvement in various biological functions, metabolic pathways, and cellular functions.

41.10 Conclusions

microRNAs are abundantly available in various organisms and affect vital functions and gene expression. Their role in beneficial processes should be explored. Using miRNA as a technology holds a lot of promise, and wide spectrum of possibilities exists which at the same time necessitate further works for better understanding of this relatively nascent field.

References

Ashraf SI, McLoon AL, Sclarsic SM, Kunes S (2006) Synaptic protein synthesis associated with memory is regulated by the RISC pathway in Drosophila. Cell 124(1):191–205. Erratum in: Cell. 2006 Aug 25;126 (4):812

Bellon M, Lepelletier Y, Hermine O, Nicot C (2009) Deregulation of microRNA involved in hematopoiesis and the immune response in HTLV-I adult T-cell leukemia. Blood 113(20):4914–4917. https://doi.org/10.1182/blood-2008-11-189845 (Epub 2009 Feb 26)

Broderick JA, Zamore PD (2011) MicroRNA therapeutics. Gene Ther. 18(12):1104–1110. https://doi.org/10.1038/gt.2011.50 (Epub 2011 Apr 28). Review

Cai X, Hagedorn CH, Cullen BR (2004) Human microRNAs are processed from capped, polyadenylated transcripts that can also function as mRNAs. RNA 10 (12):1957–1966 (Epub 2004 Nov 3)

Chhabra R, Adlakha YK, Hariharan M, Scaria V, Saini N (2009) Upregulation of miR-23a-27a-24-2 cluster induces caspase-dependent and -independent apoptosis in human embryonic kidney cells. PLoS ONE 4(6): e5848. https://doi.org/10.1371/journal.pone.0005848

Cimmino A, Calin GA, Fabbri M, Iorio MV, Ferracin M, Shimizu M, Wojcik SE, Aqeilan RI, Zupo S, Dono M, Rassenti L, Alder H, Volinia S, Liu CG, Kipps TJ, Negrini M, Croce CM (2005) miR-15 and miR-16 induce apoptosis by targeting BCL2. Proc Natl Acad Sci U S A. 102(39):13944–13949. (Epub 2005 Sep 15). Erratum in: Proc Natl Acad Sci U S A. 2006 Feb 14;103(7):2464

Elmén J, Lindow M, Silahtaroglu A, Bak M, Christensen M, Lind-Thomsen A, Hedtjärn M, Hansen JB, Hansen HF, Straarup EM, McCullagh K, Kearney P, Kauppinen S (2008) Antagonism of microRNA-122 in mice by systemically administered LNA-antimiR leads to up-regulation of a large set of predicted target mRNAs in the liver. Nucleic Acids Res 36(4):1153–1162 (Epub 2007 Dec 23)

Griffiths-Jones S, Grocock RJ, van Dongen S, Bateman A, Enright AJ (2006) miRBase: microRNA sequences, targets and gene nomenclature. Nucleic Acids Res 34 (Database issue):D140–144

Grishok A, Pasquinelli AE, Conte D, Li N, Parrish S, Ha I, Baillie DL, Fire A, Ruvkun G, Mello CC (2001) Genes and mechanisms related to RNA interference regulate expression of the small temporal RNAs that control C. *elegans* developmental timing. Cell 106 (1):23–34

Gusev Y, Brackett DJ (2007) MicroRNA expression profiling in cancer from a bioinformatics prospective. Expert Rev Mol Diagn 7(6):787–792. Review

Harfe BD, McManus MT, Mansfield JH, Hornstein E, Tabin CJ (2005) The RNaseIII enzyme Dicer is required for morphogenesis but not patterning of the vertebrate limb. Proc Natl Acad Sci U S A 102 (31):10898–10903 (Epub 2005 Jul 22)

Hatfield SD, Shcherbata HR, Fischer KA, Nakahara K, Carthew RW, Ruohola-Baker H (2005) Stem cell division is regulated by the microRNA pathway. Nature 435(7044):974–978 (Epub 2005 Jun 8)

He L, Thomson JM, Hemann MT, Hernando-Monge E, Mu D, Goodson S, Powers S, Cordon-Cardo C, Lowe SW, Hannon GJ, Hammond SM (2005) A microRNA polycistron as a potential human oncogene. Nature 435(7043):828–833

Hou X, Tang Z, Liu H, Wang N, Ju H, Li K (2012) Discovery of MicroRNAs associated with myogenesis by deep sequencing of serial developmental skeletal muscles in pigs. PLoS One 7(12):e52123. https://doi.org/10.1371/journal.pone.0052123 (Epub 2012 Dec 21)

Knight SW, Bass BL (2001) A role for the RNase III enzyme DCR-1 in RNA interference and germ line development in Caenorhabditis elegans. Science 293 (5538):2269–2271 (Epub 2001 Aug 2)

Kole R, Krainer AR, Altman S (2012) RNA therapeutics: beyond RNA interference and antisense oligonucleotides. Nat Rev Drug Discov. 11(2):125–140. https://doi.org/10.1038/nrd3625 Review

Kozomara A, Birgaoanu M, Griffiths-Jones S (2018) miRBase: from microRNA sequences to function. Nucleic Acids Res. https://doi.org/10.1093/nar/gky1141 (Epub ahead of print)

Krützfeldt J, Rajewsky N, Braich R, Rajeev KG, Tuschl T, Manoharan M, Stoffel M (2005) Silencing of microRNAs in vivo with'antagomirs'. Nature 438 (7068):685–689 (Epub 2005 Oct 30)

Kuo CH, Ying SY (2012) Advances in microRNA-mediated reprogramming technology. Stem Cells Int. 2012:823709. https://doi.org/10.1155/2012/823709 (Epub 2012 Mar 28)

Lagos-Quintana M, Rauhut R, Lendeckel W, Tuschl T (2001) Identification of novel genes coding for small expressed RNAs. Science 294(5543):853–858

Lecellier CH, Dunoyer P, Arar K, Lehmann-Che J, Eyquem S, Himber C, Saïb A, Voinnet O (2005) A cellular microRNA mediates antiviral defense in human cells. Science 308(5721):557–560

Lee RC, Feinbaum RL, Ambros V. The C (1993) Elegans heterochronic gene lin-4 encodes small RNAs with antisense complementarity to lin-14. Cell. 75(5):843–854

Li L, Wang X, Sasidharan R, Stolc V, Deng W, He H, Korbel J, Chen X, Tongprasit W, Ronald P, Chen R, Gerstein M, Deng XW (2007) Global identification and characterization of transcriptionally active regions in the rice genome. PLoS ONE 2(3):e294

Lin Q, Gao Z, Alarcon RM, Ye J, Yun Z (2009) A role of miR-27 in the regulation of adipogenesis. FEBS J 276 (8):2348–2358

Lu J, Getz G, Miska EA, Alvarez-Saavedra E, Lamb J, Peck D, Sweet-Cordero A, Ebert BL, Mak RH, Ferrando AA, Downing JR, Jacks T, Horvitz HR, Golub TR (2005) MicroRNa expression profiles classify human cancers. Nature 435(7043):834–838

Luo W, Nie Q, Zhang X (2013) MicroRNAs involved in skeletal muscle differentiation. J Genet Genomics 40 (3):107–116. https://doi.org/10.1016/j.jgg.2013.02.002 (Epub 2013 Feb 20)

Lv J, Zhang Z, Pan L, Zhang Y (2018) MicroRNA-34/449 family and viral infections. Virus Res 260:1–6. https://doi.org/10.1016/j.virusres.2018.11.001 (Epub ahead of print)

McBride JL, Boudreau RL, Harper SQ, Staber PD, Monteys AM, Martins I, Gilmore BL, Burstein H, Peluso RW, Polisky B, Carter BJ, Davidson BL (2008) Artificial miRNAs mitigate shRNA-mediated toxicity in the brain: implications for the therapeutic development of RNAi. Proc Natl Acad Sci U S A 105 (15):5868–5873. https://doi.org/10.1073/pnas.0801775105 (Epub 2008 Apr 8)

Mourelatos Z, Dostie J, Paushkin S, Sharma A, Charroux B, Abel L, Rappsilber J, Mann M, Dreyfuss G (2002) miRNPs: a novel class of ribonucleoproteins containing numerous microRNAs. Genes Dev 16 (6):720–728

Nelson PT, Hatzigeorgiou AG, Mourelatos Z (2004) miRNP:mRNA association in polyribosomes in a human neuronal cell line. RNA 10(3):387–394

Ozsolak F, Poling LL, Wang Z, Liu H, Liu XS, Roeder RG, Zhang X, Song JS, Fisher DE (2008) Chromatin structure analyses identify miRNA promoters. Genes Dev 22(22):3172–3183. https://doi.org/10.1101/gad.1706508

Reinhart BJ, Slack FJ, Basson M, Pasquinelli AE, Bettinger JC, Rougvie AE, Horvitz HR, Ruvkun G (2000) The 21-nucleotide let-7 RNA regulates developmental timing in Caenorhabditis elegans. Nature 403(6772):901–906

Schickel R, Boyerinas B, Park SM, Peter ME (2008) MicroRNAs: key players in the immune system, differentiation, tumorigenesis and cell death. Oncogene 27(45):5959–5974. https://doi.org/10.1038/onc.2008.274 Review

Stark A, Brennecke J, Bushati N, Russell RB, Cohen SM (2005) Animal MicroRNAs confer robustness to gene expression and have a significant impact on 3'UTR evolution. Cell 123(6):1133–1146

Swanton C, Caldas C (2009) Molecular classification of solid tumours: towards pathway-driven therapeutics. Br J Cancer 100(10):1517–1522. https://doi.org/10.1038/sj.bjc.6605031 (Epub 2009 Apr 14). Review

Tannenbaum E (2006) An RNA-centered view of eukaryotic cells. Biosystems. 84:217–224

The RNAcentral Constortium (2018) RNAcentral: a hub of information for non-coding RNA sequences. Nucleic Acids Res https://doi.org/10.1093/nar/gky1034

van Herwijnen MJC, Driedonks TAP, Snoek BL, Kroon AMT, Kleinjan M, Jorritsma R, Pieterse CMJ, Hoen ENMN, Wauben MHM (2018) Abundantly present miRNAs in milk-derived extracellular vesicles are conserved between mammals. Front Nutr 5:81. https://doi.org/10.3389/fnut.2018.00081 (eCollection 2018)

Volinia S, Calin GA, Liu CG, Ambs S, Cimmino A, Petrocca F, Visone R, Iorio M, Roldo C, Ferracin M, Prueitt RL, Yanaihara N, Lanza G, Scarpa A, Vecchione A, Negrini M, Harris CC, Croce CM (2006) A microRNA expression signature of human solid tumors defines cancer gene targets. Proc Natl Acad Sci U S A 103(7):2257–2261 (Epub 2006 Feb 3)

Voorhoeve PM, le Sage C, Schrier M, Gillis AJ, Stoop H, Nagel R, Liu YP, van Duijse J, Drost J, Griekspoor A, Zlotorynski E, Yabuta N, De Vita G, Nojima H, Looijenga LH, Agami R (2006) A genetic screen implicates miRNA-372 and miRNA-373 as oncogenes in testicular germ cell tumors. Cell 124(6):1169–1181

Wang W, Cai X, Lin P, Bai R (2018) Separation and determination of microRNAs by high-speed capillary sieving electrophoresis. J Sep Sci 41(20):3925–3931. https://doi.org/10.1002/jssc.201800635 (Epub 2018 Sep 5)

Wienholds E, Koudijs MJ, van Eeden FJ, Cuppen E, Plasterk RH (2003) The microRNA-producing enzyme Dicer1 is essential for zebrafish development. Nat Genet 35:217–218

Wu N, Gu T, Lu L, Cao Z, Song Q, Wang Z, Zhang Y, Chang G, Xu Q, Chen G (2018) Roles of miRNA-1 and miRNA-133 in the proliferation and differentiation of myoblasts in duck skeletal muscle. J Cell Physiol. https://doi.org/10.1002/jcp.26857

Xiong J, Lu X, Zhou Z, Chang Y, Yuan D, Tian M, Zhou Z, Wang L, Fu C, Orias E, Miao W (2012) Transcriptome analysis of the model protozoan, Tetrahymena thermophila, using Deep RNA sequencing. PLoS One. 7(2):e30630. https://doi.org/10.1371/journal.pone.0030630 (Epub 2012 Feb 7)

Xu C, Wu S, Zhao W, Mipam T, Liu J, Liu W, Yi C, Shah MA, Yu S, Cai X (2018) Differentially expressed microRNAs between cattleyak and yak testis. Sci Rep 8(1):592. https://doi.org/10.1038/s41598-017-18607-0

Genome Editing in Farm Animals 42

Abstract

Due to several reasons, animals are preferred for producing recombinant therapeutics. At present, it is possible to synthesize genes, or extract genes of interest, and introduce them into across cell membranes by means of viral vectors, electroporation or mechanically. The newly introduced enzymatic methods of gene modifications are easy, precise, and cheaper and used in human gene therapies, biomedical sciences, and livestock genome manipulation. Gene-edited livestock is poised to become a commercial reality for producing transgenic or model animals.

Highlights

- Repertoire of molecular tools allow precise modification of genome at rapid pace and precision
- Instead of yeasts and animals cells, the animals are preferred for producing recombinant therapeutic proteins
- Significant improvements are made in modifying genomes by enzyme-catalyzed methods of gene manipulation.

Keywords

Genome editing · Gene modification enzymes · Enzyme-catalyzed transgenesis · Designer nucleases · CRISPR/cas9 · ZFNs · TALENs · RNA-guided genome modification

42.1 Introduction

Development of effective and reliable ways to make precise and targeted changes to the genome of a living organism is a long-standing goal of biomedical scientists. Genome editing or gene editing has enabled scientists to manipulate genome by adding, removing, or altering the genetic material at particular locations. Earlier, the geneticists made use of chemicals or radiations to induce mutations. However, it was difficult to control how many or where the mutations would have occurred.

Several traditional gene targeting methods are used to manipulate gene functions (Tan et al. 2016). Some of the most distinguished approaches are—homologous recombination (Capecchi 2005), RNA interference (RNAi) (Fire et al. 1998), and sequence-specific designer nucleases, also called as programmable nucleases. For over a decade, RNAi was used to interrupt the genes of interest. Major limitation of this technique was that it provided only temporary inhibition of gene function and unpredictable off-target effects (Alic et al. 2012).

42.2 Enzyme-Catalyzed Transgenesis

During the past few years, transgenic animals have got popularity because of availability of annotated genome sequences of various

B. Singh et al., *Advances in Animal Biotechnology*,
https://doi.org/10.1007/978-3-030-21309-1_42

organisms for analysis, interpretation, and use. Recombinant DNA technology was most frequently used to generate genetically modified cells or organisms. Other technologies, primarily the enzyme-catalyzed methods of editing genomes via introduction of site-specific DNA double-stranded breaks, have emerged as powerful means of genome engineering or genome editing.

Major customizable DNA-binding proteins used for genome editing include (i) zinc-finger nucleases based on eukaryotic transcription factors, (ii) transcription activator-like effector nucleases (TALENs) from *Xanthomonas* bacteria (Gaj et al. 2013), (iii) meganucleases derived from microbial mobile genetic elements (Smith et al. 2006), and (iv) RNA-guided DNA endonuclease Cas9 from type II bacterial adaptive immune system, the CRISPR (Cong et al. 2013; Mali et al. 2013; Yang et al. 2014). In addition, transposon systems such as Sleeping Beauty (Ding et al. 2005, 2014; Katter et al. 2013; Ivics et al. 2014), piggyBac (Urschitz and Moisyadi 2013; Li et al. 2014), designer nucleases including zinc-finger nucleases (ZFNs) (Geurts et al. 2009; Meyer et al. 2010), and RNA-guided nucleases (Shen et al. 2013) are the prominent examples. Cre recombinase and ϕC31 integrase are of interest to modify genes in farm animals (Xu et al. 2008; Yu et al. 2013).

42.3 Recombinases

The Cre-loxP and Flp-FRT are the most extensively studied members of recombinase family (Grainge and Jayaram 1999). The LoxP recombination site comprises of two inverted 13-bp repeat sequences, flanking an 8-bp core element which decides orientation of site. If two identical target sites located on a circular and linear DNA molecules are recombined, each will be integrated into a linear molecule. Two targeted sites possessing different orientation sites on one DNA molecule will cause inversion of DNA molecule between recognition target sites (Kues 2018).

The mutated recombination sites allow the design of recombinase-mediated cassette exchange (RMCE), where host genome is formerly tagged with complementary docking sites (Grainge and Jayaram 1999). The gene of interest is delivered in a DNA flanked by target sites that are identical to their respective docking site. RMCE is one of the promising methods for precise genetic engineering of model animals such as drosophila (Schetelig et al. 2018).

42.4 Designer Nucleases

Designer or programmable nucleases are modern tools of genetic manipulation of cells and animals. They are capable of precise targeted genome modifications with broad areas of applications. Programmable nucleases include ZFNs, TALENs, and RNA-guided genome modification system CRISPR/Cas9. Indeed, these systems are combinations of DNA molecules, proteins, and RNAs, having common modes of action. They recognize a predetermined nucleotide sequence and induce double-strand break in DNA strand. Broken DNA strand induces or activates host repair mechanisms through error-prone non-homologous end joining, and or homology-directed repair. This strategy has been used to produce genetically modified cattle resistant to mastitis (Liu et al. 2014), genome edited sheep and cattle (Proudfoot et al. 2015), and myostatin mutants pigs (Wang et al. 2015, 2017).

42.5 CRISPR-Cas9—A New Approach for Genome Editing

Recently, a new technology based on a bacterial CRISPR-associated protein-9 nuclease (Cas9) from *Streptococcus pyogenes* has generated considerable excitement among scientific community (Cong et al. 2013). CRISPR-Cas9 system, i.e., clustered regularly interspaced short palindromic repeats is a simple, fast, cheap, and precise method of genetic manipulation compared to contemporary genome-editing methods. CRISPR-Cas9 was initially discovered in *Escherichia coli* (Ishino et al. 1987) at Osaka University in Japan. However, it was reported that “the biological significance of these sequences is unknown.” Barrangou et al. (2007), in their studies reported that *Streptococcus thermophilus* can acquire resistance against virus invaders much like immune system by integrating a genome fragment of the infectious virus into its CRISPR locus. In response to invasion of virus or phage, the bacterium produces two short RNAs of which one provides a sequence that matches the RNA of invading virus. By using CRISPR system, the bacteria snip out DNA from invading virus and use them to create DNA segments known as CRISPR arrays. These arrays serve as a genetic memory that helps them to recognize and destroy the virus if it again attacks the bacterium. In January 2013, Prof. Zhang and his team published the first method to engineer CRISPR to edit the genome in mouse and human cells (Cong et al. 2013).

42.6 Mechanism of Action of CRISPR/Cas9

Three types of CRISPR mechanisms have been identified, of which type II is most widely used for genome editing or engineering (Makarove et al. 2015). In CRISPR-Cas9 technique, DNA from the invading virus or bacteriophage is cleaved into small fragments and incorporated into CRISPR locus. The transcripts are then processed to generate small RNAs known as crRNA-CRISPR RNA (each about 20 bases long) which guides the Cas9, an endonuclease enzyme that binds and cuts the double-stranded DNA at a specific location in the genome of the invading bacteriophage in such a way that small fragments of DNA are either added or deleted (Jinek et al. 2012). Cas9 enzyme relies on the presence of two nuclease domains: a RuvC-like nuclease domain located at the amino terminus and a HNH-like nuclease domain that resides in the mid-region of the protein (Sapranauskas et al. 2011). In order to recognize and cleave the target DNA, Cas9 forms a complex with both a crRNA and a separate trans-activating crRNA (tracrRNA or trRNA), that is partially complementary to the crRNA (Jinek et al. 2012). TracrRNA is required for crRNA maturation from a primary transcript encoding multiple pre-crRNAs. It occurs in the presence of RNase III and Cas9 (Deltcheva et al. 2011).

During the destruction of target DNA, the HNH- and RuvC-like nuclease domains break the double-stranded DNA at a 20-nucleotide target sequence within an associated crRNA transcript. The HNH domain cleaves the complementary strand, while the RuvC domain cleaves the non-complementary strand. The double-stranded endonuclease activity of Cas9 also requires the presence of 2-5 nucleotides sequence known as protospacer-associated motif (PAM) sequence located adjacent to the 3′- of the crRNA complementary sequence (Swarts et al. 2012). In the absence of this PAM sequence (Stemberg et al. 2014), the Cas9 cannot bind with even entirely complementary sequences. Once Cas9 cuts the double-stranded DNA, the cell recognizes the damaged DNA and starts repairing it by normal repairing process. Researchers take advantage of cellular DNA repair machinery to modify target DNA by adding or deleting the genomic fragments, or by replacing an existing DNA fragment with customized nucleotide sequences.

Since CRISPR-Cas9 system can incise double-stranded DNA, CRISPRs do not need separate cleaving enzymes. They can also easily be paired with tailor-made “guide” RNA (gRNA) sequences designed to lead them to their DNA targets. The gRNA has bases that are complementary target DNA sequence in genome. Thus, gRNA will only bind to the target DNA sequence and not to other regions of the genome. Tens of thousands of such gRNA sequences have already been created and are available.

42.7 Applications in Livestock

- Gene knockout is one of the first and most widely used applications of the CRISPR–Cas9 system. Double-stranded breaks are generated by Cas9 at the single guide RNA (gRNA)-targeted locus (Cong et al. 2013; Mali et al. 2013; Jinek et al. 2012; Fu et al. 2014; Kleinstiver et al. 2016). These breaks can be repaired by homologous recombination which can be used to introduce new mutations such as deletions, insertions, inversions or translocations, and stop codons leading to functional knockout of the gene (Richardson et al. 2016).

- CRISPR-Cas9 is adopted to target many important genes in many cell lines and organisms, including human (Mali et al. 2013), bacteria (Pyne et al. 2015), zebrafish (Hwang et al. 2013), plants (Mali et al. 2013), *Xenopus tropicalis* (Jiang et al. 2013), *Drosophila* sp. (Guo et al. 2014), monkeys (Niu et al. 2014), rabbit (Yang et al. 2014), pigs (Hai et al. 2014), and yeasts (DiCarlo et al. 2013).
- CRISPR system has been widely used in industries that utilize bacterial cultures to make them resistant to phages which would otherwise impede their productivity. CRISPR immunity was originally discovered by a team of Danisco scientists who reported that certain viruses can infect *Streptococcus thermophilus*, a dairy culture used in making yogurts and cheeses (Barrangou et al. 2007; Pennisi 2013). It was discovered that CRISPR sequences imparted immunity to *S. thermophilus* against phages. Manufacturers have used this technique to improve dairy cultures.
- CRISPR-Cas9 is used in medical applications to correct disease-causing gene mutations in humans as well as animals. The technique is being widely explored on a wide variety of diseases including single-gene disorders such as cystic fibrosis, hemophilia, and sickle cell disease, hepatitis B or even high cholesterol. This technique has a potential to provide treatment and prevention of more complex diseases such as cancer, heart disease, mental illness, and human immunodeficiency virus (HIV) infection (Deng et al. 2018; Yin et al. 2018).
- Genome editing with CRISPR-Cas9 in adult mice cured a rare liver disorder by replacing the mutant form of a gene with its correct sequence, providing evidence that gene-editing technique can reverse disease symptoms in living animals. Using this technique, scientists have developed genetically engineered monkeys as models of human diseases.
- CRISPR can be applied to treat infectious diseases, by generating specialized classes of antibiotics that target only pathogenic strains, and do not affect the beneficial gut bugs.

42.8 The Future of CRISPR-Cas9

Diversity and efficacy of CRISPR-Cas9 are driving biomedical and biotechnological revolution. CRISPR-Cas9, due to its simplicity, high efficiency, and versatility, has emerged as a most user-friendly technique for cell and molecular biology research. This technique has been routinely used in humans and animal models with aim to eventually use this technique to prevent and treat diseases in humans. In addition, scientists are now working to eliminate the "off-targets" effects, wherein CRISPR-Cas9 system cuts at a different gene other than the target gene.

Studies in monkey models of muscular dystrophy show that CRISPR/Cas9 could be safe and effective in modeling genetic diseases and can be used to correct humans genetic disorders (Wang et al. 2018). Similarly, Duchenne muscular dystrophy model dogs, and cure of the disease therein raises the hope of using this novel interventions in humans (Cohen 2018). However, Mattei (2018) has raised issues concerning the precision of CRISPR/Cas9, stating that the technique is not as precise as believed earlier. It is, therefore, essential that precision should be improved to efficiently use the rapidly evolving landscape of CRISPR/Cas system.

42.9 Outlook and Challenges

Transgenic animals are now integral component of human therapeutics and regenerative medicine. Pronuclear microinjection and SCNT using genetically modified cells were the prime methods of generating transgenic offspring. The procedures were cumbersome and efficiency was low. To overcome the problems comparatively newer approaches such as ICSI-mediated transgenesis and cytoplasmic plasmid injection (CPI), as well as molecular tools such as transposons (Sleeping Beauty and piggyBac), recombinases, and designer nuclei are developed. The studies should be carried out to enhance success rate of the approaches. We foresee that CISPR/Cas9 and its refined versions would enhance precise genetic modification in animals, and debugging genetic disorders in humans. The gene-editing tools open new avenues of research in farm animal species, extending from basic research, to commercial production of animals for food and therapeutic products.

Ethical concerns arise when human genome is edited by technologies such as CRISPR-Cas9. The genome editing is limited to somatic cells as the changes done in somatic cells will affect only certain tissues and are not passed from one generation to the next. Because any changes made in germline cells will be inherited from generation to generation, it has important ethical implications. Based on concerns about ethics and safety, germline cell and embryo genome editing are currently not permissible. Approval in humans can be granted only after safety of the methods is assured. This demands comprehensive studies using cell lines and model animals.

42.10 Conclusions

Among various enzyme-catalyzed strategies, CRISPR/Cas9 has emerged as valuable tool to create genetically modified cells, transgenic embryos and cloned transgenic animals. The technique is used to correct deleterious genetic mutations in humans. The methods have direct implications for research and animal pharming aimed to improve human welfare.

References

Alic N, Hoddinott MP, Foley A, Slack C, Piper MD, Partridge L (2012) Detrimental effects of RNAi: a cautionary note on its use in Drosophila ageing studies. PLoS ONE 7(9):e45367

Barrangou R, Fremaux C, Deveau H, Richards M, Boyaval P, Moineau S, Romero DA, Horvath P (2007) CRISPR provides acquired resistance against viruses in prokaryotes. Science 315(5819):1709–1712

Capecchi MR (2005) Gene targeting in mice: functional analysis of the mammalian genome for the twenty-first century. Nat Rev Genet 6(6):507

Cohen J (2018) In dogs, CRISPR fixes a muscular dystrophy. Science 361(6405):835. https://doi.org/10.1126/science.361.6405.835 (No abstract available)

Cong L, Ran FA, Cox D, Lin S, Barretto R, Habib N, Hsu PD, Wu X, Jiang W, Marraffini L, Zhang F (2013) Multiplex genome engineering using CRISPR/Cas systems. Science 3:1231143

Deltcheva E, Chylinski K, Sharma CM, Gonzales K, Chao Y, Pirzada ZA, Eckert MR, Vogel J, Charpentier E (2011) CRISPR RNA maturation by trans-encoded small RNA and host factor RNase III. Nature 471(7340):602–607. https://doi.org/10.1038/nature09886

Deng Q, Chen Z, Shi L, Lin H (2018) Developmental progress of CRISPR/Cas9 and its therapeutic applications for HIV-1 infection. Rev Med Virol 28(5):e1998. https://doi.org/10.1002/rmv.1998 (Epub 2018 Jul 19, Review)

DiCarlo JE, Norville JE, Mali P, Rios X, Aach J, Church GM (2013) Genome engineering in Saccharomyces cerevisiae using CRISPR-Cas systems. Nucleic Acids Res 41(7):4336–4343. https://doi.org/10.1093/nar/gkt135 (Epub 2013 Mar 4)

Ding S, Wu X, Li G, Han M, Zhuang Y, Xu T (2005) Efficient transposition of the piggyBac (PB) transposon in mammalian cells and mice. Cell 122(3):473–483

Ding S, Xu T, Wu X (2014) Generation of genetically engineered mice by the piggyBac transposon system. Methods Mol Biol 1194:171–185. https://doi.org/10.1007/978-1-4939-1215-5_9

Fire A, Xu S, Montgomery MK, Kostas SA, Driver SE, Mello CC (1998) Potent and specific genetic interference by double-stranded RNA in Caenorhabditis elegans. Nature 391(6669):806–811

Fu Y, Sander JD, Reyon D, Cascio VM, Joung JK (2014) Improving CRISPR-Cas nuclease specificity using truncated guide RNAs. Nat Biotechnol 32(3):279–

284. https://doi.org/10.1038/nbt.2808 (Epub 2014 Jan 26)

Gaj T, Gersbach CA, Barbas CF III (2013) ZFN, TALEN, and CRISPR/Cas-based methods for genome engineering. Trends Biotechnol 31(7):397–405

Geurts AM, Cost GJ, Freyvert Y, Zeitler B, Miller JC, Choi VM, Jenkins SS, Wood A, Cui X, Meng X, Vincent A, Lam S, Michalkiewicz M, Schilling R, Foeckler J, Kalloway S, Weiler H, Ménoret S, Anegon I, Davis GD, Zhang L, Rebar EJ, Gregory PD, Urnov FD, Jacob HJ, Buelow R (2009) Knockout rats via embryo microinjection of zinc-finger nucleases. Science 325(5939):433. https://doi.org/10.1126/science.1172447

Grainge I, Jayaram M (1999) The integrase family of recombinase: organization and function of the active site. Mol Microbiol 33(3):449–456 (Review)

Guo X, Zhang T, Hu Z, Zhang Y, Shi Z, Wang Q, Cui Y, Wang F, Zhao H, Chen Y (2014) Efficient RNA/Cas9-mediated genome editing in Xenopus tropicalis. Development 141(3):707–714. https://doi.org/10.1242/dev.099853 (Epub 2014 Jan 8)

Hai T, Teng F, Guo R, Li W, Zhou Q (2014) One-step generation of knockout pigs by zygote injection of CRISPR/Cas system. Cell Res 24(3):372–375. https://doi.org/10.1038/cr.2014.11 (Epub 2014 Jan 31, No abstract available)

Hwang WY, Fu Y, Reyon D, Maeder ML, Tsai SQ, Sander JD, Peterson RT, Yeh JR, Joung JK (2013) Efficient genome editing in zebrafish using a CRISPR-Cas system. Nat Biotechnol 31(3):227–229. https://doi.org/10.1038/nbt.2501 (Epub 2013 Jan 29)

Ishino Y, Shinagawa H, Makino K, Amemura M, Nakata A (1987) Nucleotide sequence of the iap gene, responsible for alkaline phosphatase isozyme conversion in Escherichia coli, and identification of the gene product. J Bacteriol 169(12):5429–5433

Ivics Z, Garrels W, Mátés L, Yau TY, Bashir S, Zidek V, Landa V, Geurts A, Pravenec M, Rülicke T, Kues WA, Izsvák Z (2014) Germline transgenesis in pigs by cytoplasmic microinjection of sleeping beauty transposons. Nat Protoc 9(4):810–827. https://doi.org/10.1038/nprot.2014.010 (Epub 2014 Mar 13)

Jiang W, Bikard D, Cox D, Zhang F, Marraffini LA (2013) RNA-guided editing of bacterial genomes using CRISPR-Cas systems. Nat Biotechnol 31(3):233–239. https://doi.org/10.1038/nbt.2508 (Epub 2013 Jan 29)

Jinek M, Chylinski K, Fonfara I, Hauer M, Doudna JA, Charpentier E (2012) A programmable dual-RNA-guided DNA endonuclease in adaptive bacterial immunity. Science 337(6096):816–821. https://doi.org/10.1126/science.1225829 (Epub 2012 Jun 28)

Katter K, Geurts AM, Hoffmann O, Mátés L, Landa V, Hiripi L, Moreno C, Lazar J, Bashir S, Zidek V, Popova E, Jerchow B, Becker K, Devaraj A, Walter I, Grzybowksi M, Corbett M, Filho AR, Hodges MR, Bader M, Ivics Z, Jacob HJ, Pravenec M, Bosze Z, Rülicke T, Izsvák Z (2013) Transposon-mediated transgenesis, transgenic rescue, and tissue-specific gene expression in rodents and rabbits. FASEB J 27(3):930–941. https://doi.org/10.1096/fj.12-205526 (Epub 2012 Nov 29)

Kleinstiver BP, Pattanayak V, Prew MS, Tsai SQ, Nguyen NT, Zheng Z, Joung JK (2016) High-fidelity CRISPR-Cas9 nucleases with no detectable genome-wide off-target effects. Nature 529 (7587):490–495. https://doi.org/10.1038/nature16526 (Epub 2016 Jan 6)

Kues WA (2018) Progressive refinements in transgenesis and genetic engineering in livestock. ISSRF Newletter 22:21–24

Li Z, Zeng F, Meng F, Xu Z, Zhang X, Huang X, Tang F, Gao W, Shi J, He X, Liu D, Wang C, Urschitz J, Moisyadi S, Wu Z (2014) Generation of transgenic pigs by cytoplasmic injection of piggyBac transposase-based pmGENIE-3 plasmids. Biol Reprod 90(5):93. https://doi.org/10.1095/biolreprod.113.116905 (Print 2014 May)

Liu X, Wang Y, Tian Y, Yu Y, Gao M, Hu G, Su F, Pan S, Luo Y, Guo Z, Quan F, Zhang Y (2014) Generation of mastitis resistance in cows by targeting human lysozyme gene to β-casein locus using zinc-finger nucleases. Proc Biol Sci 281 (1780):20133368. https://doi.org/10.1098/rspb.2013.3368 (Print 2014 Apr 7)

Makarova KS, Wolf YI, Alkhnbashi OS, Costa F, Shah SA, Saunders SJ, Barrangou R, Brouns SJ, Charpentier E, Haft DH, Horvath P, Moineau S, Mojica FJ, Terns RM, Terns MP, White MF, Yakunin AF, Garrett RA, van der Oost J, Backofen R, Koonin EV (2015) An updated evolutionary classification of CRISPR-Cas systems. Nat Rev Microbiol 13 (11):722–736. https://doi.org/10.1038/nrmicro3569 (Epub 2015 Sep 28, Review)

Mali P, Yang L, Esvelt KM, Aach J, Guell M, DiCarlo JE, Norville JE, Church GM (2013) RNA-guided human genome engineering via Cas9. Science 339(6121):823–826. https://doi.org/10.1126/science.1232033 (Epub 2013 Jan 3)

Mattei TA (2018) The CRISPR-Cas9 genome editing system: not as precise as previously believed. World Neurosurg 118:377–378. https://doi.org/10.1016/j.wneu.2018.08.042 (Epub 2018 Aug 14)

Meyer M, de Angelis MH, Wurst W, Kühn R (2010) Gene targeting by homologous recombination in mouse zygotes mediated by zinc-finger nucleases. Proc Natl Acad Sci U S A 107(34):15022–15026. https://doi.org/10.1073/pnas.1009424107 (Epub 2010 Aug 4)

Niu Y, Shen B, Cui Y, Chen Y, Wang J, Wang L, Kang Y, Zhao X, Si W, Li W, Xiang AP, Zhou J, Guo X, Bi Y, Si C, Hu B, Dong G, Wang H, Zhou Z, Li T, Tan T, Pu X, Wang F, Ji S, Zhou Q, Huang X, Ji W, Sha J (2014) Generation of gene-modified cynomolgus monkey via Cas9/RNA-mediated gene targeting in one-cell embryos. Cell 156(4):836–843. https://doi.org/10.1016/j.cell.2014.01.027 (Epub 2014 Jan 30)

Pennisi E (2013) The CRISPR craze. Science 341 (6148):833–836. https://doi.org/10.1126/science.341.6148.833 (No abstract available)

Proudfoot C, Carlson DF, Huddart R, Long CR, Pryor JH, King TJ, Lillico SG, Mileham AJ, McLaren DG, Whitelaw CB, Fahrenkrug SC (2015) Genome edited sheep and cattle. Transgenic Res 24(1):147–153. https://doi.org/10.1007/s11248-014-9832-x (Epub 2014 Sep 10)

Pyne ME, Moo-Young M, Chung DA, Chou CP (2015) Coupling the CRISPR/Cas9 system with lambda red recombineering enables simplified chromosomal gene replacement in *Escherichia coli*. Appl Environ Microbiol 81(15):5103–5114. https://doi.org/10.1128/AEM.01248-15 (Epub 2015 May 22)

Richardson CD, Ray GJ, DeWitt MA, Curie GL, Corn JE (2016) Enhancing homology-directed genome editing by catalytically active and inactive CRISPR-Cas9 using asymmetric donor DNA. Nat Biotechnol 34 (3):339–344. https://doi.org/10.1038/nbt.3481 (Epub 2016 Jan 20)

Sapranauskas R, Gasiunas G, Fremaux C, Barrangou R, Horvath P, Siksnys V (2011) The Streptococcus thermophilus CRISPR/Cas system provides immunity in *Escherichia coli*. Nucleic Acids Res 39(21):9275–9282. https://doi.org/10.1093/nar/gkr606 (Epub 2011 Aug 3)

Schetelig MF, Yan Y, Zhao Y, Handler AM (2018) Genomic targeting by recombinase-mediated cassette exchange in the spotted wing drosophila, Drosophila suzukii. Insect Mol Biol. https://doi.org/10.1111/imb.12537 (Epub ahead of print)

Shen B, Zhang J, Wu H, Wang J, Ma K, Li Z, Zhang X, Zhang P, Huang X (2013) Generation of gene-modified mice via Cas9/RNA-mediated gene targeting. Cell Res 23(5):720–723. https://doi.org/10.1038/cr.2013.46 (Epub 2013 Apr 2, No abstract available)

Smith J, Grizot S, Arnould S, Duclert A, Epinat JC, Chames P, Prieto J, Redondo P, Blanco FJ, Bravo J, Montoya G, Pâques F, Duchateau P (2006) A combinatorial approach to create artificial homing endonucleases cleaving chosen sequences. Nucleic Acids Res 34(22):e149 (Epub 2006 Nov 27)

Sternberg SH, Redding S, Jinek M, Greene EC, Doudna JA (2014) DNA interrogation by the CRISPR RNA-guided endonuclease Cas9. Nature 507(7490): 62–67. https://doi.org/10.1038/nature13011 (Epub 2014 Jan 29)

Swarts DC, Mosterd C, van Passel MW, Brouns SJ (2012) CRISPR interference directs strand specific spacer acquisition. PLoS One 7(4):e35888. https://doi.org/10.1371/journal.pone.0035888 (Epub 2012 Apr 27)

Tan W, Proudfoot C, Lillico SG, Whitelaw CB (2016) Gene targeting, genome editing: from Dolly to editors. Transgenic Res 25(3):273–287. https://doi.org/10.1007/s11248-016-9932-x (Epub 2016 Feb 3)

Urschitz J, Moisyadi S (2013) Transpositional transgenesis with *piggyBac*. Mob Genet Elements 3(3):e25167 (Epub 2013 May 24)

Xu Y, Liu S, Yu G, Chen J, Chen J, Xu X, Wu Y, Zhang A, Dowdy SF, Cheng G (2008) Excision of selectable genes from transgenic goat cells by a protein transducible TAT-Cre recombinase. Gene 419 (1–2):70–74. https://doi.org/10.1016/j.gene.2008.04.020 (Epub 2008 May 13)

Yang L, Mali P, Kim-Kiselak C, Church G (2014) CRISPR-Cas-mediated targeted genome editing in human cells. Methods Mol Biol 1114:245–267. https://doi.org/10.1007/978-1-62703-761-7_16

Yin L, Hu S, Mei S, Sun H, Xu F, Li J, Zhu W, Liu X, Zhao F, Zhang D, Cen S, Liang C, Guo F (2018) CRISPR/Cas9 inhibits multiple steps of HIV-1 infection. Hum Gene Ther 29(11):1264–1276. https://doi.org/10.1089/hum.2018.018

Yu Y, Wang Y, Tong Q, Liu X, Su F, Quan F, Guo Z, Zhang Y (2013) A site-specific recombinase-based method to produce antibiotic selectable marker free transgeniccattle. PLoS One 8(5):e62457. https://doi.org/10.1371/journal.pone.0062457 (Print 2013)

Wang K, Ouyang H, Xie Z, Yao C, Guo N, Li M, Jiao H, Pang D (2015) Efficient generation of myostatin mutations in pigs using the CRISPR/Cas9 system. Sci Rep 13(5):16623. https://doi.org/10.1038/srep16623

Wang K, Tang X, Xie Z, Zou X, Li M, Yuan H, Guo N, Ouyang H, Jiao H, Pang D (2017) CRISPR/Cas9-mediated knockout of myostatin in Chinese indigenous Erhualian pigs. Transgenic Res 26(6):799–805. https://doi.org/10.1007/s11248-017-0044-z

Wang S, Ren S, Bai R, Xiao P, Zhou Q, Zhou Y, Zhou Z, Niu Y, Ji W, Chen Y (2018) No off-target mutations in functional genome regions of a CRISPR/Cas9-generated monkey model of muscular dystrophy. J Biol Chem 293(30):11654–11658. https://doi.org/10.1074/jbc.AC118.004404

Next-Generation Sequencing Vis-à-Vis Veterinary Health Management

43

Abstract

The global market for animal health products and treatments is a multibillion-dollar business. Cost of animal disease outbreak is likely to increase with urbanization and a growing demand for animal products. Combating emerging diseases caused by drug-resistant pathogenic microorganisms is challenging. The “omics” and genetic engineering and their applications in biotechnology industry have revolutionized animal healthcare management by improving rapidity, specificity, and sensitivity of diagnostic assays and decreased rates of false positive assays.

Highlights

- Genome sequencing is now an integral component of livestock management
- The technologies have provided valuable information for managing infectious diseases.

Keywords

Genome · NextGen sequencing · NGS technologies · Livestock genome · SNPs · Genome data analysis · Computational chemistry · Bioinformatics methods

43.1 Introduction

Next-generation sequencing (NGS) is a hot topic of research in commercial scientific animal farming. Development of next-generation sequencing technologies has changed the scenario to produce and analyze genomic, transcriptomic and proteomic data. In recent years, genomes of various livestock species such as cattle, buffalo, sheep, horse, swine, and chicken are partly or completely sequenced (Table 43.1). Importance of the genome sequencing, transcriptome sequencing, metagenomics to unravel population structure, and genetic diversity is highlighted in some of the previously published (Miller et al. 2013; Bayliss et al. 2017).

NGS technologies are revolutionizing the genomics research by analysis of whole genome genotyping, RNA sequencing, de novo assembling of genomes, genome-wide structural variations, detection of mutations, and complex diseases caused by pathogens (Greenwood et al. 2016; Anis et al. 2018). NGS technologies produce 100 times more data as compared to first generation genome sequencing technologies, i.e., Sanger and Maxam-Gilbert sequencing methods.

Genome sequence information is crucial for biological research. It involves different

B. Singh et al., *Advances in Animal Biotechnology*,
https://doi.org/10.1007/978-3-030-21309-1_43

Table 43.1 Various livestock genomes sequenced with their genome size, GC content, sequencing technology used, genome coverage, and sequencing submitter information

Sl. no.	Animal	Genome size (Mb)	GC content (%)	Sequencing technology	Genome coverage	Sequencing submitter	Total SRA experiments[a]
1	*Bos taurus*	2715.85	41.94	Illumina and Pacific Biosciences	80×	United States Department of Agriculture, Agricultural Research Service	32,418
2	*Sus scrofa*	2501.91	41.97	Pacific Biosciences	65×	The Swine Genome Sequencing Consortium	19,123
3	*Ovis aries*	2615.52	41.98	Illumina and Pacific Biosciences	166×	International Sheep Genome Consortium	7208
4	*Equus caballus*	2506.97	41.53	Illumina and Pacific Biosciences	88×	University of Louisville	5552
5	*Gallus gallus*	1065.37	42.30	Pacific Biosciences	82×	Washington University Genome Sequencing Center	12,732
6	*Bubalus bubalis*	2655.78	41.81	Pacific Biosciences	69×	Italian Buffalo Genome Consortium	1021
7	*Anas platyrhynchos*	1105.05	41.20	Solexa	60×	China Agricultural University	3655
8	*Canis lupus familiaris*	2410.98	41.30	Sanger	7×	Dog Genome Sequence Consortium	10,250

[a]SRA (Sequence Read Archive) contains raw sequence data from high-throughput next-generation sequencing platforms with in NCBI

techniques for handling the genome. The massive data generated by NGS technologies is a challenge for storage and data analysis, but can be managed and utilized by means of advanced bioinformatics software and in silico methods (Zhang et al. 2011; Sharma et al. 2018).

Genome sequencing of the livestock species has helped in understanding how genes govern quantitative traits at cellular and molecular levels. Application of NGS has broadened from basic to applied research, uncovering DNA–protein interactions, insight into epigenetics modification and gene expression networks. NGS incorporated in cattle research program has accelerated genetic improvement and safeguard food, health, nutritional, and environmental security (Ghosh et al. 2018).

43.2 NGS Technologies

In the past 10 years, the sequencing technologies have made progressive improvements including overall performance, cost-effectiveness, and resolving the complexity of protocols. There are different high-throughput sequencing technologies available for biological research which use different strategies and generate different outputs. NGS technologies have enabled the scientists to understand the complex biological world in a broader and meaningful perspective (Zhang et al. 2011). The individual technologies have their own advantages and capabilities which were reviewed earlier (Loman et al. 2012; Buermans and Den Dunnen 2014).

Announcement of NGS platforms in the market with focus on high-throughput data generation was initially restricted to genome sequencing. Later on NGS technologies were used in biological studies including genomics, transcriptomics, proteomics, breeding, medicine, cancer, host–pathogen interactions, and clinical microbiology (Zhang et al. 2011; Bai et al. 2012; Anis et al. 2018). 454 sequencer system by Life Sciences is the first, and Solexa 1G by Illumina is the second commercial platform for NGS (Morozova and Marra 2008). The 454 platform is based on principle of pyrosequencing, Illumina on sequence by synthesis, and ABI-SOLiD on sequence by ligation.

Uninterrupted improvement of surviving platforms and some newer technologies have been developed. Platforms based on single-molecule real-time sequencing (SMRT) are available which have promising advantages in terms of simplicity, low cost, and long read lengths. The SMRT does not involve clonal amplification of DNA; therefore, no chances of errors are associated with clonal amplification. PacBio (Pacific Biosciences) is the platform which is based on SMRT sequencing technology which sequences long single DNA in real time (Nakano et al. 2017). Other sequencing technologies such as Oxford nanopore is different from NGS platforms and does not involve amplification of DNA, but uses electric signal to detect the nucleotides (Buermans and Den Dunnen 2014; Senol et al. 2018).

43.3 NGS in Livestock Genomes

In the early era of genomics, when NGS technologies were not present, the animal breeding programs primarily relied on molecular marker-assisted selection (Williams 2005). Genome sequencing indicates the genetic makeup of an organism, and RNA sequencing displays the sequences that are actively expressed in the cell. Transcriptome sequencing (RNA-seq) is a genomic reduction analytical method which is exploited to study ecological genomics, adaptation and single nucleotide polymorphism in different species (McCormack et al. 2013).

NGS technologies are very helpful in the sequencing of animal genomes. Livestock genome sequencing information is helpful in identification of genetic markers used in animal breeding programs and detecting pathogens which may include traits involved to pathogen resistance and interaction with microbes in chicken (Ramos et al. 2009; Aslam et al. 2012; Diaz-Sanchez et al. 2013). Progress in the sequencing technologies over the past years benefited the animal research in multiple ways (Tellam et al. 2009; Williams et al. 2017). NGS has generated massive data at very low cost. Among various livestock species, dog was first species whose genome was sequenced completely in 2005 (Lindblad-Toh et al. 2005). Cattle (*Bos taurus*) genome was sequences shortly thereafter, followed by *Sus scrofa* (Table 43.1).

NGS has opened new approaches to investigate relationship between phenotypic and genetic diversity. Whole genome sequences and draft genome sequences of livestock from different species are available under various public databases, and various gene sequencing projects are underway. Data generated by NGS can lead to determine genetic markers present in the whole genome (Zhang et al. 2011). Advent of NGS technologies has pioneered a new area of research and identification of single nucleotide polymorphisms (SNPs). Ample number of SNP's has been identified in livestock species including cattle, sheep, pig, and others (Djari et al. 2013). Interpretation of NGS data is more advantageous over SNP array, as it provides better information on genomics and genomic prediction. Use of WGS data and transcriptomics data is beneficial for multi-breed prediction (Iheshiulor et al. 2016). Various types of NGS types can be used according to the need of the research project.

43.4 NGS in Livestock Infectious Diseases

Livestock infectious diseases are great threat to livestock and economic gains from livestock (Brooks-Pollock et al. 2015). Emerging infectious diseases in livestock are responsible for enormous population decline due to massive culling of livestock posing threat of zoonotic diseases (Tomley and Shirley 2009; Lefrançois and Pineau 2014). Adverse climate or climatic stress renders animals susceptible to pathogens, parasites, and pests (Gibbs 2005). Hence, it is important that livestock should be free of pathogens so that disease does not affect utilization of livestock products, and the international trade based on livestock products. Notably, there is serious concern of the presence of insecticides or antibiotic residues in livestock and agriculture products meant for export to other countries.

NGS has improved understanding of livestock genome, transcriptome, and epigenome. NGS technologies are progressively used for monitoring and diagnosis of infectious diseases including host–pathogen interactions. Diagnosing infectious diseases in laboratory is very challenging, particularly the diseases caused by mixed multiple infectious agents. In everyday laboratory work, it requires handling and processing of large number of clinical samples. Strategies have been developed to detect infectious agents utilizing NGS data (Anis et al. 2018). For instance, the metagenomics approaches are employed to amplify and discover all the infectious agents which are present in the biological sample leading to unbiased detection of pathogens therein (Miller et al. 2013). However, data generated by NGS is huge and needs processing of millions of reads to conclude detection of suspected pathogens or disease markers. Targeted NGS is another technique that involves selective amplification of defined genomic regions which are of interest (Mertes et al. 2011).

Compared to metagenomics, targeted NGS has limited prospects for discovering new pathogens. Therefore, metagenomic sequencing is the utmost important tool which is progressively being used in disease diagnosis, management and can lead to the identification of novel pathogens. It is desirable to use NGS as identification tool into animal diagnostic laboratory with cost-effective manner.

NGS is presently the leading technology in molecular genetics research. Development of NGS technologies has changed the vision of researchers to analyze the genome, transcriptome, and proteome of an organism. It has provided the biological insight and showed the potential of NGS technologies which directly have impact to manage animal infectious disease (Deurenberg et al. 2017).

NGS can be used in pathogen biology for screening transposon libraries for identifying pathways and genes that contribute to survival in different environment (van Opijnen et al. 2009; Langridge et al. 2009). RNA-Seq has revolutionized the bacterial research by providing the extensive perspective of pathogen transcriptomes (Sorek and Cossart 2010). RNA-Seq has been applied to various bacterial transcriptomes and important insights are generated (Martin et al. 2010; Camarena et al. 2010; Sharma et al. 2010). Even NGS can be used to analyze transcriptome of pathogen present in tissues of patients (Azhikina et al. 2010). Nowadays, high-throughput sequencing is combined with translational research.

High-throughput sequencing of pathogen genome has helped to identify the immunogenic proteins, based on which diagnostic immunoassays have been developed (Greub et al. 2009). NGS has also been applied to detect mutations leading to acquisition of resistance in bacteria and viral pathogens (Wang et al. 2007; Feng et al. 2009; Urbaniak et al. 2018). NGS can be used to study evolution of infectious agents. Different studies have been performed to get deep insight into the evolution of pathogenic bacteria by using NGS (Srivatsan et al. 2008; Bryant et al. 2012).

NGS facilitates sequencing of microorganisms without prior culturing. It helps in identifying novel metabolites, enzymes, microorganisms and antimicrobial peptides (AMPs) or bacteriocins for livestock or industrial applications (Singh et al. 2008; Challis 2014). Rapid sequence of bacterial genomes by NGS helps to locate genes and pathways involved in synthesis of new compounds which may be inactive in in vitro conditions, but is of utmost importance as pharmaceuticals (Gomez-Escribano et al. 2015). Complete sequence of bacteriophage was extracted which contains antimicrobial properties inhabiting chicken intestine and has potential to combat bacterial disease (Diaz-Sanchez et al. 2013). Earlier various genes were identified from gastrointestinal ecosystem of ruminants (Singh et al. 2012; Sharma et al. 2017), but metagenomics may provide better interpretations and novel genes can be identified which can be used to improve the health of livestock without culturing the gastrointestinal anaerobic bacteria which is difficult, time-consuming, labor-intensive and costs very high (Kim et al. 2017). Further, metagenomics NGS has the possible use for quality control of vaccines and various biological products.

43.5 Outlook and Challenges

NGS technologies are elite and revolutionized the research in infection biology and animal infectious diseases. Availability of high-quality genomic and transcriptomic data produced by various NGS platforms will help in the development of novel drugs and vaccine which will ultimately benefit the reducing livestock.

NGS technologies are progressively used in the study of genomics, transcriptomics, evolution, etiology, host–pathogen interactions, and epidemiology of livestock infectious diseases. Genome sequence data should be analyzed for predicting new genes and proteins as targets for therapeutic interventions.

43.6 Conclusions

Sequencing of genome has already changed the study of infectious agents. NGS can be applied to the infectious agents isolated from the pure culture which can lead to our better understanding of epidemiological infection and pathogenesis caused by the pathogen.

References

Anis E, Hawkins IK, Ilha MRS, Woldemeskel MW, Saliki JT, Wilkes RP (2018) Evaluation of targeted next-generation sequencing for detection of bovine pathogens in clinical samples. J Clin Microbiol 56(7). pii: e00399-18. https://doi.org/10.1128/jcm.00399-18 (Print 2018 Jul)

Aslam ML, Bastiaansen JW, Elferink MG, Megens HJ, Crooijmans RP, le Blomberg A, Fleischer RC, Van Tassell CP, Sonstegard TS, Schroeder SG, Groenen MA, Long JA (2012) Whole genome SNP discovery and analysis of genetic diversity in Turkey (Meleagris gallopavo). BMC Genom 14(13):391. https://doi.org/10.1186/1471-2164-13-391

Azhikina TL, Skvortsov TA, Radaeva TV, Mardanov AV, Ravin NV, Apt AS, Sverdlov ED (2010) A new technique for obtaining whole pathogen transcriptomes from infected host tissues. Biotechniques 48:139–144

Bai Y, Sarto M, Cavalcoli J (2012) Current status and future perspectives for sequencing livestock genomes. J Anim Sci Biotechnol 3(1):8

Bayliss SC, Verner-Jeffreys DW, Bartie KL, Aanensen DM, Sheppard SK, Adams A, Feil EJ (2017) The promise of whole genome pathogen sequencing for the molecular epidemiology of emerging aquaculture pathogens. Front Microbiol 8:121. https://doi.org/10.3389/fmicb.2017.00121 (eCollection 2017)

Brooks-Pollock E, de Jong MC, Keeling MJ, Klinkenberg D, Wood JL (2015) Eight challenges in modelling infectious livestock diseases. Epidemics 10:1–5. https://doi.org/10.1016/j.epidem.2014.08.005

Bryant J, Chewapreecha C, Bentley SD (2012) Developing insights into the mechanisms of evolution of bacterial pathogens from whole-genomesequences. Future Microbiol 7(11):1283 1296. https://doi.org/10.2217/fmb.12.108 (Review)

Buermans HP, Den Dunnen JT (2014) Next generation sequencing technology: advances and applications. Biochim Biophys Acta 1842(10):1932–1941. https://doi.org/10.1016/j.bbadis.2014.06.015 (Epub 2014 Jul 1. Review)

Camarena L, Bruno V, Euskirchen G, Poggio S, Snyder M (2010) Molecular mechanisms of ethanol-induced pathogenesis revealed by RNA-sequencing. PLoS Pathog 6:e1000834. https://doi.org/10.1371/journal.ppat.1000834

Challis GL (2014) Exploitation of the Streptomyces coelicolor A3(2) genome sequence for discovery of new natural products and biosynthetic pathways. J Ind Microbiol Biotechnol 41:219–232. https://doi.org/10.1007/s10295-013-1383-2

Deurenberg RH, Bathoorn E, Chlebowicz MA, Couto N, Ferdous M, García-Cobos S, Kooistra-Smid AM, Raangs EC, Rosema S, Veloo AC, Zhou K, Friedrich AW, Rossen JW (2017) Application of next generation sequencing in clinical microbiology and infection prevention. J Biotechnol 10(243):16–24. https://doi.org/10.1016/j.jbiotec.2016.12.022

Diaz-Sanchez S, Hanning I, Pendleton S, D'Souza D (2013) Next-generation sequencing: the future of molecular genetics in poultry production and food safety. Poult Sci 92(2):562–572. https://doi.org/10.3382/ps.2012-02741 (Review)

Djari A, Esquerré D, Weiss B, Martins F, Meersseman C, Boussaha M, Klopp C, Rocha D (2013) Gene-based single nucleotide polymorphism discovery in bovine muscle using next-generation transcriptomic sequencing. BMC Genom 14:307. https://doi.org/10.1186/1471-2164-14-307

Feng J, Lupien A, Gingras H, Wasserscheid J, Dewar K, Légaré D, Ouellette M (2009) Genome sequencing of linezolid-resistant Streptococcus pneumoniae mutants reveals novel mechanisms of resistance. Genome Res 19(7):1214–1223. https://doi.org/10.1101/gr.089342.108

Ghosh M, Sharma N, Singh AK, Gera M, Pulicherla KK, Jeong DK (2018) Transformation of animal genomics by next-generation sequencing technologies: a decade of challenges and their impact on genetic architecture. Crit Rev Biotechnol 38(8):1157–1175. https://doi.org/10.1080/07388551.2018.1451819

Gibbs EP (2005) Emerging zoonotic epidemics in the interconnected global community. Vet Rec. 157 (22):673–9

Gomez-Escribano JP, Castro JF, Razmilic V, Chandra G, Andrews B, Asenjo JA, Bibb MJ (2015) The Streptomyces leeuwenhoekii genome: de novo sequencing and assembly in single contigs of the chromosome, circular plasmid pSLE1 and linear plasmid pSLE2. BMC Genom 30(16):485. https://doi.org/10.1186/s12864-015-1652-8

Greenwood JM, Ezquerra AL, Behrens S, Branca A, Mallet L (2016) Current analysis of host-parasite interactions with a focus on next generation sequencing data. Zoology (Jena) 119(4):298–306. https://doi.org/10.1016/j.zool.2016.06.010

Greub G, Kebbi-Beghdadi C, Bertelli C, Collyn F, Riederer BM, Yersin C, Croxatto A, Raoult D (2009) High throughput sequencing and proteomics to identify immunogenic proteins of a new pathogen: the dirty genome approach. PLoS ONE 4(12):e8423. https://doi.org/10.1371/journal.pone.0008423

Iheshiulor OO, Woolliams JA, Yu X, Wellmann R, Meuwissen TH (2016) Within- and across-breed genomic prediction using whole-genome sequence and single nucleotide polymorphism panels. Genet Sel Evol 19(48):15. https://doi.org/10.1186/s12711-016-0193-1

Kim M, Park T, Yu Z (2017) Metagenomic investigation of gastrointestinal microbiome in cattle. Asian Aust J Anim Sci 30:1515

Langridge GC, Phan MD, Turner DJ, Perkins TT, Parts L, Haase J, Charles I, Maskell DJ, Peters SE, Dougan G, Wain J, Parkhill J, Turner AK (2009) Simultaneous assay of every Salmonella Typhi gene using one million transposon mutants. Genome Res 19 (12):2308–2316. https://doi.org/10.1101/gr.097097.109

Lefrançois T, Pineau T (2014) Public health and livestock: emerging diseases in food animals. Anim Front 4(1):4–6

Lindblad-Toh K, Wade CM, Mikkelsen TS, Karlsson EK, Jaffe DB, Kamal M, Clamp M, Chang JL, Kulbokas EJ 3rd, Zody MC, Mauceli E, Xie X, Breen M, Wayne RK, Ostrander EA, Ponting CP, Galibert F, Smith DR, DeJong PJ, Kirkness E, Alvarez P, Biagi T, Brockman W, Butler J, Chin CW, Cook A, Cuff J, Daly MJ, DeCaprio D, Gnerre S, Grabherr M, Kellis M, Kleber M, Bardeleben C, Goodstadt L, Heger A, Hitte C, Kim L, Koepfli KP, Parker HG, Pollinger JP, Searle SM, Sutter NB, Thomas R, Webber C, Baldwin J, Abebe A, Abouelleil A, Aftuck L, Ait-Zahra M, Aldredge T, Allen N, An P, Anderson S, Antoine C, Arachchi H, Aslam A, Ayotte L, Bachantsang P, Barry A, Bayul T, Benamara M, Berlin A, Bessette D, Blitshteyn B, Bloom T, Blye J, Boguslavskiy L, Bonnet C, Boukhgalter B, Brown A, Cahill P, Calixte N, Camarata J, Cheshatsang Y, Chu J, Citroen M, Collymore A, Cooke P, Dawoe T, Daza R, Decktor K, DeGray S, Dhargay N, Dooley K, Dooley K, Dorje P, Dorjee K, Dorris L, Duffey N, Dupes A, Egbiremolen O, Elong R, Falk J, Farina A, Faro S, Ferguson D, Ferreira P, Fisher S, FitzGerald M, Foley K, Foley C, Franke A, Friedrich D, Gage D, Garber M, Gearin G, Giannoukos G, Goode T, Goyette A, Graham J, Grandbois E, Gyaltsen K, Hafez N, Hagopian D, Hagos B, Hall J, Healy C, Hegarty R, Honan T, Horn A, Houde N, Hughes L, Hunnicutt L, Husby M, Jester B, Jones C, Kamat A, Kanga B, Kells C, Khazanovich D, Kieu AC, Kisner P, Kumar M, Lance K, Landers T, Lara M, Lee W, Leger JP, Lennon N, Leuper L, LeVine S, Liu J, Liu X, Lokyitsang Y, Lokyitsang T, Lui A, Macdonald J, Major J, Marabella R, Maru K, Matthews C, McDonough S, Mehta T, Meldrim J, Melnikov A, Meneus L, Mihalev A, Mihova T, Miller K, Mittelman R, Mlenga V, Mulrain L, Munson G, Navidi A, Naylor J, Nguyen T, Nguyen N, Nguyen C, Nguyen T, Nicol R, Norbu N, Norbu C, Novod N, Nyima T, Olandt P, O'Neill B, O'Neill K, Osman S, Oyono L, Patti C, Perrin D, Phunkhang P, Pierre F, Priest M, Rachupka A, Raghuraman S, Rameau R, Ray V, Raymond C, Rege F, Rise C, Rogers J, Rogov P, Sahalie J, Settipalli S, Sharpe T, Shea T, Sheehan M, Sherpa N, Shi J, Shih D, Sloan J, Smith C, Sparrow T, Stalker J, Stange-Thomann N, Stavropoulos S, Stone C, Stone S, Sykes S, Tchuinga P, Tenzing P, Tesfaye S, Thoulutsang D, Thoulutsang Y, Topham K, Topping I, Tsamla T, Vassiliev H, Venkataraman V, Vo A, Wangchuk T, Wangdi T, Weiand M, Wilkinson J, Wilson A, Yadav S, Yang S, Yang X, Young G, Yu Q, Zainoun J, Zembek L, Zimmer A, Lander ES (2005) Genome sequence, comparative analysis and haplotype structure of the domestic dog. Nature 438:803–819

Loman NJ, Misra RV, Dallman TJ, Constantinidou C, Gharbia SE, Wain J, Pallen MJ (2012) Performance comparison of benchtop high-throughput sequencing platforms. Nat Biotechnol 30(5):434–439. https://doi.org/10.1038/nbt.2198 (Erratum. In: Nat Biotechnol. 2012 Jun; 30(6):562)

Martin J, Zhu W, Passalacqua KD, Bergman N, Borodovsky M (2010) Bacillus anthracis genome organization in light of whole transcriptome sequencing. BMC Bioinform 11(Suppl 3):S10. https://doi.org/10.1186/1471-2105-11-S3-S10

McCormack JE, Hird SM, Zellmer AJ, Carstens BC, Brumfield RT (2013) Applications of next-generation sequencing to phylogeography and phylogenetics. Mol Phylogenet Evol 66(2):526–538. https://doi.org/10.1016/j.ympev.2011.12.007

Mertes F, Elsharawy A, Sauer S, van Helvoort JM, van der Zaag PJ, Franke A, Nilsson M, Lehrach H, Brookes AJ (2011) Targeted enrichment of genomic DNA regions for next-generation sequencing. Brief Funct Genom 10 (6):374–386. https://doi.org/10.1093/bfgp/elr033

Miller RR, Montoya V, Gardy JL, Patrick DM, Tang P (2013) Metagenomics for pathogen detection in public health. Genome Med 5(9):81. https://doi.org/10.1186/gm485 (eCollection 2013)

Morozova O, Marra MA (2008) Applications of next-generation sequencing technologies in functional genomics. Genomics 92(5):255–264. https://doi.org/10.1016/j.ygeno.2008.07.001

Nakano K, Shiroma A, Shimoji M, Tamotsu H, Ashimine N, Ohki S, Shinzato M, Minami M, Nakanishi T, Teruya K, Satou K, Hirano T (2017) Advantages of genome sequencing by long-read sequencer using SMRT technology in medical area. Hum Cell 30(3):149–161. https://doi.org/10.1007/s13577-017-0168-8

Ramos AM, Crooijmans RP, Affara NA, Amaral AJ, Archibald AL, Beever JE, Bendixen C, Churcher C, Clark R, Dehais P, Hansen MS, Hedegaard J, Hu ZL, Kerstens HH, Law AS, Megens HJ, Milan D, Nonneman DJ, Rohrer GA, Rothschild MF, Smith TP, Schnabel RD, Van Tassell CP, Taylor JF, Wiedmann RT, Schook LB, Groenen MA (2009) Design of a high density SNP genotyping assay in the pig using SNPs identified and characterized by next generation sequencing technology. PLoS ONE 4(8):e6524. https://doi.org/10.1371/journal.pone.0006524

Senol Cali D, Kim JS, Ghose S, Alkan C, Mutlu O (2018) Nanopore sequencing technology and tools for genome assembly: computational analysis of the current state, bottlenecks and future directions. Brief Bioinform. https://doi.org/10.1093/bib/bby017

Sharma CM, Hoffmann S, Darfeuille F, Reignier J, Findeiss S, Sittka A, Chabas S, Reiche K, Hackermüller J, Reinhardt R, Stadler PF, Vogel J (2010) The primary transcriptome of the major human pathogen Helicobacter pylori. Nature 464 (7286):250–255. https://doi.org/10.1038/nature08756

Sharma D, Mal G, Kannan A, Bhar R, Sharma R, Singh B (2017) Degradation of euptox A by tannase-producing rumen bacteria from migratory goats. J Appl Microbiol 123(5):1194–1202. https://doi.org/10.1111/jam.13563 (Epub 2017 Sep 19)

Sharma D, Sharma A, Verma SK, Singh B (2018) Targeting metabolic pathways proteins of Orientia tsutsugamushi using combined hierarchical approach to combat scrub typhus J Mol Recognit. 32(4):e2766. https://doi.org/10.1002/jmr.2766

Singh B, Gautam SK, Verma V, Kumar M, Singh B (2008) Metagenomics in animal gastrointestinal ecosystem: potential biotechnological prospects. Anaerobe 14(3):138–144. https://doi.org/10.1016/j.anaerobe.2008.03.002 (Epub 2008 Mar 26. Review)

Singh B, Bhat TK, Sharma OP, Kanwar SS, Rahi P, Gulati A (2012) Isolation of tannase-producing Enterobacter ludwigii GRT-1 from the rumen of migratory goats. Small Rumin Res 102(2–3):172–176

Sorek R, Cossart P (2010) Prokaryotic transcriptomics: a new view on regulation, physiology and pathogenicity. Nat Rev Genet 11(1):9–16. https://doi.org/10.1038/nrg2695

Srivatsan A, Han Y, Peng J, Tehranchi AK, Gibbs R, Wang JD, Chen R (2008) High-precision, whole-genome sequencing of laboratory strains facilitates genetic studies. PLoS Genet 4(8):e1000139. https://doi.org/10.1371/journal.pgen.1000139

Tellam RL, Lemay DG, Van Tassell CP, Lewin HA, Worley KC, Elsik CG (2009) Unlocking the bovine genome. BMC Genom 24(10):193. https://doi.org/10.1186/1471-2164-10-193

Tomley FM, Shirley MW (2009) Livestock infectious diseases and zoonoses. Philos Trans R Soc Lond B Biol Sci 364(1530):2637–2642. https://doi.org/10.1098/rstb.2009.0133

Urbaniak C, Sielaff AC, Frey KG, Allen JE, Singh N, Jaing C, Wheeler K, Venkateswaran K (2018) Detection of antimicrobial resistance genes associated with the International Space Station environmental surfaces. Sci Rep 8:814. https://doi.org/10.1038/s41598-017-18506-4

van Opijnen T, Bodi KL, Camilli A (2009) Tn-seq: high-throughput parallel sequencing for fitness and genetic interaction studies in microorganisms. Nat Methods 6(10):767–772. https://doi.org/10.1038/nmeth.1377

Wang C, Mitsuya Y, Gharizadeh B, Ronaghi M, Shafer RW (2007) Characterization of mutation spectra with ultra-deep pyrosequencing: application to HIV-1 drug resistance. Genome Res 17(8):1195–1201

Williams JL (2005) The use of marker-assisted selection in animal breeding and biotechnology. Rev Sci Tech 24(1):379–391

Williams JL, Iamartino D, Pruitt KD, Sonstegard T, Smith TPL, Low WY, Biagini T, Bomba L, Capomaccio S, Castiglioni B, Coletta A, Corrado F, Ferré F, Iannuzzi L, Lawley C, Macciotta N, McClure M, Mancini G, Matassino D, Mazza R, Milanesi M, Moioli B, Morandi N, Ramunno L, Peretti V, Pilla F, Ramelli P, Schroeder S, Strozzi F, Thibaud-Nissen F, Zicarelli L, Ajmone-Marsan P, Valentini A, Chillemi G, Zimin A (2017) Genome assembly and transcriptome resource for river buffalo, *Bubalus bubalis* (2n = 50). Gigascience 6(10):1–6. https://doi.org/10.1093/gigascience/gix088

Zhang J, Chiodini R, Badr A, Zhang G (2011) The impact of next-generation sequencing on genomics. J Genet Genomics. 38(3):95–109. https://doi.org/10.1016/j.jgg.2011.02.003

Computer-Aided Drug Discovery

44

Abstract

Computer-aided drug designating or drug discovery is the methodology based on computational and bioinformatics approaches to discover, develop, and analyze the drugs and similar biologically active molecules. The computer-aided drug discovery is benefited from massive genome and proteome data of pathogens and hosts accessible for analysis and interpretation. It is possible to discover potential proteins and metabolic pathways of pathogenic microorganisms and the parasites and develop novel biomolecules as drugs or therapeutics.

Highlights

- Computer-aided drug discovery is an *in silico* method of developing drugs or drug-like molecules
- The technique has important contribution to develop drugs against pathogens and parasites.

Keywords

Computational drug designing · Proteome data · Computer-aided drug discovery · Protein structure prediction · Molecular dynamics · Ligand-based drug discovery

44.1 Introduction

The drug is the small organic molecule that inhibits the activity of certain biological molecules which ultimately induces or leads to therapeutic benefits. *In silico,* drug designing is an innovative process of discovering new therapeutic molecules based on the information of biological targets. Computer-aided drug discovery (CADD) is the process of designing small molecules which are complementary in shape to the biological target sites with which they interact and bind. CADD is utilized to expedite the development of therapeutically relevant small molecules. The CADD is based on computational techniques to streamline drug discovery process, assistance of biological and chemical information of target and ligands to discover new drugs (Kapetanovic 2008; Glicksberg et al. 2019).

At present, it is possible to combine biological and chemical space with the help of computational tools and programs to reinforce drug design, development, and optimization (Sliwoski et al. 2013). Computational techniques for drug discovery and designing are gaining appreciation, as these are anticipated to play an important role in recognizing the precise molecular events and interactions of targeted protein with candidate hits, ultimately leading to identify and

B. Singh et al., *Advances in Animal Biotechnology*,
https://doi.org/10.1007/978-3-030-21309-1_44

design improved lead molecules for target (Shaikh et al. 2007). The general phases of computer-aided drug development of pharmaceuticals are depicted in Figs. 44.1, 44.2, 44.3, 44.4, 44.5, 44.6 and 44.7.

The process of computational drug designing primarily includes two broad approaches, (i) structure-based drug design (drug-target docking), and (ii) ligand-based drug design (pharmacophore and quantity structure-activity relationship (QSAR)). The CADD has been broadly engaged in lead identification and lead optimization processes against different macromolecular targets. Rational CADD when

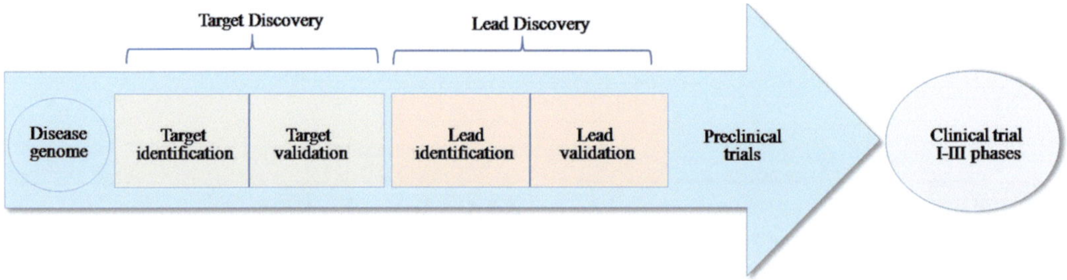

Fig. 44.1 The phases of computer-aided drug development

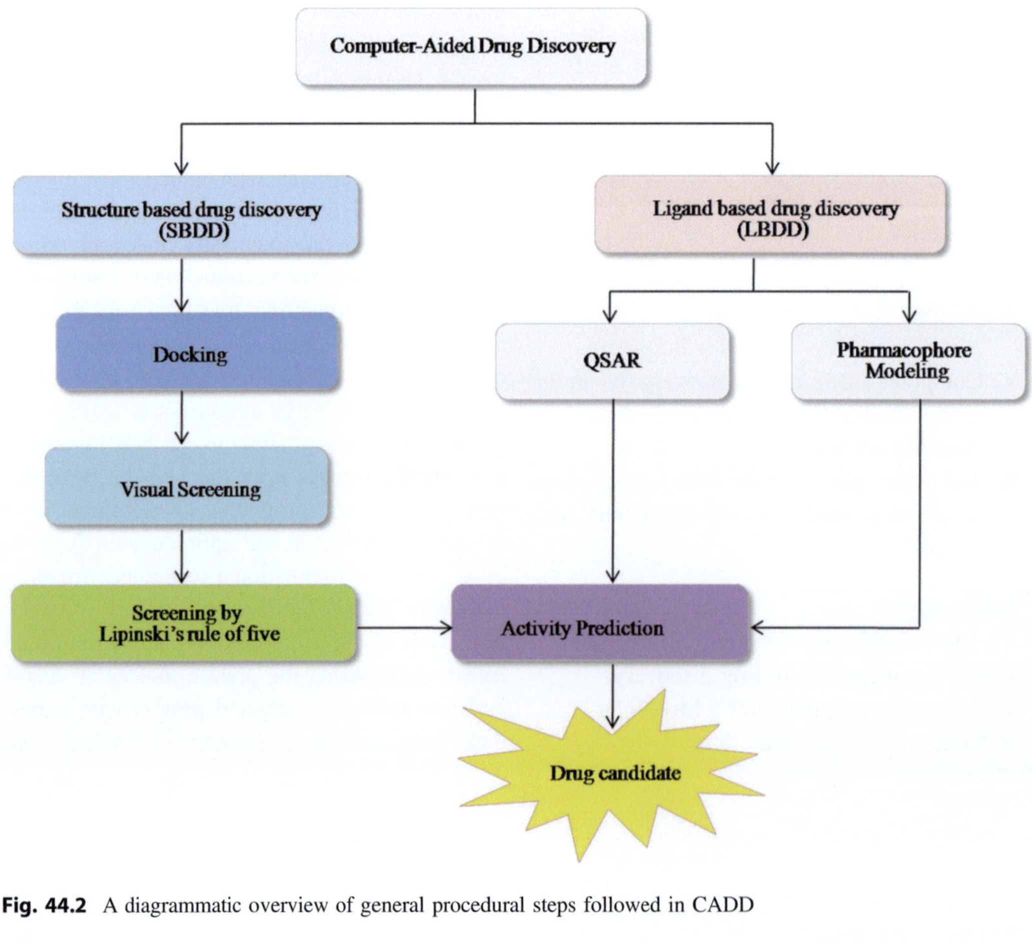

Fig. 44.2 A diagrammatic overview of general procedural steps followed in CADD

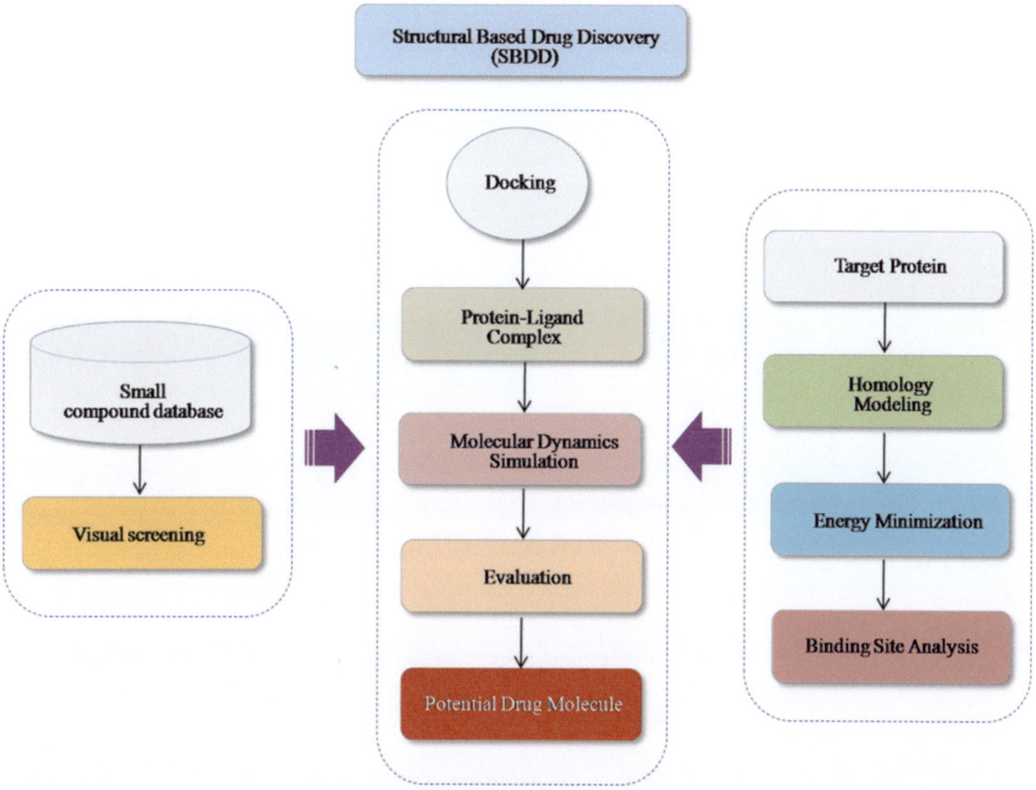

Fig. 44.3 Diagrammatic illustration of docking approach

compared to traditional drug discovery brings down the cost and time involved in discovering drug candidates. The CADD is categorized into two main types: (a) structure-based drug discovery (SBDD), and (b) ligand-based drug discovery (LBDD). The generalized overview of CADD is depicted in Fig. 44.2.

44.2 Structure-Based Drug Discovery

SBDD is the technique where three-dimensional structural (3D) information of biological target is employed for developing its inhibitors. SBDD directs the development of drug lead, which actually is not the drug product, but a chemical compound that exhibits biological activity for developing new drug (Verlinde and Hol 1994). This 3D structure of the biological target can be exploited for ligand binding which initiates drug discovery process. The SBDD approach is the most efficient approach for drug development process (Fig. 44.3).

44.3 Protein Structure Prediction

In SBDD, the first step before exploring the receptor-ligand relationship is to achieve the biological target structure. The 3D structure of the biological target protein can be achieved experimentally by NMR or X-ray crystallography (Martí-Renom et al. 2000). Majority of the three-dimensional structures and sequences are available in public domains and can be accessed or downloaded from Protein Data Bank (PDB) (Berman et al. 2006). However, when experimental structure of the target is not known, computational techniques, such as homology modeling can be used to predict 3D structure of unknown proteins (Barcellos et al. 2008).

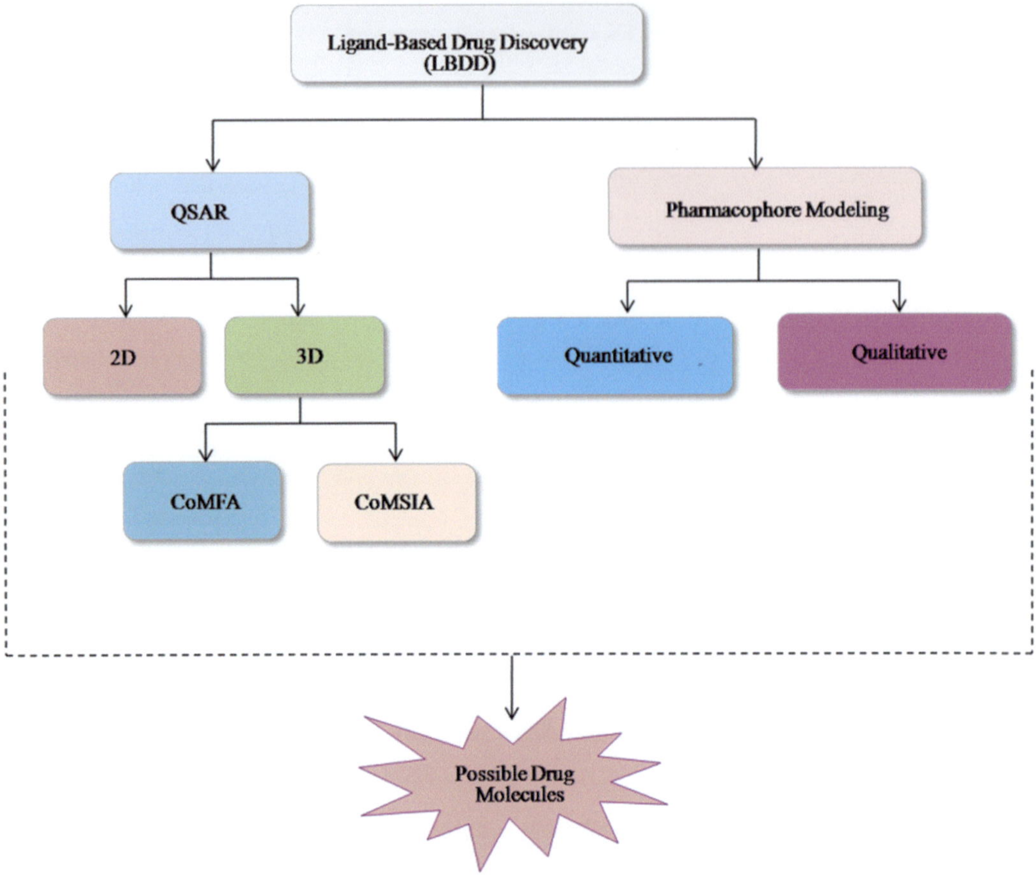

Fig. 44.4 Steps involved during Ligand-based drug discovery

44.4 Homology Modeling

Homology modeling is the most efficient and fast approach to model protein structures as a part drug development process, and protein–protein interactions (PPI) (Sujatha et al. 2009; Singh et al. 2017; Sharma et al. 2019). The proteins showing more than 30% sequence homology with their templates (homologous), can be modeled further for determining 3D structures (Martí-Renom et al. 2000). This homology modeling strategy was employed by various researchers in order to build or predict 3D structure of the proteins. Homology modeling is the knowledge-based process which involved different steps (Fig. 44.7) such as template recognition, multiple sequence alignment, protein model building, model refinement, and model validation (Joo et al. 2012). There are various programs (Table 44.1) which can be used for homology modeling of the protein.

44.5 Docking

In rational drug discovery, the role of protein-ligand docking is important to predicting the position of ligand when bound to the receptor protein. Molecular docking explores the activities of small molecules in active site of targeted protein. The electrostatic interactions, *van der* Walls interactions, Coulombic interactions and hydrogen bond formation play important role in docking. Computations of all the interactions result in the docking score, ultimately

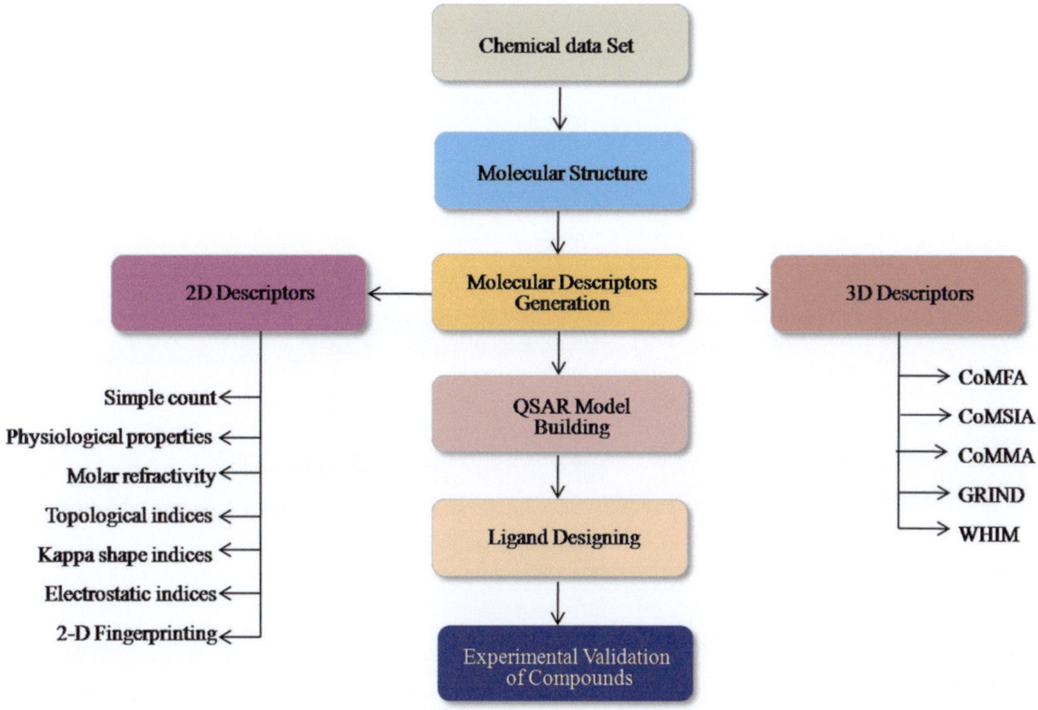

Fig. 44.5 General steps during QSAR

representing binding potentiality of ligand with protein binding site (Alberg and Schreiber 1993). The identification of ligand is performed with the help of virtual screening. The selected compounds from screening database are ranked and experimental testing for biological activity was performed for a given receptor (Shoichet 2004). The docking approach is depicted in Fig. 44.3.

Molecular docking constitutes of two sections, i.e., search algorithm and scoring function. The search algorithms are the set of parameters and rules used to predict molecular configurations. Docking makes use of various algorithms such as fast shape matching, Monte Carlo simulation, incremental construction, evolutionary programming, distance geometry, genetic algorithms, and simulated annealing (Dias and de Azevedo 2008). The scoring functions are mathematical methods used for calculation of binding affinity. The scoring function will help to find the ligand with highest-affinity to bind with the target. Scoring consists of two different expressions, (i) ranking of the generated configurations, and (ii) virtual screening (different ligands ranking against the target protein).

There are various scoring functions such as LigScore, F-Score, G-Score, D-Score, ChemScore, DrugScore, X-Score, GoldScore, and many more (Wang et al. 2003). The ligand placement is optimized by scoring functions during docking process.

Primarily, there are two types of docking, *i.e.,* rigid docking and flexible docking. Rigid docking is based upon assumption of "lock and key hypothesis" which describes both receptor and ligand are treated as rigid body. Rigid docking generates large number of docked conformations with again ranking of the docked conformations by means of free energy (Mezei 2003). Flexible docking is based on assumption of "induced-fit" hypothesis, stating that receptor and ligand are considered as flexible during docking. The flexible docking algorithms predict the binding mode of the molecules more precisely than rigid docking algorithms, and its binding affinity to other compounds (Hammes 2002).

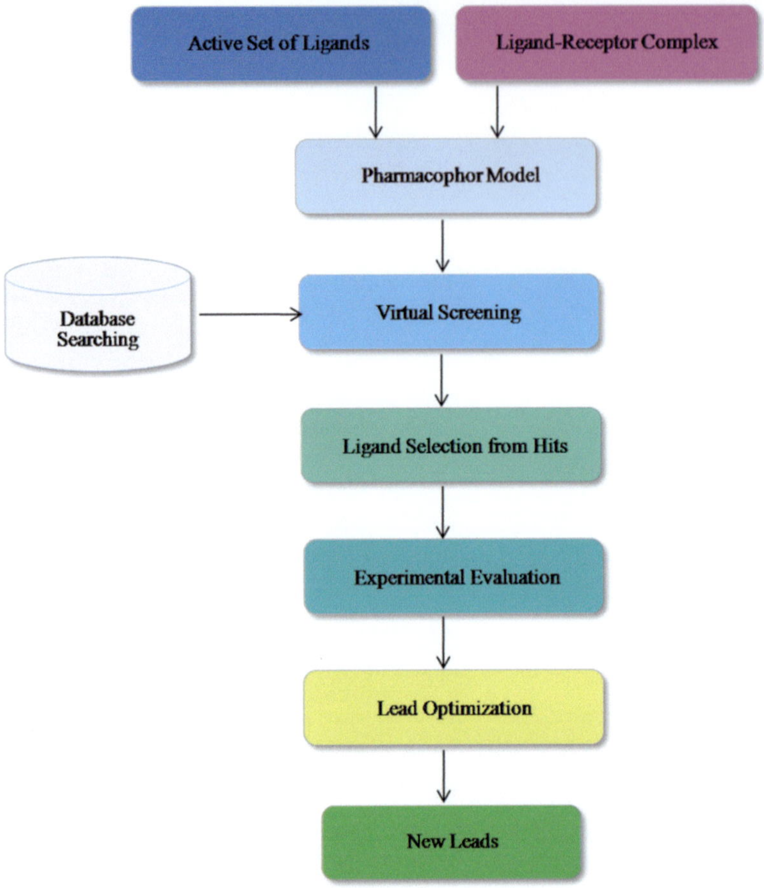

Fig. 44.6 The pharmacophore modeling was explained

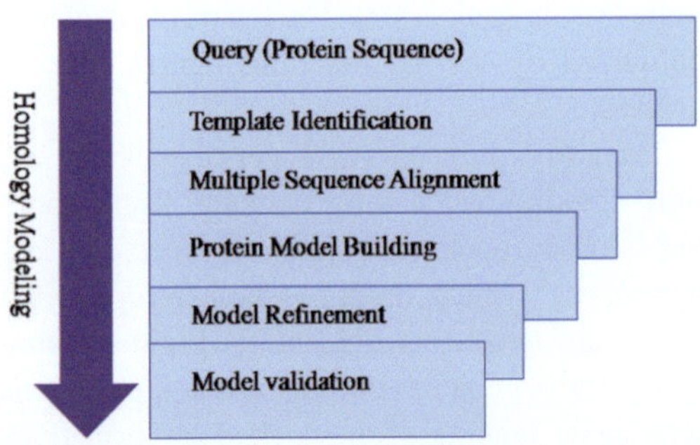

Fig. 44.7 Steps involved in homology modeling

There are more than 60 different docking programs and tools developed during the past two decades. Some of the important docking programs are summarized in Table 44.2. SBDD is the powerful technique for fast identification of small chemical molecules against the protein 3D structure available by experimental methods or computational methods.

44.6 Ligand-Based Drug Discovery (LBDD)

When 3D structure information of the target macromolecule is unidentified and cannot be identified by homology modeling, the alternative approach followed is LBDD. This method involves the investigation of ligands which are actually known to interact with macromolecular target of interest. Molecular similarity approaches such as QSAR and pharmacophore modeling are most widely and frequently used methods in LBDD. The molecular fingerprints of known ligands can be used for identification of

Table 44.1 List of some important programs for homology modeling

Sl. no.	Program	Web address	Availability	Reference
1	Phyre2	http://www.sbg.bio.ic.ac.uk/phyre2/html/page.cgi?id=index	Free	Kelley et al. (2015)
2	Modeler	Standalone	Free	Webb and Sali (2014)
3	SWISS-MODEL	https://swissmodel.expasy.org/	Free	Biasini et al. (2014)
4	ROBETTA	http://robetta.bakerlab.org/	Free	Kim et al. (2004)
5	SCHRODINGER	Standalone	Commercial	Sindhikara et al. (2017)
6	EsyPred3D	http://www.unamur.be/sciences/biologie/urbm/bioinfo/esypred/	Free	Lambert et al. (2002)
7	RaptorX	http://raptorx.uchicago.edu/	Free	Källberg et al. (2014)
8	IntFOLD	http://www.reading.ac.uk/bioinf/IntFOLD/	Free	McGuffin et al. (2015)

Table 44.2 List of some important protein-ligand docking programs

Sl. no.	Software	License	Year	Description	Reference
1	GOLD	Commercial	1995	Ligand docking	Verdonk et al. (2003)
2	GLIDE	Commercial	2004	Ligand docking and fast scoring	Friesner et al. (2004)
3	AutoDock	Open	1990	Docking with selective receptor flexibility	Goodsell and Olson (1990)
4	Autodock Vina	Open	2010	Molecular docking and virtual screening	Trott and Olson (2010)
5	FlexX	Commercial	1996	Incremental construction	Rarey et al. (1996)
6	Surflex	Commercial	2003	Docking using molecular similarity	Jain (2003)
7	ICM	Commercial	1994	Ligand docking	Abagyan et al. (1994)
8	MOE-Dock	Commercial	2008	Docks small molecules and fragments, modeling and simulation	Chemical Computing Group (2008)
9	DARWIN	Open	2000	Docking flexible small molecules	Taylor and Burnett (2000)
10	EADock	Open	2007	Small molecules docking	Grosdidier et al. (2007)
11	FlexAID	Open	2015	Docks small molecules	Gaudreault and Najmanovich (2015)
12	MOLS 2.0	Open	2016	Peptide modeling and ligand docking	Paul and Gautham (2016)
13	rDock	Open	2014	Ligand docking	Ruiz-Carmona et al. (2014)
14	DOCK 6.0	Academic	2015	Docks fragments and small molecules	Allen et al. (2015)
15	MCDock	Academic	1999	Monte Carlo simulation for ligand placement	Liu and Wang (1999)

molecules with similar fingerprints by screening the databases. The activity of novel molecules can be predicted by building the models with QSAR (Acharya et al. 2011). The most accepted approaches for LBDD are pharmacophore modeling and QSAR methods. Figure 44.4 outlines the general steps of LBDD process.

44.7 Quantitative Structure-Activity Relationships (QSAR)

QSAR is the computational technique which correlates chemical structure of compounds with a particular biological or chemical process using mathematical models. The QSAR method is based on the approach that similar physiochemical properties and structure yield related activity (Akamatsu 2002). Initially, the activity of set of chemical compounds of desired biological interest is determined. Then, the quantitative relation is commenced between physiochemical properties of active compounds and biological activity, which results in the construction of QSAR model. The constructed QSAR model describes the relationship and used to optimize the active molecules to exploit the significant biological activity. These predicted molecules/compounds are further tested for their activity by experimental methods. The QSAR method is the guiding tool for identification and characterization of compounds with improved activity (Acharya et al. 2011). There are two types of QSAR, i.e., classical QSAR or 2D-QSAR and 3D-QSAR. The 2D-QSAR used to relate the structural property molecular descriptors, *viz.* simple count, molar refractivity, topological indices, electrostatic indices, 2D-fingerprints, *etc.* to biological activity. The 2D descriptors are not able to describe accurately the relationship between 3D spatial arrangements of physiochemical properties and biological activity; therefore, 3D-QSAR approaches were developed. The 3D-QSAR comprises descriptors which explain 3D characteristics of the molecule to develop QSAR model. There are various 3D descriptors CoMFA, CoMSIA, CoMMA, GRIND, etc., but two frequently used approaches are CoMFA (comparative molecular field analysis) and CoMSIA (comparative molecular similarity indices analysis). The QSAR is explained in Fig. 44.5.

44.8 Comparative Molecular Field Analysis (CoMFA)

The CoMFA is one of the largely accepted 3D QSAR methods and based on the conception that biological activity of the molecule depends on the surrounding molecular fields (electrostatic and steric fields). The electrostatic and steric fields were computed by CoMFA with Lennard-Jones potential and Coulomb potential. This method was widely preferred, but there are certain problems associated with this method which limits its applications (Cramer et al. 1989).

44.9 Comparative Molecular Similarity Indices Analysis (CoMISA)

CoMISA is the expansion of CoMFA. Here, the biological activity of the molecule is dependent on the more molecular field properties (electrostatic, steric, hydrophobic, hydrogen bond acceptor and hydrogen bond donor).

CoMISA is improved algorithm which is least influenced by distance to van der Waals surface. CoMISA model is more precise structural activity relationship model than CoMFA (Klebe et al. 1994).

44.10 Pharmacophore Modeling

Paul Ehrlich (1909) defined the term Pharmacophore as molecular framework which bears the necessary characteristics responsible for biological activity of a drug. The currently used definition of pharmacophore was coined by Peter Gund which defines it as “a set of structural features in a molecule that is recognized at a receptor site, and is responsible for that molecules biological activity" (Gund 1977). Generally, a pharmacophore would reveal the position of the key amino acids in the active site of targeted protein (van Drie 2003). The pharmacophore modeling presents a valuable skeleton for better understanding of existing data and used as the dynamic tool in design of molecules/compounds with improved potency and selectivity. A pharmacophore model is generated by examining the structure–activity relationship and mapping common features of active analogs. The pharmacophore modeling is explained in Fig. 44.6. These pharmacophore models are broadly used to extract the precise inhibitors of disease-related proteins including enzymes, G-protein coupled receptors and ion channels (Kubinyi 2006a, b).

44.11 Applications of Computer-Aided Drug Discovery in Veterinary Sciences

Computer-aided drug discovery is equally useful in developing therapeutic interventions in veterinary and animal health. The strategy is useful for developing drugs against cancer (De et al. 2018) and infectious agents (Singh et al. 2017; Dai et al. 2018; Sharma et al. 2019).

For instance, docking calculation was performed to develop structure between receptor M2 proton channel and ligands and calculates the binding-free energies to analyze the interaction between M2 proton channel and adamantine-based inhibitors. The finally designed inhibitors of M2 proton channels were envisaged to overcome the drug-resistant menace of influenza A virus resistant to adamantane-based therapeutics in porcine (Du et al. 2010).

The shape-based computer-aided virtual screening was used for identifying novel compounds against paracoccidioidomycosis (Silva et al. 2018).

Despite the availability of prophylactic vaccines, it is necessary to go for drug-based treatment against infection and oncogenic effects of certain viruses such as HPV. In silico methods combining ADME prediction, SBVS, and MD were to identify small molecule inhibitors as anti-HPV drugs against HPV E6 protein, based on the disruption of E6-E6AP interactions (Ricci-López et al. 2019).

44.12 Opportunities and Challenges

The surge of public diseases and availability of data related to drugs has urged computational methodologists to delve into sequence data to propose candidate drugs. Computer-aided drug discovery, based on prediction of the structure of proteins is a well-established method of developing novel drugs against pathogens and preventing diseases such as cancers. Various methods are used for structure-based *in silico* drug designing. In view of development of machine learning principles accompanied by available pharmacological data, the artificial learning is suggested to be a powerful data mining tool for drug designing such as virtual screening, activity scoring *de novo* drug design, quantitative structure–activity relationship (QSAR), and evaluating absorption, distribution, metabolism, excretion and toxicity of drug molecules (Zhong et al. 2018).

44.13 Conclusions

In the era of plentiful genome and proteome data available online and accessible for analysis, the *in silico* methods have important role in identifying proteins, and lifecycles that are not yet elucidated by wet-lab methods. The genome and proteome data of human pathogens, and veterinary or zoonotic pathogens is a valuable resource for computational and bioinformatics methods to identify cryptic or uncharacterized proteins and metabolic pathways that can be used as targets to develop novel antimicrobial interventions.

References

Abagyan R, Totrov M, Kuznetsov D (1994) ICM—a new method for protein modeling and design: applications to docking and structure prediction from the distorted native conformation. J Comput Chem 15(5):488–506

Acharya C, Coop A, Polli JE, Mackerell AD Jr (2011) Recent advances in ligand-based drug design: relevance and utility of the conformationally sampled pharmacophore approach. Curr Comput Aided Drug Des. 2011 Mar;7(1):10–22 (Review)

Akamatsu M (2002) Current state and perspectives of 3D-QSAR. Curr Top Med Chem 2(12):1381–1394 (Review)

Alberg DG, Schreiber SL (1993) Structure-based design of a cyclophilin-calcineurin bridging ligand. Science 262(5131):248–250

Allen WJ, Balius TE, Mukherjee S, Brozell SR, Moustakas DT, Lang PT, Case DA, Kuntz ID, Rizzo RC (2015) DOCK 6: Impact of new features and current docking performance. J Comput Chem 36(15):1132–1156. https://doi.org/10.1002/jcc.23905

Barcellos GB, Pauli I, Caceres RA, Timmers LF, Dias R, de Azevedo WF Jr (2008) Molecular modeling as a tool for drug discovery. Curr Drug Targets 9 (12):1084–1091 (Review)

Berman HM, Westbrook J, Feng Z, Gilliland G, Bhat TN, Weissig H, Shindyalov IN, Bourne PE (2006) The protein data bank, 1999. In: International tables for crystallography volume F: crystallography of biological macromolecules. Springer, Netherlands, pp. 675–684

Biasini M, Bienert S, Waterhouse A, Arnold K, Studer G, Schmidt T, Kiefer F, Gallo Cassarino T, Bertoni M, Bordoli L, Schwede T (2014) SWISS-MODEL: modelling protein tertiary and quaternary structure using evolutionary information. Nucleic Acids Res 42 (Web Server issue):W252–W258. https://doi.org/10.1093/nar/gku340. Epub 2014 Apr 29

Chemical Computing Group, M. O. E. (2008) Molecular operating environment

Cramer RD 3rd, Patterson DE, Bunce JD (1989) Recent advances in comparative molecular field analysis (CoMFA). Prog Clin Biol Res 291:161–165

Dai J, Dan W, Li N, Wang J (2018) Computer-aided drug discovery: novel 3,9-disubstituted eudistomin U derivatives as potent antibacterial agents. Eur J Med Chem 5(157):333–338. https://doi.org/10.1016/j.ejmech.2018.08.001 (Epub 2018 Aug 4)

De B, Bhandari K, Mendonça FJB, Scotti MT, Scotti L (2018) Computational studies in drug design against cancer. Anticancer Agents Med Chem. https://doi.org/10.2174/1871520618666180911125700 (Epub ahead of print)

Dias R, de Azevedo Jr WF (2008) Molecular docking algorithms. Curr Drug Targets 9(12):1040–1047 (Review)

Du QS, Huang RB, Wang SQ, Chou KC (2010) Designing inhibitors of M2 proton channel against H1N1 swine influenza virus. PLoS ONE 5(2):e9388. https://doi.org/10.1371/journal.pone.0009388

Ehrlich P (1909) Über den jetzigen Stand der Chemotherapie. Eur J Inorg Chem 42(1):17–47

Friesner RA, Banks JL, Murphy RB, Halgren TA, Klicic JJ, Mainz DT, Repasky MP, Knoll EH, Shelley M, Perry JK, Shaw DE, Francis P, Shenkin PS (2004) Glide: a new approach for rapid, accurate docking and scoring. 1. Method and assessment of docking accuracy. J Med Chem 47(7):1739–1749

Gaudreault F, Najmanovich RJ (2015) FlexAID: revisiting docking on non-native-complex structures. J Chem Inf Model 55(7):1323–1336. https://doi.org/10.1021/acs.jcim.5b00078 (Epub 2015 Jun 24)

Glicksberg BS, Li L, Chen R, Dudley J, Chen B (2019) Leveraging big data to transform drug discovery. Methods Mol Biol 1939:91–118. https://doi.org/10.1007/978-1-4939-9089-4_6

Goodsell DS, Olson AJ (1990) Automated docking of substrates to proteins by simulated annealing. Proteins 8(3):195–202

Grosdidier A, Zoete V, Michielin O (2007) EADock: docking of small molecules into protein active sites with a multiobjective evolutionary optimization. Proteins. 67(4):1010–1025

Gund P (1977) Three-dimensional pharmacophoric pattern searching. In: Progress in molecular and subcellular biology. Springer, Berlin, pp 117–143

Hammes GG (2002) Multiple conformational changes in enzyme catalysis. Biochemistry. 2002 Jul 2;41 (26):8221–8 (Review)

Jain AN (2003) Surflex: fully automatic flexible molecular docking using a molecular similarity-based search engine. J Med Chem 46(4):499–511

Joo K, Lee J, Lee J (2012) Methods for accurate homology modeling by global optimization. Methods Mol Biol 857:175–188. https://doi.org/10.1007/978-1-61779-588-6_7

Källberg M, Margaryan G, Wang S, Ma J, Xu J (2014) RaptorX server: a resource for template-based protein structure modeling. Methods Mol Biol 1137:17–27. https://doi.org/10.1007/978-1-4939-0366-5_2

Kapetanovic IM (2008) Computer-aided drug discovery and development (CADDD): in silico-chemico-biological approach. Chem Biol Interact 171(2):165–176 (Epub 2006 Dec 16. Review)

Kelley LA, Mezulis S, Yates CM, Wass MN, Sternberg MJ (2015) The Phyre2 web portal for protein modeling, prediction and analysis. Nat Protoc 10 (6):845–858. https://doi.org/10.1038/nprot.2015.053 (Epub 2015 May 7)

Kim DE, Chivian D, Baker D (2004) Protein structure prediction and analysis using the Robetta server. Nucleic Acids Res 32(Web Server issue):W526–W531

Klebe G, Abraham U, Mietzner T (1994) Molecular similarity indices in a comparative analysis (CoMSIA) of drug molecules to correlate and predict their biological activity. J Med Chem 37(24):4130–4146

Kubinyi H (2006a) Success stories of computer-aided design. Comput Appl Pharm Res Develop 2:377

Kubinyi H (2006b) Chemogenomics in drug discovery. Ernst Schering Res Found Workshop. 58:1–19

Lambert C, Léonard N, De Bolle X, Depiereux E (2002) ESyPred3D: prediction of proteins 3D structures. Bioinformatics 18(9):1250–1256

Liu M, Wang S (1999) MCDOCK: a Monte Carlo simulation approach to the molecular docking problem. J Comput Aided Mol Des 13(5):435–451

Martí-Renom MA, Stuart AC, Fiser A, Sánchez R, Melo F, Sali A (2000) Comparative protein structure modeling of genes and genomes. Annu Rev Biophys Biomol Struct 29:291–325 (Review)

McGuffin LJ, Atkins JD, Salehe BR, Shuid AN, Roche DB (2015) IntFOLD: an integrated server for modelling protein structures and functions from amino acid sequences. Nucleic Acids Res 43(W1):W169–W173. https://doi.org/10.1093/nar/gkv236 (Epub 2015 Mar 27)

Mezei M (2003) A novel fingerprint for the characterization of protein folds. Protein Eng 16(10):713–715

Paul DS, Gautham N (2016) MOLS 2.0: software package for peptide modeling and protein-ligand docking. J Mol Model 22(10):239. https://doi.org/10.1007/s00894-016-3106-x (Epub 2016 Sep 16)

Rarey M, Kramer B, Lengauer T, Klebe G (1996) A fast flexible docking method using an incremental construction algorithm. J Mol Biol 261(3):470–489

Ricci-López J, Vidal-Limon A, Zunñiga M, Jimènez VA, Alderete JB, Brizuela CA, Aguila S (2019) Molecular modeling **simulation** studies reveal new potential inhibitors against HPV E6 protein. PLoS ONE 14 (3):e0213028. https://doi.org/10.1371/journal.pone.0213028 (eCollection 2019)

Ruiz-Carmona S, Alvarez-Garcia D, Foloppe N, Garmendia-Doval AB, Juhos S, Schmidtke P, Barril X, Hubbard RE, Morley SD (2014) rDock: a fast, versatile and open source program for docking ligands to proteins and nucleic acids. PLoS Comput Biol 10 (4):e1003571. https://doi.org/10.1371/journal.pcbi.1003571 (eCollection 2014 Apr)

Shaikh SA, Jain T, Sandhu G, Latha N, Jayaram B (2007) From drug target to leads–sketching a physicochemical pathway for lead molecule design in silico. Curr Pharm Des 13(34):3454–3470 (Review)

Sharma D, Sharma A, Verma SK, Singh B (2019) Targeting metabolic pathways proteins of Orientia tsutsugamushi using combined hierarchical approach to combat scrub typhus. J Mol Recognit 32(4):e2766. https://doi.org/10.1002/jmr.2766 (Epub 2018 Oct 21)

Shoichet BK (2004) Virtual screening of chemical libraries. Nature 432(7019):862–865 (Review)

Silva LC, Neves BJ, Gomes MN, Melo-Filho CC, Soares CM, Andrade CH, Pereira M (2018) Computer-aided identification of novel anti-paracoccidioidomycosis compounds. Future Microbiol 13:1523–1535. https://doi.org/10.2217/fmb-2018-0175 (Epub 2018 Oct 12)

Sindhikara D, Spronk SA, Day T, Borrelli K, Cheney DL, Posy SL (2017) Improving accuracy, diversity, and speed with prime macrocycle conformational sampling. J Chem Inf Model 57(8):1881–1894. https://doi.org/10.1021/acs.jcim.7b00052 (Epub 2017 Aug 8)

Singh G, Sharma D, Singh V, Rani J, Marotta F, Kumar M, Mal G, Singh B (2017) *In silico* functional elucidation of uncharacterized proteins of *Chlamydia abortus* strain LLG. Future Sci OA 3(1):FSO169. https://doi.org/10.4155/fsoa-2016-0066 (eCollection 2017 Mar. Erratum in: Future Sci OA. 2017 Oct 05;3(4):FSO66C1)

Sliwoski G, Kothiwale S, Meiler J, Lowe EW Jr (2013) Computational methods in drug discovery. Pharmacol Rev 66(1):334–395. https://doi.org/10.1124/pr.112.007336 (Print 2014. Review)

Sujatha K, Mahalakshmi A, Solaiman DK, Shenbagarathai R (2009) Sequence analysis, structure prediction, and functional validation of phaC1/phaC2 genes of Pseudomonas sp. LDC-25 and its importance in polyhydroxyalkanoate accumulation. J Biomol Struct Dyn 26(6):771–779

Taylor JS, Burnett RM (2000) DARWIN: a program for docking flexible molecules. Proteins 41(2):173–191

Trott O, Olson AJ (2010) AutoDock Vina: improving the speed and accuracy of docking with a new scoring function, efficient optimization, and multithreading. J Comput Chem 31(2):455–461. https://doi.org/10.1002/jcc.21334

van Drie JH (2003) Pharmacophore discovery–lessons learned. Curr Pharm Des9(20):1649–1664 (Review)

Verdonk ML, Cole JC, Hartshorn MJ, Murray CW, Taylor RD (2003) Improved protein-ligand docking using GOLD. Proteins 52(4):609–623

Verlinde CL, Hol WG (1994) Structure-based drug design: progress, results and challenges. Structure 2 (7):577–587 (Review)

Wang R, Lu Y, Wang S (2003) Comparative evaluation of 11 scoring functions for molecular docking. J Med Chem 46(12):2287–2303

Webb B, Sali A (2014) Protein structure modeling with MODELLER. Methods Mol Biol 1137:1–15. https://doi.org/10.1007/978-1-4939-0366-5_1

Zhong F, Xing J, Li X, Liu X, Fu Z, Xiong Z, Lu D, Wu X, Zhao J, Tan X, Li F, Luo X, Li Z, Chen K, Zheng M, Jiang H (2018) Artificial intelligence in drug design. Sci China Life Sci. https://doi.org/10.1007/s11427-018-9342-2

Part V

Animal Biotechnology in Global Perspective

Steps Toward Sustainable Livestock Development: Technologies to Boost Indigenous Livestock

45

Abstract

Livestock is an integral component of economy and livelihood of millions of people worldwide. There is a need to use sustainable methods of animal health and production in an environment challenged by climate change. The native livestock with evolutionary merits to adverse climate and low inputs offers benefits that are good for humans and the environment.

Highlights

- Growing human population and standard of living have raised the demand for livestock products
- It is imperative to conserve and expand the genetic potential of livestock through biotechnology-based inventions.

Keywords

Livestock · Animal biotechnology · Nutrition · Health · Thermal stress · Adaptive genetic merits

45.1 Introduction

Animals provide more than just food and income. The livestock sector is vastly diverse and globally dynamic sector contributing to human welfare and the national economy. In developing countries, the livestock sector has developed in response to increasing demands for farm animals products, i.e., milk, meat, and eggs. In developed countries, livestock products systems are increasing their efficiency keeping in mind the economy and environmental sustainability (Thornton et al. 2010).

The role of cattle, buffaloes, sheep, goats, and many less known livestock to human nutrition and livelihood is multifaceted. Majority of the world's milk and meat come from ruminants. Among non-ruminants, pigs and poultry are the prime contributors of meat and eggs (Eisler et al. 2014; Wanapat et al. 2015).

Low-income or landless pastoralists have less priority for high-yielding cattle and buffaloes. Small animals like goats, sheep, pig, and backyard poultry are more profitable to them as farmers get cash by selling them during the financial crisis. Paradoxically, the attitude of people to earn instant benefits from short-term investments into livestock has hampered the long-term gains.

According to an estimate, the human population is likely to increase from existing 7.3 billion in 2015 to 9.7 billion by 2050. Most growth is expected in poor countries, where income and standards of living are set to rise speedily. There will be an increase in demand for resource-intensive livestock-origin foods, i.e., milk, meat, and eggs (Global food status, 2050, accessed on Feb. 21, 2019). This can be achieved

B. Singh et al., *Advances in Animal Biotechnology*,
https://doi.org/10.1007/978-3-030-21309-1_45

only when livestock producers switch to viable and eco-friendly practices of animal rearing.

45.2 Realistic Strategies to Sustain Livestock Production

Consumer demand for animal-origin products puts pressure on small-scale as well as commercial animal husbandry to use robust scientific methods to increase the production performance of the animals. At present, around 20 species of animal are domesticated of which several breeds of seven major livestock species viz., cattle, pigs, goats, sheep, horse, cats, and dogs are found ubiquitously. Other animals (Fig. 45.1) such as yak (*Bos grunniens*), camel, llama (*Lama glama*), reindeer (*Rangifer tarandus*), mithun (*Bos frontalis*), ferret (*Mustela putorius furo*), and guinea fowl are distributed randomly and adapted utterly to specialized agroclimatic niches. For instance, mithun (Fig. 45.2) is probably the domesticated form of wild ox, the gaur (*B. gaurus*), that has been domesticated for the past 8000 years. Yak is a long-haired multipurpose domestic bovid reared throughout the high-altitude Himalayan region of Indian subcontinent, Tibet, Mongolia, and Russia.

As livestock sector comprises of multiple components including farmers, buyers, and the correlated industries, a mutual coordination is must to sustain healthy and productive livestock. Moreover, commercial livestock production is feasible only if the quality of products obtained

Fig. 45.1 Some sturdy and low-input indigenous livestock species. **a** Short hill cattle; **b** high-yielding tropical Sahiwal bull; **c** yak (India); **d** Churu, a cross between cattle and yak; **e** migratory sheep; **f** migratory cashmere goat; **g** double-humped camel of cold arid deserts of Ladakh; **h** a pony; and **i** non-descript buffalo

Fig. 45.2 A herd of mithun in Nagaland (India). Mithun is multipurpose livestock used for milk, meat, manure, skin, and agriculture operations *Image courtesy* Dr. A. Mitra, Director, ICAR-National Research Centre on Mithun, Jharnapani (India)

conforms to international standards of quality and safety. Some of the realistic strategies to sustain livestock system are summarized herein.

45.3 Animal Nutrition and Feeding

It is observed that developed countries feed around 70% of the total cereals (maize, barley, sorghum, etc.) to livestock, of which 40% is fed to ruminants, and remaining is offered to pig and poultry (Eisler et al. 2014). It is possible to save this huge amount of cereals and use it for humans. This is because ruminants have a highly specialized digestive system to efficiently utilize fibrous pasture, crop residues, and agro-industrial by-products such as oil cakes and non-protein nitrogen (NPN) for energy and proteins.

Quality of crops and forage should also be improved by genetic manipulation to increase the amount of essential amino acids, and minimizing the secondary metabolites that are toxic or anti-nutritional. Additionally, the quality of fodder and feed should be improved by removing anti-nutritional metabolites therein by physical, chemical, and biological treatments (Bhat et al. 2013).

There are plentiful anaerobic fibrolytic bacteria and fungi in GI tract of domestic ruminants (Hess et al. 2011; White et al. 2014; Gharechahi and Salekdeh 2018) and wild herbivores (Salgado-Flores et al. 2016; Svartström et al. 2017; Ungerfeld et al. 2018). In addition, GI microorganisms with specialized metabolic traits to alleviate toxic effects of ingested phytometabolites are developed in response to respective dietary ingredients in ruminants (Singh et al. 2001; Sharma et al. 2017; Leong et al. 2017) and other non-ruminants (Kohl et al. 2016; Shiffman et al. 2017).

The rumen microorganisms are of immense importance. Cellulolytic genes of rumen microbiota should be utilized to generate transgenic microorganism for use as probiotics or microbial feed supplements to efficiently degrade dietary plant biomass. Already, examples of transferring genes from rumen (*Butyrivibrio fibrisolvens*) into other bacteria (*Moraxella sp.*) to produce recombinant microorganisms of nutritional importance are existing (Gregg et al. 1998).

Futuristic progress in this area will help in minimizing the quantity of cereals fed to ruminants. China's dairy animals which initially relied on cereals have been shifted to utilize crop residues and forage. Cattle in New Zealand are made to depend for energy from grazing on pastures and rangelands (Bocquier and González-García 2010).

The rumen microorganisms and microbial enzymes should be used as inoculums to make good quality making, and processing the pig and poultry feed to improve utilization of plant structural carbohydrates, and minerals like phosphorus and nitrogen. This strategy is essentially required to improve the nutrition of livestock in developing countries where animals fail to maintain their production value due to lack of quality nutrition. To some extent, the strategy will solve the adverse effects of animals on environment. This is because roughage-based diets contribute to produce high GHG.

45.4 Promoting Native Livestock

Technologies are available to select the animals based on indicators of desirable traits. Technologies, such as artificial insemination, IVF, embryo flushing, and nuclear transfer cloning, involve manipulating the reproduction potential of superior livestock. Focus of earlier technologies was to manipulate the reproduction efficiency and propagate the desirable animals through selective breeding. In the late 1970s, the emphasis was on increasing milk production through crossbreeding of low-yielding indigenous breeds with high-yielding exotic breeds. This led to the dissemination of semen of Holstein and other breeds.

Meanwhile, abrupt alterations in climate and its impact on animal agriculture were also observed (Box 1). It was realized that high-yielding pure and crossbred females were less suitable in hot humid climate and pests, parasites, and pathogens flourishing therein. The non-humped males of exotic breeds are less suitable for agriculture, mechanical, or draft power. Moreover, farmers having low income in developing countries have to invest a lot to buy readymade feed, fodder, and medicine to maintain exotic elite animals under controlled environment. Despite this, the exotic animals fail to produce expected quantity of milk.

Due to nutritional and mineral deficiencies, genitourinary infections, and other health problems, the females become infertile. Ultimately, farmers sell males and unproductive females for culling. Most male calves due to negligible role in agriculture die due to lack of care. On the contrary, native migratory goats, sheep, and buffaloes (Fig. 45.3), being multipurpose species, are more profitable as there is no restriction in consuming their milk or meat or selling them.

Box 1 Effects and Consequences of Extreme Events on Livestock

Extreme weather events: Increased expenditure on animal housing, mortality during migration, biotic, and abiotic stress. Prevalence of insects, flies, and mosquitoes, spread of drug-resistant pathogens and parasites, risk of zoonotic diseases.

Droughts and floods: Scarcity of drinking water, reduced soil fertility, scarcity of forage and feed, risk of disease-spreading vectors, snake bites, increased the cost of nutrition and maintenance, deaths of animals.

Ecosystem changes: Soil erosion, soil mineral leaching, alteration in soil health and ecosystem, alteration in animal gut ecosystem in response to changes in dietary ingredients.

Compared to high-yielding animals, the native animals are small in body size, need fewer inputs for health, nutrition and housing, and hence, they are more suitable to poor families. Indigenous zebu cattle, for example, dwarf Vechur cow of Kerala, need less fodder; dwarf hilly cattle of

Fig. 45.3 Sturdy livestock fit for low-input management and tough agroclimatic conditions. **a** Migratory goats and sheep; **b** native buffaloes of NWHR, India. Migratory sheep and goat nibble derive nutrients from fiber-rich grasses and tree foliage that also contain anti-nutritional phytometabolites. Notably, the plants synthesize diverse phytometabolites as defensive means against predation by herbivores

Northwest Himalayan Region (NWHR) are able to graze and move in high hilly terrains. Tharparkar, Kankrej, and Rathi of tropical Indian planes are sturdy cattle that produce milk despite water shortage and thermal stress. Males are suitable to pull cart and perform agriculture operations. Ladakh (India) cattle have characteristic evolutionary adaptation to hypoxic temperate milieu (Verma et al. 2018a, b). Livestock in Africa are comparatively resistant to trypanosomiasis.

45.5 Innovative Assisted Reproduction

Reproduction biotechnology has already transformed dairy industry. Practice of AI with the sperm of elite bulls, and marker-assisted selection has produced dairy cows that yield four times more milk than the animals did in the 1940s, before the use of methods of assisted reproduction.

Reproductive biologists have come up with a new concept of surrogate male. The male is gene edited so that it loses ability to generate sperm through normal spermatogenesis. Genetically modified or SSCs of other males with desirable traits are transplanted into the male having crippled spermatogenesis. This way, the surrogate bull will produce transgenic sperms or the sperms of other valuable males (Ledford 2019). Hence, the surrogate sire is a new next-generation breeding tool in livestock-assisted reproduction. The goal is to spread the genes for desirable traits such as resistance to thermal stress or disease resistance (Giassetti et al. 2019). Park et al. (2017) have already reported a CRISPR/Cas9 gene-edited NANOS2 knockout surrogate pig which is unable to produce sperms due to ablated spermatogenesis, but are otherwise normal. NANOS2 knockout male pigs are ideal surrogate males for transplanting SSCs, thereby producing sperms from other genetically enviable sires.

If lingering technical hurdles are surmounted, the technique could prove to be highly valuable for beef cattle, pigs, equines, chickens, and other livestock species wherein AI is less successful. Of note, AI is not preferred in beef cattle as beef animals are allowed to graze freely over pastures, making it hard to track the females in estrus. Sperms are susceptible so cryopreservation in pig, while IVF is less successful in equines.

45.6 Health of Animals

Health of livestock is a vital issue as several pathogens infecting animals do infect humans. Tens of thousands of people suffer from zoonotic (viral, bacterial, protozoal, fungal, and parasitic) diseases, especially where humans and animals live closer to each other. Swine flu, avian influenza, and rabies are common zoonotic viral infections, while *Mycobacterium bovis* causes bovine tuberculosis (bTB) which is a contagious zoonotic bacterial infection. *M. bovis* has a broad range of hosts and causes significant economic loss to farmers. Zoonotic transmission of *M. bovis* to humans occurs via contaminated milk or milk products. Healthy animals are infected when they come in contact with infected animals. Children below 5 years, elder people above 65 years of age, and persons with weak immunity are more susceptible.

High densities of animals spread diseases faster. In addition, abandoned animals (cattle, equines, and pigs), stray and wild animals, such as boar (Brown et al. 2018; Fredriksson-Ahomaa 2019), bear (Scott Weese et al. 2019), bats (Rupprecht et al. 2006; Katz et al. 2016), and birds (Nga et al. 2019), spread trillions of harmful viruses, bacteria, fungi, and parasites of zoonotic nature through aerosol discharges, defecation and manure, uninhibited mating, and discharge of body fluids. Carcass of abandoned and stray animals spreads disease germs into soil and scavengers. Slaughter of animals in open pollutes water bodies and vegetation. This poses threats not only for humans, but young and weak animals also die due to infections.

Neonatal and premature deaths of animals are a loss of precious livestock genetics. Gene editing is utilized to induce heritable resistance among animals against infectious diseases. Site-specific transcription activator-like effector (TALE)-mediated knockin of *SP110*, nuclear body protein gene in newly generated Holstein Friesian cattle endowed resistance against tuberculosis (Wu et al. 2015). Cas9 nickase (Cas9n) was used to insert natural resistance-associated macrophage protein-1 (NRAMP1) gene into the bovine genome to generate gene-edited somatic cells. The cells were used to generate nuclear transfer cloned gene-edited calves having resistance to bTB (Gao et al. 2017). This approach will help to establish livestock herds that are more productive under stressful environment (Tables 45.1 and 45.2).

45.7 Precise Use of Antimicrobials

Discovery and use of antibiotics in healthcare setup in the twentieth century, together with improved hygiene and vaccination, has radically increased the life expectancy (Berkner et al. 2014). Antibiotics, vaccines, and antibodies have saved humans and animals against fatal infections.

Foot and mouth disease (FMD) virus and bTB account for heavy expenditure in many developed countries. Animals infected with bTB are hard to identify from clinical symptoms alone (Verteramo Chiu et al. 2019).

Development of resistance toward multiple antibiotics in pathogens and parasites is serious global health threat. Lax and irresponsible practice of using antibiotics has worsened the situation. Massive use of antibiotics in animal feeding as growth promoters in pig and poultry, release of residual antibiotics and metals from pharmaceutical industries, and hospitals have contributed to develop antibiotic resistance among pathogens.

Threat has aggravated as no new broad range antibiotics are developed during the past decades. Treating antibiotic-resistant pathogens is expensive and delays the cure of disease. Alternative supplementary platforms and paradigms suggested to combat antibiotic-resistant pathogens treat other perilous infectious diseases include the use of bioengineered probiotic microorganisms,

Table 45.1 Evolutionary adaptive merits of indigenous livestock to cope with environmental stress

Livestock species	Prominent adaptive features
Cattle	Smaller body size in hilly cattle to walk in mountainous terrains, and less requirement of nutrients for maintenance, higher density of sweat glands to sweat freely, reflection of radiation by short, thick and glossy hair coat, loose skin that enables the animals to withstand warm weather, minimum production of internal body heat compared to exotic cattle breeds, remarkable draught power for use in ploughing Adaptation in Ladakh (India) cattle to high altitude and low oxygen (Verma et al. 2018a), expression of Hypoxia-inducing factor-1 (HIF-1), and its associated genes *viz.*, glucose transporter 1 (GLUT1), vascular endothelial growth factor (VEGF), and hexokinase 2 (HK2) associated with homeostatic response to hypoxia at high altitude (Verma et al. 2018b)
Buffaloes	Adaptation to hot and humid climates and muddy terrain, more resistance to infectious disease compared to cattle (NRC 1981), protection from solar UV irradiation by epidermal skin melatonin, habit to keep body cool by wallowing in water and mud (Marai and Haeeb 2010), adaptation to salinity in Chilika buffaloes, the distinct buffalo breed of India (Singh et al. 2017)
Sheep	Habit to walk long distances and hilly terrains in migratory sheep reared in North West Himalayan Region, efficient fiber digestion, high prolificacy, resistance to foot rot disease, and adaptation to salinity as in Garole sheep, resistance to drought in Malpura, Chakla, and Marwari breeds
Goats	Ability to utilize fibrous forages containing anti-nutritional phytometabolites (Singh et al. 2012; Sharma et al. 2017), adaptation to high-altitude and cold deserts in *Chegu* and *Changthangi* goats (Sharma et al. 2010), production of cashmere and coarse fibre in breeds, ability of Andaman goats to utilize saline forages
Equine	Adaptation of Spiti ponies of Northwest Himalayan Region (NWHR) to hilly terrain, ability to work under low oxygen, and temperate climate of higher altitudes
Camel	Adaptation to poor-quality and unconventional roughage, less requirement of water, ability to walk and work in desert, production of milk with therapeutic attributes
Yak and mithun	Adaptation to high altitudes, cold and damp environment, utilization of regional forages, production of milk, meat and fibre, and as source of draught power

Table 45.2 A summary of consequences of climatic stress on high-yielding livestock

Attributes	Consequences (references)
Nutrition	Reduction in feed intake, and body weight gain in calves (O'Brien et al. 2010), and adult animals (Rhoads et al. 2009) Changes in biological functions, disturbances in general metabolism (Marai and Haeeb 2010) Increase in negative energy balance in high-yielding cows (Lacetera et al. 2009), changes in hepatic functioning and metabolism (Itoh et al. 1998a, b; Moore et al. 2005), thus reduced milk yield in lactating animals Increased susceptibility of the thermally stressed animals to subclinical and acute rumen acidosis (Kadzere et al. 2002)
Endocrinological functioning	Compromised steroidogenic capacity of theca and granulose cells, impaired ovarian follicular dynamics (Argov et al. 2005), depressed thyroid functioning, increase in prolactin levels, and stress-responsive hormones (glucocorticoids), decrease in aldosterone and parathyroid hormone secretion, decline in triiodothyronine (T_3) in summer in species such as buffalo and cattle (Marai and Haeeb 2010), with ultimate adverse effects on milk yield
Male reproduction	Reduction in number of total spermatozoa and motility in cattle sperm (Mathevon et al. 1998), and boar (Kunavongkrit et al. 2005), sperm morphological deformities (Nichi et al. 2006), loss of sperm viability and genome integrity (Pérez-Crespo et al. 2008); reduced developmental competence of embryos produced from fertilization by sperm exposed to thermal stress (Paul et al. 2008) Loss in fertility in poultry (Karaca et al. 2002) and equines (Sharma et al. 2010)
Female reproduction	Failure in overt expression of estrus, alterations in the ovarian follicular growth (Roth et al. 2000); recruitment of multiple ovarian follicles due to decreased levels of inhibin and elevated levels of FSH (Roth et al. 2000); repression of dominant follicle (Wolfenson et al. 2000) Reduction in fertilization due to altered physiology of oocytes, and subsequent developmental competence of (Schrock et al. 2007; Roth 2008), and embryo mortality (Sugiyama et al. 2003), activation of genes involved in apoptosis in oocytes (Soto and Smith 2009), sensitivity of winter oocytes to thermal stress, and more resistance to thermal stress in ovine oocytes in summer (Ahmadi et al. 2019), overall negative impact on outcome of pregnancy (Alhussien et al. 2018)
Embryo development	Reduced compaction of heat-stressed ova and low development of hatching blastocysts from the oocytes exposed to thermal stress (Edwards et al. 2005), increase in oxidative stress in pre-implantation embryos (Sakatani et al. 2008), reduced likelihood of embryo implantation (de Rensis and Scaramuzzi 2003)
Livestock products	Direct effect on organ and muscle metabolism due to HS that can persist after slaughter, increased risks of pale soft-exudative meat in pigs and turkeys, heat shortening in broilers, dark cutting beef in cattle and dehydration (Gregory 2010) Alterations in milk and meat quality due to the introduction of stress-tolerant livestock breeds, feeding of low-protein and high-fat finisher ration, and introducing novel adaptations in dairy and meat animals to cope with global climatic stress, increased incidences of spoilage of processed and packaged livestock products, increased costs of preservation of the animal products
Ecological imbalance	Enhanced incidences of emergence and distribution of pests, and pathogens (Vorou et al. 2007; Semenza and Menne 2009), and vector-borne diseases, reduced resistance of host to infectious agents and food-borne diseases (Khansis and Nettleman 2005; Nardone et al. 2010)
Miscellaneous	Temperature-related morbidity and mortality of animals (Nardone et al. 2010), suppression of immune functions making a stressed organism more prone to pathologies, increase in oxidative stress (Bernabucci et al. 2002)

antimicrobial peptides (AMPs), phage therapy, antibody-antibiotic conjugates, and some phytometabolites.

45.8 Nutritional Supplements

Use of microorganisms (probiotics), and microbial metabolites (postbiotics), is a common practice in animals including poultry, pigs, young calves, and pets.

The calf mortality is a matter of serious concern as it impedes benefits from animals. Improving calf survival through dietary interventions is a prerequisite to harness the benefit from small-scale as well as commercial dairy farming. Identification of animal-origin probiotics and biologically active postbiotics could serve as alternative therapies to prevent enteric infections. Animal-origin probiotic strains were found to be promising than standard probiotic strains (Dowarah et al. 2018a). *Pediococcus acidilactici* FT28 had positive effect on improving carcass quality and physiological profile of pork (Dowarah et al. 2018b).

The tiny aquatic pteridophyte, *Azolla caroliniana* found in tropical aquatic habitats such as swamps, ditches, and lakes can be grown at farmers levels in shallow ponds and tanks. Azolla is a beneficial plant which with the symbiosis of *Anabaena azollae*, cyanobacteria, fixes atmospheric nitrogen, and suppresses volatilization of NH_3 and CH_4 synthesis in paddy soil (Liu et al. 2017). *A. caroliniana* due to being rich source of proteins and vitamins is recommended for use as animal feed supplement.

Mixed fodder comprising of legumes, red or white clover, and grasses constitutes a balanced diet. Residues from transgenic crops may supplement nutrients to animals. Transgenic corn containing *Escherichia coli* phytase genes (Nyannor et al. 2009), and phytase-containing transgenic aquatic freshwater plant, the *Lemna minor*, is a useful and cost-effective feed supplement and bioremediator to broilers (Ghosh et al. 2018).

The focus should be on feeding animals the good quality forage. It is logical to explore the possibility of using phytometabolites present in native flora to improve rumen fermentation and suppress GI parasites.

45.9 Environmental Friendly Livestock Production

Livestock and environment affect each other. Ruminants are significant contributors of GHG emission, with over half of the total anthropogenic CH_4 produced. Total CH_4 produced from cattle is more than that emitted from buffaloes, sheep, and goats. Besides, methanogens are found in non-ruminants and insects . 16S rRNA analysis of fecal contents of Bactrian camels (*Camelus bactrianus*) revealed the existence of *Methanobrevibacter* sp. as dominating methanogens (Turnbull et al. 2012). Forestomach of Alpacas harbored methane-producing *M. millerae* (An et al. 2005).

Methane, being a more potent GHG than CO_2, has more role in global warming (Balcombe et al. 2018). Targeting methanogens is a big challenge. The interest in biology and control of gut methanogens is because of current need to minimize methane emissions from ruminant livestock and other sources and utilize methanogens to produce biofuel from organic wastes.

Some plant metabolites are beneficial feed supplements. Essential oils suppress CH_4 formation from rumen fermentation of dietary fiber and increase rumen microbial protein synthesis (Klop et al. 2017; Joch et al. 2018). Due to their anti-protozoal effects, the plant saponins lower rumen CH_4 synthesis (Belanche et al. 2016). Proanthocyanidins or condensed tannins have anthelmintic effects against GI nematodes (Bhat et al. 2013; Anantasook et al. 2016).

Several approaches, such as the use of probiotics and postbiotics (Jeyanathan et al. 2014), feeding biologically treated roughage (Tuyen et al. 2013), vaccination against methanogens might be the practical approaches to mitigate methane emission from ruminants (Wedlock et al. 2013; Zhang et al. 2015). Shifting to non-meat alternative is an option speculated to reduce the burden on climate to raise meat-producing animals.

45.10 Management Tactics Livestock

The traditional livestock management practices are not enough to protect the animals against climatic and biotic stress. Alternative strategies (Box 2) should be adopted to alleviate stress of the animals.

Box 2 Summary of the Management Packages to Promoting Native Livestock

Breeding management: Conserving indigenous breeds, using superior indigenous bulls for breeding, avoiding inbreeding, strengthening indigenous knowledge of animal rearing, rebuilding flocks or herds to promote genetic diversity, avoiding crossbreeding with exotic breeds.

Nutritional interventions: Avoiding overgrazing of grasslands, removing toxic weeds from crop fields and grasslands, analyzing locally available forages for presence of anti-nutritional plant metabolites, promoting cultivation of multipurpose beneficial trees such as beul (*Grewia optiva*), and tremal (*Ficus roxburghii*) (for milch buffaloes and cattle), khejri tree (*Prosopis cineraria*) for goats and sheep, and grasses and plants tolerant to abiotic stress. Other approaches include use of minerals mix in diet of animals, utilizing elite gut microbiota as feed supplements to enhance nutrient utilization and forage toxicity (Singh et al. 2008), preserving surplus native forages by ensiling (Bhat et al. 2013; Singh et al. 2015) to feed animals during scarcity of green fodder.

General maintenance: Evolving weather-based index insurance linked to measurable climate change events such as extreme heat, low rainfall, managing weather-dependency, protecting animals from pests in hot humid climate, alleviating heat stress through proper animals housing, ventilation and allowing animals wallowing, practicing an integrated system of animal farming, avoiding use of herbicides so that forage trees are not damaged.

45.11 Nanotechnology

Nanotechnology is highly diversified and dynamic field of research and applications in animal sciences. Nanotechnology is already in use in advanced agrichemicals and biological application systems.

Nanotechnology offers new solutions to many old problems. Next decades will see applications of nanotechnology applied to diagnostics, drug delivery methods, and regenerative medicine. Various applications of nanoscience and nanotechnology have started their way into biomedical sciences, nanomedicine and veterinary sector in the form of pet care products, nanovaccines and nanoadjuvants, products of veterinary medicine and farm animal disinfectants, and animal nutrition (Bai et al. 2018; El-Sayed and Kamel 2018). It is envisaged that nanotechnology will help in finding practical solution to eliminate microbial and mycotoxins from animal feeds, and specter of rising antibiotic resistance in veterinary or zoonotic pathogens and parasites.

Enhancing biogas production from animal manure and reducing odor from anaerobic digesters are other prospective areas that might be benefited from pursuit of nanotechnology.

45.12 Outlook and Challenges

Animal agriculture is an indispensable component of integrated farming systems. With increasing economy and lifestyle, the demand for high-quality livestock-origin food has also increased. In order to meet the protein requirements of human population, the livestock production has to be increased. This means number of milk- and meat-producing animals has to increase.

Biotechnology applications can only increase the productivity of animals. The emphasis should be on promoting native livestock species that perform well under stressful climate where exotic livestock is unfit. The quality of feeding and forage would play a role in supplementing nutrients.

It is expected that global warming will have negative impact on the availability of land for cultivating crops and fodder for animals. Thermal stress, emergence of new pests, parasites, and pathogens will increase that have adverse effects on animal productivity.

The metabolic and other functional capabilities of gut microbiota are necessary for nutrition and well-being of the host. Commercial and poor farmers have to follow scientific methods of rearing animals to achieve long-term gains from animals.

Field veterinary health experts and extension works will have to educate farmers for healthy practices of animal husbandry. It is imperative that awareness programs should be organized to educate and urge the farmers to deworm and vaccinate their animals.

Despite the considerable potential, there are uncertainties associated with the use of nanotechnology in veterinary health and long-term impact on the environment. Possible hazards and threats, pros and cons should be resolved based on scientific investigations in view of strict regulations and legislation.

45.13 Conclusions

The livestock sector is highly dynamic. Animals are indispensable components of economy and human evolution. The livestock serves humans by contributing high biological value proteins, fiber, draft power, and farm manure. Farm animals provide high-quality proteins, skin, wool, traction, and manure. Animal productivity has increased as a result of advances in nutrition, health augmentation, and genetic potential.

A healthy and scientifically managed livestock is important not only for economic gains, but also to prevent human and animals health. Further developments in animal breeding, nutrition, and health will contribute to increasing production efficiency and genetic gains. Emphasis should be on improving animal husbandry and productivity through education and financial aid to farmers. It is important to conserve and disseminate locally available animals under resource-poor conditions and deteriorating agro-climatic conditions.

References

Ahmadi E, Nazari H, Hossini-Fahraji H (2019) Low developmental competence and high tolerance to thermal stress of ovine oocytes in the warm compared with the cold season. Trop Anim Health Prod. https://doi.org/10.1007/s11250-019-01854-w (In press)

An D, Dong X, Dong Z (2005) Prokaryote diversity in the rumen of yak (Bos grunniens) and Jinnan cattle (Bos taurus) estimated by 16S rDNA homology analyses. Anaerobe 11(4):207–215

Anantasook N, Wanapat M, Gunun P, Cherdthong A (2016) Reducing methane production by supplementation of Terminalia chebula RETZ. containing tannins and saponins. Anim Sci J 87(6):783–790. https://doi.org/10.1111/asj.12494 (Epub 2015 Dec 21)

Alhussien MN, Kamboj A, Aljader MA, Panda BSK, Yadav ML, Sharma L, Mohammed S, Sheikh AA, Lotfan M, Kapila R, Mohanty AK, Dang AK (2018) Effect of tropical thermal stress on peri-implantation immune responses in cows. Theriogenology 1(114): 149–158. https://doi.org/10.1016/j.theriogenology.2018.03.036 (Epub 2018 Mar 29)

Argov N, Moallem U, Sklan D (2005) Summer heat stress alters the mRNA expression of selective-uptake and endocytotic receptors in bovine ovarian cells. Theriogenology 64(7):1475–1489

Bai DP, Lin XY, Huang YF, Zhang XF (2018) Theranostics aspects of various nanoparticles in veterinary medicine. Int J Mol Sci 19(11). pii: E3299. https://doi.org/10.3390/ijms19113299 (Review)

Balcombe P, Speirs JF, Brandon NP, Hawkes AD (2018) Methane emissions: choosing the right climate metric and time horizon. Environ Sci Process Impacts 20 (10):1323–1339. https://doi.org/10.1039/c8em00414e

Bernabucci U, Ronchi B, Lacetera N, Nardone A (2002) Markers of oxidative status in plasma and erythrocytes of transition dairy cows during hot season. J Dairy Sci 85(9):2173–2179

Berkner S, Konradi S, Schönfeld J (2014) Antibiotic resistance and the environment–there and back again: science & society series on science and drugs. EMBO Rep 15(7):740–744. https://doi.org/10.15252/embr.201438978

Belanche A, Pinloche E, Preskett D, Newbold CJ (2016) Effects and mode of action of chitosan and ivy fruit saponins on the microbiome, fermentation and methanogenesis in the rumen simulation technique. FEMS Microbiol Ecol 92(1). pii: fiv160. https://doi.org/10.1093/femsec/fiv160 (Epub 2015 Dec 15)

Bhat TK, Kannan A, Singh B, Sharma OP (2013) Value addition of feed and fodder by alleviating the antinutritional effects of tannins. Agric Res 2:189–206

Bocquier F, González-García E (2010) Sustainability of ruminant agriculture in the new context: feeding strategies and features of animal adaptability into the necessary holistic approach. Animal 4(7):1258–1273. https://doi.org/10.1017/S1751731110001023

Brown VR, Bowen RA, Bosco-Lauth AM (2018) Zoonotic pathogens from feral swine that pose a significant threat to public health. Transbound Emerg Dis 65(3):649–659. https://doi.org/10.1111/tbed.12820 (Epub 2018 Jan 31)

De Rensis F, Scaramuzzi RJ (2003) Heat stress and seasonal effects on reproduction in the dairy cow–a review. Theriogenology 60(6):1139–1151 (Review)

Dowarah R, Verma AK, Agarwal N, Singh P, Singh BR (2018a) Selection and characterization of probiotic lactic acid bacteria and its impact on growth, nutrient digestibility, health and antioxidant status in weaned piglets. PLoS One 13(3):e0192978. https://doi.org/10.1371/journal.pone.0192978 (eCollection 2018)

Dowarah R, Verma AK, Agarwal N, Singh P (2018b) Efficacy of species-specific probiotic Pediococcus acidilactici FT28 on blood biochemical profile, carcass traits and physicochemical properties of meat in fattening pigs. Res Vet Sci 117:60–64. https://doi.org/10.1016/j.rvsc.2017.11.011 (Epub 2017 Nov 21)

Edwards JL, Saxton AM, Lawrence JL, Payton RR, Dunlap JR (2005) Exposure to a physiologically relevant elevated temperature hastens in vitro maturation in bovine oocytes. J Dairy Sci 88(12):4326–4333

Eisler MC, Lee MR, Tarlton JF, Martin GB, Beddington J, Dungait JA, Greathead H, Liu J, Mathew S, Miller H, Misselbrook T, Murray P, Vinod VK, Van Saun R, Winter M (2014) Agriculture: steps to sustainable livestock. Nature 507(7490):32–34

El-Sayed A, Kamel M (2018) Advanced applications of nanotechnology in veterinary medicine. Environ Sci Pollut Res Int. https://doi.org/10.1007/s11356-018-3913-y (Epub ahead of print)

Fredriksson-Ahomaa M (2019) Wild boar: a reservoir of foodborne zoonoses. Foodborne Pathog Dis 16 (3):153–165. https://doi.org/10.1089/fpd.2018.2512 (Epub 2018 Nov 27)

Gao Y, Wu H, Wang Y, Liu X, Chen L, Li Q, Cui C, Liu X, Zhang J, Zhang Y (2017) Single Cas9 nickase induced generation of NRAMP1 knockin cattle with reduced off-target effects. Genome Biol 18(1):13. https://doi.org/10.1186/s13059-016-1144-4

Gharechahi J, Salekdeh GH (2018) A metagenomic analysis of the camel rumen's microbiome identifies the major microbes responsible for lignocellulose degradation and fermentation. Biotechnol Biofuels 11:216. https://doi.org/10.1186/s13068-018-1214-9 (eCollection 2018)

Ghosh M, Sharma N, Gera M, Kim N, Huynh D, Zhang J, Min T, Sodhi SS, Kim MB, Rekha VPB, Ko S, Jeong DK (2018) Insights into phytase-containing transgenic *Lemna minor* (L.) as a novel feed additive. Transgenic Res 27(2):211–224. https://doi.org/10.1007/s11248-018-0068-z (Epub 2018 Mar 24)

Giassetti MI, Ciccarelli M, Oatley JM (2019) Spermatogonial stem cell transplantation: insights and outlook for domestic animals. Annu Rev Anim Biosci 15 (7):385–401. https://doi.org/10.1146/annurev-animal-020518-115239

Gregg K, Hamdorf B, Henderson K, Kopecny J, Wong C (1998) Genetically modified ruminal bacteria protect sheep from fluoroacetate poisoning. Appl Environ Microbiol 64(9):3496–3498

Gregory NG (2010) How climatic changes could affect meat quality. Food Res Intern 43:1866–1873. https://doi.org/10.1016/j.foodres.2009.05.018

Hess M, Sczyrba A, Egan R, Kim TW, Chokhawala H, Schroth G, Luo S, Clark DS, Chen F, Zhang T, Mackie RI, Pennacchio LA, Tringe SG, Visel A, Woyke T, Wang Z, Rubin EM (2011) Metagenomic discovery of biomass-degrading genes and genomes from cow rumen. Science 331(6016):463–467. https://doi.org/10.1126/science.1200387

Itoh F, Obara Y, Rose MT, Fuse H (1998a) Heat influences on plasma insulin and glucagon in response to secretogogues in non-lactating dairy cows. Domest Anim Endocrinol 15:499–510

Itoh F, Obara Y, Rose MT, Fuse H, Hashimoto H (1998b) Insulin and glucagon secretion in lactating cows during heat exposure. J Anim Sci 76:2182–2189

Jeyanathan J, Martin C, Morgavi DP (2014) The use of direct-fed microbials for mitigation of ruminant methane emissions: a review. Animal 8(2):250–261. https://doi.org/10.1017/s1751731113002085 (Epub 2013 Nov 25, Review)

Joch M, Mrázek J, Skřivanová E, Čermák L, Marounek M (2018) Effects of pure plant secondary metabolites on methane production, rumen fermentation and rumen bacteria populations in vitro. J Anim Physiol Anim Nutr (Berl) 102(4):869–881. https://doi.org/10.1111/jpn.12910 (Epub 2018 Apr 29)

Kadzere CT, Murphy MR, Silanikove N, Maltz E (2002) Heat stress in lactating dairy cows: a review. Liv Prod Sci 77:59–71. https://doi.org/10.1016/S0301-6226(01)00330-X

Karaca AG, Parker HM, Yeatman JB, McDaniel CD (2002) Role of seminal plasma in heat stress infertility of broiler breeder males. Poult Sci 81(12):1904–1909

Katz IS, Fuoco NL, Chaves LB, Rodrigues AC, Ribeiro OG, Scheffer KC, Asano KM (2016) Delayed progression of rabies transmitted by a vampire bat. Arch Virol 161(9):2561–2566. https://doi.org/10.1007/s00705-016-2927-3 (Epub 2016 Jun 15)

Khasnis AA, Nettleman MD (2005) Global warming and infectious disease. Arch Med Res 36(6):689–696

Klop G, Dijkstra J, Dieho K, Hendriks WH, Bannink A (2017) Enteric methane production in lactating dairy cows with continuous feeding of essential oils or rotational feeding of essential oils and lauric acid. J Dairy Sci 100(5):3563–3575. https://doi.org/10.3168/jds.2016-12033 (Epub 2017 Feb 23)

Kohl KD, Connelly JW, Dearing MD, Forbey JS (2016) Microbial detoxification in the gut of a specialist avian herbivore, the Greater Sage-Grouse. FEMS Microbiol Lett 363(14). pii: fnw144. https://doi.org/10.1093/femsle/fnw144 (Epub 2016 May 29)

Kunavongkrit A, Suriyasomboon A, Lundeheim N, Heard TW, Einarsson S (2005) Management and sperm production of boars under differing environmental conditions. Theriogenology 63(2):657–667 (Review)

Lacetera N, Bernabucci U, Basiricò L, Morera P, Nardone A (2009) Heat shock impairs DNA synthesis and down-regulates gene expression for leptin and Ob-Rb receptor in concanavalin A-stimulated bovine peripheral blood mononuclear cells. Vet Immunol Immunopathol 127(1–2):190–194. https://doi.org/10.1016/j.vetimm.2008.09.020 (Epub 2008 Oct 2)

Ledford H (2019) Bull 'super dads' are being engineered to produce sperm from another father. Nature (in press)

Liu J, Xu H, Jiang Y, Zhang K, Hu Y, Zeng Z (2017) Methane emissions and microbial communities as influenced by dual cropping of Azolla along with early rice. Sci Rep 17(7):40635. https://doi.org/10.1038/srep40635

Leong LEX, Khan S, Davis CK, Denman SE, McSweeney CS (2017) Fluoroacetate in plants—a review of its distribution, toxicity to livestock and microbial detoxification. J Anim Sci Biotechnol 1 (8):55. https://doi.org/10.1186/s40104-017-0180-6 (eCollection 2017, Review)

Marai IFM, Haeeb AAM (2010) Buffalo's biological functions as affected by heat stress—a review. LIv Sci 127:89–109. https://doi.org/10.1016/j.livsci.2009.08.001

Mathevon M, Buhr MM, Dekkers JC (1998) Environmental, management, and genetic factors affecting semen production in Holstein bulls. J Dairy Sci 81 (12):3321–3330

Moore CE, Kay JK, Collier RJ, Vanbaale MJ, Baumgard LH (2005) Effect of supplemental conjugated linoleic acids on heat-stressed brown swiss and holstein cows. J Dairy Sci 88(5):1732–1740

NRC (1981) The water buffalo: new prospects for an underutilized animal. National Academy Press, Washington DC

Nardone A, Ronchi B, Lacetera N, Ranieri MS, Bernabucci U (2010) Effects of climate changes on animal production and sustainability of livestock systems. LIv Sci 130:57–69. https://doi.org/10.1016/j.livsci.2010.02.011

Nga VT, Ngoc TU, Minh LB, Ngoc VTN, Pham VH, Nghia LL, Son NLH, Van Pham TH, Bac ND, Tien TV, Tuan NNM, Tao Y, Show PL, Chu DT (2019) Zoonotic diseases from birds to humans in Vietnam: possible diseases and their associated risk factors. Eur J Clin Microbiol Infect Dis. https://doi.org/10.1007/s10096-019-03505-2 (Epub ahead of print, Review)

Nichi M, Bols PE, Züge RM, Barnabe VH, Goovaerts IG, Barnabe RC, Cortada CN (2006) Seasonal variation in semen quality in Bos indicus and Bos taurus bulls raised under tropical conditions. Theriogenology 66 (4):822–828 (Epub 2006 Mar 9)

Nyannor EK, Bedford MR, Adeola O (2009) Corn expressing an Escherichia coli-derived phytase gene: residual phytase activity and microstructure of digesta in broiler chicks. Poult Sci 88(7):1413–1420. https://doi.org/10.3382/ps.2009-00003

O'Brien MD, Rhoads RP, Sanders SR, Duff GC, Baumgard LH (2010) Metabolic adaptations to heat stress in growing cattle. Domest Anim Endocrinol 38(2):86–94. https://doi.org/10.1016/j.domaniend.2009.08.005 (Epub 2009 Sep 18)

Park KE, Kaucher AV, Powell A, Waqas MS, Sandmaier SE, Oatley MJ, Park CH, Tibary A, Donovan DM, Blomberg LA, Lillico SG, Whitelaw CB, Mileham A, Telugu BP, Oatley JM (2017) Generation of germline ablated male pigs by CRISPR/Cas9 editing of the NANOS2 gene. Sci Rep 10(7):40176. https://doi.org/10.1038/srep40176

Paul C, Murray AA, Spears N, Saunders PT (2008) A single, mild, transient scrotal heat stress causes DNA damage, subfertility and impairs formation of blastocysts in mice. Reproduction 136(1):73–84. https://doi.org/10.1530/REP-08-0036 (Epub 2008 Apr 4)

Pérez-Crespo M, Pintado B, Gutiérrez-Adán A (2008) Scrotal heat stress effects on sperm viability, sperm DNA integrity, and the offspring sex ratio in mice. Mol Reprod Dev 75(1):40–47

Rhoads ML, Rhoads RP, VanBaale MJ, Collier RJ, Sanders SR, Weber WJ, Crooker BA, Baumgard LH (2009) Effects of heat stress and plane of nutrition on lactating Holstein cows: I. Production, metabolism, and aspects of circulating somatotropin. J Dairy Sci 92 (5):1986–1997. https://doi.org/10.3168/jds.2008-1641

Roth Z (2008) Heat stress, the follicle, and its enclosed oocyte: mechanisms and potential strategies to improve fertility in dairy cows. Reprod Domest Anim 43(Suppl 2):238–244. https://doi.org/10.1111/j.1439-0531.2008.01168.x (Review)

Roth Z, Meidan R, Braw-Tal R, Wolfenson D (2000) Immediate and delayed effects of heat stress on follicular development and its association with plasma FSH and inhibin concentration in cows. J Reprod Fertil 120(1):83–90

Rupprecht CE, Hanlon CA, Slate D (2006) Control and prevention of rabies in animals: paradigm shifts. Dev Biol (Basel) 125:103–11 (Review)

Sakatani M, Yamanaka K, Kobayashi S, Takahashi M (2008) Heat shock-derived reactive oxygen species induce embryonic mortality in in vitro early stage bovine embryos. J Reprod Dev 54(6):496–501 (Epub 2008 Sep 2)

Salgado-Flores A, Bockwoldt M, Hagen LH, Pope PB, Sundset MA (2016) First insight into the faecal microbiota of the high Arctic muskoxen (*Ovibos moschatus*). Microb Genom 2(7):e000066. https://doi.org/10.1099/mgen.0.000066 (eCollection 2016 Jul)

Schrock GE, Saxton AM, Schrick FN, Edwards JL (2007) Early in vitro fertilization improves development of bovine ova heat stressed during in vitro maturation. J Dairy Sci 90(9):4297–4303

Scott Weese J, Salgado-Bierman F, Rupnik M, Smith DA, van Coeverden de Groot P (2019) Clostridium (Clostridioides) difficile shedding by polar bears (Ursus maritimus) in the Canadian Arctic. Anaerobe. pii: S1075-9964(19)30052-6. https://doi.org/10.1016/j.anaerobe.2019.03.013 (Epub ahead of print)

Semenza JC, Menne B (2009) Climate change and infectious diseases in Europe. Lancet Infect Dis 9 (6):365–375. https://doi.org/10.1016/S1473-3099(09)70104-5

Sharma D, Mal G, Kannan A, Bhar R, Sharma R, Singh B (2017) Degradation of euptox A by tannase-producing rumen bacteria from migratory goats. J Appl Microbiol 123(5):1194–1202. https://doi.org/10.1111/jam.13563 (Epub 2017 Sep 19)

Shiffman ME, Soo RM, Dennis PG, Morrison M, Tyson GW, Hugenholtz P (2017) Gene and genome-centric analyses of koala and wombat fecal microbiomes point to metabolic specialization for *Eucalyptus* digestion. PeerJ 5:e4075. https://doi.org/10.7717/peerj.4075 (eCollection 2017)

Singh B, Bhat TK, Sharma OP, Kanwar SS, Rahi P, Gulati A (2012) Isolation of tannase-producing *Enterobacter ludwigii* GRT-1 from the rumen of migratory goats. Small Rumin Res 102:172–176

Singh B, Bhat TK, Kurade NP, Sharma OP (2008) Metagenomics in animal gastrointestinal ecosystem: a microbiological and biotechnological perspective. Indian J Microbiol 48:216–227

Singh B, Bhat TK, Singh B (2001) Exploiting gastrointestinal microbes for livestock and industrial development. Asian-Australas J Anim Sci 14:567–586

Singh B, Gautam SK, Chauhan MS, Singla SK (2015) Textbook of animal biotechnology. The Energy and Resources Institute (TERI), New Delhi. Status: In Press. ISBN: 9788179933275

Singh R, Mishra SK, Rajesh C, Dash SK, Niranjan SK, Kataria RS (2017) Chilika- a distinct registered buffalo breed of India. Int J Livest Res 7(9):259–266. https://doi.org/10.5455/ijlr.20170704044822

Soto P, Smith LC (2009) BH4 peptide derived from Bcl-xL and Bax-inhibitor peptide suppresses apoptotic mitochondrial changes in heat stressed bovine oocytes. Mol Reprod Dev 76(7):637–646. https://doi.org/10.1002/mrd.20986

Sugiyama S, McGowan M, Kafi M, Phillips N, Young M (2003) Effects of increased ambient temperature on the development of in vitro derived bovine zygotes. Theriogenology 60(6):1039–1047

Svartström O, Alneberg J, Terrapon N, Lombard V, de Bruijn I, Malmsten J, Dalin AM, El Muller E, Shah P, Wilmes P, Henrissat B, Aspeborg H, Andersson AF (2017) Ninety-nine de novo assembled genomes from the moose (*Alces alces*) rumen microbiome provide new insights into microbial plant biomass degradation. ISME J 11(11):2538–2551. https://doi.org/10.1038/ismej.2017.108 (Epub 2017 Jul 21)

Thornton PK (2010) Livestock production: recent trends, future prospects. Phil Trans R Soc B 365:2853–2867. https://doi.org/10.1098/rstb.2010.0134

Turnbull KL, Smith RP, St-Pierre B, Wright AD (2012) Molecular diversity of methanogens in fecal samples from Bactrian camels (*Camelus bactrianus*) at two zoos. Res Vet Sci 93(1):246–249. https://doi.org/10.1016/j.rvsc.2011.08.013 (Epub 2011 Sep 13)

Tuyen DV, Phuong HN, Cone JW, Baars JJ, Sonnenberg AS, Hendriks WH (2013) Effect of fungal treatments of fibrous agricultural by-products on chemical composition and in vitro rumen fermentation and methane production. Bioresour Technol 129:256–263. https://doi.org/10.1016/j.biortech.2012.10.128 (Epub 2012 Nov 15)

Ungerfeld EM, Leigh MB, Forster RJ, Barboza PS (2018) Influence of season and diet on fiber digestion and bacterial community structure in the Rumen of Muskoxen (*Ovibos moschatus*). Microorganisms 6(3). pii: E89. https://doi.org/10.3390/microorganisms6030089

Verma P, Sharma A, Sodhi M, Thakur K, Kataria RS, Niranjan SK, Bharti VK, Kumar P, Giri A, Kalia S, Mukesh M (2018a) Transcriptome analysis of circulating PBMCs to understand mechanism of high altitude adaptation in native cattle of Ladakh region. Sci Rep 8(1):7681. https://doi.org/10.1038/s41598-018-25736-7

Verma P, Sharma A, Sodhi M, Thakur K, Bharti VK, Kumar P, Giri A, Kalia S, Swami SK, Mukesh M (2018b) Overexpression of genes associated with hypoxia in cattle adapted to Trans Himalayan region of Ladakh. Cell Biol Int. https://doi.org/10.1002/cbin.10981 (In press)

Verteramo Chiu LJ, Tauer LW, Smith RL, Grohn YT (2019) Assessment of the bovine tuberculosis elimination protocol in the United States. J Dairy Sci 102(3):2384–2400. https://doi.org/10.3168/jds.2018-14990

Vorou RM, Papavassiliou VG, Tsiodras S (2007) Emerging zoonoses and vector-borne infections affecting humans in Europe. Epidemiol Infect 135(8):1231–1247 (Epub 2007 Apr 20)

Wanapat M, Cherdthong A, Phesatcha K, Kang S (2015) Dietary sources and their effects on animal production and environmental sustainability. Anim Nutr 1(3):96–103. https://doi.org/10.1016/j.aninu.2015.07.004

Wedlock DN, Janssen PH, Leahy SC, Shu D, Buddle BM (2013) Progress in the development of vaccines against rumen methanogens. Animal 7(Suppl 2):244–252. https://doi.org/10.1017/S1751731113000 682 (Review)

White BA, Lamed R, Bayer EA, Flint HJ (2014) Biomass utilization by gut microbiomes. Annu Rev Microbiol 68:279–296. https://doi.org/10.1146/annurev-micro-092412-155618 (Epub 2014 Jun 16, Review)

Wolfenson D, Roth Z, Meidan R (2000) Impaired reproduction in heat-stressed cattle: basic and applied aspects. Anim Reprod Sci 60–61:535–547 (Review)

Wu H, Wang Y, Zhang Y, Yang M, Lv J, Liu J, Zhang Y (2015) TALE nickase-mediated SP110 knockin endows cattle with increased resistance to tuberculosis. Proc Natl Acad Sci U S A 112(13):E1530–E1539. https://doi.org/10.1073/pnas.1421587112 (Epub 2015 Mar 2)

Zhang L, Huang X, Xue B, Peng Q, Wang Z, Yan T, Wang L (2015) Immunization against Rumen Methanogenesis by vaccination with a new recombinant protein. PLoS One 10(10):e0140086. https://doi.org/10.1371/journal.pone.0140086 (eCollection 2015)

Biotechnology for Wildlife

46

Abstract
Wildlife is of paramount significance to welfare of humans. In modern era, the wild animals are the sources of income, food, fur, micro-organisms, and other products besides its role in maintaining ecological balance. Shrinking habitat, diseases and pest prevalence, and illegal hunting are the major threats due to which many wild species have become extinct and many are endangered. Scientific interventions are being used to not only harnessing the potential of wildlife, but conserving them through assisted reproduction, genomics, and public awareness.

Key points

- Wildlife is an essential component of natural ecosystem
- Wild natural flora and fauna are declining rapidly
- Biotechnological interventions should be used conserve and increase wild animals.

Keywords
Wildlife · Biotechnology applications · Diversity conservation · Endangered animals · Wildlife conservation · Reproduction biotechnology

46.1 Introduction

46.1.1 Wildlife in Human Welfare

The term wildlife refers to animals, plants, and other things that are not domesticated by humans. Wildlife has important role in protecting and balancing the environment by providing stability to normal processes of ecosystem. Wildlife (Figs. 46.1 and 46.2) is important source of flora for therapeutics, biological macromolecules, micro-organisms, insects, birds, fish and seafood, reptiles and mammals as source of micro-organisms with health and economic importance (Ueda et al. 2018). The insects are of immense importance in agriculture and horticulture.

Hunting of wild animals provides livelihood and economic security to people in many parts of the world. The tribes in many parts of the world depend primarily on wildlife resources for asylum, food and well-being.

Large cat species or felids are important wild animals. Being strict carnivores, the wild big cats are recognized and admired animals that remain at the top of food chain, where they have indirect effect on plant life, and play an important ecological role in structuring the animal communities and regulating the prey populations, through

B. Singh et al., *Advances in Animal Biotechnology*,
https://doi.org/10.1007/978-3-030-21309-1_46

Fig. 46.1 Some important wild animals. Wild herbivores and carnivores constitute a natural ecosystem. Herbivores are important components of agriculture as they contribute to pollination and dispersal of seeds and plants

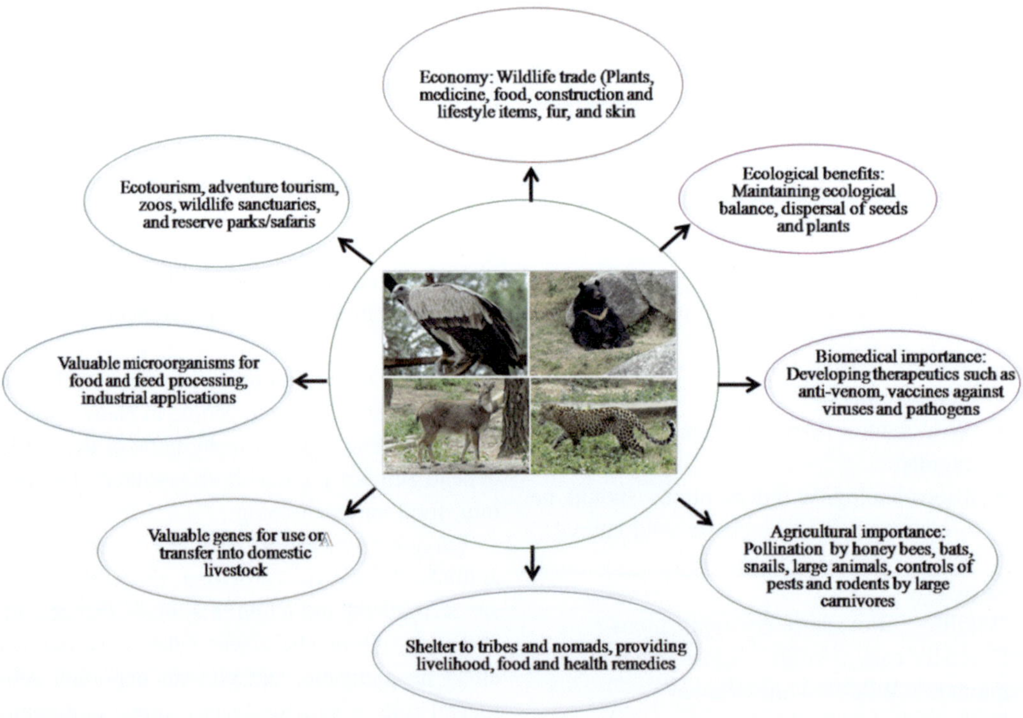

Fig. 46.2 Diagrammatic depiction of benefits of wildlife to humans and ecosystem

a process called "trophic cascade". In addition, some wild cats, for example, cougar (*Puma concolor*), a large and widely distributed feline plays important role in dispersal of seeds through their scat. When cougars eat doves, the seed predators in forests, the dove's digestive process is interrupted and seeds pass the gut of cougar. Similarly, other cats also assist long-distance dispersal of seeds.

As human population is growing, so is demand for wildlife resources including plants and animals. Importance of wildlife can be categorized as ecological importance, economic importance, investigatory importance, and conservation of flora and fauna diversity. In modern era, the wildlife is a source of economy to many countries. Educational TV shows such as National Geographic, Discovery Channel, and Animal Planet Channel shows are primarily motivated and inspired by wildlife.

Many wild animals are important for their biomedical applications. For instance, the classical mice strains are used in mapping immunological traits. As most of the classical mice strains originate from a limited number of founder stock (e.g., *Mus musculus domesticus* subspecies), their genetic diversity is finally limited which make them less suitable for exhaustive experimentation. Wild mice-derived strains with ability to breed with inbred mice strains are deemed to be more suitable for evolutionary and immunological studies (Poltorak et al. 2018).

Based on habitats, the wild animals may be terrestrial or aquatic. The animals including insects, wild herbivores (Bears, mountain goats, elk, rhinoceros, bison, deer, elephants, giraffe), carnivores (lions, tiger, leopard), and omnivorous animals (such as bear) are important (Box 1). Various programs are initiated to investigate the multiple aspects of wildlife bioresources and utilizing them for human and animal welfare and conserving them through management and biotechnological approaches.

Box 1. Benefits of Wildlife to Humans

- Maintenance of ecological balance through food habits and food chain.
- Maintenance of materials cycles such as carbon and nitrogen.
- Improvement and progress in agriculture, animal husbandry and fishery.
- Wild animals serve as valuable source of gene pool that can be introduced into domesticated animals.
- Wild animals as source of food, fur, and therapeutic products (e.g., venom for developing anti-venom).
- Role in agriculture—pollination (e.g., entomophily, ornithophily, and chiropterophily, malacophily), and dispersal of seeds and plants.
- Tourism and ecotourism, and establishing of herbal parks. Africa's ecotourism is the apt example of contribution of wildlife to human economy and livelihood.
- Economic contribution—the current education television shows.

46.2 Threats to Wild Animals

A large number of animal species are either endangered or at the brink of extinction. In addition to destruction of their habitat by human activities, loss of prey (herbivores) as food for carnivores, and accidental conflicts with humans, many of wild carnivores are culled or hunted illicitly by criminal cartels for smuggling of live animals and/or fur, bones, teeth, horns, and other body parts (Fig. 46.3). This has led to reduction in population of wildlife in many parts of the world. According to International Union for Conservation of Nature (IUCN) Red List of extinction categories, around 70 wild cat species are endangered (https://www.wildcatfamily.com/endangered-cat-species-list/, accessed on August 2, 2018). Amur leopards (*Panthera pardus orientalis*) of Primorye region of southeastern Russia and the Jilin Province of northeast China, Iberian lynx (*Lynx pardinus*) of Iberian Peninsula in southwestern Europe, Asiatic cheetahs (*Acinonyx jubatus venaticus*) in Iran, Japanese Iriomote cats (*Prionailurus bengalensis iriomotensis*), Scottish wildcats (*Felis silvestris grampia*), and South China tigers (*Panthera tigris amoyensis*) are among the most endangered

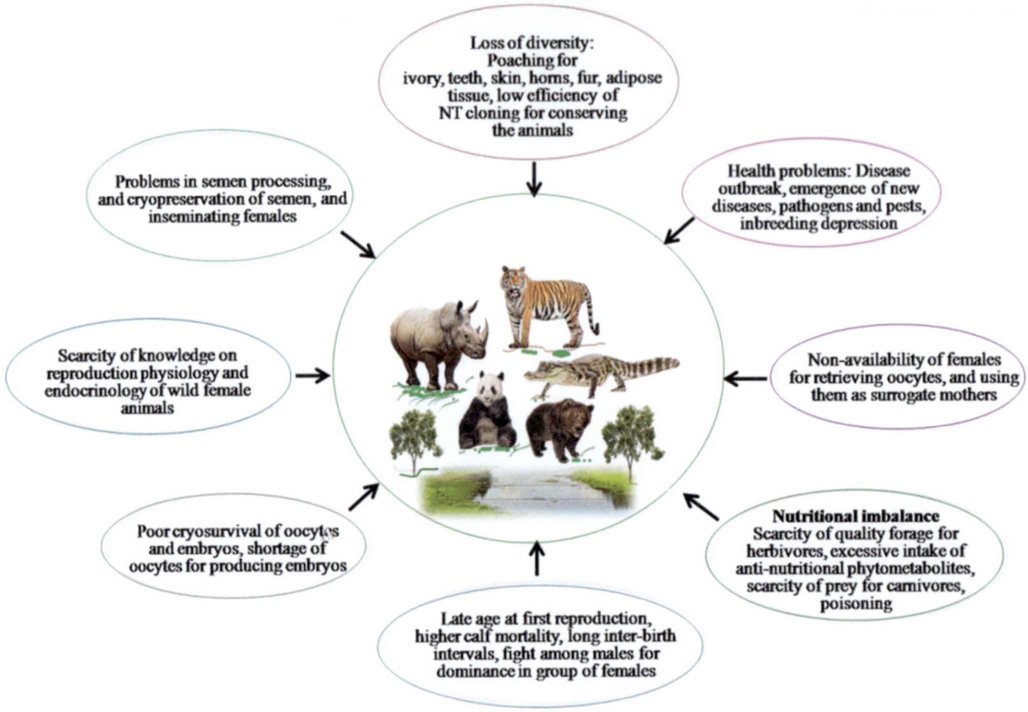

Fig. 46.3 Summary of threats to wildlife. Illegal trade, shrinking habitat, and food resources are the major factors contributing to decline in wildlife. The world is dealing with unprecedented spike in wildlife trade. A collapse in imminent in global biodiversity, especially in tropics, if stern preventive measures are not taken to reverse the species loss

wild felids (https://blogs.scientificamerican.com/extinction-countdown/6-most-endangered-feline-species/, accessed on August 2, 2018). Similarly, most large herbivores such as gorilla (*Gorilla beringei* and *Gorilla gorilla*) and Asian elephant (*Elephas maximus*) are among most endangered species. Some species such as mammoths, rhinoceros-sized marsupials and, marsupial lions are already in the list of extinct species.

46.3 Problems in Conservation of Wild Animals

Illegal hunting of wild animals for their body parts, skewing habitat due to expanding human populations and anthropological activities, high predation and break of infectious diseases are the major factors detrimental to survival of some wild animals. In some species, inbreeding depression is also a threat of serious concern. In addition, certain traits such as delayed puberty, age at first gestation, long-birth intervals, and low parturition density affect their population growth.

Assisted reproduction technologies (ARTs) have a limited success in wild animals, mostly due to non-availability of oocytes, and dearth of females to get large number of oocytes needed for producing embryos. Wherever feasible, gametes are obtained from wild and companion animals after death. In many cases, the intergenus females are used as surrogate mothers to give births to young one of other animal species.

Further, information is scarce on reproductive physiology, endocrinology, and behavior in many species of feral animals. Knowledge acquired on reproduction and endocrinological aspects in biological similar species is used to augment the reproductive success of the non-domesticated animals as well as humans (Prieto-Pablos et al. 2016; Comizzoli et al. 2018).

46.4 Wildlife Conservation

The list of endangered animal and plant species increases every day. Wildlife conservation refers to strategies aimed to maintain animals in their natural habitat, developing habitat to wild animals or conserving them through ARTs. Various nations have initiated programs to conserve their wildlife resources. For instance, University of Florida had initiated a program called Biotechnology for Ecological, Evolutionary and Conservation Sciences (BEECS) that apply biotechnological tools (DNA synthesis and genome, transcriptome and proteome sequencing, and others) to promote and conserve wildlife. The veterinary health management and biotechnological advances have been applied for conserving and repopulating endangered or threatened wild birds, reptiles, and mammals.

Endangered wild mammalian species can be repopulated by developing embryos in vivo and transferring them to estrus-synchronized surrogate mothers. Low availability of donor females and less number of oocytes recovered from females is one of the important limiting factors that impede production of embryos.

46.5 Conservation of Herbivore Feral Animals Through ARTs

Despite the difficulties in handling and getting animals to be studied, and getting the sample materials, the ARTs (semen, oocyte, and embryo cryopreservation, ICSI, IVF, in vitro production of embryos including SCNT embryos, intergenic embryo transfer, and pregnancy diagnosis) are now widely applied to breed endangered wild herbivores and carnivores. Researchers have used SCNT to disseminate wild species by nuclear transfer cloning.

Among the 200 deer subspecies are found across the globe, more than 40 are considered as endangered. At the moment, in vitro production of embryos is a promising strategy to increase and disseminate increase population of endangered feral mammalians.

Scarcity of oocytes is among the major constraints in wild mammalian species ARTs. In addition, retrieving oocytes and low availability of females for use as surrogate mothers impede implementation and success of embryo-related biotechniques. Collecting oocytes by laparoscopic ovum pick-up (LOPU) coupled with IVF and IVP may possibly assist increase the population of endangered captive or feral herbivores.

Embryo biotechnology programs were initiated in some endangered herbivores such as sika deer. The oocytes collected by using LOPU from sika deer were matured in vitro and fertilized using fresh or frozen-thawed sperm to produce embryos. The study shows that irrespective of breeding and non-breeding season, good-quality oocytes were obtained that could be used for obtaining embryos by IVF (Locatelli et al. 2006). A high developmental rate of embryos (30% of the oocytes used), their ability to withstand cryopreservation implies that LOPU–IVF might be a successful tool for establishing embryo banks from endangered sika deer species (Locatelli et al. 2006, 2012).

In 2007, eight red deer (*Cervus elaphus*) clones were produced using multipotent antler stem cells and their progeny as donor nuclei (Berg et al. 2007). Aimed at increasing the efficacy of SCNT in endangered sika deer, efforts were made for quick and accurate enucleation of oocytes to produce cytoplasts, cell–cytoplast fusion, forming cloned blastocysts (Yin et al. 2014), IVM and parthenogenetic activation to increase the embryo development efficiency. Table 46.1 summarizes the strategies used to produce embryos in vitro and their applications.

Studies are carried out to conserve and increase populations of large herbivorous animals such as rhinoceros (Hermes et al. 2018), elephants (Hermes et al. 2013; Saragusty et al. 2015; Arnold et al. 2017), giraffe (Wilsher et al. 2013), by assisted reproduction including collection and

Table 46.1 Examples of biotechnological applications applied to augment reproduction efficacy of endangered wild animals

Species	Targets	Technological interventions	Inferences and recommendations (references)
Mammalian herbivores			
Monkeys (*Macaca fascicularis*)	Establishing primate models	Using SCNT to produce NT clones animals	The first report on cloning in non-human primates, inferring that macaques could be cloned (Liu et al. 2018)
M. fuscata	Development and physiological studies using *M. fuscata* as model species	Developing iPSCs from dermal fibroblasts with Sendai virus or plasmids	The iPSCs had similarities with human iPSCs, expressed various pluripotency-specific markers. It is anticipated that cells would provide robust in vitro tools for studying mechanisms of development and physiology using *M. fuscata* as model species (Nakai et al. 2018)
Mouflon (*Ovis orientalis musimon*)	Rescue of an endangered animals	iSCNT	The SCNT explores the feasibility of reviving endangered species (Loi et al. 2001)
Rhesus macaque	Reprogramming of adult cells	SCNT	Development of pluripotent stem cell lines from adult skin fibroblasts through SCNT explore the feasibility of therapeutic cloning in primates (Byrne et al. 2007)
Rhinoceros (*Diceros bicornus*)	Genetic management	Microsatellite profiling	The study provides evidence of polygamy in black rhinoceros (Garnier et al. 2001)
	Prenatal sex determination	PCR-amplification of SRY-genes	The sex-determination envisaged to be a valuable tool for managing assisted reproduction of the managed in captivity (Stoops et al. 2018)
Sika deer (*Cervus nipon*)	Conservation	Using SCNT to produce cloned embryos	Various types of cells viz., antlerogenic periosteum, adipocytes and bone cells shown comparable reprogramming ability with no difference on development of SCNT cloned embryos, and births of cloned fawns (Berg et al. 2007)
Tapir (*Tapirus bairdii*)	Semen preservation	Cryopreservation and evaluation	The study shows feasibility of Tapir semen by cryopreservation (Pukazhenthi et al. 2011)
Vietnamese deer (*Cervus nipon*)	Collecting oocytes	Induction of super-ovulation by eFSH (0.25 and 0.5U) and collection	LOPU–IVF envisaged to be a promising method for retrieving oocytes for producing embryos from the species (Locatelli et al. 2012)
Wild ox (*Bos gaurus*)	Feasibility of repopulation by cloning	Using iSCNT to produce cloned embryos	Successful live birth of *B. gaurus* calves proved the feasibility of iSCNT to conserve NT wild mammalian cloning species (Lanza et al. 2000)

(continued)

Table 46.1 (continued)

Species	Targets	Technological interventions	Inferences and recommendations (references)
Wild sheep (*Ovis ammon*)	Salvaging endangered sheep	Using iSCNT to produce cloned embryos	The cloned embryos produced from fusion of fibroblasts from *O. amnon*, and cytoplasts prepared from IVM oocytes of domestic sheep, transferred to sheep to establish pregnancies indicating the feasibility of the iSCNT to conserve wild sheep (White et al. 1999)
Monkey (Macaca fascicularis)	Autologous cell therapy	Parthenogenetic embryos	The parthenogenetic stem cells with ability to induced differentiation into various lineages such as functional neurons, and epithelial morphologies could serve as potential source of stem cells autologous cell therapy (Vrana et al. 2003)
Mammalian carnivores/omnivores			
African wild cat (*Felis silvestris lybica*)	Conservation	Using iSCNT to produce cloned embryos	Transfer of cloned embryos to domestic cat, birth of kitten proves use of the technique to produce cloned carnivore (Gomez et al. 2004)
African wild dog (*Lycaon pictus*)	Conservation	Cryopreservation	Improving sperm cryopreservation by manipulating semen extenders (Van den Berghee et al. (2018)
Coyotes (*Canis latrans*)	Conservation	Using iSCNT to produce cloned embryos	Births of eight live offspring from transfer of embryos to dogs, indicating that despite limitations the technique can be used to conserve and multiply endangered wild canines (Hwang et al. 2013)
Gray wolf (*Canis lupus*)	Conservation	Using iSCNT to produce cloned embryos	The ability of cells collected from a dead wolf to form cloned embryos when fused with cytoplasts developed from dog oocytes, implies that iSCNT could be strategy to resurrect the endangered wild canine after death (Oh et al. 2008)
Grey wolf (*Canis lupus*)	Conservation	Using iSCNT to produce cloned embryos	The dogs used as source of oocytes, and surrogate mothers to produce cloned grey wolves implying that SCNT might be a practical approach for repopulating the endangered wild canines (Kim et al. 2007)
Jaguar (*Panthera onca*)	Collection and semen evaluation	Chemical restraining of animals and semen collection	Collecting semen by uretheral catherization found to be promising method for collecting semen from wild and captive carnivores (Araujo et al. 2017)
Jungle cat (*Felis chaos*)	Semen collection and evaluation	Chemical restraining of animals and semen collection	Collecting semen by uretheral catherization found to be promising method for collecting semen from collection by restraining wild and captive carnivores (Kheirkhah et al. 2017)

(continued)

Table 46.1 (continued)

Species	Targets	Technological interventions	Inferences and recommendations (references)
Leopard cat (*Prionailurus bengalensis*)	Feasibility of SCNT is the species	iSCNT	Development interspecies cloned nuclei from fibroblasts of *P. bengalensis* and cytoplasts from domestic cat oocytes, the cloned embryos could be used for establishing pregnancies indicating the feasibility of use of the SCNT as alternative strategy for producing offspring (Yin et al. 2006)
Black lion tamarin (*Leontopithecus chrysopygus*)	Genetic assessment and diversity species	Microsatellite-based polymorphism in different	The data envisaged to provide insights into genetic diversity in different species that could support breeding programs and prevent inbreeding depression, and conserving genetic biodiversity (Ayala-Burbano et al. 2017)
Lion (*Panthera leo*))	Diversity conservation	In vitro embryo production	Production of embryos from oocytes retrieved from lioness after death, and embryos produced through ICSI, by using homologous frozen-thawed sperm. The technique explores the feasibility of producing blastocysts in lions from IVM of oocytes and ICSI frozen-thawed sperm (Fernandez-Gonzalez et al. 2015)
Snow leopard (*Panthera uncia*)	Conservation	Inducing pluripotency in fibroblasts	The pluripotency induced by retroviral transfection with Moloney-based retroviral vectors (pMXs) encoding four factors (OCT4, SOX2, KLF4 and cMYC), led to development of iPSCs. The iPSCs might have applications in NT cloning, genetic conservation, and deriving games in vitro (Verma et al. 2012)
Tammar wallaby (*Macropus eugenii*)	Conservation	ARTs	Birth of young ones after AI (Paris et al. 2005)

Abbreviations eFSH—equine follicle stimulating hormone; ICSI—intercytoplasmic sperm injection; IVF—in vitro fertilization, IVM—in vitro maturation, IVEP—in vitro embryo production, SCNT—somatic cell nuclear transfer, iSCNT—interspecies somatic cell nuclear transfer, SRY—sex-determining region Y

cryopreservation of sperm, improving quality of semen, understanding ovarian functions, IVF, and embryo biotechnology.

A study on investigating the effects of sex-sorting and cryopreservation on post-thaw characteristics and fertility of red deer (*Cervus elaphus*) semen collected by electroejaculation from 10 mature stags during breeding season shows that Y-chromosome bearing sperm possessed adequately acceptable post-thaw sperm parameters (viz., sperm motility, chromatin stability, and fertility) and expected offspring sex ratio. It was inferred that technique has potential in wild animals though further studies should be carried out to resolve the sperm damage (Anel-Lopez et al. 2017a, b).

46.6 Conserving Felids and Other Carnivores

Reproductive biotechnological interventions are expected to conserve or repopulate wild felids (Holt et al. 2004; Yin et al. 2006; Gomez and Pope 2015; Angrimani et al. 2017). In 2003, for the first time, SCNT cloned African Wildcat (*Felis silvestris lybica*) kitten were reported from domestic cat as surrogate mothers. Embryos were developed from fusion of fibroblasts from *F. silvestris lybica* to cytoplasts from oocytes obtained of domestic cat (Gómez et al. 2004). Since then other species are also cloned.

Similarly, SCNT is used to produce clones of wild canids (Table 46.1). Other ARTs, such as sperm cryopreservation (Anel-Lopez et al. 2017), oocyte and embryo biotechniques (Singh et al. 2009), and ICSI (Salamone et al. 2017) are used to conserve and increase the population of domestic and wild mammals. It is necessary to be acquainted with fundamentals of reproductive biotechniques for their successful implementation in wild mammalian fauna (Jewgenow et al. 2017).

One of the most important benefits of freeze-drying and storage is that lyophilized sperm can be preserved when liquid nitrogen supply is interrupted in natural calamities such as Hurricane, floods or other natural or man-mediated disasters (Dickey et al. 2006). It is envisaged that semen of some wild species including chimpanzee, giraffe, jaguar, weasel, and the long-haired rat can withstand freeze-drying. Their sperm is found to remain viable after lyophilization. As pronuclei were formed after the injection of freeze-dried sperm into the mouse oocytes, the technique has potential in micro-insemination techniques, and for establishing "freeze-drying zoo" to conserve and repopulate endangered wild animals when semen availability is less (Takehito Kaneko et al. 2014).

46.7 Genomics Advances in Wildlife Conservation Management

A large number of studies are carried out in studying wild animals. Prime objective studying genomics in wild animals is assessing their genetic make-up, and diversity, comparing genetic diversity parameters between wild populations and captive groups and developing molecular markers as parameters of preventing genetic diversity and inbreeding depression. In addition, the molecular markers are indicators of their evolutionary adaptation to harsh climate.

Currently, the data is equivocal livestock domestication was a multiple or singular process. Early animal populations have contributed in a different ways to modern animal rearing and human nutrition and welfare. Wildlife genomics includes analysis of genetic materials using large-scale genome analysis tools. In addition, "omics" tools have provided a lot of valuable information on physiological adaptations, population history, behavioral aspects, health, and dynamics of wild animals. Transcriptomics studies have provided insights into molecular biological and biochemical mechanisms regulating hibernation in dwarf lemurs (*Cheirogaleus crossleyi*) (Faherty et al. 2018).

Genomics has important role in conservation of endangered species and studying endangered animal populations. More specifically, the conservation genetics has facilitated empirical insights into the effect of inbreeding and increased genetic drift leading to minimal genetic diversity in isolated populations of wild animals. The genetic information obtained on wildlife is valuable for wildlife managers and conservationists, calculating harvest rates, and managing the migration or translocation of wild animals.

Koala, the only extant species of the marsupial family Phascolarctidae, is classified as "vulnerable" due to shrinking habitat, scarcity of dietary forage, and outbreak of infectious diseases. Genome of koala is sequenced recently. Koala feeds exclusively on eucalyptus leaves that are known to contain anti-nutritional hydrolysable tannins (HTs). Genome sequence data revealed that koala's ability to detoxify eucalypt HTs might be due to expansions within a cytochrome P450 gene family, and its ability to smell, taste, and regulate intake of phytometabolites owing to expansions in the vomeronasal and taste receptors (Johnson et al. 2018). Genetically diverse population's studies require habitat corridors and translocation strategies to assist the survival of koala in its natural habitat (Johnson et al. 2018).

Metagenomic analysis of gut ecosystem of bamboo-eating giant panda (*Ailuropoda melanoleuca*) and red panda (*Ailurus fulgens*) has shown abundance of cyanide-degrading bacteria (Zhu et al. 2018). It is envisioned that gut ecosystem of other herbivores adapted to unconventional diets might yield valuable microbial resources. Integrated transcriptome sequencing, proteomics, phenotypic, and biochemical studies revealed presence of highly fibrolytic fungi from the herbivores gut and found that enzyme synthesis is triggered in response to substrate, i.e. lignocelluloses ingested by the host (Solomon et al. 2016).

Characterization and analysis of phylogenetically conserved and derived transcripts (55,910 expressed sequence tag or ESTs) contigs in *Loxodonta Africana*, the African elephant, has revealed evolution and variation of eutherian placentation showing that some candidate genes might be important for normal development and functioning of human placenta. A total of 2963 genes were found to be expressed commonly in the placentas of some eutherian mammals (mouse, cattle, and humans) studied. Further, it was inferred that dysfunction these genes might lead to complications to human fetal development (Hou et al. 2012).

46.8 Negative Impact of Wildlife

There are negative aspects of wild animals that need attention of biologists, microbiologists, and personnel involved in veterinary-public health management. Many wild animals serve as natural hosts or reservoirs of deadly viruses, pathogens, food-borne pathogens, pests, and parasites. Diseases spread from animals to humans are called zoonotic diseases.

It is noted that of the 37 new infectious diseases encountered during past 30 years, more than two-thirds are zoonotic in nature (http://needtoknow.nas.edu/id/threats/animal-carriers/, accessed August 7, 2018). Around 48 human diseases spread from pests that bite animals. Water infected with animal feces and urine is the source of several infections in humans. Wild boar is reservoir of numerous zoonotic diseases.

Some of the remarkable diseases caused by wild animals include protozoal infections (toxoplasmosis), bacterial (leptospirosis), and viral (encephalitis caused by Nipah virus).

While some wild birds are natural reservoirs of West Nile virus, fatal bird flu, the bats spread Nipah viruses, and severe acute respiratory syndrome (SARS), the rodents are carriers of deadly plague. In addition, wild animals pose threats to domestic animals by preying on them, attack occasionally humans, and devastate crops and horticulture.

46.9 Outlook and Challenges

Protecting and conserving wildlife is a major issue across the globe. The threats from smugglers and poachers are of paramount concerns. Technologies are developed to monitor wildlife and prevent unauthorized exploitation of the wildlife. However, technology and data developed for protecting endangered feral species can also be utilized by the hunters and poachers.

Reproductive techniques are used to clone many livestock species and have also been found promising to increase population of wild animals. Somatic cell banking, semen, oocyte, and embryo banking are explicitly useful to conserve wild animals. Nuclear transfer cloning has limited success when used in wild animals. The underlying technical causes should be resolved. NT cloning of endangered mammals presents practical problems, many of which stem from the paucity of knowledge about their basic reproductive biology.

In endangered species, research should focus on characterizing reproductive traits, reproductive behavior, and species-specific endocrine principals. The knowledge acquired will be useful in manipulating female reproductive cycle of females intended for use surrogate mothers. Efforts should also be made to study molecular and cellular mechanisms of gamete development, their cryopreservation, and optimizing applications of ARTs at larger scale in wild species. Also, efforts should be made for scientific control of overpopulating wild and feral animals that have posed threats to humans, crops, gardens, and domestic animals.

46.10 Conclusions

Wild animals and plants are of immense importance to humans, ecological balance, and agriculture. Regrettably, our planet is losing wild animals rapidly, many have become extinct. More than half of the total wild animals have decimated during past four decades due to habitat loss, climate change, pollution, diseases, and hunting. Clearly, protecting wildlife has become an environmental and security concern. Modern ARTs tested in domestic livestock should be optimized for use in conserving and disseminating valuable wild animals. Also strategies should be evolved for preventing reproduction in wild animals that are harmful to agriculture and humans.

References

Anel-Lopez L, Ortega-Ferrusola C, Álvarez M, Borragán S, Chamorro C, Peña FJ, Morrell J, Anel L, de Paz P (2017a) Improving sperm banking efficiency in endangered species through the use of a sperm selection method in brown bear (*Ursus arctos*) thawed sperm. BMC Vet Res 13(1):200. https://doi.org/10.1186/s12917-017-1124-2

Anel-López L, García-Álvarez O, Parrilla I, Del Olmo D, Maroto-Morales A, Fernandez-Santos MR, Ortiz JA, Soler AJ, Martínez EM, Vazquez JM, Garde JJ (2017) Effect of sex-sorting and cryopreservation on the post-thaw sperm quality of Iberian red deer spermatozoa. Theriogenology 89:206–213. https://doi.org/10.1016/j.theriogenology.2016.11.010 (Epub 2016 Nov 14)

Angrimani DSR, Barros PMH, Losano JDA, Cortada CNM, Bertolla RP, Guimarães MABV, Correa SHR, Barnabe VH, Nichi M (2017) Effect of different semen extenders for the storage of chilled sperm in Tigrina (Leopardus tigrinus). Theriogenology 89:146–154. https://doi.org/10.1016/j.theriogenology.2016.10.015 (Epub 2016 Oct 27)

Araujo GR, Paula TAR, Deco-Souza T, Morato RG, Bergo LCF, Silva LCD, Costa DS, Braud C (2017) Comparison of semen samples collected from wild and captive jaguars (Panthera onca) by urethral catheterization after pharmacological induction. Anim Reprod Sci. pii: S0378-4320(17)30552-3. https://doi.org/10.1016/j.anireprosci.2017.12.019 (Epub ahead of print)

Arnold DM, Gray C, Roth TL, Mitchell S, Graham LH (2017) A simple, field-friendly technique for cryopreserving semen from Asian elephants (Elephas maximus). Anim Reprod Sci 182:84–94. https://doi.org/10.1016/j.anireprosci.2017.05.003

Ayala-Burbano PA, Caldano L, Junior PMG, Pissinatti A, Marques MC, Wormell D, Domingues de Freitas P (2017) Genetic assessment for the endangered black lion tamarin Leontopithecus chrysopygus (Mikan, 1823), Callitrichidae, Primates. Am J Primatol 79(12). https://doi.org/10.1002/ajp.22719 (Epub 2017 Nov 2)

Berg DK, Li C, Asher G, Wells DN, Oback B (2007) Red deer cloned from antler stem cells and their differentiated progeny. Biol Reprod 77(3):384–394 (Epub 2007 May 23)

Byrne JA, Pedersen DA, Clepper LL, Nelson M, Sanger WG, Gokhale S, Wolf DP, Mitalipov SM (2007) Producing primate embryonic stem cells by somatic cell nuclear transfer. Nature 450(7169):497–502 (Epub 2007 Nov 14). Erratum in: Nature. 2014 Dec 11;516(7530):276

Comizzoli P, Paulson EE, McGinnis LK (2018) The mutual benefits of research in wild animal species and human-assisted reproduction. J Assist Reprod Genet 35(4):551–560. https://doi.org/10.1007/s10815-018-1136-2

Dickey RP, Lu PY, Sartor BM, Dunaway HE Jr, Pyrzak R, Klumpp AM (2006) Steps taken to protect and rescue cryopreserved embryos during Hurricane Katrina. Fertil Steril 86(3):732–734

Faherty SL, Villanueva-Cañas JL, Blanco MB, Albà MM, Yoder AD (2018) Transcriptomics in the wild: hibernation physiology in free-ranging dwarf lemurs. Mol Ecol 27(3):709–722. https://doi.org/10.1111/mec.14483 (Epub 2018 Jan 29)

Fernandez-Gonzalez L, Hribal R, Stagegaard J, Zahmel J, Jewgenow K (2015) Production of lion (Panthera leo) blastocysts after in vitro maturation of oocytes and intracytoplasmic sperm injection. Theriogenology 83(6):995–999. https://doi.org/10.1016/j.theriogenology.2014.11.037

Garnier JN[1], Bruford MW, Goossens B (2001) Mating system and reproductive skew in the black rhinoceros. Mol Ecol 10(8):2031–2041

Gómez MC, Pope CE (2015) Cloning Endangered felids by interspecies somatic cell nuclear transfer. Methods Mol Biol 1330:133–152. https://doi.org/10.1007/978-1-4939-2848-4_13

Gómez MC, Pope CE, Giraldo A, Lyons LA, Harris RF, King AL, Cole A, Godke RA, Dresser BL (2004) Birth of African Wildcat cloned kittens born from domestic cats. Cloning Stem Cells 6(3):247–258

Hermes R, Saragusty J, Göritz F, Bartels P, Potier R, Baker B, Streich WJ, Hildebrandt TB (2013) Freezing

African elephant semen as a new population management tool. PLoS ONE 8(3):e57616. https://doi.org/10.1371/journal.pone.0057616

Hermes R, Hildebrandt TB, Göritz F (2018) Cryopreservation in rhinoceros-setting a new benchmark for sperm cryosurvival. PLoS One 13(7):e0200154. https://doi.org/10.1371/journal.pone.0200154 (eCollection 2018)

Holt WV, Pickard AR, Prather RS (2004) Wildlife conservation and reproductive cloning. Reproduction 127(3):317–324

Hou ZC, Sterner KN, Romero R, Than NG, Gonzalez JM, Weckle A, Xing J, Benirschke K, Goodman M, Wildman DE (2012) Elephant transcriptome provides insights into the evolution of eutherian placentation. Genome Biol Evol 4(5):713–725. https://doi.org/10.1093/gbe/evs045

Hwang I, Jeong YW, Kim JJ, Lee HJ, Kang M, Park KB, Park JH, Kim YW, Kim WT, Shin T, Hyun SH, Jeung EB, Hwang WS (2013) Successful cloning of coyotes through interspecies somatic cell nuclear transfer using domestic dog oocytes. Reprod Fertil Dev 25 (8):1142–1148. https://doi.org/10.1071/rd12256

Jewgenow K, Braun BC, Dehnhard M, Zahmel J, Goeritz F (2017) Research on reproduction is essential for captive breeding of endangered carnivore species. Reprod Domest Anim. 52(Suppl 2):18–23. https://doi.org/10.1111/rda.12836 (Epub 2016 Nov 15. Review)

Johnson RN, O'Meally D, Chen Z, Etherington GJ, Ho SYW, Nash WJ, Grueber CE, Cheng Y, Whittington CM, Dennison S, Peel E, Haerty W, O'Neill RJ, Colgan D, Russell TL, Alquezar-Planas DE, Attenbrow V, Bragg JG, Brandies PA, Chong AY, Deakin JE, Di Palma F, Duda Z, Eldridge MDB, Ewart KM, Hogg CJ, Frankham GJ, Georges A, Gillett AK, Govendir M, Greenwood AD, Hayakawa T, Helgen KM, Hobbs M, Holleley CE, Heider TN, Jones EA, King A, Madden D, Graves JAM, Morris KM, Neaves LE, Patel HR, Polkinghorne A, Renfree MB, Robin C, Salinas R, Tsangaras K, Waters PD, Waters SA, Wright B, Wilkins MR, Timms P, Belov K (2018) Adaptation and conservation insights from the koala genome. Nat Genet 50(8):1102–1111. https://doi.org/10.1038/s41588-018-0153-5 (Epub 2018 Jul 2)

Kaneko T, Ito H, Sakamoto H, Onuma M, Inoue-Murayama M (2014) Sperm preservation by freeze-drying for the conservation of wild animals. PLoS One 9(11): e113381. 19. https://doi.org/10.1371/journal.pone.0113381

Kheirkhah MS, Mollapour Sisakht M, Mohammadsadegh M, Moslemi HR (2017) Sperm evaluation of Jungle Cat (Felis chaus) obtained by urethral catheterization (CT) after medetomidine administration. Theriogenology 15(91):17–20. https://doi.org/10.1016/j.theriogenology.2016.12.034

Kim MK, Jang G, Oh HJ, Yuda F, Kim HJ, Hwang WS, Hossein MS, Kim JJ, Shin NS, Kang SK, Lee BC (2007) Endangered wolves cloned from adult somatic cells. Cloning Stem Cells 9(1):130–137. Erratum in: Cloning Stem Cells. 2007 Autumn; 9(3):450

Lanza RP, Cibelli JB, Diaz F, Moraes CT, Farin PW, Farin CE, Hammer CJ, West MD, Damiani P (2000) Cloning of an endangered species (Bos gaurus) using interspecies nuclear transfer. Cloning 2(2):79–90

Liu Z, Cai Y, Wang Y, Nie Y, Zhang C, Xu Y, Zhang X, Lu Y, Wang Z, Poo M, Sun Q (2018) Cloning of macaque monkeys by somatic cell nuclear transfer. Cell 174(1):245. https://doi.org/10.1016/j.cell.2018.01.036

Locatelli Y, Vallet JC, Huyghe FP, Cognié Y, Legendre X, Mermillod P (2006) Laparoscopic ovum pick-up and in vitro production of sika deer embryos: effect of season and culture conditions. Theriogenology 66(5):1334–1342

Locatelli Y, Hendriks A, Vallet JC, Baril G, Duffard N, Bon N, Ortiz K, Scala C, Maurel MC, Mermillod P, Legendre X (2012) Assessment LOPU-IVF in Japanese sika deer (Cervus nippon nippon) and application to Vietnamese sika deer (Cervus nippon pseudaxis) a related subspecies threatened with extinction. Theriogenology 78(9):2039–2049. https://doi.org/10.1016/j.theriogenology.2012.07.025

Loi P, Ptak G, Barboni B, Fulka J Jr, Cappai P, Clinton M (2001) Genetic rescue of an endangered mammal by cross-species nuclear transfer using post-mortem somatic cells. Nat Biotechnol 19(10):962–964

Nakai R, Ohnuki M, Kuroki K, Ito H, Hirai H, Kitajima R, Fujimoto T, Nakagawa M, Enard W, Imamura M (2018) Derivation of induced pluripotent stem cells in Japanese macaque (Macaca fuscata). Sci Rep 8(1):12187. https://doi.org/10.1038/s41598-018-30734-w

Oh HJ, Kim MK, Jang G, Kim HJ, Hong SG, Park JE, Park K, Park C, Sohn SH, Kim DY, Shin NS, Lee BC (2008) Cloning endangered gray wolves (Canis lupus) from somatic cells collected postmortem. Theriogenology 70(4):638–647. https://doi.org/10.1016/j.theriogenology.2008.04.032

Paris DB, Taggart DA, Shaw G, Temple-Smith PD, Renfree MB (2005) Birth of pouch young after artificial insemination in the tammar wallaby (Macropus eugenii). Biol Reprod 72(2):451–459

Poltorak A, Apalko S, Sherbak S (2018) Wild-derived mice: from genetic diversity to variation in immune responses. Mamm Genome 29(7–8):577–584. https://doi.org/10.1007/s00335-018-9766-3

Prieto-Pablos MT, Sánchez-Calabuig MJ, Hildebrandt TB, Göritz F, Ortmann S, Eder S, Santiago-Moreno J, Hermes R, Saragusty J (2016) Cryopreservation of captive roe deer (Capreolus capreolus) semen. Theriogenology 86(3):695–703. https://doi.org/10.1016/j.theriogenology.2016.02.023

Pukazhenthi BS, Togna GD, Padilla L, Smith D, Sanchez C, Pelican K, Sanjur OI (2011) Ejaculate traits and sperm cryopreservation in the endangered Baird's tapir (Tapirus bairdii). J Androl 32(3):260–270. https://doi.org/10.2164/jandrol.110.011833

Salamone DF, Canel NG, Rodríguez MB (2017) Intracytoplasmic sperm injection in domestic and wild mammals. Reproduction 154(6):F111–F124. https://doi.org/10.1530/rep-17-0357 (Epub 2017 Dec 1)

Saragusty J, Prieto MT, Courtiol A, Potier R, Göritz F, Hildebrandt TB, Hermes R (2015) Sperm rescue in wild African elephants. Reprod Fertil Dev. https://doi.org/10.1071/rd14378

Singh B, Chauhan MS, Singla SK, Gautam SK, Verma V, Manik RS, Singh AK, Sodhi M, Mukesh M (2009) Reproductive biotechniques in buffaloes (Bubalus bubalis): status, prospects and challenges. Reprod Fertil Dev 21(4):499–510. https://doi.org/10.1071/rd08172

Solomon KV, Haitjema CH, Henske JK, Gilmore SP, Borges-Rivera D, Lipzen A, Brewer HM, Purvine SO, Wright AT, Theodorou MK, Grigoriev IV, Regev A, Thompson DA, O'Malley MA (2016) Early-branching gut fungi possess a large, comprehensive array of biomass-degrading enzymes. Science 351 (6278):1192–1195. https://doi.org/10.1126/science.aad1431

Stoops MA, Winget GD, DeChant CJ, Ball RL, Roth TL (2018) Early fetal sexing in the rhinoceros by detection of male-specific genes in maternal serum. Mol Reprod Dev Mar 85(3):197–204. https://doi.org/10.1002/mrd.22953

Ueda A, Kobayashi A, Tsuchida S, Yamada T, Murata K, Nakamura H, Ushida K (2018) Cecal microbiome analyses on wild japanese rock ptarmigans (*Lagopus muta japonica*) reveals high level of coexistence of lactic acid bacteria and lactate-utilizing bacteria. Microorganisms. 6(3). pii: E77. https://doi.org/10.3390/microorganisms6030077

Van den Berghe F, Paris MCJ, Briggs MB, Farstad WK, Paris DBBP (2018) A two-step dilution tris-egg yolk extender containing Equex STM significantly improves sperm cryopreservation in the African wild dog (Lycaon pictus). Cryobiology 80:18–25. https://doi.org/10.1016/j.cryobiol.2017.12.095

Verma R, Holland MK, Temple-Smith P, Verma PJ (2012) Inducing pluripotency in somatic cells from the snow leopard (Panthera uncia), an endangered felid. Theriogenology 77(1):220–228, 228.e1-2. https://doi.org/10.1016/j.theriogenology.2011.09.022

Vrana KE, Hipp JD, Goss AM, McCool BA, Riddle DR, Walker SJ, Wettstein PJ, Studer LP, Tabar V, Cunniff K, Chapman K, Vilner L, West MD, Grant KA, Cibelli JB (2003) Nonhuman primate parthenogenetic stem cells. Proc Natl Acad Sci U S A 100(Suppl 1):11911–6 (Epub 2003 Sep 22). Erratum in: Proc Natl Acad Sci U S A. 2004 Jan 13;101(2):693

White KL, Bunch TD, Mitalipov S, Reed WA (1999) Establishment of pregnancy after the transfer of nuclear transfer embryos produced from the fusion of argali (Ovis ammon) nuclei into domestic sheep (*Ovis aries*) enucleated oocytes. Cloning. 1(1):47–54

Wilsher S, Stansfield F, Greenwood RE, Trethowan PD, Anderson RA, Wooding FB, Allen WR (2013) Ovarian and placental morphology and endocrine functions in the pregnant giraffe (Giraffa camelopardalis). Reproduction 145(6):541–554. https://doi.org/10.1530/rep-13-0060

Yin XJ, Lee Y, Lee H, Kim N, Kim L, Shin H, Kong I (2006) In vitro production and initiation of pregnancies in inter-genus nuclear transfer embryos derived from leopard cat (Prionailurus bengalensis) nuclei fused with domestic cat (Felis silverstris catus) enucleated oocytes. Theriogenology 66(2):275–282

Yin Y, Mei M, Zhang D, Zhang S, Fan A, Zhou H, Li Z (2014) The construction of cloned Sika deer embryos (Cervus nippon hortulorum) by demecolcine auxiliary enucleation. Reprod Domest Anim 49(1):164–169. https://doi.org/10.1111/rda.12246 (Epub 2013 Oct 21)

Zhu L, Yang Z, Yao R, Xu L, Chen H, Gu X, Wu T, Yang X (2018) Potential mechanism of detoxification of cyanide compounds by gut microbiomes of bamboo-eating pandas. mSphere. 3(3). pii: e00229-18. https://doi.org/10.1128/msphere.00229-18

Non-meat Alternatives

47

Abstract

In vitro meat, seafood, and insects are viewed as alternatives to meat obtained from livestock. In vitro meat is in fact produced from cultured animal cells. The technologies seem to be moving forward, but significant concerns such as large-scale production, consumer acceptance, economy, and environmental impact should be addressed.

Highlights

- Production of meat from ruminants has certain limitations and dismal environmental concerns
- It is high time to explore alternative sources of meat to meet requirements and save the environment.

Keywords

In vitro meat · Lab-grown meat · Non-meal alternatives · Entomophagy · Environmental health

47.1 Introduction

Worldwide meat production has tripled over the past four decades and raised 20% during the last 10 years. Developed or industrial countries consume double the quantity in comparison with developing countries (http://www.worldwatch.org/global-meat-production-and-consumption-continue-rise, accessed on Nov. 12, 2018). While poultry and pig are the prime sources of meat among monogastric livestock species, cattle, goats, sheep, and buffaloes are the ruminants that contribute to meat production. In addition, horse, camel, yak, reindeer, and mithun are also used as minor sources of meat in some countries. Some tribes prefer dogs, cats, and rodent meat.

Excessive rearing of animals to meet global demand of meat has adverse effects on environment, public health, animal genetic diversity as well as on overall economy. Whereas ruminants produce methane from enteric fermentation, animal wastes release methane, nitrous oxide, and carbon dioxide. Waste animal tissues and blood from abattoir are causes of putrification and hideous stench. Besides, the abattoir waste is serious threat to water bodies and environment.

Another important negative impact of commercial meat production is that mass quantities of antibiotics are used as growth promoters in poultry and swine. In many cases, use of antibiotics is random and uncontrolled which has led to the development of antibiotic resistance among various microorganisms including spoilage and pathogenic bacteria.

The horse production is associated with low enteric methane emissions. Compared to ruminants, the equines require less feed and can be used for mechanical power and milk (Aspri et al. 2018; Liu et al. 2018). With 50% more protein, and 30% more iron contents than leanest beef,

B. Singh et al., *Advances in Animal Biotechnology*,
https://doi.org/10.1007/978-3-030-21309-1_47

the horse meat is a prospective substitute to red meat. The horse meat and its traditional products, viz. cecina, fermented sausage, liver pate, dry-cured loin, have high nutritional value (Lorenzo et al. 2017). In addition, better uptake of polyunsaturated fatty acids by horses has positive effects on meat quality (Belaunzaran et al. 2015).

In some parts of the world, dogs' and cats' meat are also consumed. Some impoverished tribes prefer roasting or stewing wild rats, and other rodents such as capybara (Nogueira-Filho and da Cunha Nogueira 2018) as cheap source of animal protein.

According to US FAO, the world would face a critical food shortage in near future as global demand for meat is likely to increase by more than two-thirds. As a strategy, alternatives to meat such as microalgae, single-cell proteins, yeasts, mushrooms, and plant-derived proteins and co-products (Box 1) should be developed.

Box 1. Some Protein-Rich Alternatives to Animal Meat for Human Consumption

Name	Source	Benefits
Quinoa (*Chenopodium quinoa*)	Plant seeds	Economical, seed are fine source of proteins, contains eight essential amino acids, gluten-free, enriched with fiber, vitamin B and iron
Tofu Nuggets	Soy seeds (*Glycine max*)	Also known as bean curd, tofu is low in fat and nutritious alternative to meat. The soy seeds are rich sources of beneficial polyphenols and phytoestrogens. The soy crop enhances soil fertility by fixing atmospheric nitrogen
Mushrooms	Fungi (Several species)	Low calorie, free of readily fermentable sugars, chewy texture can be supplemented with other protein-rich foods, the leftover substrate and waste materials can be used as manure or soil fertilizer
Cottage cheese	Milk	Good source of proteins, creamy taste, minimal fat contents
Seafood	Marine water	Several species such as prawns, lobsters, octopus, tuna, and fish are used. Most are good source of proteins, and omega fatty acids with pro-health benefits such as lowering blood pressure and cholesterolemia, and high levels of iron is present in shellfish
Lentils	Leguminous crops	Good sources of proteins and phytoestrogens, often used as vegetarian sources of proteins, lentils are useful for soil health
Quorn	Multiple s component	Quorn contains mycoproteins from a fungus known as *Fusarium venenatum* grown by fermentation
Edamane beans	Multiple beans	Rich sources of proteins, low in fat, improves soil fertility
Cultured meat	Muscle cells	The technology is emerging as alternative to meat obtained from animals
Insects	Multiple species	Several insects such as grasshoppers, cicadas, leafhoppers and bugs are rich sources of proteins (20 to 76% of DM), fat (2-20% of DM), low carbohydrates, reasonable amounts of minerals (Na, K, Ca, Fe, P and Mn) (Kouřimská and Adámková 2016)

(continued)

47.2 Cultured Meat or Tissue-Engineered Meat

The meat comprises of multiple tissues including bones, muscle, adipose, and connective tissue. In vitro-produced meat is a new concept of growing muscle cells and using them as animal-origin meat. The tissue-engineered meat is a hopeful alternative source of animal proteins, and alternative to conventional meat. To realistically mimic the multiple tissues components of meat, food compatible methods for bovine adipose tissue have to be developed.

In vitro-produced meat is the next step in food production chain. The basic methodology of in vitro meat production system (IMPS) involves culturing muscle tissue in liquid culture system on a large scale. Indeed, protocols have been developed to isolate the related cells such as pre-adipocytes, and differentiating them to adipocytes (Cui et al. 2018; Mehta et al. 2019), and allowing them to proliferate such a way that they assemble in 2D culture conditions or 3D alginate scaffolds (Mehta et al. 2019).

Despite several issues remains unsolved, production of in vitro meat has generated interest because of its potential as human food and protecting environmental and animals. One of the major problems associated with in vitro meat is its acceptance by consumers. If technology is successful, it will free up land used for growing grains and forage for producing crops and animals (Hocquette 2016).

47.3 Insects as Alternative Source of Animal Proteins

The insects serve as additional alternative environmental friendly sources of future food systems or supplements for humans (Kouřimská and Adámková 2016; Purschke et al. 2018), animals (Khan et al. 2018), and as aqua-feed (van Huis 2016). Insects are viewed as alternative sources of animal meat, and this phenomenon is known as entomophagy. Greater food conversion efficiency, shorter life cycles, different lifecycle stages, low emissions of greenhouse gases, less requirement for land and water, and ability to utilize broad range of plants, grains, seeds, and fruits, make entomophagy as attractive source of proteins, fats, vitamins and minerals. Analysis of fossils from caves in the USA and Mexico shows the evidences of use of insects as human food since prehistoric times (reviewed in Kouřimská and Adámková 2016). Mitsuhashi (2016) has enumerated a total of 2141 edible insects including their history, location and taxonomic information, and food and medicinal use. The number will increase in future with more reports on new species, and their nutritional and medicinal magnitude.

Mealworms (*Tenebrio molitor*) are new food items in Europe. The efforts are underway to preserve the nutrient quality, safety, acceptability and palatability and digestibility of insects. A recent study analyzed the effect of different cooking systems on nutritive quality of mealworm protein digestibility and fatty acids composition. It was concluded that boiling and cooking under vacuum are appropriate for reducing microbial load and maintaining high protein and PUFA of these mealworms (Caparros Megido et al. 2018).

47.4 Nutritional Relevance of Entomophagy

Certain insects such as *Tenebrio molitor* are sources of macro- and micronutrients. While insects are not routinely eaten in most Western societies, some people in the world eat various insects. Several insects such as grasshoppers, cicadas, leafhoppers, and bugs are rich sources of proteins (20 to 76% of DM), fat (2-20% of DM), low in carbohydrates, and have reasonable amounts of minerals (Na, K and Ca), and trace elements such as Cu, P, Fe, Mn, Se and Zn, water-soluble or lipophilic vitamins such as pantothenic acid and folic acid (Rumpold and Schlüter 2013; Kouřimská and Adámková 2016).

Larvae of insects are rich in fat contents. *Tenebrio molitor*, *Alphitobius diaperinus*, *Acheta domesticus* and *Blaptica dubia*, reared in the Netherlands, were found to contain C18:1 and C18:2 isomers, and trans isomers of

C16:1 and C18:1 (Tzompa-Sosa et al. 2014). Escamoles ant (*Liometupum apiculatum* M.) were found to have 505 μg/100 g of retinol, 3.31 μg/100 g of cholecalciferol, and 2.22 μg/100 g of α-tocopherol (Melo-Ruiz et al. 2013).

Insects are high in proteins containing more protein per 100 g compared to meat, fish, and egg. The wasps, bees, ants have highest protein contents. The protein contents depending on age and nutrition status of insects vary between 48 g and 74 g in aphids and pond skaters, and 23 g and 65 g in crickets (https://www.dailymail.co.uk/femail/food/article-4850314/How-insects-TWICE-protein.html, accessed on Nov. 12, 2018).

Analysis of protein contents of around 100 species including beetles, butterflies, homopterans, hemipterans, hymenopterans, orthopterans, dragon flies, etc. showed that these species had protein contents varying from 13% to 77% of the total DM contents (Xiaoming et al. 2010).

Larvae of insects such as Rhynchophorus family beetle having high lysine can compensate amino acid deficiency in human populations that thrive primarily on tubers.

47.5 Outlook and Challenges

Growing demand for meat, shrinking resources for growing animals, and the associated environment concerns have put pressure to evolve and explore strategies to develop alternative practices of meat production. The concept is impressive, but has certain important issues (Box 2) that should be resolved. Consumer acceptance of these alternative sources is a major issue. It is believed that consumption of cultured meat will depend on a conflict of values and perceptions at individual or public values.

Cultured meat might be an alternative to real meat, and a step toward reducing environmental impact of animal on climate, minimizing threat to public health, and taking care of animal welfare (https://culturedbeef.org/; accessed on Nov. 9, 2018).

Some issues are associated with techniques of producing meat in vitro. According to a recent analysis, commercial entrepreneurs have raised huge funds to promote eco-friendly methods of generating lab-grown meat. However, the industries don't have technical expertise and techniques to go ahead. It is only academic research that might play a role to advance the production of lab-grown at commercial scale (Dolgin 2019). The issues that should be resolved include developing muscle stem cell lines, developing scaffolding materials to assist the stem cell lines or combinations of cells to amass as tissue, and the bioreactors to achieve this.

The challenges of in vitro meat as alternative to real meat are on one side, convincing consumers to turn away from animal-derived meat is on other side. Both have to be satisfied. It is envisaged that commercial entrepreneurs will succeed in convincing the consumers to adopt alternative meat sources through publicity and advertisements. The beautifully packaged raw and tastefully processed alternative meat substitutes would one day make people to accept these products.

Box 2. Summary of issues related to use of insects as alternative animal proteins for humans

- Consumer acceptance of insect-origin proteins as alternatives to animal meat (Bryant and Barnett 2018, 2019)
- Off-flavor, toxicity, and allergy caused by insects (Gao et al. 2018)
- Swallowing of insects parts for use as food items
- Appropriate harvest time of insects in relation to nutrient contents
- Using genetic engineering or genome editing techniques to improve nutritional quality, feed conversion efficiency, and alleviate undesirable traits of food-grade insects

- Enhancing acquaintance with pathogens and predators of insects, and diseases when insects are produced at commercial scale
- Environmental impact assessment of commercial insect farming

47.6 Conclusions

The concept of culturing meat in vitro dates back decades. As an alternative to meat from livestock, the meat produced in vitro is humane, safe, and environment-friendly. Insects are nutritionally interesting and may be included as nutritional supplements. However, for including insects into normal diets, standard conditions for their rearing, monitoring of biological active components and associated allergenic activities should be strictly monitored.

References

Aspri M, Leni G, Galaverna G, Papademas P (2018) Bioactive properties of fermented donkey milk, before and after in vitro simulated gastrointestinal digestion. Food Chem 1(268):476–484. https://doi.org/10.1016/j.foodchem.2018.06.119

Belaunzaran X, Bessa RJ, Lavín P, Mantecón AR, Kramer JK, Aldai N (2015) Horse-meat for human consumption—current research and future opportunities. Meat Sci 108:74–81. https://doi.org/10.1016/j.meatsci.2015.05.006 (Epub 2015 May 14. Review)

Bryant C Barnett J (2018) Consumer acceptance of cultured meat: a systematic review. Meat Sci 143:8–17. https://doi.org/10.1016/j.meatsci.2018.04.008 (Epub 2018 Apr 12)

Bryant CJ, Barnett JC (2019) What's in a name? Consumer perceptions of in vitro meat under different names. Appetite 3(137):104–113. https://doi.org/10.1016/j.appet.2019.02.021

Caparros Megido R, Poelaert C, Ernens M, Liotta M, Blecker C, Danthine S, Tyteca E, Haubruge É, Alabi T, Bindelle J, Francis F (2018) Effect of household cooking techniques on the microbiological load and the nutritional quality of mealworms (Tenebrio molitor L. 1758). Food Res Int 106:503–508. https://doi.org/10.1016/j.foodres.2018.01.002 (Epub 2018 Jan 6)

Cui HX, Guo LP, Zhao GP, Liu RR, Li QH, Zheng MQ, Wen J (2018) Method using a co-culture system with high-purity intramuscular preadipocytes and satellite cells from chicken pectoralis major muscle. Poult Sci 97(10):3691–3697. https://doi.org/10.3382/ps/pey023

Dolgin E (2019) Sizzling interest in lab-grown meat belies lack of basic research. Nature 566(7743):161–162. https://doi.org/10.1038/d41586-019-00373-w

Gao Y, Wang D, Xu ML, Shi SS, Xiong JF (2018) Toxicological characteristics of edible insects in China: A historical review. Food Chem Toxicol 119:237–251. https://doi.org/10.1016/j.fct.2018.04.016 (Epub 2018 Apr 10)

Hocquette JF (2016) Is in vitro meat the solution for the future? Meat Sci 120:167–176. https://doi.org/10.1016/j.meatsci.2016.04.036 (Epub 2016 Apr 29)

Khan S, Khan RU, Alam W, Sultan A (2018) Evaluating the nutritive profile of three insect meals and their effects to replace soya bean in broiler diet. J Anim Physiol Anim Nutr (Berl). 102(2):e662–e668. https://doi.org/10.1111/jpn.12809 (Epub 2017 Nov 3)

Kouřimská L, Adámková A (2016) Nutritional and sensory quality of edible insects. NFS J (Official J Soc Nutr Food Sci) 4:22–26

Liu LL, Fang C, Liu WJ (2018) Identification on novel locus of dairy traits of Kazakh horse in Xinjiang. Gene 30(677):105–110. https://doi.org/10.1016/j.gene.2018.07.009 (Epub 2018 Jul 3)

Lorenzo JM, Munekata PES, Campagnol PCB, Zhu Z, Alpas H, Barba FJ, Tomasevic I (2017) Technological aspects of horse meat products—a review. Food Res Int 102:176–183. https://doi.org/10.1016/j.foodres.2017.09.094 (Epub 2017 Sep 30. Review)

Mehta F, Theunissen R, Post MJ (2019) Adipogenesis from Bovine Precursors. Methods Mol Biol 1889:111–125. https://doi.org/10.1007/978-1-4939-8897-6_8

Melo-Ruiz V, Quirino-Barreda T, Calvo-Carrillo C, Sánchez-Herrera K, Sandoval-Trujillo H (2013) Assessment of nutrients of Escamoles ant eggs *Limotepum apiculatum* M. by spectroscopy methods. J Chem Chem Eng7:1181–1187. https://doi.org/10.17265/1934-7375/2013.12.013

Mitsuhashi J (2016) Edible insects of the world. CRC Press. Boca Raton. ISBN 9781498756570–CAT# K27533

Nogueira-Filho SLG, da Cunha Nogueira SS (2018) Capybara meat: An extraordinary resource for food security in South America. Meat Sci 145:329–333. https://doi.org/10.1016/j.meatsci.2018.07.010

Purschke B, Tanzmeister H, Meinlschmidt P, Baumgartner S, Lauter K, Jäger H (2018) Recovery of soluble proteins from migratory locust (*Locusta migratoria*) and characterisation of their compositional and

techno-functional properties. Food Res Int 106:271–279. https://doi.org/10.1016/j.foodres.2017.12.067
Rumpold BA, Schlüter OK (2013) Nutritional composition and safety aspects of edible insects. Mol Nutr Food Res 57(5):802–823. https://doi.org/10.1002/mnfr.201200735
Tzompa-Sosa DA, Yi L, van Valenberg HJF, van Boekel MAJS, Lakemond CMM (2014) Insect lipid profile: aqueous versus organic solvent-based extraction methods. Food Res Int 62:1087–1094
van Huis A (2016) Edible insects are the future? Proc Nutr Soc. 75(3):294–305. https://doi.org/10.1017/S0029665116000069
Xiaoming C, Ying F, Hong Z (2010) Review of the nutritive value of edible insects. Edible insects and other invertebrates in Australia: future prospects. In: Proceedings of a workshop on Asia-Pacific resources and their potential for development, 19–21 Feb 2008, Bangkok 2010, pp 85–92

Career Opportunities in Animal Biotechnology

48

Abstract
Animal biotechnology involves applications of scientific, computational engineering and bioinformatics principals to animal nutrition, health and reproduction. A scientific intellect with zeal for research is must to build a career in animal biotechnology. The professionals with a sound biotechnology background can do well in animal biotechnology research and education. The marketing and business expertise is required in the marketing of the instruments, consumables, and animal products.

Key Points

- Animal biotechnology is a multidisciplinary area of research and development
- Animal biotechnology has provided career opportunities and employment to millions of professionals and "various professionals" or "specialists with different training backgrounds".

Keywords
Animal biotechnology · Entrepreneurs · Employment generation · Intellectual property

48.1 Introduction

Animal biotechnology is a well-established research and development sector. Recombinant DNA technology, genome engineering, bioinformatics, and computational engineering have advanced the field of basic and applied animal biotechnology (Fig. 48.1). Originally started as a fundamental subject of applied biological sciences, the animal biotechnology is today an important sector providing employment to a huge workforce involving researchers, principal investigators, mathematicians, technicians, and administrative persons.

The global market for protein drugs has the highest compound annual growth among the pharmaceutical class. Techniques to generate transgenic animals to produce high-quality biologically active proteins have made it possible (Herron et al. 2018).

48.2 Thrust Areas of Animal Biotechnology

Animal biotechnology encompasses various areas of biological sciences including animal healthcare, microbiology of prokaryotic and

B. Singh et al., *Advances in Animal Biotechnology*,
https://doi.org/10.1007/978-3-030-21309-1_48

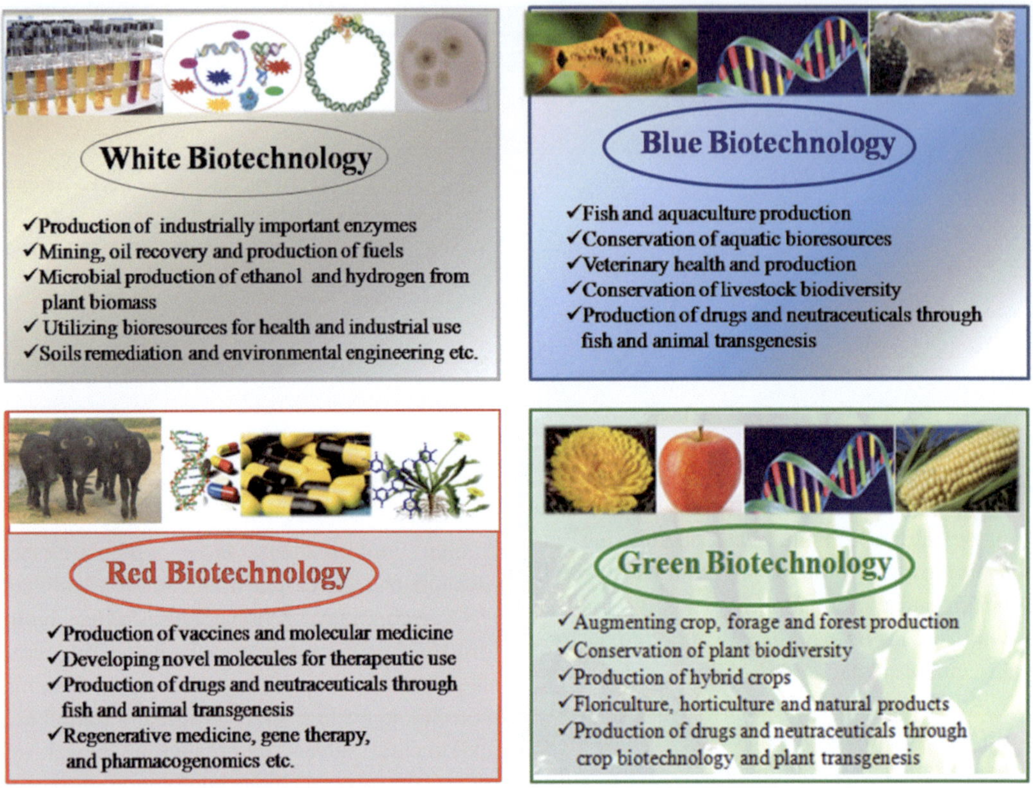

Fig. 48.1 Role of animal biotechnology in human employment. Four fields, namely, white biotechnology or industrial biotechnology, blue biotechnology (aquaculture), red biotechnology (medicines and therapeutics), and green biotechnology (crops, food, feed and fodder) are important contributions of biotechnological interventions

eukaryotic microorganisms of animal origin, assisted reproduction techniques, cryopreservation, bioengineering and genome editing of microorganisms, cells and animals, and genome, transcriptome and proteome sequencing and applications. Development of model animals to investigate human diseases (Ruan et al. 2017; Bjursell et al. 2018), and genetic engineering of animals to produce recombinant therapeutics, and genetic improvement of livestock are important contributions of animal biotechnology. Another important objective of animal biotechnology is to select animals with superior genetic merit enhanced nutrient utilization, growth, resistance to diseases, parasites and pathogens, and climatic stress (Jiang and Shen 2019). Genetic engineering and gene-editing techniques *viz.*, zinc finger nucleases (ZFN), transcription activator-like effector nuclease (TALEN), and clustered regularly interspaced short palindromic repeats (CRISPR)/CRISPR-Cas 9 (Jiang and Shen 2019) are used in genetic engineering of microorganisms originating from GI tract, stem cells, and embryos of animals.

48.3 Global Entrepreneur in Animal Biotechnology

Innovative technical growth in the dairy industry has made India the largest producer of milk in the world. The dairy industry offers seamless business opportunities in thrust areas such as growth in value-added dairy products, formulation of infants milk formula, low cholesterol milk fat products, genetic engineering of starter cultures for improved qualities of commercial value, enzymes and whole immobilization, cryopreservation or lyophilization of microbial cultures. Similarly, meat production, processing, preservation, and import have provided employment to millions of people.

Among major types of instruments, CO_2 incubators, deep freezers, cryocontainers, PCR machines of assorted kinds, micromanipulators, powerful digital microscopes, electric cell manipulators, fast performance liquid chromatography (FPLC), automated NextGen sequencers, powerful workstations and lyophilizers, etc. are more common in most laboratories. Around 123 ranked multinational commercial companies are involved in manufacturing instruments and consumables used in various facets of animal biotechnology. Several renowned institutes and academia are contributing to augment animal health, production and diversity conservation through research and teaching.

48.4 Researchers and Investigators

Researchers and scientists are the key resource persons. Scholars with a doctoral degree in microbial biotechnology, reproduction biology, genetic engineering, virology or immunology are more apposite for planning and monitoring experimentation and data management. Prime duty of researchers is to plan and execute the experiments and arrange for funds. In present scenario, multidisciplinary approaches are successful and lead to valuable outcomes. Many laboratories have international collaborations and joint research programs.

Growing biotechnology sector followed by increasing research and development (R&D) expenses will be a major market growth in the future. In 2016, the biotech industries had generated around USD 139.4 billion revenue. Increasing prevalence of lifestyle non-communicable diseases, such as cardiovascular diseases (Lena et al. 2018; Kotseva et al. 2019; Meier et al. 2019), diabetes (Afroz et al. 2018; Cuschieri 2019), obesity (Kumar and Kaufman 2018; Jaacks et al. 2019), and drug resistance in pathogens, and the expenditure involved in managing human and livestock health, and role of transgenic animals in investigating and curing the diseases will further boost the market. Some companies are interested in animals as donors of organs for human transplantation.

48.5 Mathematics, Computational Biology and Drug Discovery

Biotechnology is a multidisciplinary subject. Mathematical biology, bioinformatics, and computational approaches help foster future interactions and design novel drugs and drug targets. Indeed, mathematics opens new vistas in bioinformatics, biochemical engineering, system biology and instrumentation. Hence, the knowledge of mathematics is a strong indicator of success in the biotechnology industry. Mathematical modeling is an important tool to understand and solve biological complexities such as signal transduction processes, understanding transcriptomics (Mura et al. 2019), designing fermentation and enzyme-based processes, describing past performances, and predicting the futuristic performance of biotechnological processes. Therefore, there are ample admirable opportunities for mathematics graduating Ph.D. students and postdocs to move into biotechnology-based industry careers (Allen and Moore 2019).

48.6 Career in IPR

Like other areas of scientific interventions, protecting intellectual property (IP) by legislation is a prioritized area in animal biotechnology. Vaccines, drugs and genetically modified microorganisms and animals, and bioinformatics programs have to be protected by copyrights or patenting. Patent law is an area for professionals in biomedical sciences. Job opportunities in this area include positions of patenting agents and skills involved in drafting the applications and patent prosecution. Education in this area offers opportunities for in-house work in the biotech industry, business development, mergers, and acquisitions (Mamajiwalla 2018).

48.7 Animal Health Specialists

The animals viz., worms, fish, amphibians, birds, rodents, sheep, goat, horses, rabbits, guinea pigs, cattle, cats, hamsters, dogs, and monkeys are used as model animals for drug or pharmaceutical testing need specialized care.

Working with animals needs skilled persons. Veterinarians, pharmacists, and field supervisors are essentially required to assist in monitoring the health of model animals, transgenic animals, and collecting samples from animals. Maintaining certain model animals, such as gnotobiotic or germ-free mice used for investigating human and animal gut microbiota, fecal microbiota transplantation and drug discovery need highly controlled hygienic facilities, and experienced microbial ecologists (Fig. 48.2).

48.8 Animal Geneticists and Breeders

As a part of scientific animal husbandry practice, the genetic testing is used to select the best animals for milk or meat and resistance to disease and climatic stress. Current emphasis is selecting livestock based on transcriptomic analysis of functional genome and related molecular genetic markers. This provides opportunities of development and employment of molecular biologists in animal breeding industries.

The field of assisted reproduction biotechniques namely, nuclear transfer cloning, sperm sexing, in vitro embryo production, biobanking, and regenerative medicine are dependent on highly skilled manpower, and advanced research and analytical techniques. Some commercial companies offer cloning of pet animals.

48.9 Outlook and Challenges

Animals are important contributors of products for human nutrition, health, and livelihood. Biotechnology is an important contribution to enhance animal production and conservation. Animal biotechnology has generated opportunities to create employment to millions of professionals and semi-skilled workforce. The role of animal biotechnology will be important in future to enhance the productivity of animals to ensure food security.

There is increasing demand of organs for humans. In future, commercial companies may come to provide immunocompatible organs. This area is under progress and needs further scientific inputs.

Not only advantages are there, but also animals will face several challenges in the future. The incidences of pathogens, parasites, and pests will increase that will affect livestock productivity.

Principles of marketing should be applied for the cost-effective and creative promotion of biotechnology products. Durability of the product, attractive logo and packaging, and advertisement of the product are necessary to generate the interest of the consumers.

This is because consumers wish to see the company products many times before purchasing the same. Of course, reliability and responsible attitude of suppliers matter a lot.

In conclusion, the animal biotechnology has done a lot for the welfare of humans and animals during recent decades. Small and sharply focused entrepreneurs will continue to excel in research, development, and marketing.

48.10 Conclusions

Despite social or ethical issues, the long-term benefits of advanced areas viz., genome editing and genetic engineering are more than the bottlenecks currently associated with these techniques. With new high-throughput experimental techniques, development of tools for data

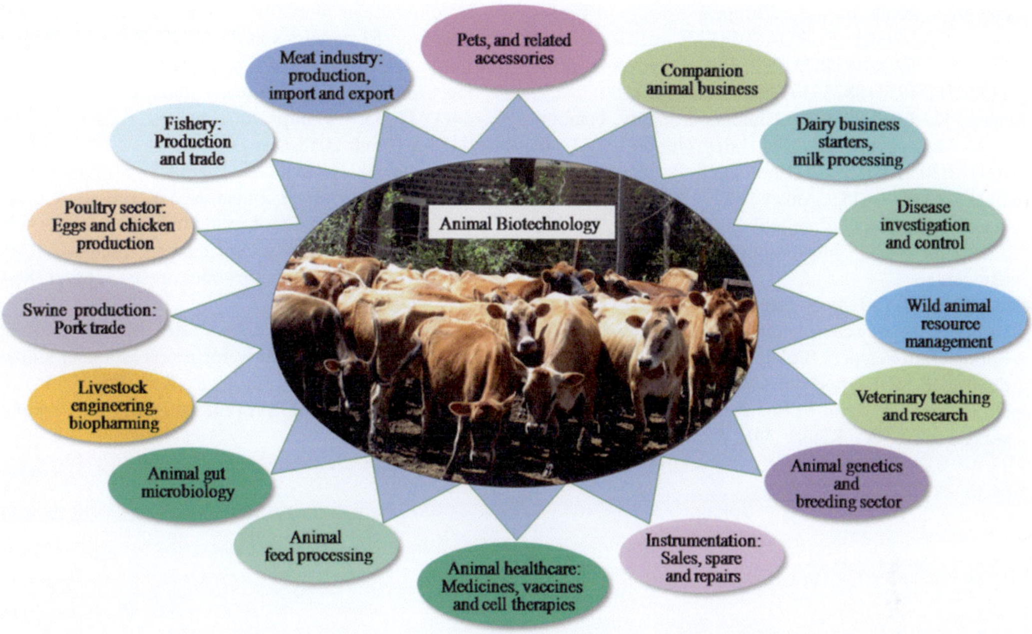

Fig. 48.2 Diagrammatic depiction of career opportunities in various sectors of animal biotechnology

analysis and predicting the success of the experimental models represent enormous opportunity for basic and applied research in animal biotechnology.

References

Afroz A, Alramadan MJ, Hossain MN, Romero L, Alam K, Magliano DJ, Billah B (2018) Cost-of-illness of type 2 diabetes mellitus in low and lower-middle income countries: a systematic review. BMC Health Serv Res. 18(1):972. https://doi.org/10.1186/s12913-018-3772-8

Allen R, Moore H (2019) Perspectives on the role of mathematics in drug discovery and development. Bull Math Biol. https://doi.org/10.1007/s11538-018-00556-y (Epub ahead of print)

Bjursell M, Porritt MJ, Ericson E, Taheri-Ghahfarokhi A, Clausen M, Magnusson L, Admyre T, Nitsch R, Mayr L, Aasehaug L, Seeliger F, Maresca M, Bohlooly-Y M, Wiseman J (2018) Therapeutic genome editing with CRISPR/Cas9 in a humanized mouse model ameliorates α1-antitrypsin deficiency phenotype. EBioMed 29:104–111. https://doi.org/10.1016/j.ebiom.2018.02.015 (Epub 2018 Feb 19)

Cuschieri S (2019) Type 2 diabetes—an unresolved disease across centuries contributing to a public health emergency. Diabetes Metab Syndr 13(1):450–453. https://doi.org/10.1016/j.dsx.2018.11.010 (Epub 2018 Nov 3. Review)

Herron LR, Pridans C, Turnbull ML, Smith N, Lillico S, Sherman A, Gilhooley HJ, Wear M, Kurian D, Papadakos G, Digard P, Hume DA, Gill AC, Sang HM (2018) A chicken bioreactor for efficient production of functional cytokines. BMC Biotechnol 18(1):82. https://doi.org/10.1186/s12896-018-0495-1

Jaacks LM, Vandevijvere S, Pan A, McGowan CJ, Wallace C, Imamura F, Mozaffarian D, Swinburn B, Ezzati M (2019) The obesity transition: stages of the global epidemic. Lancet Diabetes Endocrinol 7(3):231–240. https://doi.org/10.1016/s2213-8587(19)30026-9 (Epub 2019 Jan 28. Review)

Jiang S, Shen QW (2019) Principles of gene editing techniques and applications in animal husbandry. 3 Biotech 9(1):28. https://doi.org/10.1007/s13205-018-1563-x (Epub 2019 Jan 3)

Kotseva K, De Backer G, De Bacquer D, Rydén L, Hoes A, Grobbee D, Maggioni A, Marques-Vidal P, Jennings C, Abreu A, Aguiar C, Badariene J, Bruthans J, Castro Conde A, Cifkova R, Crowley J, Davletov K, Deckers J, De Smedt D, De Sutter J, Dilic M, Dolzhenko M, Dzerve V, Erglis A, Fras Z, Gaita D, Gotcheva N, Heuschmann P, Hasan-Ali H, Jankowski P, Lalic N, Lehto S, Lovic D, Mancas S, Mellbin L, Milicic D, Mirrakhimov E, Oganov R, Pogosova N, Reiner Z, Stöerk S, Tokgözoğlu L, Tsioufis C, Vulic D, Wood D; EUROASPIRE Investigators (2019) Lifestyle and impact on cardiovascular risk factor control in coronary patients across 27 countries: results from the European Society of Cardiology ESC-EORP EUROASPIRE V registry. Eur J Prev Cardiol 2047487318825350. https://doi.org/10.1177/2047487318825350 (Epub ahead of print)

Kumar S, Kaufman T (2018) Childhood obesity. Panminerva Med 60(4):200–212. https://doi.org/10.23736/s0031-0808.18.03557-7 (Epub 2018 Oct 5. Review)

Lena A, Coats AJS, Anker MS (2018) Metabolic disorders in heart failure and cancer. ESC Heart Fail. 5(6):1092–1098. https://doi.org/10.1002/ehf2.12389 (Review)

Mamajiwalla S (2018) A career in patent law: at the cutting edge of science, but not at the bench. Cold Spring Harb Perspect Biol 10(7). pii: a032920. https://doi.org/10.1101/cshperspect.a032920

Meier T, Gräfe K, Senn F, Sur P, Stangl GI, Dawczynski C, März W, Kleber ME, Lorkowski S (2019) Cardiovascular mortality attributable to dietary risk factors in 51 countries in the WHO European Region from 1990 to 2016: a systematic analysis of the Global Burden of Disease Study. Eur J Epidemiol 34(1):37–55. https://doi.org/10.1007/s10654-018-0473-x (Epub 2018 Dec 14)

Mura M, Jaksik R, Lalik A, Biernacki K, Kimmel M, Rzeszowska-Wolny J, Fujarewicz K (2019) A mathematical model as a tool to identify microRNAs with highest impact on transcriptome changes. BMC Genom 20(1):114. https://doi.org/10.1186/s12864-019-5464-0

Ruan J, Xu J, Chen-Tsai RY, Li K (2017) Genome editing in livestock: are we ready for a revolution in animal breeding industry? Transgenic Res 26(6):715–726. https://doi.org/10.1007/s11248-017-0049-7 (Epub 2017 Nov 1)

Intellectual Property Rights in Animal Biotechnology

49

Abstract
Intellectual property is the foundation of innovations. The science of animal biotechnology has progressed vastly during past three decades. Vaccines, antibodies, gene-editing, or transgenic animals as model for investigating human diseases and producing recombinant therapeutics, and the bioinformatics methods are the innovative intellectual outcomes.

Highlights

- Animal biotechnology is oriented toward developing products at commercial scale
- IPRs are essential components of animal biotechnology research and development.

Keywords
Intellectual property · Patents · Copyrights · Diagnostics · Recombinant vaccines · Genetically modified organisms

49.1 Introduction

Use of biotechnology is not a new concept. Humans have used biotechnology since the dawn of civilization. The practice of animal biotechnology began more than 8000 years ago with the domestication and selective breeding of animals. Invention of genetic code in the mid-1950s pioneered the present concept of animal biotechnology.

Initially, the biotechnology progressed at slow pace and remained a subject of little attention. Development of techniques viz., genetic engineering or recombinant DNA (rDNA) techniques, cell culturing and manipulation, cell fusion, production of monoclonal antibodies, and gene-editing heralded a new era of revolution in biomedical sciences.

Meanwhile, it was realized that above innovations in conjunction with reproduction technologies could modify the cells, embryos, and animals, and indeed, the outcomes were amazing. This initiated the concerns about intellectual property and its protection in biotechnology applied to animals.

At present, the animal biotechnology is a major area of basic and applied biological research aimed to develop veterinary vaccines, molecular diagnostics, transgenic or gene-edited animals for diagnosis of human diseases, using genetically engineered animals for therapeutics and organs. All this involves intellectual inputs, and outcomes are of enormous economic importance.

B. Singh et al., *Advances in Animal Biotechnology*,
https://doi.org/10.1007/978-3-030-21309-1_49

49.2 What Is Intellectual Property?

The term intellectual property (IP) is generic legal term that describes various types of intangible assets and laws which protect the innovations and applications of thoughts, ideas, and information of profitable value. In broader sense, IP protection is about the laws related to patents, copyrights, trademarks, trade secrets, and other similar rights (Cornish 1989).

Currently, a number of processes and methods applied to biotechnology are under legal protection, and most of the processed initiated in the USA. Later on, other countries also started competing in new biotechnological markets. It became important to them to amend their national laws in order to protect and boost investment in biotechnology. Despite the fact that international consensus is lacking on how biotechnology has to be treated, the biotechnology innovators opt for patenting their invention.

Webber (2003) has insisted that research outcomes of pharmaceutical sector, any biotech start-up or academia should be protected by patents. This is because enormous efforts, hard work, and financial inputs are needed to develop a novel drug or product out of hundreds and thousands of lead compounds. It is therefore prerogative for a company or institute to protect their innovative products from unauthorized duplication (Raj et al. 2015).

The patenting grants legal, time-bound monopolies to the eligible scientists or inventors (Murphy et al. 2015). Similarly, the trade secrets (information related to a formula, compilation, device or program or technique) can be protected by law that varies depending on a country and its legislation. The owner of trade secret is known as originator.

49.3 Intellectual Property Laws

Science plays key role in economic prosperity of the nation. Economy is strong when science and technology are allowed to do well. It is the technology that delivers goods and services, hence, directly or indirectly, the science is correlated to generate the revenue.

The very basic purpose of IP protection is to allow the right holder to prohibit others from using the IP (patents, copyrights, trademarks, and trade secrets) rights in well-defined ways (Brown 2003).

Activities of practicing scientist inextricably intersect with business world (Poticha and Duncan 2019). Some inventions are protected by various types of IP protection. It is important to take into consideration the business and legal factors before opting for a particular strategy (Voss et al. 2017). Some IPs are protected by simple regulatory norms, others need stringent IP protection (Voss et al. 2017).

49.4 Types of IPs in Animal Biotechnology

49.4.1 Patenting Genetically Modified Organisms

Patents cover various technologies used to modify mice or rat genome and the stem cells. For instance, the French institute CERBM and IGBMC (Centre Européen de Recherche en Biologie et en Médecine–L'Institut de génétique et de biologie moléculaire et cellulaire) have patents of describing the use of the drug tamoxifen to induce Cre recombinase activity in vivo in a transgenic mouse (https://www.taconic.com/taconic-insights/model-generation-solutions//biotechnology-patents-and-gmos.html, accessed on Feb. 26, 2019). The US Patent Office had granted a patent (No. US4736866), for a transgenic mouse model, named "Oncomouse," whose germ and somatic cells carry recombinant activated oncogene (https://patents.google.com/patent/US4736866A/en, accessed on Feb. 26, 2019). The transgenic mouse was developed at Harvard Medical School (Brown 2000).

By 2004, around 600 animal patents had been granted worldwide, 80% in the USA, and most relating to "animal models" for biomedical investigations. Besides the USA and European

Union, only three countries had allowed patents for experimental animals (Lesser 2006).

49.4.2 IP Issues in *In Silico* Biology

Bioinformatics and in silico methods are indispensable tools to analyze gigantic "omics" sequence data. The bioinformatics programs are used to predict hypothetical gene, proteins, and metabolic pathways from the sequence data of microorganism (Singh et al. 2017; Sharma et al. 2018). In silico methods are used in computational modeling in medicinal computational chemistry, predicting drug-target interactions and developing (Quantitative) Structure-Activity Relationships ((Q)SARs) (Duardo-Sánchez et al. 2008).

Potentially patentable bioinformatics programmes include lines of code, algorithms, data content, data structure, and user interfaces (Harrison 2003). Several bioinformatics algorithms are available for analyzing sequence data and predicting genes and proteins from the data. Many of them are freely accessible while others need payments. Some commercial companies provide services to analyze the sequence data. Success or failure of bioinformatics and in silico biology needs the appropriate use of legal tools for protecting and utilizing the intellectual property. Patenting of products discovered using in silico procedures is still a debatable issue and needs a due attention.

49.4.3 Therapeutics and IP

Vaccines, monoclonal antibodies, and antibiotics provide protection against infectious diseases. As there are many forms of vaccines, and components of vaccines, delivery systems, and distribution networks, a variety of IP protections are applicable to vaccines (Durell 2016). With advances in understanding molecular biology of viruses, there is increase in the application of virus sequences and viral gene expression strategies to diagnose and treat the diseases. This extends the scope of patenting multiple constituents of vaccines (Venkataraman et al. 2017). Patents have also been granted for recombinant antigen-based rapid sero-diagnosis of viral infectious diseases, synthetic peptide antigens, and multiple antigenic peptide (MAP) assays for detecting viral infectious diseases of livestock.

49.5 Outlook and Challenges

Applications of biotechnology and molecular biological tools in animal sciences have reached new horizons. In present scenario, the transgenic animals have a large role in developing pharmaceutical industries and medicinal research. Developed countries have already realized potential of biotechnology-oriented industries and the importance of patenting the genetically modified animals.

Surprisingly, the world's richest biodiversity is present in countries that are poor, and lack resources and necessary wherewithal to transform their bioresources into products and earn the revenue. Also, the deprived or developing countries are unable to invest revenue to promote the research and provide adequate patent protection for genetically modified animals.

The legal costs involved in IP can be minimized by strategic planning and diligence. Due to high-controversial nature and processes involved in legal grant of protection for animal biotechnology, the short-term developments are likely to take place at national and regional levels. We emphasize whether a person works in a biotech start-up or a university or in commercial pharmaceutical company, a sound knowledge of patent system is necessary to protect the outcomes of novel research process or the end product.

IP impinges on almost every invention a scientist does. Scientists and researchers working in the fields of animal biotechnology and genetic engineering, however, need training to deal with complex issues of IP, their rights and obligations.

49.6 Conclusions

Protecting IP by law is elemental right of an inventor to protect and utilize their invention for commercial use. The patents protect intellectual property by granting inventors the right to prevent others from making, utilizing, or commercial use of their invention. Computational and in silico biology, genome sequencing, gene synthesis, regenerative medicine, and genome engineering have enabled humans to utilize animals for qualitative and quantitative production of food and therapeutics. Although each country has its own procedures to define and evaluate the inventions and permit their patenting.

References

Brown C (2000) Patenting life: genetically altered mice an invention, court declares. CMAJ 163(7):867–868 (No abstract available)

Brown WM (2003) Intellectual property law: a primer for scientists. Mol Biotechnol 23(3):213–224 (Review)

Cornish WR (1989) Intellectual property: patents, copyright, trade marks, and allied rights. 2nd ed. London: Sweet & Maxwell

Duardo-Sánchez A, Patlewicz G, López-Díaz A (2008) Current topics on software use in medicinal chemistry: intellectual property, taxes, and regulatory issues. Curr Top Med Chem 8(18):1666–1675

Durell K (2016) Vaccines and IP rights: a multifaceted relationship. Methods Mol Biol 1404:791–811. https://doi.org/10.1007/978-1-4939-3389-1_52

Harrison R (2003) Protecting innovation in bioinformatics and in-silico biology. BioDrugs 17(4):227–231

Lesser W (2006) Living organism (animal) patents. Reviews in Cell Biology and Molecular Medicine. https://doi.org/10.1002/3527600906.mcb.200400043

Murphy A, Stramiello M, Lewis S, Irving T (2015) Introduction to intellectual property: A U.S. perspective. Cold Spring Harb Perspect Med 5: a020776. https://doi.org/10.1101/cshperspect.a020776

Poticha D, Duncan MW (2019) Intellectual property—the foundation of innovation. J Mass Spectrom. https://doi.org/10.1002/jms.4331

Raj GM, Priyadarshini R, Mathaiyan J (2015) Drug patents and intellectual property rights. Eur J Clin Pharmacol 71(4):403–409. https://doi.org/10.1007/s00228-015-1811-5 (Epub 2015 Feb 3)

Sharma D, Sharma A, Verma SK, Singh B (2018) Targeting metabolic pathways proteins of Orientia tsutsugamushi using combined hierarchical approach to combat scrub typhus. J Mol Recognit 21:e2766. https://doi.org/10.1002/jmr.2766

Singh G, Sharma D, Singh V, Rani J, Marotta F, Kumar M, Mal G, Singh B (2017) *In silico* functional elucidation of uncharacterized proteins of *Chlamydia abortus* strain LLG. Future Sci OA 3(1):FSO169. https://doi.org/10.4155/fsoa-2016-0066 (eCollection 2017 Mar. Erratum in: Future Sci OA. 2017 Oct 05;3(4):FSO66C1)

Venkataraman S, Ahmad T, Haidar MA, Hefferon KL (2017) Recently patented viral nucleotide sequences and generation of virus-derived vaccines. Recent Pat Antiinfect Drug Discov. 12(1):31–43. https://doi.org/10.2174/1574891X12666170619102506

Voss T, Paranjpe AS, Cook TG, Garrison NDW (2017) A Short Introduction to Intellectual Property Rights. Tech Vasc Interv Radiol. 20(2):116–120. https://doi.org/10.1053/j.tvir.2017.04.007

Webber PM (2003) A guide to drug discovery. Protecting your inventions: the patent system. Nat Rev Drug Discov. 2(10):823–830

Subject Index

B. Singh et al., *Advances in Animal Biotechnology*,
https://doi.org/10.1007/978-3-030-21309-1

B

F

G

O

S

U

MIX
Papier aus verantwortungsvollen Quellen
Paper from responsible sources
FSC® C105338

If you have any concerns about our products,
you can contact us on
ProductSafety@springernature.com

In case Publisher is established outside the EU,
the EU authorized representative is:
Springer Nature Customer Service Center GmbH
Europaplatz 3, 69115 Heidelberg, Germany

Printed by Libri Plureos GmbH
in Hamburg, Germany